Hans-Georg Elias
Macromolecules

Macromolecules. Hans-Georg Elias
Copyright © 2009 WILEY-VCH Verlag GmbH & Co. KGaA, Weinheim
ISBN: 978-3-527-31175-0

Related Titles

Xanthos, M. (ed.)

Functional Fillers for Plastics

2nd edition

2009

ISBN: 978-3-527-32361-6

Schnabel, W.

Polymers and Light

Fundamentals and Technical Applications

2007

ISBN: 978-3-527-31866-7

Tadmor, Zehev / Gogos, Costas G.

Principles of Polymer Processing

2nd edition

2006

ISBN: 978-0-471-38770-1

Meyer, Thierry / Keurentjes, Jos (Hrsg.)

Handbook of Polymer Reaction Engineering

2005

ISBN: 978-3-527-31014-2

Advincula, R. C. / Brittain, W. J. / Caster, K. C. / Rühe, J. (eds.)

Polymer Brushes

Synthesis, Characterization, Applications

2004

ISBN: 978-3-527-31033-3

Fakirov, S. (ed.)

Handbook of Thermoplastic Polyesters

Homopolymers, Copolymers, Blends and Composites

2002

ISBN: 978-3-527-30113-3

Urban, D., Takamura, K. (eds.)

Polymer Dispersions and Their Industrial Applications

2002

ISBN: 978-3-527-30286-4

Wilks, E. S. (ed.)

Industrial Polymers Handbook

Products, Processes, Applications

2001

ISBN: 978-3-527-30260-4

Hans-Georg Elias

Macromolecules

Volume 4: Applications of Polymers

WILEY-VCH Verlag GmbH & Co. KGaA

The Author

Prof. Dr. Hans-Georg Elias
Michigan Molecular Institute
1910 West St. Andrews Road
Midland, Michigan 48640
USA

Volume 1   Chemical Structures and Syntheses
           (Published in 2005)
Volume 2   Industrial Polymers and Syntheses
           (Published in 2007)
Volume 3   Physical Structures and Properties
           (Published in 2008)
Volume 4   Processing and Application of Polymers

Library of Congress Card No.:
applied for

British Library Cataloguing-in-Publication Data
A catalogue record for this book is available from the British Library.

Bibliographic information published by the Deutsche Nationalbibliothek
The Deutsche Nationalbibliothek lists this publication in the Deutsche Nationalbibliografie; detailed bibliographic data are available in the Internet at http://dnb.d-nb.de.

Printed in the Federal Republic of Germany
Printed on acid-free paper

Printing   Strauss GmbH, Mörlenbach
Binding   Litges & Dopf Buchbinderei,
                Heppenheim
Cover Design Gunther Schulz, Fußgönheim

ISBN: 978-3-527-31175-0

What the gentlemen cannot prove,
they call practice,
and what they cannot disprove,
they call theory.

L. Bamberger
Speeches in the German Parliament

Preface

The four volumes of the series "Macromolecules" have a long history. After completing my doctoral thesis at the Technical University of Munich in 1957, I stayed on to work on my habilitation thesis (a kind of second doctorate) which was at that time (and still is at many German universities) required for all of those who wanted to pursue an academic career. Then, I received in the Fall of 1959 a surprise invitation to join the Swiss Federal Institute of Technology (ETH) in 1960 as Head Assistant in the Department of Organic-Chemical Technology. At ETH, I presented my habilitation thesis in 1961, was appointed Privatdozent, and started lecturing on the physical chemistry of polymers as my Department Head (H. Hopff) asked me to become his co-author.

Hopff, in turn, had been approached by a publisher because there was a need for a German language textbook on polymer chemistry which was just evolving as an independent academic discipline. As a young scientist, I was, of course, deeply honored to be asked to be the co-author of a respected researcher who had in his industrial (BASF) and academic (ETH) career amassed some 150 publications and 300 patents. However, after many months, Hopff lost interest in the venture while I had joined the professorial ranks (1963) and still had no textbook on polymer science for my lecture courses.

So I expanded my lecture notes, discussed details in seminars with my graduate students and postdoctoral workers, and injected some knowledge that I had gained as a consultant to industry in Switzerland and Germany. After I had spent some 6000 hours on drafting and writing, the first edition of the German-language book "Makromoleküle" (856 pp.) was published in the Fall of 1971. I never used the book as a back-up advanced textbook, though, because in the same month I became the founding president of what was to become the Michigan Molecular Institute in Midland, Michigan.

Over the years, while the field was growing and I gained more knowledge, the German-language editions grew from one volume (1st to 4th editions, 1971-1981) to two volumes (5th edition (1990, 1992)) and then to four volumes (1999-2003). It was only thereafter that I found the time to prepare the present English edition which follows in general the format of the 6th German edition but is not a cover-to-cover translation.

Volume I of this series discusses chemical structures and principles of syntheses of synthetic and some natural macromolecules. Volume II is concerned with raw materials and energy sources for the polymer industry, monomer syntheses, industrial polymer manufacture, and general properties of individual polymers. Volume III treats physical structures and physical properties of both single macromolecules and macromolecular substances, i.e., polymers. The present, final Volume IV is concerned with applications of polymers as plastics, fibers, elastomers, thickeners, adhesives, coatings, dielectrics, etc. Like Volumes I-III, Volume IV is basically rewritten and updated. It also contains many new subchapters, tables and graphs; 18 % of the latter are new.

Volume IV consists of a very short introduction (Chapter 1; 6 pp.), three major parts (Fundamentals (4 chapters), Basic Polymer Applications (5 chapters), Special Polymer Applications (5 chapters)), and an Appendix. In all of the chapters, I have tried to relate scientific terminology to industrial practice, explain the history and driving forces behind industrial developments, and outline the scientific background of applications. The chapters contain, in part, information about specific syntheses and special physical properties but not detailed discussions about the scientific backgrounds because this information can be found in Volume I (principles of chemical syntheses), Volume II (industrial syntheses), and Volume III (principles of physical structure, solution behavior, and mechanical properties). With respect to company names and trademarks, no attempt was made to follow the recent series of company mergers, splittings, and name changes.

In Part I, Chapter 2 (12 pp.) reviews fundamental terms for the synthesis and structure of polymers in order to provide a common basis for the other chapters. Chapter 3 (56 pp.) discusses adjuvants (dyestuffs, fillers, plasticizers, processing agents, protecting agents, etc.) and their applications while Chapter 4 (56 pp.) is concerned with rheology and its effect on basic processing methods. Chapter 5 (22 pp.) contains information about important testing methods, many of which often produce data which do not translate easily to scientific properties. Of course, it is not possible to describe all or even most of the specialized tests that are in use in the various fields.

Part II outlines the basic applications of polymers as fibers (Chapter 6; 96 pp.), elastomers (Chapter 7; 53 pp.), and plastics (Chapters 8-10; 170 pp.). It may surprise the reader that I chose to present these applications in this order and not in the usual order plastics–elastomers–fibers. There are two reasons: (a) many technologies were first developed for fibers and rubbers before they were adopted for plastics and (b) fibers and rubbers are common adjuvants for plastics, which makes it expedient to treat them first.

Natural and synthetic fibers (Chapter 6; 96 pp.) are treated in greater detail, mainly because their application as textiles depends less on the chemical structure of their constituent polymers than on the physical after-treatment of fibers and the joining of fibers to fabrics. Rubbers and elastomers (Chapter 7; 53 pp.) describes not only the various types but also the economic and political factors leading to natural rubber production and the synthetic rubber industry. For convenience, plastics are discussed in three chapters: basic types in Chapter 8 (68 pp.), reinforcement in Chapter 9 (41 pp.), and polymer blends in Chapter 10 (61 pp.)

Part III discusses special applications of polymers: in Chapter 11 (39 pp.) solutions (exclusive of their role as vehicles), in Chapter 12 (40 pp) coatings and adhesives, in Chapter 13 (20 pp.) packaging materials, in Chapter 14 (40 pp.) applications as dielectrics and in the electronics industry, and in Chapter 15 (30 pp.) polymers in optics and optoelectronics. The work concludes with an Appendix (Chapter 16; 23 pp.) that lists SI quantities and units, conversion of outdated physical quantities, definitions of physical terms, and abbreviations and acronyms for fibers, rubbers, and plastics.

Industry is fond of using abbreviations, acronyms, and number and letter codes. Each field also has its own terminology. As a result, this volume contains a very large Subject Index. No attempt was made to follow name changes of companies and brands caused by recent mergers, splittings, and marketing considerations.

Like the preceding three volumes, this volume has again had the benefit of constant advice from my good friends and former colleagues at Michigan Molecular Institute (MMI), Professors Petar R. Dvornic and Steven E. Keinath, who read and checked the drafts of all 2788 pages of Volumes I-IV and made many helpful suggestions. I would also like to thank Professor Hans Schnecko, Hanau, Germany, for advice on Chapter 7 and Dr. Abhijit Sarkar, MMI, for comments on Section 15.3.

Midland, Michigan   Hans-Georg Elias
Spring 2009

List of Symbols

Symbols for physical units are strictly those of the International Standardization Organization (ISO) (see Appendix).

Symbols for physical quantities follow the recommendations of the International Union of Pure and Applied Chemistry (IUPAC) and the International Union of Pure and Applied Physics (IUPAP), except for those used in industrial practice. In particular, all symbols for physical quantities are slanted, two-letter symbols are used only for dimensionless quantities (for example, Reynolds number), and vectorial quantities are in bold letters. Specific quantities ($\equiv$ physical quantity divided by mass m) are written in small letters, using the same symbol as for the quantity itself (for example, C_p = isobaric heat capacity, $c_p = C_p/m$ = specific isobaric heat capacity). For "normalized", "reduced", etc., quantities, see the Appendix.

Indices are slanted if they refer to a *physical* quantity that is held constant (for example, C_p = heat capacity at constant pressure). They are written upright if they do not indicate a constant quantity and are just indicators, for example, numbers.

Symbols for Languages

D = German (Deutsch)	G = classical Greek
F = French	L = classical Latin

The Greek letter υ (upsilon) was transliterated as "y" (instead of the customary phonetic "u") in order to make an easier connection to written English (example: πολυς = polys (many)). For the same reason, χ was transliterated as "ch" and not as the customary phonetic "kh."

Mathematical Symbols (IUPAC; all mathematical symbols should be upright).

=	equal to	>	greater than
≠	not equal to	≥	greater than or equal to
≡	identically equal to	>>	much greater than
≈	approximately equal to	<	less than
~	proportional to (IUPAC: ~ or ∝)	≤	less than or equal to
≙	corresponds to	<<	much less than
→	approaches, tends to	±	plus or minus
d	derivative		
∂	partial derivative		
Δ	difference	sin	sine of
δ	differential	cos	cosine of
f	function of (IUPAC: f)	tan	tangent of
Σ	sum	cot	cotangent of
∫	integral	sinh	hyperbolic sine of
Π	product	arctan	inverse tangent
lg	logarithm to the base 10 (IUPAC: lg or $\log_{10}$)		
ln	logarithm to the base e (natural logarithm) (IUPAC: ln or $\log_e$)		
∴	identical with		

Symbols for Chemical Structures

R   symbol for a monovalent substituent, for example, CH_3- or C_6H_5-
Z   symbol for a divalent unit, for example, $-CH_2-$ or $-p-C_6H_4-$
Y   symbol for a trivalent unit, for example, $-C(R)<$ or $-N<$
X   symbol for a tetravalent unit, for example, $>C<$ or $>Si<$
*   symbol for an active site: radical ($^\bullet$), anion ($^\ominus$), or cation ($^\oplus$)
*p*Ph   *para*-phenylene (in text)
p-C_6H_4 *para*-phenylene = 1,4-phenylene (in line formulas)

Averages and Other Markings

‾   line above letter indicates common average, for example, $\overline{M}_n$ = number-average of molar mass (note: subscript is not italicized since it does *not* represent a physical quantity that is kept constant)
~   tilde indicates a partial quantity, for example, $\tilde{v}_A$ = partial specific volume of component A
[ ]   square brackets surrounding the symbol of the substance indicates the amount (of substance) concentration ("mole concentration"), usually in mol/L
⟨ ⟩   angled brackets surrounding a letter indicate spatial averages, for example, $\langle s^2 \rangle$ = mean-square average of radius of gyration (IUPAC)
| | two vertical lines enclosing the symbol for a vectorial quantity indicate the magnitude of that quantity. Example, $|q|$ = magnitude of the scattering vector q

Exponents and Superscripts

Symbols for exponents are slanted if they indicate physical quantities but upright if the symbol indicates a number. Example: exponent α in the intrinsic viscosity = f(molar mass) relationship, $[\eta] = K_v M^\alpha$.

°   degree of plane angle [= $(\pi/180)$ rad]
'   minute of plane angle [= $(\pi/10\ 800)$ rad]
"   second of plane angle [= $(\pi/648\ 000)$ rad]
o   pure substance
∞   infinite, for example, dilution or molecular weight
m   amount-of-substance related quantity if a subscript is inexpedient. According to IUPAC, ^{m} can be used as either a superscript or a subscript
(q)   qth order of a moment (always in parentheses since it does not represent a power)
‡   activated quantity, for example, $E^\ddagger$ = activation energy
a   general exponent in $P = K_P M^a$ (P = property)
q   general exponent
α   exponent in $[\eta] = K_v M^\alpha$
v   exponent in $\langle s^2 \rangle^{1/2} = K_s M^v$
ε   exponent in $\eta_0 = K_\eta M^\varepsilon$

Indices and Subscripts

Subscripts are slanted if they refer to physical properties that are held constant. Example: C_p = isobaric heat capacity. Letters indicating variable physical quantities or numbers of physical entities are upright (example: N_i (with i = 1, 2 or 3)).

o	standard or original state, for example, T_0 = reference temperature
0	state at time zero
1	solvent
2	solute, usually polymer
3	additional component (salt, precipitant, etc.)
∞	final state
A	substance A, for example, M_A = molar mass of substance A
a	group or monomeric unit of A, for example, mass m_a of group a
am	amorphous
B	fracture
B	substance B
bp	boiling temperature (boiling point)
br	branch, branched
c	chain (L: *catena*), for example, in networks
cr	crystalline
crit	critical
cycl	cyclic
e	entanglement
el	elastic
eff	effective
end	endgroup
eq	equilibrium
exc	excess
F	filler, fiber
fl	flexural
G	glass transformation
g	any statistical weight or statistical weight fraction, e.g., n, m, z or x, w, Z
H	hydrodynamically effective property or hydration
h	hydrodynamic average
ht	heterotactic
i	*i*th component
i	isotactic diad (IUPAC recommends m = *meso*)
inh	inherent (dilute solution viscosity)
is	heterotactic triad (IUPAC: mr)
lisl	sum of heterotactic triads, lisl = is + si
it	isotactic
j	*j*th component

k	variable
L	liquid, melt
M	melting
M	matrix (in blends or reinforcd polymers)
m	monomeric unit in macromolecules
m	molar (also as superscript)
mol	molecule
mon	monomer
mu	monomeric unit
n	number average
P	polymer
p	index for quantity at constant pressure
pol	polymer (if P is confusing)
q	index, defined differently for each section or chapter
q	number of electric charges
r	relative (only in M_r = relative molecular mass = molecular weight)
r	based on end-to-end distance, e.g., α_r = linear expansion coefficient of a coil (with respect to the end-to-end distance)
red	reduced
rel	relative
rep	repeating unit
rlx	relaxation
S	solvating solvent
s	syndiotactic diad (IUPAC recommends r = *racemo*)
s	related to radius of gyration
seg	segment
si	heterotactic triad (IUPAC recommends rm)
soln	solution
sph	sphere
T	index for quantity at constant temperature
u	monomeric unit in polymer
V	quantity at constant volume
v	viscosity average (solutions)
w	mass average (weight average if dimensionless)
x	crosslink(ed)
y	yield (stress-strain)
z	z average
η	viscosity average (melts)

Prefixes of Words (in systematic polymer names in *italics*)

alt	alternating
at	atactic
blend	polymer blend
block	block (large constitutionally uniform segment)
br	branched. IUPAC recommends sh-branch = short chain branch, l-branch = long chain branch, f-branch = branched with a branching point of functionality f
cis	cis configuration with respect to C=C double bonds
co	joint (unspecified)
comb	comb
compl	polymer–polymer complex
cyclo	cyclic
ct	cis-tactic
g	graft
ht	heterotactic
ipn	interpenetrating network
it	isotactic
net	network; μ-net = micro network
per	periodic
r	random (Bernoulli distribution)
sipn	semi-interpenetrating network
st	syndiotactic
star	star-like. f-star, if the functionality f is known; f is then a number
stat	statistical (unspecified distribution)
trans	trans configuration with respect to C=C double bonds
tt	trans-tactic

Other Abbreviations

AIBN	N,N'-azobisisobutyronitrile
BPO	dibenzoylperoxide
Bu	butyl group (iBu = isobutyl group; nBu = normal butyl group (according to IUPAC, the normal butyl group is not characterized by n, which rules out Bu as an unspecified butyl group); sBu = secondary butyl group; tBu = tertiary butyl group)
Bz	benzene or benzyl
C	catalyst; C* = active catalyst or active catalytic center
cell	cellulose residue
Cp	cyclopentadienyl group
DMF	N,N-dimethylformamide
DMSO	dimethylsulfoxide
Et	ethyl group
G	gauche conformation
Glc	glucose
GPC	gel permeation chromatography

I   initiator
IR   infrared
L   solvent (liquid)
LC   liquid-crystalline
LS   light scattering
MC   main chain
Me   methyl group
Mt   metal atom
Np   naphthalene
NMR   nuclear magnetic resonance
P   polymer
Ph   phenyl group
Pr   propyl group
SANS   small-angle neutron scattring
SAXS   small-angle X-ray scattering
SC   side chain
SEC   size exclusion chromatography
THF   tetrahydrofuran
UV   ultraviolet

Quantity Symbols (unit symbols: see Chapter 16, Appendix)

Quantity symbols follow in general the recommendations of IUPAC: quantity symbols are always slanted, and vectorial quantities are given in bold letters.

A area; A_c = cross-sectional area of a chain
A numerical aperture
$A^{\ddagger}$ pre-exponential constant (in $k = A^{\ddagger} \exp(- E^{\ddagger}/RT)$)
a specific surface area (in $m^2\ g^{-1}$)
a thermodynamic activity
a areal fraction
a_T shift factor in the WLF equation

b bond length; b_{eff} = effective bond length
bbl   barrel petroleum (= 42 US gallons = 0.158 987 m^3)

C number concentration (number of entities per total volume, $C = cN_A/M$)
$[C]$ amount-of-substance concentration of substance C = amount of substance C per total volume = "molar concentration of C"
C heat capacity (usually in J/K); C_p = isobaric heat capacity (heat capacity at constant pressure p); C_V = isochoric heat capacity (heat capacity at constant volume V); C_m = molar heat capacity (heat capacity per amount-of-substance n)
C electrical capacitance ($C = Q/U$)
c crystallographic bond length = crystallographic length of a repeating unit (usually crystallographic c axis)
c specific heat capacity (usually in J/(g K)); c_p = isobaric specific heat capacity; c_V = isochoric specific heat capacity. Formerly: specific heat

c concentration = mass concentration (= mass-of-substance per total volume) = "weight concentration." IUPAC calls this quantity "mass density" (quantity symbol ρ). The quantity symbol c has, however, traditionally been used for a special case of mass concentration, i.e., mass-of-substance per volume of solution and the quantity symbol ρ for another special case, the mass density ("density") = mass-of-substance per volume of substance.

         The mass concentration of a solute 2 is related to its density ρ and volume fraction by $c_2 = \rho_2 \phi_2$ if volumes are additive.

$\hat{c}$ velocity of light or sound (depends on chapter)

D diffusion coefficient; D_{rot} = rotatory diffusion coefficient

DP   often used in literature as the symbol for "degree of polymerization". This book uses X instead since slanted (!) two-letter symbols of physical quantities are reserved for dimensionless *transport* quantities (ISO).

d diameter

d dimensionality

E energy

E tensile modulus (= modulus of elasticity, Young's modulus); E_f = flexural modulus

E electric field strength (vectorial quantity)

e elementary charge

e cohesion energy density

e component of elongation or shearing (tensor)

F force (vectorial quantity)

F electrical field (vectorial quantity)

f fraction (unspecified); see also x = amount fraction ("mole fraction"), w = mass fraction ("weight fraction"), ϕ = volume fraction

f fineness = linear density = $m/L = \rho/A$

f_o functionality of a molecule

G Gibbs energy ($G = H - TS$); formerly: free enthalpy

G shear modulus (in J m^{-3}), G' = shear storage modulus (real modulus, in-phase modulus, "elastic modulus"), G'' = shear loss modulus (imaginary modulus, 90° out-of-phase modulus, viscous modulus), G_N^o = plateau modulus

G statistical weight fraction ($G_i = g_i/\Sigma_i\, g_i$)

G electrical conductance

G gloss

G_{IC} fracture toughness (in J m^{-2})

g acceleration (due to gravity)

g statistical weight (for example: n, x, w). IUPAC recommends k for this quantity which is problematic because of the many other uses of k. Similarly, K cannot be used for the statistical weight fraction because of the many other meanings of K.

H height

H enthalpy; ΔH_{mix} = enthalpy of mixing, $\Delta H_{mix,m}$ = molar enthalpy of mixing

H haze

h	Planck constant ($h = 6.626\ 075\ 5 \cdot 10^{-34}$ J s)
h	height
I	electric current
I	light intensity
i	radiation intensity of a molecule
i	variable (*i*th component, etc.)
J	flux (of mass, volume, energy, etc.)
J	shear compliance
K	general constant; equilibrium constant
K	compression modulus
k_B	Boltzmann constant ($k_B = R/N_A = 1.380\ 658 \cdot 10^{-23}$ J K^{-1})

L — length (always geometric); L_{chain} = true (historic) contour length of a chain (= number of chain bonds times length of valence bonds); L_{cont} = conventional contour length of a chain (= length of chain in all-trans macroconformation); L_K = length of a Kuhn segment (Kuhnian length); L_{ps} = persistence length; L_{seg} = segment length

M — molar mass of a molecule in g mol^{-1} (= physical unit of mass of molecule divided by amount of molecule). $\overline{M}_n$ = number-average molar mass; $\overline{M}_w$ = mass-average molar mass. Physical methods (osmometry, light scattering, etc., determine molar masses whereas chemical methods (end-group determinations, etc., usually lead to molecular weights, i.e., relative molar masses M_r (dimensionless, physical unit "1").

This book uses the same symbol for molar mass and molecular weight since M is used mostly in a descriptive manner and literature data usually do not distinguish between molar mass and molecular weight.

M	milkiness
m	mass; m_{mol} = mass of molecule
N	number of entities
N_A	Avogadro constant ($N_A = 6.022\ 136\ 7 \cdot 10^{23}$ mol^{-1})
n	amount of substance (in mol); formerly: mole number
n	refractive index in medium; n_1 = refractive index of solvent; n_2 = refractive index of solute
P	permeability coefficient ($P = DS$)
P	power, electric power
$\boldsymbol{P}$	dielectric polarization (= electric dipole moment per volume) (vector)
p	pressure
$\boldsymbol{p}$	dipole moment (vectorial quantity)
Q	electric charge = quantity of electricity
Q	heat
Q	intermediate variable or constant, usually a ratio; varies with section
q	intermediate variable or constant, usually a ratio; varies with section
q	charge of an ion

R	molar gas constant ($R = 8.314\ 510$ J K^{-1} mol^{-1})
R	electrical resistance
R	Pockels coefficient (electrooptics)
R	reflectivity
R	radius: R_d = Stokes radius (from diffusion coefficient), R_{sph} = radius of equivalent sphere, R_v = Einstein radius (from dilute solution viscosity)
RH	relative humidity
r	radius
r	shift of electrons (vectorial quantity)
S	entropy; ΔS_{mix} = entropy of mixing, $\Delta S_{mix,m}$ = molar entropy of mixing
S	solubility coefficient
S	electric strength
S	Kerr coefficient (electrooptics)
s	radius of gyration (IUPAC), shorthand for $\langle s^2 \rangle^{1/2}$ (IUPAC); in the literature often as R_g
T	temperature (always with units). In physical equations always as thermodynamic temperature with unit kelvin; in descriptions, either as thermodynamic temperature (unit: kelvin) or as Celsius temperature (unit: degree Celsius). Mix-ups can be ruled out because the physical unit is always given. IUPAC recommends for the Celsius temperature either t as a quantity symbol (which can be confused with t for time) or θ (which can be confused with Θ for the theta temperature). T_c = ceiling temperature, T_d = decomposition temperature, T_G = glass temperature, T_M = melting temperature
T	transparency
t	time
U	internal energy
U	electric potential difference (voltage drop, voltage)
U	tracking force
u	fractional conversion of monomer molecules (p = fractional conversion of groups; y = yield of substance)
u	excluded volume
V	volume; V_h = hydrodynamic volume, V_m = molar volume; $\tilde{V}_m$ = partial molar volume
V	electrical potential difference (voltage drop, voltage)
v	specific volume; $\tilde{v}$ = partial specific volume
v	linear velocity ($v = dL/dt$)
W	work
w	mass fraction = weight fraction. For example, mass fraction of component 2 in a mixture with component 1: $w_2 = m_2/m = x_2/[x_2 + x_1(M_1/M_2)]$
X	degree of polymerization of a molecule with respect to monomeric units (not to repeating units!); $\overline{X}_n$ = number-average degree of polymerization of a substance; $\overline{X}_w$ = mass-average degree of polymerization of a substance
x	mole fraction (amount-of-substance fraction)

Y refractive index increment ($= \mathrm{d}n/\mathrm{d}c$)
Y degree of polymerization with respect to repeating unit
y yield of substance

Z tear strength
z z-statistical weight
z coordination number, number of neighbors

α angle
α linear thermal expansion coefficient of materials or random coils ($\alpha = (1/L)(\mathrm{d}L/\mathrm{d}T)$). Note: in literature often as β!
α degree of crystallinity (with index for method: X = X-ray, d = density, etc.)
α electrical linear polarizability of a molecule

β angle
β compressibility coefficient
β cubic thermal expansion coefficient [$\beta = (1/V)(\mathrm{d}V/\mathrm{d}T)$]; in literature often as α
β electrical second (or quadratic) hyperpolarizability of a molecule

Γ_{H} parameter of preferential solvation (preferential hydration)
γ angle
γ surface tension, interfacial energy; γ_{I} = impact strength, $\gamma_{\mathrm{I,N}}$= notched impact strength
γ electrical third (or cubic) hyperpolarizability of a molecule
γ_{e} shear strain
$\dot{\gamma}$ shear rate (velocity gradient)

δ loss angle
δ solubility parameter

ε linear extension [$\varepsilon = (L - L_0)/L_0$] = nominal strain (Cauchy strain); $\varepsilon_{\mathrm{H}} = \ln(L/L_0)$ = Hencky strain (true strain); ε_{B} = fracture elongation; $\dot{\varepsilon}$ = elongational strain rate = tensile strain rate; η_{e} = extensional viscosity, $\eta_{\mathrm{e,0}}$ = Troutonian viscosity
ε energy per molecule, cohesive energy
ε_{r} relative permittivity. Formerly: dielectric constant

η dynamic viscosity. η_0 = viscosity at rest (first Newtonian viscosity) = inverse of the stationary fluidity ϕ_0, $\eta_0 = 1/\phi_0$. η_∞ = (apparent) second Newtonian viscosity of non-Newtonian fluids at very large shear rates.

η_1 = viscosity of solvent
η_{r} = η/η_1 = relative viscosity
η_{i} = $(\eta - \eta_1)/\eta_1$ = relative visc. increment   (= specific viscosity η_{sp}) *
η_{inh} = $(\ln \eta_{\mathrm{r}})/c$ = inherent viscosity   (= logarithmic visc. number)
η_{red} = $(\eta - \eta_1)/(\eta_1 c)$ = reduced viscosity   (= viscosity number η_{sp}/c)
$[\eta]$ = $\lim \eta_{\mathrm{red},c \to 0}$ = limiting visc. number   (= intrinsic viscosity)

* IUPAC recommends "relative viscosity increment" and the symbol η_{i}. However, the symbol η_{i} is easily confused with the symbol η_i for the viscosity of the substance i.

Θ	characteristic temperature, especially theta temperature
θ	torsional angle (conformational angle in macromolecular science)
ϑ	angle, especially scattering angle or torsional angle (organic chemistry)
κ	isothermal (cubic) compressibility
κ	dielectric constant (microelectronics industry only)
$\kappa^{(i)}$	first (i = 1), second (i = 2), third (i = 3) susceptibility
Λ	aspect ratio = axial ratio of rods (length/diameter) or rotational ellipsoids (main axis/secondary axis)
λ	wavelength in medium, $\lambda = \lambda_0/n$ (λ_0 = wavelength of incident light)
λ	thermal conductivity
λ	strain ratio = draw ratio ($\lambda = L/L_0$)
μ	chemical potential
μ	mobility of an electrical carrier
μ	electric dipole moment
μ	Poisson ratio
v	kinematic viscosity, $v = \eta v$ (η = dynamic viscosity; v = specific volume)
v	frequency
v	Abbé number
v	speed
v	velocity
π	mathematical constant pi
ρ	density (= mass/volume of the same matter), for example, mass of substance A per volume of substance A. ρ is also used by IUPAC for other densities, for example, for the number density (= number of entities per volume of matter)
ρ	electric resistivity (= volume resistivity, volume resistance)
σ	nominal mechanical stress; σ_{11} = normal stress, σ_{21} = shear stress, σ' = true stress; σ_B = tensile strength at break. "Tensile strength" denotes the strength at mechanical failure; it may refer to the upper yield strength σ_{UY} of ductile materials or the tensile strength at break, σ_B, for brittle ones
σ	electrical conductivity
τ	relaxation time
τ	shear stress (= σ_{21})
τ_i	light transmission; τ_{it} = internal transmission; τ_{et} = external transmission
ϕ	volume fraction; ϕ_f = free volume fraction
ϕ	angle
$\Delta\phi$	electrical potential difference (voltage drop, voltage)
χ^∞	reduced chemical potential
χ	Flory-Huggins interaction parameter
Ω	angle
ω	angular velocity *or* angular (circular) frequency

Table of Contents

1 Introduction

1.1 Materials

Humans need air, water, and food for sheer survival; materials for clothing, housing, transportation, and many goods to make life enjoyable; and energy to convert materials into goods and for heating/cooling, transportation, and communication. The amounts of consumed materials are considerable as can be seen, for example, for the United States of America, a highly industrialized and wealthy country (Table 1-1). Note that materials are sold by weight but applied per volume so that volume per capita is more important than mass per capita.

Table 1-1 Density and consumption of some materials in the US (1999) [1]. Population: $272 \cdot 10^6$ (1999), $300 \cdot 10^6$ (2006). Abbreviations and acronyms: LI = low-molecular weight inorganics, Mo = modified natural product, Mt = metal, Na = natural product, PI = polymeric inorganics, PO = polymeric organics, Sy = synthetic product.

Material	Type	Source	Density in g/cm^3	Consumption in 10^6 t/a	kg/capita	L/capita
Non-metallic minerals						
Stones	LI, PI	Na	≈ 3	2600	9580	3190
Sand, gravel	LI, PI	Na	≈ 2.65	1108	4080	1540
Cement	PI	Mo	3.2	86	317	100
Clays	PI	Mo	≈ 2.4	42.2	155	65
Limestone	LI	Mo	≈ 3	21	77.3	26
Gypsum	LI	Na	2.32	19	70	30
Pumice		Na	0.3	0.62	2.3	7.7
Talc		Na	2.7	0.95	3.5	1.3
Feldspar		Na	2.6	0.90	3.3	1.3
Metals						
Raw steel	Mt	Sy	7.87	13	480	61
Aluminum, primary	Mt	Sy	2.70	6.0	22	8.1
Copper	Mt	Sy	8.94	1.7	6.3	0.70
Lead	Mt	Sy	11.34	0.52	1.9	0.17
Zinc	Mt	Sy	7.13	0.78	2.9	0.41
Organic polymers						
Wood and wood products						
Lumber	PO	Na	≈ 0.8	190	700	875
Paper, cardboard	PO	Mo	≈ 0.65	87.9	323	500
Plywood, veneers	PO	Mo	≈ 0.8	25.2	93	116
Fibers (without import/export of fabrics and clothing)						
Synthetic fibers	PO	Sy	≈ 1.25	5.41	19.9	15.9
Cotton	PO	Na	1.54	2.15	7.90	5.1
Viscose, acetate fibers	PO	Mo	1.54	0.13	0.48	0.31
Wool, silk	PO	Na	1.25	0.063	0.23	0.18
Rubbers (without import/export of rubber products)						
Synthetic rubbers	PO	Sy	≈ 0.94	3.58	13.2	14.0
Natural rubber	PO	Na	0.91	1.23	4.53	5.0
Plastics raw materials (without import/export of plastic products)						
Synthetic resins	PO	Sy	≈ 0.91	32	116	127

The data of Table 1-1 indicate that in 1999 each US resident consumed per year ca. 15 000 kg of stone and other non-metallic minerals, ca. 800 kg of lumber and wood products, and ca. 500 kg of metals. Each inhabitant also used 323 kg of paper, cardboard, and other cellulosic products.

In comparison, US per-capita consumption of synthetic polymers was relatively small: 116 kg of plastics, ca. 23 kg of synthetic fibers, and 14 kg of synthetic rubbers. The picture does not change greatly if one adds natural fibers, natural rubber, and net imports of finished goods such as plastics products, textiles, rubber tires, etc.

The first semi-synthetic polymers became commercial a mere 150 years ago and the first synthetic polymers only ca. 100 years ago. Very often, it took many decades before a newly discovered polymer became a commercial product (Fig. 1-1). A major obstacle was the unknown structure of these new materials. 150 years ago, chemists knew how one could determine the elemental composition of an organic material but still had to learn how chemical elements are arranged in groups such as substituents (see Volume I, Section 1.3). About 100 years ago, "chemical structure" was synonymous with "chemical composition" but it was baffling why, for example, natural rubber with the composition [CH_2–$C(CH_3)$=CH–CH_2] behaved very differently than the compound with the structure H–CH_2–$C(CH_3)$=CH–CH_2–H: the former is a rubber that formed colloidal solutions whereas the latter is a liquid. In 1910, these differences were correctly explained by S.S.Pickles as being caused by the chain-like structure of natural rubber. In 1920, a thorough survey of the existing literature then led Hermann Staudinger to postulate that organic colloids consist of large chain molecules, i.e., "macromolecules" (see also p. 7). The macromolecular hypothesis was subsequently confirmed by his systematic experimental investigations; much later (1953), he earned the Nobel prize for his work.

Science and technology now had a working hypothesis for the effects of chemical structure on physical properties of these macromolecules (or "polymers", see Chapter 2) and, as a result, the time span between the discovery of new polymers and their commercial production shortened considerably (Fig. 1-1), and many new polymers were introduced to the market. As a result, polymer production grew exponentially (Fig. 1-2).

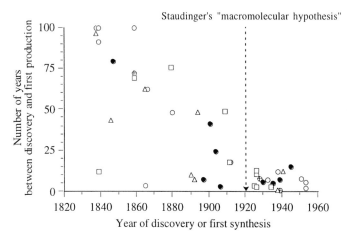

Fig. 1-1 Scientific insight reduced the time span between the synthesis of synthetic and semi-synthetic polymers and their commercial production (data of Table 1-1, Volume II).
□ Elastomers, Δ fibers, O thermoplastics, ● thermosets, ⊕ thickeners.

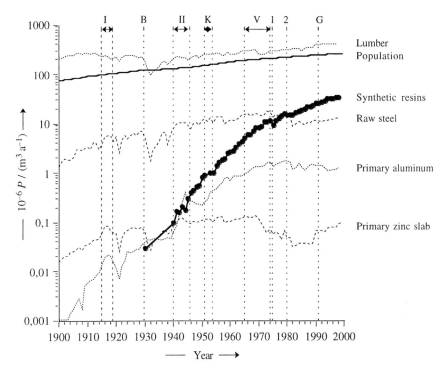

Fig. 1-2 Production P of important materials in the United States (in million cubic meters per year) [2]. The data are given on a volume basis and not on the customary weight basis since materials are used by volume although they are sold by weight. The axis of the ordinates is on a logarithmic scale: the slopes of the curves thus refer to changes in the growth *rates*. The population is given in millions. Synthetic polymers are historically known as "resins." See also Chapter 8.

 I = First World War, B = "Black Friday" 1929, II = Second World War, K = Korean War, V = Vietnam War, 1 = first oil crisis, 2 = second oil crisis, G = Gulf War.

Natural fibers such as wool, silk, and cotton have been used for thousands of years but semi-synthetic fibers have only been produced since 1889 (cellulose nitrate for gun cotton), regenerated natural fibers since 1899 (viscose fibers via the xanthate process), synthetic industrial fibers since 1934 (poly(vinyl chloride) (Pe–Ce fiber; IG Farben)), and synthetic textile fibers since 1938 (nylon 6.6; DuPont) (see Chapter 6).

Native Americans knew about natural rubber and its properties long before their first contacts with Europeans. They used natural rubber not only for balls but also cast it in clay molds to form shaped articles such as bottles and primitive rubber boots. However, natural rubber became a versatile material only after the 1839 discovery of vulcanization by sulfur. The German rubber shortage during World War I (1914-1918) led to the development of methyl rubber (no longer produced). Modern synthetic rubber technology started in 1929 with the production of poly(butadiene) (Buna, no longer a commercial product), poly(chloroprene), and polysulfide rubbers (Thiokol®) (see Chapter 7).

The first synthetic thermosets were modified natural products. The milk protein casein has been converted with lye to artificial horn for intarsia (inlaid work) since at least 1530 and with formaldehyde to Galalith® for haberdashery (commercial since 1904). The first synthetic thermoset was Bakelite® which has been commercially produced from phenol and formaldehyde since 1909 (see Chapter 8).

The first thermoplastic (Celluloid®, commercial since 1869) was also based on a natural polymer, cellulose, that was nitrated and then made pliable by addition of camphor. The first commercial synthetic thermoplastics were poly(methyl methacrylate) (since 1928) and poly(styrene) (since 1930).

The polymer industry first grew relatively slowly, in part, because raw materials for synthetic polymers were expensive (many monomers were based on coal and acetylene chemistry); in part, because polymers were not fine-tuned with adjuvants; and, in part, because processors and polymer users had to get acquainted with these new materials. About 1950, acetylene chemistry was replaced world-wide by petrochemicals from petroleum (crude oil) such as ethene, propene, and other hydrocarbons. Low raw material prices led to a stronger demand for polymers which required larger production units. The resulting economy of scale led to lower polymer prices, etc.

Between 1955 and 1970, the cost of one barrel petroleum (bbl) hovered around 1 US- dollar ($) but the price of high-density poly(ethylene) fell from ca. 110 cts/kg to ca. 30 cts/kg and that of poly(vinyl chloride) from ca. 80 cts/kg to ca. 20 cts/kg. The nationalization of oil fields by OPEC nations led to drastic increases in the price of petroleum which rose from ca. 2 $/bbl in 1973 to ca. 40 $/bbl in 1981 (see Volume II, p. 47 ff.). The oil price then decreased in two steps to ca. 14 $/bbl in 1986, fluctuated drastically, and saw a new round of more or less continuous price increases since 2002 which peaked at ca. 150 $/bbl in 2008 and then fell to ca. 40 $/bbl.

During the last fifty years, industrialized nations developed many new types of polymers, many of which remained niche products whereas some became bulk chemicals. There are several reasons for these developments. Companies will introduce new products only if the products and processes leading to them are well protected by patents, the flow of raw materials and energy is secured, and a certain sales volume is guaranteed. Important are also customers that are willing and able to process the new materials to products that are better and/or less expensive than the existing ones. For example, a polymer producer without experience in the textile business will have a hard time to introduce new fiber-forming polymers. Old experiences with amorphous polymers cannot be transferred to new semicrystalline polymers, and so on.

Therefore, new developments are often not revolutionary but evolutionary. Even if a new polymer is accepted by the market, it may take five to ten years to get established. For example, it cost DuPont almost 20 years (1940-1959) and ca. 42 million US dollars to develop Delrin®, a poly(oxymethylene) $-\!\!-\!\!OCH_2-\!\!\!\!\frac{}{}_n$, although so-called paraformaldehyde ($n \approx$ 8-100) had been known since 1897 (M.Delépine), its chemical structure since 1925 (H.Staudinger and M.Lüthi), and its physical structure since 1932 (E.Sauter) (see Volume III, p. 234). Since costs of developing new polymers may range into the tens and hundreds of millions of dollars, it is clear that the development of additional grades of known polymers is often more financially rewarding and less risky than the introduction of polymers with completely new chemical structures.

The considerations above apply to polymeric materials with known applications. The situation is completely different if a new polymer has revolutionary properties and/or allows new production techniques. Some of these small volume polymers are included in this book but most of them are not because the field is much too diversified. Examples are membranes for fuel cells, antifouling polymers, polymers in agriculture, the whole biomedical sector, and the like.

1.2 Polymeric Materials

Polymers are chemical substances composed of large polymer molecules that are also called macromolecules (see Section 2.1). **Natural polymers** are found in nature (example: cellulose) whereas **synthetic polymers** are synthesized by man from small molecules (**monomers**) (example: poly(ethylene) from the polymerization of ethene) whereas **semisynthetic polymers** are obtained from the chemical conversion of natural polymers (example: cellulose acetate from the reaction of cellulose with acetic acid).

Early synthetic polymers resembled **natural resins**, which are organic solids that break with a conchoidal fracture in contrast to the plane surfaces created upon the fracture of crystalline materials or the drawn-out zones formed upon the breaking of gums and waxes. The designation "natural resin" now mainly refers to oleoresins from tree saps; these resins are not composed of polymers.

Because of this resemblance to natural resins, early synthetic polymers were called **synthetic resins**. The latter term (or just **"resin"**) is still used in industry for all materials that serve as raw materials for plastics, coatings, inks, etc., but not for elastomers and fibers. Such resins may by high-molecular weight materials such as poly(ethylene) or so-called **prepolymers** which are fairly low-molecular weight substances (molecular weights usually smaller than 1000) that can be converted to true polymers.

Polymers are mainly used as *raw materials*, for example, for plastics, elastomers, fibers, coatings, and adhesives, and, in smaller amounts, directly, for example, as thickeners, dielectrics, resists, ion-exchange resins. **Plastics** (Chapter 8) are usually subdivided into two groups according to their hardening processes: thermoplastics and thermosets.

Polymers are called **thermoplastic** if they can be repeatedly heated above their highest physical transformation temperature and then solidified by cooling without losing their solubility. The highest physical transformation temperatures are the melting temperature T_M of (semi)crystalline polymers and the glass temperature T_G of amorphous polymers. For practical applications, these characteristic temperatures must be substantially above the intended use temperature.

Thermoplastic polymers are usually of high molecular weight. They solidify (harden) by freezing-in thermal motions, either ordered as crystalline regions in semicrystalline polymers or unordered in amorphous ones.

Thermosetting resins are mostly prepolymers. They harden by chemical crosslinking reactions between their molecules, usually at elevated temperatures, whereupon they become **thermosets** since their shapes and properties are "set" by these processes. Upon renewed heating, they do not melt or soften but decompose chemically. Once "set," they can be neither remelted and reshaped nor dissolved; their formation is irreversible.

Fibers (Chapter 6) are characterized by large aspect ratios, i.e., ratios of lengths to diameters, and considerable flexibilities. Most synthetic fibers are thermoplastics but natural fibers are not. In any case, the highest physical transformation temperature (T_M or T_G) must be considerably higher than the intended use temperature.

Rubbers (see Chapter 7 for various definitions of "rubber") are uncrosslinked, elastic polymers above their glass temperatures. On fairly light chemical crosslinking, they convert to **elastomers**. Strong chemical crosslinking leads to thermosets. Thermoplastic elastomers owe their elasticity to physical crosslinking. Some authors reserve "rubber" for natural products and "elastomer" for synthetic ones.

Literature to Chapter 1

HISTORICAL DEVELOPMENT
R.Olby, The Macromolecular Concept and the Origin of Molecular Biology, J.Chem.Educ.
   4 (1970) 168
J.H.DuBois, Plastics History-U.S.A., Cahners Books, Boston (MA) 1972
J.K.Craver, R.W.Tess, Eds., Applied Polymer Science, Am.Chem.Soc., Washington (DC) 1975
F.M.McMillan, The Chain Straighteners: Fruitful Innovation. The Discovery of Linear and Stereo-
   regular Polymers, MacMillan, London 1981
R.B.Seymour, Ed., History of Polymer Science and Technology, Dekker, New York 1982
   (reprint of papers in J.Macromol.Sci.-Chem. **A 15** (1981) 1065-1460)
R.Friedel, Pioneer Plastic. The Making and Selling of Celluloid, University of Wisconsin Press,
   Madison (WI) 1983
W.H.Stockmayer, B.H.Zimm, When Polymer Science Looked Easy, Ann.Rev.Phys.Chem.
   35 (1984) 1
H.Morawetz, Polymers: The Origins and Growth of a Science, Wiley-Interscience, New York 1985
R.B.Seymour, G.A.Stahl, Eds., Genesis of Polymer Science, Am.Chem.Soc., Washington (DC)
   1985
R.B.Seymour, G.S.Kirshenbaum, Eds., High Performance Polymers: Their Origin and Develop-
   ment, Elsevier, New York 1986
D.A.Hounshell, J.K.Smith, Jr., The Nylon Drama, Americn Heritage of Invention and Technology
   4/2 (1988) 40
R.B.Seymour, Ed., Pioneers in Polymer Science, Kluwer, Boston (MA) 1989
J.Alper, G.L.Nelson, Eds., Polymeric Materials: Chemistry for the Future, Amer.Chem.Soc.,
   Washington (DC) 1989
S.T.I.Mossman, P.J.T.Morris, Ed., The Development of Plastics, Royal Society of Chemistry,
   Cambridge 1994
J.L.Meikle, American Plastic: A Cultural History, Rutgers Univ. Press, New Brunswick (NJ) 1995
S.Fenichell, Plastic-The Making of a Synthetic Century, Harper Business, New York 1996
Y.Furukawa, Inventing Polymer Science. Staudinger, Carothers, and the Emergence of Macro-
   molecular Chemistry, University of Pennsylvania Press, Philadelphia (PA) 1998
T.M.Brown, A.T.Dronsfield, P.J.T.Morris, In the Beginning ... was Ebonite, Educ.Chem.
   39/1 (2002) 18
H.-G.Elias, Macromolecules, Volume I: Chemical Structures and Syntheses, Section 1.3:
   Development of the Macromolecular Hypothesis, Wiley-VCH, Weinheim 2005
   Volume II: Industrial Polymers and Syntheses, Wiley-VCH, Weinheim 2007
E.A.Campo, Industrial Polymers, Hanser, Munich 2007

References to Chapter 1

[1]   U.S. Statistical Abstracts 2001
[2]   H.-G.Elias, An Introduction to Plastics, Wiley-VCH, Weinheim, 2nd ed. 2003, Fig. 1-2

2 Polymers

2.1 Chemical Structure

2.1.1 Constitution

The words matter, material, and substance are not officially defined. Not only do they have various meanings in common language, physics, and chemistry but the meaning may also vary within the common language or physics. In chemical parlance, **matter** (L: *materia* = lumber) always consists of a chemical substance or more than one. Matter is defined here as a space-occupying entity that has a mass; it differs from antimatter, vacuum, and energy.

A **chemical substance** (L: *substantia* = substance, essence), as defined by chemistry, is matter that consists of either the many atoms of a **chemical element** or the many molecules (composed of two or more atoms) of a **chemical compound**. The physical quantity of a chemical substance is the **amount of substance**, measured in the physical unit "mol", which is independent of the physical state, shape, etc., of the chemical compound.

No official definition seems to exist for **material** (in the physics sense) (L: *materia* = matter) but in general "material" designates a solid "of which a thing is or may be constructed" (The American Heritage Dictionary of the English Language). A material is therefore a subclass of matter; it usually comes in various **grades**.

In recent years, it has become custom in materials science to distinguish **soft matter** (polymers, liquid crystals, colloids, gels, foams, etc.) from **hard matter** (metals, ceramics, semiconductors, etc.). Soft and hard matter differ in their hardness, i.e., the mechanical resistance against the penetration by another body. The examples and the common meaning of these terms indicate that "soft matter" in this sense is really a misnomer for "soft materials" and "hard matter" a misnomer for "hard materials." Gases and liquids are certainly "soft matter" but definitely not "soft materials."

Constitution-based Polymer Terms

The meaning of the word "polymeric" (G: *polys* = several, many; *meros* = part) has changed many times since the term was introduced in 1831 by Jøns Jacob Berzelius (see Volume I, p. 7). Originally, "polymer" referred to two or more chemical substances that had the same relative composition (in the modern meaning of the term). Historic examples are ethene, C_2H_4, and 1-butene, C_4H_8, both of which have the relative composition of two hydrogen atoms per one carbon atom. These "polymeric" substances had low molecular weights, and when truly high-molecular weight substances were discovered much later they were therefore called "**high polymers**."

In 1920, Hermann Staudinger reviewed the status of so-called organic colloids and concluded that these kinds of matter consist of large molecules whose "units" are covalently interconnected. Because of the various meanings of the word "polymer" at that time (including non-covalently bound entities), he proposed to call these large molecules "macromolecules" (L: *makros* = large; G: *molecula* = diminutive of *moles* = mass). The designation "polymer" survived and prospered albeit mostly as a synonym for "macromolecule" and not, more correctly, in the sense of a "macromolecular substance."

According to the Commission on Macromolecular Nomenclature, Macromolecular Division, International Union of Pure and Applied Chemistry (IUPAC), "polymer molecule" and "macromolecule" are synonyms as are also "polymer" and "macromolecular substance." However, "polymer" is used in general only for those large molecules that consist of many polymeric units of the same type or very few types. On the other hand, protein molecules that consist of up to 22 types of encoded amino acid units and are true macromolecules are rarely called "polymer molecules."

In the simplest case, a polymer molecule consists of identical **chain units** which may be of the same type as the methylene unit $-CH_2-$ in poly(methylene), $+CH_2+_n$, of two types $-CH_2-$ and $-CHCl-$ as in poly(vinyl chloride), $+CH_2-CHCl+_n$, and of three types $-O-$, $-CH_2-$, and $-CH(CH_3)-$ as in poly(oxypropylene), $+O-CH_2-CH(CH_3)+_n$.

The smallest **constitutional unit** of such polymer molecules is the **repeating unit**. Examples of repeating units are $-CH_2-$ in poly(methylene), $-CH_2-CHCl-$ in poly(vinyl chloride), and $-O-CH_2-CH(CH_3)-$ in poly(oxypropylene), all with one monomeric unit each. The repeating unit $-NH(CH_2)_6NH-CO(CH_2)_4CO-$ of poly(hexamethylene adipamide) consists of two monomeric units. **Systematic names** of polymeric substances are based on the chemical constitution of repeating units (Volume I, p. 28); for example, nylon 66 has the systematic name poly(iminoadipoyliminohexamethylene).

Systematic names are cumbersome and rarely used, except for complex chemical structures. Commonly used are **generic names** which are the names of the **monomer**(s) (or **monomeric unit**(s) derived therefrom) that are prefixed by "poly." Repeating units, monomeric units, and monomers are therefore not necessarily identical (Table 2-1). "Monomeric unit" (a unit of a polymer molecule) is not identical with "monomer" (a chemical substance from which a polymer is prepared). A monomer in a polymer is an unwanted impurity and not a unit of a polymer molecule!

This volume is concerned with industrial aspects of polymers and uses therefore common names of polymers and not their systematic names. In a few cases, old names of monomeric units have been retained although monomers themselves are referred to by their new names. For example, the polymer of ethene (IUPAC Rule A-3.1; trivial name: ethylene) is still called poly(ethylene) because "ethylene" is the name of the constituting monomeric unit, $-CH_2CH_2-$.

For systematic reasons, names of *all* monomeric units are written in parentheses. Examples are poly(ethylene) instead of the more common polyethylene and poly-(ethylene terephthalate) instead of polyethylene terephthalate. Names of chemical groups are *not* written in parentheses, for example, polyamide and not poly(amide).

Table 2-1 Examples of polymer names, the constitution of their repeating units and monomeric units, and the constitution of monomers from which the polymers are derived.

Polymer name	Repeating unit	Monomeric unit(s)	Monomer(s)
Poly(methylene)	$-CH_2-$	$-CH_2-$	CH_2N
Poly(ethylene)	$-CH_2-$	$-CH_2CH_2-$	$CH_2=CH_2$
Poly(propylene)	$-CH_2CH(CH_3)-$	$-CH_2CH(CH_3)-$	$CH_2=CH(CH_3)$
Poly(hexamethylene adipamide)	$-NH(CH_2)_6NHCO(CH_2)_4CO-$	$-NH(CH_2)_6NH-$ + $-CO(CH_2)_4CO-$	$H_2N(CH_2)_6NH_2$ + $HOOC(CH_2)_4COOH$

Macromolecular substances may be **homochain polymers** with a single type of **chain atoms** or **heterochain polymers** with two or more types of chain atoms. Examples of homochains are poly(sulfur), $+S+_n$, with sulfur atoms as **chain atoms**, and poly-(ethylene), $+CH_2CH_2+_n$, with carbon atoms as chain atoms in **chain units** $-CH_2-$. Examples of heterochain polymers are poly(oxymethylene), $+OCH_2+_n$, with $-O-$ and $-CH_2-$ as chain units, and poly(ε-caprolactam), $+NH(CH_2)_5CO+_n$, with $-NH-$, $-CH_2-$, and $-CO-$ as chain units. The structure-related terms "homochain polymer" and "heterochain polymer" should not be confused with the process-related terms "homopolymer" and "heteropolymer" (see below).

The simplest polymer molecules have **linear chains** which are chains without branches such as those of poly(methylene) molecules (Fig. 2-1). In polymer science, **branches** are defined as side chains of linear chains that are produced by side reactions during polymerization, i.e., they are not defined topologically as in organic chemistry. For example, $CH_3-CH_2-CH(C_4H_9)-CH_3$ is a branched molecule in organic chemistry but $CH_3+CH_2-CH(C_4H_9)+_nCH_3$ is not a branched molecule in polymer science because the butyl group $-C_4H_9$ is already present in the monomer, $CH_2=CH(C_4H_9)$.

The designation "linear chain" has historic roots: the founder of macromolecular science, Hermann Staudinger, assumed that chain molecules such as poly(methylene) exist in solution as "totally stretched" chains, i.e., as zigzag chains (Fig. 2-1, bottom left). In reality, such macroconformations (see Section 2.2) are rare: they are only present as relatively short segments in certain semicrystalline polymers (Section 2.2) and not at all in dilute solution where such chains form random coils (Fig. 2-1, right). A "non-linear polymer", by the way, is not a branched polymer but a shorthand notation for a polymer with so-called non-linear optical properties (Section 15.3.2).

Linear chains may be present as open chains like the poly(ethylene) of Fig. 2-1 or as ring-shaped molecules (Fig. 2-2). **Ring polymers** are also known as **cyclic polymers.** However, a **cyclopolymer** is not a substance containing large ring-shaped molecules but a polymer with cyclic main chain groups that resulted from alternating intramolecular and intermolecular functional group reactions (see Volume I, p. 161).

Macromolecules with *one* electrical charge per molecule are called **macroions** and those with several charges per molecule, **polyions**. In a similar manner, one distinguishes **macrocations** and **polycations**, **macroanions** and **polyanions**, **macroacids** and **polyacids**, **macrobases** and **polybases**, and **macroradicals** and **polyradicals**.

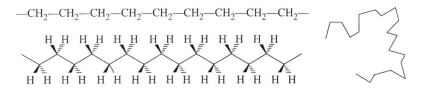

Fig. 2-1 Top left: constitution of poly(methylene) from the polymerization of diazomethane, CH_2N. Bottom left: section of the conformational structure of poly(methylene) in the all-trans macroconformation (present in the stems of chain-folded crystals). Right: projection of the three-dimensional macroconformation of the same chain segment on the paper plane (present in solutions and melts).

Process-based Terms

Polymerizations are chemical reactions that convert **monomer molecules** to **polymer molecules**. There are four basic types of polymerizations (see Volume I, p. 153 ff.):

chain polymerization $P_i{}^* + M \rightarrow P_{i+1}{}^*$; often called "addition polymerization" or "chain-growth polymerization." Example: polymerization of styrene by anions, cations, or free radicals

polyaddition $R_m + R_n \rightarrow R_{m+n}$; no traditional English name. Example: synthesis of polyurethanes

polycondensation $R_m + R_n \rightarrow R_{m+n} + L$; often called "step-growth polymerization". Example: synthesis of PA 6.6

polyelimination $P_i{}^* + M \rightarrow P_{i+1}{}^* + L$;   no traditional English name. Example: polymerization of *N*-carboxy anhydrides of α-amino acids with release of CO_2 $(= L)$

In **chain polymerizations**, an initiator molecule I* (radical I$^\bullet$, anion $A^\ominus$, cation $C^\oplus$, etc.) adds to a monomer molecule to form an active molecule IM*. In a chain reaction (hence: *chain* polymerization), the monomeric *end* of the active IM* *adds* more monomer molecules to yield an active polymer chain, $P_i{}^*$, consisting of *i* monomeric units. In **insertion** (chain) **polymerizations**, monomer is *inserted* between the catalyst and the growing chain. In these types of polymerization, no leaving molecules L are released.

Another type of chain reaction proceeds with the formation of leaving molecules L. IUPAC proposes to call this type of polymerization "condensative chain polymerization" because it resembles a chain polymerization on one hand (chain reaction) and a polycondensation on the other (formation of leaving molecules L). But there is no condensation between polymer molecules of various sizes. This book thus uses the term **polyelimination** for this rare type of polymerization.

In **polycondensations**, reactant molecules (monomers, dimers, trimers, etc., react with each other to form larger reactant molecules while leaving a by-product such as H_2O in the reaction of diamines and dicarboxylic acid to polyamides. The reacting molecules must be at least bifunctional, i.e., carrying at least two reactive functional groups.

Polyaddition is similar to polycondensation with respect to participating reactant molecules albeit without formation of leaving molecules. The only large application of this type of polymerization is for the syntheses of polyurethanes and related polymers. Unfortunately, the term "polyaddition" sounds too similar to "addition polymerization."

Polymerizations are also distinguished according to the medium in which they take place: **bulk polymerization (polymerization in mass, neat polymerization), solution polymerization**, **dispersion polymerization** (in W/O emulsion), **emulsion polymerization** and **suspension polymerization** (both in O/W emulsion), **solid-state polymerization**, and **gas-phase polymerization** (see Volumes I and II). Polymers from these polymerizations are known as **bulk polymers, solution polymers, suspension polymers**, etc.

Polymers are called **homopolymers** if they are generated from a single type of monomer and **copolymers** if they are produced by **copolymerization** of two *or more* types of monomers. "Homopolymer" and "copolymer" are thus process-oriented terms and not structure-related ones. The term "copolymerization" is used only for chain polymerizations; AABB polycondensations of monomers AA (such as hexamethylenediamine) and BB (such as adipic acid) are *not* called copolycondensations.

Bi-, Ter-, Quater-, and **Quinterpolymerizations** are copolymerizations of 2, 3, 4, or 5 types of monomers. However, literature often uses "copolymer" as a synonym for "bipolymer" and "copolymerization" as a synonym for "bipolymerization."

Periodic copolymers are copolymers with regular successions of sequences with two or more types of monomeric units a, b, c, etc., for example ...(ab)$_n$..., ...(abc)$_n$..., and ...(abb)$_n$ ... The special case ...(ab)$_n$... is an **alternating bipolymer.** Another special case are **block polymers** in which long "blocks" of the same type of monomeric units are interconnected via their ends; examples are a$_n$b$_m$, a$_n$b$_m$a$_n$, and a$_n$b$_m$c$_p$. IUPAC defines **block *co*polymers** as block polymers that result from the copolymerization of two or more types of monomers whereas **block polymers** are prepared by other means, for example, by connecting molecules of two independently prepared homopolymers. **Multiblock (co)polymers** with many short block lengths n, m, p, etc., are known as **segment (co)polymers** or **segmented polymers**.

Gradient copolymers consist of copolymer molecules with a systematic variation of the composition of segments from one end to the other. **Statistical copolymers** are copolymers with statistical distributions of the length of sequences of like monomer units along their chains (Markov chains of zeroth, first, second, ... order). Statistical bipolymers with a Bernouillian distribution (Markov zeroth order) of lengths of sequences of monomer units are truly **random bipolymers**.

Polymer molecules may not only be linear but also cyclic (ring-shaped) or branched (Fig. 2-2). **Comb polymers** are composed of branched molecules that consist of molecules with a main chain to which longer side chains of equal or different lengths are attached at equal or variable distances. They are usually produced by grafting of monomeric units onto or from an existing polymer chain (**graft (co)polymers**). A special type of comb polymer results from the homopolymerization of monomer molecules with very long side chains, such as $CH_2=CH(COO(CH_2)_{50}CH_3)$.

In **star molecules,** three or more arms of equal or unequal length radiate from a central core. The arms may consist of different types of monomeric units (**miktoarms**). **Dendrimers** are star molecules with regular branches upon branches, called generations. Dendrimers with up to nine generations of branches have been synthesized.

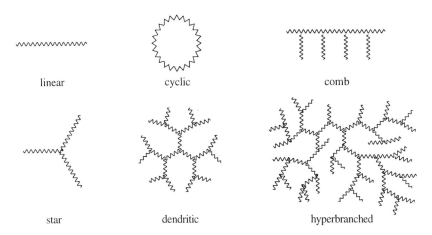

linear	cyclic	comb

star	dendritic	hyperbranched

Fig. 2-2 Structures of linear and branched macromolecules (schematic). Hyperbranched polymers have random branches-upon-branches whereas random ones rarely have secondary branches (not shown).

Statistically branched polymers (randomly branched polymers) consist of molecules with irregular branches, i.e., different lengths of chain segments between branching points; they rarely have secondary branches and vary rarely tertiary ones.

Hyperbranched polymers are similar albeit with chain segments of equal length between branching points and random branch-upon-branch structures (Fig. 2-2).

Statistically crosslinked polymers result from either post-polymerization intermolecular reactions of uncrosslinked polymer molecules (for example, the crosslinking of natural rubber by sulfur) or from the polymerization of polyfunctional monomer molecules, i.e., molecules with three or more functional groups. Examples are the free-radical polymerization of the tetrafunctional *o*-diallyl phthalate, o-$C_6H_4(COOCH_2CH=CH_2)_2$, and the polyaddition of tetrafunctional polyols, $X(OH)_4$, to difunctional diisocyanates, $Z(NCO)_2$.

The crosslinking polycondensation of low-molecular weight compounds to highly crosslinked **thermosets** (Section 8.5) is a technology that was developed empirically long before the macromolecular character of the resulting crosslinked polymers was recognized. For historical reasons, the starting materials are here called **thermosetting resins** (see Volume II, p. 78 ff.). The conversion of soluble resins to insoluble thermosets by condensation proceeds in three steps which are distinguished as A–, B–, and C–stages.

The vast majority of uncrosslinked polymers are soluble and usually meltable without decomposition. Such polymers can be processed in their molten states, solidified on cooling, and molten again: they are **thermoplastics.**

Molecular Weights

Monomers are low-molecular weight chemical substances while polymers are those with high molecular weights. **Oligomers** are low-molecular weight polymers, usually those with less than ca. 20 monomeric units per molecule. In industry, crosslinkable oligomers (and sometimes also monomers) are known as **prepolymers**.

The determination of molecular masses with chemical methods such as the number of endgroups per mass of *linear* macromolecules delivers **relative molecular masses** M_r. These dimensionless quantities are traditionally known as **molecular weights**. Physical methods such as membrane osmometry or static light scattering do not deliver molecular weights but **molar masses** with the physical unit mass per amount-of-substance, for example, g/mol (see Volume I, p. 84 ff., and Volume III, p. 16 ff).

The **degree of polymerization**, X, indicates the number of *repeating units* (*not* monomeric units!) per polymer molecule. This theoretical quantity cannot be determined experimentally. In literature, it is often symbolized by *DP* but such two-letter symbols are reserved by ISO for dimensionless quantities such as the Reynolds number.

With few exceptions (such as enzymes and some oligomers), macromolecular substances are **molecularly non-uniform** since their molecules differ with respect to the number of monomeric units and thus also in their molecular masses. Hence, the polymer (= macromolecular substance) has a **molecular weight distribution** and **molar mass distribution**, respectively.

Molecularly non-uniform polymers are often called "polydisperse" in the literature. However, "disperse" is used exclusively in natural science for the distribution of matter in another matter, i.e., for *multiphase* systems (L: *dispergare* = to distribute). "Monodisperse" is an oxymoron since "disperse" relates to "more than one" and it cannot be "mono" as it is already multiple ("disperse").

Because of molar mass distributions of polymers, various experimental methods deliver different **molar mass averages**. Membrane osmometry leads to **number-average molar masses**, $\overline{M}_n$, whereas static light scattering furnishes **mass-average molar masses**, $\overline{M}_w$ (but neither *mass*-average molecular *weights* nor *weight*-average molar *masses*).

These two molar mass averages are defined as

$$(2\text{-}1) \qquad \overline{M}_n \equiv \frac{\sum_i n_i M_i}{\sum_i n_i} = \frac{\sum_i N_i M_i}{\sum_i N_i} = \sum_i x_i M_i$$

$$(2\text{-}2) \qquad \overline{M}_w \equiv \frac{\sum_i m_i M_i}{\sum_i m_i} = \sum_i w_i M_i = \frac{\sum_i n_i M_i^2}{\sum_i n_i M_i}$$

where N_i = number, n_i = amount (in moles), x_i = mole fraction, m_i = mass, w_i = mass fraction (weight fraction) of *molecularly uniform* molecules with the molar mass M_i (for a detailed discussion, see Volume III, Section 2.3). Endgroups are usually neglected.

Industrial polymers are usually not characterized by absolute molar masses but by physical quantities that depend on molar masses. Most often used are viscosity-related quantities (Section 4.2). Intrinsic viscosities $[\eta]$ of dilute solutions are in general use for all types of polymers. Thermoplastics are usually characterized by the melt-flow index MFI or the melt-volume index MVI and elastomers by so-called Mooney viscosities.

2.1.2 Chemical Configuration

The word "configuration" has different meanings in physics and chemistry. In physics, the "configuration" of a polymer denotes the general spatial arrangement of atoms and groups of atoms of a chain. Chemistry restricts "configuration" to high energy barriers for the interconversion of spatial arrangements of groups (this Section) and "conformation" to low ones (Section 2.2) (for details, see Volume I, Chapters 4 and 5; Volume III, Section 2.2 and Chapter 3).

Adjacent chain units with carbon atoms as chain atoms display **stereoisomerism** because of the tetrahedral arrangement of valence bonds around the carbon atom: (chemical) configurational isomerism around a carbon atom, torsional isomerism around a carbon-carbon double bond, and conformational isomerism with respect to two or more carbon chain atoms.

Polymer chemistry is interested in the *relative* configuration about stereogenic centers, the so-called **tacticity** (G: *taktikos* = ordered). For example, poly(1-olefin)s with the constitutional repeating unit $-$*CHR$-$CH$_2-$ have one stereogenic center * per repeating unit. Chains with the same relative configuration of all stereogenic centers are called **isotactic** (G: *isos* = equal) and chains with alternating relative configurations, **syndiotactic** (G: *syn* = together; *dios* = two) (Fig. 2-3). In *Fischer projections* of chains, all substituents are either on the same side (isotactic) or opposite sides (syndiotactic).

isotactic poly(1-olefin) syndiotactic poly(1-olefin)

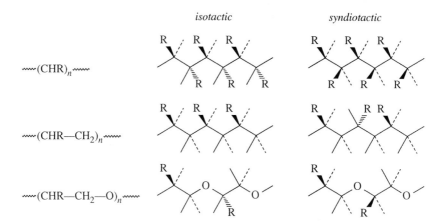

Fig. 2-3 Segments of isotactic (left) and syndiotactic (right) polymer chains in hypothetical all-trans-conformations. From top to bottom for R = CH_3: poly(methylmethylene), poly(propylene), and poly-(oxypropylene). Compare the relative positions of substituents R in chains with all-trans conformations with those in all-cis conformation (bottom of page 15).

For *stereoprojections* of isotactic chains in all-trans conformation (i.e., zigzag chains), the "same side" rule applies only to monomeric units with even numbers of chain atoms, for example, $+CHR–CH_2+_n$ or $+CHR–O+_n$ (Fig. 2-3). However, substituents R of isotactic chains are on opposite sides of chains if the chains consist of monomeric units with of odd numbers of chain atoms such as $+CHR+_n$ (one chain atom per monomeric unit), $+CHR–CH_2–O+_n$ (three chain atoms per monomeric unit,), etc.

Industrial tactic polymers are usually not completely tactic. In science, degrees of tacticity are usually characterized by the mole fractions of, for example, isotactic diads (two stereocenters), triads (three stereocenters), etc., which are obtained from NMR spectroscopy. In industry, isotactic poly(propylene)s are characterized by their **isotacticity index**, which is the fraction of polymer that is insoluble in boiling xylene. This index thus does not measure isotacticity *per se* but the fraction of polymer molecules in insoluble crystallites (which is related to tacticity). Polymers are never truly 100 % tactic; the highest stereoregularities of industrial polymers are usually 96 % or less.

Truly **atactic** polymers have a Bernoullian (= Markov zeroth order) distribution of tactic J-ads, i.e., mole fractions $x_i = x_s = 0.5$ of isotactic (i) and syndiotactic (s) diads, $x_{ii} = x_{is} = x_{si} = x_{ss} = 0.25$ of triads (i.e., $x_{is} + x_{si} \equiv x_{ht} = 0.5$ of heterotactic triads), etc. However, the literature uses the term "atactic" usually as "not predominantly tactic."

Torsional stereoisomers (= geometric isomers) have tactic structures around carbon-carbon double bonds (Fig. 2-4). Poly(butadiene)s (R = H) may have 1,4-structures in cis (ct) or trans (tt) configurations of carbon-carbon double bonds as well as 1,2-structures with either isotactic (it) or syndiotactic (st) diads. Poly(isoprene)s (R = CH_3) may have the same 1,4 and 1,2 structures as well as 3,4 ones.

Despite the presence of centers of stereoisomerism and helical chain structures produced by them (see below), isotactic polymers are not optically active since polymer chains are present in equal amounts of left-handed and right-handed helices. "Optical activity" means either "rotation of the plane of polarized light" (chemistry; Volume III, Section 3.4) or "non-linear optical properties" (photonics, this Volume, Section 13.3).

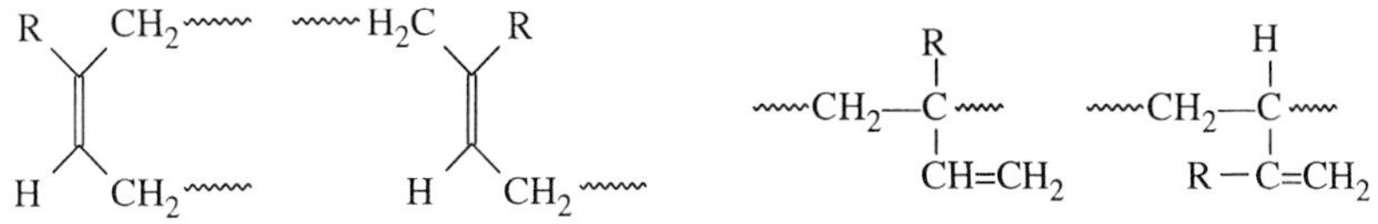

R   CH₂ ⌇⌇⌇   ⌇⌇⌇ H₂C   R

 ⟩—⟨   ⟩—⟨

H   CH₂ ⌇⌇⌇   H   CH₂ ⌇⌇⌇

1,4-cis-   1,4-trans-
  (ct)   (tt)

              R   H
              | |
      ⌇⌇⌇CH₂—C ⌇⌇⌇   ⌇⌇⌇CH₂—C ⌇⌇⌇
              | |
          CH=CH₂   R—C=CH₂

            1,2-   3,4-
         (it or st)   (it or st)

Fig. 2-4 Constitution and configuration of monomeric units from $CH_2=CR–CH=CH_2$.

2.2 Physical Structure

2.2.1 Conformation

Substituents R in chain units –CHR–, –NR–, etc., can rotate around chain bonds and adopt time-averaged preferred energetic conformations (organic chemistry) and **micro-conformations** (macromolecular chemistry), respectively, relative to adjacent substituents (Volume I, Chapter 5; Volume III, Chapter 3). These preferential conformations carry different names in organic and macromolecular chemistry (Fig. 2-5); names of angles are also different.

In crystalline states, polymer segments may be present in regular sequences of micro-conformations, for example, in the all-trans **macroconformation** (zigzag chain) of poly-(ethylene) (Figs. 2-1 and 2-6). Because of strong repulsion between the bulky methyl groups of isotactic poly(propylene), every second methyl group is in a gauche micro-conformation instead of trans, leading to a sequence ...TGTGTGTG.... For steric reasons, all subsequent gauche positions must have the same sign in a given chain. Since G^+ and G^- are energetically equal, a crystalline it-poly(propylene) contains equal amounts of chains in ...TG$^+$TG$^+$TG$^+$TG$^+$... and ...TG$^-$TG$^-$TG$^-$TG$^-$... sequences.

These regular sequences force a poly(propylene) chain to adopt a certain screw sense, resulting in helical structures with opposite screw directions for the two types of macro-conformations, $(TG^+)_i$ and $(TG^-)_i$. Isotactic poly(propylene) forms 3_1 helices (3 propy-lene units required for a complete turn), isotactic poly(1-butene) exists in 4_1 helices, poly(oxyethylene) in 7_2 helices, and so on.

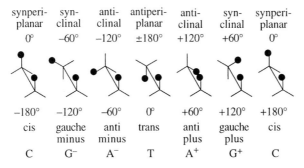

synperi-planar	syn-clinal	anti-clinal	antiperi-planar	anti-clinal	syn-clinal	synperi-planar
0°	−60°	−120°	±180°	+120°	+60°	0°

−180°	−120°	−60°	0°	+60°	+120°	+180°
cis	gauche minus	anti minus	trans	anti plus	gauche plus	cis
C	G^-	A^-	T	A^+	G^+	C

Fig. 2-5 Names of and conventions for microconformations of molecules $RCH_2–CH_2R$ (● = R). Top: organic chemistry, bottom: macromolecular chemistry.

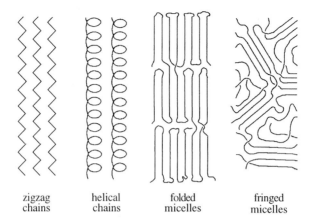

zigzag   helical   folded   fringed
chains   chains   micelles   micelles

Fig. 2-6 Macroconformations of chain sections of linear macromolecules. The straight lines of folded and fringed micelles represent both zigzag chains and helices.

In real polymers, the lengths of straight zigzag and/or helical segments are limited by constitutional, configurational, and/or conformational mistakes, endgroups, non-equilibrium states, and the like. Real polymers are therefore always semicrystalline. The degree of (semi)crystallinity varies with the experimental method for their determination (Volume III, Section 7.5) because each method detects various degrees of regularity.

In solution, helices and/or helical sections survive only if either strong attraction forces are present between substituents or such severe steric hindrances exist that the solvent has no chance to interact with the backbone of the chain. Most dissolved chain molecules consist therefore of irregular sequences of microconformations that sometimes may or may not include short helical sections. As a result, such dissolved polymer chains adopt the shape of a so-called random coil (Volume III, Chapter 4).

Random coils exist in various types, depending on the range of interactions between segments of the same polymer molecule. These interactions may be short-range if they exist between nearby segments (such as hydrogen bonds between adjacent monomeric units) or long-range (such as between monomeric units that are separated from each other by many monomeric units along the chain). Larger short-range interactions stiffen chain segments, which then have a certain persistence with respect to the spatial direction of segments: the chain becomes wormlike. Repulsive long-range interactions exclude the space taken by these segments to all other segments: the chain becomes perturbed (as compared to an infinitely thin chain) and the molecule has an internal excluded volume. These excluded volumes disappear if repulsive and attractive forces of long-range interactions cancel each other; the coil is then unperturbed (see Volume III, Chapter 4). Such coils exist in melts and in so-called theta solvents.

2.2.2 Morphology, Transitions, and Relaxations

In crystalline polymers, chain segments may be in zigzag (all-trans) or helical (trans-gauche) conformations. Segments of the same chain and/or other chains are bundled in folded or fringed micelles (Fig. 2-6) which, in turn, form larger aggregates such as

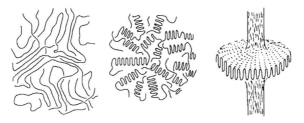

Fig. 2-7 Schematic representation of some polymer morphologies. Left: fringed micelle; center: cross-section of a spherulite formed in bulk; right: shashlik structures may form on crystallization from flowing polymer solutions). Straight lines symbolize either zigzag chains or helices.

spherulites (Fig. 2-7). Depending on molecule parameters (molar mass, stiffness of chains), molecule interactions) and external conditions (cooling rate, flow, etc.), solid polymers may have very different morphologies which then lead to very different polymer properties. An example is the formation of so-called shashliks or shish kebabs (Fig. 2-7).

On heating above their **melting temperatures** T_M, long-range ordered structures of semicrystalline polymers dissolve. The resulting melt consists of interpenetrating random coils (Section 5.2.2 and Volume III, Section 13.3). The space within one coil is filled with segments of other coils. Segments of sufficiently long chains can therefore entangle and form temporary physical networks above a critical **entanglement** molar mass of M_{ent}. The entanglements impede the viscous flow of polymer melts so that the melt viscosity η of entangled polymers is much more strongly dependent on the molar mass ($\eta \sim M^{3.4}$) than that of non-entangled polymers ($\eta \sim M$) (p. 88 ff.).

Solid non-crystalline (amorphous) polymers are glass-like without long-range order of polymer segments. On heating such polymers, chain segments begin to move. At the so-called **glass temperature** T_G, glassy polymers convert to melts. The conversion glass → melt is a relaxation process but is neither a thermodynamic first-order *transition* like the melting of crystallites nor a second-order thermodynamic transition like that of smectic liquid-crystalline phases to nematic liquid-crystalline ones (Volume III, p. 448). T_G is therefore not a glass *transition* temperature but may be called a glass *transformation* temperature or just a glass temperature (see Volume III, Section 13.5). On cooling, melts of non-crystallizing polymers solidify, usually without morphological changes. Like the melt, the resulting amorphous polymers consist of networks of entangled chains.

Semicrystalline polymers may have both crystalline and non-crystalline regions. Therefore they have both glass and melting temperatures ($T_G < T_M$). Both semicrystalline and amorphous polymers may also exhibit other transformation temperatures. The segmental, molecular, and/or supramolecular processes leading to these transformations are mostly unknown.

Liquid crystalline polymers exhibit thermodynamic transition temperatures between the various liquid-crystalline phases (smectic, nematic, cholesteric) and an isotropization temperature between the nematic structure and the melt (Volume III, p. 450).

Literature to Chapter 2

HANDBOOKS
G.Allen, J.C.Bevington, Eds., Comprehensive Polymer Science, Pergamon, Oxford, 7 vols. (1989),
    First Supplement (1992), Second Supplement (1996)
J.E.Mark, Ed., Physical Properties of Polymers Handbook, AIP Press, Williston (VT), 1996
J.C.Salamone, Ed., Polymeric Materials Encyclopedia, CRC Press, Boca Raton (FL) 1996, 12 vols.;
    short version: Concise Polymeric Materials Encyclopedia, CRC Press, Boca Raton (FL) 1998
    (1 volume)
E.S.Wilks, Ed., Industrial Polymers Handbook (relevant chapters of Ullmann's Encyclopedia of
    Industrial Chemistry, 5th ed.), Wiley-VCH, Weinheim 2001 (5 vols.)
H.F.Mark, Encyclopedia of Polymer Science and Engineering, Wiley, Hoboken (NJ), 3rd ed. (2003-
    2007), 12 volumes
J.L.Atwood, J.W.Steed, Eds., Encyclopedia of Supramolecular Chemistry, Dekker, New York 2004
-, Encyclopedia of Polymer Science and Engineering, Concise 3rd ed., Wiley-Interscience, Hoboken
    (NJ) 2007

DATA COLLECTIONS
J.E.Mark, Ed., Physical Properties of Polymers Handbook, AIP Press, Woodbury (NY) 1996
N.A.Waterman, M.F.Ashby, Eds., The Materials Selector, Chapman & Hall, Boca Raton (FL) 1999
    (CD-ROM)
J.E.Mark, Ed., Polymer Data Handbook, Oxford Univ. Press, New York 1999
N.A.Waterman, M.F.Ashby, Eds., The Materials Selector, Chapman & Hall, Boca Raton (FL) 1999
D.J.David, A.Misra, Relating Materials Properties to Structure. Handbook and Software for Polymer
    Calculations and Materials Properties, Technomic Publ., Lancaster (PA) 2000 (databases,
    computational programs)
B.Ellis, Polymers. A Property Database, CRC Press, Boca Raton (FL), 2nd ed. 2008 (CD-ROM)
-, The Wiley Database of Polymer Properties, Wiley, Hoboken (NJ) 2002 (selected data from
    J.Brandrup, E.H.Immergut, E.A.Grulke, Eds., Polymer Handbook, Wiley, New York, 4th ed.,
    1 volume (1999) = 2 volumes (2004)
-, CAMPUS®, diskettes with properties of plastics that are produced by ca. 30 international compa-
    nies belonging to the CAMPUS system
-, PLASPEC, Fachinformationszentrum Chemie, Berlin (data bank, ca. 7000 plastics)
-, Plastics Databases, ASM International, Materials Park (OH)
POLYMERIS™ (includes calculations of polymer properties by the methods of Bicerano and
    van Krevelen), Scivision, Burlington (MA)

DICTIONARIES
T.Whelan, Polymer Technology Dictionary, Chapman and Hall, London 1994
M.S.M.Alger, Polymer Science Dictionary, Chapman and Hall, London, 2nd ed. 1997
V.Beddoes, The Polymer Lexicon, RAPRA, Shawbury, UK, 1998 (abbreviations)
H.Keller, U.Erb, Dictionary of Engineering Materials, Wiley-Interscience, Hoboken (NJ) 2004

PREDICTION OF POLYMER PROPERTIES
D.W. van Krevelen, Properties of Polymers–Correlation with Chemical Structure, Elsevier,
    Amsterdam, 1st ed. 1969, 2nd ed. 1976, 3rd ed. 1990
J.Bicerano, Prediction of Polymer Properties, Dekker, New York, 3rd ed. 2002
A.A.Askadskii, Computational Materials Science of Polymers, Cambridge Int.Sci.Publ.,
    Cambridge 2003

POLYMER SYNTHESES
    For PRINCIPLES, see Volume I (2005); for INDUSTRIAL PROCESSES, see Volume II (2007).
H.R.Kricheldorf, O.Nuyken, G.Swift, Eds., Handbook of Polymer Synthesis, CRC Press, Boca
    Raton (FL), 2nd Ed. 2005
K.Matyjaszewski, Y.Gnanou, L.Leibler, Eds., Macromolecular Engineering. Precise Synthesis,
    Materials Properties, Applications, Wiley-VCH, Weinheim, 4 vols. 2007

POLYMER PROPERTIES
    For the PHYSICS and PHYSICAL CHEMISTRY, see Volume III (2008)

3 Polymer Adjuvants

3.1 Overview

3.1.1 Introduction

Solid polymers are usually not used as such but practically always in combination with **adjuvants**. Rare examples of polymeric materials without adjuvants are packaging films for foodstuffs (poly(ethylene)) and water purification membranes (aramides).

The vast majority of polymers serves as raw materials for polymer grades that contain adjuvants such as fillers, plasticizers, antioxidants, and so on. The term **polymer grade** is not officially defined. Like the also used terms "polymer type" and "polymer quality", it denotes a polymeric material that differs from other polymeric materials with the same monomeric units by molar mass, molar mass distribution, degree of branching, type and amount of adjuvants, and so on.

For the purpose of this book, a polymer **adjuvant** is defined as an agent that is added to a polymer in order to assist in the processing of the polymer to semi-finished goods and final articles and/or to modify or improve its properties (from *adjuvare* (L) = to assist, aid)). The term "adjuvant" has been long used in biomedicine but is relatively new to the polymer field. Terms similar to "adjuvant" have been used in several subfields (plastics, elastomers, fibers, coatings, adhesives, etc.) where one encounters terms like additives, auxiliaries, modifiers, etc., albeit with different meanings from subfield to subfield. The term "adjuvant", as used here, comprises **additives** that do not markedly change mechanical properties and **modifiers** that do. Additives are usually added in amounts of less than 5 % and modifiers in larger amounts.

Adjuvants comprise large groups of chemical compounds such as dyestuffs (Section 3.2), fillers and reinforcing agents (Section 3.3), processing aids (Section 3.5), and adjuvants for improving chemical (Section 3.6) and physical properties (Section 3.4).

Adjuvants may affect mechanical, electrical, and/or optical properties of polymers, provide polymer articles with a pleasant appearance, and/or extend the lifetime of polymers. For example, fillers, plasticizers, nucleation agents, and blowing agents affect bulk mechanical properties; antioxidants, metal deactivators, light stabilizers, and flame retardants the chemical ageing; lubricants, release agents, and nucleating agents the processing; and dyestuffs, pigments, and brightening agents the appearance.

The group of adjuvants sometimes includes also peroxidic crosslinking agents for unsaturated polyesters and similar prepolymers, activators for crosslinking agents for elastomers, and chemicals for wash-and-wear finishes of textiles.

Not included in the group of adjuvants are polymers that are added to other polymers in larger proportions such as elastomers for rubber-reinforced plastics (Section 10.4.7) or the combination of two polymers in multicomponent fibers (see Chapter 6). Also not included are extenders and diluting agents for lacquers (see Chapter 12).

The field of adjuvants is very large and diversified. Types and percentages of of additives, modifiers, etc., used vary from industry to industry. The percentage of adjuvants in polymers is largest in elastomers (on average 40 %), lesser in plastics (ca. 10-15 %) and least in textile fibers (ca. 3-5 %) (Table 3-1). However, the cost of adjuvants per mass of plastic may be similar to that of the polymer!

Table 3-1 Types, commonly used weight fractions w, and annual global consumption of adjuvants for plastics, elastomers, and fibers in 1988. The proportion of adjuvants per raw polymer does not change much from year to year. 0: not used; empty space: unknown data, if any.

Use and type of adjuvant	$w/\%$	Consumption in 1000 t a^{-1} in		
		Plastics	Elastomers	Fibers
Processing				
Processing stabilizers				
Heat stabilizers	0.1 - 1	120	0	0
Processing aids				
Lubricants, factices, etc.	0.5 - 5	200	258	250
Polymeric adjuvants		2		0
Release agents		2		0
Nucleation agents		2	0	0
Vulcanizing and crosslinking agents				
Initiators, accelerators		50	251	0
Sulfur		0	215	0
Metal oxides (ZnO, etc.)		0	599	0
Fatty acids		0	196	0
Application				
Stabilizers				
Antioxidants	0.1 - 2	80	157	
Metal deactivators		2		
Light stabilizers	0.1 - 1	12		
Flame retardants		300	45	
Biocides		4		
Modifiers				
Dyestuffs	0.02 - 3	9	0	1 000
Color pigments (without carbon black)		450		125
Carbon black		100	5 000	0
Fillers, mineral	≈ 30	3 000	2 400	0
organic		500	0	0
Fibers, glass		1 300	0	
organic (including textile fibers)	4	4 600	0	
Plasticizers	< 60	3 100	1 427	0
Resins		-	105	
Blowing agents (without gases)		35	25	0
Impact modifiers		130	0	0
Antistatics		10		
Other		15		
Total Consumption of Polymer Adjuvants		*14 023*	*10 878*	
Total Consumption of Polymers		*93 800*	*15 100*	*39 450*

The process of mixing adjuvants with polymers is referred to differently in different fields. In the plastics industry, the combining of polymers and fillers, reinforcing agents, plasticizers, catalysts, colorants, antioxidants, etc., is called **compounding** and the resulting admixture, a (physical) **compound** (not to be confused with "chemical compound").

The same process is called **formulating** in the rubber, paint, and adhesive industries; a "compound" in the paint industry is a lacquer that is liquefied by melting and not by addition of solvents. Rubber compounding is done by rubber processors whereas compounding of polymers for plastics with adjuvants is usually done by specialized companies (**compounders**) and neither by polymer producers nor by plastics processors.

Additives (adjuvants) are sold as individual species or as **additive systems** which are adjusted combinations of the various effectors. Additive systems may be simple mixtures of adjuvants or concentrates of such mixtures in a polymer as carrier, so-called **master batches**. A master batch of a colorant in a polymer is called **color concentrate**. Master batches and color concentrates allow easy dosage of their additives by calculating the **let-down ratio** = weight of resin divided by weight of master batch (concentrate).

The following sections describe properties of adjuvants and also very often their effect on the properties of polymeric materials. In other cases, it is more efficient to treat the action of adjuvants in separate chapters such as the effect of fillers and reinforcing agents in plastics and elastomers in Chapter 9 and that of polymer blends, plasticizers, and blowing agents in Chapter 10. Supplementing information can also be found in the main chapters on plastics (Chapter 8), elastomers (Chapter 7), fibers (Chapter 6), and in the special chapters on thickeners (Chapter 11), coatings and adhesives (Chapter 12), packaging materials (Chapter 13), electrical/electronic uses (Chapter 14) and optical and optoelectronic applications (Chapter 15).

3.1.2 Compounding

Compounding of polymers comprises the working in of adjuvants as well as the homogenization and granulation or pelletizing of the resulting products.

Discontinuous polymerization of monomers leads to polymers that differ somewhat from **batch** to batch. Powders of such batches are then mixed in order to deliver **grades** with constant specification to the customer. This **microhomogenization** is distinguished from the **macrohomogenization** of granulates and similar prefabricates.

Microhomogenization of batches of polymers with different molar masses and molar mass distributions delivers grades with broader molar mass distributions. The ratio of mass-average and number-average molar masses of a binary mixture,

(3-1)

$$\frac{\overline{M}_w}{\overline{M}_n} = w_A \left(\frac{\overline{M}_w}{\overline{M}_n}\right)_A \left[\frac{w_B + w_A\{\overline{M}_{n,B}/\overline{M}_{n,A}\}}{\{\overline{M}_{n,B}/\overline{M}_{n,A}\}}\right] + w_B \left(\frac{\overline{M}_w}{\overline{M}_n}\right)_B \left[w_B + w_A\{\overline{M}_{n,B}/\overline{M}_{n,A}\}\right]$$

is controlled by three quantities: the mass fraction w_A (or $w_B = 1 - w_A$) of the polymers of the two batches A and B, the ratios $(\overline{M}_w/\overline{M}_n)_i$ of mass-average and number-average molar mass of each batch, and the ratio $\overline{M}_{n,B}/\overline{M}_{n,A}$ of the number-average molar masses of the two batches.

A 1:1 mixture of two polymers (batches) with unimodal Schulz-Flory distributions of molar masses, i.e., $(\overline{M}_w/\overline{M}_n)_i = 2$, delivers for $\overline{M}_{n,B}/\overline{M}_{n,A} = 2$ a polymer grade with a bimodal molar mass distribution and a mass-number molar mass ratio ("polydispersity") of $\overline{M}_w/\overline{M}_n = 2.25$.

Adjuvants may be powders or pellets, tough elastomers, or liquids. Depending on their viscosities and that of the polymeric material, they are worked in by agitators, mixers, kneaders, or calenders. The working in of adjuvants in polymeric raw materials is called **mixing, blending, dispersing**, or **compounding** in plastics technology; **mixing** or **blending** in the rubber field; **dyeing** (dyestuffs) or **finishing** in the fiber and textile business; and **diluting, extending**, or **adulterating** in lacquer manufacturing.

Most of these terms are not defined exactly and their use varies also from field to field. In physical chemistry, "mixing" refers to processes which lead to distributions of molecules. In technology, "mixing" also refers to the uniting of powders which are particles consisting of, for example, molecules. "Blending" in the narrow sense refers to the removal of the air layer around polymer particles but "blending" in the wider sense is the mixing of two types of matter with similar rheological properties, such as two rubbers, a rubber and a thermoplastic, or two whiskies ("adulterating" if expensive whisky is diluted by a cheaper one; the same term is used in the lacquer industry). Correspondingly, a mixture of polymers that are molecularly distributed is called a "homogeneous blend" in the polymer industry.

Dispersing refers to the distributing of powders or liquid droplets in liquids. **Compounding** is therefore sometimes defined as very intensive dispersing that removes voids in polymers.

Powdery or liquid adjuvants can be worked into powdery or liquid polymers by paddle mixers, turbine impellers, ball mills, and the like; the mixing process requires 35-100 kJ/kg. Much more energy per mass is used to work in adjuvants in thermoplastics (ca. 700 kJ/kg) or rubbers (ca. 1500 kJ/kg). The latter processes require heavy machinery such as kneaders, double-screw extruders, Banbury mixers, or calenders.

The working in of fibers into polymers depends on the type of polymer and the length of the fibers. Short glass fibers (US: fiberglass) are either **chopped strands** with mean fiber lengths of $L = 4.5$ mm and aspect ratios of $L/d = 320$ or **milled fibers** with $L = 0.2$ mm and $L/d = 20$. They are mixed with powdery thermoplastics or resins and the mixture is extruded and granulated. These processes and the subsequent fabrication of articles by molding, extruding, etc., reduces the length of chopped fibers to ca. 0.65 mm and the aspect ratio to ca. 47 whereas the dimensions of milled fibers remain the same.

Long glass fibers are first impregnated with the liquid thermoplastic. The impregnated fibers are then cut to 6-12 mm length and worked into the thermoplastic. Mats of long fibers are impregnated with liquid thermosetting resins and then cured (Section 4.4.3).

Fibers are worked into rubber tires differently. Parallel steel fibers are coated on both sides with the rubber mixture in a calender whereas textile fibers are used as fabrics.

Mixing of poly(vinyl chloride) powders and adjuvants in a conventional mixer delivers a **heterogeneous compound** (**premix**) that cannot be processed directly to PVC articles. High-intensity mixers deliver free-flowing (pulverulent) PVC mixtures; these **dry blends** can be processed directly by injection molding or extrusion.

3.1.3   Drying and Devolatilization

After the working in of adjuvants (especially fillers and pigments), volatiles have to be removed from melts and granulates, either at atmospheric pressure or, more effectively, under a vacuum of 5-10 kPa (= 50-100 mbar). Devolatilization removes not only occluded gases such as air but also moisture and other volatiles such as residual monomers, volatile oligomers, volatile sizes, etc. (see also p. 78).

All these impurities would interfere with the wetting of dispersed particles (fillers, fibers, etc.) by the melt. The remaining volatile components would also generate insufficient mechanical properties by incomplete interactions between solid adjuvants and polymer molecules as well as forming voids and optically unacceptable surface structures. The removal of residual monomer in plastics for food packaging is especially important since legally admissible monomer contents are often in the few parts per million range.

3.1.4 Diffusion of Adjuvants

Terminology

Adjuvants and polymeric matrices are often not in thermodynamic equilibrium. Hence, the system tries to demix. The speed of demixing depends on the polymer and the adjuvant, the interaction between these two, the viscosity of the system, the environment (air, surrounding liquids, contacting solids), and the temperature. Low-molar mass adjuvants travel to the surface layer if the polymer is surrounded by air or non-dissolving liquids or through the surface into contacting solids or dissolving liquids.

Diffusion of adjuvants through or out of the polymeric matrix is mostly undesirable (see below) but desirable in a few cases. Antistatics and lubricants can only act on surfaces but not in the interiors of polymers. The diffusion of poly(methyl methacrylate) and other incompatible thermoplastics out of their blends with glass fiber reinforced unsaturated polyester resins leads to exceptionally smooth (low-shrinkage), easily printable surfaces of articles (**low-profile**) (p. 356).

The diffusion of adjuvants from the interior to the surface of the matrix is known by various names. That of fluid adjuvants (plasticizers, etc.) to the surface of plastics in contact with air is called **exudation** or **bleeding through** and that of solid adjuvants, **efflorescence** or **blooming**. In the latter case, one distinguishes the deposition of white fillers as **chalking** from the **floating** of pigments and the **bleeding** of dyestuffs. The deposition of efflorescing solids on the surface of the mold is known as **plate-out**.

The diffusion of solids from the specimen into a surrounding liquid is called **bleeding** (mainly for colorants) and that of liquids, **extraction**. The transport of plasticizer molecules from one plastic into a contacting other one is known as **migration** but this term is also used in the sense of bleeding, blooming, or chalking. A **bleed-through** is the transport of an adhesive through thin solids such as paper, films, or thin sheets.

Processes

Adjuvants move from the interior of a material to the surface and/or surrounding air only if the adjuvant is insoluble in the polymeric host. However, transport of adjuvants from one polymer to another polymer or to another solid material or liquid also occurs for thermodynamically stable polymer–adjuvant systems. The reason for the bleeding, extraction, or migration is the difference in chemical potentials of adjuvant and polymer which is not present for the exudation, efflorescence, chalking, floating, or bleeding from the materials to the surface against air.

The rate of escape of adjuvants from the polymer matrix is controlled by the diffusion of adjuvants within the polymer. Kinetic factors can slow down these processes but do not prevent them: polymeric plasticizers migrate despite their high viscosities, and barrier layers may slow down migrating plasticizers but do not block them completely.

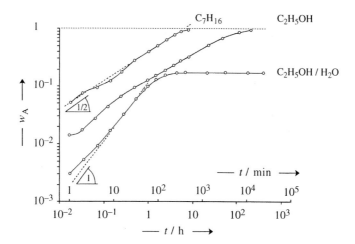

Fig. 3-1 Time dependence of the weight fraction w_A of octadecane, $C_{18}H_{38}$, from an octadecane swollen high-density poly(ethylene) that is extracted at 60°C by a hundred-fold excess of the surrounding liquids C_7H_{16}, C_2H_5OH , or 1:1 C_2H_5OH/H_2O (1:1 by volume) [1].

For example, if an octadecane-swollen poly(ethylene) (PE) is submerged in a vast surplus of heptane, a thermodynamically good solvent for octadecane, octadecane will be extracted and practically completely replaced by hexane. The highly swollen poly-(ethylene) relaxes fast; it behaves like a viscous liquid. The weight fraction w_A of the extracted octadecane is proportional to the square root of time (Fig. 3-1), which corresponds to Case I of permeation (Fickian behavior; see Volume III, p. 486).

However, if the surrounding liquid (C_2H_5OH/H_2O) is a precipitant for the polymer (PE) and a thermodynamically bad solvent for the adjuvant ($C_{18}H_{38}$), the adjuvant will travel out of the specimen but C_2H_5OH/H_2O will not move into the polymer. The slowly relaxing polymer behaves like an elastic body (Case II of permeation). Because of the high Deborah number (>10), extracted weight fractions w_A are initially directly proportional to the time. Finally, a thermodynamic distribution equilibrium is established, i.e., only a part of the adjuvant is extracted (in Fig. 3-1: $w_A \approx 0.16$).

The distribution coefficient K of the adjuvant in the polymer and in the surrounding non-absorbable liquid (such as C_2H_5OH/H_2O in Fig. 3-1) is controlled by the solubility coefficient S and the volume-related activity coefficient γ^∞ at infinite dilution:

(3-2) $\ln K = \ln S + \ln \gamma^\infty$ or $\gamma^\infty = K/S$

The natural logarithm of the activity coefficient is related by $\ln \gamma^\infty = 1 + \chi^\infty$ to the reduced chemical potential χ^∞ which describes the thermodynamic interaction of the polymer with the infinitely small proportion of the adjuvant.

The distribution coefficient K has to be replaced by K/ϕ_q if the adjuvant can enter only a fraction ϕ_q of the polymer host, for example, only amorphous regions ($\phi_q = \phi_a$).

The reduced chemical potential can be determined by inverse gas chromatography. It is always positive and adopts values of $0 < \chi^\infty < 2$ for apolar adjuvants in apolar polymers, while values of $\chi^\infty > 2$ are observed for combinations of apolar and polar substances and/or large molecules of the adjuvant.

Properties

The migration (exudation, etc.) of adjuvants is toxicologically harmless if the amount does not exceed the legal limit. However, migration is always aesthetically unpleasant, technologically dubious because of changed properties, and often simply annoying.

Efflorescence of solid adjuvants makes surfaces unsightly and leads to a partial loss of the effect of antioxidants, light stabilizers, etc. It can be prevented by using polymers with incorporated small fractions of comonomeric units that are able to physically bind the adjuvant in the interior of the material.

Exudation of plasticizers and lubricants and their subsequent condensation on other parts (such as condensation of plasticizers from plasticized PVC) can be prevented by "internal plasticization" through the use of appropriate comonomeric units (Chapter 10) or slowed down by using polymeric plasticizers.

3.2 Dyestuffs

3.2.1 Introduction

Materials are colored to protect them from radiation, to indicate various functions, or to make them look pleasant. Color is provided by **colorants**, i.e., dyestuffs *in* materials or paints or inks *on* surfaces). **Dyestuffs** are dyes (Section 3.2.2) or pigments (Section 3.2.3). Optical appearance is furthermore affected by **brighteners** (Section 3.2.4).

Dyes consist of chromophore molecules that are molecularly dissolved in fibers, plastics, and elastomers. In fibers, dye molecules may also be molecularly adsorbed on fiber surfaces or bonded chemically to fiber molecules (so-called **reactive dyes**). Color by dyes is exclusively determined by the electronic properties of dye molecules.

Pigments are dispersed in matrices. Whether a dyestuff behaves as a dye or as a pigment depends on its chemical structure, its interaction with the polymeric matrix, and the temperature (Fig. 3-2). An organic dyestuff may be insoluble in the matrix at low temperatures and behaves as a pigment but may dissolve at higher temperatures that are required for polymer processing such as injection molding, extrusion, or fiber spinning.

The different behavior manifests itself in the (optical) **absorptance**, $A = I_{abs}/I_0$, (formerly: **absorptivity**) where $A + T + R = 1$ (T = transmittance, R = reflectance). The absorptance of dyes increases linearly with the dye concentration whereas that of pigments first increases and then becomes independent of the pigment concentration (Fig. 3-2). A further increase of the pigment concentration does not change the hue but increases the opacity (Section 15.2.2).

Textiles are practically exclusively colored by dyestuffs on the surface of fibers; the bulk coloration of fibers and yarns is relatively rare. The coloring of plastics and elastomers is predominantly by pigments (98 %) in polymers and less by dyestuffs (2 %) because pigments possess a greater **lightfastness** and migrate less easily.

Controlled amounts of pigments are worked into plastics as a masterbatch or color concentrate with plastics, as a paste with a plasticizer, as a mixture with a filler, or on the surface of plastics granulates where they are bound by electrostatic charging of pigments. Granulates may surface-bind up to 1 % pigment.

Table 3-2 True densities ρ, apparent densities ρ_{app}, and ratios $w = \rho_{app}/\rho$ of pigment densities [3].

Pigment	$\dfrac{\rho}{g\ cm^{-3}}$	$\dfrac{\rho_{app}}{g\ cm^{-3}}$	$10^2\ w$	Pigment	$\dfrac{\rho}{g\ cm^{-3}}$	$\dfrac{\rho_{app}}{g\ cm^{-3}}$	$10^2\ w$
Organic Pigments				*Inorganic Pigments*			
Toluidine yellow	1.40	0.70	50.0	Ultramarine blue	2.30	0.43	18.6
Hansa yellow	1.48	0.67	45.2	Chalk	2.70	0.37	13.7
Para red	1.50	0.67	44.7	Siena earth (brown ochre)	3.30	0.30	9.1
Lithol red	1.60	0.64	40.0	Iron oxide, natural	3.65	0.29	7.9
Phthalocyanine blue	1.55	0.65	41.9	Cobalt blue	3.80	0.26	6.8
Carbon black	1.80	0.55	30.6	TiO_2 (manufactured rutile)	3.90	0.25	6.4
				Lithopone	4.30	0.23	5.3
				Zinc oxide	5.60	0.18	3.2
				Chromium yellow	5.80	0.17	2.9
				Zinc dust	7.00	0.14	2.0
				Bronze powder	8.00	0.12	1.5

3.2.4   Effect Pigments

Effect pigments consist of small light-reflecting platelets of 5-35 µm diameter that usually lie plane parallel to the surface of plastics. Effect pigments such as TiO_2 coated mica platelets have a pearly shine if they are colorless or show only the iridescent colors of thin platelets; this shine is caused by multiple reflections (Section 13.1.3). A mother-of-pearl effect is produced if such platelets are oriented parallel.

Before World War II, pearlescent effects were obtained by grinding up fish scales. They are now produced from thin aluminum platelets or from mica that is coated with layers of TiO_2 or iron oxides (results in green, silver, gold, red, copper, violet, and blue hues).

Interference pigments shift colors on viewing under different angles (e.g., green $\rightarrow$ blue $\rightarrow$ purple $\rightarrow$ black). They are produced by vaporizing metals in a vacuum chamber by an electron beam. The metal atoms are then deposited as a thin layer on a film of poly(ethylene terephthalate). The polymer film is dissolved and the metal film cut to flakes of 17-20 µm diameter and ca. 1 µm thickness.

Iridescent pigments with a fish-scale look are obtained by combining very thin layers of MgF_2, Al, and Cr (all by vaporization). The color is controlled by keeping the allowable variation (tolerance) of the thickness of each layer within a few atoms.

3.2.5   Properties

Dyestuffs must of course have all the required optical properties such as hue (Section 14.3.2), transparency (Section 13.1.4), hiding power (Section 13.2.3), etc., and also certain processing and use properties. Since polymers are usually processed at elevated temperatures, dyestuffs must be resistant against heat and oxidation. They should also distribute fast and completely in polymer matrices and/or deposit homogeneously on fibers and fabrics.

Dyes require a special affinity to the polymer matrix but pigments do not. However, pigments should be wettable, which can be achieved by treating them with surface-active compounds. Pigments should also not form lumps. Lumping is often caused by inclusion of air, which can be prevented by applying a vacuum to the pigment. Pigments may also act as nucleating agents for crystallizable polymers.

Some inorganic pigments are so hard that they act as abrasives and remove metal particles from processing machinery. This undesirable effect can be prevented by embedding pigment particles in waxes with melting temperatures just below the highest transformation temperature of the polymer (melting temperature or glass temperature); the waxes will act as lubricants.

Pigments for lacquers and printing inks are often coated with carrier resins such as hydrogenated colophony (pine resin), ethyl cellulose, vinyl acetate–vinyl chloride copolymers, etc. All these pretreatment processes are known as **conditioning**.

Properties of pigments also depend on the size of pigment particles. In general, inorganic pigments should have diameters between 0.3 µm and 0.8 µm, which allows the coloration of films and fibers with thicknesses down to 20 µm. Carbon blacks for ink-jet printers have diameters of 0.1 µm. Colored plastics are transparent if the diameter of pigment particles is smaller than one-half of the wavelength λ_0 of the incident light (visible light: 0.190 µm < $\lambda_0/2$ < 0.375 µm).

Inorganic pigments with diameters of 0.3-0.8 µm cause fibers and films with thicknesses of less than 20 µm to break at pigment-matrix interfaces because polymers do not adhere well to these pigments and pigment diameters, and matrix thicknesses become comparable. Such very thin fibers and films require organic pigments which can be ground to smaller particles than inorganic ones (specific surfaces of (400-700) m²/g *versus* (10-70) m²/g). The finer grinding produces lighter colors and a greater swellability of organic pigments.

Depending on the chemical structures of polymers, dyestuffs, and other adjuvants (antioxidants, light stabilizers, etc.), dyestuffs may either protect polymers from degradation or accelerate their deterioration. The effects also depend on the environment (temperature, relative humidity) and the physical structure of the system (dispersity, surface treatment, etc.).

These effects are especially pronounced in fibers. An example is the effect of relative humidity on various fibers that had been dyed with different dyes (Table 3-3).

Table 3-3 Effect of the chemical structure of anthraquinone dyes and the relative humidity (RH = 0 % or 100 %) on the fracture strength of fibers that were exposed to sunlight [4].

Dye	Loss of fracture strength in percent							
	Silk		PA 6.6		Cotton		Rayon	
RH in percent →	100	0	100	0	100	0	100	0
Cibanone Yellow R	100	66	85	73	-	-	80	38
Cibanone Orange R	100	66	67	65	69	45	52	30
Cibanone Orange 6R	68	34	20	34	-	-	31	22
Caledon Gold Orange G	97	34	80	80	44	26	16	16
Caledon Red BN	92	25	15	15	18	18	11	7
Caledon Jade Green G	21	6	8	8	-	-	8	8

There are several reasons for the strong effects of dyes on fibers. Dyes are molecularly dissolved in fibers and fibers have far larger specific surfaces than plastic parts. Textiles would not be comfortable to wear if they did not absorb humidity but absorbed water swells fibers so that polymer chains are more accessible to degradation. The resulting loss of fracture strength is especially large for silk and polyamide 6.6 because the amide groups of these polymers complex better with anthraquinone dyes than the hydroxyl groups of celluloses (cotton, rayon). Such degradations may also occur in certain pigmented plastics.

3.3   Fillers

3.3.1   Overview

Fillers are solid inorganic or organic, natural or synthetic materials that are dispersed in polymers. The shape of filler particles may be spheroidal, fibrous, or platelet-like. The chemical structure of fillers is usually not important except for so-called active ones.

Fillers are used in considerable proportions in plastics, elastomers, fibers, papers, coatings, adhesives, etc. In thermoplastics, fillers are used mainly in poly(vinyl chloride), it-poly(propylene), poly(ethylene), and polyamides and in thermosetting polymers such as phenolic resins and unsaturated polyester resins.

Inactive fillers dilute expensive polymers without significantly changing their use properties. They reduce costs and are therefore also called **extenders**.

Active fillers improve certain mechanical properties. They are therefore also known as **reinforcing fillers**, and for molding materials (compounds) as **resin binders**. The activity of such fillers depends on the concentration and chemical structure of atomic groups on particle surfaces. "Reinforcement" is not well defined: it may denote the increase of tensile strength, notched impact strength, or flexural strength, or the decrease of abrasion, etc. (Chapter 9).

Many different materials serve as fillers (Table 3-4). Most of the commercially used fillers are not pure in the chemical sense. They contain impurities from their origin and/or their manufacture (for example, minerals) or from their after-treatment (for example, surface coatings).

Most often used are spheroidal *inorganic* materials such as stone powders, chalk, kaolin, talc, mica, heavy spar, diatomite, pyrogenic silicic acid, etc.; most of these fillers are inactive. Reinforcing fillers are often fibers, either short fibers of glass, metals, or metal oxides or long fibers of glass or various inorganics.

Organic fillers are not so common. They include particulates such as wood meal, nut shells, and cellulose flakes (all for thermosetting resins), starch (for poly(ethylene) and poly(vinyl alcohol)), carbon black (elastomers, plastics, inks, coatings), and also various organic fibers, diced paper or plastic foam, and paper and fabric webs.

Carbon black (p. 34) and some other fillers provide additional properties. Carbon black is also a prominent black pigment (p. 27). Zinc oxide is a filler, a white pigment, and a vulcanization agent. Water is a plasticizer but may also serve as a filler (Section 3.3.2) and air is not only a blowing agent but also a "pigment" (Section 15.2.3).

Table 3-4 Average main dimensions L (diameter of spheres, fiber lengths, diameter of platelets), ratios L/d of lengths L and thicknesses d, specific surface areas a, densities ρ, tensile moduli E, fracture strengths σ_B, linear thermal expansion coefficients α, and decomposition temperatures T_d of fillers. All fillers have distributions of dimensions and properties that vary with the type, grade, and supplier. * Beginning of decomposition. For carbon blacks, see Table 3-6.

Type and Name	$\dfrac{L}{\mu m}$	$\dfrac{L}{d}$	$\dfrac{a}{m^2\,g^{-1}}$	$\dfrac{\rho}{g\,cm^{-3}}$	$\dfrac{E}{GPa}$	$\dfrac{\sigma_B}{GPa}$	$\dfrac{10^6\,\alpha}{K^{-1}}$	$\dfrac{T_d}{°C}$
Spheres, spheroids, and cubes								
Glass spheres, s.a. Table 3-7	25	1	0.3	2.54	60		8.6	
Glass spheres, hollow	50	1	0.6	≤ 1.1	0.2		8.8	
Silica, fumed (pyrogeneous)	0.8	≈ 1	225±175	2.2	20		0.5	2000
, dried gel	6		450±350	2.2				
, precipitated	9		180±120	2.0			0.5	2000
Diatomite				2-2.5			0.5	
Quartz, sand				2.65	30		10	1726
Chalk, natural, milled	120	1-5	0.3	2.71	26		10	550
	14	1-5	0.9	2.71	26			
	3	1-5	2.2	2.71	26			
, "synthetic"	1.0	≈ 1	5.3	2.71	26			
	0.07	≈ 1	23	2.65	26			
Dolomite				2.87	35		10	500
Magnesite				3.0	35		10	350
Gypsum				2.32			10	65*
Heavy spar	3		0.6	≤ 4.5	30		10	1580*
Wood meal				0.6	10		50	200
Nutshell meal	30		0.45	≤ 1.32	10			
Fibers								
Glass fibers, E glass, short	200	20	0.8	2.54	73	3.5	5	
, A glass, short	200	20		2.48	45	2.5	8.6	
, S glass, short	200	20		2.49	87	4.9	5.9	
Franklin fiber	2	40						
Asbestos (chrysotile)	≈ 100	1000	40	2.5	145	3		
Wollastonite	2	60	1.3	≤ 2.9	30		6.5	1540
Carbon fibers (Section 6.6.9)	7	25		1.55	63	0.4		
Graphite fibers				1.7	390	2.5		400
Aluminum oxide, whiskers				3.96	420	20		
Silicon carbide, whiskers				3.19	850	11		
Steel	250			7.75	210	4.1		
Aramid (Kevlar 29®)	12	62		1.44	62	2.8	–3.5	
Platelets								
Feldspar	15		0.85	2.6	30		6.5	
	5		2.1	2.6	30		6.5	
Kaolin	1-10	≤ 10	10-40	2.60	20		8	450
, calcined				2.5-2.6	200		4.5	1100
Clay				2.6				
Talc(um)	7	9	7	2.8	20		8	750
	1.8	9	17	2.8	20			
Mica, muscovite	0.5-300	40	30	2.85	172	860	2.5	900
Aluminum hydroxide	40		0.19	2.4		30	5	200
	0.7		0.19	2.4		30		
Magnesium hydroxide				2.38				300

Many properties of filled polymers can be predicted if the properties of the polymers and fillers are known (Chapter 9). For fillers, one needs to know their geometric properties (Section 3.3.6), including their specific surfaces, as well as their densities, elastic moduli, fracture strengths, and decomposition temperatures (Table 3-4).

For inactive fillers, many mechanical properties can be calculated from these data using mixing rules (Section 9.2 ff.). Since all fillers have considerably larger moduli of elasticity and lower elongations than conventional plastics, filled polymers are usually (but not always) more rigid than unfilled ones (Table 3-5). These more or less additive effects are moderated by filler-polymer interactions and the type of stress.

Fracture strengths of particulate-filled polymers are generally lower than those of unfilled ones whereas those of short fiber-reinforced ones are higher (Table 3-5). Heat distortion temperatures are increased by fibers but not much affected by particulates.

It is therefore expedient to discuss here only the structures and properties of fillers themselves but not the properties of filled and reinforced polymers (see Chapter 9). It should also be noted that many fillers are not used as such but are coated with ca. 1 % stearic acid as a lubricant (Section 3.5.1).

Table 3-5 Mechanical properties of unfilled polymers and of thermoplastics with 30 wt% fillers [5]. HDPE = high-density poly(ethylene), LDPE = low-density poly(ethylene), PA 6 = poly(ε-caprolactam), PBT = poly(butylene terephthalate), PS = atactic poly(styrene).

Polymer		Properties after filling with 30 wt% filler				
Filler →	unfilled	Glass	Chalk	Talc	Asbestos	Glass
Shape →	-	Spheres	Spheroids	Platelets	Fibers	Fibers
Tensile modulus E in MPa						
LDPE	210	290	900	600	670	1200
PA 6	1200	3000	3300	-	8000	5500
HDPE	1400	-	1900	-	-	7800
PBT	2700	3500	4700	-	-	8500
PS	3800	-	2000	5600	8200	11000
Tensile strength in MPa						
LDPE	10	10	16	16	16	24
HDPE	27	-	20	-	-	60
PS	55	-	15	39	71	95
PBT	60	50	62	-	-	130
PA 6	64	65	50	-	123	148
Fracture elongation in %						
PS	4	-	2	2	-	2
PBT	120	4	4	-	-	4
PA 6	220	20	30	-	3	4
LDPE	500	73	220	40	40	65
HDPE	>550	-	-	9	-	2
Heat distortion temperature in °C						
LDPE	35	-	-	-	-	-
HDPE	50	-	-	-	-	108
PBT	67	70	67	-	-	205
PA 6	80	-	60	-	193	208
PS	86	-	-	-	91	93

3.3.2 Spheroidal Fillers

Calcium carbonate is the most often used filler for plastics (>85 % of the total consumption of all fillers). Industrial products contain 90-99.5 % $CaCO_3$, 0.5-10 % $MgCO_3$, and less than 0.2 % Fe_2O_3. These materials are mainly obtained by grinding natural products such as chalk, limestone, or marble. Dissolution of such natural products and subsequent precipitation generates very fine particles of a "synthetic" calcium carbonate (CCP = *calcium carbonicum praecipitatum*).

Chalk and limestone form compact particles that do not absorb as much plasticizer as CCP (9-21 % *versus* 35-40 %). They are also less hard than CCP (less abrasion of machine parts) and require less energy during processing because of their smaller specific surfaces. Use of the more expensive CCP, on the other hand, leads in general to better mechanical and optical properties of the filled polymers.

Similar to calcium carbonates are calcium magnesium carbonates, magnesium carbonates, calcium sulfates, and barium sulfates. These mineral fillers are all easy to disperse; they all lead to high degrees of whiteness.

Industrially used *calcium magnesium carbonate* is obtained by grinding dolomite, a mineral containing 54 % $CaCO_3$, 45.6 % $MgCO_3$, and <0.2 % Fe_2O_3. Magnesite is a relatively pure *magnesium carbonate* consisting of 90-99.5 % $MgCO_3$, 0.5-10 % $CaCO_3$, and < 0.2 % Fe_2O_3.

Water-free *calcium sulfate* is found in nature as the mineral anhydrite; it is also a by-product in the production of hydrofluoric acid from fluorspar (fluorite, florspar), CaF_2. Dry $CaSO_4$ is used industrially as a binder and construction material but not as a filler for polymers because it readily absorbs humidity (laboratory use as drying material (Drierite®)). The filler for polymers is rather the naturally occuring dihydrate of $CaSO_4$ (gypsum, also known as terra alba or alabaster). The naturally occurring, glass-clear crystalline form is not used technically. Ground industrial gypsum contains 90 % $CaSO_4 \cdot 2\ H_2O$, 10 % $Ca/MgCO_3$, and <0.2 % Fe_2O_3. The industrial dihydrate is used in polymeric foams; the hemihydrate is the so-called Franklin fiber (see p. 35).

Barium sulfate, $BaSO_4$, has an extraordinarily high density (Table 3-4) which led to the mineral names heavy spar and baryte (G: *barus* = heavy). Industrial products contain 85-99 % $BaSO_4$, 0.5-2 % $SrSO_4$, and <2 % SiO_2. Pigments are also called blanc fixe or permanent white because they do not react with H_2S (no darkening of paints).

Feldspar (D: *Feld* = old form of *Fels* = rock; D: *spar* (outdated) = easily cleavable) is a series of naturally occurring aluminum silicates, $Mt[AlSi_3O_8]$, that are either spheroidal or platelet-like. Granite delivers potassium feldspar (orthoclase) whereas nepheline with the approximate composition $KNa_3[AlSiO_4]_4$ is obtained from the igneous rock, syenite. Feldspars are the least reactive inactive fillers. They have low specific surfaces.

Naturally occurring crystalline *silicon dioxides*, SiO_2, are used relatively little (*quartz* with its modifications α-quartz (low quartz), β-quartz (high quartz), tridymite, and cristobalite). Industrially used are various amorphous silicic acids, $SiO_2 \cdot x\ H_2O$, such as kieselguhr (diatomite) which consists of 70-90 % SiO_2, 3-12 % H_2O, and various proportions of metal oxides. Also used are perlites which are volcanic glasses from quartz or quartz plus alkali feldspars that include pockets of water. On heating perlites to 1200°C, water evaporates and the mineral expands to a fluffy material with cellular structure (density ca. 0.03 g/cm^3).

Fumed silicas (pyrogenic silicas) are produced by pyrolysis of $SiCl_4$ (giving Aerosil®) or calcining ethyl silicate (giving silica white, also known as carbon white in incorrect analogy to carbon black). Fumed silicas are highly active fillers (and whitening agents) that can be used in small proportions to decrease the Barus effect on procesing (see p. 94), reduce the shrinkage, and increase the dimensional stability.

Glass beads are available in many sizes, both compact with diameters between 4 µm and 5000 µm and hollow with external diameters between 10 µm and 250 µm. On processing, polymers filled with glass beads shrink evenly in all directions, in contrast to polymers filled with short glass fibers that tend to orient in the flow direction.

Spheroidal fillers also comprise wood meal (wood flour) and similar products from nut shells. These wood products serve as fillers for phenolic resins, poly(ethylene)s, and poly(vinyl chloride).

Carbon blacks are produced by thermal-oxidative processes or thermal processes (see Volume II, p. 213) from methane, natural gas, aromatic hydrocarbons, acetylene, and the like by more than one hundred different processes. Accordingly, properties vary widely (examples in Table 3-6). Carbon blacks are used predominantly for the reinforcement of rubbers and as active fillers and pigments in plastics.

Table 3-6 Properties of some carbon blacks [6]. CB = carbon black, DBP = dibutyl phthalate, I_2 = iodine, LSO = linseed oil; n.a. = not applicable. * N_2 adsorption.

	Physical unit	Thermal-oxidative processes			Thermal processes	
		Channel blacks	Gas blacks	Furnace blacks	Thermal blacks	Acetylene blacks
Particle diameter	nm	110-120	10-30	10-80	120-500	32-42
Specific surface	m^2/g CB	16-24	90-500	15-450	6-15	ca. 65
Absorption	mg I_2/g CB	23-33	n.a.	15-450	6-10	ca. 100
	mL DBP/100 g CB	100-120	n.a.	40-200	37-43	150-200
	g LSO/100 g CB	2.5-4.0	2.2-11.0	2.0-5.0	0.65-0.90	4.0-5.0

3.3.3 Fibrous Fillers

Fibrous fillers are always reinforcing agents. The most important reinforcing fibers for plastics are *short glass fibers* (US: fiberglass) so-called E glass, an alkali-poor, water-resistant boron silicate glass that was originally developed for the *e*lectrical industry (hence: E glass; for the chemical composition, see page 230). Far less used are R glasses (*r*esistant against acids; good strengths) and S glasses (good *s*trengths). Properties of quartz fibers and some glass fibers are shown in Table 3-7.

Short glass fibers are produced by the chopping or milling of long fibers. Subsequent compounding and processing reduces fiber lengths further. For example, the average fiber lengths of glass fibers in a compounded and processed it-poly(propylene) is reduced from 4.5 mm to 0.66 mm and the axial ratio from 320 to 50.

The American glass fiber industry characterizes average diameters *d* of glass fibers by letters whereas the European industry uses fineness (= linear density (Section 6.1.3)). Glass fibers for thermoplastics are mainly type G (*d* = 8.9-10.2 µm) and type K (*d* =

Table 3-7 Properties of quartz fibers and short glass fibers of types A (alkali rich), E (electrical), HM (high modulus), R (resistant against acids), S (strength), and X.

Property	Physical unit	Quartz	A	E	HM	R	S	X
Density (of fiber glass)	g/cm^3	2.20	2.45	2.59	2.58	2.53	2.48	2.49
Softening temperature	°C	1720	1000	1320				
Expansion coefficient, linear	10^{-6} K^{-1}	1	8.6	5.0			5.9	
Tensile modulus	GPa	66	74	73	115	86	86	118
Tensile strength (freshly drawn)	GPa	3.2	3.1	3.4	3.4	4.4	4.6	6.9
Fracture elongation	%	1.0	1.5	2.5	3		2.8	
Hardness (Mohs' scale)	-	7	6.5	6.5			6.5	
Relative permittivity (at 10 kHz)	1		5	5.8			5.2	
Electric surface resistance	Ω		10^7	10^7			10^7	
Fiber diameter	µm	9	10	3-13	9	9	9	
Fiber length, cut fibers	mm			4.5				
, milled fibers	mm			0.2				

12.7-14 µm) whereas those for thermosets are types H (d = 10.2-11.4 µm), M (d = 15.2-16.5 µm), and T (d = 22.9-24.1 µm). Some special glass fiber mats for thermosets consist of type DE (5.1-6.4 µm to 6.4-7.6 µm) and those for roof coverings of type M.

Before processing, mats of glass fibers are treated with **sizes**. Such sizes should protect the chafe-sensitive glass fibers during processing and promote the adhesion between glass fiber and polymer matrix. Sizes for glass fibers thus differ from textile sizes which should temporarily glue protruding fibers to fabrics (Section 6.4.2). Sizes for glass fibers consist of coupling agents, polymeric film formers, and wetting agents.

The content of organic material in treated glass fibers is characterized in the United States by a so-called LOI value. This value indicates the *loss on ignition* and has nothing to do with the LOI (= *limiting oxygen index*) of the polymer industry, which is a measure of the flammability of the material (Section 3.6.6).

Another synthetic inorganic fiber is the so-called *Franklin fiber* which consists of the hemihydrate of calcium sulfate. The fiber has lengths between 150 µm and 300 µm and is therefore called a "microfiber" (the same term is used in the textile industry, albeit with a different meaning (Section 6.1.3)). Another mineral fiber is called PMF. This *processed mineral fiber* is produced from blast furnace slag. PMF has a length of 275 µm and a diameter of 5 µm.

The mineral *wollastonite*, CaSiO$_3$, forms pure white small needles that compete as a filler with the platelets of talc and mica.

Whiskers are very fine fibers with diameters of 1-25 µm and axial ratios between 50 and 15 000. They are produced from carbon, boron, ceramics (alumina, beryllia, boron carbide or nitride, silicon carbide or nitride), or metals (aluminum, cobalt, iron, nickel, rhenium, tungsten). They have exceptional strengths and moduli and good temperature stabilities but command very high prices which restricts their use mainly to materials for space, deep sea, and weapons (see also pp. 230, 234).

Also used are inorganic long fibers, for example, fabrics and mats of long glass fibers (Section 6.8.1) and wires from steel or beryllium. Reinforcing organic long fibers comprise those of aromatic polyamides, graphite, and carbon (see Chapter 6).

3.3.4   Platelet-like Fillers

Kaolins (Volume II, p. 569) are hydrated aluminum silicates with the average composition $Al_2O_3 \cdot SiO_2 \cdot 2\ H_2O$ that were formed by the weathering of granites and feldspars.

More than 50 % of kaolins are used in the paper industry. They are also the second most important filler in the rubber industry. This industry also distinguishes between "hard" and "soft" kaolins which has nothing to do with hardness. In hard kaolins, 75 % of the particles have diameters of less than 2 μm; reinforcement of rubber with these grades produces large moduli of elasticity (hence: "hardness," meaning rigidity). "Soft" kaolins consist of larger platelets that lead to lower moduli of elasticity (hence: "softness").

In plastics, kaolins improve not only mechanical properties but also chemical and electrical ones. They reduce the water uptake and, because the platelets are more or less parallel to the surface, also the surface hardness.

Micas are complex potassium/aluminum silicates with phyllo structures. The light-colored muscovite (common mica) has the composition $KAl_2(OH,F)_2[AlSi_3O_{10}]$ whereas in the dark-colored phlogopite (amber mica), $K(Mg,Fe^{2+})_3(OH,F)_2[AlSi_3O_{10}]$, aluminum ions are replaced by magnesium and iron. The reinforcing action of mica depends on the axial ratio of the platelets, which is similar to the action of talc(um).

Talc(um), $Mg_3[(OH)_2Si_4O_{10}]$, is a phyllo silicate that comprises ca. 10 % of all fillers for plastics. Like all platelet fillers, it leads to larger moduli of elasticity and greater surface hardnesses of plastics where it also reduces fracture strengths, fracture elongations, and notched impact strengths. Small talc particles are good nucleation agents.

Aluminum (tri)hydroxide, $Al(OH)_3$ (ATH; gibbsite), and *magnesium hydroxide*, $Mg(OH)_2$ (brucite), are used mainly as flame retardants. Both compounds also improve (di)electric strengths and tracking resistances. $Al(OH)_3$ begins to dehydrate at temperatures above ca. 200°C, which may lead to difficulties in processing.

3.3.5   Fluid Fillers

Fillers are not necessarily solids. Incorporated air acts as a "pigment" and a "filler" in coatings (Section 15.2.2) and as "plasticizer" and "blowing agent" in foamed materials (Section 10.5).

Water is not only a plasticizer for hydrophilic polymers (p. 79, 425) but also a filler for unsaturated polyester resins (UP), silicones (SI), poly(vinyl chloride)s (PVC), and latex rubbers; in the latter, water can be dispersed in up to 2 mm large droplets as a type of water-in-oil emulsion. Unsaturated polyesters filled with 50 % water still have good mechanical properties, for example, moduli of elasticity of ca. 800 MPa and fracture strengths of ca. 17 MPa.

3.3.6   Geometric Properties of Fillers

Properties of filled polymers are most importantly affected by the geometric properties of fillers: the length and length distribution of the largest particle axis and the average shape of particles.

Table 3-8 Aspect ratios and circularities of some simple geometric figures. Aspect ratios are given as the ratio of the lengths of the longest and shortest symmetry axes (dotted lines) in circles, hexagons, squares, triangles, and various rectangles. Circularities are calculated with Eq.(3-3).

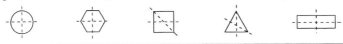

Shape	Aspect ratio	Circularity	Shape	Aspect ratio	Circularity
Circle	1	1	Rectangle	5	0.661
Hexagon	$2/3^{1/2} \approx 1.155$	0.952	Rectangle	10	0.510
Square	$2^{1/2} \approx 1.414$	0.866	Rectangle	20	0.377
Triangle	2	0.778	Rectangle	100	0.175

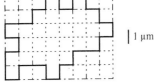

| 1 μm

Particle with an area of $A = 28$ μm^2, a perimeter of $P = 34$ μm, and a circularity of $C = 0.552$.

The shape of filler particles is usually described by the ratio of the geometric main axis to the length of one of the secondary axes, the so-called **aspect ratio**. This ratio is 1 for spherical fillers, ca. 1.1-1.5 for spheroidal ones, larger than 10 for rodlike fillers (fibers), and usually between 5 and 100 for platelets.

Important but almost never used is also the **circularity** C of a particle, which can be calculated from the area A and the perimeter P of particles:

(3-3) $C = 2\,(\pi A)^{1/2} P^{-1}$

A circle with the radius R, the area $A = \pi R^2$, and the circumference (= perimeter) $P = 2\pi R$ has a circularity of unity. Circularities and aspect ratios of some other simple geometric shapes and an irregular shape are shown in Table 3-8.

Because of the strong effect of axial ratios of filler particles on the properties of filled polymers, distribution curves or at least averages of characteristic geometric properties should be known. Examples are the distribution curves and averages of diameters of spherical particles and the lengths or aspect ratios of rods.

The g-average of a property P is defined as

(3-4) $\overline{P}_g \equiv \sum_i g_i P_i / \sum_i P_i$

where g may be the number, mass, z, ..., etc., fraction of entities and P their molar mass, length, surface area, volume, ..., etc. The length, surface, and volume averages of particles can be correlated with the number, mass, ... averages as the following examples show.

The masses m_i and numbers N_i of particles of size i can be calculated from their masses m_i, volumes V_i, radii R_i, surfaces A_i, and densities ρ_i. Assuming constant densities ρ of particles regardless of sizes, the mass of a sphere with the radius R_i is calculated from

(3-5) $m_i = N_i m_i = N_i V_{part,i} \rho = N_i (4\,\pi/3) R_i^3 \rho = N_i(\rho/3) A_i R_i$ (sphere)

that of circular discs with surface $A_i = 2\,\pi R_i^2 + 2\,\pi R_i h_i$ (with $A_i \approx 2\,\pi R_i^2$ for very thin platelets) and constant heights h from

(3-6) $m_i = N_i \rho(\pi R_i^2)h = N_i \rho h[(A_i/2) - \pi R_i h_i] \approx N_i \rho h A_i/2$ (circular disc)

and that of circular cylinders with constant radii R and varying lengths L from

(3-7) $m_i = N_i \rho(\pi R^2)L_i = N_i(\rho R/2)A_i$ (circular cylinder)

Table 3-9 contains the primary and secondary definitions of number, surface, and mass averages of spheres, circular discs, and circular cylinders. Primary definitions are indicated by vertical lines at the left of the definitions. For example, the mass average diameter of spheres is given by $\overline{d}_w = \Sigma_i m_i d_i / \Sigma_i m_i$ (right data column, seventh line) if the mass m itself is used as a statistical weight. However, if the mass average is written for numbers N_i of spheres as statistical weights, the expression becomes $\overline{d}_w = \Sigma_i N_i d_i^4 / \Sigma_i N_i d_i^3$ (left data column, seventh line) because the mass is proportional to the third power of the diameter (or radius) of a sphere (Eq.(3-5)).

Conversion of statistical weights of dimensions of particles thus cannot be done "in analogy" to that of molar masses as the following example shows:

Molar mass $\overline{M}_w = \Sigma_i m_i M_i / \Sigma_i m_i \quad \rightarrow \quad \overline{M}_w = \Sigma_i N_i M_i^2 / \Sigma_i N_i M_i$

Diameter of spheres $\overline{d}_w = \Sigma_i m_i d_i / \Sigma_i m_i \quad \rightarrow \quad \overline{d}_w = \Sigma_i N_i d_i^4 / \Sigma_i N_i d_i^3$

Table 3-9 Number, surface, and mass averages of diameters d of spheres and thin circular discs of constant thickness and lengths L of circular cylinders with constant diameter d. N = number, A = surface, m = mass. Primary definitions are marked by a vertical line at the left of the expression.

	Number N	Applied statistical weight Area A	Mass m
Number averages			
Spheres	$\Sigma_i N_i d_i / \Sigma_i N_i$	$\Sigma_i A_i d_i^{-1} / \Sigma_i A_i d_i^{-2}$	$\Sigma_i m_i d_i^{-2} / \Sigma_i m_i d_i^{-3}$
Circular discs	$\Sigma_i N_i d_i / \Sigma_i N_i$	$\Sigma_i A_i / \Sigma_i A_i d_i^{-1}$	$\Sigma_i m_i d_i^{-1} / \Sigma_i m_i d_i^{-2}$
Circular cylinders	$\Sigma_i N_i L_i / \Sigma_i N_i$	$\Sigma_i A_i / \Sigma_i A_i L_i^{-1}$	$\Sigma_i m_i / \Sigma_i m_i L_i^{-1}$
Surface averages			
Spheres	$\Sigma_i N_i L_i^3 / \Sigma_i N_i d_i^2$	$\Sigma_i A_i d_i / \Sigma_i A_i$	$\Sigma_i m_i / \Sigma_i m_i d_i^{-1}$
Circular discs	$\Sigma_i N_i d_i^2 / \Sigma_i N_i d_i$	$\Sigma_i A_i d_i / \Sigma_i A_i$	$\Sigma_i m_i / \Sigma_i m_i d_i^{-1}$
Circular cylinders	$\Sigma_i N_i L_i^2 / \Sigma_i N_i L_i$	$\Sigma_i A_i L_i / \Sigma_i A_i$	$\Sigma_i m_i L_i / \Sigma_i m_i$
Mass averages			
Spheres	$\Sigma_i N_i d_i^4 / \Sigma_i N_i d_i^3$	$\Sigma_i A_i d_i^2 / \Sigma_i A_i d_i$	$\Sigma_i m_i d_i / \Sigma_i m_i$
Circular discs	$\Sigma_i N_i d_i^3 / \Sigma_i N_i d_i^2$	$\Sigma_i A_i d_i^2 / \Sigma_i A_i d_i$	$\Sigma_i m_i d_i / \Sigma_i m_i$
Circular cylinders	$\Sigma_i N_i L_i^2 / \Sigma_i N_i L_i$	$\Sigma_i A_i L_i / \Sigma_i A_i$	$\Sigma_i m_i L_i / \Sigma_i m_i$

3.4 Other Adjuvants for Application Properties

3.4.1 Nucleating Agents

Crystallization of polymer melts is usually initiated heterogeneously by dust particles, container walls, filler and/or pigment particles, or left-over nuclei (Volume III, Section 7.3.1). Concentrations of such crystallization nuclei vary between 1 and 10^{12} nuclei per cubic centimeter. Homogeneous formations of nuclei by spontaneous clustering of chain segments are rare.

Crystallization leads to spherulites if all sides of nuclei add chain segments at the same rates. The smaller the concentration of nuclei, the larger are the spherulites and the worse is the resistance of the polymer to fracture. High concentrations of nuclei lead to faster crystallizations and therefore also to shorter cycle times in processing.

The concentration of nuclei in polymers can be increased by the addition of **nucleating agents** which improve both end-use properties of polymers (fracture behavior) and processing properties (shorter cycle times). Nucleating agents that lead to improved transparencies of polymers are known as **clarifiers**; these adjuvants strongly increase the concentration of nuclei and may also change the habitus of the resulting crystallites.

Nucleating agents are used predominantly in poly(1-olefin)s and here especially in isotactic poly(propylene)s. The efficiency of nucleating agents can be measured by the magnitude of the crystallization temperature, i.e., the maximum of the heat effect during controlled cooling of the melt. For example, cooling a melt of it-poly(propylene) from 200°C at a rate of 12.5 K/min leads to a crystallization temperature of 106°C. Addition of 0.25 wt% of nucleating agents increases the crystallization temperature to 110°C (calcium oxide), 118°C (sodium caproate), and 131°C (sodium benzoate).

The efficiency of a nucleating agent depends predominantly on its solubility in the polymer melt. Insoluble additives either increase the crystallization rate (if they are wetted by the polymer melt) or do not change it (if they are not wetted). Soluble inert additives act only as diluents: they decrease crystallization rates but do not nucleate crystallizations. However, such additives may react with the polymer to form new species that are the true nucleating agents.

An example is sodium benzoate, C_6H_5COONa, which is insoluble in it-poly(propylene) (iPP) but soluble in poly(ethylene terephthalate) (PET). In the latter case, PET molecules, $\text{+OCH}_2\text{CH}_2\text{O–CO(1,4-C}_6\text{H}_4)\text{CO+}_n$, are transesterified by sodium benzoate which leads to polyester molecules with ionic endgroups and sodium (4-carboxymethyl)benzoate, $CH_3OOC(p\text{-}C_6H_4)COONa$. The latter compound then dismutates to $CH_3OOC(p\text{-}C_6H_4)COOCH_3$ and $NaOOC(p\text{-}C_6H_4)COONa$, the true nucleating agent.

Nucleation can also be initiated inadvertently by crystalline pigments and other adjuvants. Inorganic pigments are insoluble in polymer melts but are probably wetted and then act as nucleating agents. Organic pigments sometimes lead to epitaxial growth of polymers on pigment particles. Such pigments may produce other common crystal modifications, for example, the less common ß-modification of it-poly(propylene).

Organic pigments are usually somewhat soluble in the polymer, which leads to slower nucleations and crystallizations. Cooling such a melt during processing causes an after-crystallization of the melt, a shrinkage of the article, and, because of the anisotropy of heat conductivity, also to a distortion of the part.

3.4.2   Coupling Agents

Non-reactive Coupling Agents

Coupling agents are used to improve the adherence of polymers to fillers. For reinforcing fibers, they are known as **finishes**.

Non-reactive coupling agents adsorb on the surface of filler particles and bind to polymer molecules by physical bonds (hydrogen bonds, dipole-dipole interactions, van der Waals bonds).

For stearic acid, $C_{17}H_{35}COOH$, as a coupling agent for mineral fillers in poly(1-olefin)s, these interactions are between the $C_{17}H_{35}$ groups and the $-CH_2-CHR-$ groups of the polymer on one hand and the COOH groups and the surface groups of fillers on the other. At small concentrations of coupling agents, bound mass fractions w_b of the coupling agent per specific area A/m_F of the filler are directly proportional to the original concentration w_0 of stearic acid in CCl_4 (Fig. 3-3). At higher initial concentrations w_0, saturation is observed.

For a given filler, the saturation concentration of the coupling agent is higher, the greater the specific area of the filler (see data for $CaCO_3$ in Fig. 3-3). Of course, saturation concentrations also depend on the chemical nature of the filler surfaces and coupling agents. For example, SiO_2 has the largest specific surface but the lowest saturation concentration for stearic acid.

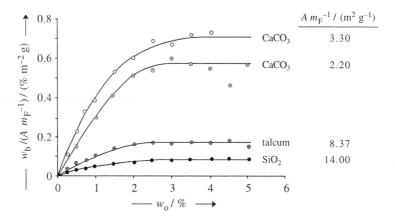

Fig. 3-3 Effect of the type and specific surface area, A/m_F, of fillers on the mass fraction m_b of stearic acid that is bound to the specific surface area of the filler as a function of the original mass fraction w_0 of stearic acid in CCl_4 [7a]. Stearic acid was worked into the fillers for 30 min with a Haake kneader. With kind permission of Academic Press, San Diego (CA).

The achievable saturation concentration is also controlled by the type of mixer (and presumably by the mixing speed and mixing time). For the system $CaCO_3$–SiO_2 in Fig. 3-3, it was larger for a Haake sigma kneader than for a Brabender planetary mixer (Fig. 3-4). In these experiments, it was not checked whether the Sigma kneader breaks calcium carbonate particles more strongly than the planetary mixer. Such an action would lead to greater filler surfaces at (probably) constant specific surfaces and hence to a higher saturation concentration.

In sigma kneaders, two arms or blades, separated by a saddle, move in opposite directions. In planetary mixers, an agitator moves in a circular path while rotating.

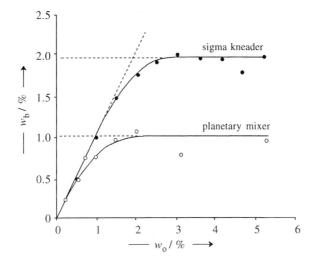

Fig. 3-4 Effect of the type of mixing device on the mass fraction w_b of stearic acid boound to calcium carbonate as function of the original mass fraction w_o of stearic acid in CCl_4 [7b].
    With kind permission of Academic Press, San Diego (CA).

Reactive Coupling Agents

Reactive coupling agents form chemical bonds with surface groups of filler particles. For glass fibers and certain mineral fillers such as clays, micas, kaolin, and trihydroxy aluminum, coupling agents are mostly silicon derivatives $(RO)_3SiZR''$ where R'' is a group that reacts with the polymer, for example, $CH_2=CH-$, $Cl-$, NH_2-, or $HS-$. Z is a spacer group, for example, $-(CH_2)_3-$.

Commonly used are vinyl triethoxysilane $CH_2=CHSi(OC_2H_5)_3$ (for UP, PE, PP, and PVC), γ-methacryloxypropyl trimethoxysilane $CH_2=C(CH_3)COO(CH_2)_3Si(OCH_3)_3$ (for UP, EP, PE, PP, PS, ABS, SAN, and PMMA), and γ-aminopropyl triethoxysilane $NH_2(CH_2)_3Si(OC_2H_5)_3$ (for EP, PF, MF, PC, PE, PP, PC, PA, PMMA, and PSU).

Also used are titanates $(RO)_mTi(OR'R''Y)_n$; they are known as monoalkoxy types (m = 1, n = 3), coordination types (m = 4, n = 2), and chelate types (m = 1, n = 2).

Silanes hydrolyze in humid air or aqueous solutions, for example, $(RO)_3SiR'R'' \rightarrow (HO)_3SiR'R''$. The resulting silanol groups bind to the surface of fillers whereas the remaining functional groups bind to the polymer matrix. The latter bonds are certainly physical for most of the polymeric groups but chemical for thermosetting resins such as UP, EP, PF, and MF and possibly also for thermoplastics with reactive chemical groups such as polyamides and poly(butylene terephthalate).

Whether or not reactive coupling agents also bind *chemically* to some thermoplastics is discussed controversially in the literature. A chemical reaction of the coupling agent with reactive endgroups or incorporated "wrong" chain units has been suggested but not proven.

Bonds between silanol groups and glass fibers or mineral fillers are probably reversibly hydrolyzable. Such reactions could equilibrate stresses if filled polymers are subjected to mechanical forces. For the possible formation of interlayers between fillers and polymeric matrices, see Section 9.2.3.

3.4.3   Plasticizers

Types

Technologically, **external plasticizers** are substances that are added to polymers in order to improve deformabilities. Scientifically, they are chemical compounds that lower the glass temperature (Section 10.3). Both effects can also be achieved by **internal plasticizing** through incorporation of flexibilizing groups in polymer chains.

Plasticizing always involves the external addition or internal incorporation of chemical substances in polymer chains. *Plastifying* is the softening of a plastic by contact heat and *plastication* that by contact heat and friction.

The first industrially used plasticizer was camphor (1,7,7-trimethylbicyclo[2.2.1]-2-heptanone ($C_{10}H_{16}O$); T_M = 178°C); it was used to plasticize cellulose 2 1/2-acetate. Solid low-molecular weight plasticizers are practically not used any more; essentially all external plasticizers are now low-molecular weight liquids.

Suppliers offer about 500 different plasticizers. Elastomers are predominantly plasticized by mineral oils; they may contain between 10 and 40 wt% plasticizers. Ca. 90 wt% of all plasticizers for plastics are used in plasticized poly(vinyl chloride); these plasticizers are usually phthalic acid esters (Table 3-10) which are also used for polyurethanes, unsaturated polyesters, and phenolic resins. The plasticizer of choice was DOP (DEHP) which, like DBP, is now banned in the European Union for use in toys and all child care articles and will be banned in California (if >0.1 %), starting in 2009.

Phosphoric acid esters are good plasticizers for poly(vinyl acetate), poly(vinyl butyral), cellulose 2 1/2-acetate, and phenolic resins. Sulfone amides are specialty plasticizers for amino resins, phenolic resins, unsaturated polyesters, polyamides, and cellulose acetate that cannot be used for poly(vinyl chloride). Specialty plasticizers for poly(vinyl butyrate) are di(*n*-hexyl) adipate and the di(*n*-heptanoic acid) esters of tri- and tetraethylene glycol.

Literature rarely uses chemical names of plasticizers but usually abbreviations and acronyms. Unfortunately, different organizations (ASTM, BS, DIN, ISO, IUPAC, etc.) recommend different symbols for the same substance. Di(2-ethylhexyl) phthalate (DEHP) is commonly called dioctyl phthalate (DOP); for tricresyl phosphate, both TCP and TCF are used because TCP® is a registered trademark in the United Kingdom.

Table 3-10 Properties and use of important plasticizers. ρ = density (20°C), η = viscosity (20°C), T_S = solubility temperature for poly(vinyl chloride). For polymer abbreviations, see Appendix.

Symbol	Name	$\dfrac{\rho}{\text{g cm}^{-3}}$	$\dfrac{\eta}{\text{mPa s}}$	$\dfrac{T_S}{°C}$	Use in
DIDP	di(*i*-decyl) phthalate	0.965	123	141	PVC, CAB, CN, PMMA, PS
DEHP, DOP	di(2-ethylhexyl) phthalate	0.983	80	120	PVC, CN, CAB, VC/VAC, PS
DINP	di(*i*-nonyl) phthalate	0.968	47	135	PVC, CAB, VC/VAC
DINA	di(*i*-nonyl) adipate	0.927	29	152	PVC, CN, PS
DBP	dibutyl phthalate	1.047	21	93	PVC, CN, CAB
DOA	di(2-ethylhexyl) adipate	0.924	14	149	PVC, PVB, PS, VC/VAC
TOF	tri(2-ethylhexyl) phosphate	0.92	14	126	CN, CAB, PVAC
TCP, TCF	tricresyl phosphate	≈1.18	≈ 70	101	CN, CAB, VC/VAC

Polymeric plasticizers are not very common; they are usually polyesters or poly-ethers. Polyesters are synthesized by polycondensation. They have thus fairly broad molar mass distributions and therefore also contain monomers and oligomers.

Action

Plasticized polymers can be considered special homogeneous blends. They are thus discussed in greater detail in Chapter 10, together with polymer blends. This section discusses only some properties of plasticizers themselves.

External plasticizers are subdivided into primary and secondary ones. **Primary plasticizers** interact directly with polymer molecules, for example, via interactions between polymer and plasticizer dipoles. **Secondary plasticizers** are diluents for primary plasticizers. They are also called **extenders** (not to be confused with inactive fillers that are also called extenders).

Depending on the polymer, a plasticizer may be either primary or secondary. For example, heavy oils are apolar and therefore extenders for the polar poly(vinyl chloride) but primary plasticizers for the apolar diene elastomers. Because the plasticizer action in diene elastomers is via dispersion forces and not by polar groups, one also talks of **oil-extended rubbers**.

Good plasticizer actions are generally produced by low-viscosity plasticizers since low viscosities indicate small interactions between molecules (Section 10.3). Unfortunately, low viscosities go hand in hand with high volatilities and thus with a tendency to exudation. An example is the fogging of the interior sides of windows of new cars that is caused by exuded plasticizers from PVC parts when the car is heated by the sun.

The exudation of plasticizers in air or foodstuffs has caused concern because of toxicity, not only for small children (see above) but also for adults. Floor coverings of DOP plasticized PVC lead to concentrations of ca. 1 μg DOP per m^3 of air which is four decades smaller than the maximum permissible workplace concentration of 10 mg/m^3. DOP saturated air was inhaled lifelong by hamsters without tumor formation.

Animal fodder enriched with DOP and DOA led to extremely low *acute* toxicities but very high doses during long periods of time showed increased tumor formations in rodents. In the European Union, DOP and DOA are therefore classified as carcinogenic with a maximum dose of 80 μg of DOP per day for a non-significant risk.

PVC Plastisols

PVC plastisols are dispersions of PVC particles in plasticizers. Such particles are produced by emulsion or microemulsion polymerizations (see Volume II, p. 163); the dispersions also contain thermal stabilizers, pigments, fillers, antistatics, etc. On heating, particles "melt" (i.e., soften) and the material "gels" to a homogenous mass that can be processed to films, sheets, molded articles, etc.

PVC plastisols should have viscosities as low as possible. In the past, this was obtained by dilution with glycol ether, white spirits, etc. However, volatile diluents cause formation of bubbles whereas non-volatile ones tend to exude. Better effects are achieved by the addition of wetting agents in proportions of 0.5-2 phr (definition p. 652) that work best at low shear rates. Combinations of surfactants and oligomeric hydrocarbons work best.

3.4.4   Compatibilizers

For thermodynamic or kinetic reasons, many mixtures or blends of two different polymers A_p^* and B_q^* are heterogenous (Chapter 10). The microphases of these blends do not adhere well to each other and separate easily on mechanical stress, especially on impact. These undesirable effects can be reduced or even prevented by using so-called compatibilizers. Unfortunately, compatibilizers are expensive and they also require considerable energy to work them into the polymer melts.

Compatibilizers are bipolymers with long sequences A_n and B_m of each type of monomeric units A and B (Volume III, Section 8.5.2). They may be diblock polymers A_n–B_m, segmented (co)polymers A_n–B_m–A_r–B_s...A_t–B_u..., graft polymers of A_n segments on B_m chains or vice versa, triblock polymers A_n–B_m–A_n, etc. Segments A_n of compatibilizer molecules must be compatible with polymer segments A_p^* and compatibilizer segments B_m with polymer segments B_q^*.

Monomeric units A of compatibilizers need not be chemically identical with monomeric units A* of polymers and neither need B and B* be the same. For this reason, chemical structures of compatibilizers vary widely and so do their actions (see Section 10.4.6). Compatibilizers may also act as processing aids (Section 3.5) or impact modifiers; the latter action is mentioned in various other sections.

3.4.5   Impact Modifiers and Tougheners

The impact strength of thermoplastics is improved by mixing them with rubbers or elastomers (see Chapter 10). In the strict meaning of the words, these added elastic materials are impact modifiers. However, the term "impact modifier" is usually used only for certain elastomers that are added to unplasticized (hard) poly(vinyl chloride)s (U-PVC). The impact modification of other plastics is called toughening (see Section 10.4.7).

U-PVC from suspension polymerization consists of smaller primary particles that are aggregated to larger secondary particles (Section 8.2.4). Impact modifiers act on suspension U-PVC in either of two different ways.

So-called predefined impact modifiers are dispersed as spherical particles in the PVC matrix. This group comprises methyl methacrylate–butadiene–styrene copolymers (MBS), acrylic (ACR) and modified acrylic rubbers (MACR), acrylate–styrene–acrylonitrile terpolymers (ASA), and acrylonitrile–butadiene–styrene polymers (ABS).

Non-predefined impact modifiers encapsulate primary particles, which leads to an elastomeric network in the PVC matrix. This network is much more sensitive to shear than the dispersion of predefined modifiers which, in turn, leads to a smaller processing window. This group consists of chlorinated poly(ethylene) (CPE), ethylene–vinyl acetate copolymers (EVAC = EVA (USA)), acrylic elastomers (ACR), and polyurethane (PUR) and polyolefin (PO) elastomers.

3.4.6   Antifogging Agents

Antifogging agents prevent the condensation of water on the internal surface of packaging films. These water droplets are caused by the moisture released from packaged humid goods (fresh vegetables, meat). Their appearance depends on the relative humidities and the temperature difference between the good and the exterior air. If water droplets are formed on the packaging film, packaged goods can no longer be inspected.

Antifogging agents create transparent layers on the films. These layers are either not wettable by water so that no water droplets are produced or completely wettable so that a continuous water layer is produced. An example of the first type is silicone oil, an example of the second type, polyglycols.

3.4.7 Antistatic Agents and Conductivity Improvers

Most polymers are electrical insulators (Chapter 14), which makes them useful for many electrical and electronic parts and articles. However, the very same property also leads to electrostatic charging of plastics, elastomers, and fibers which may be aesthetically unpleasant (dust on surfaces, wavy trouser legs in contact with socks of other fibers) or disturb the acoustic pleasure of listening to vinyl records (dust). Electrostatic charging may also interfere with the processing of polymers to films and fibers, destroy parts by spark discharges, cause dust explosions, and/or lead to medical problems by deposition of hygienically questionable impurities.

Polymers are therefore often provided with antistatic agents (hydrophilic substances, often low-molecular weight) or conductivity improvers (electrically conducting fillers). **Antistatic agents** lower the electrical surface resistivity of conventional polymers from ca. 10^{14}–10^{16} Ω to ca. 10^8 Ω. External antistatics are only present on surfaces; their total concentrations are therefore small (ca. 10^{-4} %). They are usually applied only for temporary purposes such as for processing. Internal antistatics are distributed throughout the bulk of the polymer; they need to be applied in larger concentrations of 0.02–2.5 %.

Antistatics for plastics are amphiphilic compounds such as ethoxylated fatty amines, aliphatic sulfonates, polar fatty acid esters, quaternary ammonium compounds, etc. The hydrophobic part either adheres to the surface of the polymer (external antistatics) or is anchored in the polymer (internal antistatics). The hydrophilic part always faces the environment where it attracts humidity which in turn leads to a conducting "aqueous" surface. The same effect is produced in textiles by applying a silicone layer.

Conductivity improvers (metal powders, graphite, carbon black, organic semiconductors) are applied in much higher concentrations of 1–40 % They not only reduce the electrical surface resistivity but also decrease the volume resistivity ("specific resistivity") to ca. 0.1 Ω cm, i.e., increase the volume conductivity to ca. 10 S/cm.

Electrical conductivity requires direct contact of the dispersed conducting filler particles of micrometer size or larger (nano-"particles" behave differently). For the same filler fraction, the proportion of contacts increases with the aspect ratio of the particles (Figs. 3-5 and 9-6). Rodlike particles are therefore more efficient than spherical ones: rods need to be applied in smaller concentrations in order to achieve the same effect.

Each particle needs to be in contact with at least two other particles in order to provide conductivity. Electrical conductivities of polymers with electrically conducting spherical filler particles therefore increase with increasing weight fraction of spheres and become constant for $N \geq 2$ contacts (Fig. 3-5). Since the conductivity is provided by surface contacts, electrically conducting surfaces of filler particles suffice: ii is not necessary that the whole filler particle is electrically conducting.

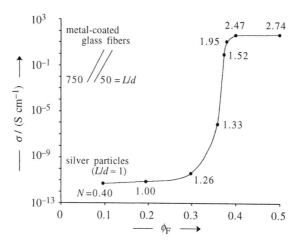

Fig. 3-5 Electrical conductivity σ of filled phenolic resins as function of the volume fraction ϕ_F of fillers (metal-coated glass fibers with aspect ratios of $L/d = 50$ or 750 [8] or spheroidal silver particles with $L/d \approx 1$ [9]). N = average numbers of particle-particle contacts per filler particle.

3.5 Processing Aids

Processing of polymers is improved by various processing aids which may be either low-molecular weight chemical compounds (such as waxes) or high-molecular weight polymers (such as silicones or fluoro polymers). The same additive can work in various ways so that its designation may not cover all its actions.

3.5.1 Lubricants

Lubricants are processing aids for polymer melts or polymeric powders that are applied in concentrations between 0.01 % and 0.1 %. External lubricants effect the rheological processes at the interface polymer/processing tools whereas internal lubricants improve the flow properties and the homogeneity of the melt. Both external and internal lubricants reduce frictional heat and thus act as heat regulators and physical thermostabilizers. Lubricants may also act as mold release agents (Section 3.5.3) and slip agents (Section 3.5.4).

Ca. 60 % of all lubricants are used for the processing of poly(vinyl chloride)s, especially for the production of sheets and bottles. Lubricants for poly(olefin)s are always combined with fillers, especially with chalk and talc.

Internal Lubricants

Phenomenologically, internal lubricants reduce viscosities and dissipate heat in high-speed processes. Molecular processes leading to these actions are not well understood.

Thermal processing of polymer powders or granulates, especially those from poly-(vinyl chloride) and poly(1-olefin)s, causes polymer particles to melt to smaller "flow units" of ca. 0.1-1 μm diameter and not to homogeneous liquids. These flow units are very stable agglomerates that are probably already formed during polymerization and dissolve completely only at very high temperatures. However, heating to higher temperatures requires more energy, leads to longer cycle times, and may cause polymer degradation. Hence, internal lubricants are used.

These agents reside predominantly in the surface layers of polymer agglomerates, similarly to surfactants at the water-air interface or diblock polymers at phase boundaries of heterogeneous polymer blends. The interpenetration of the surface region of the agglomerates by internal lubricants causes the surface regions to swell and merge with the already liquid regions. The resulting homogenization of the melt decreases both the Barus effect and the melt fracture (p. 94).

Internal lubricants thus act as external plasticizers for interior surface areas, hence the designation as "internal lubricants" or "slip additives" (not to be confused with "slip agents" (see below)).

The action of an internal lubricant is controlled by its affinity to the polymer. Typical internal lubricants for poly(olefin)s and poly(vinyl chloride) are partially soluble, modified esters of long-chain fatty acids, fatty alcohols, wax acids and esters, and metal soaps.

External Lubricants

External lubricants affect processes at interfaces: the friction between machinery walls and polymer powders and polymer melts, respectively, the friction between polymer particles themselves, and the adherence of articles to the tools.

External lubricants can act only if they form unstable yet controllable blends with the polymer. On one hand, they should not mix well with the polymer so that they can exude and enrich at the polymer surface. On the other hand, they should not be completely insoluble in the polymer so that they are not expelled from it. If the latter were the case, the external lubricant would congregate completely at the surface of the tool. The polymer melt would then stick at and turn around with the extruder screw and would not be extruded at all. Similarly, without adhesion of polymers to rolls, calenders would not function.

External lubricants should thus form a lubricating layer between the wall and the polymer melt. They must contain polar groups so that they can form dipole-dipole interactions with the oxide and hydroxide groups on the steel surface of the tools. This, in turn, would screen the polar metal surface by apolar groups of the external lubricant which improves the gliding of the polymer melt.

However, this interaction between lubricant molecules and tool surface should not be so strong that all lubricant molecules assemble on the walls of the tool. Some lubricant molecules must also reside in the polymer melt itself. External lubricants for polar polymers such as poly(vinyl chloride) are therefore always amphiphilic compounds such as metal stearates or fatty acid amides whereas those for apolar polymers are waxes. Yet, film blowing of linear-low density poly(ethylene) (LLDPE) employs fatty acid amides, especially erucic acid amide, as external lubricants, and also fluoropolymers.

Lubricant molecules on the surface of processing machinery are continuously removed with the processed polymers. They are replaced by new lubricant molecules from the interior of the polymer melt. It takes time to establish this dynamic equilibrium, and polymer processing thus requires a certain conditioning time.

3.5.2   Plasticating Agents

Plasticating agents are copolymers of methyl methacrylate and ethyl or butyl meth-acrylate, usually of very high molecular weight, that are added to poly(vinyl chloride)s in amounts of $(1\text{-}3)\cdot10^{-4}$ %. These very effective processing aids promote the disintegration of PVC particles and improve the melt strength (p. 99) by increasing the elongational viscosity. These actions are probably caused by the ability of the very long chains of the agents to provide entanglements to the molecules of the relatively low-molecular weight PVC molecules.

3.5.3   Parting Agents

Parting agents (release agents®) prevent the adhesion of molded parts at the walls of molds and improve their release. They act similarly to lubricating agents but should reside mainly on the surface of the mold and not in the surface of the melt.

Such parting agents are poorly wetting polymers with low critical surface tensions such as silicones or fluoropolymers that are applied directly to the surface of the mold. They are used in the processing of poly(1-olefin)s, polyurethanes, polycarbonates, polyesters, and epoxy resins. For polar resins, fatty acids and waxes are also used.

3.5.4   Antiblocking Agents

Cold flow or static electricity often causes polymer films to stick together, which can be prevented by application of adjuvants that are called **antiblocking agents** (PVC), **slip agents** (poly(1-olefin)s), or **slip depressants**. Their action causes films to lose their wavy surfaces; they are therefore also known as **flatt(en)ing agents**.

Antiblocking agents act similarly to external lubricating agents or parting agents. Such agents for poly(vinyl chloride)s are waxes, metal salts of fatty acids, silicones, fluoropolymers, or poly(vinyl alcohol). Thin polyolefin films are kept separate by fine particles of 20-30 μm diameter, for example, pyrogenic silicic acid (for LDPE, LLDPE, PP, PET), natural silicic acid (for LDPE), talcum (for LDPE, LLDPE), limestone (for PP), or zeolites (for LLDPE, PP, and also PVC).

3.6   Chemically Acting Adjuvants

3.6.1   Aging

The aging of a polymer is the unwanted change of its physical and chemical structure and the resulting change of its use properties with time (Table 3-11). Chemical aging of polymers is mostly caused by atmospheric agents during processing (heat, oxidation) and use (oxidation, photochemical reactions, hydrolysis). Physical aging results from mechanical forces that act over long times, even without chemical forces.

Table 3-11 The resistance of some commercial polymers to various aging effects [10]. ++++ High, +++ moderate, ++ fair, + poor, - very poor.

Polymer	Resistance against			
	Hydrolysis	Oxidation, thermal	Oxidation, photochemical	Radiation, gamma
Poly(tetrafluoroethylene)	++++	++++	++++	-
Poly(styrene), at-	++++	++	+	+++
Poly(ethylene)	++++	+	+	++
Poly(propylene), it-	++++	-	-	+
Poly(vinyl chloride), at-	++++	-	+	+
Poly(dimethylsiloxane)	+++	++++	++++	
Poly(*p*-phenylene terephthalamide)	+++	++++	-	-
Poly(methyl methacrylate), at-	+++	+++	++++	++
Poly(ethylene terephthalate)	+++	+++	+++	+++
Polycarbonate, bisphenol A-	+++	++	+	+++
Poly(2,6-dimethyl-1,4-oxyphenylene)	+++	-	++	+++
Poly(acrylonitrile), at-	+++	+	+++	++
Polyamides 6 and 6.6	++	++	++	++
Poly(oxymethylene)	++	+	+	-

Chemical aging by oxidation can be prevented or at least reduced by added antioxidants (Section 3.6.4) and that by photooxidation by light stabilizers (Section 3.6.5). Poly(vinyl chloride) is often provided with heat stabilizers (Section 3.6.3).

Processes leading to combustion are similar to those caused by thermal and oxidative degradation (Section 3.6.6). For protection against ignition, flame propagation, and smoke development, polymers are provided with flame retardants (Section 3.6.6).

Combustion, on the other hand, is also used for the thermal recycling to low-molecular weight raw materials in low-temperature pyrolysis and for complete incineration for the generation of energy. Polymers may also be disposed of by "biological" degradation by the action of air, light, and water, and, less often, by microbes.

3.6.2 Degradation

"Degradation" is not well defined. In general, it refers to undesirable and often uncontrolled chemical reactions that may lead to a decrease of molecular weight, a change of the constitution of the polymer by side reactions, or a combination of both.

The decrease of molecular weight without side reactions may be caused by chain scission or depolymerization or a combination of both types of reactions. **Chain scission** is a random splitting of chains that generates degraded polymers with lower degrees of polymerization. **Depolymerization** is an unzipping reaction that leads to a mixture of monomer molecules and unreacted polymer molecules.

Chain Scission

Chain scissions are reverse polyadditions if they proceed without participation of small molecules QL, schematically,

(3-8) $M_{i+j} \rightarrow M_i + M_j \rightarrow M_{i-k} + M_k + M_{j-m} + M_m$ etc.

and reverse polycondensations if they do:

(3-9) $M_{i+j} + QL \rightarrow M_iQ + LM_j$ etc.

In both types of chain scissions, molecular weights decrease with increasing extent of chain scissions. Molecular non-uniformities ("polydispersities") of the resulting polymers vary with the initial non-uniformity.

For example, chain scission of linear polymers from the equilibrium polycondensation of AB monomers leads to a continuous decrease from $(\overline{X}_w/\overline{X}_n)_0 = 2$ to $(\overline{X}_w/\overline{X}_n)_\infty = 1$. Chain scission of uniform polymers with $(\overline{X}_w/\overline{X}_n)_0 = 1$ leads first to non-uniform polymers with $\overline{X}_w/\overline{X}_n > 1$, then to a maximum of $\overline{X}_w/\overline{X}_n$, and finally to a decrease of $\overline{X}_w/\overline{X}_n$ until complete scission to a mixture of monomers with $(\overline{X}_w/\overline{X}_n)_\infty = 1$.

The rate of chain scission depends on the concentration [C] of the catalyst and the proportion of the remaining chain bonds. For polymers with Schulz-Flory distributions of degrees of polymerization (Volume I, Section 13.2) and only one type of cleavable chain bonds, inverse degrees of polymerization increase linearly with time,

(3-10) $1/\overline{X}_g = 1/\overline{X}_{g,0} + Ak \, [C] \, t$

where $\overline{X}_{g,0}$ = initial degree of polymerization and k = rate constant. Eq.(3-10) applies to both number-average (g = n; A = 1) and mass-average (g = w; A = 1/2) degrees of polymerization.

Depolymerization

Depolymerizations are unzipping reactions that start at active chain ends *,

(3-11) $\sim M_i^* \rightarrow \sim M_{i-1}^* + M; \quad \sim M_{i-1}^* \rightarrow \sim M_{i-2}^* + M;$ etc.

or proceed by intramolecular back-biting reactions (example: certain silicones). Activated chain ends (* = $^\ominus$, $^\oplus$, $^\bullet$) are formed directly in living polymerizations with macroanions $\sim M_i^\ominus$ or macrocations $\sim M_i^\oplus$. Depolymerizations via macroradicals, require a homolytic scission of chains, $\sim M_i{-}M_j\sim \rightarrow \sim M_i^\bullet + {}^\bullet M_j\sim$, as the start reaction.

Depolymerizations proceed without side reactions if substituents and their bonds to the main chain are much more stable than chain bonds of the main chain. Theory predicts a linear decrease of the number-average degree of polymerization, $\overline{X}_n$, with the fraction of generated monomer molecules, f_M, if the depolymerization is living and termination and transfer reactions can be neglected:

(3-12) $\overline{X}_n = [1 - f_M] \, \overline{X}_{n,0}$

Eq.(3-12) is usually found for *high* degrees of polymerization and small fractions of generated monomer molecules. At larger degrees of depolymerization, termination and transfer reactions cannot be neglected which leads to dead polymer chains that can depolymerize only after homolysis.

Termination reactions decrease the probability of the formation of monomer molecules. This probability can be described by the **zip length** Ξ which is the number of monomer molecules that is generated per *kinetic* chain. The actual monomer yield depends on the type and extent of side reactions.

Hence, small values of Ξ do not necessarily indicate low monomer yields. At 300°C, poly(ethylene) thermally degrades with a zip length of 0.01 and generates 21 mol% ethene whereas poly(styrene) proceeds with a zip length of 3.1 to a yield of 65 mol% styrene. Poly(α-methylstyrene), on the other hand, has a low thermodynamic ceiling temperature (Volume I, p. 199), a zip length of more than 200, and generates 100 % monomeric α-methyl styrene because of the absence of side reactions.

Thermal Degradation, Thermo-oxidative Degradation, and Pyrolysis

At higher temperatures, chain scissions and depolymerizations are always accompanied by side reactions. This thermal degradation by heat alone can be reduced by heat stabilizers (Section 3.6.3). Thermo-oxidative degradations can be prevented by anti-oxidants (Section 3.6.4). Thermo-oxidative reactions in flames lead to pyrolysis which can be battled with flame retardants (Section 3.6.6)

Pyrolyses occur relatively easily if polymer molecules possess long sequences of methylene groups, electron-attracting groups, branching points, or segments that can react to five-membered or six-membered rings. Conversely, thermal stability is increased for those polymers that contain aromatic rings or ladder structures and/or are rich in fluorine as substituents and poor in hydrogen substituents (see also p. 69).

3.6.3   Heat Stabilization

At temperatures above ca. 120°C, homopolymers and copolymers of vinyl chloride split off hydrogen chloride and change color from off-white to yellow, brown, and black. Simultaneously, mechanical properties deteriorate.

Poly(vinyl chloride)s are stabilized against this degradation by added **heat stabilizers** which are inorganic and organic derivatives of lead or organic compounds of barium, cadmium, zinc, and tin. These primary stabilizers are often combined with secondary ones such as organophosphites, dicyanodiamides, or epoxidized vegetable oils.

The mechanism of thermal dehydrochlorination is complex and still not fully understood. Investigations of model compounds have found that poly(vinyl chloride) with the ideal chain constitution $\sim$CH$_2$–CHCl–CH$_2$–CHCl....CH$_2$–CHCl$\sim$ is thermally stable. A spontaneous dehydrochlorination according to

(3-13)   – HCl

can thus be excluded. Hence, dehydrochlorinations must start at "wrong" chain structures that were formed during polymerization.

Newer investigations have also ruled out that dehydrochlorination starts at ketoallyl structures, Eq.(3-14), or at tertiary (allylic) chlorine atoms, Eq.(3-15):

(3-14)

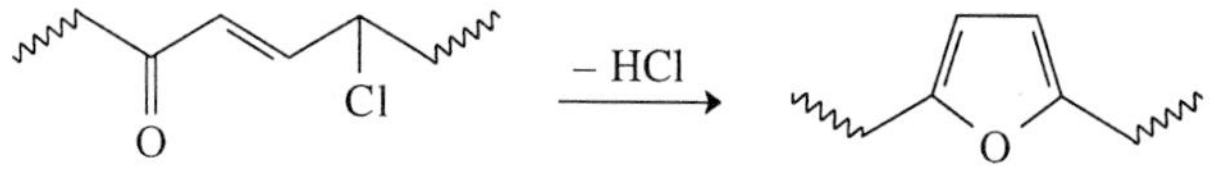

(3-15)

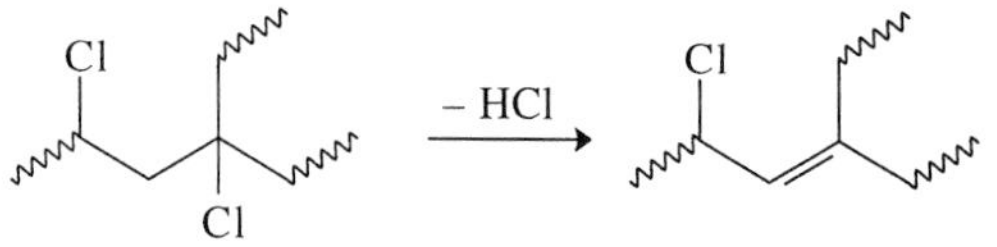

Poly(vinyl chloride)s do contain ca. 0.5-1.5 double bonds per 1000 carbon atoms; these double bonds may be endgroups or within the chain itself. During polymerization double bonds at chain ends are generated by chain transfer reactions to monomer molecules and double bonds within the chain by chain transfer to polymer molecules.

Dehydrochlorinations are assumed to start at these double bonds and to proceed in a zip reaction:

(3-16) $\sim\sim$CH$=$CH$-$CHCl$-$CH$_2\sim\sim$ $\longrightarrow$ $\sim\sim$CH$=$CH$-$CH$=$CH$\sim\sim$ + HCl, etc.

The color becomes yellowish for sequences of six to seven conjugated double bonds and intensifies with increasing length of conjugation.

HCl develops faster in air than in nitrogen, probably because radicals are generated from the decomposition of hydroperoxides that were formed by reaction of polymer chains with oxygen. These radicals react with PVC chains and generate reactive sites at which dehydrochlorination can start. Reactions in air also lead to the formation of other oxygen-containing groups, such as carbonyl groups

Discoloration also seems to be caused by charge-transfer complexes of double bonds with hydrochloric acid which can explain the often postulated "autocatalytic" action of HCl, i.e., the deepening of color in the presence of HCl caused by extensions of the length of polyene sequences:

(3-17)

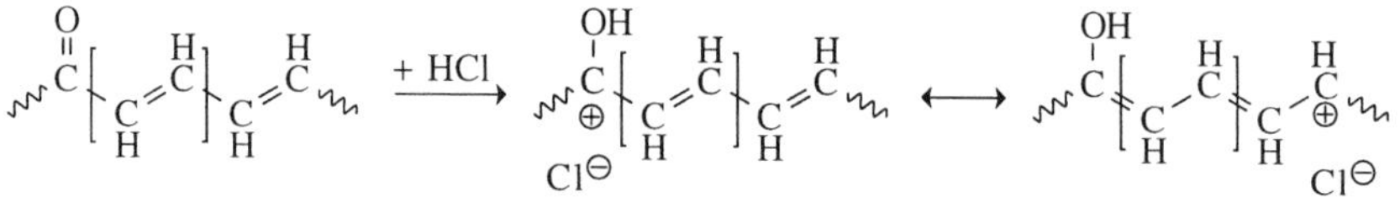

Both free-radical and ionic mechanisms have been proposed for dehydrochlorinations. According to some publications, ionic mechanisms dominate at lower temperatures and free-radical ones at higher temperatures.

The effect of primary stabilizers is not known in detail. They certainly do not prevent dehydrochlorination but will decrease effects generated by the production of HCl. All primary stabilizers can react with HCl, for example, metal carboxylates according to

(3-18)   Zn(OOCR)$_2$ $\xrightarrow[- \text{RCOOH}]{+ \text{HCl}}$ ZnCl(OOCR) $\xrightarrow[- \text{RCOOH}]{+ \text{HCl}}$ ZnCl$_2$

The resulting zinc chloride then forms a coordinatively unsaturated zinc complex with the generated carboxylic acid:

(3-19) $RCOOH + ZnCl_2 \longrightarrow$

$$R-C \overset{O}{\underset{O}{<}} Zn \overset{Cl}{<} + HCl$$

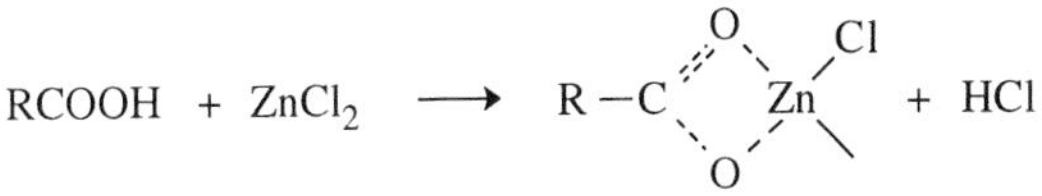

This complex reacts with the polyene sequences of dehydrochlorinated poly(vinyl chloride) chains. The polyene sequence is interrupted and the discoloration decreases:

(3-20)

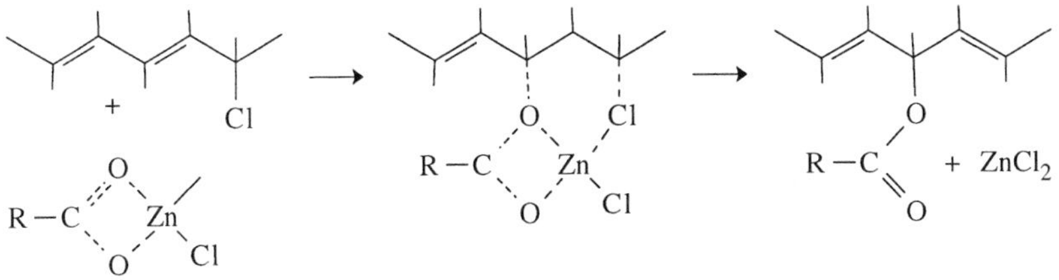

Poly(vinyl chloride) discolors not only thermally at elevated temperatures but also by photochemical oxidation which results in the formation of conjugated systems and the release of hydrochloric acid:

(3-21)

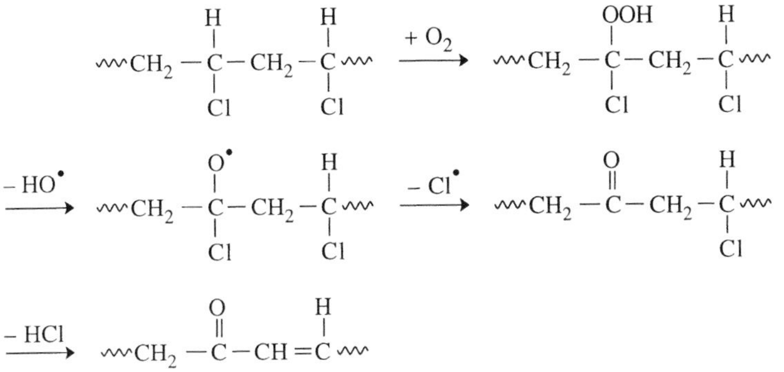

3.6.4 Thermo-oxidative Stabilization

Oxidation Processes

Polymer molecules RH react with oxygen O_2, ozone O_3, and nitrogen oxide NO in many ways. The direct formation of free radicals according to

(3-22) $RH + O_2 \rightarrow R^\bullet + {}^\bullet OOH$

(3-23) $RH + O_3 \rightarrow RO^\bullet + {}^\bullet OOH$

is very slow for oxygen, unproven for ozone, and not present for nitrogen oxide. Oxygen rather reacts with polymer molecules first to hydroperoxides which subsequently decompose to oxy and hydroxy radicals:

(3-24) $RH + O_2 \rightarrow ROOH \rightarrow RO^\bullet + {}^\bullet OH$

Hydroperoxides ROOH decompose in reactions that are induced and accelerated by transition metal ions Mt^{n+} and may by oxidizing or reducing, depending on the redox potential:

(3-25) $ROOH + Mt^{N\oplus} \rightarrow RO^{\bullet} + OH^{\ominus} + Mt^{(N+1)\oplus}$

(3-26) $ROOH + Mt^{N\oplus} \rightarrow ROO^{\bullet} + H^{\oplus} + Mt^{(N-1)\oplus}$

Peroxy radicals attack the polymer and generate hydrocarbon radicals that add triplet oxygen in a very fast reaction:

(3-27) $ROO^{\bullet} + RH \rightarrow ROOH + R^{\bullet}$

(3-28) $R^{\bullet} + {}^3O_2 \rightarrow ROO^{\bullet}$

The oxidation therefore leads to peroxy radicals $ROO^{\bullet}$ (hydroperoxy radicals $HOO^{\bullet}$ if R = H) by reactions 3-26 and 3-28; oxy radicals $RO^{\bullet}$ and hydroxy radicals $HO^{\bullet}$, respectively, by reactions 3-24 and 3-25; and alkyl radicals $R^{\bullet}$ by reaction 3-27. These radicals react further, depending on the substrate.

The oxidation of hydrocarbons is accompanied by a weak light emission in the blue-violet range. The quantum yield is very low, only 10^{-8} to 10^{-10}. It is assumed that this chemoluminescence results from a deactivation of activated carbonyl groups. Such groups result from the combination of two peroxy radicals (Russell mechanism):

(3-29) $HC{-}O{-}O^{\bullet} + {}^{\bullet}O{-}O{-}CH \longrightarrow C{=}O^* + O_2 + {}^{\bullet}O{-}CH$

Chemoluminescence is directly proportional to the oxidation rate. Because its sensitivity is much greater than that of other methods, it allows one to measure very small degrees of oxidation. The method showed that oxidations of poly(propylene), polydienes, ethylene-propylene-diene polymers, etc., are induced by stress.

According to Eq.(3-29), *peroxy radicals* form not only keto groups, >C=O, but also oxy radicals >CH–O$^{\bullet}$ that, in turn, generate alkyl radicals and hydroxy groups according to >CH–O$^{\bullet}$ + RH → >CH–OH + R$^{\bullet}$. With saturated hydrocarbons, peroxy radicals also lead to hydroperoxide (Eq.(3-26)). The rate of reaction is highest for tertiary hydrogen atoms, smaller for secondary ones, and smallest for primary ones. Peroxy radicals in poly(olefin)s lead to neighboring group effects:

(3-30)

In poly(acrylonitrile), $+CH_2{-}CH(CN)+_n$, the directing effect of the nitrile group leads to an attack on the methylene group instead of the tertiary hydrogen atom.

Peroxy radicals react with olefins by either addition to the olefinic double bond or by formation of an epoxy group:

(3-31) $ROO^\bullet$ + C=C → $ROO-C-C^\bullet$

 $RO^\bullet$ + C—C (with O bridge)

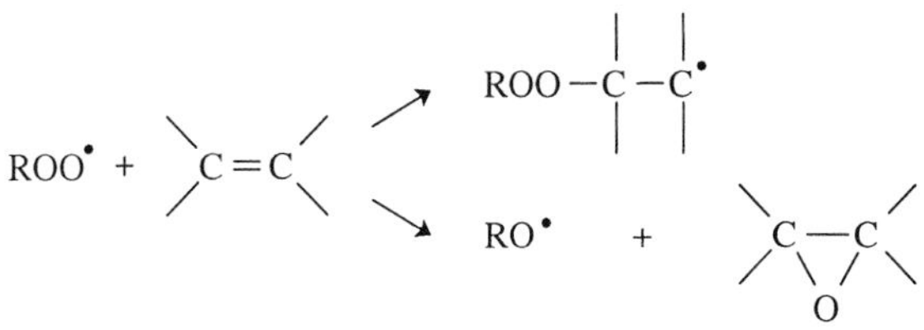

Peroxy radicals also add to conjugated double bonds:

(3-32) $ROO^\bullet$ + ⌇⌇CH=CH−CH=CH⌇⌇ → ⌇⌇CH−CH=CH−$\overset{\bullet}{C}$H⌇⌇

 OOR

Oxyradicals $RO^\bullet$ (or hydroxy radicals $HO^\bullet$) lead to hydrogen transfer (Eq.(3-33)), induced decomposition of hydroperoxides (Eq.(3-34)), or addition to double bonds (Eq.(3-35)), depending on the substrate:

(3-33) $RO^\bullet$ + RH ⟶ ROH + $RO^\bullet$

(3-34) $RO^\bullet$ + ROOH ⟶ ROH + $ROO^\bullet$

(3-35) $RO^\bullet$ + ⌇⌇CH=CH⌇⌇ ⟶ ⌇⌇CH−CH⌇⌇

 RO

Polymer molecules are crosslinked by combination of two radicals: from reaction 3-35 to >CH–CH< bridges, from radicals of reactions 3-33 and 3-35 to >CH–O–R bridges, and from reaction 3-34 with release of oxygen to peroxide groups, 2 ~CH($OO^\bullet$)~ → >C(H)–O–O–C(H)< + O_2. However, such combinations are not very probable because they require the diffusion of radicals which is very slow in solid polymers. Indeed, cross-linking was never observed in the autoxidation of truly saturated poly(ethylene)s, i.e., poly(ethylene)s in which vinyl groups from side reactions during polymerization were removed by after-hydrogenation.

Antioxidants

Oxidation of polymers by oxygen from air is diminished if (a) oxidizable groups are less accessible and/or (b) oxidation is supressed or reversed by antioxidants. For the first reason, semicrystalline polymers are oxidized much more slowly than amorphous ones since crystalline regions are less accessible to oxygen than amorphous ones. The diffusion of oxygen into polymers can also be reduced by addition of surface protecting agents such as waxes that are added in proportions of ca. 1 %.

In general, protection against oxidation is achieved by addition of antixoxidants, a subgroup of **stabilizers**. The world consumption of antioxidants is now ca. 250 000 t/a. In the plastics industry, more than 90 % of antioxidants are used for polymeric hydrocarbons such as poly(ethylene)s, poly(propylene)s, poly(styrene)s, and ABS polymers (polymers from styrene and poly(acrylonitrile-*co*-butadiene)).

Antioxidants are subdivided into primary and secondary ones. **Primary antioxidants (chain terminators, radical scavengers)** interfere with the kinetic chain reaction and destroy radicals. **Secondary antioxidants** are also known as **deinitiators**.

Secondary Antioxidants prevent the formation of hydroperoxides or regulate their decomposition is such a way that fewer radicals are formed. Deinitiators include peroxide deactivators, metal deactivators, and UV absorbers (which belong to the group of light stabilizers (Section 3.6.5)).

Peroxide deactivators decompose hydroperoxides and peroxides before they can form radicals. The main groups of peroxide deactivators comprise organophosphorus compounds (global: ca. 65 000 t/a (1997)) and thioethers (global: ca. 20 000 t/a (1997)). Organophosphorus compounds are either triaryl phosphites $P(OR)_3$ or phosphonites $(RO)_2PR'$:

P-1, P-3   P-4, P-5   P-2

with R:

P-1, P-2, P-5   P-3   P-4

Both organophosphites and thioethers as peroxide deactivators contain large hydrocarbon groups in order to improve the compatibility of their active groups with polymer matrices. Such thioethers R"–S–R" are termed S-1 with R" = $-CH_2CH_2COOC_{18}H_{37}$, S-2 with $-C_{18}H_{37}$, and S-3 with $-CH_2CH_2COOC_{12}H_{25}$,

Hydroperoxides oxidize organic phosphites to esters of orthophosphoric acid R_3PO_4 (Eq.(3-36)) and thioethers to sulfoxides (Eq.(3-37)) and further to sulfones (Eq.(3-38)):

(3-36) $(RO)_3P + R'OOH \longrightarrow (RO)_3PO + R'OH$

(3-37) $(R")_2S + R'OOH \longrightarrow (R")_2SO + R'OH$

(3-38) $(R")_2SO + R'OOH \longrightarrow (R")_2SO_2 + R'OH$

Sulfoxides from Eq.(3-37) decompose thermally to sulfenic acids, R"–S–OH, which are then oxidized to sulfonic acid $R"SO_3H$ and finally decompose to various compounds, for example, the sulfoxide from S-3:

(3-39)

A single thioether molecule can therefore destroy many hydroperoxide molecules (Eqs.(3-37)-(3-39)). This **thiosynergism** makes thio compounds good deinitiators.

Primary antioxidants are **radical scavengers** that terminate *kinetic* chains. This group comprises mainly hindered phenols (1997: ca. 120 000 t/a) and aromatic amines.

A large number of phenolic chain terminators are based on the double butylated hydroxy toluene (BHT) = 2,6-di(*t*-butyl)-*p*-cresol (DBMC) [A-1], for example:

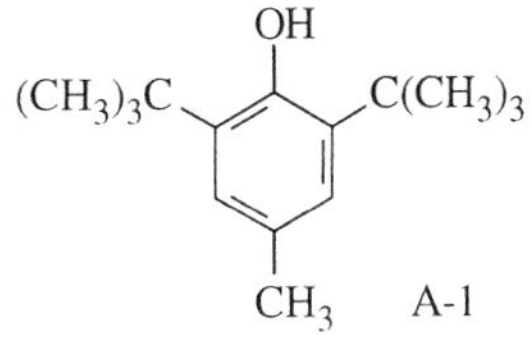
A-1

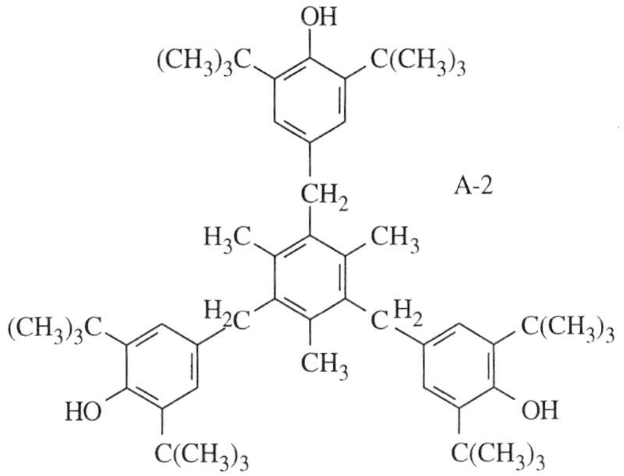
A-2

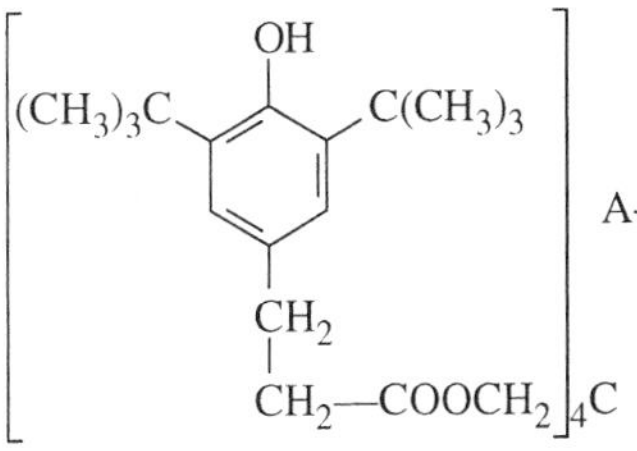
A-3

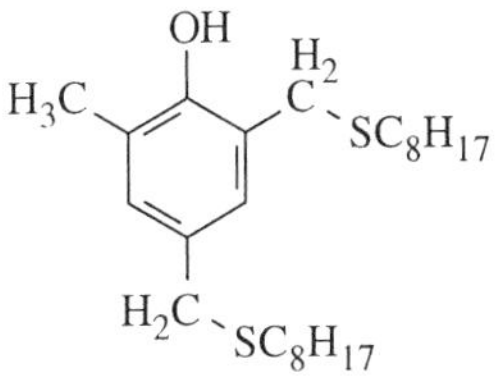
A-4

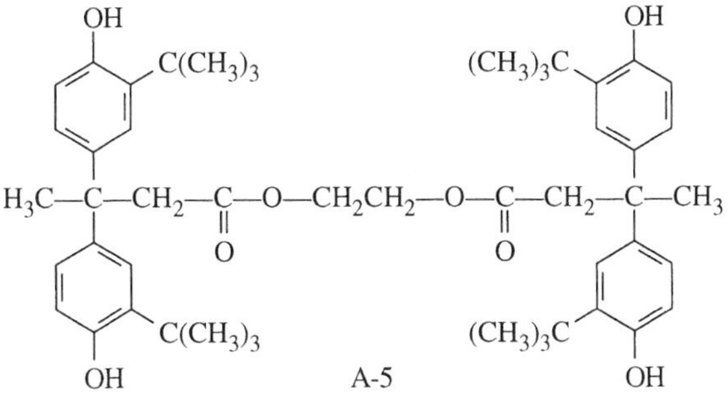
A-5

Sterically hindered phenols act as hydrogen donors and with suitable substitution also as radical scavengers that destroy two radicals per phenolic group, for example, BHT:

(3-40)

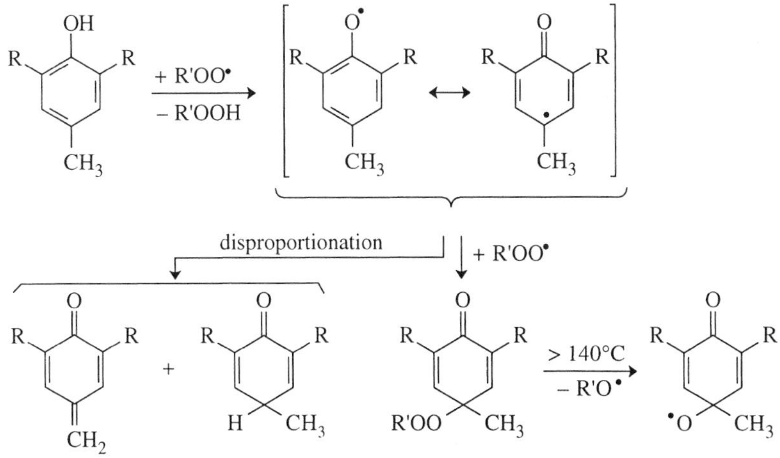

Reactions (3-40) proceed probably via a primary π complex between BHT and radicals ROO˙. BHT is used for poly(olefin)s, poly(styrene)s, vinyl polymers, and diene elastomers but no longer for food packaging in Europe. Substituted phenols become inefficient at temperatures above ca. 140°C since they form new radicals by homolysis.

Secondary aromatic, non-sterically hindered amines such as

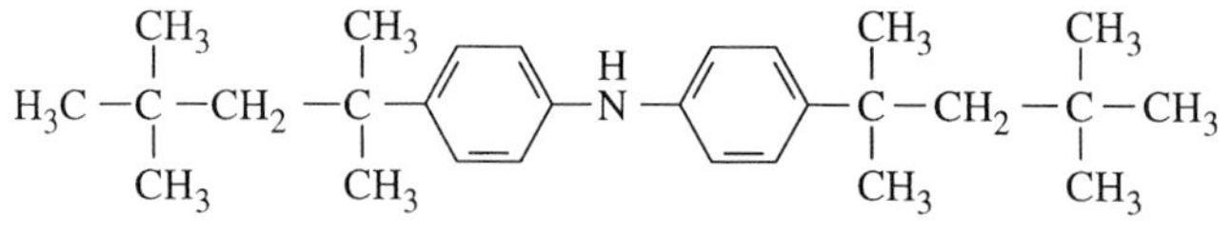

are also proton donors. They are more effective than sterically hindered phenols but lead to discolored polymers. For this reason, they are used mainly for dark-colored articles such as carbon black-filled elastomers and styrene polymers, tire cords from polyamide fibers, flexible polyurethane foams, and melt adhesives based on polyamides.

Sterically *hindered amine stabilizers* (HAS) were originally developed as *hindered amine light stabilizers* and are therefore also known as HALS. All industrially used HALS contain 2,6-dimethyl substituted piperidine groups as effective groups:

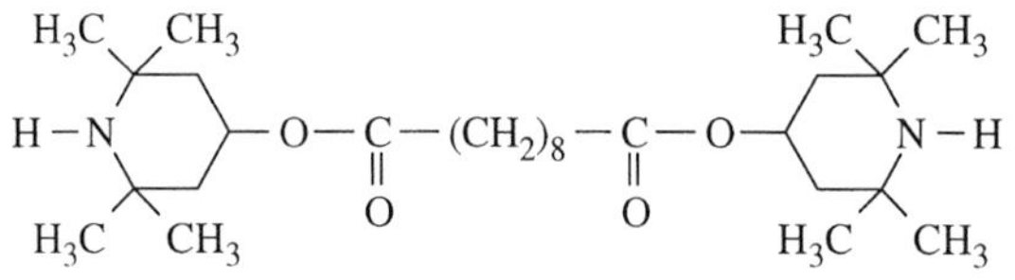

HALS-1

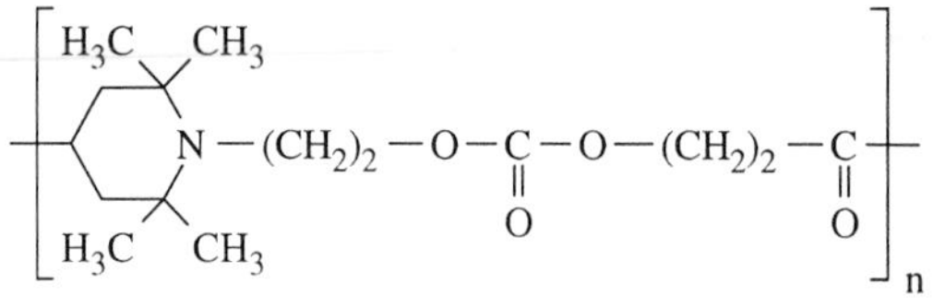

HALS-2

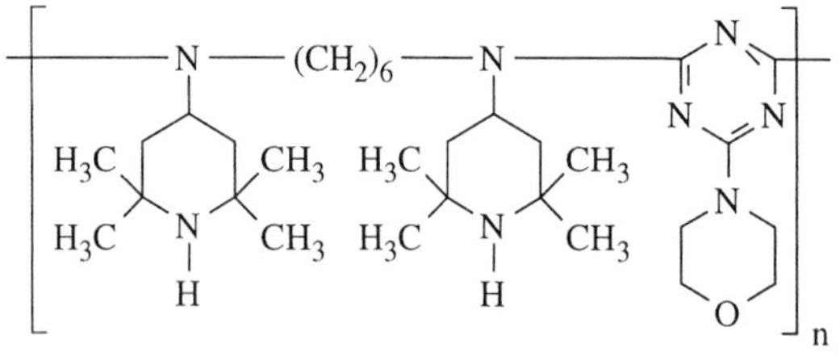

HALS-3

These amines do not react directly with radicals. As shown in Eq.(3-41), they rather form radical cations such as I from HALS-1 by direct photolysis or electron transfer. The radical cation I deprotonates to II that reacts with oxygen to form radicals >NOO˙ (III) which add radicals R˙ to give IV. Heat and light during the use of the polymer articles causes groups >N–O–O–R to dissociate into nitroxyl radicals (aminoxyls) >N–O˙ and radicals RO˙.

HALS-1 is too volatile at higher processing temperatures. It is therefore combined with peroxide deactivators such as organophosphites or organophosphonates.

(3-41)

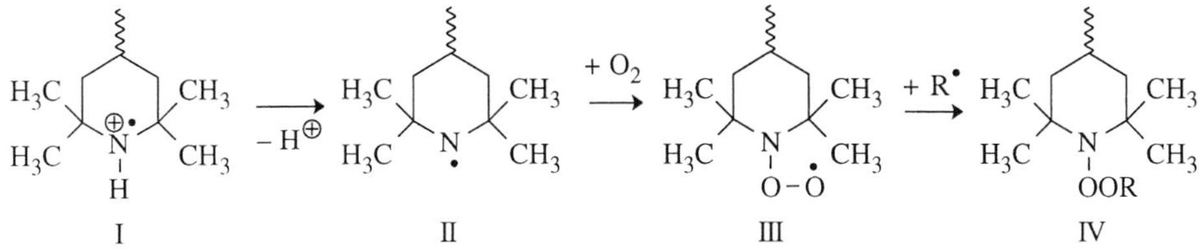

Several other types of stabilizers are used for special purposes, for example, *zinc dibutyl dithio carbamate* V that reacts in a very complex way. It is first oxidized by hydroperoxides (Eq.(3-42)), probably to an unstable sulfonate VI which decomposes thermally and releases SO_2:

(3-42)
$$\underset{\text{V}}{(Bu_2N-C(=S)-S)_2Zn} \xrightarrow[-\,2\,ROH]{+\,2\,ROOH} \underset{\text{VI}}{(Bu_2N-C(=S)-S(=O)_2-O)_2Zn}$$
$$\xrightarrow[-\,2\,SO_2]{} (Bu_2N-C(=S)-O)_2Zn \xrightarrow[-\,Zn(OBu)_2]{} 2\ Bu-N{=}C{=}S$$

Such a decomposition to an isothiocyanate and SO_2 is known for the model compound dialkylbenzthiazole-2-sulfonic acid.

The sulfur dioxide from this decomposition catalyzes the decomposition of hydroperoxides, for example, that of cumene hydroperoxide:

(3-43)
$$C_6H_5-C(CH_3)_2-OOH \xrightarrow{+\,SO_2} C_6H_5-C(CH_3)_2-O^{\oplus} + HSO_3^{\ominus}$$
$$\xrightarrow[-\,SO_2]{} C_6H_5OH + (CH_3)_2CO$$

The combination of deinitiators and kinetic chain terminators often increases inhibition times and reduces the rates of oxygen uptake more than the sum of the two single reactions. These synergistic effects occur because both compounds react successively.

Antagonistic effects are also known. An example is the adsorption of certain antioxidants by carbon black which renders the antioxidants ineffective.

3.6.5 Light Stabilization

Processes

Plastics, polymer fibers, polymer coatings, etc., are attacked by the combined action of sunlight and oxygen (**photo-oxidation**). The sun emits radiation with wavelengths greater than 100 nm. Sunlight with wavelengths between 100 nm and 175 nm (ultraviolet) is absorbed by the oxygen of the upper atmosphere (ca. 100 km above earth) which generates ozone that forms an ozone layer in the stratosphere at heights between ca. 15 km and 40 km. This layer absorbs radiation mainly between 185 nm and 290 nm (maximum at 253 nm), but also somewhat between 290 nm and 330 nm.

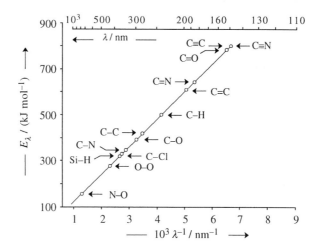

Fig. 3-6 Dissociation energy E_λ of chemical bonds as a function of the inverse wavelength λ of incident light.

The remaining ultraviolet radiation can be absorbed directly by chemical compounds. However, light is also scattered by the atmosphere. This effect is stronger, the shorter the wavelength. Since light is scattered in all directions, chemical compounds can also absorb indirect light.

Depending on the type of chemical bond, absorbed light may have sufficient energy to dissociate covalent bonds (Fig. 3-6). In the ultraviolet region, only multiple bonds can absorb light and one should therefore expect that polymers with such bonds will be especially prone to degradation by ultraviolet light.

For example, UV light causes homolytic chain scissions in *cis*-1,4-poly(isoprene):

$$
\begin{array}{c}
\text{(3-44)} \quad \text{wwCH}_2\!-\!\overset{\overset{\displaystyle CH_3}{|}}{C}\!=\!CH\!-\!CH_2\!-\!CH_2\!-\!\overset{\overset{\displaystyle CH_3}{|}}{C}\!=\!CH\!-\!CH_2\text{ww}\\[2mm]
\xrightarrow{h\nu}\quad \text{wwCH}_2\!-\!\overset{\overset{\displaystyle CH_3}{|}}{C}\!=\!CH\!-\!\overset{\bullet}{C}H_2 \;+\; \overset{\bullet}{C}H_2\!-\!\overset{\overset{\displaystyle CH_3}{|}}{C}\!=\!CH\!-\!CH_2\text{ww}
\end{array}
$$

The resulting macroradicals have radical sites at chain ends. Subsequent chain transfer reactions lead to resonance-stabilized radical sites within the chains

$$
\begin{array}{c}
\text{(3-45)} \quad \text{wwCH}_2\!-\!\overset{\overset{\displaystyle CH_3}{|}}{C}\!=\!CH\!-\!\overset{\bullet}{C}H_2 \;+\; \text{wwCH}_2\!-\!\overset{\overset{\displaystyle CH_3}{|}}{C}\!=\!CH\!-\!CH_2\text{ww}\\[2mm]
\longrightarrow \text{wwCH}_2\!-\!\overset{\overset{\displaystyle CH_3}{|}}{C}\!=\!CH\!-\!CH_3 \;+\; \text{wwCH}_2\!-\!\overset{\overset{\displaystyle CH_3}{|}}{C}\!=\!CH\!-\!\overset{\bullet}{C}H\text{ww}
\end{array}
$$

The combination of such radicals leads to crosslinked polymers.

Crosslinking is not observed if radicals from the action of UV light are not at all or only a little resonance stabilized. An example is the formation of radicals in poly(methyl methacrylate) which leads to chain disproportionation reactions and therefore to chain scission and not to crosslinking:

$$\underset{\substack{| \\ \text{COOCH}_3 \ \ \text{COOCH}_3}}{\text{wwCH}_2\text{—}\overset{\overset{\displaystyle \text{CH}_3}{|}}{\text{C}}\text{—CH}_2\text{—}\overset{\overset{\displaystyle \text{CH}_3}{|}}{\text{C}}\text{ww}} \xrightarrow{\text{– }^{\bullet}\text{COOCH}_3} \text{wwCH}_2\text{—}\overset{\overset{\displaystyle \text{CH}_3}{|}}{\underset{\underset{\text{COOCH}_3}{|}}{\text{C}}}\text{—CH}_2\text{—}\overset{\overset{\displaystyle \text{CH}_3}{|}}{\underset{\bullet}{\text{C}}}\text{ww}$$

(3-46)

$$\longrightarrow \ \text{wwCH}_2\text{—}\overset{\overset{\displaystyle \text{CH}_3}{|}}{\underset{\underset{\text{COOCH}_3}{|}}{\text{C}}}{}^{\bullet} \ \ + \ \text{CH}_2\text{=}\overset{\overset{\displaystyle \text{CH}_3}{|}}{\underset{\underset{\text{COOCH}_3}{|}}{\text{C}}}\text{ww}$$

Hence, one would expect that saturated polymers such as poly(ethylene) would not absorb ultraviolet light and therefore neither crosslink nor degrade to lower molecular weights. However, industrial polymers always absorb UV light, for example, poly-(methyl methacrylate)s at (290-315) nm, poly(ethylene)s at 300 nm, poly(vinyl chloride)s at 310 nm, and poly(styrene)s at 318 nm.

These absorptions are caused by structural mistakes from polymerization and/or processing, or stem from impurities. The nature of the absorbing groups is usually not known with certainty since they are often present in very small proportions.

Examples are hydroperoxide, peroxide, and carbonyl groups as well as carbon-carbon double bonds in industrial poly(ethylene)s, it-poly(propylene)s, and poly(vinyl chloride)s. In industrial poly(styrene), stilbene groups, $-C_6H_4-CH=CH-C_6H_4-$, were found in the main chain and phenyl ketone groups, $-CH_2-CO-C_6H_5$, as endgroups. Ultraviolet absorption of poly(ethylene)s and it-poly(propylene)s is also caused by catalyst residues and that of poly(vinyl chloride)s by initiator residues and fragments. More than 80 % of all UV absorbers are therefore used for poly(olefin)s.

On irradiation by UV light, polymers with hydroperoxide groups are degraded by free-radical reactions

(3-47)

$$\underset{\underset{\text{OOH}}{|}}{\overset{\overset{\displaystyle R}{|}}{\text{ww}\overset{|}{\text{C}}\text{ww}}} \xrightarrow[\text{– HO}^{\bullet}]{hv} \underset{\underset{\text{O}^{\bullet}}{|}}{\overset{\overset{\displaystyle R}{|}}{\text{ww}\overset{|}{\text{C}}\text{ww}}} \overset{\nearrow}{\underset{\searrow}{}} \begin{array}{l} \text{ww}\overset{\overset{\displaystyle}{||}}{\underset{\text{O}}{\text{C}}}\text{ww} \ \ + \ \ \text{R}^{\bullet} \\[2em] \text{ww}\overset{\overset{\displaystyle}{||}}{\underset{\text{O}}{\text{C}}}\text{—R} \ + \ {}^{\bullet}\text{ww} \end{array}$$

which can be stopped by radical catchers. However, in the degradation of poly(ethylene terephthalate), only the top reaction is a radical process whereas the bottom one is a molecular process which cannot be prevented by common antioxidants but only by special light stabilizers:

(3-48)

$$\text{ww}\!\!\left\langle\!\!\bigcirc\!\!\right\rangle\!\!\overset{\overset{\displaystyle}{||}}{\underset{\text{O}}{\text{C}}}\text{—OCH}_2\text{CH}_2\text{ww} \ \overset{\nearrow}{\underset{\searrow}{}} \begin{array}{l} \text{ww}\!\!\left\langle\!\!\bigcirc\!\!\right\rangle\!\!\overset{\overset{\displaystyle}{||}}{\underset{\text{O}}{\text{C}}}{}^{\bullet} \ + \ {}^{\bullet}\text{OCH}_2\text{CH}_2\text{ww} \\[2em] \text{ww}\!\!\left\langle\!\!\bigcirc\!\!\right\rangle\!\!\overset{\overset{\displaystyle}{||}}{\underset{\text{O}}{\text{C}}}\text{—OH} \ + \ \text{CH}_2\text{=CH ww} \end{array}$$

The extent of degradation depends on the weathering which can occur in the open-air atmosphere in various parts of the world or be produced artificially by instruments such as the weatherometer. An example is the open-air exposure of sheets of it-poly-

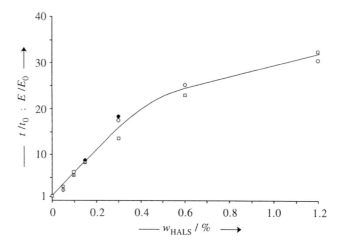

Fig. 3-7 Normalized times t/t_0 or normalized energies E/E_0, respectively, that are needed to reduce the tensile strength of the polymer to 50 % of the original value as function of the mass fraction w_{HALS} of HALS-1 (p. 58) as a light stabilizer for an isotactic poly(propylene) [11]. The curve is independent of natural (outdoor) or artificial weathering and also of the intensity of irradiation.

In the absence of HALS, a 50 % strength reduction required $t_0 = 495$ h in the weatherometer (□) and energies of $E_0 = 110$ kJ/cm^2 in Florida (●) and $E_0 = 165$ kJ/cm^2 in Basel (○) outdoors.

The polymer contained 0.1 wt% calcium stearate + 0.05 wt% A-3 (p. 57) + 0.05 wt% S-1 (p. 56) and various mass fractions of HALS-1. Test specimens were bands of 50 μm thickness.

(propylene) that were facing South at an angle of 45°. A 50 % decrease of tensile strength was observed after 69 days in Florida (irradiation energy: 580 kJ/cm^2 per year) but 117 days in Basel (irradiation energy: 340 kJ/cm^2 per year). However, relative degradations are the same whether artificial weathering is used or specimens are exposed to the elements in various parts of the world (Fig. 3-7).

As indicated by the initial slope in the range $0 \leq w_{HALS} \leq 0.3$ %, the gain of stability is directly proportional to the mass fraction w_{HALS} of HALS. At higher mass fractions of $0.5 \% \leq w_{HALS} \leq 1.2$ %, the stability is greater but the gain in stability per mass stabilizer is smaller as shown by the smaller proportionality coefficient. The two slopes differ by a factor of ca. 5.

Degradation processes do not stop on irradiation but continue in the dark. A measure of the degradation is the discoloration which can be measured by so-called B values. These values increase with the number of days in the dark until they approach a constant value as shown in Fig. 3-8 for a poly(vinyl chloride). It is speculated that new carbon-carbon double bonds are formed by dehydrochlorination and by isomerizations of longer polyene sequences without HCl formation. The initial increases and the final values both depend on the type and duration of natural and artificial weathering.

Renewed irradiation of the species leads to a decrease of the discoloration (compare the two bottom curves of Fig. 3-8 which shows the data after 18 h and 500 h weathering in the weather-O-meter). It seems that this decrease of discoloration is caused by a shortening of polyene sequences by oxidation, probably by addition of OH radicals to carbon-carbon double bonds. It has been experimentally shown that hydroxyl radicals are indeed formed on the surface of titanium dioxide particles that had been irradiated in the presence of humidity.

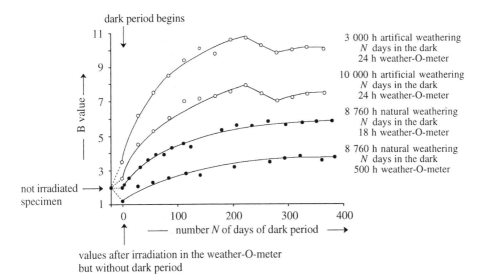

dark period begins

3 000 h artifical weathering
N days in the dark
24 h weather-O-meter

10 000 h artificial weathering
N days in the dark
24 h weather-O-meter

8 760 h natural weathering
N days in the dark
18 h weather-O-meter

8 760 h natural weathering
N days in the dark
500 h weather-O-meter

not irradiated specimen

values after irradiation in the weather-O-meter
but without dark period

Fig. 3-8 B values of impact-resistant poly(vinyl chloride) as a function of the number of days, N, in the dark [12]. B values are a measure of discoloration the paper did not provide a definition of B. The B value of the non-irradiated specimen is ca. 2.

Specimens were first subjected to 8760 h (= 1 year) in the Mediterranean or 3000 h and 10 000 h, respectively, in the weather-O-meter (Xenon irradiation). B values after these treatments are indicated for $N = 0$. Subsequently, specimens were stored N days in the dark and then weathered again in the weather-O-meter for 18, 24, or 500 hours.

Light Stabilizers

Prevention of UV absorption is the first line of defence against light-induced degradation. Absorption of ultraviolet light can be reduced by use of ideal polymer constitutions without light-absorbing "wrong" groups, exclusion of light-absorbing impurities, and application of adjuvants that are stable against irradiation or that reflect light. However, use of special UV absorbers is almost always essential. Other defensive measures include deactivation of excited states by energy tranferring agents (quenchers), the use of radical catchers, and the addition of antioxidants.

Light absorption can be reduced by certain pigments which reflect light so strongly that only a few radicals can be formed. An example is carbon black which is not only strongly UV absorbing but also a good radical catcher since it itself contains many carbon radicals. The white pigment titanium dioxide (manufactured rutile form), on the other hand, is a sensitizer that promotes degradation.

Transparent polymers always need to be protected by UV absorbers against degradation. These adjuvants should absorb light (Fig. 3-9) but should not form radicals. Because absorption maxima of many polymers for plastics are between 290 nm and 360 nm, UV absorbers for these polymers must absorb light below 420 nm. UV absorbers for cosmetics, on the other hand, must absorb below 320 nm because human skin has a sharp sensitivity maximum at 297 nm.

UV absorbers such as *o*-hydroxybenzophenone absorb ultraviolet light via the hydrogen bridge, become excited, and convert the absorbed energy to infrared radiation, i.e., heat Δ:

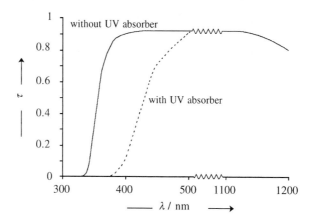

Fig. 3-9 Transmission τ of a poly(methyl methacrylate) with and without addition of a UV absorber as a function of the wavelength λ.

(3-49)

Phenyl salicylate is not a light absorber but converts on irradiation to the UV absorber 2,4-dihydroxybenzophenone. Other effective UV absorbers are 2-(2'-hydroxyphenyl)-benzotriazole and its derivatives. Benzophenones are universally applicable UV absorbers whereas benzotriazoles are effective for polar polymers. Nickel compounds (see below) are used for poly(olefin)s.

phenyl salicylate	2,4-dihydroxy- benzophenone	2-(2'-hydroxyphenyl)- benzotriazole

UV absorbers for polymers should not only absorb ultraviolet light and protect the polymers but must be also compatible with the substrate, have high lightfastness, good thermal stability during processing, and, for fibers, good color stability.

Excited states of polymer molecules (Volume I, p. 370) can be deactivated by **quenchers** that absorb the energy of the excited sensitizer and become excited themselves. Quenchers for the protection of polymers against degradation by UV light must also dissipate their accumulated energy without harming the polymer. Presently, only some nickel compounds are suitable for this purpose.

Nickel dibutyl dithiocarbamate (NiDBC) and nickel acetophenodioxime (NiOx) have long been thought of as quenchers. However, they are not quenchers for carbonyl groups or singlet oxygen. At high temperatures, they rather convert hydroperoxides to non-radical products fast, i.e., they remove the primary photo-initiating groups. These actions are catalytical for NiDBC but stoichiometric for NiOx.

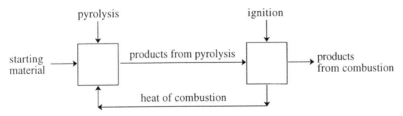

NiDBC   NiOx

3.6.6   Flame Retardation

Combustion

At room temperature and above, all organic compounds are thermodynamically unstable against oxygen. For biochemical processes, the activation energy for oxidation is provided at room temperature but the direct combustion of organics by oxygen itself usually requires much higher temperatures.

Burning of organic compounds is a complex process that can be subdivided into heating-up, pyrolysis, ignition, combustion, and flame propagation (Fig. 3-10).

Heating-up brings the material to the temperature that is required for burning. It is faster, the smaller the specific heat capacity and the higher the thermal conductivity. *Pyrolysis* degrades the polymer and delivers gases, liquids, and carbon-like residue. The flammable gases are *ignited* either spontaneously or by a spark or a flame leading to *combustion*, a free-radical chain reaction. The burning produces heat which promotes further pyrolysis and the propagation of fire.

Fig. 3-10 Burning of organic compounds (schematic).

Limiting Oxygen Index

The limiting oxygen index (*LOI*) is a quantitative measure of inflammability and combustion. It is defined as the limiting value of the volume fraction of oxygen in a mixture of oxygen and nitrogen, ϕ_{O_2}, that is just sufficient to maintain the burning of matter after ignition by an extraneous flame:

(3-50) $LOI = \phi_{O_2} = V_{O_2} / (V_{O_2} + V_{N_2})$

Table 3-12 Upper (U) and lower (L) limiting oxygen indices LOI of various materials; ignition times at heat flows of 5.8 and 10.5 W/cm^2, respectively; temperatures T_I for spontaneous (S) and extraneous (E) ignitions; indices FI for the propagation of flames; and indices SI for the density of smoke.

Ignition times for poly(ethylene terephthalate), wool, polyamide 6.6, and poly(acrylonitrile) refer to carpets from these polymers with their conventional backings (latex foam for wool and PA 6.6, jute for acrylic fibers); ignition times for poly(vinyl chloride) are for floor coverings from filled PVC.

Material	$10^2\,\phi_{O_2}$ U	L	t/s 5.8	10.5	T_I/°C E	S	FI/%	SI/%
Hydrogen	5.4	5.4						
Formaldehyde	7.1	7.1						
Poly(oxymethylene)	14	12						
Benzene	13.1							
Poly(styrene)	17.8		108	31	296	491	355	94
Poly(methyl methacrylate)	17.3		115	31				1
Poly(ethylene), low density	17.4				341	349		7
, high density	17.4							15
Paper	17.5				230	230		
Cotton, fiber	18.0							
, upholstery			9	3				
Spruce wood (1.9 cm thick)	21		46	11	260	260	143	
Oak wood (1.9 cm thick)	23		55	17				
Poly(ethylene terephthalate), fiber	20	16		485		43		
, carpet	21		79	28				
, upholstery			58	11				
Polycarbonate (from bisphenol A)	24.9							88
Poly(acrylonitrile), fiber	18.2				465			416
, carpet	18.2		39	9				
Wool, fiber	25.2				570			
, carpet			26	6				
Polyamide 6.6, bulk	24.3		29	12	421	424		
, carpet	20.1		29	12				5
Poly(m-phenylene isophthalamide)	28.5	17						
Poly(oxy-2,6-dimethyl-1,4-phenylene)	30.0							
Poly(1,4-phenylene sulfide)	>40							
Polybenzimidazole	41.5	29						
Polyurethane, foam					310	416	>1500	
Poly(vinyl chloride), fiber	37.5	20			346	454	10	
, unplasticized plastic	41.5		269	78				
Linoleum, 3 mm thick			59	12				
Poly(vinylidene chloride)	60							46
Poly(tetrafluoroethylene) (Teflon)	95				>600			

LOI values are usually determined by igniting the upper part of the specimen and letting it burn like a candle (LOI-U). More relevant to actual conditions is to ignite the lower part of the specimen (LOI-L) since this would heat up a greater proportion of the specimen and lead to a stronger pyrolysis and more intensive burning of the specimen because of the feedback of heat. LOI-L values are therefore always lower than LOI-U ones (Table 3-12).

LOI values also depend on the geometric dimensions of the specimen and the temperature of the igniting flame. Because oxygen diffusion is impeded, they increase in general by 15 % if the ratio of mass and area is doubled. As a consequence, fabrics, fibers, and plastic articles of the same polymer all have different LOI values.

Specimens with LOI-U > 22.5 are usually called **flame-retardant** and those with LOI-U > 27, **self-extinguishing**. Such terms can be misleading if they are applied to materials instead of specimens and if the testing procedures are not mentioned.

For example, poly(tetrafluoroethylene) has a very large LOI-U value of 95 (Table 3-12) which would put it into the self-extinguishing category. It usually does not burn in and with oxygen because (a) too much energy is required to attain the degradation temperature of more than 600°C and (b) degradation products cannot be oxidized easily. However, PTFE does continue to burn once it is ignited.

Inflammabilities depend not only on local oxygen concentrations but also on extraneous and spontaneous ignition temperatures, rates of heat take-up, types of pyrolysis products, and the melting behavior of polymers. Ignition temperatures vary with the extent of heat flow (Table 3-12).

Spontaneous and extraneous ignition temperatures are often identical but the latter can also be much lower than the former (Table 3-12). For example, on extraneous ignition of thermoplastic fibers, material melts away from the flame, solidifies, and the flame extinguishes if no flammable gases are produced. In blended fabrics, on the other hand, unmeltable fibers such as cotton may act as wicks and also prevent the flow of meltable components.

Flame propagation varies widely. On burning, some polymers release "extinguishing" gases such as HCl from poly(vinyl chloride), CO_2 from polymers with ester groups, and H_2O from cellulosics. Other polymers produce many flammable gases which cause a rapid spreading of flames, especially in specimens with large surface areas that allow gases to escape rapidly. For this reason, foamed polymers burn especially fast.

The higher the value of LOI, the greater the fraction f_C of coal-like residues that is usually produced by burning (Fig. 3-11). As a good approximation, one finds

$$(3\text{-}51) \qquad LOI = 0.17 + 0.43 \, f_C$$

Little is known about the interrelationships between inflammability, combustibility, and formation and toxicity of smoke. Smoke contains gases and dispersed solid particles; it may be toxic and/or suffocating. Strong smoke can blur the view and impede or even prevent rescue operations.

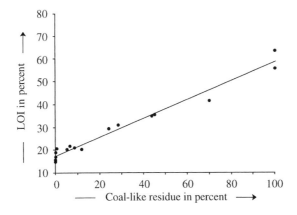

Fig. 3-11 Interrelationship between LOI values and the percentage of coal-like residues.

Table 3-13 Mass fraction w of emitted gases during combustion of various materials. Ald = aldehydes (formaldehyde, acetaldehyde, butyraldehyde, acrolein, etc.).

Material	CO	CO$_2$	O$_2$	$10^2\,w$ HCN	NO$_2$	HCl	Ald
Woolen carpet	1.9	18	5.0	1.0	0	0	
Oak wood	1.4	9.3	10	0.004	0.001	0	0.043
PVC floor covering	0.15		0	0	0.1		
Poly(acrylonitrile) carpet	0.11	3.3	15	0.01	0.0001	0	
Cotton							0.037

Smoke formation depends strongly on both heat flow and the supply of oxygen. Polymers with aromatic groups burn strongly and with strong smoke. Halogen-containing polymers are not very combustible but emit strong smoke. Poly(olefin)s, on the other hand, burn well but emit little smoke. The smoke from burning poly-(ethylene)s, natural rubber, and celluloses contains free radicals but the smoke from polyamides, poly(vinyl chloride)s, or poly(tetrafluoroethylene) does not.

With respect to smoke formation, woolen carpets are much more dangerous than carpets from acrylic fibers (Table 3-13) although wool has a higher LOI than poly(acrylonitrile) (Table 3-12). MoO$_3$, Zn/Mg complexes, and Al(OH)$_3$ are good smoke suppressants; Al(OH)$_3$ is also a good flame retardant. ZnO cannot be used in poly(vinyl chloride)s since it reduces the thermal stability of PVC.

Flame Retardants

The fire resistance of polymers can be improved by application of coatings that reduce the access of oxygen and/or decrease the flammability. Non-burning or reflecting coatings increase the pyrolysis temperatures. The access of oxygen is also prevented if the polymers themselves or the added flame retardants emit non-burnable gases.

For example, polycarbonates emit CO$_2$; they are self-extinguishing. Celluloses can be protected by addition of ZnCl$_2$, which promotes the formation of carbon and H$_2$O during burning: carbon forms a protecting layer and water vapor extinguishes fires.

Fire protection can also be improved by use of a flame retardant. All industrial flame retardants contain one or more than one of the following elements: P, Sb, Cl, Br, B, and/or N (Table 3-14). Since 2004, most bromine-containing flame retardants are no longer allowed in the European Union. It must be noted that all added flame retardants impair other use properties of polymers.

Flame retardants are usually chosen empirically since their ways of action are little understood. Two effects are proven: the prevention of oxygen access by release of non-burnable gases and the "poisoning" of flames by radicals.

The so-called thermal theory of action of flame retardants is disputed. According to this hypothesis, flame retardants are endothermally degraded on heating. The loss of energy by this process is said to decrease the surface temperature of polymers so much that the release of energy from oxidation is no longer sufficient to degrade polymer molecules to easily burnable fragments.

Table 3-12 Use of flame retardants in poly(ethylene)s (PE), poly(propylene)s (PP), poly(styrene)s (PS), acrylonitrile-butadiene-styrene copolymers (ABS), poly(vinyl chloride)s (PVC), unsaturated polyesters (UP), epoxies (EP), and polyurethanes (PUR).

Flame retardant	PE	PP	PS	ABS	PVC	UP	EP	PUR
Diantimony trioxide	+	+	+	+	+	+		
Brominated hydrocarbons	+	+	+	+		+	+	+
Chlorinated hydrocarbons	+	+			+			
Organophosphorus compounds, halogen-containing			+	+		+		+
, other					+			
Zinc diborate					+			
Aluminum trihydroxide						+		

On combustion of flame retardant-containing polymers, many flame retardants split off water and form glass-like coatings, for example, B_2O_3 from $B(OH)_3$. The endothermic release of H_2O from $Al(OH)_3$ at temperatures just above 200°C reduces the heat flow to the polymer; $Mg(OH)_2$ acts in the same manner.

Combustion causes halogen-free phosphorus-containing flame retardants to oxidize to phosphorus oxides which form glass-like surface coatings on the substrate; these coatings decrease the escape of burnable gases and lower the access of oxygen. Phosphorus oxides also react with water to phosphorus-containing acids which catalyze the release of water from organic polymers, thereby producing larger char contents and less burnable gases.

Phosphoric acid from these flame retardants splits off methanol from the ester groups of poly(methyl methacrylate). The reaction leads to cyclic anhydride groups which, in turn, reduces the zip length in depolymerizations (see Volume I, p. 596) and lowers the proportion of flammable methyl methacrylate monomer.

Antimony(III) oxide itself is not a flame retardant. With halogen-containing additives, Sb_4O_6 reacts to $SbOCl$ and $SbCl_3$. These compounds then react further with polymer radicals from the decomposition of polymers which terminates the kinetic chain.

Flame retardants are most often chlorine- or bromine-containing compounds that are used in proportions of 2-4 wt% (bromine) or 20-40 wt% (chlorine). Bromine-containing flame retardants are very effective but less stable against light than chlorine compounds and also much more expensive. All halogen-containing flame retardants generate halogen radicals that combine with radicals from the combustion of polymers. These reactions terminate the kinetic chain which in turn "poisons" the flame.

For polyamides and poly(butylene terephthalate)s for electrical appliances, water-releasing flame retardants are unusable and halogen-containing ones undesirable. These polymers are rather provided with red phosphorus. On combustion, it converts polymers into unsaturated compounds that cyclize and then carbonize. Phosphorus acids resulting from these reactions form a glassy surface coating.

Heat-resistant Polymers

High-temperature-resistant polymers usually contain uncommon monomeric units (Table 3-13). Because of their high prices, they are used only for very special purposes.

Table 3-15 Repeating units and decomposition temperatures T_d of poly(ethylene) and some high-temperature resistant polymers. Decomposition temperatures correspond to the inflection points in thermogravimetric curves (see Volume I, p. 598).

Repeating unit	$T_d/°C$ in nitrogen	$T_d/°C$ in air
—CH$_2$—CH$_2$—	415	
(p-phenylene)	660	
(p-phenylene)—S—	555	
(p-phenylene)—S—(p-phenylene)—SO$_2$—	505	490
(p-phenylene)—CH$_2$—CH$_2$—	450	420
(p-phenylene)—CF$_2$—CF$_2$—	535	
—C(O)—(p-phenylene)—C(O)—O—(p-phenylene)—O—	465	465
(tetrafluoro-p-phenylene)	720	
(2,6-dimethylphenylene)—O—	420	330
(tetramethylphenylene)—CH$_2$—CH$_2$—	475	420
(polybenzothiazole structure)	650	500

All high temperature-resistant polymers contain cyclic structures, either isolated aromatic moieties or ladder structures. Suitable polymers must require high activation energies for the scission of bonds: polymer chains should therefore have few (if any) hydrogen substituents and no branching points, long methylene sequences, or electron-attracting groups. C–F bonds are always preferred over C–H bonds.

3.6.7 Biostabilizers

Biocides are chemical substances that destroy life (G: *bios* = life; L: *caedere* = to kill). Major groups of biocides are **pesticides** (including insecticides, herbicides, fungicides, rodenticides, etc.) and **antimicrobials** (antibiotics, bactericides, antivirals, etc.). The classification of biocides by the European Union comprises 23 product categories in a total of 4 groups: disinfectants and general biocides, preservatives, pest controls, and other.

In polymer technology, one is interested in those biocides that prevent or at least slow down the deterioration of materials caused by microorganisms (bacteria, fungi, yeasts). The chemical compounds of this group of biocides are also known as **biostabilizers**.

Most synthetic polymer molecules are resistant to attacks by microorganisms because the organisms are not biologically programmed to sever carbon-carbon bonds, carbon-hydrogen bonds, and the like. Microorganisms can attack the heteroatom-containing bonds of polyurethanes, polyesters, and polyethers. Biodegradation is especially severe for commercial polymers that contain certain adjuvants, especially plasticizers and lubricants; examples are plasticized poly(vinyl chloride) and poly(ethylene)s. Such polymers are used for swimming pool lining, floor and wall coverings, upholstery, shower curtains, refrigerator seals, water beds, roof coverings, and the like.

Industrial biocides include 10,10'-oxybisphenoxyarsine (I), *N*-(trichloromethylthio)-phthalimide (II), and tributyltin compounds such as III. These chemicals are used in proportions between 0.025 wt% and 1 wt%. The military also uses copper-*bis*(8-hydroxyquinoline) (IV) which also acts as a colorant (leads to yellow-green polymers).

I II III IV

Literature to Chapter 3

3.1 OVERVIEWS
J.S.Dick, Compounding Materials for the Polymer Industries, Noyes, Park Ridge (NJ) 1987
Radian Corp., Chemical Additives for the Polymer Industry, Noyes, Park Ridge (NJ) 1987
J.T.Lutz, Ed., Thermoplastic Polymer Additives, Dekker, New York 1988

J.Edenbaum, Ed., Plastics Additives and Modifiers Handbook, Van Nostrand Reinhold, New York 1992
J.Murphy, Additives for Plastics Handbook, Elsevier, Amsterdam 1996
S.Al-Malaika, Ed., Reactive Modifiers for Polymers, Chapman and Hall, New York 1997
G.Pritchard, Ed., Plastics Additives: An A-Z Reference, Chapman and Hall, London 1998
J.T.Lutz, Jr., R.F.Grossman, Eds., Polymer Modifiers and Additives, Dekker, New York 2000
J.H.Bieleman, Ed., Additives for Coatings, Wiley-VCH, Weinheim 2000
H.Zweifel, Ed., Plastics Additives Handbook, Hanser, Munich, 5th ed. 2001
J.J.Florio, D.J.Miller, Handbook of Coatings Additives, Dekker, New York, 2nd ed. 2004
M.Ash, I.Ash, Handbook of Plastic and Rubber Chemical Additives. An International Guide to more
    than 22 000 Plastic and Rubber Additives, Synapse Information Resources, Endicott (NY) 2004
C.Bart, Additives in Polymers, Wiley-VCH, Weinheim 2005
M.Bolgar, J.Hubball, S.Meroneck, J.Groeger, Handbook for the Chemical Analysis of Plastic and
    Polymer Additives, CRC Press, Boca Raton (FL) 2008

3.1.2 COMPOUNDING

G.A.R.Matthews, Ed., Polymer Mixing Technology, Appl.Sci.Publ., London 1982
J.A.Biesenberger, Ed., Devolatilization of Polymers, Hanser, Munich 1983
E.J.Wickson, Ed., Handbook of PVC Formulation, Wiley, New York 1993
C.Rauwendaal, Polymer Mixing. A Self-Study Guide, Hanser Gardner, Cincinnati (OH) 1998
R.H.Wildi, C.Maier, Understanding Compounding, Hanser Gardner, Cincinnati (OH) 1998
D.B.Todd, Ed., Plastics Compounding. Equipment and Processing, Hanser, Munich 1998
J.White, A.Coran, A.Moet, Polymer Mixing: Technology and Engineering, Hanser, Munich 2001

3.2 DYESTUFFS (see also Section 15.4 COLOR)

D.H.Solomon, D.G.Hawthorne, Chemistry of Pigments and Fillers, Wiley, New York 1983
K.McLaren, The Colour Science of Dyes and Pigments, Hilger, Bristol, 2nd ed. 1986
P.Lewis, Ed., Pigment Handbook, Vol. 1, Properties and Economics, Wiley, New York, 2nd ed. 1988
H.Zollinger, Color Chemistry. Synthesis, Properties and Applications of Organic Dyes and Pig-
    ments, VCH, Weinheim, 2nd ed. 1991
A.Reife, H.S.Freeman, Eds., Environmental Chemistry of Dyes and Pigments, Wiley, New York
    1996
R.M.Harris, Ed., Coloring Technology for Plastics, Plastics Design Library, Bristol 1999
I.Wheeler, Metallic Pigments in Polymers, RAPRA, Shropshire, UK, 2000
H.Macdonald Smith, Ed., High Performance Pigments, Wiley-VCH, Weinheim 2001
K.Hunger, Ed., Industrial Dyes, Wiley-VCH, Weinheim 2002
A.Müller, Coloring. Fundamentals–Colorants–Preparation, Hanser, Munich 2003
R.A.Charvat, Ed., Coloring of Plastics. Fundamentals, Wiley, Hoboken (NJ) 2003
W.Herbst, K.Hunger, Industrial Organic Pigments, Wiley-VCH, Weinheim 2004
G.Buxbaum, G.Pfaff, Eds., Industrial Inorganic Pigments, Wiley-VCH, Weinheim, 3rd ed. 2005
E.Gutoff, E.D.Cohen, Coating and Drying Defects, Wiley, Hoboken (NJ), 2nd ed. 2006
P.Wissling, Metallic Effect Pigments, W.Andrew, Norwich (NY) 2006

3.3. FILLERS (see also Chapter 9)

H.G.Barth, Ed., Modern Methods of Particle Size Analysis, Wiley, New York 1984
J.V.Milewski, S.Katz, Eds., Handbook of Reinforcements for Plastics, Van Nostrand, New York,
    2nd ed. 1987
T.Allen, Particle Size Measurement, Chapman and Hall, New York, 4th ed. 1990
J.T.Lutz, Jr., D.L.Dunkelberger, Impact Modifiers for PVC, Wiley, New York 1992
J.-B.Donnet, R.C.Bansal, M.-J.Wang, Eds., Carbon Black, Dekker, New York, 2nd ed. 1993
G.Wyprych, Handbook of Fillers, Plastics Design Library, Toronto, 2nd ed. 1999
R.N.Rothen, Mineral Fillers in Thermoplastics: Filler Manufacture and Characterisation, Adv.Polym.
    Sci. **139** (1999) 68
T.Sugimoto, Ed., Fine Particles. Synthesis, Characterization, and Mechanisms of Growth, Dekker,
    New York 2000
J.-F.Gerard, I.Meisel, C.S.Kniep, S.Spiegel, K.Grieve, Fillers and Filled Polymers, Wiley, New
    York 2001
M.Xanthos, Ed., Functional Fillers for Plastics, Wiley-VCH, Weinheim 2005
H.E.Bergna, W.O.Roberts, Eds., Colloidal Silica. Fundamentals and Applications, CRC Press,
    Boca Raton (FL) 2006

3.4.2 COUPLING AGENTS
D.E.Leyden, W.Collins, Eds., Silylated Surfaces, Gordon and Breach, New York 1980

3.4.3 PLASTICIZERS (see also Chapter 10)
J.Kern, J.R.Darby, The Technology of Plasticizers, Wiley, New York 1982

3.4.4 COMPATIBILIZERS
S.Datta, D.J.Lohse, Polymeric Compatibilisers, Uses and Benefits in Polymer Blends, Hanser, Munich 1996

3.4.7 ANTISTATIC AGENTS AND CONDUCTIVITY IMPROVERS (see also Chapter 14)
S.K.Bhattacharya, Ed., Metal-Filled Polymers - Properties and Applications, Dekker, New York 1986

3.5.2 PLASTICATING AGENTS
G.Buttlers, Ed., Particulate Nature of PVC, Elsevier, Amsterdam 1982
H.A.Sarvetnick, Ed., Plastisols and Organosols, Van Nostrand Reinhold, New York 1972

3.6.2 DEGRADATION
G.Geuskens, Ed., Degradation and Stabilization of Polymers, Appl.Sci.Publ., Barking, Essex 1975
H.H.G.Jellinek, Ed., Aspects of Degradation and Stabilization of Polymers, Elsevier, Amsterdam1977
W.Schnabel, Polymer Degradation, Hanser, Munich 1981
M.T.Gillies, Ed., Stabilizers for Synthetic Resins, Noyes, Park Ridge (NJ), 1981
A.Davis, D.Sims, Weathering of Polymers, Elsevier Appl. Sci., Amsterdam 1983
W.Lincoln Hawkins, Polymer Degradation and Stabilization, Springer, Berlin 1984
N.M.Emanuel, A.L.Buchachenko, Chemical Physics of Polymer Degradation and Stabilization, VSP, Utrecht 1987
G.Scott, D.Gilead, Ed., Degradable Polymers: Principles and Applications, Chapman and Hall, New York 1995
H.Zweifel, Stabilization of Polymeric Materials, Springer, Berlin 1998
L.K.Massey, The Effect of UV Light and Weather on Plastics and Elastomers, W.Andrew, Norwich (NY), 2nd. ed. 2006
G.Wyprych, Handbook of Material Weathering, W.Andrew, Norwich (NY), 4th ed. 2007

3.6.3 HEAT STABILIZATION
Z.Mayer, Thermal Decomposition of Poly(vinyl chloride) and of Its Low-Molecular-Weight Model Compounds, J.Macromol.Sci.-Revs. **C 10** (1974) 263
G.Ayrey, B.C.Head, R.C.Poller, The Thermal Dehydrochlorination and Stabilization of Poly(vinyl chloride), Macromol.Revs. **8** (1974) 1

3.6.4 THERMO-OXIDATIVE STABILIZATION
J.Pospíšil, P.P.Klemchuk, Eds., Oxidation Inhibition in Organic Materials, CRC Press, Boca Raton (FL), 2 vols. 1989
G.Scott, Ed., Atmospheric Oxidation and Antioxidants, Elsevier, Amsterdam 1993, 3 vols.
Yu.A.Shlyapnikov, S.G.Kiryushkin, A.P.Mar'in, Antioxidative Stabilisation of Polymers, Taylor and Francis, London 1997
E.T.Denisov, T.Denisova, Handbook of Antioxidants, CRC Press, Boca Raton (FL), 2nd ed. 1999

3.6.5 LIGHT STABILIZATION
B.Rånby, J.F.Rabek, Photodegradation, Photo-Oxidation and Photostabilization of Polymers, Wiley, New York 1975
V.Ya.Shlyapintokh, Photochemical Conversion and Stabilization of Polymers, Hanser, Munich 1984
J.F.Rabek, Photodegradation of Polymers. Physical Characterization and Applications, Springer, Berlin 1996
L.K.Massey, The Effect of UV Light and Weather on Plastics and Elastomers, W.Andrew, Norwich (NY), 2nd ed. 2006
W.Schnabel, Polymers and Light, Wiley-VCH, Weinheim 2007

3.6.6a COMBUSTION

W.C.Kuryla, A.J.Papa, Eds., Flame Retardancy of Polymeric Materials, Dekker, New York 1978 (6 vols.)

C.F.Cullis, M.M.Hirschler, Eds., The Combustion of Organic Polymers, Oxford University Press, Oxford 1981

R.M.Aseeva, G.E.Zaikov, Combustion of Polymer Materials, Hanser, Munich 1986

-, Fire Toxicity of Plastics, RAPRA, Shawbury, Shropshire, UK, 1990

G.Pal, H.Macskasy, Plastics - Their Behavior in Fires, Elsevier, Amsterdam 1991

C.J.Hilando, Flammability Handbook for Plastics, Technomic, Lancaster (PA), 5th ed. 1998

J.Troitzsch, Ed,, Plastics Flammabiliy Handbook, Hanser, Munich, 3rd ed. 2004

F.H.Pragr, H.Rosteck, Polyurethane and Fire, Wiley-VCH, Weinheim 2006

3.6.6b FLAME RETARDANTS

M.Lewis, S.M.Atlas, E.M.Pearce, Eds., Flame-Retardant Polymeric Materials, Plenum, New York 1976

G.C.Tesoro, Chemical Modification of Polymers with Flame-Retardant Compounds, Macromol.Revs. **13** (1978) 283

A.Granzow, Flame Retardation by Phosphorous Compounds, Acc.Chem.Res. **11** (1978) 177

A.E.Grand, C.A.Wilkie, Eds., Fire Retardancy of Polymeric Materials, Dekker, New York 2000

S.M.Lomakin, G.E.Zaikov, Modern Polymer Flame Retardancy, VSP, Utrecht, The Netherlands, 2004

A.B.Morgan, C.A.Wilkie, Flame Retardant Polymer Nanocomposites, Wiley, Hoboken (NJ) 2007

3.6.6c HEAT-RESISTANT POLYMERS

A.H.Frazer, High-Temperature Resistant Polymers, Interscience, New York 1968

V.V.Korshak, Heat-Resistant Polymers, Israel Progr.Sci.Translations, Jerusalem 1971

P.E.Cassidy, Thermally Stable Polymers, Dekker, New York 1980

J.P.Critchley, G.J.Knight, W.W.Wright, Heat-Resistant Polymers, Plenum, New York 1983

3.6.7 BIOSTABILIZERS

R.E.Klausmeier, C.C.Andrews, Plastics, Econ.Microbiol. **6** (1981) 431 (microbiol. degradation)

E.W.Flick, Fungicides, Biocides and Preservatives for Industrial Applications, William Andrew, Norwich (NY) 1988

S.J.Huang, Ed., Biodegradable Polymers, Hanser, Munich 1990 (pharmaceuticals, agriculture)

References to Chapter 3

[1] I.C.Sanchez, S.S.Chang, L.E.Smith, Polymer News **6** (1980) 249, Figs. 1 and 4

[2] G.Kaufmann, Plastverarbeiter **20** (1969) 457, several data

[3] J.Jousset, Matières plastiques, Dunod, Paris, Bd. I, S. 48

[4] G.S.Egerton, J.Soc. Dyers Colour. **65** (1949) 764

[5] H.P.Schlumpf, in R.Gächter, H.Müller, Eds., Plastics Additives Handbook, Hanser, Munich, 3rd ed. (1990), p. 525 ff.

[6] –, Was ist Russ?, company literature of Degussa AG, Frankfurt/Main (no year given) ("What is Carbon Black?")

[7] E.Fekete, B.Pukánzhky, A.Tóth, I.Bertóti, J.Colloid Interface Sci. **135** (1990) 200, (a) Fig. 2, (b) Fig. 1

[8] D.E.Davenport, Polymer News **8** (1982) 134, data of Fig. 1

[9] J.Garland, Trans.Met.Soc. (AIME) **235** (1966) 642

[10] C.E.Carraher, Jr., Polymer News **14** (1989) 243, Table 1

[11] F.Gugumus, in R.Gächter, H.Müller, Eds., Plastics Additives Handbook, Hanser, Munich, 3rd ed. (1990), p. 215, Table 12

[12] D.Hepp, Angew.Makromol.Chem. **171** (1989) 39, Table 1-3

4 Processing

4.1 Introduction

4.1.1 Overview

Polymeric endproducts such as plastic articles, rubber parts, textile fibers, coatings, adhesives, etc., can be manufactured from various raw materials such as monomers, prepolymers, or polymers in various ways, mostly via the liquid state. Depending on the raw material, five different types of processes are possible for the conversion of raw materials to desired goods:

A. Direct polymerization of monomers, possibly as admixtures with adjuvants, to polymers with simultaneous shaping;

B. Polymerization of monomers to polymers, compounding, processing;

C. Polymerization of monomers to prepolymers, compounding, curing with simultaneous shaping;

D. Polymerization of monomers to polymers, compounding, processing to semi-finished goods, further processing;

E. Polymerization of monomers to prepolymers, compounding, curing, processing to semi-finished goods, further processing.

The number of processing steps increases from type A to type E. In principle, type A processes should therefore be the most economical. However, these types of processes present the technical problem that a low-viscosity liquid must be converted to a very high-viscosity, solid final product. For this reason, type A processes are used mainly for the preparation of adhesives and some coatings (Chapter 12) but relatively little for shaped articles such as fibers (Chapter 6), films and sheets (Chapter 13), and three-dimensional articles (Chapters 7-10).

Hence, type B processes are the main processes for the manufacture of plastic articles and thermoplastic fibers. Type B processes are also used for the manufacture of rubber goods albeit with simultaneous crosslinking during shaping.

Type C processes are characteristic for thermosets. Sometimes, type D processes are more advantageous for thermosets; these processes may combine a type C process with subsequent machining of the semifinished goods. Type E processes are usually avoided; they combine type A-D processes with, for example, flash removal.

Very many different processing methods exist for each of the types A-E, especially for the processing of raw materials to thermoplastics and thermosets. Methods of processing rubbers to elastomers and polymers to fibers are in general very similar to those of thermoplastics and thermosetting materials; for example, extrusion can be used for both thermoplastics and fibers, and calendering for both rubbers and plastic films.

The principles of major processing methods are therefore discussed collectively in this chapter for all polymer applications (plastics, elastomers, fibers, coatings, etc.). Special types of processing can be found in the respective chapters on fibers, elastomers, films, coating, etc.

The choice of a processing method depends *technically* on the rheological properties of the material that is to be processed and on the desired shape of the final product. *Economically*, one has to consider the costs of materials, energy, processing machinery, work force, selling, and capital; some of these costs depend heavily on the throughput.

Processing methods can be subdivided according to the processing technology or the type of shaping process. For plastics (Chapters 8-10, 12-15) processes can be subdivided into groups according to the *rheological states* of raw materials during shaping (for fibers, see Chapter 6; for elastomers, see Chapter 7):

viscous:	casting, molding, spraying, coating;
elastoviscous:	injection molding, extrusion, calendering, rolling, kneading;
elastoplastic:	drawing, blowing, foaming;
viscoelastic:	sintering, welding;
solid:	machining, joining, glueing;

or according to the type of *shaping*:

forming:	casting, dipping, press molding, compression molding, injection molding, extrusion, foaming, sintering;
transforming:	calendering, embossing, bending, deep drawing;
joining:	welding, glueing, rivetting, shrink coating or wrapping;
coating:	laminating, coating, flame spraying, fluidized bed coating, lining;
separating:	cutting, machining.

Shaping methods may also be subdivided according to the applied pressure, the type of procedure (continuous, discontinuous), the type of the resulting product (formed article, semi-finished goods, profile, surface coating, expanded material, etc.). From the technical point of view, no *large* principal differences exist between the production of plastic articles and fibers (for example, extrusions *versus* melt spinning) but there are many different technical terms that are based on phenomenology (see Chapter 6-8).

4.1.2   General Factors

Plastics are processed as **powders** (L: *pulvis* = dust) or compacted powders such as **granulates (granules)**, i.e., asymmetric aggregates of powder particles (L: *granulum* = diminutive of *granum* = grain), or **pellets**, i.e., spherical particles (L: *pila* = ball).

Density

In discontinuous shaping processes, polymeric materials must be measured in exact amounts because otherwise molds will not be filled completely or material will be wasted. For measuring, the *relevant* density must be known.

In chemistry and technology, density ρ is defined as the ratio of mass m (not weight!) to occupied volume V, $\rho = m/V$. Physicists use another definition: density = quantity A per quantity B where the two quantities may be anything. Examples are the mass density = mass per volume, the number density = number of entities per volume, the surface density = mass per surface area, etc. (see Volume III, p. 168).

The **pure density** (**true density**) of a solid is the ratio of mass m and volume V of *matter*. The measured density of solids, usually just called "density", is not the true density of matter but a *material* **density** (**gross density**) because it also depends on the degree of crystallinity of the solid (if any) as well as the volume of pores within the material.

Solids are usually present as **particulates** (powders, granulates, pellets, short fibers, etc.) that contain separate **particles**. The maximum packing of regularly shaped particles (**packing fraction**, a volume fraction) depends on the shape of particles, the type of particle distribution (if any), and the kind of packing (Table 4-1).

In general, particles are not maximally but irregularly packed. Hence, the volume of a particulate is the sum of the volumes of particles and the interspaces between them. The density of a particulate is thus a **bulk density** (called **powder density** for powders). It may be a **filling density** (loosely filled), a **press density** (filled under pressure), etc.

Table 4-1 Packing fractions = maximum volume fractions ϕ_{max} of space that can be occupied by an infinite number of like particles (where relevant: with length L and diameter d). Numbers in parentheses indicate the numbers of nearest neighbors.
$\phi_{max,exact}$ = mathematically exact value, $\phi_{max, approx.}$ = approximate value (n = numerical approximation of exact value; t = theoretical approximation of unknown exact value; e = experimental).

Type (shape) and distribution	Packing		$\phi_{max, exact}$	$\phi_{max, approx.}$	
Spheres in a plane	hexagonal close		$\pi / 12^{1/2}$	≈ 0.9069	n
Spheres, unimodal, 3-dimens.	hexagonal close (12)		$\pi / 3 \cdot 2^{1/2}$	≈ 0.7405	n
	cubic close (face-centered) (12)		$\pi / 3 \cdot 2^{1/2}$	≈ 0.7405	n
	body-centered cubic (8)		$\pi / (8/3^{1/2})$	≈ 0.6802	n
	primitive cubic (6)		$\pi / 6$	≈ 0.5236	n
	diamond cubic			≈ 0.34	t
	random, densest			≈ 0.637	t
	random, loose			≈ 0.601	t
Spheres, bimodal, 3-dimens.	random, densest,	$d_2/d_1 = \quad 3.8$		≈ 0.684	t
		$d_2/d_1 = \quad 47$		≈ 0.814	t
		$d_2/d_1 = 100$		≈ 0.868	t
Spheres, tetramodal, 3-dimens.				≈ 0.98	t
Tetrahedra	random (4)		$\pi \, 3^{1/2} / 16$	≈ 0.3026	n
Cubes	random (6)		$\pi / 6$	≈ 0.5236	n
Octahedra	random (8)			≈ 0.6053	t
Dodecahedra	random (12)			≈ 0.754	t
Icosahedra	random (20)			≈ 0.8387	t
Rods, cylindrical, unimodal	hexagonal, parallel			≈ 0.907	t
	cubic, parallel			≈ 0.785	t
	random, parallel			≈ 0.82	t
	random, three-dimensional, $L/d = 1$			≈ 0.704	t
	random, three-dimensional, $L/d = 2$			≈ 0.704	t
	random, three-dimensional, $L/d = 4$			≈ 0.704	t
	random, three-dimensional, $L/d = 8\text{-}70$			*)	t
Ellipsoids			$[\pi(24 \cdot 2^{1/2} - 6 \cdot 3^{1/2} - 2\,\pi)]/72$	≈ 0.7533	n
Glass fibers, broken, unimodal	experience values for common sizes			0.476-0.500	e
, bimodal	experience value for common sizes			0.566	e
Chalk, precipitated	experience values for commercial products			0.36-0.50	e
Aluminum, powder	experience values for commercial products			0.25-0.32	e

*) $1/\phi_{max} = 1.052 + 0.123 \, (L/d) + 0.00111 \, (L/d)^2$

For foamed polymers, one distinguishes the **overall density** of the material (= **gross density**) that includes the usually more dense surface layers from the **core density** of the center parts of the expanded polymer.

In general, one finds

pure density > gross density > bulk density > sinter density

The ratio of gross density to bulk density is known as **bulk factor A**, f_A, which indicates the ratio of the volume of the material to the volume of the mold. The inverse bulk factor increases with increasing pressure and approaches unity if the bulk density equals the gross density (Fig. 4-1) since pressure reduces free volume *between* particles. The inverse bulk factor can exceed unity if pressure reduces pores *within* particles.

The density of compacted molding material is called **compacted bulk density** or, for compacted powders, **tablet density.** The ratio of gross density to compacted bulk density (or tablet density) is known as **bulk factor B**.

Part weights and materials costs are sometimes evaluated with the help of the so-called **specific gravity** which is defined by ASTM as the ratio of the density of water to the density of the material, both at $73°F = 22.78°C$. The term is a misnomer because it is neither a "gravity" (= intensity of gravitational forces between two sufficiently massive bodies) nor "specific" (= ratio of a physical property of matter to its mass).

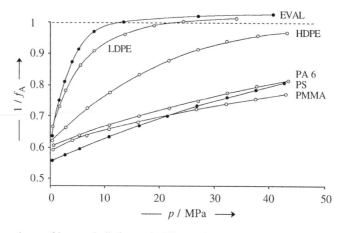

Fig. 4-1 Dependence of inverse bulk factor A, $1/f_A$, on the compacting pressure p, of granules of a poly(ethylene-*co*-vinyl alcohol) (EVAL), a low density poly(ethylene) (LDPE), a high-density poly-(ethylene) (HDPE), a poly(ε-caprolactam) (PA 6), an atactic poly(styrene) (PS), and an atactic poly-(methyl methacrylate) (PMMA) [1].
With kind permission by Industrial Media, Inc., Denver (CO).

Devolatilization

Industrial polymers contain residual monomers, oligomers, other low-molecular weight compounds (including solvents), catalyst residues, and also usually moisture. Some of these impurities are volatile and leave the polymer at processing temperatures; others become entrapped in the polymer as bubbles, voids, or pinholes. For health reasons, concentrations of impurities must be below legal limits which are usually in the ppm range (10^{-4} %) and sometimes in the ppb range (10^{-7} %).

Volatile components are sometimes removed by steaming with inert gases. In general, they are eliminated by applying vacuum, usually by extrusion. In special cases, one lets polymer particles fall in towers at very low pressures. Powders, granulates, and pellets are also predried before processing.

Moisture

Polymers take up moisture by absorption, adsorption, and capillary condensation. A pure absorption should follow **Henry's law**, i.e., the mass fraction w_w of water in the polymer should be proportional to the pressure p or the relative humidity ϕ_{rh},

(4-1) $w_w = S_p \cdot p = S_{rh} \cdot \phi_{rh}$

where S_p and S_{rh} are solubility coefficients. Henry's law is well obeyed for isotactic poly-(propylene), except for very high relative humidities of more than ca. 90 %, and other apolar polymers (not shown). It is not followed by hydrophilic polymers such as poly-amide 6.6, poly(ethylene-*co*-vinyl alcohol), and poly(vinyl alcohol) (Fig. 4-2).

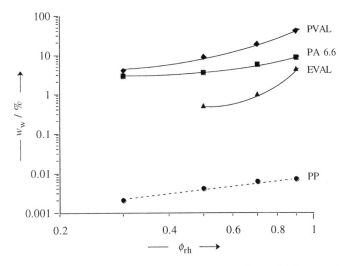

Fig. 4-2 Logarithm of mass fraction w_w of moisture in polymers in equilibrium as a function of the logarithm of relative humidity, ϕ_{rh} [2]. For log-log plots, Henry's law predicts straight lines with slopes of unity (broken line for poly(propylene) (PP)). EVAL = poly(ethylene)-*co*-vinyl alcohol), PA 6.6 = poly(hexamethylene adipamide), PVAL = poly(vinyl alcohol).

Moisture acts as a plasticizer in hydrophilic polymers. It can also add considerably to the weight of polymeric materials. For these reasons, many polymers such as polyamides have to be "conditioned" before they can be shipped as pellets or processed to articles, i.e., they must be exposed to humidity for a specified time period.

Moisture uptake can take days to weeks before equilibrium is reached, even by relatively thin parts or pellets of hydrophilic polymers such as polyamide 6.6 at relative humidities of 100 % (Fig. 4-3). For shipping, polymers are therefore conditioned, usually by exposure to 50 % relative humidity for 24 hours. They also have to be conditioned before processing.

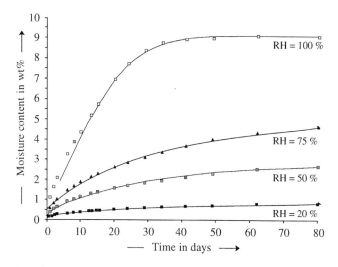

Fig. 4-3 Time dependence of moisture uptake of 1.6 mm thick polyamide 6.6 discs of 50 mm diameter at room temperature [3]. By kind permission of Hanser Publishers, Munich.

Warehousing and Transporting

Polymers are stocked and transported in various shapes, depending on their elasticity,
hydrophobicity, and electrostatic behavior. Also important are densities, hardnesses, and
chemical stabilities, against air, light, and humidity (Table 4-2). Powders and grinding
stocks of plastics may get entangled and clogged on warehousing, which leads to unfavorable bulk properties. A tendency to become charged electrostatically (Chapter
14.2.7) may lead powders to cling to container and silo walls; it may also lead to dust
explosions. Sensitive polymers are therefore stored in wet states.

Elastomer and plastic refuse are shipped and stored in bales. Particulate thermoplastics are transported either mechanically by conveyor belts, conveying chutes, or
screw conveyors; pneumatically by air; or hydraulically by water. They are shipped in
bags or as tank loads.

Size separation may be by screening for coarser dry goods or by classifying for finer
separations, using fluid-dynamic separation devices.

Table 4-2 Effect of various properties on warehousing, transport, and processing of powders, etc.

Process	Electro-statics	Hydro-phobicity	Density	Elasticity	Hardness	Chemical stability
Warehousing	+					
Transport, conveyance	+	+	+	+	+	
Size reduction	+			+	+	+
Classifying, sorting, etc.		+	+			
Washing	+	+	+			
Drying		+				+
Melting						+
Agglomeration, regranulation	+					+

Table 4-3 Processing windows ΔT_{proc} of injection-molded polymers [4]. T_G = glass temperature, T_M = (crystalline) melting temperature, $T_{d,x}$ = temperature of onset of decomposition in air (x = O) or nitrogen or helium (x = N). Processing temperatures may vary somewhat with the polymer grade.

Polymer	T_G/°C	T_M/°C	ΔT_{proc}/°C	$T_{d,O}$/°C	$T_{d,N}$/°C
Poly(ethylene), low density	−103	110	160 - 260	265	290
Poly(vinyl chloride)	80		180 - 210	200	250
Poly(styrene), atactic	100	-	180 - 280	300	
Poly(ethylene), high density	−123	135	200 - 300		
Poly(propylene), isotactic	20	163	230 - 270	120	328
Polyamide 6	56	220	238 - 270		
Polyamide 6.6	48	256	280 - 305	185	
Polycarbonate, bisphenol A	150	-	280 - 320		

The size reduction of particles of thermosetting resins is possible by impact in impact mills, pulverizers, mixers, crushers, etc. Impact and pressure are ineffective for thermoplastic particles because of their viscoelasticity; they are reduced in size by cutting.

Heating and Melting

Most processing methods require the heating of materials to processing temperatures T_{proc} that are well above the highest transformation temperature of the material (glass temperature T_G, clearance temperature nematic $\rightleftarrows$ isotropic T_{NI}, or crystalline melting temperature T_M; see Volume III, Chapter 13). Industrially, this temperature is often called the melting temperature, regardless of the type of transformation (Table 4-3).

The lower limit of the processing window is given by the dissolution of agglomerates (amorphous polymers) and the melting of highly ordered crystallites (semicrystalline polymers) as well as the shift of melting temperatures caused by the effect of processing pressure; it is at least ca. 20 K higher than the highest transformation temperature. The upper limit of the processing temperatures is usually at least 30 K below the decomposition temperature of the polymer.

Heat is usually supplied externally and internally: externally mainly by contact (hot metal parts, immersion into baths) or less commonly by convection (blowing hot air, infrared radiation), and internally usually by friction through shearing (kneaders, screw plasticators, etc.) or seldom by vibration of groups (ultrasonic radiation). Rheological properties can also be reduced by repeated extrusion (called **shear working, shear refining**, or **shear modification**). This treatment decreases extensional viscosities (see Section 4.2.6) and increases melt flow indices (Section 4.2.2) but barely modifies steady-state shear viscosities and dynamic properties. It is probably caused by alignments of chain segments or entire chains because it can be reversed by heating or dissolution.

Cooling and Annealing

Warm articles must be solidified by cooling which is possible only by thermal conductivity, a slow process that leads to temperature gradients and, in turn, to inhomogeneities in thick-walled articles. Because of the slow process, cycle times (= number of articles produced per hour) are therefore very often controlled by cooling times.

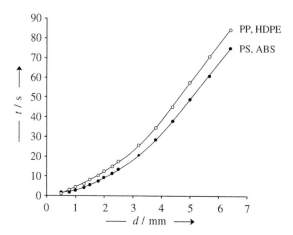

Fig. 4-4 Cooling time *t* as a function of wall thickness *d* of injection molded parts from highly crystalline isotactic poly(propylene) (PP) and high-density poly(ethylene) (HDPE) an from amorphous poly(styrene) (PS) and high-impact ABS polymers from acrylonitrile, butadiene, and styrene [5].

Cooling rates strongly depend on the thickness of the walls of parts (Fig. 4-4). Amorphous polymers (PS, ABS, etc.) cool fastest whereas highly crystalline polymers (PP, HDPE, etc.) cool slowest.

Cooling times should be as short as possible in order to achieve high cycle times and therefore economic use of processing equipment. Short cooling times are especially important for highly crystalline polymers since large spherulites would form otherwise; this would make polymers more brittle.

Crystallization does not stop after the article has reached ambient temperature. Instead, one rather observes after-crystallizations that continue for long times (Fig. 4-5). Such after-crystallizations lead to more dense polymers and therefore to post-shrinkages and different use properties. Shrinkage is smallest in cold molds but polymers from such molds also have the largest post-shrinkages (after-shrinkages) on storage. Industrially, one therefore works with as high as possible mold temperatures.

Flow processes orient chain segments during molding of amorphous polymers. The resulting anisotropy of properties (Table 4-4) causes stresses which can be removed by **annealing**, i.e., by keeping the polymers for extended times below their glass temperatures. Annealing allows chains to return to the thermodynamically preferred shape of random coils which reduces stresses.

Table 4-4 Effect of a 50-60 % orientation of chain segments on tensile moduli *E* and fracture strengths σ_B of injection molded parts longitudinal (∥) and transverse (⊥) to the direction of orientation; n.o. = not oriented specimen. ABS contains units from acrylonitrile, butadiene, and styrene.

Polymer	*E*/MPa			σ_B/MPa		
	∥	n.o.	⊥	∥	n.o.	⊥
ABS polymer	3450	2400	2070	124	48.3	41.4
Poly(ε-caprolactam) (PA 6)	3450	1240	1240	172	69.0	75.8
Isotactic poly(propylene) (PP)	2760	1240	1170	138	34.5	103
High-density poly(ethylene) (HDPE)	2070	760	620	103	27.6	20.7

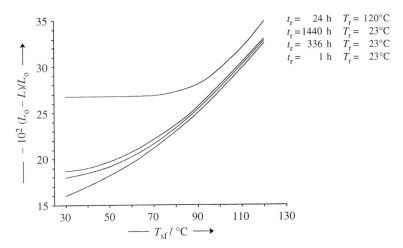

Fig. 4-5 Shrinkage $(L_0 - L)/L_0$ of an injection molded poly(oxymethylene) $(T_M = 181°C)$ as a function of the surface temperature T_{sf} of molds for an annealing temperature of 120°C and various residence times t_r and temperatures T_r [6]. The post-shrinkage is the difference of shrinkage between two residence times at T_{sf} = const. and T_r = const. By kind permission of Hanser Verlag, Munich.

Highly crystalline polymers cannot be annealed because this would lead to secondary crystallizations and larger spherulites. Annealing temperatures T_{an} are very often identical with the temperatures for maximum crystallization rates, i.e., $T_{an} = (0.80\text{-}0.87)\, T_M$.

4.2   Flow Properties

4.2.1   Viscosities

Processing of polymeric materials to shaped articles is largely controlled by the flow properties of polymers, i.e., their **viscosities** (L: *viscos* = tough, sticky; from L: *viscum* = mistletoe, bird lime made from mistletoe berries).

The **dynamic viscosity (shear viscosity)** η of a liquid is given by the ratio of **shear-(ing) stress**, $\tau = \sigma_{21} = FA$ (force times area), and **shear rate (velocity gradient)**, $\dot\gamma = dv/dy$ (= change of flow rate v with distance y perpendicular to the flow direction), where η may be a function of $\dot\gamma$ (**non-Newtonian**) or not (**Newtonian** (η_0)):

(4-2) $\eta = FA/(dv/dy) = \sigma_{21}/\dot\gamma$; $\eta = f(T, p)$

Newtonian viscosities η_0 are material constants that depend on temperature and pressure but not on the shear rate or the duration of stress. For this reason, they are also called **viscosity at rest**, **stationary viscosity**, or **zero-shear viscosity**.

Dynamic viscosities are now reported in pascal seconds, Pa s = kg m^{-1} s^{-1}, and no longer in Poise, 1 P = 1 g cm^{-1} s^{-1} = 0.1 Pa s.

The product of dynamic viscosity η and specific volume, $v = 1/\rho$, of the liquid is known as **kinematic viscosity**, $\nu = \eta v = \eta/\rho$, where ρ = density of the liquid in mass per volume. It is now measured in m^2 s^{-1} and not in Stokes, 1 St = 1 cm^2 s^{-1}.

Shear stress, $\sigma_{21} = G\gamma_e$, is the product of **shear modulus** G and **shear strain** (= **elastic shear deformation**), $\gamma_e = (\sigma_{11} - \sigma_{22})/\sigma_{21}$, where σ_{11} = tensile stress and σ_{22} = stress normal to the tensile stress (**normal stress**). The shear modulus of Newtonian liquids, G_0, is independent of the shear deformation.

Melts and concentrated solutions of polymers have very high Newtonian viscosities of $\eta_0 = 10^2 \text{-} 10^6$ Pa s. At moderately high shear rates $\dot\gamma$, they no longer behave in a Newtonian manner and their dynamic viscosity, Eq.(4-2), becomes an **apparent viscosity**, η, that depends on the shear rate $\dot\gamma$ (Section 4.2.4).

Shear viscosities are important for many polymer processes such as extrusion and injection molding. However, polymer flow can also be obtained by elongating instead of shearing since polymeric liquids can be substantially drawn without breaking (Section 4.2.6). Such extensibilities allow fiber spinning, extrusion blowing, and vacuum forming of thermoplastic materials.

Extensional viscosities (elongational viscosities, tensile viscosities) η_e are obtained from the ratio of **tensile stress** σ_{11} to **tensile strain rate (elongational rate)** $\dot\varepsilon$ of fluids:

$$(4\text{-}3) \quad \eta_e = \sigma_{11} / \dot\varepsilon$$

Extensional viscosities usually depend on elongational rates. They become true **Troutonian viscosities,** $\eta_{e,0}$, at very low values of $\dot\varepsilon$ where $\eta_{e,0} \neq \text{f}(\dot\varepsilon)$ (Section 4.2.6).

In general, viscosities of polymers depend on molar masses and are therefore used to characterize polymers with respect to molar mass and/or dimensions. The preferred method of academic laboratories is dilute solution viscometry in which the viscosity η of various polymer solutions of concentration c_2 (in g/cm^3) in solvents of viscosity η_1 is determined at low shear rates. Extrapolation of $\eta_{red} = (\eta - \eta_1)/(\eta_1 c_2)$, the so-called **reduced viscosity**, to zero polymer concentration delivers the **intrinsic viscosity** $[\eta]$,

$$(4\text{-}4) \quad [\eta] \equiv \lim_{c_2 \to 0} [(\eta - \eta_1)/(\eta_1 c_2)] \quad ; \quad [\eta] = K_v \overline{M}_v^\alpha$$

that depends on a power α of the so-called viscosity-average molar mass, $\overline{M}_v$ (see Volume I, p. 93 ff., and Volume III, Chapter 12).

Constants K_v and α have to be determined empirically for each system polymer-solvent-temperature. For flexible macromolecules, the exponent α adopts values of 0.500 for so-called theta solvents and 0.764 for thermodynamically good solvents. Values of viscosity-average molar masses of such polymers are between those of number-average and mass-average molar masses if $\alpha < 1$. They are identical with mass-average molar masses if α is unity which corresponds to the hypothetical case of freely draining coils.

The preparation of polymer solutions is labor-intensive. Hence, industrial laboratories prefer to measure viscosity-related properties in melts. Newtonian melt viscosities (Section 4.2.3) depend on molar masses in a similar way as intrinsic viscosities do, i.e., $\eta_0 = K_\eta \overline{M}_\eta^\varepsilon$ where K_η and ε are empirical constants (see Section 4.2.3); $\overline{M}_\eta$ is usually thought to be the mass-average molar mass, $\overline{M}_w$. Application of this equation requires calibration with a series of polymers with known molar masses and molar mass distributions.

For this reason and because of the direct correlation with processing parameters, industry prefers to determine melt flow indices (MFI) or melt volume indices (MVI) for thermoplastics and Mooney viscosities for elastomers (next Section).

4.2.2 Viscometry and Rheometry

Viscosities can be measured by a multitude of instruments (see also Volume III, Chapters 12 and 15). **Capillary viscometers** (Fig. 4-6) serve mainly for low viscosity liquids but are also used for melts. Shear rates are varied by changing the applied pressure (fill height or external pressure) or the diameter of the capillary. Shear gradients are not linear but parabolic because liquids adhere to the capillary walls (where the flow rate is zero) and flow with maximum speed at the center of the capillary. As a result, flow profiles can become very complicated for non-Newtonian liquids. Additional effects arise from entry effects at short capillaries. Calculation of shear rates, shear stresses, elongational rates, and tensile stresses may thus become very difficult.

Shear viscosities of high viscosity liquids can be determined more easily by rotational or cone-and-plate viscometers (Fig. 4-6). In **rotational viscometers**, a hanging rotor turns in a stator that is filled with the liquid to be measured (**Searle principle**). Conversely, a cylinder wall rotates around a stationary stator (**Couette principle**). **Cone-and-plate viscometers** for very high viscosities employ a broad, rotating cylinder on a stationary plate. In all cases, gaps between rotor and stator and cone and plate must be very narrow in order to generate linear shear gradients.

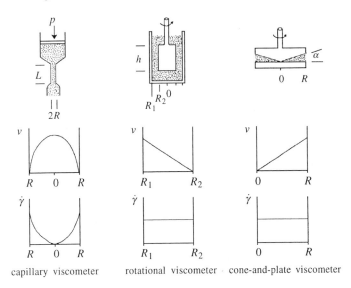

capillary viscometer   rotational viscometer   cone-and-plate viscometer

Fig. 4-6 Top from left to right: schematic representations of a capillary viscometer, a Couette rotational viscometer, and a cone-and-plate viscometer. Center: velocities v as a function of radii, R, R_1, and R_2, respectively. Bottom: shear rates $\dot{\gamma}$ at R, R_1, and R_2.

Industry uses more practical quantities as indicators of flow properties. Thermoplastic polymers are characterized by their melt flow indices or melt volume indices. Both quantities are directly connected to the rheological behavior of melts during processing.

The **melt flow index (MFI)** indicates the mass of polymer that is extruded within 10 minutes from a standard plastometer by a standard load F at a temperature T. MFI is a measure of fluidity (not viscosity) that also depends on the (unspecified) shear gradient in the plastometer. The lower the MFI, the greater the molar mass of the polymer.

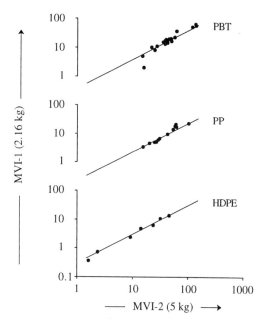

Fig. 4-7 Dependence of MVI-1 on MVI-2. All slopes are unity.
HDPE = unfilled high-density poly(ethylene)s at 190°C (Eltex® grades of Solvay, S.A.);
PP   = isotactic poly(propylene)s filled with (20-40) wt% mineral fillers, 30 wt% glass fibers,
        and/or impact improvers, all at 230°C (Moplen® grades of Himont);
PBT   = poly(butylene terephthalate)s at 260°C, either unfilled or filled with glass fibers, glass
        beads, white pigments, or elastomers (Pocan® grades of Bayer AG).

Multiplication of MFI by the density of the melt delivers the **melt volume index**
(**MVI**). MVIs are determined by the length that a plunger moves in 10 min under a load
of 2.16 kg (MVI-1) or 5 kg (MVI-2) where (MVI-1) = K(MVI-2) (Fig. 4-7).

Mooney viscosities of elastomers are measured by standard cone-and-plate viscometers. Their values ML-t/T indicate the retracting force that is experienced by elastomers on deformation for t min at constant temperature T and rotational speed.

Special practical viscometers are used in the paint industry. **Ford cups** are calibrated
vessels with a hole at the base through which the liquid runs out under its own pressure.
The flow time for a standard quantity of liquid indicates the apparent viscosity.

With **Höppler viscometers**, the time is measured for a sphere to roll through the
liquid in an inclined tube. In **Cochius tubes**, the time taken by an air bubble to rise in the
liquid is a measure of the viscosity. All these instruments deliver indicators that are not
easy to convert to true rheological quantities.

4.2.3 Newtonian Melt Viscosities

Molar Mass Dependence
Stationary shear viscosities η_0 of polymer melts increase with a power ε of an average
molar mass $\overline{M}$:

(4-5) $\eta_0 = K_\eta \overline{M}^\varepsilon$

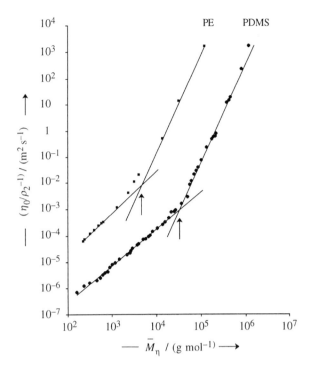

Fig. 4-8 Dependence of kinematic shear viscosities at rest, η_0/ρ_2, on the viscosity-average molar mass $\overline{M}_\eta$ of polymers with narrow molar mass distributions. Arrows indicate critical molar masses. For clarity, PE viscosities were shifted upwards by two decades. ρ_2 = density of melt.
PE: alkanes and poly(ethylene)s, H-[CH$_2$CH$_2\text{-}$]$_n$, with narrow distributions at 175°C [7].
PDMS: poly(dimethylsiloxane)s, (CH$_3$)$_3$Si-[OSi(CH$_3$)$_2\text{-}$]$_n$OSi(CH$_3$)$_3$ at 25°C [8].
Critical degrees of polymerization, $\uparrow$, are 656 for monomeric units –CH$_2$CH$_2$– (assuming endgroups H–) and 465 for monomeric units –OSi(CH$_3$)– (endgroups: CH$_3$– and (CH$_3$)$_3$Si–).

$\overline{M}$ is a viscosity average $\overline{M}_\eta$ (Volume III, p. 496 ff.) but is generally assumed to be the mass average molar mass $\overline{M}_w$ (weight-average molecular weight). The zero-shear viscosity η in Eq.(4-5) is usually the dynamic viscosity but occasionally the kinematic viscosity (Fig. 4-8) because the latter corrects for the variation of polymer densities of low-molar mass polymers with molar mass.

Plots of lg η_0 = f(lg $\overline{M}_w$) or lg (η_0/ρ_2) = f(lg $\overline{M}_\eta$) show two regions that are separated by a fairly sharp transition. The dependence of η_0 on M can be explained qualitatively for the low-molar mass region as follows.

Viscosities with the physical unit Pa s = kg/(m s) = N/m^2 = J/m^3 measure the mass that is moved per length and time, i.e., force per area = energy per volume. During the flow of compact spherical molecules, each with the same friction coefficient, only mutual friction forces have to be overcome and the viscosities are small. However, the movement of one chain unit of a chain molecule also causes all other chain units to move. Since each chain segment has the same friction coefficient and there are X chain units per molecule, kinematic viscosities as a measure of friction should be proportional to the degrees of polymerization and molar masses, respectively (Rouse theory, see Volume III, p. 497). Rouse theory predicts exponents of $\varepsilon = 1$ in Eq.(4-5) but experiments often show $\varepsilon \neq 1$, for example, $\varepsilon = 1.37$ (PDMS) and $\varepsilon = 1.64$ (PE) (Fig. 4-8).

Above a certain critical molar mass, chain molecules start to entangle (Volume III, p. 180, 605). The number of entanglements per chain does not vary with time for low shear gradients and the chain thus behaves as a physically crosslinked network. The higher the molar mass, the greater the number of entanglements and the larger the viscosity.

Because of entanglements, chains cannot move freely in all directions but through a kind of tube formed by the chains (Volume III, p. 477, 500 ff.) or like a snake through underbrush. The theory of this **reptation** predicts an exponent of $\varepsilon = 3.0$ whereas experimentally found exponents are higher, most often reported as $\varepsilon = 3.4$, but, for example, $\varepsilon = 3.72$ for poly(ethylene) and $\varepsilon = 3.51$ for poly(dimethylsiloxane) (Fig. 4-8). These deviations from theory are explained by a "breathing" of the tube.

Newtonian (zero-shear) melt viscosities of coil-like macromolecules increase so strongly with increasing molar mass that high-molar mass polymers cannot be processed as melts. For example, casting and injection molding processes require melt viscosities of less than ca. 10^3 Pa s which corresponds to molar masses of ca. 100 000 g/mol for poly-(ethylene)s with narrow molar mass distributions and a density of $\rho = 1$ g/cm^3.

Melts of liquid-crystalline polymers have two critical molar masses, an example of which is the polymer of Fig. 4-9. This polymer forms isotropic melts with an exponent of $\varepsilon = 3.0$ for molar masses of $\overline{M}_w \leq 29\ 000$ g/mol and anisotropic ones at molar masses of $\overline{M}_w \geq 74\ 000$ g/mol. Between these critical molar masses, a two-phase region was observed in which the melt viscosity is approximately independent of the molar mass. It seems that this intermediate region is caused by the formation of spheroidal domains that become more compact with increasing molar mass (hence the slight decrease of η_0).

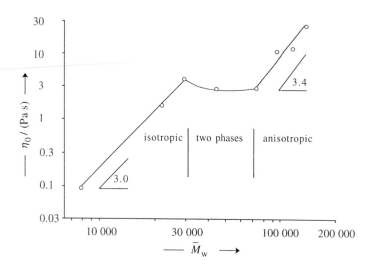

Fig. 4-9 Zero-shear viscosities of liquid-crystalline poly(2,5-didodecyl-1,4-phenylene)s with critical molar masses of $\overline{M}_w \approx 29\ 000$ g/mol and $\overline{M}_w \approx 74\ 000$ g/mol [9].

Temperature Dependence

Polymers are usually processed in a relatively narrow temperature range between their glass temperatures (if amorphous) or their melting temperatures (if semicrystalline) and their decomposition temperatures (Table 4-3). In this range, the temperature dependence of Newtonian melt viscosities η can be described by a type of **Williams-Landel-Ferry equation (WLF equation)** in which the shift factor a_T of the WLF equation for glass temperatures (Volume III, p. 467) is replaced by the ratio η/η_R where η_r = melt viscosity at a reference temperature T_o:

$$(4\text{-}6) \qquad \lg \frac{\eta}{\eta_R} = \frac{-K[T-T_o]}{K'+[T-T_o]}$$

The WLF equation is based on the Doolittle assumption that the probability of segmental movements is larger, the greater the volume fraction $\phi_{f,o}$ of free volume in the melt. It is assumed further that all polymers have the same volume fraction $\phi_{f,o}$ at the glass temperature T_G; empirically, $\phi_{f,o} \approx 0.025$ for many polymers.

The adjustable parameters K, K', and T_o are assumed to be universal, at $T_o = T_G + 50$ K formerly $K = 17.44$ K and $K' = 51.6$ K, and now $K = 8.86$ K and $K' = 101.6$ K (improved data base). More precisely, individual values have to be used for each polymer, for example, $T_o = 423$ K, $K = 13.5$ K, and $K' = 487$ K for atactic poly(styrene) with $T_G = 373$ K. For non-Newtonian viscosities, see next Section.

4.2.4 Non-Newtonian Shear Viscosities

Basic Phenomena

Shear viscosities are defined as non-Newtonian if shear stresses σ_{21} are not directly proportional to shear rates $\dot{\gamma}$ (Fig. 4-10). Shear viscosities η calculated with Eq.(4-2) are therefore apparent viscosities (Fig. 4-10, left) that may be independent of the duration of the experiment or not. They may also decrease or increase with shear rate or become "infinitely" high below a critical shear rate (Fig. 4-10, right).

The behavior of polymer melts may be Newtonian below and non-Newtonian above a certain critical shear rate that is usually in the range $0.1 \leq \dot{\gamma}/s \leq 10$ (see below). Processing often leads to very high shear rates, for example, up to 20 000 s^{-1} upon brush coating and up to 70 000 s^{-1} in the gate area of injection molding.

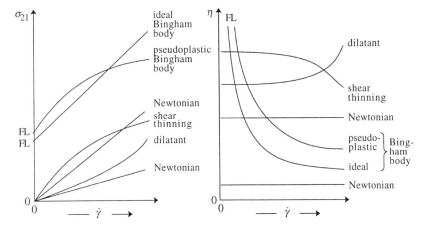

Fig. 4-10 Shear stress σ_{21} and viscosity η as a function of shear rate $\dot{\gamma}$ (schematic). FL = flow limit.

At zero shear rate, Newtonian liquids and many non-Newtonian ones have zero shear stresses (Fig. 4-10). **Bingham bodies**, on the other hand, show **flow limits**, i.e., no apparent flow below a critical shear rate above which they may behave Newtonian (ideal Bingham bodies) or non-Newtonian (pseudoplastic Bingham bodies), an example of which is tomato ketchup. Flow limits are appreciated for paints which should flow at the high shear rates of painting but should not drop from the brush under their own weight.

Polymer melts and concentrated solutions rarely behave as Bingham bodies but usually show **shear thinning** above certain molar masses, i.e., a decrease of apparent viscosities with increasing shear rate above a critical shear rate. This behavior resembles that of pseudo-ideal Bingham bodies and is therefore also called **pseudoplastic** although there is no yield value. Shear thinning is advantageous for processing because it reduces the energy that is necessary for processing.

Shear thickening is rare for polymer melts but is found for some dispersions and so-called ionomers (copolymers of ethene with, e.g., a few acrylic acid units) in apolar solvents. Such systems show **dilatancy** (L: *dilatare* = to enlarge, expand).

A material is **rheopectic** if the viscosity *increases* with time at constant shear rate (G: *rheos* = low, *pectous* = solidified, curdled). The apparent viscosity of **thixotropic** materials *decreases* with time at constant shear rate (G: *thixis* = touch, movement; *tropos* = to change); on removal of the shear, viscosity increases with time. Examples of thixotropic materials are suspensions of plate-like silicates and certain polymer melts.

Dispersions and gels frequently show **wall effects**, for example, a squeezing out of liquids near walls where the fluids act as lubricants (plug flow of toothpaste). An additional complication may be **turbulence**, which usually occurs at much lower **Reynolds numbers**, $Re \equiv \rho(dV/dt)d/\eta$, in non-Newtonian liquids than in Newtonian ones.

Flow Curves

The flow behavior of shear-thinning liquids is often represented in flow curves, either as $\lg \dot{\gamma} = f(\lg \sigma_{21})$ (Fig. 4-11) or as $\lg \eta = f(\lg \dot{\gamma})$ (Fig. 4-12). For Newtonian liquids, the slopes of the function $\lg \dot{\gamma} = f(\lg \sigma_{21})$ are unity whereas those of highly non-Newtonian liquids are S-shaped with a Newtonian range at low shear stresses σ_{21}. This range is followed by the shear-thinning range, which is larger the higher the concentration.

At high shear stresses, one often finds a second Newtonian range where $\lg \dot{\gamma} \neq f(\lg \sigma_{21})$. It is highly questionable whether this range indeed shows a Newtonian behavior of the *original* fluid and not some artefact from aggregation and/or degradation of the polymer, and/or turbulence of the fluid.

The flow behavior in the shear-thinning range is often described by empirical **flow laws** that are valid only for relatively small ranges of shear stress or shear rate. Most often used is the **Ostwald–de Waele** equation

$$(4\text{-}7) \qquad \dot{\gamma} = K_1 \sigma_{21}^{m}$$

The exponent m is known as the **flow exponent** or **pseudo-plasticity index**. It equals unity for Newtonian liquids and is smaller than unity for shear-thinning ones. Both m and K_1 must be evaluated for each *grade* and not just for every polymer type.

In the United States, $\dot{\gamma} = K_1 \sigma_{21}^{m}$ is replaced by its inverse function, the **power law**, where K_6 = **consistency index** and m = **power-law index**:

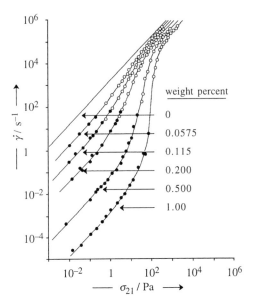

Fig. 4-11 Flow curves of butyl acetate solutions of a cellulose nitrate at 25°C as measured with capillary viscometers (O) and rotational viscometers (●) [10]. By kind permission of American Physical Society, Melville (NY).

$$(4\text{-}8) \qquad \sigma_{21} = (1/K_1)^{1/m} \dot\gamma^{1/m} = K_6 \dot\gamma^n \; ; \quad (1/K_1)^{1/m} \equiv K_6 \; ; \quad 1/m \equiv n$$

Other empirical equations are the **Prandtl-Eyring equation**, $\dot\gamma = K_2 \sinh(\sigma_{21}/K_3)$ and the **Rabinowitsch-Weissenberg equation**, $\dot\gamma = K_4\sigma_{21} + K_5(\sigma_{21})^3$.

The whole flow range from the first to the (apparent) second Newtonian range is described by several more recently proposed flow laws that are all of the type

$$(4\text{-}9) \qquad \eta = \eta_\infty + (\eta_0 - \eta_\infty)[1 + K(Q\dot\gamma)^a]^{-1/b}$$

where η_0 = first Newtonian shear viscosity, η_∞ = second Newtonian shear viscosity, $\dot\gamma$ = shear rate, and K, Q, a, b = constants with various meanings, depending on the particular theory (Table 4-5). The CAMPUS@ system uses the **Carreau equation** which corresponds to Eq.(4-9) with $K = 1$, $a = 2$, $1/b = (n - 1)/2$, and Q = relaxation time, t_{rlx}.

Dimensional analysis of Eq.(4-9) shows that for $K = 1$, Q must have the unit of time since $\dot\gamma$ has the unit of inverse time. Q may therefore be identified as the inverse critical shear rate, $Q = 1/\dot\gamma_{crit}$, at which the Newtonian behavior converts to shear thinning.

Table 4-5 Constants in Eq.(4-9) according to various authors.

Author	η_∞	K	Q	a	$1/b$
Cross	η_∞	K	1	2/3	1
Carreau	η_∞	1	t_{rlx}	2	$(n - 1)/2$
Gaidos and Darby	η_∞	$4\,q(1 - q)$	t_{rlx}	2	1

Ideally, a sharp transition should exist from Newtonian behavior, $\eta_0 = \sigma_{21}/\dot\gamma$, to shear-thinning where $\eta = f(\sigma_{21})$ or $\eta = f(\dot\gamma)$, respectively. At that ideal transition point, one must therefore have $\eta_0 \dot\gamma_{crit} = \sigma_{21,crit}$. Furthermore, the quantity η_∞ can be assumed to be zero because a second Newtonian range is very unlikely for a well-behaved system (absence of turbulence, degradation etc.). Setting a $\equiv$ by, Eq.(4-9) becomes

$$(4\text{-}10)\quad \eta/\eta_0 = \left[1+\left(\frac{\eta_0\dot\gamma}{\sigma_{21,crit}}\right)^{by}\right]^{-1/b} \quad ; \quad \lg\frac{\eta}{\eta_0} = -\frac{1}{b}\lg\left[1+\left(\frac{\eta_0\dot\gamma}{\sigma_{21,crit}}\right)^{by}\right]$$

At $\dot\gamma \to 0$, Newtonian behavior is recovered ($\eta/\eta_0 \equiv 1$). In **Vinogradov-Malkin** plots of $\lg(\eta/\eta_0) = f[\lg(\eta_0\dot\gamma)]$ for shear-thinning liquids, Eq.(4-10) delivers a zero slope and $\lg(\eta/\eta_0) = 0$ for small values of $\lg(\eta_0\dot\gamma)$ and $(\eta/\eta_0) = -y\lg(\eta_0\dot\gamma/\sigma_{21,crit})$ for very large values. The two linear sections of $\lg(\eta/\eta_0) = f[\lg(\eta_0\dot\gamma)]$ intersect at a critical value of $\eta_0\dot\gamma = \sigma_{21,crit}$. The transition range between Newtonian behavior and perfect shear thinning has a width of about one decade of pascal for poly(ethylene)s (PE) but is very narrow for a glass-fiber filled poly(styrene) (PS-GF) (Fig. 4-12).

Remarkably, these two polymers and also a plasticized poly(vinyl butyral) (PVB) have the same critical transition at $\eta_0\dot\gamma = \sigma_{21,crit} = 10^4$ Pa. The same value was observed for poly(isobutylene)s (PIB), regardless of temperature (–20°C to 80°C) (Volume III, p. 507). The critical shear stress, $\sigma_{21,crit}$, thus appears to be universal for *linear* polymers, regardless of the polymer constitution (PE, PS, PIB, PVB), the molar mass (PE), and the temperature (PIB) (see also Volume III, p. 507). This value is not affected by the presence of a filler (PS-GF).

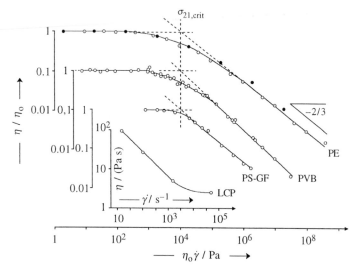

Fig. 4-12 Normalized melt viscosities η/η_0 as a function of $\eta_0\dot\gamma$ for three poly(ethylene)s PE with different molecular weights at 150°C, a glass-fiber reinforced poly(styrene) PS-GF (no temperature given), and a plasticized poly(vinyl butyral) (PVB) at 125°C (see also Volume III, Fig. 5-11). For clarity, ordinate axes were shifted as indicated.
Insert: shear viscosities $\eta/(\text{Pa s})$ of a liquid crystalline polyester with 70 % $-O(p\text{-}C_6H_4)CO-$, 15 % $-CO(p\text{-}C_6H_4)CO-$, and 15 % $-OCH_2CH_2O-$ units as a function of shear rate $\dot\gamma/s^{-1}$.

Since $\sigma_{21,crit}$ has the physical unit of pressure, this behavior seems to indicate that shear thinning sets in at the same critical free volume of melts of linear polymers. Indeed, high-generation dendrimers with a branch-upon-branch constitution showed a far higher critical shear stress of $\sigma_{21,crit} \approx 6 \cdot 10^5$ Pa (Volume III, Fig. 15-12) because they are much more difficult to compress.

The curves for the shear viscosities of poly(ethylene)s, the glass-fiber filled poly(styrene), and a liquid-crystalline polymer (all Fig. 4-12) as well as a poly(isobutylene) at various temperatures, and the high-generation polyamidoamine dendrimers (Volume III, p. 507) also have the same slope of $-2/3$ for $\eta/\eta_0 = f(\eta_0 \dot\gamma)$ in the shear-thinning range. The slope for plasticized poly(vinyl butyral) is more negative (ca. $-3/4$) (Fig. 4-12).

For rotational viscometers, the Newtonian range at small values of $\eta_0 \dot\gamma$ is observed only after a certain time lag (Fig. 4-13, see also Fig. 4-176). The dependence of apparent viscosities on shear rate *and* temperature is described by CAMPUS® by a combined Carreau-WLF equation with five empirical constants K_1-K_5 (for the physical meaning of these constants, see Volume III, p. 467):

$$(4\text{-}11) \quad \eta = \frac{K_1 a_T}{(1 + K_2 a_T \dot\gamma)^{K_3}} \quad ; \quad \lg a_T = \frac{8.86(K_4 - K_5)}{101.6 + K_4 - K_5} - \frac{8.86(T - K_5)}{101.6 + T - K_5}$$

4.2.5 Melt Elasticity

Coil molecules deform on shearing (Volume III, Fig. 15-4). At short experimental times, very large coil molecules can also not disentangle fast enough to adopt a new equilibrium shape. As a result, a tensile stress σ_{11} arises normal to the flow direction (Fig. 4-13): the poly(ethylene) melt is thixotropic. The tensile stress σ_{11} is much larger than the shear stress σ_{21}; it decreases only slowly (Fig. 4-13).

This behavior of polymer melts cannot be observed in capillary viscometry but shows up in cone-and-plate viscometers because these viscometers allow one to determine shear stresses and tensile stresses separately. The shearing during rotation here generates a torsional momentum which is transferred to the rotor; this process allows one to calculate the shear stress. Tensile stresses push the cone and plate apart which can be prevented by applying a force that is proportional to the tensile stress.

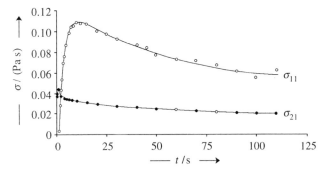

Fig. 4-13 Time dependence of stresses σ (shear stress σ_{21} or tensile stress σ_{11}) of a poly(ethylene) melt at 150°C and a shear rate of 8.8 s^{-1} as measured by a cone-and-plate viscometer) [11].

The flow through tubes (as in capillary viscometers or extruders) also generates a normal stress; however, this stress is not counteracted by a pressure but rather is stored as elastic energy per volume. At the exit of the tube, the flowing material is no longer confined by the walls of the tube. The coils return to their equilibrium macroconformations (shapes), the melt expands perpendicularly to the flow direction (Fig. 4-14), and the diameter of the exiting jet becomes larger than the tube diameter. However, it may also be smaller than the tube diameter if molecules with rodlike segments crystallize on exiting.

The enlargement of the jet is known in different industries by various names: **Barus effect** or **memory effect** in melt viscometry, **parison swell** in extrusion, **swelling** in injection blow molding, and **onion formation** in fiber spinning. The effect arises for the same reason as the **rod climbing effect (Weissenberg effect)** where a liquid creeps up an immersed rotating rod.

The effect of normal stress can be determined quantitatively for the flow in capillaries or nozzles by the **Bagley diagram** (Fig. 4-14). The flow in tubes of length L and radius R by a pressure p generates a shear stress $\sigma_{21} = \eta_0 \dot\gamma$ at the tube wall. For both Newtonian and non-Newtonian liquids, this shear stresss is given by $\sigma_{21} = (pR)/(2\,L)$ (Volume III, p. 514). A plot of pressure p as a function of the nozzle geometry L/R at constant shear rate $\dot\gamma$ according to

$$(4\text{-}12) \qquad p = 2\,\sigma_{21}(L/R) = 2\,\eta_0\dot\gamma(L/R)$$

delivers a straight line for sufficiently small values of L/R (Fig. 4-14). This line intersects the ordinate axis at a certain pressure p (**pressure correction**) and the abscissa at a certain value of L/R (**Bagley correction**).

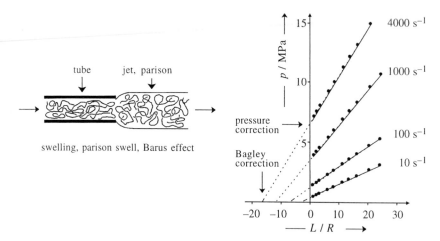

Fig. 4-14 Parison swell (left; F: *paraison*, from *parer* = to prepare) and pressure p as a function of the nozzle geometry L/R for measurements at various shear rates. Data for a high-impact poly(styrene) at 190°C [12]. The pressure correction (= value p_0 of the ordinate at $L/R = 0$) and the Bagley correction (= value $(L/R)_0$ of the abscissa at $p = 0$ MPa) are indicated for the data at 4000 s^{-1}.

The pressure correction at $L/R = 0$ indicates the loss of pressure that is caused by the elastically stored energy of the flowing liquid and the formation of a stationary flow profile at both ends of the tube. The larger the nozzle geometry L/R, the greater the pressure that has to be applied. The tube radius R cannot be changed since it is controlled by the shape of the desired specimen. In order to keep the pressure correction as small as possible, one therefore needs to work with tubes as short as possible.

The parison swell can be reduced by repeated extrusions, possibly because more and more entanglements are dissolved. Such a treatment lowers the extensional viscosity (next Section) but does not change the shear viscosity at rest. It also increases the melt volume index, but this does not come from a decrease of the molecular weight because the effect can be reversed by a heat treatment of the melt or a dissolution and subsequent recovery of the polymer.

Entanglements increase the elasticity of polymer melts and concentrated polymer solutions. The flow of such elastic liquids produces elastic vibrations which are less and less dampened the higher the shear rate or shear stress. The surface layer of the melt becomes detached from the wall of the tube, which leads to critical turbulence and a roughness of the surface of the melt (Fig. 4-15).

Rapid cooling of the extruded strand freezes the roughness and the surface appears fractured. The phenomenon is therefore called **melt fracture, surface roughness, shark-skin, orange peel**, or **matte**. "Melt fracture" thus does not indicate a breaking of the strand. The formation of melt fracture is sometimes accompanied by a cyclic thickening and narrowing of the diameter of the strand, which is called **draw resonance**.

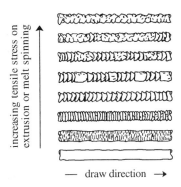

Fig. 4-15 Formation of melt fracture with increasing shear stress during extrusion or melt spinning of thermoplastics [13]. With kind permission of the ACS Rubber Division, Akron (OH).

4.2.6 Extensional Viscosity

Melts and concentrated solutions of high-molar mass polymers can be stretched considerably without breaking. This extensibility allows the spinning of fibers from melts and concentrated solutions and the blow molding of hollow articles. Without the extensibility, vacuum forming of films and sheets would be impossible and injection molding considerably more difficult.

The elastic behavior of melts and concentrated solutions is controlled by the **extensional viscosity** (**elongational viscosity**) of these fluids and the melt strength (p. 99) caused thereby. It is strongly affected by the temperature, the molar mass and molar mass distribution of molecules, the stiffness and architecture of polymers, and, for solutions, also by the polymer concentration.

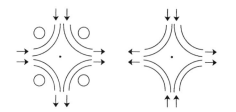

Fig. 4-16 Measurement of elongational viscosity of liquids using rotating rolls (left) or suction nozzles (right).

→ Flow, —— flow lines

Phenomena

The extensional viscosity of polymeric liquids is probably the most important property for the industrial processing of polymers. However, it is the least understood rheological property because it is very difficult to measure. It can be studied by elongating specimens vertically or horizontally, using clamps or rolls in the latter case (see Volume III, Fig. 15-13). One can also let fluids flow against each other and then divert the flow 90° to the initial flow direction by rolls or by sucking it off with nozzles (Fig. 4-16). All these procedures are instrumentally challenging, time consuming, and unfit for routine measurements. More promising are acoustical rheometers that use sound waves to extend and contract fluids. Knowledge of extensional viscosities is highly desirable because the processing of polymers depends much more strongly on elongational viscosities than is commonly assumed.

The variation of elongation ε with time, the elongational rate, $\dot{\varepsilon} = d\ln(L/L_0)/dt$, is given by the time dependence of the **Hencky elongation**, $\varepsilon_H = \ln(L/L_0)$, and not by that of the Cauchy elongation, $\varepsilon = (L - L_0)/L_0$ (see Volume III, Chapter 16). The three main deformation rates in the 11, 22, and 33 directions are set as $\dot{\varepsilon}_{11} \geq \dot{\varepsilon}_{22} \geq \dot{\varepsilon}_{33}$ where the ratio $K = \dot{\varepsilon}_{22}/\dot{\varepsilon}_{11}$ defines the type of elongational flow. This ratio assumes values of $K = -1/2$ for uniaxial elongations, $K = 1$ for biaxial ones with equal extensions in both directions, and $K = 0$ for planar ones (so-called pure shearing). In contrast to shear viscosities, elongational viscosities must always be characterized by the type of deformation.

Shear viscosities $\eta_{s,0}$ at zero shear rate (Newtonian viscosities) and elongational viscosities $\eta_{e,0}$ at zero elongational rates (Troutonian viscosities) are related by $\eta_{e,0} = 3\,\eta_{s,0}$ for uniaxial elongations ($K = -1/2$) and by $\eta_{e,0} = 6\,\eta_{s,0}$ for equally biaxial ones ($K = 1$). These simple relations do not apply to non-Newtonian (η_s dependent on shear rate) and non-Troutonian viscosities (η_e dependent on elongational rate) (Fig. 4-17).

At small deformation rates $\dot{\gamma}$, shear and extensional viscosities may increase with time, i.e., show thixotropy like the branched poly(ethylene) of Fig. 4-17. At long times and small rates of $\dot{\varepsilon}$ and $\dot{\gamma}$, both lg η_e and lg η_s finally become independent of time. For the whole time span, uniaxial Trouton viscosities are always 3 times larger than zero-shear Newtonian viscosities, as required by theory.

At higher deformation rates, logarithms of shear viscosities of this thixotropic poly(ethylene) melt (see also Fig. 4-13) first pass through a maximum with the logarithm of time until lg σ_s finally becomes a linear function of lg t (Fig. 4-17, bottom). The logarithm of shear viscosity, on the other hand, increases rapidly with the logarithm of time to a maximum (Fig. 4-17, top). The reversible elongation is small at small elongational rates (typical of a viscous liquid) but increases strongly at high deformation rates. At these rates and not too large deformations, polymer melts behave as rubbers. This polymer melt is thixotropic for all but the smallest deformation rate.

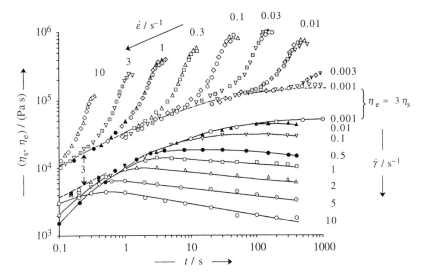

Fig. 4-17 Time dependence of elongational viscosities η_e (top) and shear viscosities η_s (bottom) of a melt of a branched poly(ethylene) at 150°C and various elongational rates $\dot{\varepsilon}$ and shear rates $\dot{\gamma}$ [14].
By kind permission of the Schweizerische Chemische Gesellschaft, Basel.

Effect of Polymer Structure

The dependence of elongational viscosity on elongational rate is strongly affected by the chemical and physical polymer structure and the state of the fluid (Fig. 4-18).

With increasing shear rate, the longitudinal axes of *rod-like* (or stiff) *macromolecules in dilute solutions* orient themselves more and more in the flow direction. The resistance against flow becomes smaller, and the shear viscosity decreases (Fig 4-18, top, right). In elongational flow, on the other hand, rods are elongated, which increases the resistance against more elongation. Elongational viscosities increase and finally become constant because bond angles between chain atoms can only widen to a certain extent before the chain breaks which indeed happens at large deformation rates (for an application of this behavior, see Section 11.4.2).

Rods in concentrated solutions cannot reside in random positions. They rather form domains in which the longitudinal axes of rods lie more or less parallel to each other whereas the main orientation axes of the domains themselves are spatially random (see Volume III, Section 8.2). This feature leads to the following flow behavior.

In deformations like elongation or shearing, only the main axes of domains have to be oriented in the flow direction, which requires much less frictional energy than the orientation of the rods themselves because domains have a much smaller surface/volume ratio than rods. As a result, there will be little, if any, initial increase of viscosity with increasing deformation rate but then a strong decrease of lg η with lg $\dot{\varepsilon}$ or lg $\dot{\gamma}$.

Dilute solutions of coil molecules behave quite differently on shearing or elongation. Above a certain critical deformation rate, shear viscosities *de*crease with increasing shear rate whereas elongational viscosities *in*crease with increasing elongation rate. The reason for this behavior has to be sought in the structure of random coils which are kidney-like with axial ratios of ca. 12:2.7:1 (Volume III, p. 84).

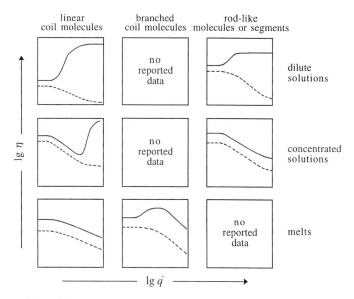

Fig. 4-18 Logarithm of shear viscosities $\eta = \eta_s$ (- - -) and elongational viscosities η_e (——) as a function of the logarithm of the deformation rate $\dot{q}$ (= $\dot{\gamma}$ on shearing, $\dot{\varepsilon}$ on elongation) [15]. The second Newtonian range (left, top and middle) is probably an experimental artefact caused by degradation of molecules, turbulent flow, and the like.

Shear flow causes coils to become more oriented with respect to their longest axis, which reduces their resistance to flow: shear viscosities decrease with increasing shear rate. Uniaxial elongational flow tries to extend the coils in the flow direction, which is resisted for entropic reasons. The extensional viscosity remains constant up to a critical elongation rate, whereupon it increases dramatically until it finally becomes constant and much larger than the shear viscosity (Fig. 4-18, top left). At even higher shear rates (not shown in Fig. 4-18), coil molecules become so extended that they break in the center (Volume III, p. 513).

For *coils in melts*, entanglements between molecules are more important than elongations of molecules themselves. The number of entanglements decreases with increasing extensional rate and so does the extensional viscosity (Fig. 4-18, bottom left).

Coils of linear polymers *in semi-concentrated solutions* combine the features of melts and dilute solutions (Fig. 4-18, left center). At low extensional rates, elongational viscosities decrease because of a (partial?) dissolution of entanglements. At higher extensional rates, elongational viscosities increase (and finally become constant?) because molecules are more stretched (Fig. 4-18, left center).

Branched polymer molecules in melts can entangle less with other polymer molecules than linear ones. With increasing elongational rate, the elongational viscosities of these polymers first increase to a maximum and then decrease.

Fillers act as physical crosslinking agents; they increase elongational viscosities of polymers (Fig. 4-19). At very small Hencky elongations ε_H, extensional viscosities η_e of filled polymers at constant elongational rate $\dot{\varepsilon}$ increase somewhat with increasing ε_H but then do not vary with the Hencky elongation ε_H over a large range of ε_H. This behavior contrasts with the behavior of the corresponding unfilled polymers where the viscosity function, $\lg \eta_e = f(\varepsilon_H)$, passes through a weak maximum with increasing ε_H.

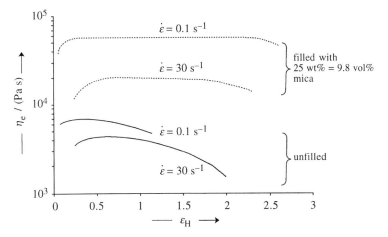

Fig. 4-19 Effect of extensional rate $\dot{\varepsilon}$ on the dependence of uniaxial extensional viscosity η_e on the Hencky elongation ε_H for an unfilled and a filled high-density poly(ethylene) at 150°C [16].

Melt Strengths

Polymers melts should have high strengths for processing, for example, for melt spinning and thermoforming. Melt strengths are forces, $F = A_0 \eta_e \dot{\varepsilon} / \exp(\dot{\varepsilon} t)$, that can be calculated from the time dependence of extensional viscosities η_e at constant extensional rate $\dot{\varepsilon}$ and the initial cross-section of the specimen, A_0.

These time dependences can be quite different for various grades of the same polymer (Fig. 4-20). In both cases, extensional viscosities increase with time but the increase is concave to the time axis for the standard grade and convex for the high-impact polymer. The curves from the standard grade are also practically independent of the extensional rate whereas those of the latter deviate earlier, the greater the extensional rate.

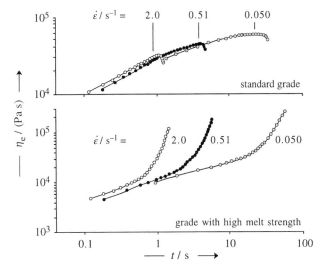

Fig. 4-20 Effect of extensional rates $\dot{\varepsilon}$ on the time dependence of uniaxial extensional viscosities η_e of two grades of isotactic poly(propylene) at 180°C [17].

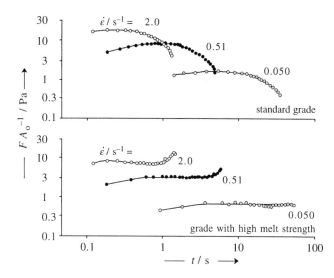

Fig. 4-21 Time dependence of melt strength F per initial cross-sectional area A_0 of the poly(propyl-ene) grades of Fig. 4-20 at various extensional rates $\dot{\varepsilon}/s^{-1}$.

Melt strengths per area, F/A_0, of the two types of it-poly(propylene) grades of Fig. 4-20 are both the higher, the greater the elongational rate (Fig. 4-21). However, the values of F/A_0 of the standard it-poly(propylene) decrease with time whereas those of the grade with high melt strength remain constant. The chemical and physical differences between the two grades were not revealed by the authors.

4.3   Processing Fundamentals

4.3.1   Types of Processes

"**Processing**" usually refers to those treatments of materials that lead to articles with the desired microstructures and shapes. Processing can thus include chemical processes such as the fabrication of ceramic parts by sol-gel processes, the formation of alloys in metallurgy, as well as crosslinking reactions leading to thermosets. For thermoplastics, "processing" only refers to shaping.

Types of processes can be subdivided in various groups, depending on the discipline. Engineers are fond of distinguishing procedures according to the machinery (for example, pressure-less *versus* pressure-requiring processes), the process (continuous *versus* discontinuous processes, forming *versus* transforming, etc.), or the state of matter (thermoplastic *versus* thermosetting materials, melt spinning *versus* solution spinning, etc.). For materials scientists, the most important property for processing is the stability during shaping, a property that is mainly controlled by the thermal and rheological properties of starting materials and the resulting articles. The two main groups of processes for plastics here are molding processes that use a support (external stabilization) and extrusion processes that do not (no external stabilization) (Table 4-6).

Table 4-6 Very often (++) and often (+) used processes for plastics. Types A-E, see p. 75

Plastics	External stabilization by molds or supports			No external stabilization			
	Molding, casting, dipping	Laminating, filament winding	Press molding	Injection molding	Extrusion	Blow molding	Thermo-forming
Thermosets from monomers							
Diallyl phthalate	+						
Polyurethanes	++						
Epoxides	+	+	+				
Thermosets from prepolymers							
Unsaturated polyesters	+	++	++	+			
Phenol-formaldehyde			++	+			
Melamine-formaldehyde			+	+			
Semicrystalline thermoplastics							
Poly(oxymethylene)				++	+		
Polyamides 6 and 6.6				++	+		
Poly(ethylene terephthalate)				+	+	++	
Poly(ethylene), poly(propylene)			+	++	++	++	+
Amorphous thermoplastics							
Polycarbonate				+	+	+	+
Poly(styrene)				++	+	+	++
Poly(methyl methacrylate) (as monomer)				+	++		+
Poly(vinyl chloride) (plasticized)				+	++	++	+
% of all plastics by weight	?	?	3	32	36	10	?

Other processes: calendering 6 %, coating 5 %, powder processing 2 %, all others 6 %.

Processes for fibers, elastomers, and lacquers are similar to (and older than!) those for plastics but less varied and more specific. Because of their special nature, they will be discussed in Chapters 6 (fibers), 7 (elastomers), and 12 (coatings and adhesives).

Type A processes for plastics comprise the direct conversion of monomers to final polymeric products on supports; an example is the coating of wires. However, the formation of shaped articles from monomers or oligomers leading to thermoplastics (for example, polyamides or polycarbonates) or thermosets (for example, to polyurethanes and unsaturated polyesters or from epoxides or phenolic resins) by coating, casting, extrusion, or injection molding often causes problems because low-viscosity monomers or oligomers are converted to highly viscous polymers. For economic reasons, short cycle times are required for mass-produced items; such processes necessitate high rates of polymerization. An example is reaction injection molding (RIM) (Sections 4.5.3, 8.5.4).

Type B processes are typical for thermoplastics. After polymerization of the monomer(s), the polymer is isolated and warehoused (as pellets, granulate, etc.) before it is shipped to compounders, processors, and/or fabricators. Pellets, granulates, and powders have to be dried and compounded before they are processed by casting, dipping, coating, laminating, spraying, rolling, calendering, extruding, injection molding, blow molding, thermoforming, vacuum forming, and the like (see below). A few special articles are also manufactured by press molding of thermoplastics, a process that is common for crosslinkable polymers.

Type C processes are used for thermosetting polymers. Crosslinkable monomers (A stage) are polymerized to oligomers (**prepolymers**) and the polymerization stopped just before crosslinking at the so-called B stage. Adjuvants are added and the mixture is then polymerized further with simultaneous shaping (C stage). Type C processes with molds comprise molding, press molding, compression molding, transfer molding, and injection molding. Type C processes without molds include extrusion (only certain systems) and the production of semifinished goods from expanded polymers.

Difficult to prepare articles are sometimes produced from semifinished goods by Type D processes such as welding, embossing, cutting, sawing, drilling, machining, etc. Surfaces are prepared by polishing, painting, metallizing, coating, etc., for technical reasons (surface hardness, friction, etc.) and/or aesthetic ones (gloss, color, etc.).

4.3.2   Energy Consumption

Energy is a major cost factor in polymer processing since many processes require the heating of monomer, oligomers, and/or polymers and/or heating of semifinished goods.

External heat is absorbed by the materials either by bodily *contact* with hot metal surfaces, by immersion into a hot bath, or by *convection* (infrared radiation, blowing hot air). The heat is then transported to the interior by heat conduction.

Internal heat is produced by vibration of atomic groups (for example, triggered by ultrasound or microwaves) or by shearing (kneaders, screw mixers, etc.). **Plastication** combines friction and contact heat whereas **plastifying** is restricted to contact heating. Both processes should not be confused with **plasticizing**, the lowering of glass temperatures by addition of plasticizers (p. 424 ff.) or incorporation of flexibilizing chemical groups into polymer chains (see p. 42 ff.).

Contact heat generates in the surface region of polymer particles (powders, granules, pellets) a plastified ("molten") zone that may or may not adhere to the walls of the extruder or mold. Extrusion shaves off a part of the adhering melt which interpenetrates the voids between polymer particles and causes the still solid particles to melt by contact heat. However, the melt cannot penetrate cold plugs of powder particles that were compressed by the extrusion process (see Section 4.5.2). Such cold plugs are always formed if polymer melts do not adhere to extruder walls.

For this reason, contact heat is usually supplemented by heat from friction. The adiabatic change of temperature is calculated from the density ρ and the specific heat capacity c_p of the polymer and the pressure difference $\Delta p = \eta\dot{\gamma}$, i.e., the kinematic viscosity η and the shear rate $\dot{\gamma}$:

(4-12) $\Delta T = \Delta p/(\rho c_p)$

Injection molding with $\Delta p = 100$ MPa and $\dot{\gamma} = 10^4$ s^{-1} of a melt of viscosity $\eta = 10^4$ Pa s and $c_p = 2$ J/(g K) of a polymer with $\rho = 1$ g/cm^3 will thus increase the processing temperature by $\Delta T = 50$ K. The increased temperature lowers the melt viscosity, reduces the shear energy that has to be put into the system, and increases the processing speed.

Rheological properties of thermoplastics can also be reduced reversibly by **shear modification (shear working, shear refining)**, i.e., by repeated extrusion of the same polymeric material. This procedure decreases the elongational viscosity (and thus the parison swell) but practically neither the stationary shear viscosity nor the dynamic properties. The effect is probably caused by an orientation of chain segments since the original rheological properties can be reconstituted by heating the polymer.

After processing, articles are still warm. Heat is removable only by heat transfer, i.e., the cooling rate depends on the heat conductivity of the polymer. This process is slow, especially for thick-walled articles, and thus often controls cycle times.

4.4   Processing of Viscous Materials

Liquid monomers, molten prepolymers, low-concentration solutions, polymer suspensions and dispersions, and polymer powders have relatively low viscosities. They flow under their own weight, and no pressure has to be applied for all or at least the main stage of processing. However, end stages of processing sometimes do need application of pressure in order to remove voids from the very viscous material.

Polymeric materials with low viscosities can be processed by casting methods (Section 4.4.1), coating methods (Section 4.4.2), and molding processes without (Section 4.4.4), or with applied pressure (Section 4.4.5).

4.4.1   Casting

In casting processes, liquid masses are poured into a mold where they are chemically "hardened", i.e., polymerized by polycondensation (phenolic resins, amino resins, epoxies) or chain polymerization ("addition polymerization": methyl methacrylate, styrene, ε-caprolactam, N-vinyl carbazole). The casting of gelable masses, for example, plastisols of poly(vinyl chloride), is called **gelling** or **fusion**.

Molds for **monomer casting** (Fig. 4-22) are relatively inexpensive because they can be made from sheet metal, glass, ceramics, or wood. Monomer casting is also an easy way to incorporate metal parts in polymers, for example, blades into handles of knives.

These easy, low-cost processes have two disadvantages. Chemical reactions are relatively slow, especially polycondensations, so that casting is slow and not very economical for mass production. Many polymerizations are also exothermic which makes casting difficult to control.

For this reason, casting of unsaturated polyester resins is not very popular since it consists of the strongly exothermic free-radical crosslinking polymerization of unsaturated polyesters from maleic anhydride and ethylene glycol with styrene (or methyl methacrylate). Hence, thermoplastics are processed by casting in only a few cases, for example, methyl methacrylate for glasses or dentures.

A related processing method is **film casting** to (endless) sheetings which are more homogeneous than the ones produced by calendering (Section 4.5.1). The process is used for sheetings of cellulose acetate, polyamides, and polyesters.

Dip molding (dipping) is used for the manufacturing of thin-walled, hollow articles such as surgery gloves and condoms. In this process, a dipping mandrel (dip mold) is dipped so often or so long into a fluid which may be a polymer solution (for example,

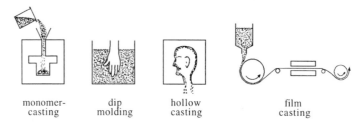

monomer-   dip   hollow   film
casting   molding   casting   casting

Fig. 4-22 Casting methods. Films, sheets, and sheetings are plane products with thicknesses that are much smaller than their lengths and widths. The definitions of these three products vary from country to country. According to ASTM, a film is a sheet that is less than 400 μm thick. A sheeting is an endless sheet, a "foil" is always a metal sheet.

of silicones), a latex (for example, of natural rubber, poly(chloroprene), or a plastisol (of poly(vinyl chloride)) until the layer has the desired thickness (Fig. 4-22). The viscosity of the liquid should be less than ca. 12 Pa s and the flow limit as small as possible. The fluid-covered mandrel is pulled out of the fluid and heated or dried to solidify the coating. Processed plastisols require an additional gelling.

Hollow casting (flow casting, slush casting) produces hollow articles with thicker walls by pouring solutions or dispersions of resins, PVC pastes, or molten plastics into a two-part hollow mold. The polymer forms a layer on the inner wall and the form is reversed to let excess polymer flow out. The deposited polymer is hardened by cooling (melts) or heating (solutions, dispersions, pastes) and the article is removed from the opened mold. Hollow casting from dry, sinterable powders are known as **slush molding**. However, "slush molding" and "slush casting" are also used interchangeably.

Centrifugal casting of melts leads to inner coatings or thick-walled rotationally symmetrical articles by centrifuging a partially filled mold (rotation around *one* axis). It is used for the manufacture of bushings, sockets, pipes, etc., by polymerization of ε-caprolactam and for inner coatings from meltable powders.

Rotational casting (rotational molding, rotomolding, slush molding, "casting") allows automatic casting of hollow articles from powdered thermoplastic resins (PE, it-PP, PA, PC, PET) with particle sizes of ca. 500 μm or PVC plastisols. The polymer is metered into the mold, and the molds are closed and then rotated around *two* (or more) axes for a homogeneous distribution of the material on the walls (Fig. 4-23). Subsequent heating melts thermoplastics and gelatinates (gelifies, gelatinizes, gels, fuses, fluxes) plastisols. The thickness of the walls of articles is determined by the amount of resin and the rotational speed. The mold is opened, and the article is removed after cooling.

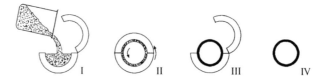

I   II   III   IV

Fig. 4-23 Rotational casting [18]: (I) A metered amount of, for example, a poly(vinyl chloride) plastisol (or a powdered thermoplastic) is poured into the mold. (II) The mold is closed and rotated about two or more axes to distribute the material homogeneously on the inner wall. Simultaneous heating gels the plastisol (or melts the thermoplastic). (III) The mold is opened, and the articles is cooled. (IV) The articles is removed. – A plastisol is a suspension of a finely divided polymer in a plasticizer.

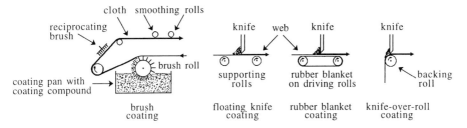

Fig. 4-24 Spread coating. From left to right: brush coating with impregnation of fabric (cloth) by a brush roll, reciprocating brush, and smoothing rolls; floating knife coating with floating doctor knife; rubber blanket coating with doctor knife; knife-over-roll coating with doctor knife and backing roll.

4.4.2 Coating

Coating processes are used to deposit solutions, melts, or dispersions of polymers permanently onto a support by spreading, roll coating, or spraying. The support may consist of the same polymer as the coating or be a different material. The support may be permanently combined with the coating product or later removed from it.

Spreading

Painting is a coating process in which solutions, suspensions, or latices of film-forming polymers are spread by brushes, rollers, spray guns, etc., on supports (walls, sheets, etc.). After evaporation of the liquid (solvent), permanent films are formed (see Section 12.4).

Spread coating is used to coat "endless", thin ("two-dimensional") articles (paper, sheets, foils, fabrics, etc.) with polymeric fluids (solutions, dispersions, pastes, melts) (Fig. 4-24). Three-dimensional articles are coated by "spray processes" (see below). Coating by direct dipping of articles into polymer melts is not a commercial process.

In **brush coating**, the fluid coating compound is transferred from a coating pan to the fabric (cloth) by a brush roll (Fig. 4-24). The coating is spread by a reciprocating brush and then smoothed by smoothing rolls.

The various **knife coatings** are the most important spread coating methods. In **floating knife coating**, a doctor knife (doctor blade) removes the surplus paste from the web (fabric) that is supported only by two supporting rolls. The doctor knife may also be supported by a rubber blanket on driving rolls (**rubber blanket coating**) or by just a backing roll (**knife-over-roll coating**).

Roll Coating

Roll coating is an umbrella term for more than 30 processes (examples in Fig. 4-25):

(I) In **roll coating with squeezing-off excess resin**, a lower roll dips into a fluid in the coating pan and transfers the fluid to the underside of the fabric which is supported by a second (upper) roll that rotates in the counter direction. The upper roll supports the fabric and squeezes off excess resin from the impregnated fabric. The upper roll may be a gravure roll.

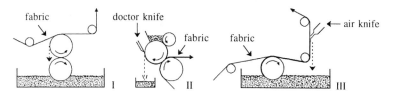

Fig. 4-25 Some roll coating processes (see text).

(II) **Reverse roll coating** works with three rolls. The gap between the upper and central roll controls the thickness of the fluid layer on the center roll which continuously coats the fabric. Excess fluid is removed by a doctor knife and falls into a catch pan.

(III) Another version of roll coating removes excess fluid by an air knife.

Spray Processes

In spray processes, matter (fluids, foams, powders) is deposited on surfaces through nozzles. Many variations are known.

Spraying of paints with spray guns is similar to painting with brushes or rolls and requires similar properties of paints (Section 12.4).

In **spray coating**, fluid matter is sprayed continuously through spray nozzles onto moving fabrics or films (Fig. 4-26). The process works well with low-viscosity fluids but delivers less even coatings than knife coating if molten resins or other highly viscous matter is used. A special version of this process is the **spray foaming** of foams on sheets. Another special type is **spray-up molding (fiber-spray gun molding)** in which short fibers are mixed in air with sprayed resin and then sprayed onto the surface.

In **flame spraying**, polymer granulates (poly(ethylene), poly(vinyl chloride), cellulose esters, epoxy resins) are heated in a flame-spraying gun (Fig. 4-26) by combustion of a combustible gas with air or oxygen and sprayed onto heated metal surfaces that have been roughened or treated with primers.

A special version of flame spraying is **hot spraying** onto cold metal surfaces. Heating metal surfaces has often to be avoided because it will change the texture of the surface. In this process, polymer powders (polyamides, epoxy resins) in argon, helium, or nitrogen are blown onto the article by a spray gun while being heated in an arc to ca. 1600°C. In this process, metal surfaces do not get hotter than 50-60°C.

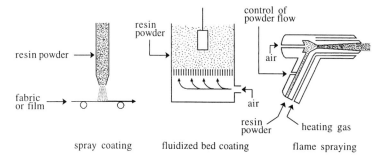

Fig. 4-26 Spray coating and flame spraying as examples of spray processes. Fluidized bed coating has the same *physical* mechanism but is not a spraying process.

A related physical process (though not a spray process) is **fluidized bed (dip) coating**. Metallic articles are degreased and mechanically roughened or treated with primers for better adhesion. The heated articles are then immersed in a fluidized (i.e., turbulent) bed of thermoplastic powders with particle sizes of ca. 200 µm diameter in inert gases (Fig. 4-26). The polymer particles adhere to the metal surface where they melt and form a dense film of ca. 200-400 µm thickness.

Suitable polymers must possess unique combinations of melt viscosities, heat stabilities, and mechanical properties. Polymers are mainly polyamides, poly(ethylene)s, and poly(vinyl chloride), and, to a lesser content, cellulose acetobutyrate. Fluidized bed coating is sometimes called **fluidized bed sintering** although it is a bulk melting process and not sintering, i.e., surface melting.

4.4.3 Laminating

Laminating involves the combining of two or more two-dimensional materials (plastic sheets, fabrics, wood panels, etc.) by a laminating resin (L: *lamina* = thin plate). The process is also called **bonding** for the combination fabric + fabric or fabric + expanded materials (plastic or rubber), **doubling** for the combination of thermoplastic films, sheets, or sheetings, and **quilting** if a filler material is used.

Laminating processes also include, in principle, the application of thermosetting resins to glass fiber mats (hand lay-up) or glass fiber strands (filament winding). However, these two processes use molds and are therefore usually considered molding operations (Section 4.4.4).

4.4.4 Pressureless Molding

The term **molding** is applied to most (but not all) processes requiring a mold. It is also used as a generic term for operations on fiber mats that are impregnated by thermo-setting resins.

Molds can be negative (female mold, cavity, impression) or positive (male mold, plug) or can have negative and positive parts. A female mold was formerly called a matrix (L: *matrix* = womb; from *mater* = mother); the corresponding term *patrix* (L: *pater* = father) was never used in English for male molds. "Matrix" and "female mold" are now considered sexist but "negative mold" also has a negative connotation. Politically correct would be "concave" and "convex" but that is not as vivid although it is certainly graphic.

Hand Lay-up Molding

The simplest discontinuous, pressureless molding process is hand lay-up molding (**hand lay-up spray molding**, **hand lay-up**, **lay-up molding**, **wet lay-up**, **contact molding**, **impression molding**) of resin-impregnated mats, usually from glass fibers but also from aramid or carbon fibers. Molds are first treated with a release agent or a sealing film upon which a 300-600 µm thick layer gel coat (= fiber-free resin) is applied with a brush or spray gun; alternatively, a resin-impregnated surfacing mat may be used. One ply of fiber mat (or several of them) is put on the gel coat, impregnated with the mixture of catalyst and resin (unsaturated polyester resin (Section 8.5.5) or epoxy resin (Section 8.5.4)), and pressed against the gel-coat by rollers. Usually, a second gel-coat is applied. The composite is cured (= crosslinked) in place, for example, by infrared radiation. Hand lay-up is especially suited for small numbers of large parts, for example, hulks of boats and rotor blades.

The procedure can be simplified by using *preimpregn*ated fibers or mats (**prepregs**). The fibers or mats are impregnated continuously with resin and partially cured in an oven before they are placed in the mold. Prepregs may be

– **preimpregnated rovings** that contain parallel strands or filaments of textile glass fibers that are assembled without intentional twist;
– **preimpregnated fabrics** from glass, carbon or aramid fibers with phenolic or epoxy resins, mainly used for thin-walled, high-load articles;
– **sheet molding compounds** (**SMCs**) composed of glass fiber mats that are impregnated with a mixture of unsaturated polyester resin (UP) and mineral fillers (chalk, talc, clay);
– **bulk molding compounds** (**BMCs**) which are compounds of UP resins, chopped glass fibers, and mineral fillers.

Fiber-spray Gun Molding

This process (**spray-up molding**) is a semi-automatic hand lay-up molding process in which the unsaturated polyester resin is mixed with the initiator ("catalyst") in a spray gun. The resulting fluid and short glass fibers (cut, not milled) are then sprayed simultaneously on the mold by pressurized air through separate nozzles. Fluid and fibers mix in air before they contact the mold.

Advantages are long shelf-lives of resin and catalyst (both stored separately), short cure times (use of a fast acting catalyst), and no loss of fibers (for hand lay-up, mats, rovings, etc., have to be cut to size; the scrap is useless). However, fiber-spray gun molding needs more skill for an even application of the fiber-filled resin and extensive start-up and clean-up operations for the equipment.

Filament Winding

Filament winding is a semi-continuous process that serves for the manufacture of large hollow articles (Fig. 4-27). Glass fiber rovings from roving reels are impregnated by resin/initiator/catalyst in an impregnating bath, and led by a reciprocating fiber guide under tension onto a winding mandrel (Provençal: *mandre* = axle, crank; from *L*: *mamphur* = bow-drill) in an exact geometric pattern. Curing (= crosslinking) of the resin is in an oven (small articles) or in a heated mold or by ultraviolet radiation (large articles).

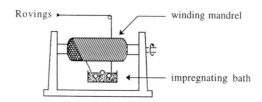

Fig. 4-27 Filament winding.

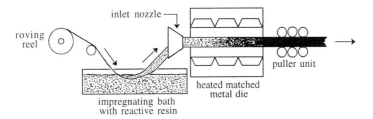

roving reel

inlet nozzle

puller unit

heated matched metal die

impregnating bath with reactive resin

Fig. 4-28 Pultrusion (schematic).

The cured molding shrinks onto the mandrel which is therefore made collapsible and removable. The process is used to prepare large, high-strength, hollow bodies from glass fiber reinforced epoxy resins. A newer modification of this process makes use of centrifugal forces by placing the rovings on the interior wall of a rotating hollow cylinder.

Pultrusion

Pultrusion is a continuous process for the production of uniaxial, "endless" semi-finished products from fiberglass-reinforced resins, especially unsaturated polyesters (Fig. 4-28). In this process, the glass fiber mat is pulled through a bath with the impregnating resin through an inlet nozzle which "extrudes" the mat with the desired shape (profile, tube, etc.), hence the name "pultrusion," although the material is not extruded by pressure as in a regular extrusion process (Section 4.5.2).

Simultaneously, surplus resin is removed, and air bubbles are squeezed out. The resin-filled fiber bundle is then pulled through a heated matched metal die where the resin polymerizes and the article hardens.

Careful adjustment of resin reactivity, heating temperature, and pulling speed is essential in order to prevent premature gelation (causes adhesion to the mold) and dimensionally unstable articles (produces deformation of profiles).

4.4.5   Pressure-assisted Molding

More compact thermosets are obtained if stronger pressures are applied to reinforced resins during curing. These pressure-assisted moldings use either single-component rigid molds or flexible rubber bags as countermolds (Fig. 4-29).

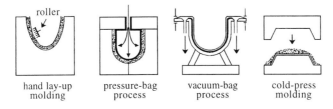

roller

| hand lay-up molding | pressure-bag process | vacuum-bag process | cold-press molding |

Fig. 4-29 Molding processes for reinforced plastics. From left to right:
Hand lay-up: a roller presses a resin-impregnated mat against a gel coat in a mold.
Pressure bag process: a rubber bag is pushed by pressure against the resin-impregnated fabric.
Vacuum bag process: vacuum causes a rubber bag to be pressed against the impregnated fabric.
Cold-press molding: a mat (fabric) is placed over a male mold. Resin is applied to the center of the
                                  mold. Closure of the mold evenly impregnates the mat; excess resin runs off.

Table 4-8 Comparison of processing temperatures T (RT = room temperature), times t to produce an article (h = several hours, d = several days), investment costs per mold, production costs per article (+ low, ++ medium, +++ high, ++++ very high), and number of parts that can be manufactured with a mold from unsaturated polyesters (UP), epoxy resins (EP), phenolic resins (PF), and vinyl ester resins (VE) [20]. f(M) = depends on mold size. BMC, SMC: see p. 108.

Process	Resin	Processing		Costs		Parts
		$T/°C$	$t/\min$	Invest.	Work	per mold
Spray-up molding	UP, EP	RT	30 – d	++	+++	≤ 200
Cold press molding	UP, EP	20 – 60	5 – 30	++	++(+)	≤ 200
Hand lay-up	UP, EP	RT	30 – d	+	+++	≤ 1000
Bag process, vacuum	UP, EP	RT	30 – h	+	+++	≤ 200
Bag process, pressure	EP	80 - 200	h	++	+++	> 1 000 000
Transfer molding	UP	RT	30 – h	++	++	
-, reinforced plastics				+++		≤ 1 000 000
Casting	EP, PF					≤ 3500
Rotational molding	UP, EP, VE	RT	10 – h	+++	+	100 000
Compression molding	UP, EP	80 – 120	2 – 10	+++	++	
-, composites, BMC	UP, EP, PF	120 – 160	2 – 5	+++	+	≤ 120 000
-, composites, SMC	UP, EP			++++		≤ 400 000
Centrifugal molding						> 1 000 000
Stamping				++++		≤ 3 000 000
Pultrusion	UP, (EP)	≤ 160		+++	+	practically ∞
Lamination	UP	≥ RT			+	practically ∞
Filament winding	UP, EP, VE	RT	f(M)	+++	++(+)	practically ∞

during shaping by extrusion and injection molding. Plastication processes are commonly used to work-in adjuvants.

The compacting of these viscoelastic materials requires application of pressure, which may lead to high shear gradients during processing. This, in turn, may result in chain cleavage, formation of radicals, and grafting/crosslinking.

4.5.1 Plasticating

Kneading

Kneading serves to mix two rubbers (called **blending**), to work in adjuvants (called **formulating**), and to plasticate materials. It is performed in special kneaders or in machines without incorporated plastication units such as presses or calenders. Kneading may be discontinuous (kneaders, presses) or continuous (kneaders, calenders).

Kneaders are heavy machines that come in extraordinarily many designs. Some employ a moving screw and stationary teeth, whereas others use overlapping kneading blades, kneading disks, sigma blades (Z type blades), etc.

Rolling and Calendering

Rolling is a discontinuous process for the blending of rubbers with other rubbers and the compounding, plastication, and/or homogenization of rubbers and plasticized thermoplastics. Roll mills consist of 2 rolls (plastics, rubbers) or 3-5 rolls (pastes) that

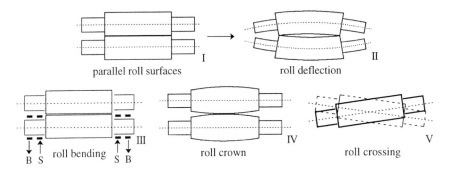

Fig. 4-32 Rolls with parallel roll surfaces (I) will cause roll deflection (II) that can be eliminated by one of the three methods III-V. B = roll bending force, S = roll supporting force. See text.

rotate inwardly with different speeds; small rolls are known as rollers. Rolls have lengths of 85-300 cm (production) or 20-45 cm (laboratory) and diameters of 20-55 cm (production) or 8-26 cm (laboratory). Rolls with parallel surfaces (Fig. 4-32, I) will bend under their own weight (Fig. 4-32, II) and by the presence of the rolled material which leads to uneven processing. Roll deflection (II) is eliminated by using convex rolls (roll crown, roll camber (IV)). One can also increase the separation of roll ends (V) or use forces that counteract roll deflection (III). Method V is commonly used for calenders.

The heated rolls counterrotate at different speeds, which forces the kneading stock to circulate and form a hide (= rough rubber sheet around the roll) only around the faster roll if temperatures and roll speeds are adjusted correctly (Fig. 4-33). The material builds up at the roll mouth where the resulting kneading stock is plasticated by the combined action of shear, elongation, and bending forces, usually with degradation of the polymer to lower molecular weights. For crossblending and better working-in of adjuvants, hides are cut and refed to the roll mouth, either by hand or automatically.

Roll mills are mainly used for the blending and compounding of rubbers but also as as part of calendering trains (see below) for the production of poly(vinyl chloride) films (very thin) and sheetings (endless sheets) as well as in combination with discontinuously working internal mixers for making thermoplastic sheets.

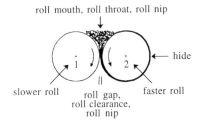

Fig. 4-33 Roll mill. Note that the terms "roll gap" and "roll nip" are used with two different meanings (the distance 1–2 between the two roll axes is also called the "roll gap").

Calendering (G: *kylindros* = cylinder, *kylindein* = to roll) is used for the shaping of rubbers and thermoplastics to films and sheetings, the reinforcement of rubbers by steel mesh, nylon fabrics, etc., and the coating of paper, textiles, etc.

Calenders have 2-7 large rolls that are heated by steam, hot water, or electricity. Rolls of calenders can be stacked vertically, in part horizontally, or at an angle of 45°C (Fig. 4-34). The arrangement of rolls resembles capital letters and one distinguishes therefore F, I, L, S, and Z calenders as well as inverted F and L and stretched Z types.

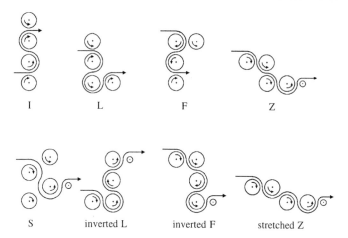

I   L   F   Z

S   inverted L   inverted F   stretched Z

Fig. 4-34 Some types of 4-roll calenders.

For example, the first and third rolls of an F-type calender may be offset (V in Fig. 4-32) to compensate for roll deflection. The thickness of the sheet is determined by the roll gap of the last pair of rolls, the gaging rolls. The last roll is the master roll that determines the speed of the subsequent train (cooler, winder, etc.), if any.

Calender lines may be quite large. For plastics films, they consist of a mixer, roll mill, conveyor, extruder, calender, embosser, cooler, edge trimmer, and winder.

Speeds of calenders range between 10 and 50 revolutions per minute. Residence times of materials decrease in the order L > F > Z. Shear rates and shearing stresses during calendering are higher than those in cold press molding but lower than those during extrusion (Table 4-9).

Calendering produces films and sheets with 60-400 μm thickness and sheetings with 400-600 μm. Thinner films and thicker sheetings are obtained by extrusion. The application of powders to films and sheeting by calendering is called **powder coating**. The deposition of a second film/sheeting of a thermoplastic on a film/sheeting of the same type is known as **doubling**, or, if it is of a different type, as **laminating**. In rubber technology, a second rubber layer may be deposited on a rubber sheet with the same rotational speed of the calender or with a different one.

4.5.2 Extrusion Processes

In extrusion processes, a melt or solution is continuously pushed out through orifices (L: *extrudere* = to thrust out). Extrusion processes are the most important processing methods for fibers, rubbers, thermoplastics, and even thermosets.

Fiber Spinning

This oldest extrusion process is used in nature by spiders and the mulberry spinner (natural silk) and in industry to prepare textile fibers and certain industrial fibers. Fibers can be spun from polymer melts (**melt spinning**), from polymer solutions into air (**dry spinning**) or into a coagulating bath (**wet spinning**), from polymer gels (**extrusion spin-**

Table 4-9 Shear rates $\dot{\gamma}$, shearing stresses σ_{21}, and viscosities η in important processing methods.

Processing method	$\dot{\gamma}/s^{-1}$	σ_{21} / MPa	η / (Pa s)
Cold press molding	1 – 10	0.01 – 0.03	1 000 – 100 000
Calendering	10 – 100	0.02 – 0.05	100 – 10 000
Extrusion	100 – 5 000	0.05 – 0.12	10 – 1 000
Injection molding	1 000 – 100 000	0.1 – 1	1 – 1 000

ning), from polymer dispersions (**dispersion spinning**), and even from monomers (**reaction** or **polymerization spinning**) (see Chapter 6). Fiber spinning uses pumps to push the fluid through spinnerets with very many orifices, and bobbins or reels are used to wind up the exiting fiber.

Extrusion

Extrusion is now the most important processing method for rubbers and thermoplastics (Table 4-6); it is also used for thermosets. In extruders for plastics, the prewarmed material is pushed through a die into air or a cooling bath (Fig. 4-35). In fiber spinning, the die is called a spinneret (Section 6.2.2). The extruded material forms tubes, profiles, sheetings, cables, or filaments.

In modern extruders, plastics are almost exclusively transported by single or double screws; pistons are still used for some thermosetting resins. The design of the screw depends on the desired task. Extruders are characterized by the ratio L/d of length and diameter of the screws. Melt-fed extruders have ratios of $L/d \approx 8$ and solid-fed ones, $L/d \approx 20$-40; the compression ratio is usually 2-4. Large commercial single-screw extruders have screw diameters of ca. 45 cm; they can extrude up to 20 tons per hour.

Extruders are universal machines that transport, degas, compact, mix, shape, and finally extrude the material as a shaped mass, the **parison** (Fig. 4-35, see also p. 94)). The parison is not externally supported and must therefore have a high viscosity η in order to maintain its shape until it is solidified by cooling.

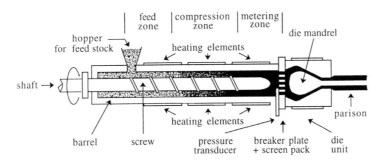

Fig. 4-35 Screw extruder for tubes (schematic). A hopper feeds plastics pellets, powders, beads, flakes, or reground material into the extruder. A turning screw transports the plastics through feed, compression, and metering zones where it is molten by friction and contact heat. The melt is extruded through a pressure transducer, a breaker plate, a screen pack, and a die. The inner dimensions of the extruded tube are controlled by the die mandrel, the outer ones by the diameter of the nozzle and the memory effect (p. 94). Parisons (pipes, tubes, etc.) are cut to the desired length.

Self-support is provided by entanglements if molar masses are high enough. For this reason, polymers with lower molar masses are sometimes slightly crosslinked. Extrusion thus demands polymers with high molecular weights ($\eta \sim M^{3.4}$) or, to be more exact, polymers with high-molecular weight *fractions*. Such polymers allow high extrusion speeds because their low-molecular weight fractions act as lubricants. They are also less prone to melt fracture (p. 95). However, higher extrusion speeds lead to greater parison swell, which can be reduced if grades with less elastic properties are used.

Because of the Barus effect (p. 94), parisons are not only larger than the die orifices but also have different shapes if they are not circular (Fig. 4-36), depending on the stress at the edges of multi-sided orifices. The effect is made use of in fiber spinning for the manufacture of trilobal or more complex cross-sections of fibers (Section 6.2.3).

Prepolymers can be simultaneously reacted and extruded, usually in piston-type extruders in order to avoid non-homogeneous crosslinking. Extrusion of monomers to thermoplastics is rarely used (an example is methyl methacrylate) as are grafting reactions in extruders. Crosslinking of rubbers proceeds in a separate process after the extrusion (Chapter 7).

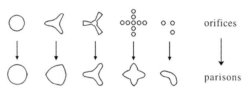

Fig. 4-36 Effect of orifice shape on parison shape.

Coextrusion

Films and sheetings for packaging and semi-finished goods are produced with two or more layers of the same or different material. An example of the former are pore-free films of low-density poly(ethylene) for general packaging (Chapter 13). Different polymers are used for the coextrusion to packaging films with aroma or gas barriers, usually in combination with primers; an example is poly(propylene)-primer-poly(ethylene-*co*-vinyl alcohol)-primer-poly(propylene). Thick sheets (boards) for walls of refrigerators may be produced in two layers with a thermoplastic styrene-butadiene copolymer (SB polymer) as substrate and poly(styrene) (PS) as glossy surface layer. Such refrigerator walls may also consist of four coextruded layers: coolant-resistant ABS + expanded SB polymer + SB masking layer + PS glossy surface layer.

Netlon® Process

Thermoplastics can be extruded to nets with permanently connected weft and warp (see Chapter 6). The extrusion head consists here of two counterrotating strainers, i.e., dies with a number of circular holes arranged in a circle. If the holes of the two dies face each other, only one strand is produced per pair of holes. On rotation of the holes, each strand separates into two strands until the holes unite again. With counterrotating strainers and constant rotational speed, tubes with diamond-like net structures are produced that, when slit lengthwise, become a flat net. Very different net structures are possible by varying the shape, thickness, and arrangement of holes, rotational speed, etc.

Film Blowing

Film blowing is the most important method for the production of polymer films. In this process, a thermoplastic is extruded from a circular die to a tubular film, usually upwards (Fig. 4-37). The resulting film tubing is cooled by an air-cooling ring, and the bubble is collapsed and laid flat by pinch rolls (nip or squeeze rollers). The collapsed tube can be converted into flat film by slitting it open or by trimming both sides.

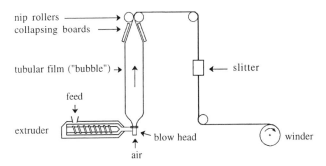

Fig. 4-37 Film blowing. A tube is extruded with a blow head directed upwards. Air expands the tube to a tubular film ("bubble"), which is collapsed by bubble-collapsing boards or rolls. The film is flattened further by nip rollers (squeeze rollers) and slit by a slitter before it is wound by a winder.

Cast Film Extrusion

A special type of extrusion is the extrusion through flat sheet dies (slot dies) to films with thicknesses of 20-100 μm. The exiting film is then cooled by cooling cylinders (**chill-roll extrusion**) or quenching baths. Slot dies are also used for **extrusion coating** of paper or cardboard with, for example, poly(ethylene), to films that can be hot sealed. Extrusion also serves for the sheathing of metal cables.

Extrusion Blow Molding (Extrusion Blowing)

This one-step process for the manufacture of hollow bodies (bottles, containers) (Fig. 4-38) requires polymer grades of PE, PVC, PC, PA, and HIPS with high melt strengths.

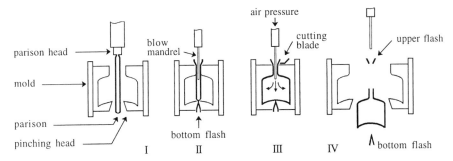

Fig. 4-38 Extrusion blow molding. I. A hollow parison (tube) is extruded from the parison head into the open mold. The blow mandrel is lifted (not shown). II. Upon closing of the mold, the pinching head generates a bottom flash; the blow mandrel is inserted. III. Air pressure causes the parison to adopt the shape of the mold. After the blow mandrel is lifted, a cutting blade cuts off the top flash. IV: The mold is opened, and the container removed.

The many known machine designs differ in the type of blowing (from the top, the bottom, or the side), and the taking and transporting of the parison, mold, and mandrel (shot-in of mandrel, sliding mold, rising table, etc.). In the example of Fig. 4-38, the extruder head points downwards. The extruded endless tubes are "blown" into a mold by air pressure; upon mold closure, they form hollow bodies with a seam at the bottom. Constant wall thicknesses are difficult to achieve by extrusion blow forming and flash (cut-off) may amount to 20 % of the material in bottle manufacturing.

Such hollow bodies can also be prepared by other processes. Filament winding (p. 108) and rotational casting (p. 104) also require only one step; they are, however, much more expensive. Injection blow molding (next Section) needs two steps since halves are separately molded and have to be welded or glued together. However, constant wall thicknesses are much more easy to obtain.

4.5.3   Injection Molding

Injection molding is an extrusion of hot, plasticated polymers by screws or double screws into cold or preheated molds under pressure (Fig. 4-39); torpedoes are now used less often. Screws and pistons (torpedoes) serve simultaneously as plastication, metering, and injection devices. Compact articles are mostly produced by single-screw machines with shot weights up to 30 kg and high clamping forces of ca. 3000 t for parts with cross sections up to 1 square meter.

Pistons are no longer used in standard injection molding machines because screws permit higher production speeds and create new surfaces by shearing; this allows for better degassing and increased flow of the melt into the interstitial zones between pellets, granulates, etc. Large articles from expanded plastics are still manufactured by special injection molding units using torpedoes instead of screws.

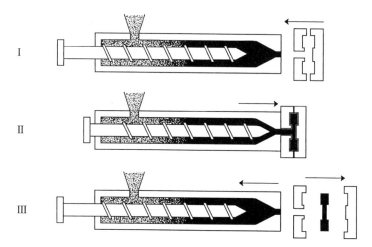

Fig. 4-39 Injection molding (here with reciprocating (moving back and forth) screw). (I) Pellets are fed into the machine through a feed hopper and transported by the rotating screw through the feed, compression, and metering zones where they are heated externally by heating elements and internally by friction. (II) The plasticated material is injected into the cold mold where it solidifies almost instantaneously. (III) The specimen is demolded; the reciprocation of the screw plasticates new material.

Table 4-10 Injection molding of thermoplastics [21]. T_G = glass temperature, T_{HDT} = heat distortion temperature, T_M = melting temperature, $T_{material}$ = temperature of the melt of the plasticated plastic, T_{mold} = temperature of mold, p = injection pressure. cp = copolymer. * Conflicting data.

Polymer	T_G °C	T_{HDT} °C	T_M °C	$T_{material}$ °C	T_{mold} °C	p MPa
Amorphous polymers						
Polycarbonate of bisphenol A	150	132	–	280 – 320	85 – 120	80 – 150
Styrene-acrylonitrile copolymer	120	103	–	200 – 260	30 – 85	
Acrylonitrile-butadiene-styrene cp.	100	98	–	200 – 280	40 – 80	60 – 180
Poly(styrene), at-	100	104	–	170 – 280	50 – 70	
Poly(methyl methacrylate)	105	90	–	150 – 200	50 – 90	70 – 120
Poly(vinyl chloride), hard	82	68	–	180 – 210	20 – 60	100 – 180
Semicrystalline polymers						
Poly(ethylene terephthalate)	70	85	265	270 – 280	120 – 140	120 – 140
Poly(chlorotrifluoroethylene)	40	66	220	220 – 280	80 – 130	150
Poly(ε-caprolactam)	50	65	215	230 – 290	40 – 60	90 – 140
Poly(oxymethylene)	–82	136	181	180 – 230	60 – 120	80 – 170
Poly(propylene), it-	–15	56	176	200 – 300	20 – 60	80 – 180
Poly(ethylene), high density	–123*	49	135	240 – 300	20 – 60	60 – 150
Poly(ethylene), low density	–103*	37	115	180 – 260	20 – 60	60 – 150

Injection molding is the most important processing method for thermoplastics (Table 4-10), including cellulose acetate; it is also used for polyurethanes, phenolic resins, amino resins, and unsaturated polyesters (for processing conditions, see Table 4-7). It is also suitable for the processing of starch that is plasticized by water.

Thermoplastics are used as pellets because powders would be compacted and form an undesirable plug in the center of the machine. Crystallizable thermoplastics must crystallize fast, which can be achieved by using nucleating agents. Slow crystallization will lead to spherulites or after-crystallizations that will cause shrinkage and warping.

Injection molding machines with screws (Fig. 4-39) are especially effective for plastication because the action of the screw lets melts flow easily into the interparticle space (good deaeration). The shearing action of the screw(s) also generates constantly new surfaces which leads to a better compounding.

Processing temperatures must be high enough to allow for good flow into the mold. These temperatures are considerably higher than the glass temperature of amorphous polymers and at least just above the melting temperature T_M of semicrystalline ones (Table 4-10), yet lower than decomposition temperatures. Melt viscosities should be low in order to obtain good mold filling at the lowest possible injection pressure (lowest energy consumption!).

The plasticated polymer is injected into the mold where it forms a ca. 0.05-mm thick layer at the cold walls that separates the plastic interior from the cold wall. The center material flows and solidifies more slowly; after-pressure (dwell pressure, holding pressure) is subsequently applied in order to prevent void formation.

Shape stability of molded *amorphous* articles is achieved if the mold temperature T_{mold} is lower than the heat distortion temperature T_{HDT} (Section 5.2.2) of the thermoplastic. The articles solidify especially fast if the heat is extracted rapidly from the mold by cooling water, thus allowing short cycle times. *Semicrystalline* polymers often attain shape stability at $T_{HDT} > T_{mold}$ because they solidify at $T_M > T_{HDT}$ (Table 4-10).

Injection molded parts thus have skin-core structures: a 100–600 μm thick skin surrounds a lower-density core with different morphology. The flow also creates radial orientations of polymer segments and filler particles, especially near sharp edges and bends. Injection-molded articles are thus always anisotropic; molds need careful design to avoid stress zones.

Sandwich Molding

This process (short for **sandwich injection molding**) allows consecutive injection of two polymers into the same mold from two different injection units. The injected second polymer balloons the first one and presses it against the wall. The process is used for the sheathing of inexpensive polymers by more expensive ones or the forming of strong skins around foamed polymers (**foam sandwich molding, sandwich foam process**).

Injection Blow Molding

In blow molding, parisons are either extruded into an open mold (Fig. 4-38) or injected into a closed mold. In injection blow molding, the core of the injection mold acts as the blow mandrel. Injection blow molding also needs two steps (instead of one in extrusion blow molding, Fig. 4-38) since halves are separately molded and have to be welded or glued together. However, constant wall thicknesses are much easier to obtain than be extrusion blow molding (p. 117).

Reaction Injection Molding (RIM)

RIM processes are fast simultaneous polymerization and injection molding processes with cycle times of a few minutes (faster than vacuum casting). They are used for the manufacture of articles from crosslinked polyurethanes and from interpenetrating networks of polyurethanes and polyacrylates. Reaction injection molding of other fast polymerizing monomers has not been commercially successful. Examples of the latter processes are the anionic polymerization of dicyclopentadiene and the formation of block polymers by anionic polymerization of ε-caprolactam in the presence of poly-(oxyethylene)s, poly(oxypropylene)s, or poly(butadiene)s with hydroxyl endgroups.

4.6   Forming

Forming is the general term for the changing of shapes of solid semi-finished goods by external tensile forces that are generally assisted by vacuum or pressure. These processes require ductile plastics; they are performed near or in the ductile region of the stress–strain curve (Fig. 5-3). Plastics are not "melted" at the applied temperatures; they remain "solid" at these "cold" processing conditions.

Two groups of forming processes can be distinguished: **strengthening** for one-dimensional and two-dimensional semifinished goods (fibers, rods, tapes, films, sheets) and **transformation** for the manufacture of three-dimensional articles from sheets.

4.6.1 Strengthening

Strengthening comprises several types of processes (Fig. 4-40). Drawing, rolling, and hydrostatic extrusion are processes that are performed between room temperature and up to temperatures near the glass temperature of amorphous polymers or just below the melting temperature of semicrystalline polymers. The applied tensile forces partially orient mobilized chain segments and crystalline lamellae, and the resulting orientation is frozen-in by cooling under tension. The finished articles are thus under considerable internal stress because molecule segments are not in their thermodynamically preferred random coil conformations. Oriented films with internal stresses are utilized as **shrink films**. These films shrink upon fast heating just above the softening temperature and cover packaged goods tightly.

Tempering (**annealing**) below the heat distortion temperature increases segmental mobilities. The molecules return to their random coil states and unwanted stresses are relieved (**thermofixing**).

The strengthening of fibers, rods, and tapes by **cold drawing** may be performed without assistance or may be assisted by mandrels or guides (Fig. 4-40, top). Cold drawing of tapes and sheets with the help of rolls is called **rolling** (Fig. 4-40, bottom, left).

The cold drawing of cables, rods, and tapes is also called **stretching**, whereas the term **orientating** is used for the cold drawing of sheets and films. These engineering terms refer to macroscopic processes; molecular orientation on the microscopic level occurs with all processes. Unidimensional cold drawing of films leads to monoaxial orientation, that is used for the manufacture of fibrillated films and fibers (Section 6.7.3).

Cold extrusion of rods and tapes is a pushing process whereas cold drawing refers to pulling. In **hydrostatic extrusion**, an all-sided pressure is applied to the specimen.

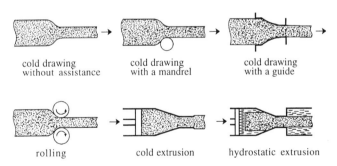

cold drawing without assistance	cold drawing with a mandrel	cold drawing with a guide
rolling	cold extrusion	hydrostatic extrusion

Fig. 4-40 Solid-state forming of one-dimensional and two-dimensional structures.

4.6.2 Transformation

Transformation processes comprise processing methods that convert two-dimensional semi-finished products (films, sheets, sheetings) into three-dimensional articles at temperatures near the maximum of the ductile region of the stress-strain curve (**thermoforming**). Such regions exist only in amorphous polymers and in semi-crystalline polymers with relatively broad melting ranges. Polymers with sharp melting ranges such as poly(4-methyl-1-pentene) cannot be thermoformed by vacuum forming.

In order to avoid orientation and to enable stress relaxation, thermoforming is performed with heated sheets or films at higher temperatures than those that are used in strengthening processes. Thus, processing temperatures are above the glass temperatures of amorphous polymers and at the melting temperatures of semicrystalline ones. Upper thermoforming temperatures are limited by the onset of flow processes.

Thermoforming processes are all **drawing processes** that are known by many names that vary not only with the type of the process but also for the same process from one machine manufacturer to the next. Machines for such processes can produce up to 8000 articles per hour.

Simple shapes are prepared by **deep drawing** from preheated sheet stock with a clamping ring and a forming plug without a negative (female mold) (Fig. 4-41, I). Excess material has to be cut off from the resulting article.

Very many other thermoforming processes also use molds and/or vacuum. Forming by sucking a preheated sheet into a cavity (female mold) with a help of a vacuum is called **vacuum thermoforming** or **vacuum forming** (Fig. 4-41, II). Again, flash has to be removed by one of the many deflashing (deburring) methods.

Vacuum forming is known in many variations, for example, as a negative process (as in Fig. 4-41, II) with mechanical or pneumatic prestretching, with an air cushion, etc. The simplest process is **straight-vacuum forming** (with or without mechanical prestretching) but this process produces articles with walls that are thinner at the lowest position than anywhere else. The manufacture of complicated articles by this process thus requires the prestretching of the prewarmed blank (billet) by compressed air before the forming by vacuum.

Male molds often allow more uniform thicknesses and better dimensional stabilities. In **drape-and-vacuum forming**, the heated billet is drawn over a positive mold with suction inlets. This process may produce wrinkles between the drawn depressions which can be avoided by **draw grid forming** in which the positive mold and its frame is replaced by several positive molds with a draw grid (drape). This process is used to produce inserts for boxes of chocolates.

Stretch forming (**stretching**) is a drawing of a preheated disc into a heated cavity (pot) with the help of a plug (Fig. 4-41, III) or a sliding hold-down plate. Molded articles do not have to be deflashed because the mold is completely closed.

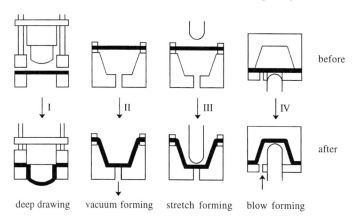

deep drawing   vacuum forming   stretch forming   blow forming

Fig. 4-41 Some thermoforming processes (I-III: negative, IV: positive).

Blow forming (blowing) of sheet materials is a special type of stretch forming in which a blank is pushed into a cavity by air. The process may be assisted by a plug or a rubber blanket. It differs from **film blowing** (p. 117) which is an extrusion process.

4.7 Fabricating

Fabricating is the process of modifying preformed articles or their assemblies to final products; this term should not be used for the manufacture of the preformed articles themselves. Three main groups of operations can be distinguished: transforming, machining, and bonding. In addition, the surface of articles may be treated.

4.7.1 Transforming

Transforming is the **conversion** of semi-finished goods into final goods without formation of shavings or addition of other materials. In the narrow sense, transforming includes stamping and forging, and in the broader sense, also calendering (Section 4.5.1), deep drawing (Section 4.6.2), and bending and creasing (folding) of rigid sheets.

Surface patterns can be created by several types of molding operations that are called **embossing** (creates regular patterns on hot sheets, usually by calenders), **stamping** (produces irregular patterns such as letters on cold sheets by sudden pressure or blow of a hot stamp or die), or **coining** (leading to a relief).

Forging is used for the transformation of poly(p-hydroxybenzoate) (PHB) and ultrahigh-molar mass poly(ethylene) (PE-UHMW) (molar mass greater than ca. 3 million). These polymers cannot be processed by, for example, injection molding, because of extremely high viscosities that are caused by high entanglement densities of ultrahigh-molar mass molecules (PE-UHMW) or frozen-in liquid-crystalline structures (PHB).

4.7.2 Detachment Operations

Plastic articles sometimes need to be detached, either to remove gates, sprues, etc., or to receive their final shapes. Detaching operations are always performed below the melting temperature (semicrystalline polymers) or below the glass temperature (amorphous polymers). They may or may not produce shavings.

No shavings are produced by cutting, shearing, or punching. Mechanical **cutting** is used for films of celluloid and poly(tetrafluoroethylene); newer systems employ high-pressure air jets (**airjet cutting**), water jets (**waterjet cutting**, up to 4000 bar), or lasers. **Shearing** is a cutting by knives with lateral motion. **Punching** is used for special parts such as O-rings.

Detachment operations with the production of shavings are called **machining**. They include **sawing, drilling, turning, milling**, and **threading**. In machining, high speeds have to be avoided because the low heat conductivity causes plastics to heat up, become viscoelastic, and start smearing. High speeds thus require small cross-sections of shavings.

4.7.3 Bonding and Joining

Bonding and joining refers to the interconnection of parts or semifinished goods to greater entities.

Plastics parts can be joined **mechanically** by screws, rivets, or bolts from plastics or metals. Such joints are sufficient for small stresses since larger stresses would loosen the joint because of the viscoelastic properties of plastics. The deformability of plastics is used to join plastic parts by snap-in joints and clamp connectors that consist of hooks or other protrusions that interlock on contact with undercuts, recesses, and the like.

Welding

Welding is the physical joining of the top layers of plastic parts by heat and pressure, usually under nitrogen, without any adhesive. It is restricted to parts of the same polymer since the polymers must be miscible. Compatible polymers of different chemical structures are rarely welded (for the difference between "miscible" and "compatible", see Section 10.2 and also Volume III, p. 320).

On welding, chains of polymer molecules interdiffuse, which requires a certain mobility of polymer segments that is provided by heating above the softening temperature by thermal conduction, convection, radiation, friction, or induction). Polymers cannot be welded if they are chemically crosslinked (thermosets) or highly entangled (for example, ultrahigh-molar mass poly(ethylene)), or if they decompose under welding conditions.

The softened top layers are pressed together. Mutual diffusion of segments across the interface then leads the coils to interpenetrate. This interpenetration is *not* affected by pressure, whose rôle is not to promote diffusion but rather to provide good surface contact and to smooth surface irregularities.

Welding is used to join tubes (flexible) and pipes (rigid), mainly from poly(ethylene) or poly(vinyl chloride). In **autogeneous welding**, the welding seam is delivered by the polymer itself, usually by overlapping the ends of the parts (**lap welding**). In **heterogeneous welding**, parts are just pressed against each other (**butt welding**). In both types of welding, heating is by hot gases, usually nitrogen.

In **friction welding** (**spin welding**), two parts are rotating fast against each other, which leads to friction and heating of parts. The heat causes the temperature to increase above the glass temperature (amorphous polymers) and melting temperature (semicrystalline polymers), respectively, which leads to self-diffusion of polymer chains. This "melting" is promoted by applied pressure.

In **induction welding**, a metal band is put into the slot between the two parts. The parts are pushed together and electricity with ca. 50 Hz is applied. The resulting induction heat softens the surface and welds the parts.

Solution welding (**solvent welding, solvent adhesion**) is a welding process and not an adhesive joining (see Section 12.3). The top layers of parts are swollen by solvents that have approximately the same solubility parameter (Volume III, p. 299) as the polymer parts to be joined. The swelling provides polymer segments with the mobility necessary for mutual diffusion. The joints are initially weak because of the long times required for the solvent to diffuse out of the joint through the polymer parts into the atmosphere. This problem can be circumvented if the solvent is a polymerizable monomer.

Sintering

Sintering is used for surface treatment, the manufacture of porous materials, or the fabrication of large hollow bodies. Small plastics particles are pressed together by high pressure and subsequently heated to such high temperatures that the upper layers of particles begin to melt and stick together. The resulting porous parts with open pores are used as filters, separators for material exchangers, or aerators. Polymers for sintering include poly(ethylene), poly(propylene), poly(tetrafluoroethylene), poly(methyl methacrylate), and poly(styrene).

In **centrifugal molding**, dry fusible powders are heated and sintered in a centrifuge to hollow bodies with volumes up to 10 000 liters. This process differs from centrifugal casting which uses fluids (p. 104).

4.8 Surface Finish

Surfaces of plastic parts sometimes need to be modified by other materials for aesthetic (color, gloss) or technical reasons (hardness, abrasion, corrosion). These other materials are either deposited as such on the surface by physical or chemical means or they react chemically with the surface. One can thus distinguish between surface deposition and surface modification processes. Surfaces should be clean and dry; they often need to be pretreated (see below).

The following is a short review of treatment processes some of which will be discussed in greater detail in other sections.

4.8.1 Coloring

Plastics and elastomers are usually colored in bulk by pigments (Section 3.2; for fibers, see Section 6.5.4). However, some plastic parts require **painting** since the color and gloss of the parts must match that of painted parts made from metal; an example is automobile front and rear parts. Color and gloss never match if articles consist of different materials and different colorants because of differences in hue, lightness, and chroma of the colorants (Section 6.6.4) and reflection, scattering, etc., of light by the parts.

The paint must wet the polymer surface but should not modify the physical structure of the plastic (no swelling, recrystallization, stress corrosion, etc.). The plastics parts need to be carefully designed because paints recede from sharp edges, leaving only a thin coating.

Colored surfaces of mass-produced parts can also be obtained by chemically bonding a **colored film** (skin) to the main parts, for example, bumpers. The colored film is produced by coextrusion of four layers of an ionomer-based polymer: a clear coat finish, a color layer, a tie layer for the adhesive bond between the coating and the backing layer, and the backing layer. The skin is thermoformed to a preform that is loaded into the injection molding press and chemically bonded to the main part.

Printing. All printing techniques (Section 12.5.1) can be applied to plastics. Hard and non-absorbent plastic surfaces require indirect techniques, which use elastic intermediates to transfer the printing ink to the substrate.

4.8.2   Metallizing

All metallizing processes demand prior degassing, degreasing, and drying of surfaces.

Vacuum deposition. Practically all plastic surfaces can be metallized in vacuum with layers of up to 1 μm thickness. An example is aluminum covered films of poly(ethylene terephthalate) for packaging purposes. Advantageous is the high gloss and the barrier properties, disadvantageous is the modest adhesive strength.

Adhesive strengths can be improved if vacuum deposition is combined with **chemical metallizing**. The plastic part is first covered with a paint containing zinc, cadmium, or lead oxides. The oxides are then reduced to a strongly adhering, electrically conducting metal layer which is then coated with silver.

Better adhesion is achieved by **electroplating**, especially of ABS polymers. Etching by a chromic acid–sulfuric acid bath oxidizes the surface of dispersed elastomer particles leaving submicroscopic pores at the surface of the polymer. The pores become the anchoring sites for chemically deposited metals such as nickel or copper. The resulting film is then reinforced by electroplating with, for example, chromium.

4.8.3   Oxide Coatings

Glazing is used both as a general term for the generation of high-gloss surfaces of polymers by infrared radiation or polishing rolls immediately after processing or for the specific process of depositing an inorganic glass on the polymer surface by electron beams in high vacuum. The latter proces is economical because high evaporation rates lead to short glazing times; too fast a rate produces cracks, however. The direct (thermal) evaporation of glasses with the help of cathode rays is too slow. Glasses may consist of borosilicates or SiO_{1-2}. Coatings with SiO_2 are not temperature resistant because plastics and SiO_2 have very different thermal expansion coefficients (Volume III, p. 434).

Sol-gel processes form polymer networks by utilizing the evaporation of alcohol from alcoholic solutions of hydrolyzable alcoholates of multivalent metals such as Ti, Si, or Al. Coatings generated at low temperatures contain many metal hydroxide groups; the resulting surfaces are hydrophilic and antistatic. At higher temperatures, metal oxides result, which form scratch-resistant surface coatings.

Scratch-resistant coatings can also be generated from 50:50 mixtures of poly(silicic acid) and poly(tetrafluoroethylene-*co*-hydroxyalkyl vinyl ether). It is not known why this process delivers surface coatings with the scratch resistance of glass.

4.8.4   Plasma Coating

Coatings can also be produced by **plasma polymerization** (Volume I, p. 377). The term refers to the vacuum polymerization of low molar mass compounds by glass plasmas (partially ionized gases consisting of ions, electrons, and neutral species). Plasma polymerization is a polymerization of molecule fragments and not a polymerization of monomers to polymers with the same monomeric units. For example, the plasma polymerization of ethene C_2H_4 in air leads to to $(C_2H_{2.6}O_{0.4})_n$ and that of ethane C_2H_6 delivers $(C_2H_3)_n$. Industrially used "monomers" are ethene and hexamethyltrisiloxane.

Literature to Chapter 4

4.1.1.a POLYMER PROCESSING, GENERAL SURVEYS

S.Middleman, Fundamentals of Polymer Processing, McGraw-Hill, New York 1977
D.C.Miles, J.H.Briston, Polymer Technology, Chem.Publ.Co., New York 1979
J.L.Throne, Plastics Process Engineering, Dekker, New York 1979
R.J.Crawford, Plastics Engineering, Pergamon, Oxford 1981 (= Progr.Polym.Sci. **7**)
G.Astarita, L.Nicolais, Eds., Polymer Processing and Properties, Plenum, New York 1984
J.R.A.Pearson, Mechanics of Polymer Processing, Elsevier, New York 1985
M.J.Folkes, Processing, Structure and Properties of Block Copolymers, Elsevier Appl.Sci.,
    London 1985
N.G.McCrum, C.P.Buckley, C.B.Bucknall, Principles of Polymer Engineering, Wiley, New York
    1988
D.H.Morton-Jones, Polymer Processing, Chapman and Hall, New York 1989
C.L.Tucker III, Fundamentals of Computer Modeling for Polymer Processing, Dekker, New York
    1991
J.-F.Agassant, P.Avenas, J.-Ph.Sergent, P.J.Carreau, Polymer Processing. Principles and
    Modeling, Hanser, Munich 1991
J.-M.Carrier, Polymeric Materials and Processing, Hanser, Munich 1991
A.I.Isayev, Ed., Modeling of Polymer Processing, Hanser, Munich 1991
P.J.Corish, Ed., Concise Encyclopedia of Polymer Processing and Applications, Pergamon Press,
    Tarrytown (NY) 1991
W.Michaeli, Plastics Processing. An Introduction, Hanser, Munich 1995
H.E.H.Meijer, Ed., Processing of Polymers (= R.W.Cahn, P.Haasen, E.J.Kramer, Eds., Materials
    Science and Technology, volume 18), Wiley-VCH, Weinheim 1997
D.G.Baird, D.I.Collias, Polymer Processing: Principles and Design, Wiley, New York 1998
T.A.Osswald, Polymer Processing Fundamentals, Hanser Gardner, Cincinnati (OH) 1998
R.S.Davé, A.Loos, Processing of Composites, Hanser Gardner, Cincinnati (OH) 2000
I.M.Ward, P.D.Coates, M.M.Dummoulin, Eds., Solid Phase Processing of Polymers, Hanser,
    Munich 2001
C.A.Harper, E.M.Petrie, Plastic Materials and Processes. A Concise Encyclopedia, Wiley-
    Interscience, Hoboken (NJ) 2003
C.A.Harper, Handbook of Plastics Processes, Wiley, Hoboken (NJ) 2006
Z.Tadmor, C.G.Gogos, Principles of Polymer Processing, Wiley, Hoboken (NJ), 2nd ed. 2006
T.A.Osswald, J.Pablo Hernandez, Polymer Processing. Modeling and Simulation, Hanser, Munich
    2006
M.Chanda, S.K.Roy, Plastics Fabrication and Recycling, CRC Press, Boca Raton (FL) 2008

4.1.1.b POLYMER PROCESSING, EQUIPMENT
J.M.Chabot, The Development of Plastics Processing - Machinery and Methods, Wiley, New York
    1992

4.1.2 GENERAL FACTORS
N.P.Cheremisinoff, Polymer Mixing and Extrusion Technology, Dekker, New York 1987
J.-M.Vergnaud, Drying of Polymeric and Solid Materials, Springer, London 1992
I.Manas-Zlaczower, Z.Tadmor, Eds., Mixing and Compounding of Polymers, Hanser,
    Munich 1994
R.J.Albalak, Ed., Polymer Devolatilization, Dekker, New York 1996
D.B.Todd, Plastics Compounding, Hanser, Munich 1998
C.Rauwendaal, Polymer Mixing. A Self-Study Guide, Hanser, Munich 1998
R.H.Wildi, C.Maier, Understanding Compounding, Hanser Gardner, Cincinnati (OH) 1998
J.White, A.Coran, A.Moet, Polymer Mixing: Technology and Engineering, Hanser, Munich 2001

4.2 FLOW PROPERTIES
H.Janeschitz-Kriegl, Polymer Melt Rheology and Flow Birefringence, Springer, Berlin 1983
J.Ferguson, N.E.Hudson, Extensional Flow of Polymers, in R.A.Pethrick, Ed., Polymer
    Yearbook 2, Harwood Academic Publ., Chur (GR), Switzerland 1985, p. 155
S.W.Churchill, Viscous Flows: The Practical Use of Theory, Butterworths, Stoneham (MA) 1988
H.A.Barnes, J.F.Hutton, K.Walters, An Introduction to Rheology, Elsevier, Amsterdam 1989

J.L.White, Principles of Polymer Engineering Rheology, Wiley, New York 1990
J.M.Dealy, K.F.Wissbrun, Melt Rheology and Its Role in Plastics Processing, Van Nostrand
    Reinhold, New York 1990
C.W.Macosko, Rheology. Principles, Measurements, Applications, Wiley, New York 1994
D.Acierno, A.A.Collyer, Rheology and Processing of Liquid Crystalline Polymers, Chapman and
    Hall, New York 1996
A.V.Shenoy, D.R.Saini, Thermoplastic Melt Rheology and Processing, Dekker, New York 1996
P.J.Carreau, D.C.R. De Kee, R.P.Chhabra, Rheology of Polymeric Systems. Principles and
    Applications, Hanser Gardner, Cincinnati (OH) 1997
P.R.Hornsby, Rheology, Compounding and Processing of Filled Thermoplastics, Adv.Polym.Sci.
    139 (1999) 155
F.A.Morrison, Understanding Rheology, Oxford Univ. Press, New York 2001
J.Furukawa, Physical Chemistry of Polymer Rheology, Springer, Berlin 2003
R.K.Gupta, Polymer and Composite Rheology, Dekker, New York, 2nd ed. (2003) (this is the second
    edition of the book of L.E.Nielsen, Polymer Rheology, Dekker, New York 1977)
A.Ya.Malkin, A.I.Isayev, Rheology: Concepts, Methods, and Applications, William Andrew,
    Norwich (NY) 2006
T.G.Mezger, The Rheology Handbook, William Andrew, Norwich (NY) 2006
J.M.Dealy, R.G.Larson, Structure and Rheology of Molten Polymers, Hanser Gardner, Cincinnati
    (OH) 2006

4.4.1 CASTING

R.J.Crawford, Rotational Moulding of Plastics, Wiley, New York 1992
M.Narkis, N.Rosenzweig, Eds., Polymer Powder Technology, Wiley, New York 1995
J.F.Stevenson, Ed., Innovation in Polymer Processing: Molding, Hanser, Munich 1996
G.L.Beall, Rotational Molding, Hanser, Munich 1998
R.J.Crawford, J.L.Throne, Rotational Molding Technology, William Andrew, Norwich (NY) 2002

4.4.4 PRESSURELESS MOLDING

D.V.Rosato, C.S.Grove, Jr., Filament Winding, Interscience, New York 1968
R.W.Meyer, Handbook of Pultrusion Technology, Chapman and Hall, London 1985
R.E.Wright, Molded Thermosets, Hanser, Munich 1992
H.G.Kia, Sheet Molding Compounds - Science and Technology, Hanser, Munich 1993
H.Potente, H.P.Heim, Specialized Molding Techniques, William Andrew, Norwich (NY) 2001

4.4.5 PRESSURE-ASSISTED MOLDING

A.I.Isayev, Ed., Injection and Compression Molding Fundamentals, Dekker, New York 1987
M.Narkis, N.Rosenzweig, Ed., Polymer Powder Technology, Wiley, New York 1995
B.A.Davies, P.Gramann, A.Rios, T.A.Osswald, Compression Molding, Hanser, Munich 2003

4.5.1 PLASTICATING

R.E.Elden, A.D.Swan, Calendering of Plastics, Iliffe, London 1971
C.I.Hester, R.L.Nicholson, M.A.Cassidy, Powder Coating Technology, Noyes Publ., Park Ridge
    (OH) 1990
M.Narkis, N.Rosenzweig, Ed., Polymer Powder Technology, Wiley, New York 1995

4.5.2.a EXTRUSION PROCESSES

J.L.White, Twin-Screw Extrusion - Technology and Principles, Hanser, Munich 1990
M.Xanthos, Reactive Extrusion. Principle and Practice, Hanser, Munich 1992
K.Y.O'Brien, Ed., Application of Computer Modeling for Extrusion and Other Continuous
    Polymer Processes, Hanser, Munich 1992
M.J.Stevens, J.A.Covas, Extruder Principles and Operation, Chapman and Hall, London,
    2nd ed. 1995
F.Hensen, Ed., Plastics Extrusion Technology, Hanser, Munich, 2nd ed. 1997
C.I.Chung, Extrusion of Polymers. Theory and Practice, Hanser Gardner, Cincinnati (OH) 2000
C.Rauwendaal, Polymer Extrusion, Hanser, Munich, 4th ed. 2002
L.P.B.M.Janssen, Reactive Extrusion Systems, Dekker, New York, and CRC Press, Boca Raton (NJ)
    2004

H.F.Giles, Jr., E.M.Mount, J.R.Wagner, Jr., Extrusion: The Definitive Processing Guide and Handbook, William Andrew, Norwich (NY) 2004

4.5.2.b BLOWING PROCESSES
E.G.Fisher, Blow Moulding of Plastics, Butterworth, London 1971
N.Lee, Plastics Blow Molding Handbook, Van Nostrand Reinhold, New York 1990
D.V.Rosato, D.V.Rosato, D.P.DiMattia, Blow Molding Handbook. Technology, Performance, Markets, Economics, Hanser, Munich, 2nd revised edition 2004
J.L.Throne, Thermoplastic Foam Extrusion, Hanser, Munich 2004
K.Cantor, Blown Film Extrusion. An Introduction, Hanser, Munich (OH) 2006
S.L.Belcher, Practical Guide to Injection Blow Molding, CRC Press, Boca Raton (FL) 2007

4.5.2.c THERMOFORMING
J.Florian, Practical Thermoforming. Principles and Applications, Dekker, New York, 2nd ed. 1996
J.L.Throne, Technology of Thermoforming, Hanser, Munich 1996
A.Illig, Ed., Thermoforming: A Practical Guide, Hanser, Munich 2001

4.5.3.a INJECTION MOLDING
A.I.Isayev, Ed., Injection and Compression Molding Fundamentals, Dekker, New York 1987
L.T.Manzione, Ed., Applications of Computer Aided Engineering in Injection Molding, Hanser, Munich 1987
F.Johannaber, Injection Molding Machines, Hanser, Munich, 3rd ed. 1994
H.Rees, Understanding Injection Molding Technology, Hanser, Munich 1994
G.Pötsch, W.Michaeli, Injection Molding. An Introduction, Hanser, Munich 1996
D.V.Rosato, D.V.Rosato, M.G.Rosato, Injection Molding Handbook, Springer, Berlin, 3rd ed. 2000
T.A.Osswald, L.-S.Turng, P.J.Gramann, Injection Molding Handbook, Hanser, Munich 2001

4.5.3.b REACTION INJECTION MOLDING
J.E.Kresta, Ed., Reaction Injection Molding and Fast Polymerization Reactions, Plenum, New York 1982
F.M.Sweeney, Reaction Injection Molding Machinery and Processes, Dekker, New York 1987
C.W.Macosko, RIM: Fundamentals of Reaction Injection Molding, Hanser, Munich 1989

4.6 FORMING
B.J.Jungnickel et al., Solid State Forming of Plastics, Mechanical Eng.Publ., Suffolk, UK, 1993 (translation of the 1987 German edition)
I.M.Ward, P.D.Coates, M.M.Dumoulin, Eds., Solid Phase Processing of Polymers, Hanser, Munich 2000

4.7.2 DETACHMENT OPERATIONS
A.Kobayashi, Machining of Plastics, McGraw-Hill, New York 1981

4.7.3 BONDING AND JOINING
M.N.Watson, Joining Plastics in Production, The Welding Institute, Cambridge, UK, 1989
PDL Staff, Handbook of Plastics Joining, Plastics Design Library, Norwich (NY) 1996
D.A.Grewell, A.Benatar, J.B.Park, Plastics and Composites Welding Handbook, Hanser, Munich 2003
J.Rotheiser, Joining of Plastics, Hanser, Munich, 2nd. ed. 2004

4.8 SURFACE FINISH
J.M.Margolis, Ed., Decorating Plastics, Hanser, Munich 1986
W.G.Simpson, Ed., Plastics. Surface and Finish, Royal Soc.Chem., Cambridge, UK, 2nd ed. 1996
E.B.Gutoff, E.D.Cohen, Coating and Drying Defects, Wiley, Hoboken (NJ), 2nd ed. 2006

References to Chapter 4

[1]   C.Y.Cheng, Plastics Compounding (March/April 1981) 29, Fig. 5
[2]   L.E.Gerlowski, ACS Polymer Preprints **30/1** (1989) 15, data of Table 1
[3]   H.A.Scheetz, in M.I.Kohan, Ed., Nylon Plastics Handbook, Hanser, Munich 1995, Fig. 12-5
[4]   D.V.Rosato, D.V.Rosato, Injection Molding Handbook, Chapman and Hall, London 1986, p. 366 (does not contain melting temperatures and glass temperatures)
[5]   F.L.Burkett, in H.F.Mark, Encyclopedia of Polymer Science and Engineering, Wiley, New York, 2nd ed., **16** (1989) 218, Table 2
[6]   G.W.Ehrenstein, Polymer-Werkstoffe, Hanser, Munich 1978, Fig. 39
[7]   D.S.Pearson, G. Ver Strate, E. von Meerwall, F.C.Schilling, Macromolecules **20** (1987) 1133, Tables I and II
[8]   P.R.Dvornic, J.D.Jovanovic, M.N.Govedarica, J.Appl.Polym.Sci. **49** (1993) 1497, Table I
[9]   G.Wegner, Macromol.Symp. **101** (1996) 257, Fig. 1
[10]   W.Philippoff, F.H.Gaskins, J.G.Brodnyan, J.Appl.Phys. **28** (1957) 1118, Fig. 2
[11]   -, Kunststoff-Physik im Gespräch, BASF AG, Ludwigshafen, 7th ed. (1988), p. 109
[12]   J.Meissner, in R.Vieweg, G.Daumiller, Eds., Kunststoff-Handbuch **5** (1969) 162, Fig. 10
[13]   J.L.Leblanc, Rubber Chem.Technol. **54** (1981) 905, Fig. 6
[14]   J.Meissner, Chimia **38/3** (1984) 35, Fig. 18
[15]   H.A.Barnes, Polym.Mat.Sci.Eng. **61** (1989) 37, Fig. 11
[16]   L.A.Utracki, J.Lara, Polym.Comp. **5** (1984) 44, selected data of Figs. 1 and 2
[17]   E.M.Phillips, K.G.McHugh, K.Ogale, M.B.Bradley, Kunststoffe **82** (1992) 671, selected data of Figs. 2A and 2B
[18]   Union Carbide, company literature
[19]   S.Artmeier, K.-H.Seemann, G.Zieschank, G.Spur, W.Brockmann, P.Berns, K.Wiebusch, Kunststoffe, Verarbeitung; Ullmanns Enzyklopädie der technischen Chemie, Verlag Chemie, Weinheim, 4th ed., **15** (1978) 281, Table 5
[20]   G.Burkhardt, U.Hüsgen, M.Kalwa, G.Pötsch, S.Schwenzer, Ullmanns Enzyklopädie der technischen Chemie, Verlag Chemie, Weinheim, 5th ed., **A 20** (1992) 663, Table 1
[21]   H.Saechtling, International Plastics Handbook, Hanser, Munich 1983, supplemented data of Table 18

5   Testing

5.1   Overview

5.1.1   Introduction

Processing and the application of polymers as working materials, elastomers, fibers, coatings, etc., require a good knowledge of polymer properties and their variation with temperature, applied stress, and other physical effects. In contrast to polymer science, polymer technology is less interested in the properties of single molecules than in the properties and behavior of assemblies of polymer molecules. Properties of interest are usually not molecular parameters such as molecular weights, thermal properties such as melting and glass temperatures, or fundamental rheological properties such as Newtonian viscosities but rather characteristic parameters that are directly connected to processing and use, such as melt flow indices (p. 85) and heat distortion temperatures (p. 136).

Characteristic parameters of a polymer often vary not only with the molecular weight, the molecular weight distribution, and the temperature (Fig. 5-1) but also with the specimen itself because polymers are usually not in their equilibrium states. Reported melting temperatures of "crystalline" polymers are usually neither those of perfect crystals nor those of crystalline regions in equilibrium with non-crystallizable ones (Volume III, p. 441). Glass (transition) temperatures of amorphous polymers are not thermodynamic transitions but rather relaxation processes that depend not only on the speed of testing but also on the preparation of the specimen (Volume III, p. 453). Both the preparation of the test specimen and the testing procedures have to be standardized in order to obtain comparable results (Section 5.1.2). Standardization is also necessary for the comparison of mechanical, electrical, optical, and other properties.

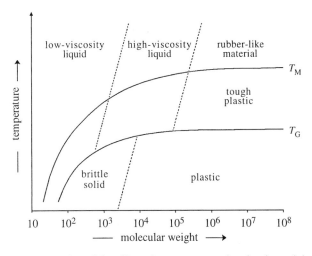

Fig. 5-1 Schematic representation of the effect of temperature and molecular weight on macroscopic properties of plastics from semi-crystalline polymers with the same constitution and configuration. The molecular weight range is only for orientation; onset of brittleness, toughness, etc., varies widely with polymer structure. T_M = melting temperature, T_G = glass temperature. Amorphous polymers have no melting temperature.

5.1.2   Standards

Industrially important polymer properties often differ from those of interest to polymer science because of more complex use conditions and the presence of time effects. Basic research often concentrates on equilibria or steady states but such states are seldom approached during processing: solid polymers usually represent frozen-in non-equilibrium conditions. The resulting "states" change with the type and duration of stresses, which makes it impossible to compare test results unless testing methods and conditions are standardized with respect to the preparation, size, and shape of test specimens and the testing conditions such as temperature, humidity, duration and speed of tests, etc. The resulting data thus represent snapshots of properties at defined conditions; they rarely allow one to draw conclusions about equilibrium properties or long-term behavior.

For a certain application, polymers are very often not chosen because of *one* important characteristic but because of a favorable combination of many properties. Since there are many possible applications, it would be a hopeless undertaking to measure all relevant properties for a host of different conditions (various temperatures, types and duration of stresses, humidities, etc.), and that not only for different grades of the same type of polymer but also for various specimens of the same grade.

Instead, polymers are characterized by some especially important properties that are measured at standardized conditions which are established by national and/or international committees. These standards should have practical relevance, allow quality controls, establish sales specifications, permit comparisons with competitors' products, let one check customer's complaints, and let one follow government laws and regulations.

The first standards were established by voluntary national organizations such as ASTM (*A*merican *S*ociety for *T*esting and *M*aterials (1898)), BSI Group ("*B*ritish *S*tandards *I*nstitution" (1901)), and DIN (*D*eutsches *I*nstitut für *N*ormung (= "German Institute for Standardization") (1917)). There are also standards by various trade organizations such as VDE = Verband Deutscher Elektriker (= Association of German Electricians). These organizations proposed many standards, for example, about 5000 for plastics by DIN. Unfortunately, the various national standards were and are not comparable.

As a result, international and supranational standards were established. 1947 saw the founding of the International Organization for Standardization (*Organisation internationale de normalisation*), a non-governmental organization located in Geneva, Switzerland, known as ISO (neither an acronym nor an abbreviation; from G: *isos* = equal). This was followed in Europe by the *Comité Européen de Normalisation* (CEN; European Committee for Standardization (1961)) that issues standards except for electrotechnical data which are issued be CENELEC = *Comité Européen de Normalisation Electrotechnique* (1973) and telecommunication (issued by ETSI = European Telecommunication Standards Institute (1988)).

High costs forbid producers, fabricators, and users of polymers to apply all tests covered by these standards. For thermoplastic bulk or film properties, many European, Japanese, and some American polymer producers have therefore settled on a set of ca. 50 mechanical, thermal, electrical, optical, processing, etc., basic data that are obtained from standard specimens using uniform testing conditions (CAMPUS® = *C*omputer *A*ided *M*aterial *P*reselection by *U*niform *S*tandards). The testing conditions differ for bulk materials and film grades. Examples of the former are shown in Table 5-1.

Table 5-1 Some CAMPUS® single-point tests of thermoplastic bulk materials [1].
    Specimen as molded material (M), sheet S, or ISO rectangular (flat) bar (I) or dumbbell (D) ("dog bone") with length *L*, width *W*, and thickness *d*, all in mm.
    Testing conditions to be reported include the testing temperature *T*, the applied load L, the solvent S used (if any), and the relative humidity (RH). For electrical tests, see Table 14-1.
    unnotch. = unnotched, notch. = notched.

Property	Type	Specimen L	W	d	Testing conditions	Standards ISO	DIN
General Properties							
Density		10	10	4		1183	53479
Indicative density (PE only)	I, D	10	10	4	23°C		
Water absorption		80	80	1	23°C, saturated		53495
Humidity absorption		80	80	1	23°C, 50 % RH		53495
Flammability UL 94		80	10	4		4589	
Chemical Structure							
Isotacticity index (p. 14)	M					6427 Annex B	
Viscosity number	M				*T*, S	1628	
Thermal Properties							
Linear thermal exp. parallel	I, D	≥10	10	4	(23-80)°C	11359	53752
normal	I, D	≥10	10	4	(23-80)°C	11359	53752
Melting temperature T_M					10°C/min	11357	
Glass (transition) temperature T_G					10°C/min	11357	
(Heat) deflection temp. T_f (1.8)	I, D	≥110	10	4	1.8 N/mm^2	75	53461
T_f (0.45)	I, D	≥110	10	4	0.45 N/mm^2	75	53461
T_f (8.0)	I, D	≥110	10	4	8.0 N/mm^2	75	53461
Vicat softening temp. A/50		≥10	10	4	50°C/h; 10 N	306	53460
B/50		≥10	10	4	50°C/h; 50 N	306	53460
Mechanical Properties							
Tensile modulus	I	80	10	4	1 mm/min	527	53457
Yield stress	I	80	10	4	50 mm/min	527	53455
Stress at 50 % strain	I	80	10	4	50 mm/min	527	53455
Stress at break	I	80	10	4	50 mm/min	527	53455
Yield strain	I	80	10	4	50 mm/min	527	53455
Nominal strain at break	I	80	10	4	50 mm/min	527	53455
	I	80	10	4	5 mm/min	527	53455
Tensile creep modulus	I	80	10	4	1 h	899	53444
	I	80	10	4	1000 h	899	53444
Charpy impact strength, unnotch.	I, D	80	10	4	+23°C	179/1eU	
	I, D	80	10	4	−30°C	179/1eU	
Charpy impact strength, notch.	I, D	80	10	4	+23°C	179/1eA	
	I, D	80	10	4	−30°C	179/1eA	
Tensile impact strength	I, D	80	10	4	+23°C	8256/1B	53448
Optical Properties							
Refractive index	S			1		489	53491
Luminous transmittance	S	60	60	2		13468	5036 T3
Rheological Properties							
Melt volume rate, MVR	M				cm^3/(10 min)	1133	
Mold shrinkage, parallel		60	60	2		294-4 (thermoplast.)	
normal		60	60	2		294-4 (thermoplast.)	

Companies store these standard data of all their polymer grades on compact discs which can be obtained from each company. Product data from various companies are collected on special commercial software. These data are now directly comparable whereas in the past testing conditions varied from company to company. It is now also possible to compare results of different tests of the same grade since the same type of test bar is now used for various tests. This was not possible in the past since companies used different test specimens, e.g., rectangular bars and dumbbells, for different tests such as tensile and impact strength because this would allow one to maximize property data (see next paragraph). Similar data banks also exist for products of companies that are not members of the CAMPUS® system, for example, DuPont (EREMIS), ICI (EPOS), and the former Rhône-Poulenc (RP3L).

It is very important that test specimens are of the same type with the same dimensions and have been produced in the same manner, for example by molding *or* injection molding. Molding delivers isotropic test specimens whereas injection molding causes orientation of molecule segments and, in semicrystalline polymers, also orientation of and in crystalline regions. Isotropic test specimens are preferred for comparing different grades and types because orientation effects are difficult to quantify. Injection molded test specimens, on the other hand, simulate much better "real" conditions since most thermoplastics are processed by injection molding, extrusion, etc.

The previous discussion was concerned only with single-point data of molded thermoplastics which are the largest group of polymers by volume. Multiple point testing will be mentioned frequently in the following sections. There are also different testing conditions for molded thermoplastics, polymer films, and expanded polymers as well as for fibers, elastomers, coatings, adhesives, and the like. Some of these data will be discussed in the appropriate chapters and sections.

5.2   Thermal Properties

5.2.1   Thermal Expansion

Bodies expand on heating because molecules, segments, and atoms become increasingly mobile and require more space. The change dV/dT of volume V with temperature T can be related to the volume to give the **cubic thermal expansion coefficient at constant pressure**, $\beta = V^{-1}(\partial V/\partial T)_p$. Isotropic bodies expand equally in all three spatial dimensions so that the **linear expansion coefficient** α is just 1/3 of the cubic one, $\alpha = \beta/3$. Anisotropic bodies expand unequally so that $\beta \neq 3\,\alpha$.

Thermal expansion is largely controlled by forces between atoms. These forces increase in the order covalent bonds > hydrogen bonds > other dipole-dipole interactions > electron-deficient bonds > dispersion forces so that thermal expansion coefficients increase in the opposite order with the approximate values of

apolar liquids	$(20\text{-}40)\cdot10^{-5}$ K^{-1}
polymers	$(5\text{-}16)\cdot10^{-5}$ K^{-1}
metals	$(1\text{-}2.5)\cdot10^{-5}$ K^{-1}
covalent solids	$\approx 0.1\cdot10^{-5}$ K^{-1}

Differences between expansion coefficients of various materials can cause trouble, for example, if polymers are bonded to metal parts.

Chain segments of semicrystalline and/or oriented polymers are preferentially oriented in one or two spatial directions, which leads to different expansions in different directions. An after-crystallization can therefore lead to different shrinkage and distortion of parts in three directions and thus to different dimensional stabilities and properties. However, amorphous polymers become dimensionally less stable if they absorb moisture.

5.2.2 Thermal Transitions and Relaxations

At the **melting temperature** T_M, crystalline regions become disordered and fuse together; melting temperatures are therefore also known as **fusing temperatures**. At T_M, the material expands and becomes considerably less viscous although the viscosity of the melt may still be very high.

Melting is a thermodynamic 1st order transition. At the melting temperature, a (semi)crystalline solid and a liquid are in equilibrium on both sides of the transition temperature (Volume III, pp. 428 and 452). In addition to 1st order transitions, liquid-crystalline polymers may also exhibit thermodynamic 2nd order transitions between two liquid-crystalline states (for example, smectic $\rightleftarrows$ nematic). The transitions crystalline $\rightleftarrows$ smectic and nematic $\rightleftarrows$ liquid are of thermodynamic 1st order (Volume III, p. 448 ff.).

Thermodynamically controlled **thermal transitions** differ from kinetically controlled **thermal relaxations** that depend on the speed or frequency of the applied method whereas true transitions do not. However, some experimental methods are so slow that a thermal relaxation may *appear* to have characteristics similar to a true 2nd order transition.

An example is the **glass** (transformation or "transition") **temperature** T_G at which a hard solid mass transforms to a softer, rubber-like material (the "melt") (Volume III, p. 452). Industrially, glass temperatures are therefore also called "softening temperatures" or even "melting temperatures" which are then distinguished from true melting temperatures T_M, the "crystallite melting temperatures."

In polymer science, thermal transitions and relaxations are usually determined at static or quasistatic conditions without external load, which allows one to correlate the resulting data with the chemical and physical structures of polymers. The melting and glass temperatures listed in this book refer in general to industrial polymers and conventional measurement conditions, i.e., they apply neither to ideal states (= 100 % crystalline or 100 % amorphous) nor to infinite molecular weights.

Crystalline polymers are usually processed above their melting temperatures and amorphous polymers above their glass temperatures. Upper service temperatures of plastics are below T_M (crystalline polymers) or below T_G (amorphous polymers).

Industry always tests the thermal behavior of polymers under external loads. These tests can be done quickly with simple instruments to deliver **Martens temperatures**, **Vicat temperatures**, or **heat distortion temperatures** (HDT), the latter also known as **heat deflection temperatures, deflection temperatures**, or **flexural softening temperatures** (Table 5-2). All three methods also reflect the elasticity of specimens. In addition, Vicat temperatures and HDTs are also affected by surface hardness.

Table 5-2 Working ranges of Martens, Vicat (ISO), and heat distortion temperatures HDT (ISO).

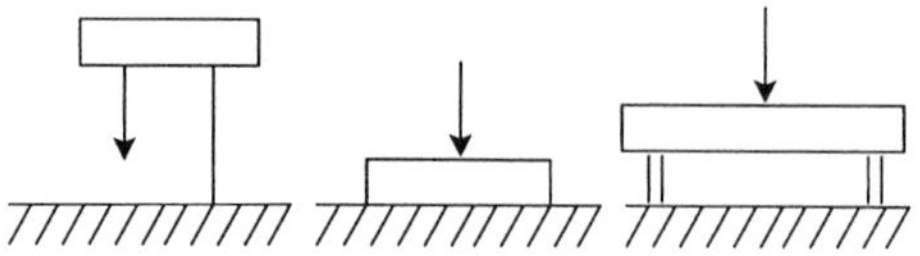

			Martens	Vicat	HDT
Standard	DIN		53458, 53462	53460	53461
	ISO			306	75
Load	flexural stress σ_F	method A	4.9 MPa		1.80 MPa
		method B			0.45 MPa
		method C			8.0 MPa
	pressure F	method A		10 N [1)]	
		method B		50 N [2)]	
Temperature change dT/dt		method A	50 K/h	50 K/h	120 K/h
		method B		120 K/h	120 K/h
		method C			120 K/h
Target deformation ΔL		method A	6 mm	1 mm	0.21 mm
		method B		1 mm	0.33 mm
		method C		1 mm	

ASTM D 648: [1)] 9.81 N, [2)] 49.05 N.

- The **Martens temperature** (Martens number) measures the temperature at which the testing specimen is lowered by 6 mm if it is heated at a rate of 50 K/h at a load of 4.9 MPa.
- The two types of **Vicat temperatures** indicate the temperatures at which a needle under a load of 1 N or 50 N goes 1 mm deep into the testing specimen that is heated at a rate of 50 or 120 K/h (Table 5-2).
- The three types of **heat distortion temperatures** employ three different flexural stresses and various target deformations at the same heating rate of 120 K/h.

The resulting thermal indicators are always lower than melting temperatures of semi-crystalline polymers (Table 5-3). In general, one finds HDT-B ≥ Vicat B ≥ HDT-A > Martens temperature.

Table 5-3 Comparison of melting, glass, Martens, Vicat, heat distortion (HDT, methods A or B), and upper service (UST) temperatures of crystalline (c) and amorphous (a) polymers. Bisphenol A polycarbonate (PC) does not crystallize under the usual processing conditions (T_M = 235°C). The crystallinity of poly(vinyl chloride) PVC is less than 5 %.

Polymer		Temperatures in °C						
		T_M	HDT-B	Vicat B	HDT-A	Martens	T_G	UST
PA 6.6	c	265	246	230	75	55	50	100
POM	c	178	150	155	101	65	−50	90
PP	c	176	165	105	120	45	−15	100
PSU	a	-	181		174		190	
PC	a	-	138	150	132	120	147	100
PS	a	-	97	101	90	74	100	80
SAN	a	-	95	99		77	106	85
PVC (hart)	a	-	82	80	70	65	90	60

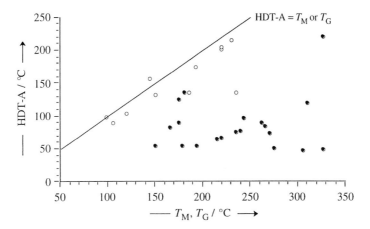

Fig. 5-2 Heat distortion temperature, method A, of unfilled polymers as a function of melting temperature of crystalline polymers (●) or glass temperature of amorphous ones (O). The solid line indicates the identities HDT-A = T_M and HDT-A = T_G, respectively. Measurements at 1.82 MPa. Data from Tables 8-18 and 8-20 through 8-24 of this Volume.

A clearer picture is obtained if heat distortion temperatures HDT-A and HDT-B are compared with the glass and melting temperatures of amorphous polymers. For amorphous polymers, one finds $T_G >$ HDT-B > HDT-A but the differences are not large (Table 5-3) and practically independent of the values of the glass temperature. HDT-Bs of amorphous polymers are also little affected by the presence of fillers (Chapter 9).

Differences are much larger for semicrystalline polymers, $T_M \gg$ HDT-B $\ggg$ HDT-A, (Table 5-3). In contrast to amorphous polymers, HDT-As of semicrystalline polymers do not vary systematically with T_M (Fig. 5-2); they rather form more or less a broad band between ca. 50°C and 185°C. The reason for this behavior is a plateau in the temperature dependence of shear moduli between T_G and T_M.

5.3   Tensile Testing

5.3.1   Experiments

Stress-strain Diagrams

Mechanical properties are most often evaluated by tensile testing in which a specimen such as a molded dumbbell or rectangular slab, a film, a fiber bundle, etc., of length L_0 and cross-sectional area A_0 is clamped on both ends and then drawn with constant speed (see Volume III, p. 526). The force F normal to the cross-sectional area is recorded as a function of time t and/or length L of the specimen.

The calculated **tensile stress**, $\sigma \equiv \sigma_{11} = F/A_0$, of plastics is plotted against the **draw ratio (strain ratio)**, $\lambda = L/L_0$, or the **nominal elongation (elongation, Cauchy elongation, tensile strain, engineering strain)**, $\varepsilon = \lambda - 1 = (L - L_0)/L_0 = \Delta L/L_0$ (Fig. 5-3). Drawing by a factor of 3 thus results in a tensile strain of 200 %. For fibers, see Chapter 6.

On drawing, elastomers such as SBS taper homogeneously so that the actual cross-sectional area A becomes smaller with time albeit to the same extent at each length

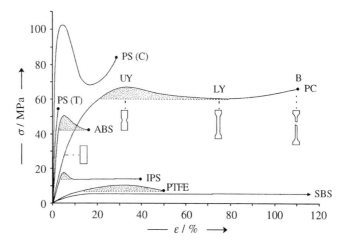

Fig. 5-3   Nominal tensile stress σ as a function of the nominal elongation ε of various polymers at room temperature. UY = upper yield point, LY = lower yield point, B = break (fracture) ($\bullet$). → additional elongation. Dotted areas: ductility ranges for tension = f(elongation).

ABS = high-impact strength "blend" from acrylonitrile+butadiene+styrene (Chapter 10), IPS = impact-resistant poly(styrene), PC = bisphenol A polycarbonate, PS = atactic poly(styrene), PTFE = poly(tetrafluoroethylene), SBS = elastomeric triblock copolymer with styrene and butadiene blocks. For comparison, stress/strain curves of PS are shown for tension (T) and compression (C).

position of the specimen. For volume-invariant deformations, the **true tensile strength** σ' is therefore greater than the nominal stress σ, i.e., $\sigma' = F/A = (F/A_0)(L/L_0) = \sigma(L/L_0)$. Hence, true stress-strain curves of elastomers differ from nominal ones although they are still very similar.

Many plastics do not taper homogeneously but rather exhibit **necking** followed by a **telescope effect** as indicated in Fig. 5-3 for rectangular test specimens of PC. Stress-strain diagrams of, for example, PC, PS (C), ABS, IPS, and PTFE (Fig. 5-3) then show a maximum which is caused by the use of nominal stresses and strains. The maximum disappears if the true stress σ' is plotted as a function of the **true strain** ε' (**Hencky strain**) (Volume III, p. 521) which is obtained as $\varepsilon' = \ln (L/L_0) = \ln (A/A_0)$ by integration over all length changes. The literature practically never reports true stress-strain curves.

The initial slope of nominal curves $\sigma = f(\varepsilon)$ indicates the **tensile modulus** E (**modulus of elasticity, Young's modulus**) which is one of the three moduli that describe elasticities (the others are the shear modulus and the bulk modulus, see Volume III, p. 523). The true initial slope is difficult to measure, even for thermoplastics, because it occurs at very small elongations. The reported values of E are therefore not obtained from tangents but from secants, for *thermoplastics* for example from the origin of $\sigma = f(\varepsilon)$ to the value of σ at elongations of 0.05 % $\leq \varepsilon \leq$ 0.25 % in the CAMPUS® system. For *elastomers*, one usually reports 100 % moduli from tangents at or secants to 100% nominal elongation and correspondingly also 300 % moduli. The moduli of *fibers* do not refer to stresses at all but rather to ratios of forces per so-called linear densities; they may be obtained as tangents, secants, or at defined elongation intervals (see Section 6.3.1).

Nominal stress-strain curves of thermoplastics (example: PC in Fig. 5-3) may show an **upper yield point** UY with a yield stress σ_{UY} at an elongation ε_{UY}, a **lower yield point** LYP with coordinates σ_{LY} and ε_{LY}, and a fracture B with **fracture strength** σ_B at the frac-

Table 5-4   Types of polymers according to their stress-strain behavior.

Designation		Values of			Examples
Correct	Conventional	E	σ_Y	ε_B	
rigid-brittle	hard-brittle	large	-	small	PS, PF
rigid-strong	hard-strong	large	large	small	PMMA
rigid-ductile	hard-tough	large	large	large	POM, PC
soft-strong	soft-strong	small	small	small	PTFE
soft-ductile	soft-tough	small	small	large	LDPE
soft-elastic	soft-weak	small	-	large	SBS

ture elongation ε_B. Other thermoplastics show only UY and B (ABS) or only B (PS-T). All polymers with an upper yield point have a ratio of upper yield stress and tensile modulus of $\sigma_{UYP}/E \approx 0.025$. The area under the curve defined by the UY and the horizontal through the lower yield point LY is the **ductility range** (see Fig. 5-3).

Tensile stresses often decrease at elongations greater than ε_{UY}, This **stress softening** occurs without the specimen being heated and is therefore also called **cold flow**. After passing through the maximum at the UY, some thermoplastics fracture at a lower value of σ (PTFE in Fig. 5-3), some extend a little (ABS) or a lot more (IPS), and some pass through a lower yield point LY with the coordinates σ_{LY} and ε_{LY} before they fracture at σ_B (PC), the **tensile strength at break (fracture strength)**, with an **elongation at break (fracture elongation)**, ε_B.

Terminology and Temperature Dependence

For fast qualitative descriptions, plastics are often described by adjectives such as "hard", "soft", "brittle", "strong", "tough", and/or "weak." Some of these epithets are justified but some are not because of other established meanings of the words. In all cases, these terms can be used only to compare tensile properties that have been obtained under the same experimental conditions such as constant temperature and strain rate.

A large initial slope in the $\sigma = f(\varepsilon)$ diagram is characteristic of a **rigid** polymer with a large tensile modulus such as atactic poly(styrene) (Fig. 5-3). Such polymers should not be called "hard" because hardness is a surface property (Section 5.5). Conversely, a polymer with a small initial slope in the function $\sigma = f(\varepsilon)$ is **soft**, i.e., easily deformable; examples are PTFE and SBS.

Polymers without a yield point do not take up much energy on deformation; they break easily and are **brittle**. Poly(styrene) PS is therefore a rigid-brittle polymer. **Ductile** polymers, on the other hand, are easily deformable (have large ductility ranges) and have large fracture elongations. Polycarbonate PC is thus a rigid-ductile polymer while low-density poly(ethylene) LDPE is a soft-ductile one. However, ductile polymers are not "tough" because "toughness" characterizes resistance to impact (Section 5.4). The correct and conventional epithets are compared in Table 5-4.

Plastics may therefore fail under different conditions: brittle polymers by fracture (σ_B, ε_B) and ductile polymers by yielding (σ_{UYP}, ε_{UYP}). Correspondingly, "**tensile strength**" without any other specification may refer to either fracture of brittle polymers or yielding of ductile ones since "strength" here only refers to "failure."

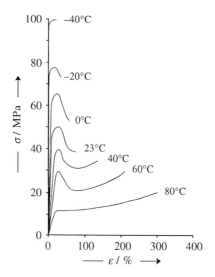

Fig. 5-4 Stress-strain diagrams of a rigid (= non-plasticized) poly(vinyl chloride) PVC at various temperatures. PVC is rigid-brittle at –40°C but soft-ductile at $T_G \approx 80$°C.

All of these designations only refer to the behavior at a defined temperature with a constant elongation rate, for thermoplastics typically 5 mm/min at 23°C. At low temperatures, all thermoplastics are rigid-brittle: there is no stress-softening and no telescope effect (Fig. 5-4). At temperatures near their highest transformation temperature T_M (semicrystalline polymers) or T_G (amorphous polymers), all thermoplastics show rubber-like behavior because of strong flow effects. In intermediate temperature ranges, stress-strain behavior ranges from rigid-ductile to soft-ductile.

Tensile tests are short-term tests which deliver data that depend on the strain rate. At small strain rates, molecule segments may still be able to relax and flow whereas at large strain rates they cannot. For the testing of plastics, elastomers, and fibers, elongation rates $\dot{\varepsilon}$ are standardized, however, unfortunately differently in different countries.

5.3.2 Tensile Moduli

During *tensile testing*, applied loads are transferred by shearing from the surface to the interior of the specimen. For isotropic plastics, this will be the full load. However, anisotropic specimens do not receive the full load since clamping produces additional stresses that balance out differently in different regions. The effect is larger, the shorter the specimen and the smaller the aspect ratio L/d.

The apparent tensile modulus thus increases with increasing ratio of length L and diameter d of the testing specimen until it finally becomes constant, for example, at $L/d \approx$ 100 for a semi-crystalline poly(ethylene) (Volume III, Section 16.2.4). Such large ratios L/d are never used in commercial testing where they are typically $L/d \leq 11$. It must be emphasized therefore that all industrially reported tensile moduli of plastics are those of the testing specimens and conditions and neither the true elastic moduli of the materials nor the lattice moduli of 100 % crystalline polymers.

5.3.3   Flexural Moduli

In principle, moduli can be determined not only by elongating a specimen, but also by bending, pressurizing, compressing, and twisting or by combining these simple types of deformation with elongating or shearing. For example, forces can act only on the free end of a specimen that is clamped at the other end (1-point measurements similar to the Martens test, Table 5-2) or on the center of a specimen that is clamped at both ends (3-point measurements similar to HDT in Table 5-2).

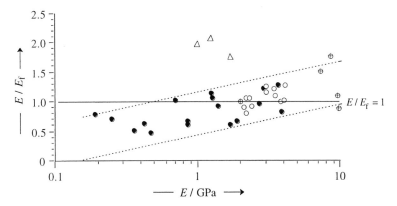

Fig. 5-5 Ratios E/E_F of tensile moduli E and flexural moduli E_F as a function of the tensile moduli of amorphous polymers (O), semicrystalline polymers (●), conditioned semicrystalline polymers (aliphatic polyamides at 50 % relative humidity) (Δ), and liquid-crystalline polymers (⊕). Data of [2].

For 3-point bendings, deformations are more complicated than for elongations or compressions because the convex side will be elongated and the concave side will be compressed. In this case, the **flexural modulus** of a rectangular specimen with length L, width b, and thickness d can be calculated from $E_f = (FL^3)/(4\,bd^3\delta)$ (Volume III, p. 528) where F = force and δ = depth of bend.

Tensile and flexural moduli should be identical, which is indeed approximately observed for amorphous polymers (Fig. 5-5). However, the ratio E/E_f of semicrystalline and liquid-crystalline polymers is smaller than unity for small tensile moduli but greater than unity for large ones. These variations are caused by orientations of molecule segments and crystalline regions. Plasticizing polymers by humidity leads to values of E/E_f far greater than unity, i.e., flexural moduli are much more affected than tensile moduli.

The **stiffness** of a body is defined by the force F that is required for bending. For elongations of isotropic bodies, it is given by $F = EA$: the stiffnes of an isotropic body (e.g., a beam) is twice as large if the cross-sectional area is doubled.

In a similar manner, one can calculate the stiffness against shearing, compression, or bending. In bending, the so-called neutral axis of a slab or beam is neither compressed nor elongated (Volume III, p. 528) if tensile moduli equal compression moduli, i.e., if Poisson's ratio is 1/3 (Volume II, p. 523). The **flexural stiffness** F_f is usually defined as that flexural moment M which deforms a body to the sector of a ring with the radius R ($F_f = MR$). In order to obtain true segments of a circle, the span L should therefore be much larger than the thickness (ca. 16:1 for 3-point measurements).

Theory delivers $F_f = MR = E_f A^{(2)}$ where $A^{(2)}$ = 2nd moment of the cross-sectional area with respect to the neutral axis which is $A^{(2)} = bd^3/12$ for a rectangular cross-section with width b and thickness d. In this case, the flexural stiffness is $F_f = E_f bd^3/12$, i.e., doubling the thickness delivers an 8-fold flexural stiffness but only a 2-fold elongational or compression stiffness.

5.3.4   Tensile Strengths

Types of Failure

On deformation, polymers and polymeric systems can fail in various ways (Fig. 5-6). Thermoplastics and thermosets may form shear bands or crazes; thermoplastics may break in a brittle or tough manner; fibers may deform by shearing, bending, or kink formation, and fiber reinforced plastics may deform in a variety of ways.

Failure is defined as the point beyond which the material can no longer perform its intended function. In *tension*, failure of plastics occurs at points of the stress-strain curve at which the specimen gives way, which may be either the upper yield point of ductile polymers or the fracture point of brittle ones (Fig. 5-3). The stress at these points is known as **tensile strength** (= energy per volume = force per area).

For fibers, failure is measured as **tenacity** (= energy per mass = force per linear density (Section 6.3.1). One distinguishes here the **force at break** at the maximum of the $F = f(\varepsilon)$ curve from the subsequent (lower) **force at rupture** (Section 6.3.1).

Shearing occurs in the direction of stress. It leads to shear bands if the stress is localized. However, the specimen may also give way perpendicular (normal) to the applied stress, which leads to the formation of **crazes** that may be up to 100 μm long and up to 10 μm wide. Crazes are not hollow holes but filled with amorphous microfibrils.

The deformation of crazes is the main mechanism for the dissipation of stress energy because it is much more efficient than dissipation by shear flow. Continued application of stress converts crazes to microcracks that are no longer filled with matter. The microcracks grow and unite until the specimen fractures catastrophically.

Fracture of plastics can therefore occur in two ways. In **brittle fracture**, the specimen does not absorb energy and fractures normal to the stress direction. Polymers are called brittle if the fracture elongation is less than 10 % (United States) or 20 % (Europe).

Ductile failure of plastics is caused by the mutual sliding of chain segments and crystalline regions (semicrystalline polymers only). The specimen fractures along microcracks in the amorphous regions, i.e., spherulitic polymers fracture between spherulites. For a more detailed discussion of fracture, see Volume III, Chapter 18.

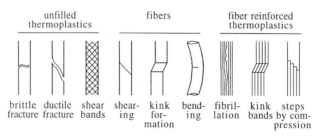

unfilled thermoplastics			fibers			fiber reinforced thermoplastics		
brittle fracture	ductile fracture	shear bands	shear- ing	kink for- mation	bend- ing	fibril- lation	kink bands	steps by com- pression

Fig. 5-6 Types of failure.

Effect of Preparation and Molar Mass of Specimens

Fracture strengths increase with increasing molecular weights M_r of polymers. Significant fracture strengths are only observed beyond the onset of chain entanglements which occurs for atactic poly(styrene)s at molar masses of $M = 13\ 600$ g/mol (from shear storage plateau modulus) and $M_r = 35\ 000$ g/mol (from Newtonian viscosity) (Volume III, p. 605). Significant strengths require even higher molecular weights, for the atactic poly(styrenes of Fig. 5-7 at $(\overline{M}_w + \overline{M}_n)/2 = 65\ 000$ g/mol which corresponds to $\overline{M}_n = 60\ 000$ g/mol (if compression molded) and $\overline{M}_n = 50\ 000$ g/mol (if injection molded) for narrowly distributed polymers.

Significant fracture elongations and fracture strengths are obtained only for *high* molar mass polymers as shown by the data for poly(styrene) (Fig. 5-7). Increasing molar masses lead to more entanglements per molecule, and the fracture strength increases until it becomes practically independent of the molar mass.

Entanglement and tensile test data are not directly comparable because of different experimental conditions. Tensile strengths $\sigma_B \geq 0$ were obtained at 23°C for molar masses of $M \geq 65\ 000$ g/mol whereas the onset of entanglement was observed at 190°C for $M \geq 35\ 000$ g/mol. However, entanglement molar masses *decrease* with decreasing temperature (as judged from the temperature dependence of entanglement data by shear storage plateau moduli (see Volume III, p. 605); the effect of temperature on the onset of entanglements has not been reported for melt viscosity as a function of molar mass). Hence, the gap between onset of entanglements and onset of fracture strength should be even greater for the same temperature.

Properties are also affected by the method of preparation of test specimens. Injection molded bars have greater fracture strengths and elongations than compression molded ones since molecule segments orient in the flow direction during the process. Tensile moduli are not affected because they are obtained at low stresses/elongations.

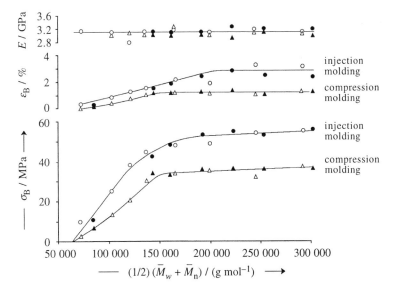

Fig. 5-7 Tensile moduli E, fracture elongations ε_B, and fracture strengths σ_B of anionically polymerized poly(styrene)s with narrow molar mass distributions ($\overline{M}_w / \overline{M}_n = 1.1 \pm 0.1$) (O,Δ) and of free-radically polymerized poly(styrene)s with broad molar mass distributions ($\overline{M}_w / \overline{M}_n = 2.2 \pm 0.4$) (●,▲) as functions of the arithmetical mean of number-average and weight-average molar masses [3]. All data at 23°C and 50 % relative humidity. Test specimens were obtained by compression molding (Δ,▲) or injection molding (O,●).

Statistical Effects

Test specimens from the same material show a distribution of properties even if they are processed in the same manner. For example, test bars do not break at the same time and the same load because specimens are not homogeneous, forces do not act in the same manner because of small differences in clamping, etc. The distribution of properties can be described by a **Weibull distribution** of the fraction f of properties with, for example, strength σ_B where σ^* and a are empirical parameters (Eq.(5-1); Fig. 5-8):

$$(5\text{-}1) \qquad \int_0^\infty f \, d\sigma = \exp[(\sigma/\sigma^*)^a]$$

in which σ^* = **characteristic strength** and a = factor for the shape of the distribution. The same material can therefore have widely varying strengths, for example, tensile strengths between 80 MPa and 114 MPa in Fig. 5-8.

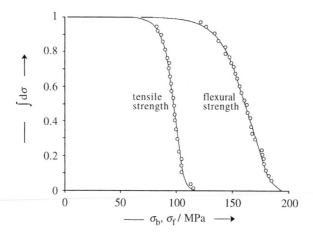

Fig. 5-8 Integral distribution of tensile strengths and flexural strengths (4-point method) of a glass fiber-reinforced sheet molding compound that was pre-impregnated with an unsaturated polyester resin [4]. Curves were calculated for Weibull distributions (Eq.(5-1)) with σ^* = 100 MPa (tensile) and 167 MPa (flexure) and a = 12.9 (tensile) and 100 (flexure), respectively. By kind permission of the Society of Plastics Engineers, Brookfield Center (CT).

5.4 Impact Strength

In conventional tensile tests, polymers are subjected to drawing speeds between 1 mm/min and 50 mm/min which corresponds to 0.017-0.83 mm/s or 0.06-3 m/h. These speeds are much lower than the ones that objects experience in daily life: doors are slammed shut at ca. 3 m/s and a car crash at 75 miles per hour (= 120.6 km/h) corresponds to 33.5 m/s.

Such high speeds can be generated with high-speed testing machines: hydraulic instruments achieve $\Delta L/t$ = 4 m/s and pneumatic ones up to 250 m/s. For a test bar with an initial length of L_0 = 10 cm, these speeds correspond to extension rates of $\dot{\varepsilon} = \Delta L/(L_0 t) = $ 40 s^{-1} and 2500 s^{-1}, respectively.

However, such machines are much too expensive for routine tests. Far less expensive are tests that measure the effect of impacts, for example, by a falling dart on an unnotched specimen or a pendulum on a notched one (Izod, Charpy) (Fig. 5-9). In these tests, unnotched specimens experience impact speeds of up to 4 m/s and extension rates of up to 60 s^{-1}. In notched specimens, extension rates may even approach 5000 s^{-1} which corresponds to those generated by pneumatic high-speed tensile testing machines.

The most important impact tests are the Izod method, the Charpy method (known as the "European method" in the United States), and the tensile impact test (Table 5-1). Izod and Charpy tests can be conducted with either unnotched or notched specimens.

In **Charpy tests**, the test specimen lies horizontally on two supports and is struck from above by a pendulum (Fig. 5-9). The specimen experiences a compression on the hit side, a bending stress in the center, and strong tensile stress at the bottom. A notch, if any, is therefore placed at the bottom.

In **Izod tests (dynostat method)**, a vertically positioned specimen is clamped at the bottom and struck at the free upper end by a pendulum (Fig. 5-9). The specimen is subjected to bending stresses (major) and shear stresses (minor). Notches are in the center of the specimen, below the impact point of the pendulum. The smaller the radius of the notch, the higher the stress concentration at the tip and the lower the impact strength at constant temperature.

The impact strength of unnotched specimens is predominantly controlled by the effect of "natural cracks" with submicroscopic dimensions; truly macroscopic cracks are probably never present. Much more revealing are impact tests with notched specimens since the notch is the biggest "crack."

Impact strengths are affected by the thickness of the specimen. Impact strengths of infinitely thick specimens are governed by the stress field; the deformation of the whole specimen is not very important. An infinitely thin specimen, on the other hand, is strongly deformed but there is no stress field.

A notch is a stress concentrator. If the notched impact strength is calculated as fracture energy per width of notch, then it is mainly controlled by the energy for fracture initiation. A fracture energy per width of notch *and* thickness of specimen (i.e., per area) on the other hand, is more indicative of crack propagation.

Since U.S. data are reported as energy per length (1 ft lbf/in. ≈ 53.38 N = 53.38 J/m) and European data as energy per area (kJ/m^2), the former are measuring crack initiations and the latter, crack propagations. However, European data average better with respect to effects of morphology and various proportions of fracture initiation and propagation.

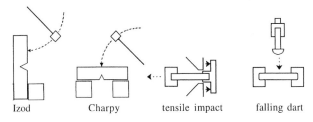

Izod   Charpy   tensile impact   falling dart

Fig. 5-9 Schematic representation of tests of notched impact strengths (Izod, Charpy) and impact strength (tensile impact, falling dart). Izod and Charpy tests may also be performed with unnotched apecimens.

Table 5-5 Impact strengths γ_i, notched impact strengths $\gamma_{i,N}$, and notched tensile impact strengths $\gamma_{i,N,t}$ of some grades of European industrial thermoplastics at 23°C and –30°C, arranged according to increasing notched impact strengths at 23°C. nf = no fracture, - not reported.

Polymer			$\dfrac{\gamma_i}{\text{kJ m}^{-2}}$	$\dfrac{\gamma_{i,N}}{\text{kJ m}^{-2}}$	$\dfrac{\gamma_{i,N,t}}{\text{kJ m}^{-2}}$	$\dfrac{\gamma_i}{\text{kJ m}^{-2}}$	$\dfrac{\gamma_{i,N}}{\text{kJ m}^{-2}}$
No.	Symbol	Trade name	23°C	23°C	23°C	–30°C	–30°C
1	PS	Polystyrol 143 E	9	2	-	9	2
2	HDPE	Hostalen GA 7260	-	2.1	45	-	3.0
3	PMMA	Degalen 6	14	2.1	-	14	2.0
4	PA 6.6	Ultramid A5, high M, dry	38	3.5	-	30	2.5
5	PA 6.6	Ultramid A5, high M, air dry	64	3.5	-	30	2.5
6	PBT	Vestodur 1000 nf	135	4.5	55	130	4.3
7	PA 6.6	Ultramid A3, low M, dry	nf	6	-	280	6
8	ABS	Magnum 3153	nf	11	47	89	9
9	POM	Delrin 100 NC-10	250	12	-	200	7
10	PA 6.6	Ultramid A3, low M, air dry	nf	12	-	300	5.5
11	HDPE	Lupolen 6031 H	nf	15	120	nf	9

Tests of **tensile impact strength** use a dumbbell type of specimen that is clamped on both sides. The pendulum then strikes against both sides of the clamp. The absorbed energy is reported per area. Different types of specimens lead to different results since the absorbed energy increases with the volume of the specimen.

Impact strengths from Charpy and Izod tests are not intrinsic material properties because they depend on external conditions such as specimen thickness and notch shape, depth, and radius. They are helpful for the comparison of various materials but are useless for design calculations. For engineering purposes, impact strengths are therefore characterized by one of two other quantities, the critical stress intensity factor K_{IC} or the fracture toughness G_{IC} (Volume III, p. 629 ff.).

Table 5-5 shows various impact strengths of grades of thermoplastics, arranged according to increasing notched impact strengths at 23°C. In general, the same ranking is also obtained for other impact strengths and other temperatures although there are some deviations. The table also shows that impact strengths of the same type of polymer can vary significantly with the molecular weight (numbers 4 *vs.* 7; 5 *vs.* 10) and that the idealized chemical structure of a polymer is not a good indicator of the performance of grades of various polymer manufacturers (numbers 2 and 11).

5.5   Hardness

The hardness of a body is defined as its resistance to deformation or fracture. The measured hardnesses are (macroscopic!) **surface hardnesses** that cannot be easily related to fundamental material properties.

Hardness is a fairly complex quantity that depends on the tensile modulus, the upper yield strength, and the stress hardening of the specimen. Yield strength and stress hardening are viscoelastic properties that vary with the type, magnitude, and duration of the load. For thin layers, all quantities also depend on the type of support. Hardnesses also vary with the thickness of the specimen if the sample has a structure gradient.

Consequently, "hardness" cannot be defined universally and there also cannot be a hardness test that is suitable for *all* types of materials. As a result, many different hardness tests and hardness scales exist that differ from industry to industry and also from material to material. Table 5-6 compares some hardness scales.

Table 5-6 Comparison of some hardness scales [5, 6]. There are many other Rockwell scales (A, B, C). Shore A corresponds approximately to IRHD (p. 151).

Material Mohs	Modified Mohs	Mohs orig.	Mohs mod.	Vickers	Brinell	Rockwell M	Rockwell α ≈ R	Shore D	Shore C	Shore A
Minerals, metals, and other inorganics										
Diamond	{diamond}	10	15	2500	1000	500				
	{boron carbide}		14							
	{silicon carbide}		13							
Corundum	{fused alumina}	9	12	1650	785	450				
	{fused zirconia}		11							
	{garnet}		10							
Topaz	(chromium)	8	9	1030	600	400				
Quartz		7	8	600	440	350				
	(vitreous silica)		7							
Orthoclase	(tungsten)	6	6	325	310	300				
Apatite	(cobalt)	5	5	155	200	250				
Fluorite	(iron)	4	4	63	120	200				
Calcite	(copper)	3	3	20	65	150				
Hard plastics										
Gypsum	(finger nail)	2	2	4	25	100				
					16	80				
					12	70	100	90		
					10	65	97	86		
					9	63	96	83		
					8	60	93	80		
					7	57	90	77		
					6	54	88	74		
Talc(um)		1	1		5	50	85	70		
					4	45			65	95
Soft plastics										
					3	40	50	60	93	98
					2	32		55	89	96
					1.5	28		50	80	94
					1	23		42	70	90
Elastomers										
					0.8	20		38	65	88
					0.6	17		35	57	85
					0.5	15		30	50	70
								25	43	75
								20	36	70
								15	27	60
								12	21	50
								10	18	40
								8	15	30
								6.5	11	20
								4	8	10

5.5.1 Scratch Hardness

There are three major groups of hardnesses: scratch hardness, indentation hardness, and rebound (or dynamic) hardness.

The oldest hardness scale is a scratch scale: in the **Mohs scale**, materials with higher Mohs numbers can scratch materials with lower Mohs numbers but not the other way around. The gaps between higher Mohs numbers are fairly wide and it is for this reason that the upper numbers of the Mohs scale have been recently expanded in order to accommodate modern hard materials such as boron carbide (Table 5-6).

Mohs numbers of polymers are generally smaller than 2 so that the Mohs scale is not sensitive enough to distinguish between the hardnesses of various polymers (Table 5-6). Mohs hardnesses are therefore not suitable for polymers. Instead, one uses so-called pencil hardnesses, especially for coatings. In these tests, pencils with "leads" of various hardnesses (as indicated by a code, see below) are used to scratch surfaces. The ASTM **scratch pencil hardness** is defined as the code of the hardest pencil that will not scratch or rupture the surface of the coating. The ASTM **gouge surface hardness** refers to the hardest pencil that will not make a cut of at least 3 mm in length into the coating.

5000 years ago, Egyptians filled hollow plant stems (reeds, papyrus, bamboo) with liquid lead and used the solidified rods as writing instruments. The Roman *stylus* was also often made from lead. At the end of the medieval ages, alloys of lead and silver were used for the same purpose. At the end of the 16th century, lead started to be replaced by graphite (from L: *graphein* = to write) which was thought to be a lead ore; "lead" therefore became the name of the writing substance of a pencil (from L: *penicellus*, diminutive of *penis* = tail). Present pencil leads consist of heat-treated mixtures of graphite and clay in a hollow wooden shaft.

In the European and United States pencil codes, "H" denotes "hardness" and "B" "blackness" since leads are darker and softer the more graphite they contain. The codes range from 9B (softest) to 9H (hardest) on the European scale (ISO 15184) and from 6B (softest) to 6H (hardest) on the US scale (ASTM D3363). In the United States, most pencil manufacturers also use a simple number code for household pencils (the Faber-Castell code differs from the common one). The Russian code is different.

Europe (ISO)	9B	8B	7B	6B ...	2B	B	HB	F	H	2H ...	6H	7H	8H	9H
U.S. (ASTM)				6B ...	2B	B	HB	F	H	2H ...	6H			
U.S. (common)						1	2	$2^1/_2$	3	4				
Faber-Castell					1	2	$2^1/_2$		3	4				
Russia						M	TM		T	2T				

Pencil hardnesses of various countries are not directly comparable. For example, the pencil hardness HB is softest in Japan, harder in Europe, and hardest in the United States.

ASTM regulates the flattening of the lead and the length to which it is exposed. The pencils are usually drawn by hand or with a machine at an angle of 45° across the surface. The results of these procedures heavily depend on the scratching angle, the pressure exerted on the lead, the shape and length of the lead, etc.

Scratch pencil hardnesses increase with the thickness of the coating (Fig. 5-10) and should become independent of the thickness of the coating and thus of the nature of support at very thick coatings. The pencil hardnesses of Fig. 5-10 scatter at higher filler contents which may be due to wrong assignments (3H instead of 4H for $d_{ind} = 20.5$ µm or 4H instead of 5H for $d_{ind} = 25.1$ µm?).

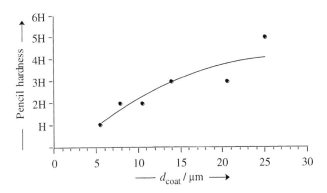

Fig. 5-10 Dependence of pencil hardness on the thickness d_{coat} of a sol-gel coating on a polycarbon-ate support [7a]. The sol-gel coating was prepared from 38 mol% GLYMO (= γ-glycidyloxypropyltri-methoxysilane) and 62 mol% TEOS (= tetraethoxysilane); it contained 43.8 wt% colloidal silica.

Pencil hardnesses depend on the elasticity of the coating and, for very thin coatings, also on the elasticity of the support. They should relate therefore to the tensile modulus E and should also correlate with the indentation hardness which is indeed found (Fig. 5-11). With increasing proportion of the hard filler from 0 % to 50 %, tensile moduli of the coating material increased 3.8-fold whereas the indentation hardnesses of the coatings were only 1.7 times greater because the latter are also effected by flow and crack processes. Indeed, these particular coatings cracked behind the indenter. Tensile cracking was the dominant failure mechanism.

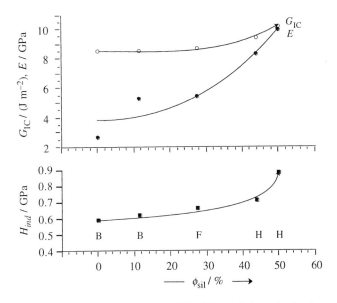

Fig. 5-11 Fracture toughness G_{IC} (top), tensile modulus E (top), indentation hardness H_{ind} (bottom), and pencil hardnesses B, F, and H as a functions of the volume fraction ϕ_{sil} of colloidal silica in a sol-gel coating from GLYMO and TEOS (see caption to Fig. 5-10). The coating had a thickness of 5.1 ± 0.3 μm. Data of [7b]. For a discussion of fracture toughness (critical strain release rate), see Volume III, Section 18.3.3.

In addition to those mentioned above, very many other hardness scales are also used. The hardness of hard plastics is often reported in **Barcol units** from HB50 (or 50B) to HB90 (or 90B) that are determined similarly to Shore D. In Europe, one often uses **ball indentation hardnesses** which measure the indentation of spheres of 5 mm diameter under a load of 358 N; hardnesses are then reported as mechanical stresses. There are also several types of hardness tests that use pendulums.

For an infinitely thick, rigid specimen, the measured hardness is indeed that of the surface region of the specimen. For soft specimens, "hardness" may include a contribution from the interior, especially, if the surface is plasticized by the humidity of air. The interior may also affect the hardness of the specimen if the melt of a crystallizable polymer is injected into a cold mold because the surface may solidify without much crystallization whereas the interior crystallizes while cooling slowly.

Literature to Chapter 5

(for literature on weathering, etc., see also Chapter 3)

P.Han, Tensile Testing, ASM International, Materials Park (OH) 1992
M.Ezrin, Plastics Failure Guide. Cause and Prevention, Hanser Gardner, Cincinnati (OH) 1996
V.Shah, Handbook of Plastics Testing Technology, Wiley-VCH, Weinheim, 2nd ed. 1998
R.P.Brown, Ed., Handbook of Polymer Testing: Physical Methods, Dekker, New York 1999
B.P.Saville, Physical Testing of Textiles, CRC Press, Boca Raton (FL) 1999
H.Chandler, Ed., Hardness Testing, ASM International, Materials Park (OH) 1999
J.Scheirs, Ed., Compositional and Failure Analysis of Polymers, Wiley, New York 2000
F.A.Morrison, Understanding Rheology, Oxford Univ. Press, New York 2001
G.W.Ehrenstein, G.Riedel, P.Trawiel, Thermal Analysis of Plastics. Principles and Practice, Hanser, Munich 2004
D.C.Hylton, Understanding Plastics Testing, Hanser Gardner, Cincinnati (OH) 2004
W.Grellmann, S.Seidler, Plastics Testing, Hanser, Munich 2007
V.Shah, Handbook of Plastics Testing and Failure Analysis, Wiley, Hoboken (NJ), 3rd ed. 2007
W.A.Woishnis, Chemical Resistance of Plastics and Elastomers, W.Andrew, Norwich (NY) 2007
LE.A.Campo, Selection of Polymeric Materials: How to Select Design Properties from Different Standards, W.Andrew, Norwich (NY) 2007
L.W.McKeen, The Effect of Temperature and Other Factors on Plastics and Elastomers, W.Andrew (NY), 2nd ed. 2008

References to Chapter 5

[1] http://www.campusplastics.com/iso.htm, accessed 2001-01-03 and 2007-12-12. Lists are constantly updated.
[2] H.-G. Elias, Makromoleküle, Vol. 2, Technologie, Hüthig & Wepf, Basel, 5th ed. (1992), data of Tables 15-8, 15-11, 15-13, 15-14, 15-15, 15-16, and 15-18
[3] H.W.McCormick, F.M.Brower, L.Kin, J.Polym.Sci. **39** (1959) 87, data of Tables 1-3
[4] C.D.Shirrell, Polymer Composites **4** (1983) 172, Fig. 5
[5] E.Rabinowicz, Friction and Wear of Materials, Wiley, New York, 2nd ed. (1995), Table 7.3
[6] D.W. van Krevelen, Properties of Polymers, Elsevier, Amsterdam, 2nd ed. (1976), Fig. 25.5 and Table 25.6
[7] E.Chwa, L.Wu, Z.Chen, Key Eng.Mat. **312** (2006) 339, (a) Table 2, (b) Table 1

6 Fibers and Fabrics

6.1 Introduction

6.1.1 Overview

Fibers (L: *fibra*) are flexible entities with large aspect ratios, i.e., ratios of length to diameter (see Section 6.1.3). They are usually subdivided according to their origin into **natural fibers** and **man-made fibers**. Most fibers consist of linear polymer molecules with high molecular weights (examples: cotton, polyester fibers) whereas others are composed of oligomers (example: glass fibers) or crosslinked polymer molecules (examples: wool, spandex fibers, phenolic resin fibers). Fibers may be therefore thermoplastics, thermosets, elastomers, or thermoplastic elastomers. Many fibers are semi-crystalline, especially those for textile and industrial applications but there are also non-crystalline ones (example: poly(styrene)).

Natural Fibers

Natural fibers may be of animal, plant, or mineral origin. They are used directly (example: some cotton types) or after clean-up operations (example: wool). All presently used animal fibers consist of protein molecules (wool, silk), all plant fibers of celluloses (cotton, flax, hemp, sisal, etc.). The best known natural mineral fiber is asbestos.

Animal fibers are subdivided into secretion fibers and hair fibers. The main representatives of **secretion fibers** are **true silks** which are secretion products of worms and spiders (Section 6.3.2). Artificial silks are man-made polymers that have properties resembling those of natural silks, especially strength and gloss. Protein-based silks have also been obtained from genetically engineered bacteria; these silks are presently at the laboratory stage.

Hair fibers are from animals or plants. **Wools** are hair fibers that can be processed to textiles; examples are the hairs of sheep, sheep camels (alpaca, guanaco, llama, vicuña), camel, goats (cashmere, mohair), and *Leporidae* (hare, rabbit, Angora rabbit). The hairs of beavers, otters, horses, humans, etc., are not converted to textiles. Glass wool, rock wool, and steel wool are socalled because they are fleecy like a wool fleece. They are not true wools because they are not processed to textiles.

Vegetal fibers (**vegetable fibers, plant fibers** (UK)) are subdivided in stem fibers, hard fibers, and seed hairs (Section 6.6.2). **Stem fibers** (also called **bast fibers**) are obtained from stems of plants; examples are hemp, jute, flax, and ramie. **Leaf fibers** are gotten from leaves or fruits, for example, from abaca, henquen, and sisal. The most important **seed hair** is cotton (Section 6.3.3); other seed hairs are coir (coconut fiber) and kapok.

Natural fibers are known by their trivial names (silk, wool, cotton, flax, etc.), and, more precisely, by the name of the animals or plants from which they are obtained (Tussah silk, sheep wool, etc.). In international trade, they are also known by two-letter symbols. Unfortunately, these symbols are not uniform: for example, silk from the domesticated moth *Bombyx mori* carries the symbol SE according to the European law for textile characterization but the symbol Ms according to DIN (German standards) (for a list of symbols, see Appendix).

Man-made Fibers

Man-made fibers are also called **manufactured fibers**. They are subdivided according to their composition into inorganic and organic fibers, with respect to their origin into synthetic and semi-synthetic fibers, and after their use into textile and industrial fibers.

By composition: Most man-made fibers are **organic fibers** such as polyesters, poly-amides, and rayon. **Inorganic fibers** comprise carbon, glass, ceramic, and metal fibers (whole metal only: metal-coated organic fibers are "metallized fibers").

By synthesis: Fully **synthetic fibers** are obtained by polymerization of monomers; examples: polyester and nylon fibers. **Semi-synthetic fibers** are based on natural fibers; an example is the esterification of cellulose to cellulose acetate. A subgroup of semi-synthetic fibers is **regenerated fibers** that result from processes in which natural fibers are converted chemically to intermediate products from which the original chemical composition was regenerated (L: *regeneratus* = born again). An example is rayon (viscose).

Commercially traded man-made fibers carry protected trade names such as Perlon®, Diolen®, Trevira®, Orlon®, Lycra®, etc. These fibers belong to any of the classes of fibers that have code names which are based on chemical names of classes of polymers (Table 6-1, Table 16-8) but in reality refer only to specific subclasses and not to general classes. An example is "polyamide fiber" which does not refer to the general class of fibers from polymers with amide groups (aliphatic, aromatic, cycloaromatic) but only to fibers from certain aliphatic polyamides. These fibers are also known as "nylons" in the United States (nylon was a registered trade name but is now a free name). There are also other differences: "spandex" in the United States is called "elastane" in Europe (formerly: elasthane) and US "rayon" is European "viscose" (see Table 6-1 and also Table 16-8).

Differences also exist between various two-, three-, and four-letter codes for man-made fibers (Table 6-1). This book uses codes and generic names as defined by ISO 2076 and BISFA (*Bureau International pour la Standardisation des Fibres Artificielles* = International Bureau for the Standardization of Man-Made Fibers). These codes and names differ in part from codes and names of other fibers by other organizations and also in part from those for plastics (see Appendix).

Use of Fibers

Fibers are used for textile purposes (Sections 6.3-6.5), in industrial applications (Section 6.6), and in the paper industry (Section 6.7); they are also the main component of true leathers (Section 6.8). In industry, **textile** refers to cloth or fabric *and* the fiber or yarn for weaving or knitting into a fabric whereas in everyday use it is usually associated with clothing and apparel. Hence, the term "textile" is an umbrella term for textile fibers as well as for semi-finished textiles (threads, yarns, fleeces, etc.), textiles (cloths, fabrics, etc.), and textile products (dresses, garments, home textiles, carpets, mats, filter cloths, etc.) that are derived from textile fibers.

Textile fibers are defined as fibers that can be woven or knitted to textiles (L: *textus* = woven thing). They comprise natural and synthetic products as well as organic and inorganic ones (Section 6.3). Examples are cotton, wool, silk, polyester, nylon, acrylics, olefin, and glass. **Technical fibers (industrial fibers)** are for non-textile purposes; they include natural, man-made organic, and inorganic fibers (Section 6.6). Examples are sisal, jute, olefin, aramid, carbon, graphite, aluminum oxide, and boron carbide.

Table 6-1 Synthetic organic fibers: BISFA codes, BISFA and FTC names (*in italics*), and main chemical structure elements (see also Appendix, Table 16.8). FTC = US Federal Trade Commission.

Code	BISFA name	FTC name	Polymers or monomeric units	wt% of units BISFA	FTC
Synthetic organic fibers					
-	-	*anidex*	$-CH_2-CH(COOR)-$ (homo or co)		>50
AR	aramid	*aramid*	aromatic amide units (may vary)	>85	>85
CLF	chlorofiber	*vinyon*	$-CH_2-CHCl-$	>50	>85
		saran	$-CH_2-CCl_2-$	>50	>80
			$-CH_2-CCl_2-$ (+ <35 wt% AN)	>65	
ED	elastodiene	*rubber*	crosslinked natural or synthetic poly-(isoprene) *or* crosslinked other diene (co)polymers with or without one or more vinyl monomers		
-	-	*lastrile*	copolymer of acrylonitrile (and a diene such as butadiene)	1	10-50
-	-	*rubber*	$-CH_2-CCl=CH-CH_2-$ (homo or co)	>35	>35
EL	elastane	*spandex*	segmented polyurethane units	>85	>85
EME	elastomultiester	*elastoester*	see Table 16-8 (FTC only)		
MAC	modacrylic	*modacrylic*	$-CH_2-CH(CN)-$ (+ $-CH_2-CHCl-$ or $-CH_2-CCl_2-$)	50-85	35-85
MF	melamine	*melamine*	melamine + formaldehyde monomers	>50	
PA	polyamide	*nylon*	aliphatic or cycloaliphatic amide units	>85	>85
PAN	acrylic	*acrylic*	$-CH_2-CH(CN)-$	>85	>85
-	-	-	polybenzimidazole		
PE	polyethylene	*olefin*	$-CH_2-CH_2-$ (FTC: also PP, etc.)	100	>85
-	-	-	poly(ethylene-2,6-naphthalate)		
PES	polyester	*polyester*	$-CO-(1,4-C_6H_4)-CO-O(CH_2CH_2)O-$	>85	>85
PI	polyimide	-	aromatic imide units		
PLA	polylactide	-	$-O-CH(CH_3)-CO-$	100	
PP	polypropylene	*olefin*	$-CH_2-CH(CH_3)-$, isotactic	100	>85
-	-	*sulfar*	$-S-(1,4-C_6H_4)-$		>85
PTFE	fluorofiber	*fluoropolym.*	aliphatic fluorocarbon monomers		
PVAL	vinylal	*vinal*	acetalized $-CH_2-CH(OH)-$ (* non-acetalized and acetalized units)		>85*
-	-	*novoloid*	crosslinked novolac		>85
-	-	*nytril*	$-CH_2-C(CN)_2-$		>85
Semi-synthetic organic fibers					
ALG	alginate	-	metal salts of alginic acid		
CA	acetate	*acetate*	acetylated cellulose units	74-92	<92
CLY	lyocell	*rayon*	regenerated cellulose (from org. solvents)	100	
CMD	modal	*rayon*	regenerated cellulose (high strength)	100	
CTA	triacetate	*triacetate*	acetylated cellulose units	>92	>92
CUP	cupro	*cupra*	regenerated cellulose (cupro process)	100	
CV	viscose	*rayon*	regenerated cellulose (viscose process) (* if <15 wt% substituted)	100	100*
-	-	*azlon*	regenerated natural protein		
Synthetic inorganic fibers					
CF	carbon	carbon	from thermal carbonization of organic fiber precursors	>90	
CEF	ceramic	-	textile fiber from ceramic materials		
GF	glass	*glass*	fiber from drawn molten glass		
MTF	metal	*metallic*	whole metal	100	
-	metallized	*metallic*	fibers coated with metals		

The textile industry has changed dramatically during the last 50 years. In the past, the chemical composition and physical structure of polymers dictated their uses: wool for warmth, cotton for water absorption, and silk for appearance. However, the modern textile industry can overcome many of these properties by changing the shape and surface structure of fibers as well as the arrangement of fibers in textiles and the combination of textiles to textile products. For example, polyester fabrics can be made to resemble silk, cotton, or wool. It is for this reason that the major subsections of this chapter are not arranged according to the chemical synthesis and structure of the fiber materials (see Volume II of this series) but with respect to their applications. For textile fibers, this comprises the main uses as silk(-like), wool(-like), or cotton(-like) fibers.

6.1.2 History and Production

The wool of wild animals is probably the oldest fiber used by man, since impressions of strings have been found on 35 000 year old string ceramics. Strings must also have been used very early to stitch together hides and skins because of the discovery of very old sewing needles from ivory. A ca. 20 000 year old cave painting shows ropes as means to harvest wild honey from bees' nests in cliffs. Sheep and goats were domesticated for milk, meat, skins, and fibers ca. 9000 BC. About 8000 BC, lake dwellers possessed woolen fabrics. The Babylonians owned woolen clothing about 4000 BC.

Cotton seems to have been known to the Egyptians since ca. 12 000 BC. Pile dwellers used flax and linen about 8000 BC. In Mexico, cotton fabrics were found that date back to 5700 BC. Looms were operated in Egypt in 4400 BC.

Hemp was utilized in Southeast Asia in 6000 BC and silk by the Chinese in 2640 BC. Asbestos was used for lamp wicks in 500 BC. In times of hardship many other plants were harvested for their fiber content, for example, stinging nettle.

Man-made fibers are much younger. The first man-made fibers seem to have been carbon fibers for electric light bulbs (1854 Heinrich Goebel; 1879 Sir Joseph Wilson Swan (dinitration of cellulose nitrate fibers); 1879 Thomas Alva Edison (commercial realization)). The spinning of cellulose nitrate to form fibers and the subsequent conversion of these fibers was discovered in 1884 by Hilaire Bernigaud Comte de Chardonnay de Grange who searched for a substitute for natural silk because the French silk industry was threatened by a disease of silk worms (mulberry spinners disease). In 1892, the xanthogenate process for viscose fibers (US: rayon) was invented by the Englishmen Charles Frederic Cross, Edward John Bevan, and Clayton Beadle.

The first fully synthetic fibers were spun in 1931 as PC fibers from poly(vinyl chloride) by IG Farbenindustrie, Germany; production ceased after several years. The first successful textile fiber, poly(hexamethylene adipamide) = PA 6.6, was invented by the American Wallace Hume Carothers in 1935 and produced since 1939 by DuPont de Nemours (USA) as Nylon® (nylon is now a free name). In the same year, IG Farbenindustrie began to produce poly(ε-caprolactam) = PA 6 (Perlon®) which was discovered by Paul Schlack in 1937.

The Carothers-DuPont patents were umbrella patents that covered the polycondensation of a very large number of dicarboxylic acids with either diamine or diols. However, these patents did not list terephthalic acid. Poly(ethylene terephthalate) was patented in

1946 by J.R.Whinfield and J.T.Dickson, Calico Printers, England. Imperial Chemical Industries PLC (ICI), United Kingdom, secured the rights to this invention and then started to cooperate with DuPont because it lacked experience in fiber spinning.

The third big synthetic fiber, poly(acrylonitrile), was prepared in 1893 by the Frenchman C.Moreau. It received little interest because it could neither be spun from the melt (degradation) nor from solution (no known solvent). In 1938, H.Rein of IG Farbenindustrie succeeded in spinning fibers from poly(acrylonitrile) in aqueous solutions of quaternary ammonium compounds or metal salts such as LiBr. The breakthrough came in 1942 as H.Rein in Germany and R.C.Latham and R.C.Houtz (DuPont) in the US simultaneously (April and June!) patented *N,N*-dimethylformamide as a solvent for PAN.

The arrival of synthetic fibers changed the textile landscape considerably (Fig. 6-1). During the last 100 years, the production of cotton increased approximately in unison with the increase in population, especially since 1950, that of wool grew more slowly until 1990 and then fell, and that of natural silk gyrated wildly. The increasing world demand for textile fibers was first satisfied with rayon (viscose) but rayon production stagnated ca. 1965 and then fell, in part because of competition by synthetic fibers, and in part because of environmental concerns with respect to their synthesis. The production of synthetic fibers, on the other hand, continues to increase unabated although the growth rate (logarithmic scale in Fig. 6-1!) decreased somewhat since 1975.

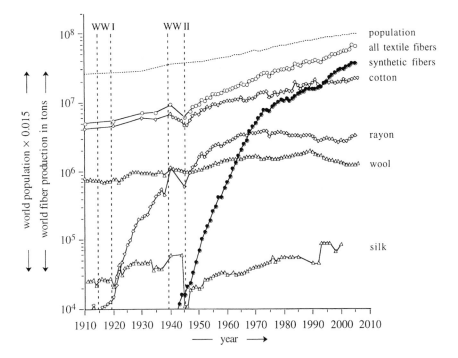

Fig. 6-1 World population (times 0.015) and world production of textile fibers [1a, 2-5]. "All textile fibers" does not include flax (for linen), ramie (China grass), and other polysaccharide fibers.

Cotton data were recalculated from reported shipments of bales; they represent seed cotton (Section 6.3.3). Wool data refer to raw wool (wool and wool fat); the conversion is weight of raw wool = 1.75·(weight of degreased wool). It is unclear whether the silk production data refer to raw silk, debasted silk, or charged silk (see Section 6.3.2). WW I and WW II = World Wars I and II.

Table 6-2 World production of textile fibers m_{tex}, world population N_{pop}, and world consumption of textile fibers per capita [1-5].

Year	1910	1920	1930	1940	1950	1960	1970	1980	1990	2000	2006
m_{tex} in 10^9 kg	5.01	5.46	7.08	9.71	9.38	14.93	21.84	29.63	29.44	53.33	66.00
N_{pop} in 10^9	1.75	1.81	1.91	2.35	2.55	2.96	3.70	4.41	5.25	6.09	6.61
kg per capita	2.90	3.01	3.71	4.13	3.67	5.05	5.91	6.71	7.52	8.76	10.00

The world-wide annual consumption of textile fibers per capita is now more than three times as large as it was 100 years ago. In 2006, it averaged 10 kg per capita (Table 6-2) with wide variations from country to country. The consumption of textiles is especially high in the United States: according to the U.S. Council for Textile Recycling, each U.S. person annually throws away about 31 kg of textiles of which ca. 4.5 kg are recycled and mainly sold as textiles to Africa and some Asian countries.

Globally, the production of cotton and synthetic fibers will continue to increase (Fig. 6-1). For individual countries and regions, the outlook is quite different (Fig. 6-2). Europe (France, Germany, United Kingdom) and the United States were the pioneers in developing man-made fibers but have fared quite differently during the last 1-2 decades. In Europe, annual productions changed relatively little (some noticeable decreases for PA and PES). The United States, on the other hand, experienced strong declines of acrylics and cellulosics since 1981 and polyesters since 2000, a moderate decline of nylons since 2000, and a stagnation of polyesters in 2000-2005.

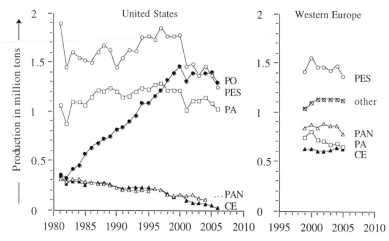

Fig. 6-2 Annual production of man-made textile fibers in the United States [6] and in Western Europe [1b]. CE = regenerated cellulosics, PA = polyamide (*nylon*), PAN = acrylics, PES = polyester, PO = polyolefins, other (not specified). United States production excludes cellulose acetates for cigarette filters; it is unclear whether cigarette filters are included in the data for Western Europe. Apparently, PES data refer just to the production of textile fibers exclusive of the production of bottle-grade and film-grade poly(ethylene terephthalate) and related aromatic polyesters.
   [1b] does not specify "Western Europe" which includes different countries in the 2002 and 2007 reports. Data for "Europe" by [6] are not comparable: older data are not specified (European Union only?) whereas newer data include the enlarged European Union (since 2003), Russia and the Confederation of Independent States (since 2005), and Turkey (2003-2004 but not 2005 ff.).

Table 6-3 2006 global production of polyester fibers (PES), polyamides (nylons) (PA), acrylics (PAN), and cellulosics (CE) without acetate tow for cigarette filters [7]. Production data for polyolefins and specialties were not reported by this source or by any other source. Data in thousands of tons.

Country or Region	PES	PA	PAN	CE	Total	% of Total
China	16 046	852	839	1 435	19 172	51.8
Western Europe	1 155	544	735	449	2 883	7.8
ASEAN	2 162	129	81	330	2 702	7.3
Taiwan	1 798	415	149	132	2 494	6.7
India	1 959	89	102	310	2 460	6.6
USA	1 253	1 023	4	27	2 307	6.2
South Korea	1 250	165	48	6	1 469	4.0
Japan	483	123	243	66	915	2.5
Other	1 632	561	313	107	2 613	7.1
World Total	27 738	3 901	2 514	2 862	37 015	100

These strong downturns in man-made fiber production were caused by the increased competition of Asian countries, first Japan, then South Korea, and now many others, especially China. However, the Japanese man-made fiber production peaked in 1997 at 1.82 million tons and is now down to 1.21 million tons, including polyolefins (2006). Most dramatic is the progress of China's man-made fiber industry: China is now (2007) the country with the largest synthetic fiber industry (ca. 52 % of the global production in 2006, exclusive of polyolefins) (Table 6-3). From 2000 to 2005, its polyester capacity rose from $5.95 \cdot 10^6$ t in 2000 to $20.6 \cdot 10^6$ t in 2005 and its production from $5.3 \cdot 10^6$ t (2000) to $13.9 \cdot 10^6$ t (2005) and $16.0 \cdot 10^6$ (2006). The rapid expansion had its price, though: capacity utilization *decreased* from 89 % in 2000 to 67 % in 2005.

Prognoses about the further development of the man-made fiber industry are difficult since the situation changes from fiber type to fiber type and from region to region as shown by Fig. 6-2. Statistics are hard to find in the open literature, and data, if published at all, are not easy to reconcile since terms are usually not defined. For example, "cotton" may refer to "seed cotton", "lint plus linters", or "raw cotton" (Section 6.3.3); "wool" to "raw wool" or "degreased wool" (Section 6.3.4); and "silk" to "raw silk", "debasted silk", or "charged silk" (Section 6.3.2). "Regenerated cellulosics" may just denote "rayon" ("viscose") or include cellulose acetates (2 1/2 acetates and/or triacetates), and if, either include or exclude cellulose acetates used for cigarette tow (Section 6.4.1), a major market in Western countries. "Polyester" may or may not include other aromatic polyester fibers besides poly(ethylene terephthalate) and may or may not exclude "film fiber" and "film tape" (Section 6.6.1), etc.

The global production for **natural textile fibers** is dominated by cotton ($25.1 \cdot 10^6$ t/a in 2006 and increasing) whereas the production of wool is considerably less ($1.2 \cdot 10^6$ t/a in 2006 and falling) (Fig. 6-1)). Natural silk had a remarkable comeback (133 000 t/a in 2006); it is now mainly produced in China.

The world production of **regenerated celluloses** for textile purposes (rayon = viscose, modal, cupro, lyocell, acetate, triacetate) is about constant ($2.9 \cdot 10^6$ t/a in 2006) (Fig. 6-1) but is falling in the United States (Fig. 6-2). Other regenerated natural fibers such as the polysaccharide fiber alginate or the protein fiber azlon are apparently no longer produced or produced only in very small quantities.

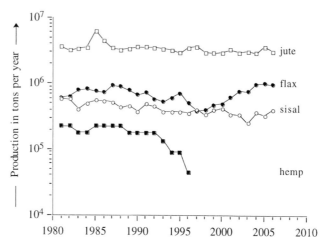

Fig. 6-3 World production of jute and jute-like fibers [8], flax fiber and tow [8], hemp [2], and sisal and other agave fibers [8].

Several other natural polysaccharide fibers serve for textile purposes. Ramie from China grass is often used to stiffen cotton fabrics. Flax is used for linen.

All of these fibers belong to the group of **vegetable fibers (vegetal fibers)** with the subgroups of **seed fibers** (cotton, kapok, etc.), **bast fibers** (jute, flax, hemp, etc.), **leaf fibers** (sisal, manila hemp, etc.), and miscellaneous (Section 6.6.2).

A great many of the vegetable fibers do not serve in textiles but are used as **industrial fibers** for ropes, filter cloth, etc., especially those with large moduli of elasticity (**hard fibers**). Of other major fibers in this group, global productions of jute and sisal decreased slightly and hemp strongly whereas flax had a remarkable comeback (Fig. 6-3). These fibers are grown in many countries, but in each group only one country dominates the market (Table 6-4).

The 2006 world production of **synthetic textile fibers** is governed by four types: polyester ($27.7 \cdot 10^6$ t/a), polyolefin (not reported, probably ca. $6 \cdot 10^6$ t/a), polyamide ($3.9 \cdot 10^6$ t/a), and acrylics ($2.5 \cdot 10^6$ t/a). Glass fibers (US: fiberglass) are mainly used for insulation and as reinforcing fibers in plastics although some are used for home textile such as curtains (world production not reported, probably $2 \cdot 10^6$ t/a) as as wave guides (p. 618). Production figures for the new polylactide fiber were not disclosed.

Table 6-4 2006 percentages of the annual world production of cotton ($24.8 \cdot 10^6$ t), jute ($3.1 \cdot 10^6$ t), flax ($0.98 \cdot 10^6$ t), and sisal ($0.43 \cdot 10^6$ t) by the five largest producer countries in each category [8].

Cotton lint		Flax fiber and tow		Jute and jute-like		Sisal and other agave fibers	
China	27.1 %	China	73.5 %	India	65.6 %	Brazil	57.9 %
United States	18.1 %	France	9.2 %	Bangladesh	25.7 %	Tanzania	6.5 %
India	14.3 %	Russ. Federation	3.7 %	China	2.8 %	Mexico	6.2 %
Pakistan	8.8 %	Belarus	3.0 %	Côte d'Ivoire	1.3 %	Kenya	5.8 %
Brazil	4.9 %	United Kingdom	2.9 %	Thailand	1.0 %	Colombia	5.0 %
79 others	26.8 %	17 others	7.7 %	14 others	13.6 %	23 others	18.6 %

Elastic textile fibers include *rubber fibers* (elastodiene fibers), *spandex fibers* (elast(h)ane fibers), and elastomultiester fibers (see Table 6-1).

Regional specialties contributed ca. 4 % to the world synthetic fiber market. A few serve as textile fibers such as acrylonitrile-vinyl(idene) chloride (*modacrylic*), homo or copolymers of acrylic esters (*anidex*), and acetalized poly(vinyl alcohol) (*vinylal, vinal*) but most of them are **technical fibers** (Section 6.6). Examples are aramids, chlorofibers (*vinyon, saran*), fluorofibers, poly(ethylene-2,6-naphthalate) (PEN), polybenzimidazole (PBI), poly(1,4-phenylene sulfide) (*sulfar*, PPS), polyimides (PI), and the thermoset fibers based on melamine or novolacs (*novoloid*). Poly(vinylidene cyanide) fibers (*nytril*) and the regenerated protein fiber *azlon* are no longer produced.

6.1.3 Designations of Fibers

Dimensions

Fibers are "one-dimensional" entities that have much larger lengths L than diameters d (Table 6-5), i.e., large **aspect ratios**, $\Lambda = L/d$. In order to spin fibers into yarn or thread, fibers must have aspect ratios of at least 800. Bagasse (residue from sugar cane), kapok, etc., can therefore not be converted to yarns or threads. Very short fibers are often used to make paper (Section 6.9) or fleece (Section 6.7).

Spinning has two meanings. Traditionally, it means the twisting of several fibers into a yarn. In the synthetic fiber industry, it refers to the production of fibers from polymer melts, solutions, dispersions, or gels that are extruded through spinnerets, i.e., plates with (usually) very many holes.

Textile fibers usually have diameters of 10-50 μm. Average diameters of such fibers are difficult to determine, even under the microscope, since fiber diameters may vary not only from fiber to fiber but also within one individual fiber. Hence, thicknesses of fibers are not characterized by diameters but by the ratio of fiber mass m and fiber length L (or density ρ and cross-sectional area A), the **linear density** or **fineness**, $f = m/L = \rho/A$, which is measured in **tex** = g/(1000 m). The United States still uses the older unit **denier**, 1 den = g/(9000 m) (from L: *denarius* (a small silver coin); from *deni* = by tens).

Table 6-5 Densities ρ, lengths L, diameters d, aspect ratios Λ, and finenesses f of dry natural fibers (rayon and synthetic fibers for comparison). All data are averages or guidelines. [1] Ultimates (p. 217).

Fiber	$\rho/(\text{g cm}^{-3})$	L/cm	d/μm	Λ	f/dtex
Silk (from *Bombyx mori*)	1.34	<120 000	8 - 15	$100\cdot10^6$	1
Wool (sheep)	1.30	<13	20	6500	4
Ramie [1]	1.50	2.5 - 25	16 - 126	2500	24
Cotton	<1.54	2.7 - 7.5	20	1400	4
Flax [1]	1.50	0.44 - 7.7	5 - 76	1100	6
Kapok [1]		1.5 - 3.0	10 - 30	1100	
Hemp [1]	1.48	0.5 - 5.5	10 - 51	1000	7
Sisal [1]		0.08 - 0.75	7 - 47	150	9
Jute [1]	1.50	0.07 - 0.7	15 - 25	200	2
Coir (from coconuts) [1]	1.15	0.03 - 0.1	12 - 24	40	
Abaca [1]		0.3 - 1.2	6 - 46	300	
Rayon (as staple fiber)	1.52	3 - 8	10 - 50	1800	11
Synthetic fibers (as staple fibers)	various	3 - 8	10 - 50	1800	various
Bagasse		0.1 - 0.4	40 - 100	35	60

The size of a "fiber" is not defined exactly. With respect to fineness, the textile field distinguishes **monofils** or **bristles** (>30 dtex) from **fibers** (1-30 dtex), and **microfibers** (0.1-1 dtex). Fibers with less than 0.3 dtex are sometimes called **supermicrofibers** and those with <0.1 dtex, **ultrafibers**, **superfine fibers**, **fibrils**, or **whiskers** (if they are single crystals). **Nanofibers** have finenesses greater than 10^{-7} dtex. However, the "microfibers" of the plastics industry are very short fibers of 0.15-0.3 mm length and 3-5 μm diameter which would make them **fine fibers** (1-2.4 dtex) in the textile industry.

Filaments and Staple Fibers

All synthetic fibers exit spinnerets as "endless" fibers, called **filaments** by the synthetic fiber industry (L: *filum* = thread). A spinneret with just one hole produces **monofilaments (monofils)** whereas spinnerets with up to 80 000 holes lead to **multifilaments**. Filaments can be processed directly to spunbonded sheet products (Section 6.4.3) but are mainly united parallel and without twist to **filament tows** that are then spun to yarns.

Staple fibers are obtained by combining numerous filament bundles to thick filament tows that are then crimped and cut. Staple fibers are sometimes subdivided into short ones (1.25-3.8 cm for cotton and cotton-type fibers) and long ones (>7.6 cm for wool and wool-type fibers). Staple fibers of regular length are then spun to yarns (Section 6.4.1) whereas very short ones of 0.05-1.5 cm length are combined to a loose fiber material, the **flock** (flake), which is used to make paper, nonwovens, and the like.

A third type of fiber is **film fiber** (Section 6.6.3) from the lengthwise division (cutting, splitting, fibrillating, etc.) of poly(propylene) or polyester films (Table 6-6). Film fibers are not used for textiles, but for strings, ropes, carpet backing, and the like.

The words "fiber", "filament", "fibril", etc., have different meanings in different fields. In the material world, "fiber" commonly denotes an elongated, slim structure except in nutrition science where "fiber" refers to indigestibles which may be soluble or insoluble and are often not fibers at all.

The use of the word "filament" for endless synthetic fibers conflicts with the much older use of the same word by botany where it refers to fibers with 5-50 nm diameter regardless of their length. Filaments of synthetic fibers have much larger diameters of 10 000-50 000 nm (Table 6-5). The "filament" of the electrical industry is a thread of metal without any specified size. For fishing lines, "monofilament" means a string made from a single type of fiber. In protein chemistry, a "microfilament" is a very thin fiber (the microfilament of the protein actin has a diameter of 7 nm).

"Fibrils" are generally very slim fibers (L: *fibralla* = diminutive of *fibra* = fiber). In botany, "fibrils" have diameters of 100-800 nm whereas the "fibrils" of protein chemistry have diameters of 1 nm.

The use of fibers is not only controlled by their length but also by their fineness (Fig. 6-4): finer ones for woven and knitted textiles (clothing, home textiles, etc.), thicker ones for industrial uses (ropes, tire cords, filter cloths, plastics reinforcement, etc.). Each use demands other properties such as appearance, comfort, washability, strength, flammability, etc. (Section 6.4).

Table 6-6 Percentage of world production of staple and filament type man-made fibers in 2005/2006.

Fiber type	Acrylics	Cellulosics	Polyesters	Polyamides	Polyolefins
Staple fiber	100.0 %	86 %	41 %	20 %	22 %
Filament	0 %	14 %	59 %	80 %	43 %
Film fiber	0 %	0 %	0 %	0 %	35 %

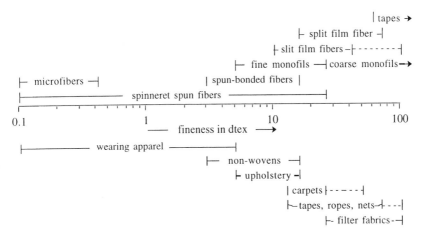

Fig. 6-4 Uses of fibers according to fineness and type of manufacture [9].

6.2 Manufacture of Fibers and Threads

6.2.1 Overview

The ability to form long thin threads and fibers is not an exclusive property of macromolecular compounds since long threads cannot only be pulled out of melts and concentrated solutions of polymers but also out of honey (aqueous solution of low-molecular weight sugars) and aqueous soap solutions (from low-molecular weight surfactants). Threads from concentrated sugar solutions may even be spun to cotton candy (spun sugar, cotton fluff, candy floss).

In these processes, polymer chains and the rod-like molecular associates of sugars and soaps orient themselves in the draw or flow direction and form lateral physical bonds. In low-molecular weight substances, the number of these physical bonds per molecule is small and the mechanical strength of the fiber is correspondingly low. The higher the degree of polymerization, the more physical bonds can be formed and the higher the mechanical strength. For polymer chains, noticeable strengths are observed only above a certain minimum degree of polymerization X (see Fig. 5-7) which is $X \approx 50$ for polyamide 6 with its strong intermolecular hydrogen bonds but $X \approx 600$ for atactic poly(styrene) with dispersion forces between molecules. Sufficient mechanical strengths require much higher degrees of polymerization such as $X \approx 120\text{-}150$ for polyamide 6 (a nylon fiber from poly(ε-caprolactam)), $X \approx 80\text{-}110$ for polyester fibers (PES) from poly(ethylene terephthalate) (PET), $X \approx 125\text{-}190$ for PET bottles, $X \approx 210\text{-}260$ for PET technical applications (Volume II, Table 7-12), and $X \approx 1500$ for acrylic fibers (from poly(acrylonitrile). Ultradrawn fibers from ultra-high molecular weight poly(ethylene) ($X \approx 125\,000$) have mechanical strengths that surpass even that of steel.

In general, polymers for textile fibers have lower degrees of polymerization than those for industrial fibers of the same polymer constitution. The type of spinning process depends on the polymer properties (thermal decomposition, solubility, dispersibility) and the desired fiber properties.

6.2.2   Fiber Spinning

Polymer synthesis and fiber formation are simultaneous processes in natural fibers such as silk, wool, and cotton but separate processes for synthetic fibers where the polymer is usually isolated before fiber spinning. There is only one known process (albeit a non-industrial one) where fibers are formed spontaneously during polymerization: the formation of aramid fibers from *p*-phenylenediamine and terephthaloyldichloride in *N,N*-dimethylacetamide in the presence of pyridine and LiCl at room temperature. Usually, the polymer just precipitates but under special conditions, such as very intense stirring, a gel is formed. After a few hours, the molecular weight increases from 15 000 to 57 000 and fibers precipitate. The fibers orient themselves lengthwise during stirring but the fiber cross-sections are not circular but ellipsoidal.

Practically all man-made fibers are spun from polymers or prepolymers. Spin methods are usually subdivided according to the formation of filaments and/or the type of wind-up. Filaments are formed from melts, solutions (dry or wet), gels, dispersions, or by reaction spinning. According to the wind-up of filaments, one distinguishes bobbin spinning, centrifugal spinning, cylinder spinning, funnel spinning, pot spinning, reel spinning, ribbon spinning, and tow spinning.

Reaction Spinning

In principle, reaction spinning (**polymerization spinning**) is the most economical process for the production of fibers since it avoids the costly separation and remelting/dissolution of polymers before fiber spinning. For reaction spinning, monomers together with initiators or catalysts, fillers, pigments, flame retardants, etc., are polymerized and the resulting polymer immediately spun at rates of up to 400 m/min. However, reaction spinning is suitable only for very fast polymerizing monomers and is therefore used only for the preparation of elastane (spandex) fibers.

Melt Spinning

Spinning fibers from melts is another economical method for most man-made fibers since there is no solvent that has to be recovered and worked-up. Melt spinning also allows spinnerets with more nozzles per area which increases the capacity of the subsequent processing train (winders, etc.).

For melt spinning, polymer pellets are molten under a nitrogen blanket. The melt is pumped through a filter and then through a spinneret with up to several hundreds of holes of 50-400 μm diameter (Fig. 6-5). The exiting filaments solidify in air or nitrogen (if prone to oxidation), are drawn with speeds of up to 4000 m/min, and are simultaneously conditioned (subjected to humidity, if necessary), smoothed, and stretched.

Stretching reduces the cross-section of the fiber by a factor of 20-100. The drawn filaments have diameters of 4-60 μm and finenesses of ca. 1.4-15 dtex.

Melt-spinnable are only those polymers which have (a) sufficiently broad processing windows between their melting and decomposition temperatures and (b) sufficiently low melt viscosities in this window. Such polymers are poly(ethylene)s with spinning temperatures of 225-230°C, it-poly(propylene) at 250-300°C, polyamide 6 at 270-280°C (and also polyamide 6.6 and some other aliphatic polyamides), poly(ethylene terephtha-

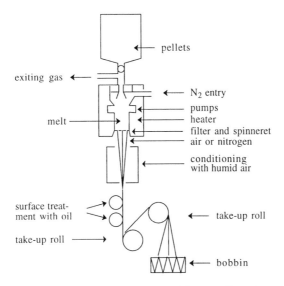

Fig. 6-5 Melt spinning of polyamides [10a]. Polymer chips or pellets from a hopper are molten under nitrogen and the melt is filtered and pressed through spinnerets. After exiting the spinneret, filaments are cooled by air or nitrogen and, in the case of polyamides, conditioned, i.e., brought to a specified relative humidity. The filament surfaces are smoothed with oil (= fiber finishing) to ease spooling, stretched, and finally wound on bobbins (F: *bobine* = roll).

late) at (280-300)°C, saran, sulfar, glass (see Section 6.6.7), basalt, and aluminum oxide (Section 6.6.8). Some polymers partially degrade under melt processing conditions, which results in monomers, oligomers, and other low-molecular weight decomposition products that are deposited on the machinery.

Dry Spinning

Dry spinning and wet spinning are used to spin fibers from polymer solutions, the former into air and the latter into a precipitating bath. These two spinning methods are used for polymers that would decompose at temperatures above their melting temperatures, if any. Dry spinning requires volatile solvents that can be easily recovered. The solvent recovery makes dry spinning more expensive than melt spinning.

Dry spinning uses polymer solutions with fairly high polymer concentrations of 15-35 wt%. On exiting the spinneret, filaments are highly swollen. The solvent is removed by blowing warm air or nitrogen counter to the direction of the moving filament (Fig. 6-6) whereupon the solvent evaporates and the filament solidifies. Since solvent evaporation is a diffusion-controlled process that proceeds from the exterior of the filament to the interior and the gas is blown in from one side only, the accompanying contraction leads to bean-like cross-sections of filaments, even if circular spinneret holes are used.

Spinning speeds are ca. 300-400 m/min, i.e., smaller than those for melt spinning but greater than those for wet spinning. Fast dry spinning processes reach speeds of up to 5000 m/min if the solvent in removed by vacuum or with very hot air. However, the resulting filaments are too uneven for textile purposes. Machinery for dry spinning is much more expensive than that for wet spinning; however, operating costs are significantly lower.

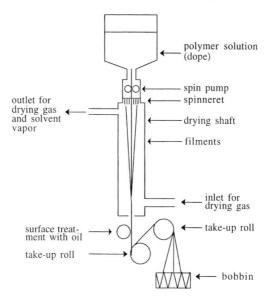

Fig. 6-6 Dry spinning [10b]. The dope is pressed through the spinneret into a shaft. The solvent of the resulting gel-like filaments is removed by warm air. The solidified filaments are further processed like melt-spun filaments.

Dry spinning is used for solutions of 25-32 wt% poly(acrylonitrile) in *N,N*-dimethyl-formamide, 25-30 wt% modacryl in DMF or acetone, 15-20 wt% aramids in DMF + 5 wt% LiCl, 20-22 wt% cellulose triacetate in CH_2Cl_2 + 5 wt% methanol, 35 wt% poly(vinyl chloride) in CS_2/acetone (40/60 = V/V), and 20-35 wt% elastane (spandex). Dry spinning is also used for cellulose 2 1/2-acetate, chlorofibers (vinyon), and poly-benzimidazole (PBI).

Wet Spinning

Wet spinning is used if polymers degrade on melting, dissolve only in a non-volatile solvent, or dissolve only after a chemical transformation. In this type of spinning process, concentrated polymer solutions are pumped through spinnerets into a bath with a precipitating liquid (Fig. 6-7). The liquid in the precipitating bath must be miscible with the polymer solvent in all proportions.

The exiting solvent-swollen, gel-like filaments coagulate by phase separation. This process is not very fast and the spinning speeds of wet-spinning processes are therefore much lower than those of dry spinning, usually only 50-100 m/min. Solvents and precipitating agents need to be recovered which makes wet spinning less economical than melt spinning and dry spinning.

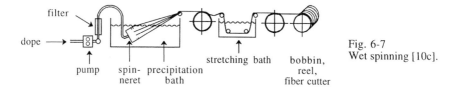

Fig. 6-7
Wet spinning [10c].

The classic wet spinning process is that of rayon (viscose) in which 7-10 wt% solutions of cellulose xanthogenate in dilute aqueous NaOH solutions (+ Na and Zn acetates) are spun into a coagulating bath of sulfuric acid + sodium acetate + zinc acetate) (Volume III, p. 399). Wet spinning is also used for 10-18 wt% aqueous solutions of poly(vinyl alcohol) into aqueous sodium sulfate solutions, 10-25 wt% solutions of poly(acrylonitrile) in DMF, NaSCN, DMA, and other solvents, 15-25 wt% acetone solutions of modacrylics into aqueous acetone, and 20 wt% sulfuric acid solutions of aramids into water.

Gel Spinning

Gel spinning (**gel extrusion spinning, extrusion spinning**) is a special variant of wet spinning which uses polymer "solutions" of much higher concentrations up to 80 wt%. The greater shape stability of the exiting gel-like filaments allows much higher spinning rates of up to 500 m/min. Gel spinning is used for 25-55 wt% solutions of poly(acrylonitrile) or poly(vinyl alcohol). It is of special interest for the dry spinning of high-modulus poly(ethylene) fibers from 1-2 wt% decalin solutions of ultra-high molecular weight polymers at 120-130°C. The exiting filaments are then hot drawn at temperatures of 70-150°C.

Dispersion Spinning

Dispersion spinning is a special process in which an insoluble and unmeltable polymer is dispersed in a molecular solution of another soluble polymer. The latter polymer serves to increase the viscosity of the polymer dispersion and to stabilize the exiting filaments.

An example is the fiber spinning of particles of poly(tetrafluoroethylene) that are dispersed in aqeous poly(vinyl alcohol) solutions. The exiting filaments are heated which evaporates the water. The poly(vinyl alcohol) is then burned off which sinters the poly(tetrafluoroethylene) particles to stable filaments. The process is also used to produce inorganic fibers from aluminum oxide, magnesium oxide, or calcium oxide.

In a variation of the process, the supporting polymer is not burned off but crosslinked. This process is used for the fiber spinning of poly(vinyl chloride) dispersions in aqueous poly(vinyl alcohol) solutions. After evaporation of the water, poly(vinyl alcohol) is crosslinked by formaldehyde.

Special Processes

Inorganic fibers can also be produced by a **soaking process** in which preformed fibers of organic polymers, for example, rayon, are soaked with solutions of inorganic salts. The soaked fibers are heated to high temperatures which causes the organic material to pyrolyse and the remaining inorganic material to sinter. The process is used for the manufacture of fibers from aluminum oxide and zirconium oxide.

Specialty fibers can also be obtained by **fiber transformation** in which a preformed fiber is converted to the desired fiber by pyrolysis or by reaction with other substances. These processes are used for carbon, graphite, boron nitride, and boron carbide fibers (Section 6.8.3).

6.2.3   Cross-sections of Fibers

Spinning processes lead to various types of cross-sections of filaments, depending on the type of orifices (holes) in spinnerets and the diffusion processes during the solidification of the exiting filaments (from L: $\bar{o}s$ (stem: *or*) = mouth; *facere* = to make).

In melt spinning, but not in other spinning processes, filaments from circular orifices are circular, too. Because of the Barus effect (p. 94), cross-sections of filaments from shear thinning polymer melts are larger than the cross-sections of the orifices (**onion formation**); they are smaller if filaments are spun from liquid crystalline polymers.

Dry spinning delivers kidney-shaped cross-sections because of temperature gradients in filaments during solidification (p. 162). Non-circular cross-sections of filaments can also be produced by varying the shape of the orifices. Because of the Barus effect, cross-sections of the resulting filaments differ from those of the orifices (Fig. 6-8).

Filaments with non-circular cross-sections are called **profile fibers**. Triangular orifices with concave lateral faces lead to filaments with concave ones and *vice versa*. Depending on the cross-section of the orifice, resulting the filaments may have trilobal, star-like, kidney-like, serrated, etc., cross-sections.

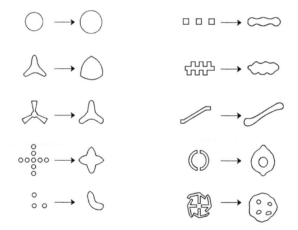

Fig. 6-8 Orifices of spinnerets and resulting cross-sections of filaments. Orifices may deliver circular filaments (top left) or profile filaments (all others) and hollow filaments (right, first and second from bottom) or compact ones (all others).

The various cross-sections lead to quite different fiber and textile properties. Profile fibers have larger surface areas than circular ones of the same cross-sectional area which leads to better dyeabilities. Textiles from profile fibers look better and have a better feel (see also Section 6.4). Triangular profile fibers have a better gloss (see Section 6.3.1), cross-shaped ones are more stiff than circular ones, etc.

Melt spinning, and less often also wet, dry, and dispersion spinning, can also produce **bicomponent fibers (bico fibers, composite fibers, heterofils)** that are also called **conjugated fibers, biconstituent fibers,** or **bilaminar fibers**. These fibers combine two different polymers in one filament, for example, two different polyamides or a poly(vinyl chloride) and a poly(vinyl alcohol). Bicomponent fibers differ from **mixed yarns** which consist of two or more types of filaments.

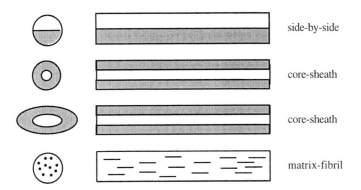

side-by-side

core-sheath

core-sheath

matrix-fibril

Fig. 6-9 Schematic representation of cross-sections (left) and longitudinal sections (right) of bicomponent fibers (top three) and matrix-fibril fibers (islands-in-a-sea) (bottom).

Bicomponent fibers are manufactured by feeding two melts or spin dopes with different compositions separately to the spinneret where they combine immediately before the spinneret mouth. In the filament, the two components may have side-by-side, core-sheath, or matrix-fibril structures (Fig. 6-9). In the United States, matrix-fibril fibers are not included in the class of bicomponent fibers.

Each subcategory of bicomponent fibers can have many variants. For example, side-by-side fibers can have compositions other than 50:50, core-sheath types may be not only circular but ellipsoidal, and/or off-center (not shown), and the outer shapes may not be circular or ellipsoidal (see Fig. 6-8).

Bicomponent fibers are part of the group of **synthetic fiber alloys** that also include bistructural fibers and composite yarns. In **bistructural fibers**, both components have the same chemical composition but different physical structures. **Composite yarns**, on the other hand, are mixtures of monofilaments with different chemical and/or physical structures (for yarns, see Section 6.4.1).

6.2.4 Spinnability

The liquid jets exiting from the spinneret produce filaments of length L that depend on the viscosity η of the liquid and the speed v of the spinning processes (Fig. 6-10). The filament length increases with the product $v\eta$, passes through a maximum and then decreases before the filament ruptures. The maximum of the function $L = \mathrm{f}(\lg v\eta)$ is called **spinnability** L_{max}.

The presence of maxima in the $L = \mathrm{f}(v\eta)$ curves for the filament producing jets that exit the spinneret indicates that at least two processes are operating in fiber spinning. Fluids suitable for fiber spinning are melts, concentrated polymer solutions, and polymer dispersions in polymer solutions. All these fluids are viscoelastic which means that they store a certain amount of elastic energy. The magnitude of this amount depends not only on the viscosity of the fluid and the speed of the spinning process but also on the elastic modulus of the fluid and the cohesion energy of the material.

The jet breaks when a critical amount of stored elastic energy is surpassed. This **cohesive failure** is a brittle fracture which severs the jet with a smooth break (Fig. 6-11).

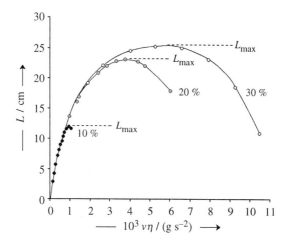

Fig. 6-10 Filament length L as a function of the product $v\eta$ of spinning speed v and fluid viscosity η in the spinning of various concentrations of a cellulose acetate in an 85/15 mixture of acetone and water [11]. Areas under the curves indicate the regions of spinnability.

However, the jet can also disintegrate by **capillary wave failure**. The latter type of fracture is a tough fracture which depends not only on the flow rate and the elastic modulus of the not yet solidified filament but also on the ratio of surface tension to the viscosity of the jet stream. Capillary wave fracture causes the fluid to separate into droplets.

Cohesive and capillary forces work simultaneously on the jet and the resulting filament, respectively. At low viscosities and small flow speeds, jets will disintegrate into droplets because of the dominating effect of surface tension, i.e., the filament fractures by capillary wave failure. High speeds of high viscosity liquids, on the other hand, cause cohesive failure because they lead to large relaxation times. High viscosities are caused by high polymer concentrations, high molecular weights, fast gel formations, and low temperatures.

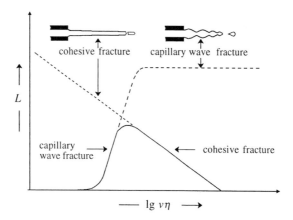

Fig. 6-11 Observed filament length L as a function of the logarithm of the product of spinning rate v and shear viscosity η (——) [11]. - - - Lengths L in the absence of capillary fracture, — — — lengths in the absence of cohesive fracture. The maximum filament length L_{max} results from the interplay of these two types of fracture processes.

6.2.5   Flow Processes

The production of filaments and fibers is a very sophisticated process. For example, in the melt spinning to filaments, the melt is transported by gear pumps at spinning pressures of ca. 5-40 MPa to the spin packs where it is filtered through metal or silicic filters and then homogenized. It is then transported to the spinneret which has 3-350 capillaries with diameters of 0.15-0.8 mm each and lengths of 1-5 times their diameter. The processing speeds of spinning processes range from 500-1800 m/min for undrawn filaments to 5500-7500 m/min for highly oriented ones.

All spinning processes proceed schematically in four steps. The first step is the extrusion of the melt or dope from the spinneret where the length of the resulting gel-like filament is controlled by the spinnability (previous section). In the second step, internal stresses are relaxed and the gel-like filament starts to solidify. In these two steps, the filament gains its shape.

The still semi-liquid filament then extends under its own weight which leads to a low level of orientation of the polymer segments. In the fourth step, the filament is lengthened by cold drawing. Steps three and four are part of the stretching process.

These four steps are not clearly separated; each step is also affected by various other phenomena which are mainly caused by the viscoelasticity of polymer melts and spinning dopes. For example, the passage of fluids from the broader prechannel to the very narrow capillaries of the spinneret gives rise to circular flow which may cause pigment particles to agglomerate. Since capillaries are short, Hagenbach-Couette type disturbances occur at the capillary entrances (Volume III, p. 396). In the capillary, melts or spinning dopes are deformed elastically which causes molecule segments to partially orient in the flow direction. A part of the deformation energy is stored as elasticity.

The axial flow *in* the capillaries is controlled by the non-Newtonian shear viscosity which decreases with increasing shear rate. At the capillary exit, both axial and radial flow is present. The axial flow is mainly controlled by the drawdown of the exiting filament whereas the radial flow is governed by the elongational viscosity.

The Barus effect (p. 94) causes the exiting strand to expand (onion formation) because the deformed polymer coils try to reestablish their equilibrium shapes. The expansion is usually 1.2-2.5 times the diameter of the capillary orifice but may be as large as 8 times. It is the greater, the larger the length/diameter ratio of the capillary, the greater the flow speed in the capillary, and the higher the drawdown speed.

The expansion zone is followed by a zone in which the still liquid strand is lengthened which causes a pre-orientation of molecule segments and, in semi-crystalline polymers, also some crystallization (see below). A transition zone with a strong viscosity increase due to the decreasing temperature is then followed by a zone in which the filament is lengthened by cold drawing.

6.2.6   Fiber Structure

Structure Formation During Fiber Spinning

The properties of fibers depend not only on the chemical structure of their polymers but also on the physical structures such as orientations of molecule segments and the proportions, sizes, and types of crystalline regions. These physical structures are controlled by the processes occurring during fiber spinning and fiber drawing.

During fiber spinning, molecule segments are oriented by three effects: flow orientations within and outside the capillaries and orientations caused by filament deformations. Flow orientations are effective for fiber orientations if the speed of solidification of the gel-like filament is larger than the inverse relaxation time of the polymer coils in the melt. Since relaxation times of coils are ca. $(0.1-1)\cdot10^{-3}$ s, solidification speeds should be greater than $(1-10)\cdot10^{3}$ s^{-1} which is only found for the surface of filaments but not for the interior. Hence, orientations of molecule segments in the capillaries does not much influence the orientation of segments in the filaments.

However, segments *are* oriented in the zone immediately below the capillary exit. In this zone, optical birefringence increases with increasing distance from the exit in an S-like manner. The limiting value is controlled by the solidification of the filament which reduces drastically the movement of molecule segments.

Hence, for conventional spinning rates, the largest proportion of segmental orientations is caused by flow processes in the liquid zone near the exit of the capillaries and a smaller one in the subsequent solid zone. Another small contribution comes from a third process: orientation of segments from a deformation of the generated physical network of crystalline regions.

Crystallinities of filaments vary widely. Fast solidifying, slowly crystallizing polymers practically do not crystallize if the spinning speed is fairly low; an example is melt-spun poly(ethylene terephthalate). The same polymer exhibits considerable crystallinities if the spinning speed exceeds 3500 m/min.

Slowly solidifying, fast crystallizing polymers usually have optimal crystallinity; examples are filaments from the wet spinning of poly(vinyl alcohol)s or rayon. Melt-spun aliphatic polyamides or it-poly(propylene)s have medium crystallinities.

The highest crystallinities are usually obtained from wet spinning since molecule segments here are mobile for long times because of the plasticizing action of the solvents. In melt spinning, on the other hand, melts are quickly cooled below the glass temperature of polymers, which drastically reduces the mobility of molecule segments and considerably lowers the degree of crystallinity of the polymer. During spinning, segments are already oriented in the flow direction. Spherulite formation is therefore practically absent in wet and dry spinning and very rare in melt spinning. Instead, molecules crystallize in lamellar forms (see Volume III, Section 7.2.1).

Stretching

On stretching the exiting filaments, lamellae are displaced from their original positions and reoriented anew so that more and more chain segments are now oriented parallel to the direction of stretching.

The resulting filaments of crystallizable, flexible chain molecules consist essentially of practically parallel microfibrils of more than 100 nm length and more than 20 nm diameter. Each microfibril is composed of lamellae with folded polymer chains that are separated by thinner amorphous layers (Fig. 6-12). The lamellae are interconnected by bundles of stretched polymer chains, so-called crystal bridges, that sometimes connect twenty or more lamellae and amorphous regions. These crystal bridges are responsible for the high mechanical strength in the filament direction.

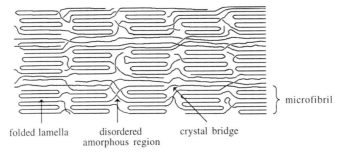

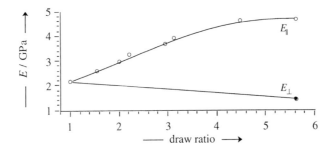

folded lamella   disordered   crystal bridge
                  amorphous region

Fig. 6-12 Structural model of stretched microcrystalline filaments.

In conventional fiber spinning, stretching orients the chain axes of folded lamellae in the draw direction until a limiting value is reached that is determined by the draw condition. As a result, tensile moduli $E_{\parallel}$ in the draw direction increase with increasing draw ratio and finally become constant whereas tensile moduli $E_{\perp}$ normal to the draw direction decrease (Fig. 6-13). Stretching also increases the strength at break and decreases the elongation at break (Table 6-7). In addition, fibers are often annealed and then set under stress, which changes the ratio of extended and folded chains and the resulting mechanical properties.

Fig. 6-13 Dependence of tensile moduli parallel ($E_{\parallel}$) and perpendicular ($E_{\perp}$) to the longitudinal filament axis on the draw ratio [12]. For fiber spinning, the draw ratio is given by the ratio of the surface speeds of the last and the first stretching roll, v/v_0.

Table 6-7 Tensile moduli $E_{\parallel}$ in the fiber direction, strength σ_B at break, and elongation ε_B at break as functions of the draw ratio v/v_0 (see text of Fig. 6-13).

Fiber	v/v_0	$E_{\parallel}$/MPa	σ_B/MPa	ε_B/%
Poly(oxymethylene)	1	3 700	70	45
	7	20 600	900	8
	22	38 400	2 000	5
Poly(propylene), isotactic	1	1 400	35	700
	8	10 700	680	6
	25	26 800	1 700	4
Poly(ε-caprolactam)	1	2 400		
	5.4	5 500		
Poly(styrene), atactic	1		35	
	6		110	

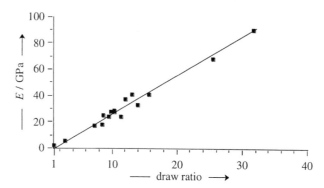

Fig. 6-14 Tensile modulus E in the draw direction of solution-spun/drawn filaments of a poly(ethylene) as a function of the draw ratio [13]. The high-molecular weight poly(ethylene)($\overline{M}_w$ = 1 500 000, $\overline{M}_n$ = 200 000) had a melting temperature of 145.5°C at a heating rate of 10 K/min.

Very high elastic moduli and tensile strengths are obtained if high-molecular weight poly(ethylene)s are spun from dilute solution and drawn (Fig. 6-14). The highest observed modulus is ca. 1/4 of the lattice modulus (354 GPa).

Completely different fiber properties are obtained if melt-spun poly(ethylene) fibers are first drawn to ratios of v/v_0 = 2.0-2.5, then relaxed to v/v_0 = 1.8-2.0, and finally set. This process does produce a certain proportion of crystal bridges but not all of these bridges are laterally oriented since some of them fold back to lamellae during the relaxation process. The final heat-setting then leads to a structure with amorphous regions and folded and extended polymer chains.

Such fibers do not have large tensile moduli and tensile strengths but they can be greatly extended before they break. On elongation of these **hard-elastic fibers,** lamellar regions are assumed to be deformed whereas crystal bridges are thought to remain in place (Fig. 6-15). Such a mechanism would increase the potential energy: the fibers will be energy-elastic and not entropy-elastic like rubber bands. According to other authors, it is not the spring-like deformation of crystalline lamellae that is important but the formation of microfibers by crazing.

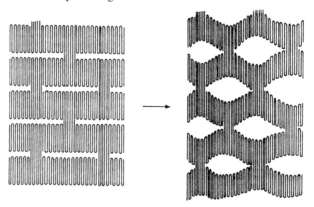

Fig. 6-15 Assumed deformation of a hard-elastic fiber: lamellar regions are deformed but cannot separate from each other because of the many crystal bridges. Removal of the tension causes lamellae to jump back to their original position.

Above a certain concentration, rigid chain molecules form lyotropic liquid crystals (Volume III, p. 244 ff.). The resulting lateral self-association of polymer segments remains in place after precipitation of the polymer which leads to very high tensile moduli and good breaking strengths of such fibers. These properties depend on the draw ratio, i.e., the ratio of fiber cross-sections after exiting the spinneret holes and fiber cross-sections after the first winding. The ratio is the larger, the greater the modulus and the smaller the elongation at break.

6.3 Textile Fibers

6.3.1 Introduction

Overview

The properties of the three classical natural fibers are still the yardstick to which man-made fibers are held: wool for warm apparel in colder regions, cotton for underwear and for garments in tropical areas, and silk as a luxury fiber because of its hand and look.

Textile properties depend only in part on the chemical and physical structure of the polymer molecules of these fibers. They also depend on the shape of the fibers, the combination of fibers in yarns, and the arrangement of yarns in wovens, knits, fleeces, and the like. The same chemical structure may therefore lead to many different textile properties. For example, polyester fibers can not only be processed to silk-like textiles but also to cotton-like and even wool-like ones.

It is therefore expedient to discuss fibers not according to their chemical structures but to their applications. Thus, this Section will treat textile fibers including spandex, Section 6.7 technical organic fibers including hard fibers and high performance fibers, Section 6.8 inorganic fibers, Section 6.9 paper and related products, and Section 6.10 leathers. For convenience, this Section on textile fibers will be followed by sections on yarns (Section 6.4), textiles (Section 6.5), and textile finishing (Section 6.6).

Textile applications of fibers depend mainly on three types of porperties:

- *Perception*: gloss, color, hiding power, luster, hand, drape, bulk, comfort, etc.;
- *Performance*: colorfastness, water repelling and take-up, static electrification, elasticity, resistance to deformation, creasing, pilling, abrasion, scouring, etc.;
- *Care*: washability, dry cleaning, drying, wrinkle resistance after laundering, ironing, pressing.

These properties depend in a complex manner on the chemical structures of polymers, additives, and finishes; on the physical structure of the polymers and the shape, texture, and combination of fibers; and the type, intensity, and duration of the demand. For example, cellulose fibers take up plenty of water but the same effect can also be obtained by making fibers microporous, even hydrophobic ones.

Perception and (in part) performance are controlled strongly by the cross-sectional shapes and surface structures of fibers (Fig. 6-16). *Silk fibers* (Section 6.3.2) have triangular to ellipsoidal cross-sections and a fairly smooth surface. Such properties cause the silky hand and the rustle of silk fibers, and, because of the increased light refraction

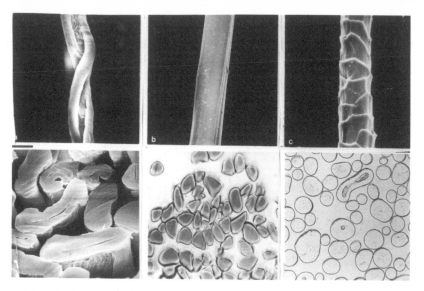

Fig. 6-16 Longitudinal (top) and cross-sectional (bottom) photographs of natural fibers [14].
   Left: cotton with a hollow, bean-like structure. The lumen is compressed because of drying.
   Center: natural silk with triangular cross-section and smooth surface.
   Right: sheep's wool with scaly surface and bicomponent structure.

by the triangular shape (similar to the effects of cut diamonds), also the high gloss. Because of their compactness, silk fibers will only take up a little water or moisture.

Cotton fibers (Section 6.3.3), on the other hand, are bean-like hollow cellulose fibers (Fig. 6-16). Their water uptake is not only caused by the hydrophilic nature of cellulose but also by the hollow center of the fibers as well as the vacuoles between cellulose chains (for a more detailed description of cotton fibers, see Volume II, p. 395).

Wool fibers have a scaly surface which causes the typical woolly touch. Their bicomponent nature (Section 6.3.4) produces the crimp and bulk of these fibers.

To some extent, these properties of natural fibers can be reproduced in synthetic fibers by texturizing and/or use of bicomponent structures. Gloss, hand, and bulk of man-made fibers can by varied through different cross-sectional shapes (Figs. 6-8, 6-17), etc., so that a wide variety of textile properties can be obtained without changing chemical structures.

Mechanical Properties

Many use and some care characteristics of textiles are controlled by the mechanical properties of fibers. These properties are not only controlled by the chemical structure and the degree of polymerization of the polymer and the physical structure of the fibers but also by their length and diameter.

The greater the diameter and length of fibers, the more probable is the presence of flaws. Industrial fibers should possess not more than 1 flaw per 2000 km fiber. Laboratory fibers have many more flaws per length than industrial fibers. Hence, only the best data on laboratory fibers can be compared with those of industrial fibers because otherwise one would compare the effect of weak points and not the effects of chemical and physical structures.

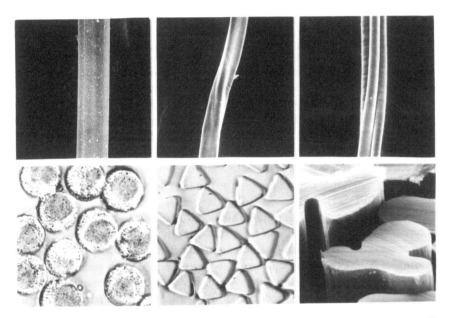

Fig. 6-17 Longitudinal (top) and cross-sectional (bottom) photographs of various synthetic fibers [14]. Left: bicomponent fiber from polyamide 6 and polyamide 6.6 with core-sheath structure. Center: polyester fiber with triangular cross-section. Right: trilobal bicomponent fiber from two acrylics.

Tensile data of textile fibers are obtained from force-elongation curves, $F = f(\varepsilon)$, (Fig. 6-18) and not from the stress-elongation data, $\sigma = f(\varepsilon)$, of technical fibers and plastics. Forces are not reported as such (see p. 161) but as force (in newton N) per linear density (in dtex), i.e., as **titer** (outdated) or **tenacity** (L: *tenere* = to hold).

Textile moduli are reported as tangent, secant, or chord moduli (Fig. 6-18). Forces F sometimes pass through a weak maximum with the **force at break**, (**breaking strength**), F_{max}, at the **elongation at break**, ε_{max} (Fig. 6-18). The maximum is sometimes followed by the lower value of the **force at rupture**, F_R, at the **elongation at rupture**, ε_R.

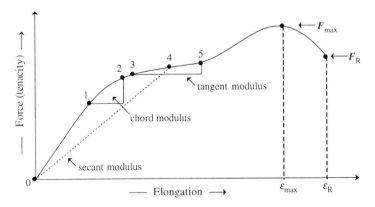

Fig. 6-18 Force-elongation diagram of textile fibers with forces at break (F_{max}) and rupture (F_R). Textile moduli are reported as **tangent moduli** between two points (e.g., 0-1 or 3-5), as **secant moduli** from the origin to a defined value of the elongation (here: 0-4), or as **chord moduli** between two defined elongations (here: 1-2).

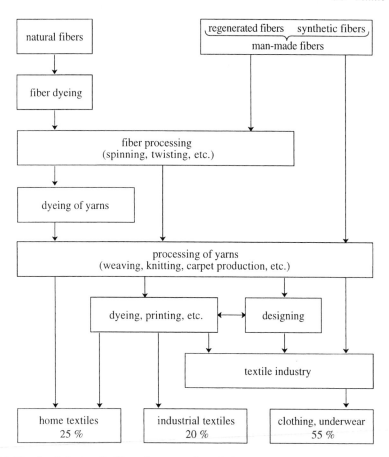

Fig. 6-20 Textile chain (textile fibers (home textiles, clothing, underwear): 54 % man-made, 35 % cotton, 4 % wool, 7 % other plant fibers).

Textile raw materials (natural, regenerated, and synthetic fibers) and textiles are separated by many processing steps, the **textile chain** (Fig. 6-20). The **textile industry** comprises practically all steps from fibers to design whereas the **textile trade** also includes the retail of clothing, underwear, and home textiles.

6.3.2 Silk Types

Overview

Natural silk was always a specialty fiber because of the high cost of production and the luxurious look. Silk was formerly the fiber of choice for expensive ladies' dresses and gentlemen's summer suits. In Western countries, it is now used practically only for accessories such as neckties and scarfs. In Japan, very expensive, classic kimonos are still tailored from natural silk.

Silk was a desirable fiber. Early man-made fibers were thus called "silks" although their mechanical properties did not resemble those of natural silk (Fig. 6-19 and Table 6-9). These **artificial silks** comprised three types of cellulosic fibers:

Table 6-9 Average properties of natural silk and some silk-like filaments. RH = relative humidity.

Property	RH (%)	Physical unit	Natural silk	Copper silk	Acetate silk	Tri-acetate	PA 6.6	PES
Density	65	g/cm^3	1.25	1.54	1.30	1.28	1.14	1.38
Textile modulus	0	N/tex	6.0					
	65	N/tex	8.9	4.1	2.9	3.3	3.5	7.9
	100	N/tex		3.3				
Strength at break	65	N/tex	0.35	0.17	0.11	0.15	0.56	0.44
	100	N/tex	0.30	0.11	0.08	0.095	0.49	0.44
Elongation at break	0	%	20			24		
	65	%	23	14	33		23	23
	100	%	30	23	38	35	30	23
Textile torsional modulus	65	N/tex	1.9	0.8	0.7	0.7	0.45	0.65
Moisture uptake	65	%	11	11	6.4	4.5	4.3	0.4
Water retention	100	%	42	110	20	14	12	4
Melting temperature	-	°C	-	-	260	300	250	248
Softening temperature	-	°C	-	-	184	250	-	240
Upper ironing temperature	-	°C	121	210	180	185	149	180

- from regenerated cellulose by the viscose process (**viscose, rayon**);
- from regenerated cellulose by the cuprammonium process (**cuprammonium rayon** ("copper silk");
- from the partial saponification of cellulose triacetate to cellulose 2 1/2-acetate (**acetyl cellulose, secondary acetate, acetate silk**).

None of these "silks" ever had the properties of natural silks, except for some luster.

The annual production of natural silk dropped drastically at the end of World War II but has recently recovered (Fig. 6-1). China is now the major producer.

The world production of regenerated cellulosic fibers has been stagnant since 1965, in part for environmental concerns, in part for economics, and in part fbecause of the better properties of some synthetic fibers. The production of these cellulosic fibers is constant in Western Europe but fell drastically in the United States. However, cellulose acetates are also used as plastics and for cigarette filters: in 1982, the annual world production of cellulose acetates was 264 000 t for textiles but already 339 000 t for cigarette tow.

Natural Silks

Natural silks are produced by the larvae of certain insects and by spiders. Commercially important are the silks from the larvae of the order *Lepidoptera* (Class *Insecta*). These consist of the protein secretions of the domesticated moth *Bombyx mori Linné* (Japan, China). Increasing in importance are the silks of the wild tussah moths *Antheraea pernyi* (China) and *Antheraea mylitta* (India) whereas the silk from the wild communal caterpillar *Anaphe moloneyi* (Africa) is used mainly for local consumption.

The silk from the drag lines of the tropical spider *Nephila clavipes* (golden orb weaver) is presently of considerable scientific interest since it has five times the strength of steel. This spider silk has also been produced on the laboratory scale by genetically engineered bacteria.

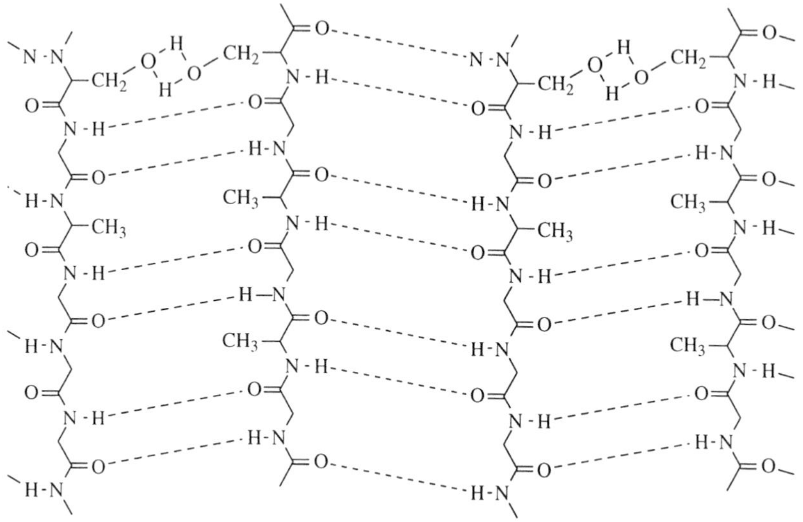

Fig. 6-21 Pleated sheet structures (β structures; Voleume III, p. 200) of the crystalline regions of natural silk from *B. mori*. Directions of chains and fibers are identical.

Composition of silks. All natural silks are proteins. *Nephila* silk consists of alanine blocks (25 % of all α-amino acid units) that are separated by glycine-rich (42 % of all α-amino acid units) blocks of glycine, arginine, and tyrosine units. The protein chains of the silk from *B. mori* contain regularly structured segments with 10 blocks of 6 α-amino acid units each and an irregular part with 33 amino acid units. The regular segments with the constitution H(ser-gly-ala-gly-ala-gly)$_{10}$OH lie parallel to the fiber direction and are interconnected by hydrogen bonds (Fig. 6-21). This structure is responsible for the high strength of silk since any strong elongation of the fiber necessitates the cleavage of hydrogen bonds, the deformation of bond angles, and the stretching of chain bonds. The chains of the irregular part are present as random coils that are easily deformable and thus responsible for the significant elongation at break (Table 6-8).

The silkworms of *B. mori* spin cocoons consisting of 78 % silk fibroin (raw silk) and 22 % sericin (a gum protein from serine (37 %), aspartates (26 %), glycine (17 %), and other α-amino acids (20 %)). The protein chains of silk fibroin have molecular weights of ca. 400 000. They are assembled in microfibrils of 10 nm width that form fibrils which are up to 2000 nm wide. Physical properties of silk fibroins of various silks (Table 6-10) depend strongly on the content of α-amino acids with short side chains.

Isolation of silk fibers. Larvae of *B. mori* feed on mulberry leaves and those of *Antheraea* on leaves from oak or castor. They then start to spin cocoons by extruding two fibroin filaments from a "spinneret" in the silkworm's head. The exiting filaments are glued together by sericin to form single threads with average diameters of 15-25 μm. The threads are bonded to a cocoon in which the larva metamorphoses to a chrysalid.

The cocoons are collected before the chrysalids can metamorphose further to moths. The chrysalids in the cocoons are then killed by heating with hot air or water vapor. The outer parts of the cocoons are brushed off to give **flock silk** which is less than 15 mm long and cannot be spun. Dipping the cocoons into hot water softens the sericin so that rotating brushes can grab the ends of the silk fibers and wind 4-10 of them on a rotating

Table 6-10 Mole fraction x_S of amino acid units with short side chains S and properties of various silk fibroins at 23°C and 65 % relative humidity. E = tensile modulus, σ_B = strength at break, ε_B = elongation at break, ΔR_{10} = elastic recovery from a 10 % elongation.

Source	x_S/%	E/GPa	σ_B/MPa	ε_B/%	ΔR_{10}/%	
					in air	in water
Anaphe moloneyi	95.2			13	50	50
Bombyx mori	87.4	12	470	23	50	60
Antherea mylitta	71.1			34	30	70
Nephila clavipes		22	1000	35		

reel to give spinnable **silk fibers** that constitute about 500-800 m of the total of 1000-3000 m long fibers. The remaining fibers are degummed which causes some fibers to break. These short fibers (**silk noils**) and those from damaged cocoons are united and then spun to threads.

The remaining, long silk fibers are made soft by dipping them into oil; they are then treated with alkali-free soaps in order to remove the sericin. The silk becomes more glossy and supple but also loses up to 25 % of its weight. It is thus treated with aqueous solutions of $SnCl_4$ and Na_2HPO_4 that react on the fiber to tin phosphate which is then reacted with water-glass. This treatment improves hand and gloss and also the balance sheet. The silk fibers are then bleached with sulfur dioxide, perborates, alkali peroxides, etc. Silk fibers for black-colored yarns are instead treated with vegetal tanning agents.

Silk fabrics are traditionally sold in **momme**, which is the weight of the fabric per square meter (1 momme = 3.75 g/m^2). Most silk charmeuse fabrics have 16-22 mommes. The lightest silk fabric is Chinese silk with 4-8 mommes.

Polyamides (Nylons)

Up to the middle of the last century, natural silk was the material for the hosiery of the elegant lady and cotton the fiber for the common folks. The situation changed with the invention of the first *successful* synthetic fiber, polyamide 6.6 = PA 6.6 = PA 66 = poly(hexamethylene adipamide), sold as Nylon® in 1938 (see Volume II, p. 456). One year later, Perlon® (= poly(ε-caprolactam) = polyamide 6 = PA 6) was introduced.

The new, silk-like "nylons" completely replaced silk in cotton stockings. In the United States, "nylon" became the generic name for aliphatic polyamides (for textiles only if the textiles contain 85 % or more of these polyamides). Fibers from PA 6.6 and 6 are still the most important polyamides. To this day, their sales still reflect the original inventor nations and sales territories: PA 6.6 in the United States and the United Kingdom and PA 6 in Germany and Western Europe. During the last ten years, the U.S. production of nylons has decreased whereas that of Western Eurpe has stagnated (Fig. 6-2).

Polyamide 6.6 (PA 66 = six-six; not "sixty-six") is produced by polycondensation of hexamethylenediamine, $H_2N(CH_2)_6NH_2$, and adipic acid, $HOOC(CH_2)_4COOH$, whereas polyamide 6 is obtained from the hydrolytic polymerization of the 7-membered cyclic amide, ε-caprolactam, C_6H_5NO (Volume II, p. 456 ff.). Fibers from PA 6.6 and PA 6 are used and are expected to continue to be used for women's hosiery, intimate apparel, carpets, certain stretch fabrics, and industrial applications. However, they do not have the moisture uptake and water retention of natural silk. Since they feel "synthetic", men's nylon dress shirts disappeared years ago despite their good mechanical properties.

Over the years, many other aliphatic, cycloaliphatic, and aromatic polyamides appeared, both of the AB type (like PA 6) and the AABB type (like PA 6.6) (see p. 185). These fibers are produced by polycondensation (all AABB types plus PA 7, 9, 11) and by hydrolytic ring-opening polymerization (PA 6, 8, 12) (see Volume II, p. 456 ff.). In addition, there are many copolymers (for example, 6/6.6, 6.6/6.12), blends (for example, PA 6/PA 6.6, PA 6.6/PA 6.12), aromatic polyamides (= aramids), etc. (see Table 6-11):

PA 4.2 is produced in a pilot plant. It is not used for textile purposes.

PA 4.6 is very highly crystalline and more temperature-resistant than PA 6.6 (Table 6-11). The main use is in the automotive and electrical/electronic industries.

PA 6.6 is not only used as a textile fiber and for tire cords but also as a thermoplastic.

PA 6.9 is produced only in the hundreds-of-tons range. It serves for industrial fibers (monofilament, bristles) and industrial parts (electrical connectors, bushings, etc.).

PA 6.10 has about the same properties and production volume as PA 6.9. It is used for the same purposes and also in sport and leisure wear.

PA 6.12: the largest application for unreinforced PA 6.12 is for toothbrushes, followed by other brushes, and uses in the automotive and electrical industries.

PA 12.12 looked like an excellent candidate for the fiber and the plastics business in the 1970s. It was abandoned because of production problems.

PA 13.13 never made it to the market.

Qiana® is a luxury silk-like textile fiber but was withdrawn from the market in the early 80s for economic reasons.

PA 2: two types have limited commercial interest; two other ones were explored:

Poly(γ-benzyl-L-glutamate) with R = $CH_2CH_2COOCH_2C_6H_5$) is an encapsulation agent for pharmaceutically active liquids.

Poly(γ-methyl-L-glutamate) with R = $CH_2CH_2COOCH_3$ has excellent silk-like properties that could not be realized with economical fiber spinning technology. The polymer is used in Japan as a coating on synthetic leathers.

Poly(L-leucine) with R = $CH_2CH(CH_3)_2$ can be spun from helicogenic solvents to wool-like fibers with α-helical polymer molecules. On stretching, they convert to silk-like fibers with β-pleated sheet structures of the polymer molecules.

Poly(γ-D-glutamic acid) with R = CH_2CH_2COOH, behaves like poly(L-leucine) but is also not economical.

PA 3 with R = H is not a commercial polymer but its dimethyl substituted (R = CH_3) representative is spun to industrial sewing yarns.

PA 4 can be dyed easily but is not significant economically.

PA 6 is one of the two most important polyamide fibers (the other one is PA 6.6), not so much because of its fiber properties but because of its favorably priced raw materials (benzene, cyclohexane, or phenol).

PA 7 is produced in small quantities in the former countries of the USSR.

PA 8 failed economically because raw materials were too expensive.

PA 9 is too expensive. It was produced in the countries of the former Soviet Union.

PA 11 for industrial applications is produced in France (based on castor oil) and in Russia (based on telomers from ethylene and carbon tetrachloride).

PA 12 has properties similar to PA 11 and is also used in the same way.

Table 6-11 Chemical structures, chemical names, short names, and codes of aliphatic and semi-aliphatic polyamides.

Structure	Name
$\left[-NH-(CH_2)_4-NH-\underset{O}{\underset{\|}{C}}-\underset{O}{\underset{\|}{C}}-\right]_n$	poly(tetramethylene oxalamide), polyamide 4.2, nylon 42, PA 4.2
$\left[-NH-(CH_2)_4-NH-\underset{O}{\underset{\|}{C}}-(CH_2)_4-\underset{O}{\underset{\|}{C}}-\right]_n$	poly(tetramethylene adipamide), polyamide 4.6, nylon 46, PA 4.6
$\left[-NH-(CH_2)_6-NH-\underset{O}{\underset{\|}{C}}-(CH_2)_4-\underset{O}{\underset{\|}{C}}-\right]_n$	poly(hexamethylene adipamide), polyamide 6.6, nylon 66, PA 6.6
$\left[-NH-(CH_2)_6-NH-\underset{O}{\underset{\|}{C}}-(CH_2)_7-\underset{O}{\underset{\|}{C}}-\right]_n$	poly(hexamethylene azelaamide), polyamide 6.9, nylon 69, PA 6.9
$\left[-NH-(CH_2)_6-NH-\underset{O}{\underset{\|}{C}}-(CH_2)_8-\underset{O}{\underset{\|}{C}}-\right]_n$	poly(hexaamethylene sebacamide), polyamide 6.10, nylon 610, PA 6.10
$\left[-NH-(CH_2)_6-NH-\underset{O}{\underset{\|}{C}}-(CH_2)_{10}-\underset{O}{\underset{\|}{C}}-\right]_n$	poly(hexamethylene dodecanoamide), polyamide 6.12, nylon 612, PA 6.12
$\left[-NH-(CH_2)_{10}-NH-\underset{O}{\underset{\|}{C}}-(CH_2)_8-\underset{O}{\underset{\|}{C}}-\right]_n$	poly(decamethylene sebacamide), polyamide 10.10, nylon 1010, PA 6.10
$\left[-NH-(CH_2)_{12}-NH-\underset{O}{\underset{\|}{C}}-(CH_2)_{10}-\underset{O}{\underset{\|}{C}}-\right]_n$	poly(dodecamethylene dodecanoamide), polyamide 12.12, nylon 1212, PA 12.12
$\left[-NH-(CH_2)_{13}-NH-\underset{O}{\underset{\|}{C}}-(CH_2)_{11}-\underset{O}{\underset{\|}{C}}-\right]_n$	poly(tridecanomethylene tridecanoamide), polyamide 13.13, nylon 1313, PA 13.13
$\left[-NH-\bigcirc-CH_2-\bigcirc-NH-\underset{O}{\underset{\|}{C}}-(CH_2)_{10}-\underset{O}{\underset{\|}{C}}-\right]_n$	poly[(4,4'-biscyclohexylene)methylene-dodecanoamide), Qiana®
$\left[-NH-CHR-\underset{O}{\underset{\|}{C}}-\right]_n$	poly(α-amino acid)s polyamides 2, see text, p. 184
$\left[-NH-CH_2-CR_2-\underset{O}{\underset{\|}{C}}-\right]_n$	poly(β-amino acid)s polyamides 3, see text, p. 184
$\left[-NH-(CH_2)_3-\underset{O}{\underset{\|}{C}}-\right]_n$	poly(γ-butyrolactam), polyamide 4, nylon 4, PA 4
$\left[-NH-(CH_2)_5-\underset{O}{\underset{\|}{C}}-\right]_n$	poly(ε-caprolactam), polyamide 6, nylon 6, PA 6, Perlon®
$\left[-NH-(CH_2)_7-\underset{O}{\underset{\|}{C}}-\right]_n$	poly(8-aminocaprylic acid), polyamide 8, nylon 8, PA 8
$\left[-NH-(CH_2)_{10}-\underset{O}{\underset{\|}{C}}-\right]_n$	poly(11-aminoundecanoic acid), polyamide 11, nylon 11, PA 11
$\left[-NH-(CH_2)_{11}-\underset{O}{\underset{\|}{C}}-\right]_n$	poly(laurolactam), polyamide 12, nylon 12, PA 12

So-called **shingosen fibers** were commercially successful. These polyester fibers are produced by a combination of chemical, physical, and textile modifications. Chemically, they consist of copolymers of terephthalic acid and ethylene glycol with isophthalic acid and bisphenol A as well as some sodium 5-sulfoisophthalate. Physically, they are micro and ultrafine fibers with triangular cross-sections and core-sheath, side-by-side, etc., structures. Textile modifications include false-twisting (Section 6.4). Various combinations of these modifications deliver supersoft fabrics ("peach skin"), superbulk fabrics ("new silk"), worsted wool types ("new weave"), or rayon-type textiles ("super drape").

Superfilaments

New processes made it possible to produce superfilaments from polyester, polyamides, and other polymers. These **superfils** range from microfibers (0.1-1 dtex) to finer structures that are known under various names, depending on the manufacturer (ultrafiber, superfine fiber, fibril, etc.). Unofficial designations are ultrafine fibers (ca. 10^{-3} dtex) and ultra superfine fibers (ca. 10^{-4} dtex). These fibers are much thinner than those of natural silk (ca. 1.3 dtex).

Microfibers can be produced mechanically by spinning from spinnerets with many holes and two sideward airjets near the orifices of the spinneret. On blowing hot air at high velocity through the jets, the exiting filaments are torn into many short pieces which are then bundled like spaghetti. The bundles are embedded into a tangle of microfibers with pore diameters that are ca. 3000 times smaller than the diameter of rain drops. Such materials do not allow water droplets to enter the fabric, especially, if the fabric consists of hydrophobic polymers such as poly(tetrafluoroethylene) (**Gore-Tex**®). Water vapor can pass through, though, so that such fabrics from microfilaments with ca. 0.5 dtex can breathe although they repel water.

Other processes employ polymer blends of two immiscible polymers such as a water-soluble matrix (for example, poly(vinyl alcohol)) and a water-insoluble fiber-forming polymer (for example, poly(ethylene terephthalate)). On extrusion, flow lines generate a fibrillar morphology. The water-soluble polymer is extracted by water, and the resulting foam-like slurry is filtered off and then spun to microfibers for filter cloths. These filaments are so fine (finenesses as low as 10^{-4} dtex) that a filament extended from the earth to the moon would weigh only 4.2 grams!

Very fine fibers are also obtained from film fibers (Section 6.6.3). Examples are blends of polyamide 6/poly(propylene), poly(ethylene terephthalate)/poly(styrene), and polyamide/poly(ethylene terephthalate).

6.3.3   Cotton Types

Cotton

Cotton fiber production is now the second largest after polyester ($25.1 \cdot 10^6$ t/a *versus* $27.7 \cdot 10^6$ t/a in 2006). The largest producers are China, the United States, and India (Table 6-4). Egypt delivers the best cotton but produces only 2 % of the total.

Cotton (Arabic: *qutn*) is the seed hair of the fruits (bolls) of a treelike perennial plant of the genus *Gossypium* (family *Malvaceae*) that is grown commercially as a bushy annual with the species *arboreum*, *barbadense*, *herbaceum*, and *hirsutum* (see also Volume II, p. 393). 90 % of the world production of cotton is from the species *G. hirsutum* (includes all varieties of American Upland Cotton) and another 8 % from *G. barbadense* (also known as extra-long staple (ELS)). Cotton fibers are normally white but there are also ochre and green varieties.

Raw cotton contains seeds. This **seed cotton** is internationally traded in bales. An Egyptian bale weighs 340.5 kg (750 lbs), the old American bale 227 kg (500 lbs), the new American bale 217.9 kg (480 lbs), and the Chinese bale I, 85 kg.

Seed hairs are sheared off by saw gins (Upland cotton) or roller gins (ELS) which have blades that reach through a grid into the chamber with the seeds ("gin" is short for "engine"). The first ginning removes long seed hairs (**lints**) and the 2nd and 3rd ginning the shorter **linters**. Both lint and linters are blown away and collected separately. Seed cotton typically delivers 34 wt% lint + linters, 57 wt% seeds, and 9 % waste and humidity. Lints from *G. barbadense* are 33-36 mm long and those from *G. hirsutum*, 26-30 mm; they are processed into textiles. First-cut linters (<15 mm) serve in mattresses and upholstery and second-cut ones (2-3 mm) as feedstock for rayon, gun cotton, specialty papers, etc. Seeds are pressed to cotton seed oil; the remaining cake serves as animal fodder.

Cotton fibers consist of a multilayer core of cellulose molecules that is surrounded by the cuticle composed of lignin, pectin, fat, and waxes (Table 6-15; for the complex fine structure, see Volume II, p. 395). Cellulose is a linear polysaccharide composed of D-glucose units that are interconnected in the β-$(1{\rightarrow}4)$ position. Commercial cotton fibers also contain about 1 COOH group per 500-1000 glucose units. Number-average degrees of polymerization of cotton harvested in the dark under N_2 are 14 000-18 000, that of conventionally harvested cotton ca. 7000, and that of industrial wood pulp 300-2000.

Idealized constitution of cellulose chains of cotton and rayon consisting of cellobiose repeating units

The many hydrophilic OH groups and the large lumen of the fiber (Fig. 6-16) give cotton a large moisture uptake and water retention (Table 6-16). Cellulose fibers are therefore well suited for underwear and, in warm countries, for apparel.

Table 6-15 Chemical composition of natural cellulose fibers. WS = water solubles.

Fiber	Composition of dry fibers in wt%					
	Cellulose	Polyoses	Pectins	Lignins	Fats, waxes	WS
Cotton	92.7	4.7	1.0	0	0.6	1.0
Ramie	68.8	13.1	1.9	0.6	0.3	5.3
Hemp	67.0	16.1	0.8	3.3	0.7	2.1
Sisal	65.8	12.0	0.8	9.9	0.3	1.2
Jute	64.4	12.0	0.2	11.8	0.5	1.1
Flax. roasted	64.1	16.7	1.8	2.0	1.5	3.9
Abaca (Manila hemp)	63.2	19.6	0.5	5.1	0.2	1.4

Table 6-16 Average properties of cotton and some other staple fibers. RH = relative humidity; na = not applicable.

Property	RH in %	Physical unit	Cotton	Rayon (viscose)	Linen (flax)	Ramie	PA 6.6	PES
Average fiber length		mm	25	25	30	150	25	25
Average fiber diameter		µm	18	20	20	70	20	20
Cellulose content		%	92.9	95.5	71.2	76.2	na	na
Degree of polymerization, $\bar{X}_n$		1	7000	320		1900		
Density	65	g/cm³	1.54	1.52	1.50	1.51	1.14	1.38
Textile modulus	0	N/tex	9.7					
	65	N/tex	5.7	4.8	19	15	3.5	3.5
	100	N/tex	4.0	2.3				
Strength at break	0	N/tex	0.32	0.29				
	65	N/tex	0.35	0.23	0.46	0.58	0.39	0.38
	100	N/tex	0.40	0.14		0.69	0.36	0.38
Elongation at break	0	%		17		2.3		
	65	%	7	20	3	4	38	35
	100	%	9	28		2.4	40	35
Moisture uptake	65	%	8	14	12	6	4.3	0.4
Water regain	100	%	40	63			6	4
Heat conductivity	65	W/(m K)	0.45	0.58			0.25	0.25
Melting temperature	-	°C	-	-	-	-	250	248
Upper ironing temperature	-	°C	182	220	193	193	149	180
Decomposition temperature	-	°C	240	250	260	260	230	

The glass temperature of cellulose is estimated as 225°C in the dry state and below 20°C in the wet state. Pressure causes wet cotton and rayon fabrics to crease and look shoddy; it is for this reason that Western apparel is not made from cotton fabrics. Conversely, moistened cotton fabrics can be easily flattened by a hot iron which causes the water to evaporate and the wrinkles to disappear.

The bad crease resistance of cotton and rayon fabrics can be improved but not eliminated by applying textile finishes (Section 6.6). It is for this reason that underwear and men's shirts are usually not made from pure cotton but consist of mixed fibers, especially with polyester yarns. These mixed fabrics are much more abrasion resistant and longer lasting than cotton itself (Table 6-17). They also require considerably less energy per wear cycle (from freshly washed to used, including a proportionate amount for the manufacture of fibers and shirts). The energy consumption can be reduced considerably if Western standards (automatic washer and dryer, electric ironing) are replaced by Third World methods (hand washing, air drying), provided one prefers the crumpled look and does not rub the fabric on washing boards (increases the abrasion and shortens the life of the fabrics, especially of those made from pure cotton).

Untreated cotton looks dull but this can be remedied by treating it with 20-26 % aqueous sodium hydroxide solutions. This **mercerization** (John Mercer, 1844) provides cotton with an attractive shine because it converts the cellulose I modification with parallel polymer chains into the cellulose II modification with antiparallel chains (Volume II, p. 388 ff.). Concurrent application of mechanical stress during this process provides an improved breaking strength of fibers and a softer hand of fabrics. Absence of stress during mercerization leads to shrunk, elastic yarns and fabrics.

Table 6-17 Energy consumption during manufacture and care of men's dress shirts from cotton (CO), polyester fibers (PES), or mixed fabrics of CO + PES according to Western and Third World standards (automatic washer and dryer + electric iron *versus* hand washing and air drying) [16].

	Physical unit	Composition of shirts in %			
		100 CO 0 PES	50 CO 50 PES	35 CO 65 PES	0 CO 100 PES
Manufacture of fibers	MJ/kg	49			179
Manufacture of shirts from fibers	MJ/shirt	95	115	117	78
Number of uses per shirt	-	50	50	75	75
Energy consumption per wash					
- automatic washer	MJ/wash	2.3	1.1	1.1	1.1
- automatic dryer	MJ/wash	2.9	1.7	1.3	0.6
- electric iron	MJ/wash	1.2	0.4	0.4	0
- hand washing, air drying	MJ/wash	0.6	0.3	0.3	0.2
Energy consumption per lifetime of shirt					
- industrialized countries	MJ/lifetime	415	275	330	203
- Third World countries	MJ/lifetime	125	130	143	90
Energy consumption per use					
- industrialized countries	MJ/use	8.3	5.5	4.4	2.7
- Third World countries	MJ/use	2.5	2.6	1.9	1.2
Energy consumption per 50 uses					
- industrialized countries	MJ/(50 uses)	415	275	220	135
- Third World countries	MJ/(50 uses)	125	130	95	60

The comfort of cotton is also improved by treating dry cotton fabrics with aqueous solutions of poly(ethylene glycol) (PEG), $H(OCH_2CH_2)_{20-25}OH$. Heating the fabrics causes PEG to bind chemically to the cotton. On wearing such treated fabrics, body warmth causes the PEG to melt. The energy required for the melting process is drawn from the body which cools. The reversible effect lasts ca. 30 minutes.

Rayon (Viscose)

In the mid-to-end 1800s, accidental discoveries, men's curiosity, the mulberry spinner's disease (p. 156), and the high price of cotton all came together and led to such remarkable inventions as **Chardonnet silk** (nitrocellulose), the cuoxam process (Bemberg process) for **cuprammonium rayon**, and the viscose process for **rayon (viscose)**.

The Chardonnet process fell out of favor because cellulose nitrate ("nitrocelulose") is flammable. Both viscose and cuprammonium rayon are still produced but the production has either stagnated [world (Fig. 6-1)), Western Europe (Fig. 6-2)] or is falling (United States (Fig. 6-2)).

In the **cuprammonium process** (Volume II, p. 398), linters are dissolved in an ammoniacal copper solution (Schweizer's reagent; commonly mispelled as "Schweitzer's reagent." Schweizer was a professor at what is now the Swiss Federal Institute of Technology (ETH), Zurich). The solution is then wet spun into water to give **cupro silk (CUP)**. The process is simpler than the viscose process but also more expensive.

In the **viscose process**, cellulose (mostly wood pulp) is converted to alkali cellulose and then to cellulose xanthogenate (Volume II, p. 398). Solutions of cellulose xanthogenate are then wet spun into a sulfuric acid bath which regenerates the cellulose. The

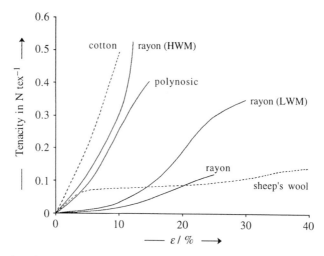

Fig. 6-22 Tenacity of cotton, sheep's wool, and various rayons as a function of elongation ε. Rayon (standard type; viscose), HWM = high wet modulus, LWM = low wet modulus,.

resulting fibers are very glossy, which led to their American name **rayon** (F: *rayon* = light beam); the international name (BISFA) is **viscose (CV)**.

Regular rayons have fairly low degrees of polymerization (ca. 260 *versus* ca. 7000 for cotton) which lead to tenacities that are much lower than those of cotton (Fig. 6-22). Changes in the viscose process first led to rayons with low wet moduli (**LWM**) and then to **modal fibers** with higher degrees of polymerization of up to 800.

According to BISFA, modal fibers are characterized by a high breaking force and a high wet modulus (Table 6-18). In the wet state, the highest tenacity should be at least 22.5 cN/tex at an elongation of not more than 15 %. The BISFA value of the wet modulus is not defined. **Wet moduli** are given by the tensile force at wet fibers that are elongated by 5 %. The values are then extrapolated to an elongation of 100 %. Hence, wet moduli are smaller than common textile moduli which are always obtained for 0 % elongation regardless of the relative humidity. Fibers with wet moduli of more than 0.1 N/tex practically do not shrink.

Modal fibers come in two types. So-called **polynosics** (F: *poly*mère *non* synthé*ti*que) are prepared by using increased CS_2 concentrations for the dope and smaller acid concentrations in the coagulating bath. These fibers not only have higher degrees of polymerization but also a higher degree of orientation and thus shorter elongations at break. **High wet modulus fibers (HWM fibers)** are obtained from xanthogenate solutions with added amines and poly(ethylene glycol)s, coagulation baths with added zinc ions, and changed spinning conditions.

Fibers with high moduli can also be prepared by other processes. In one process, cellulose acetates from linters are saponified to reconstituted (not "regenerated"!) celluloses with relatively high degrees of polymerizations. Much more efficient is the **organosolv process** which delivers celluloses that are called **lyocell (CLY, Tencel®)** and not rayon.

The lyocell process makes use of the fact that cellulose dissolves directly in some organic solvents without intermediate formation of substituted cellulose as in the viscose process. The industrial solvent of choice for the lyocell process is *N*-methylmorpholine-*N*-oxide (NMMO) with nitrosamine contents of less than 30 ppb = $3 \cdot 10^{-5}$ %.

Table 6-18 Average properties of cotton and regenerated or reconstituted cellulosic fibers. CDA = cellulose reconstituted from cellulose 2 1/2-acetate, CLY = lyocell, CMD = modal fiber (high wet modulus), CUP = from the cuprammonium process, CV = regular viscose (rayon) from the viscose process, PN = polynosic.

Property	RH in %	Physical unit	Regenerated cellulosic fibers.				Reconstituted		Cotton
			CUP	CV	PN	CMD	CLY	CDA	
Degree of polymerization		1	500	320	600	750			7000
Density	65	g/cm³	1.53	1.52	1.51	1.53		1.52	1.54
Textile modulus	65	N/tex	4.1	4.8	4.8		13.7		5.7
	100	N/tex	3.3	2.3	3.5		9.5		4.0
Wet modulus		N/tex	0.014	0.017	0.12	0.12	0.10		
Textile strength	65	N/tex	0.17	0.23	0.29	0.64	0.37	0.62	0.35
	100	N/tex	0.11	0.14	0.25	0.51		0.50	0.40
Wet strength		N/tex					0.30		
Elongation	65	%	14	20	8.5	7.5	13	6.0	7.0
	100	%	29	28	10.5	8.5		7.0	9.0
Elongation at wet state		%					15		
Moisture uptake	65	%	11	14	12	12		11	7.5
Water regain	100	%	117	95	62	65		22	40

Highly absorbent cellulose fibers are obtained if Na_2CO_3-containing dopes enter acidic precipitation baths. The developing CO_2 blows up the fiber and creates voids. Fabrics from these fibers have water regains of up to 200 %, which makes them very suitable for sports clothing.

Cotton Type Polyesters

Polyester filaments with circular cross-sections are silk-like (Section 6.3.2). However, most polyester fibers are not used as filaments but as staple fibers for clothing and home textiles (Table 6-6). Since polyester fibers from melt spinning are straight, they need to be crimped for many applications and texturized for apparel (Section 6.4).

Polyester fibers can only be dyed by dispersion dyes because they do not contain ionizable groups as nylon fibers do. The dye uptake is increased by incorporating aliphatic dicarboxylic acids, isophthalic acid, or poly(ethylene glycol)s into the polyester chains. Such comonomers decrease the crystallinity and thus increase the access to amorphous regions where the dye molecules can be deposited. However, such comonomers also lower the glass and melting temperatures.

Shrinkage can be improved by copolycondensation with aliphaitc dicarboxylic acids or long-chain diols and also by a suitable choice of drawing temperatures and draw ratios. The latter, physical processes are not favored by industry since they are partially reversible on further processing.

Polyester fibers can be made more cotton-like or more wool-like with respect to their stress-strain behavior (Fig. 6-23). The tenacity = f(elongation) curve of cotton-type PES for sewing yarns is practically identical with that of cotton itself up to $\varepsilon \approx 11$ %, upon which PES becomes plastic. Similarly, the wool-type PES fibers mimic the stress-strain behavior of wool well at small elongations but have much larger tenacities at elongations greater than ca. 5 %. For the use of other polyester fibers, see p. 186, bottom.

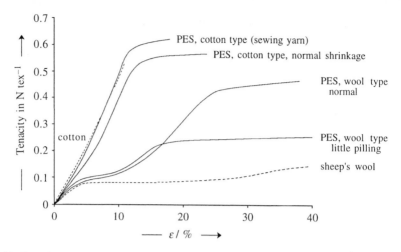

Fig. 6-23 Tenacity as a function of elongation for various polyester fibers (PES) as compared to cotton (- - -) and sheep's wool (- - -) [17].

Flax and Ramie

For thousands of years, mankind has used many other vegetable fibers in addition to cotton, most of them for non-textile applications (Section 6.7.2). **Flax** (p. 218) is used to make linen which now serves for home textiles (bed sheets, kitchen towels). The use of linen for men's shirts and tropical suits has fallen out of favor, however, because linen is very stiff and not very pliable, i.e., not very comfortable.

Ramie (p. 218) is a very strong and stiff fiber which rules out its use as a stand-alone textile fiber. However, recent fashion favors very loosely knitted pullovers which would not fit nicely if they were not reinforced by stiff ramie fibers.

6.3.4   Wool Types

Natural Wools

Natural wool is the cut hair of sheep, goats, llamas, rabbits, etc. Commercially, the wool of domesticated sheeps is the most important wool with respect to volume and value. Raw wool (greasy wool) contains 40-80 % wool fibers (on average: 60 %) and wool fat, suint (Old French: *suer* = to sweat), yolk (= grease of sheep), and plant residues.

The worldwide annual production of raw wool peaked in the early 1990s and is now slowly declining (Fig. 6-1). In 2005, it was $1.23 \cdot 10^6$ t/a of which Australia produced ca. 25 %, China 18 %, and New Zealand 11 %. The annual production of wool per sheep varies widely: 5.3 kg in Australia, New Zealand, and Argentina; 4.0 kg in South Africa, Uruguay, and the United States; 2.5 kg in Eastern Europe and China; and only 1.5 kg in the United Kingdom, Germany, and Turkey.

Virgin wool is sheep's wool before the first spinning whereas **shoddy** is recycled wool from worn garments or shredded unfelted woolen or worsted rags. **Worsted wool** has long staple fibers. **Woolen** is a short, carded (= brushed) wool for knitting and **ragg** a sturdy wool. Wool fibers have diameters from 17 μm (grade AAAA) to 60 μm (grade F).

Besides sheep's wool and the finer **lamb's wool**, several specialty wools are traded:

- **Alpaca**, a fine, long fiber from the sheep camel *Lama pacos* of the Andes.
- **Angora wool** (short: **angora**) from the angora goat (see mohair). In analogy, other animals with fine hairs have also been named "angora", for example, angora rabbit, angora cat. The hair of the angora rabbit is also called angora wool. Angora is the old name of Ankara where the angora goat came from.
- **Guanaco** from the wild *Lama guanicoe* (resembles the domesticated *Lama pacos*).
- **Yak** from the Central Asian bovine *Bos grunniens* (both wild and domesticated).
- **Camel hair** is the downy hair of the two-humped Bactrian camel (*Camelus bactrianus*) or, less often, the one-humped domesticted Arabian camel (dromedary, *Camelus dromedarius*) (G: *dromas* = runner).
- **Cashgora** from the crossbreeding of a nanny cashmere goat with a billy angora goat.
- **Cashmere** (Indian: *pashmina*) is a very fine wool (d = 13-15 µm) from the cashmere goat (*Capra hircus laniger*) of the high mountains of China (Tibet), Inner Mongolia, India, Iran, and Afghanistan. The annual yield is 150 g per billy goat and 50 g per nanny goat. The world production is ca. 8000 t/a, of which 2/3 is from China.
- **Llama wool** from the domesticated Peruvian camel (*Lama peruana*).
- **Merino** (from the Berber tribe *Beni merin* that developed this breed of sheep), a very fine wool (d = 11.7 µm), originally from Spain, now mainly from Australia.
- **Mohair** (d = 25-30 µm) from the Angora goat (*Capra hircus angorensis*).
- **Vicuña**, the finest wool ($d \approx$ 12-14 µm) known from the humpless camelid *Lama vicugna* which lives at 4000 m in the Andes, probably a wild form of the alpaca.

The names of wools from the various kinds of sheep were used to identify the textiles therefrom and are today applied to distinguish textiles regardless of the origin of the wool. **Merino** is now the name for all fine, strongly creased, soft wools of sheep, **crossbreed** a less creased wool of medium length, and **cheviot** (originally from the Cheviot sheep) the coarse twill weave for carpets and covers (also a fiber for damming materials).

Structure of Natural Wools

Hairs are dead cell tissues that consist of α-keratins which are insoluble scleroproteins with high sulfur content (see also Volume II, p. 547). Keratins consist of many types of proteins and not of a single type as silk fibroins do. The chemical structure of wool fibers is therefore more complicated than that of silk fibers, which is also reflected in their complex morphology.

Wool fibers (Fig. 6-16) are composed of the exterior cuticle and the interior cortex (Fig. 6-24) which both consist of peptide chains. The cystine-rich cuticle has a scale-like structure and consists of three layers that are connected by intercellular membranes: the ca. 10 µm thick outer epicuticle is strongly crosslinked by cystine S-S bridges. It is followed by the exocuticle and the innermost, sulfur-free endocuticle.

The cuticle surrounds the cortex which is composed of many cortex cells that are ca. 100 µm long and 3-5 µm thick. In wool from sheep, there are two types of cortex cells: one half of the cortex houses orthocortex cells and the other half, paracortex cells. Orthocortex cells contain relatively many acidic peptide units and few cystine units whereas paracortex cells have relatively more cystine units.

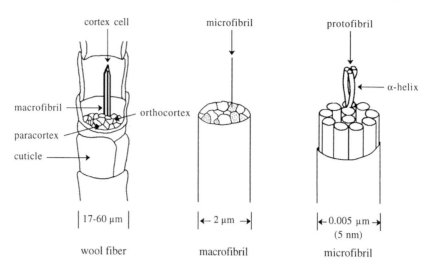

Fig. 6-24 Schematic representation of fibers from merino wool (left) with the exterior cuticle and the interior paracortex and orthocortex cells [18]. Between cuticle and cortical cells is a thin layer of a continuous intercellular material (not shown) that is part of the cell membrane structure of the fiber (Volume I, p. 512). Each cortex cell contains many macrofibrils (center) which each hold many microfibrils (right) (see text). Each microfibril consists of 11 protofibrils and each protofibril of 2-3 α-helices of peptide chains.

The proportion and arrangement of orthocortex and paracortex cells determines the textile character of the various types of wool. Fibers from merino wool are bicomponent fibers because of the side-by-side structure of paracortex and orthocortex cells. The two types of cells have different expansion coefficients; therefore, the bilateral arrangement of paracortex and orthocortex cells causes merino wool to crimp.

In the wool of the Lincoln sheep, on the other hand, paracortex and orthocortex cells are arranged core-sheath (Fig. 6-9): the fiber structure is not asymmetrical and the fibers do not crimp. Mohair contains practically only orthocortex cells and fibers from black-face sheep practically only paracortex cells.

The cortex of sheep wool is composed of several bundles of long, spindle-shape, polyhedral cortical cells called macrofibrils (Table 6-19). Each macrofibril contains ca. 70 microfibrils that are embedded in a sulfur-rich protein matrix. A microfibril consists of 9 outer protofibrils that surround 2 inner ones (Fig. 6-24). Each protofibril incorporates 2-3 polypeptide chains with molecular weights between 200 000 and 1 000 000 that are present in the macroconformation of an α-helix (Volume I, p. 533).

Table 6-19 Structure of wool fibers from sheep.

Type	Length (nm)	Diameter (nm)	Composition
Cortex cells	≈ 100 000	3000-5000	several macrofibrils
Macrofibril	10 000	200-300	ca. 70 microfibrils
Microfibril	1 000	7.3	11 protofibrils
Protofibril	1 000	2	2-3 peptide chains
α-Helix of peptide chains	50	0.5	1 polypeptide chain

The complex physical structure of wool fibers is a consequence of its complex chemical composition. Wool consists of approximately 200 different macromolecular compounds of which ca. 80 % are keratins belonging to the groups of scleroproteins (Volume I, p. 541), 17 % non-keratin proteins, 1.5 % polysaccharides and nucleic acids, and 1.5 % lipids and inorganic compounds.

Keratins of wools consist of three groups: sulfur-poor helix-forming proteins (ca. 20 types), sulfur-rich proteins (ca. 100 types), and glycine-tyrosine rich proteins (ca. 50 types). Non-keratin proteins comprise the proteins of the nuclei of cells and those of the cytoplasm.

Most peptide units of wool proteins have very voluminous side groups that require much space and prevent the formation of folded sheet structures in α-keratins. The resulting helices are crosslinked by bridges from disulfide units –S–S– of cystines and from N_ε-(γ-glutamyl)lysine units $-CH_2CH_2CO-NH(CH_2)_4-$. Because of its crosslinked structure, wool is insoluble in all solvents, in contrast to all other natural fibers.

Work-up of Wool

Fibers of raw wool are glued together by wool fat, suint, and yolk which have to be removed. Wool is first **scoured**, i.e., washed with warm water, or industrially with nonionic detergents. Alkali treatment causes the wool to yellow, degrades its mechanical properties, and impairs the hand of wool.

Plant residues (seeds, etc.) are then removed by **carbonizing** which consists of treating the wool with 4-7 % sulfuric acid, drying it at 100-120°C, and finally beating it for the mechanical removal of cellulosic impurities.

During carbonizing, several chemical processes take place in some of the α-amino acid units of wool (Volume II, p. 547): in some serine units, an N/O peptidyl shift or an esterification with subsequent decomposition by β-elimination; a sulfidation of some tyrosine units; and, to a small extent, the reaction of free amino groups of peptide units with sulfuric acid to sulfamic acid groups.

For unknown reasons, wool is slightly yellow. For many applications, it is therefore **bleached** with SO_2 or H_2O_2. Bleaching requires great care because a too harsh treatment would cleave peptide bonds.

The felting of wool is assumed to be caused by its scaly surface. Felting can be prevented by a slight **chlorination** which modifies the surface of fibers. Mild chlorination is also required for shrink-proofing which consists of treating wool with solutions or emulsions of polymers that contain reactive side-chains such as acetidinium groups.

Properties of Wool

The scaly surface of wool fibers (Fig. 6-16) causes the characteristic hand and bulkiness of wool: wool fabrics can contain up to 85 % air which retains heat in cold climates and keeps it out in warm ones (wool clothing of bedouins!). Wool is a viscoelastic material since it consists of crystalline fibrils which are embedded in an amorphous matrix (Fig. 6-24) that is plasticized by water.

The paracortex and the orthocortex of merino wool have different chemical compositions and take up different proportions of water. The resulting crimping is fairly stable at

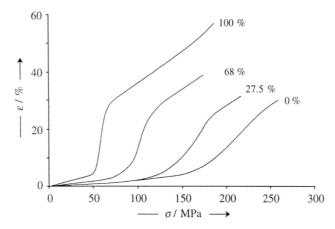

Fig. 6-25 Elongation ε of sheep wool as a function of stress σ at various relative humidities [19].

not too large forces since small loads produce only small elongations. However, elongations increase steeply above certain critical stresses which are smaller the larger the relative humidity (Fig. 6-25).

Strain-stress curves such as the ones shown in Fig. 6-25 exhibit hysteresis if loads are not too large. For this reason, it is sufficient to hang creased woolen textiles in steaming bathrooms in order to remove creases. Conversely, woolen fabrics have to be ironed with very heavy pressure if one wants creases in trousers or one desires smooth fabrics.

Ironing of wool fabrics or preparation of a permanent wave of hairs causes crosslinks to sever and subsequently form again at another position. Especially effective is the treatment of heated wool or hair with alkali which splits cysteine bridges and leads to thiol groups that then catalyze an exchange of –S–S– bonds (Eq.(6-1)).

The reaction is promoted by elongating the fibers. Pressing fabrics briefly with a steam iron (or hair with a heated curler) will only cause –S–S– bonds to snap. But doing so for longer times will allow the –S–S– bridges to rearrange and the fabric or hair to "set" in the desired position with the wanted elongation. Correspondingly, hair becomes "permanently set."

(6-1)

$$\begin{array}{c} \text{S—S} \\ \text{SH} \quad \text{S} \\ \quad\quad \text{S} \end{array} \xrightarrow[\text{– H}_2\text{O}]{+\ \text{OH}^{\ominus}} \begin{array}{c} \text{S—S} \\ \text{S—S} \\ \quad \text{S}^{\ominus} \end{array}$$

Wool has the greatest moisture uptake and the smallest water retention of all natural fibers (Table 6-8). The moisture uptake increases strongly with increasing relative humidity, showing some hysteresis (Fig. 6-26). Textile moduli therefore decrease with increasing moisture uptake. This effect is much greater than predicted by the mixing rule (Section 9.2) for the diminishing small modulus of water. As a consequence, wool becomes softer and more deformable at high relative humidities (Figs. 6-25 and 6-26).

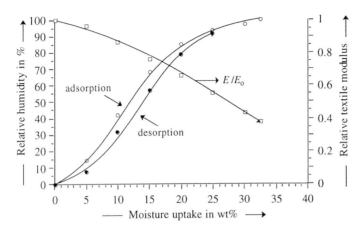

Fig. 6-26 Left: relationship between relative humidity and moisture uptake of wool on adsorption and desorption of water. Right: dependence of relative textile moduli E/E_0 on moisture uptake, set as unity for zero moisture uptake. Data of [20].

Despite the large moisture uptake, wool textiles behave "hydrophobically" because of the relatively small water retention (Table 6-8). It is for this reason that people in rainy Northern climates prefer to wear felt hats and loden coats (loden: a water-repelling, dense felt from sheep's wool). Because the body temperature is greater than the outdoor temperature, wool begins to take up water vapor from the humid air. The fibers start to swell which makes the fabric less porous so that it cannot be penetrated any more by rain drops. Simultaneously, passage of cold air becomes more constricted so that the body feels warmer. This effect cannot be achieved by normal wool pullovers because they are too loosely knitted.

In contrast to all other known fibers, wool may lead to allergic reactions in some people. Such reactions may also occur with other natural and man-made fibers but they are not caused by the fibers themselves but by dyestuffs, sizes, and the like.

Acrylics

Acrylic fibers are somewhat wool-like in character (Fig. 6-19). These fibers are always copolymers of acrylonitrile, $CH_2=CH(CN)$, with other monomers since fibers from pure poly(acrylonitrile) are difficult to spin and dye. Regular **acrylics** contain at least 85 wt% of acrylonitrile units, $-CH_2-CH(CN)-$, whereas **modacrylics** (= modified acrylics) consist of 35-85 wt% acrylonitrile units.

Comonomers for acrylics are methyl acrylate, $CH_2=CH(COOCH_3)$, or vinyl acetate, $CH_2=CH(OOCCH_3)$. Most modacrylics contain high proportions of vinyl chloride units, $-CH_2-CHCl-$, or vinylidene chloride units, $-CH_2-CCl_2-$. The incorporation of these units reduces the crystallinity of acrylics and modacrylics and lowers their glass temperature which, in turn, increases the solubility of acrylics in dopes and improves the diffusion rate of dyestuffs into the fiber. The uptake of dyestuffs can also be improved by incorporating monomeric units from sulfonated monomers such as sodium styrene sulfonate or sodium methallyl sulfonate. Acrylics and cotton have similar flammabilities (LOI = 18.2 *versus* 18.0) but those of modacrylics are greater (27-37).

Table 6-20 Average properties of sheep's wool and some wool-like fibers. Azlon = regenerated protein fiber, αPLL = α-form of poly(L-leucine), $+\text{NH}-\text{CH}(\text{CH}_2\text{CH}(\text{CH}_3)_2)+_n$.

Property	RH (%)	Physical unit	Sheep's wool	Azlon	αPLL	Mod-acrylics	Acrylics staple	filaments
Density	65	g/cm³	1.30	1.25	1.04	1.35	1.18	1.18
Textile modulus	65	N/tex	2.7	2.2	1.9	1.8	2.9	5.8
Textile strength at rupture	65	N/tex	0.12	0.09	0.055	0.20	0.24	0.42
	100	N/tex	0.11	0.045		0.18	0.22	0.38
Elongation at rupture	65	%	20	35	55	38	40	28
	100	%	38	55	97	40	45	30
Moisture uptake	65	%	16	13		1.7	1.2	1.2
Water retention	100	%	35			15	8.5	8.5

Properties. (Mod)acrylics have properties similar to those of wool with respect to elongation but are stiffer and stronger and have less moisture uptake and water retention (Table 6-20). They are characterized by high bulk, good heat retention, very good light and weather stability, simple dyeability, and good care properties. Both wool and acrylics are suitable for home textiles and for apparel that is worn only for a short time such as fashionable ladies' apparel and children's wear.

The reason for not using acrylonitrile copolymers in long-lived, demanding applications such as men's suits is their tendency to **pilling** which is greater than that of wool. Rubbing of fabrics from wool or acrylics pulls out fibrils that curl up into little balls called pills which are connected with the fabric by only a few fibrils. The pills break off relatively easily from wool fabrics because of wool's low rupture strength (Table 6-20) but remain much longer on the surface of acrylic fabrics with their higher strength. As a result, fabrics from acrylics look shabby on longer use.

Pilling can be reduced by lowering tensile strengths, for example, by using polymers with smaller molecular weights but this leads to lower viscosities and thus to problems during fiber spinning. Other means are to provide the polymers with branches or weak links in chains by using suitable comonomers. Polymers with weak links can still have high molecular weights and melt viscosities but the links can be broken after fiber spinning, for example, by hot water vapor if the links are hydrolyzable. Pilling is also affected by the cross-sectional shape of fibers. Wet-spun standard acrylics have circular to kidney-like cross-sections; they pill easily. Dry-spun acrylics are provided with trilobal or pentalobal cross-sections (see Fig. 6-17); their pilling tendency is lower.

(Mod)acrylics lack the moisture uptake and water retention of wool because they are far less hydrophilic. This deficiency has led to the development of absorbent acrylic fibers which are produced by dry spinning. The fibers have a core-sheath structure (Fig. 6-27) and apparent densities of 0.85-0.95 g/cm³. The fine capillaries of the core take up water fast and retain it up to 30-40 % of the fiber weight, similarly to sheep's wool, and much higher than for common acrylics (5-12 %) and modacrylics (10-20 %).

In contrast to wool and cotton, the water uptake does not swell the fiber: fabrics are not compacted and the circulation of humid air through the empty spaces in the fabric is not reduced. The compact fiber surface prevents one from feeling the humidity despite the high water content: sweat is rapidly transported to the fiber surface where it evaporates. Hollow fibers for the desalination of water have similar structures (Fig. 6-27).

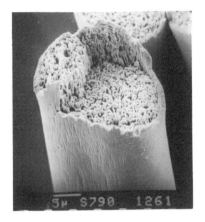 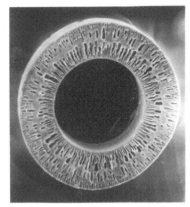

Fig. 6-27 Left: Absorbent acrylic fiber for textile applications (courtesy of Bayer AG); right: hollow poly(acrylonitrile) fiber for water desalination (courtesy of Asahi Chemical).

Such fibers are of interest for textiles that are worn close to the skin (underwear, socks, sports clothes) (for outer wear, see Gore-Tex®, p. 188). Water-repellent membranes are obtained by sintering additive-containing micropowders of poly(tetrafluoroethylene) under pressure just below their melting temperature followed by an extraction of the additives.

Wool Type Polyesters

Textile methods can also produce wool-like fibers from polyesters. Such fibers have stress-strain curves that resemble those of wool (Fig. 6-23). There are also wool-type polyesters that are not very prone to pilling.

Other Wool-like Fibers

Proteins from milk, corn (maize), peanuts, beans, and soy can be solubilized and then spun to so-called **azlon** fibers. The solubilization breaks up quaternary and tertiary structures that are not regenerated on fiber spinning to these **regenerated fibers**. Azlon fibers therefore have lower tensile strengths and higher fracture elongations than wool (Table 6-20) despite similar chemical structures. For economic reasons, azlon fibers are no longer produced.

Wool-like fibers are also formed by the helical α-modification of poly(L-leucine). These fibers are not produced industrially. Their textile properties are shown in Table 6-20 in order to demonstrate the effect of macroconformation on physical and textile properties (compare the properties of poly(L-leucine) in pleated sheet conformation (Table 6-14).

6.3.5 Elastic Fibers

Types

Elastic fibers are desirable for many textiles such as socks or sports outerwear. Such textiles can be obtained by six different approaches: chemical synthesis of elastofibers, chemical after-treatment of fibers, yarns, or fabrics, special processes for fiber spinning, modification of physical properties of fibers, mechanical after-treatment of fibers, and special manufacturing processes for fabrics.

Textile fabrics can be produced by weaving or knitting yarns, or by production of non-woven fabrics from staple fibers or spunbonded sheet products from filaments (Section 6.4). Knitted fabrics are more elastic than woven ones; this was first utilized for woolen jerseys and later for so-called double-knits from polyester fibers (now out of fashion). Fibers can also be made elastic by crimping or texturing.

More drastic changes in the elastic properties of fibers are achieved by chemical synthesis. **Elastic fibers** comprise three groups: elastodiene fibers, elastane fibers, and elastomultiesters (BISFA; see Table 6-1). These entropy-elastic fibers have glass temperatures below room temperature, very large elongations at break, and small textile moduli. So-called **hard-elastic fibers** are energy-elastic and have glass temperatures above room temperature. Their elasticity is not caused by a special chemical structure but by their physical structure and mechanism of deformation. They do not seem to be commercial.

Elastodiene Fibers

Elastodiene fibers, commonly called **rubber fibers**, are the oldest group of elastic fibers. They consist of vulcanized rubbers (see Chapter 7), usually natural rubber but also nitrile or chloroprene rubbers. Elastodiene fibers are obtained by two different methods. In the first process, conventionally produced formulated rubbers are calendered to thin sheets that are vulcanized and subsequently cut to small bands. In the second process, formulated latices are extruded into acidic baths. The coagulated fibers are then continuously washed, dried and vulcanized.

Elastane Fibers

The second large group of elastic fibers comprises elastanes (formerly elasthane; in the United States known as **spandex**, originally a trademark). These fibers are composed of at least 85 % by mass of a polyurethane that contains "soft segments" (blocks with glass temperatures below room temperature) and "hard segments" (blocks with glass temperatures above room temperature) (see Volume II, p. 495).

In the first of such fibers, Lycra® from the DuPont company, soft blocks contain macroglycol units with $P = [CH_2CH_2(OCH_2CH_2)_i]$ whereas hard blocks have hydrazine units $(E = 0)$; both types of blocks contain 1,4-phenylene units $(Ar = -C_6H_4-)$. Elastane fibers from other industrial companies have different chemical compositions such as different macroglycol and chain extender units:

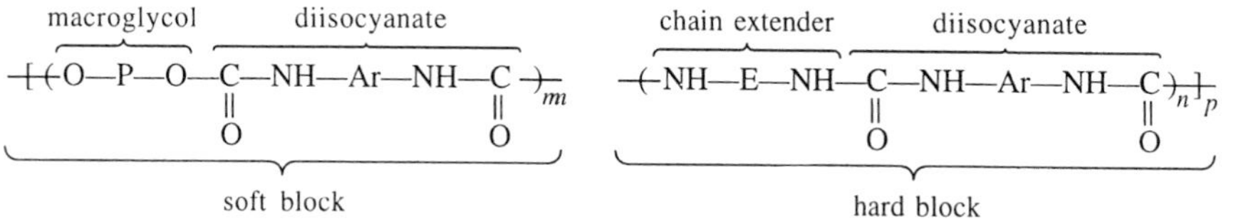

These polymers are processed to fibers by melt or solution spinning. In elastane fibers, the "hard" polyurethane segments serve as physical crosslinkers for the rubbery matrix of "soft" polyether segments.

Table 6-21 Average properties of elastodiene fibers from natural rubber (NR), dry spun (D) or reaction-spun (R) elastane fibers (EA), and hard-elastic fiber (HE) from poly(oxymethylene) (POM) or isotactic poly(propylene) (PP). RH = relative humidity.

Property	RH (%)	Physical unit	NR	EA D	EA R	POM HE	PP HE
Density	65	g/cm^3	0.91	1.21	1.21	1.40	0.91
Textile modulus	65	N/tex	0.0018	0.0044	0.0044	2.1	2.8
Tenacity	65	N/tex	0.079	0.090	0.042	0.24	0.13
Elongation at break	65	%	500	460	520	400	250
Moisture uptake	65	%	~ 0	1	1		~ 0
Water retention	100	%	~ 0	9	9		~ 0
Tenacity	-	tex	-	4.4	7.8	-	-

Elastane fibers have about the same elongation at break as elastodiene fibers but greater tenacity, about twice as high textile moduli (Fig. 6-28) and greater moisture uptake and water retention (Table 6-21). They are usually combined with other fibers.

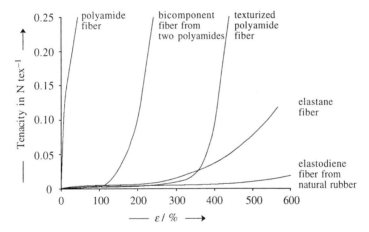

Fig. 6-28 Tenacity as a function of elongation of elastic fibers and a conventional polyamide [21].

Elastomultiesters and Other Elastic Fibers

Elastomultiesters are a new class of fibers consisting of polyester-based bicomponent fibers that will stretch up to 50 % and can return to their original length. The Directive of the European Union defines an elastomultiester fiber as

"a fiber formed by interaction of two or more chemically distinct linear macromolecules in two or more distinct phases (of which none exceeds 85 % by mass) which contains ester groups as dominant functional unit (at least 85 %) and which, after suitable treatment when stretched to one and one half times its original length and released, recovers rapidly and substantially to its initial length."

Apparently, details about the chemical structure of such fibers, their manufacure, and their properties have not been released.

Elastic fibers also comprise poly(butylene terephthalate) (PBT) (p. 186) and **Lastrol**, a new elastic core-sheath fiber with a cotton sheath wrapped around a polyolefin fiber with 99 % ethylene units). Little is known about poly(trimethylene terephthalate) (PTT) and the new fiber **Aquafil**®, a blend of a polyamide and poly(propylene). Fig. 6-28 compares the stress-strain behavior of elastodiene and elastane fibers with a conventional polyamide fiber, a texturized polyamide fiber, and a polyamide bicomponent fiber.

It is unclear whether any of these fibers is a hard elastic fiber. Such fibers, if based on poly(oxymethylene), are less extendable than elastane fibers but have larger elongations at break and far higher textile moduli (Table 6-21).

6.4 Yarns

Silk, cotton, and selected man-made fibers are used to a small extent as threads, for example, as sewing yarns.. In most cases, natural fibers and filaments and staple fibers from man-made fibers are processed further, either directly to nonwovens or spun-bonded materials or first to yarns (this section) and then to textiles (Section 6.5).

6.4.1   Terms

Fibers are flexible materials of great fineness (p. 161) and high aspect ratio (= ratio of length to diameter). A **filament** is a practically continuous fiber. Filaments with some crimp (waviness) are known as **bulked continuous filaments (BCF)**.

A **yarn** is a textile product of substantial length and relatively small cross-section, containing fibers with or without twist, that is spun with speeds of 8-30 m/s (conventional) to 177 m/s (superhigh). **Filament yarns** are composed of one or more filaments. **Monofil yarns** consist of a single filament; they are called **monofils** if they have diameters of more than 0.1 mm. Rigid monofils of limited length are known as **bristles**.

Multifil yarns are filament yarns composed of two or more filaments, often ca. 100 per cross-section. Multifilament yarns are said to be **interlaced** (or intermingled or **tangled**) if the filaments are entwined with random interlacing points. **Tows** are multifil yarns with finenesses of more than ca. 30 000 dtex, assembled from very many non-twisted filaments. Such tows are either cut or stretch broken to staple fibers.

Staple fibers may be cut to bundles of essentially constant length (**square cut**), as tow (see below) **stretch-broken** to a defined limit, or **bias cut** to variable length. **Flock** are very short fibers that are not used for yarn spinning.

The staple fibers are then either used directly for yarn spinning or assembled in "infinitely long" entities in which the staple fibers lie parallel without **twist** (**sliver** or **top**), with a slight twist (**roving**), or bonded by a twist (**spun yarn**). The direction of a twist can be to the left or to the right (Fig. 6-29).

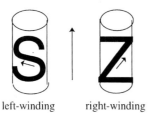

left-winding   right-winding

Fig. 6-29 A twist is a helical disposition of the components of a yarn. It is left-winding (minus; S type) if it winds to the left (small arrow) of the viewing direction (large arrow) and right winding (plus; Z type) if it winds to the right (small arrow).

Single yarns are composed of staple fibers, a single filament, or multiple filaments. Two or more single yarns may be combined by a single twisting operation to a **folded yarn** (also called a **plied yarn**). The combined yarns are called **cabled yarn** if at least one of the single yarns is a folded one and all components are combined by one or more twisting operations. An **assembled yarn** (**multiple wound yarn**) is a yarn from two or more single, folded or cabled yarns that are assembled without twist.

Folded or cabled yarns are called **cords** for certain industrial applications, for example, steel cords for rubber tires. A **compact cord** comprises several filaments that are twisted in the same direction and have the same **lay length** which is the axial distance between the ends of a 360° turn.

A linear assembly of fibers and filaments is commonly called a **strand**. This term is especially common for yarns that are used to make ropes and cordage (thick ropes). "Strand" is also used for assemblies of glass fibers that are used for reinforcement.

Yarns are usually also **textured** (crimped, twisted, etc.) in order to obtain certain textile properties such as **bulk** (the "volume" of a fabric) and **hand** (the feel of a fabric) (next Section). Cotton and wool yarns are combed to remove the shortest fibers. The resulting **worsted wool** is a compacted yarn where the long staple fibers are paralleled (Fig. 6-30, right, bottom) (name from the village of Worthstede in Norfolk, England, where such yarns were first made). Cotton and wool yarns are also **carded**, i.e., the surfaces of yarns are roughened by brushes (*L: carduus* = thistle) (Fig. 6-30, right, top).

6.4.2   Texturing

Spinning of *filaments* of man-made fibers leads to **multifilament yarns** in which the filaments are practically all parallel to the long axis of the yarn (Fig. 6-30, top left). Such yarns are smooth and compact and do not have the properties that one usually associates with textile products. Compact twisting leads to **twisted yarns** (Fig. 6-30, left, second). A **wick** is a twisted yarn of a larger diameter.

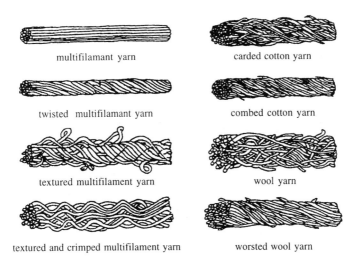

multifilamant yarn   carded cotton yarn

twisted multifilamant yarn   combed cotton yarn

textured multifilament yarn   wool yarn

textured and crimped multifilament yarn   worsted wool yarn

Fig. 6-30 Types of yarns [22]. Left: from man-made fibers; right: from cotton and wool.

Natural fibers have "texture", i.e., they are crimped, but twisted synthetic yarns are smooth and do not stick to each other. Filaments of multifil yarns become irregularly entwined by **texturing** in two or three directions, a method that got its name from the (regularly!) intertwined yarns of wovens (see Fig. 6-32, p. 207) (L: *textus* = woven thing). Textured yarns are bulky and elastic (Fig. 6-30, left, third from above).

Before texturing, filaments must be **heat-set**. The reason for this operation is the following. Filaments arrive from fiber spinning processes (Section 6.2.2) as drawn materials with high strength and elasticity. On heating above a certain temperature (washing, drying, ironing), they would contract, especially if water is present as the swelling agent. Fabrics from such fibers would therefore not be dimensionally stable.

Filaments are therefore heat-set, i.e., heated to a certain temperature which may be done with or without tension and with or without a swelling agent. Heat-setting relieves tensions (and also changes the physical structure of the filament) and the filament remains stable during all subsequent textile operations as long as these are milder than the heat-setting conditions.

More than 90 % of all polyester and polyamide fibers with textile titers of 22-330 dtex are textured by false-twist texturing (see below) which can be used for both multifilaments and staple fibers. The twisting interlocks the fibers which therefore cannot slide past each other if the yarn is stretched.

Similar properties can be obtained by **air-jet texturing** of multifilament yarns (where turbulent air forms loops and curls) and **stuffer box texturing** (where yarn is stuffed by pressure into a heated, pressure-controlled box). Bistructural texturing, an integrated process, has not caught on, in part, because yarns from this process had another hand and crimp than the established friction textured yarns.

In **false-twist texturing**, tows of heat-set twisted *multifil* yarns are continuously drawn under tension, twisted with ca. 2000-3000 twists per meter, heat-set, and then detwisted (Fig. 6-31, bottom; Fig. 6-32).

Yarns from *staple fibers* are produced similarly (Fig. 6-31, top). Thousands of continuous filaments are bundled to a **tow** which is usually crimped by passing the hot tow between intermeshing gear wheels or stuffing it into a confined space before cutting it normal to its long axis. The resulting **flock** is drawn to a band or fleece which is twisted, heat-set, cooled, and detwisted. Such yarns consist of ca. 1000 fibers per cross-section.

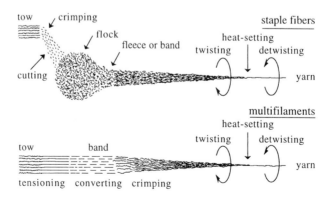

Fig. 6-31 Spinning of yarns from tows of man-made fibers [23]. Top: spinning of staple fibers. Bottom: spinning of multifilaments. By kind permission of Hanser Publishers, Munich, Germany.

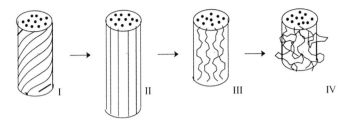

Fig. 6-32 False-twist texturing of multifilaments [24]. (I) The tow of the multifilaments is twisted and heat-set. (II) Tensioning untwists the "natural" helical state of the filaments of I and puts the filaments under torsional stress. (III) The stresses are partially relieved if one lets the untwisted tow II contract by 10-20 % whereupon the filaments form alternating helices.(IV) The helices are stabilized by heat-setting whereupon the filaments relieve torque and coil up to highly snarled forms (see Fig. 6-30, left, third from top).

There are many types of heat-setting and texturing processes. In the Helanca® process, for example, two polyamide multifils are twisted counterrotating which makes the texture permanent. On stretching such yarns, both types of multifils extend to different extents which makes Helanca® yarns highly elastic.

Texturing considerably improves the bulkiness of multifilament yarns. The resulting improvement of wear properties (warmth, hand, etc.) allows such yarns to enter textile applications that are not accessible to yarns that are not textured.

6.5 Textiles

6.5.1 Wovens and Knits

Properties of textiles are not only affected by the chemical and physical structures of fibers and the properties of yarns but also by the way in which fibers or yarns are combined to fabrics, nonwovens, etc. Fibers can be combined to nonwoven or spunbonded materials without the intermediate conversion to yarns; these products are used mainly for non-textile applications such as diapers, filters, carpet backing, etc., and also for medical textiles (surgical gowns, caps, etc.) (Section 6.4.3).

For textiles, yarns are converted to fabrics by weaving, knitting, or crocheting. Weaving is almost completely mechanized; hand-weaving is a rare handicraft. For commercial use, knitting is also done by machines but there is still a lot of hand knitting. Both weaving and knitting may not only be two-dimensional for outerwear and underwear but also three-dimensional for reinforced plastics (Chapter 9). Crocheting is done exclusively by hand; it involves only one active loop instead of several loops as in knitting.

Wovens

Weaving is the oldest process for the manufacture of fabrics. The resulting wovens are two-dimensional materials in which yarns cross each other at constant angles (90°, 60°, etc.) (Fig. 6-33). Depending on the bonds between the yarns, one distinguishes plain weaves (tabby or taffeta), twills and satin weaves.

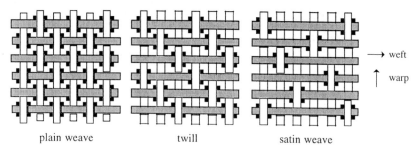

plain weave   twill   satin weave

Fig. 6-33 The three basic weaves. The points ■ contact each other at yarn crossings in plain weaves and twills (for clarity, not shown as contacting) but not in satin weaves.

In **plain weaves**, horizontal, parallel running yarns (**weft, filling yarn, woof**) are woven alternately above and below at right angles through vertically arranged, parallel yarns (**warp**) (Fig. 6-33). In plain weaves and twills, weft and warp contact each other at each intersection which makes the resulting wovens dense and strong.

Plain weaves from natural fibers (wool, silk, cotton) have different appearances and heavinesses. The corresponding textiles were therefore given special names that were later carried over to plain weaves from man-made fibers that looked and behaved similarly: cloth (originally from wool; "cloth" now designates any fabric from natural or man-made fibers, whether woven, knitted, felted, etc.), taffeta (originally from silk; now also from rayon or nylon; glossy and alike on both sides), batiste (the finest plain weave from cotton, now for clothing from any type of fiber), cambric (a fine plain weave from cotton or linen), calico (a coarse cotton cloth), cretonne (a heavy plain weave from cotton), chiffon (sheer silk or rayon), organza (a stiff fabric from silk or man-made fibers), etc.

In **twills**, one weft thread runs over one warp thread and then under 2-12 warp threads (in Fig. 6-33, only 2). The contact points touch each other on diagonals to weft and warp. Such twill bonds can be varied considerably, either via the number of bound warp threads or via the slope of the diagonal warp/weft bonds. Twills are the weakest of the three plain weaves. Twills include chino, denim, gabardine, serge, and tweed.

In **satin weaves**, four or more weft yarns are floating over a warp yarn or vice versa; there are no contact points at the crossing of weft and warp. A fabric with satin weave is called satin if it shows the warp threads and sateen if the weft threads are visible.

In these weaves, weft and warp can not only be arranged to each other at angles of 90° (called biaxial) as shown by Fig. 6-33 but also at angles of 60° (called triaxial). One can weave tightly or loosely, vary the type and color of yarns, etc. These differences affect the strength, hand, bulk, gloss, etc. of the fabric. For example, plain weaves look relatively dull whereas satin weaves are glossy.

Since warp threads are subjected to high mechanical stresses during weaving, fibers of these threads should not stick out of the surfaces of yarns. They are therefore held down temporarily by glues called (textile) **sizes** which make the yarn more compact, smooth, and soft.

Sizes are aqueous solutions of starches, phosphate groups containing starches, carboxymethyl celluloses, poly(galactomannane)s, polyacrylates, and poly(vinyl alcohol) with 10-20 % acetate groups. Sizes are usually removed after weaving: most of them are washed out but starch must be removed enzymatically or by oxidation.

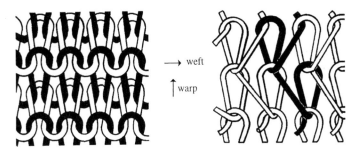

Fig. 6-34 Two knits. Left: weft knitted fabric with horizontally arranged meshes (crochet type). Right: warp knitted fabric (with one warp thread shown in black).

Knits

Yarns in knits are intertwined to give double-S type loops. In hand knitting, crocheting, and simple machine knitting, meshes are formed from one thread that runs horizontally and is continuously interlocked with another thread (Fig. 6-34, left). Machine knitting allows many threads to unite to meshes simultaneously (Fig. 6-34, right). Knits can be very loose (typical for crochets (Old French: diminutive of *croc(he)* = hook)) or very dense like a plain weave. All knits are considerably flexible.

The mechanical wear comfort of textiles is thus controlled not only by the tenacity = f(elongation) characteristics but also by the type of yarn spinning and fabric manufacture. In general, knitted fabrics are more flexible and extendable than woven ones. They are therefore more suitable for sports wear and non-iron, crease-resistant clothes. But they are also more prone to runs (ladders) and running threads.

Other properties of textiles can also not be traced directly to fiber properties. An example is the burning of textiles which depends not only on the limiting oxygen index, the melting temperature, the flow behavior, etc., of polymers but also on the fiber and textile construction (especially the weight per area) and the applied dyestuffs and finishes. For example, the time to ignition increases with the area weight if the textile chars but decreases if the polymer melts.

6.5.2 Nonwovens

Production speeds of woven and knitted textiles are not very high (Table 6-22) and one therefore tries to reduce the number of steps that are necessary to produce textiles from fibers (Section 6.4). In general, one distinguishes between **staple nonwovens** from staple fibers and **spunlaid nonwovens** from filaments.

Table 6-22 Machine production speeds of textiles in area per hour.

Process	$v/(m^2\ h^{-1})$	Process	$v/(m^2\ h^{-1})$
Weaving	5	Staple nonwovens, dry process	5 000
Knitting, crochet-type	20	Staple nonwovens, wet process	30 000
Knitting	80	Paper production	240 000

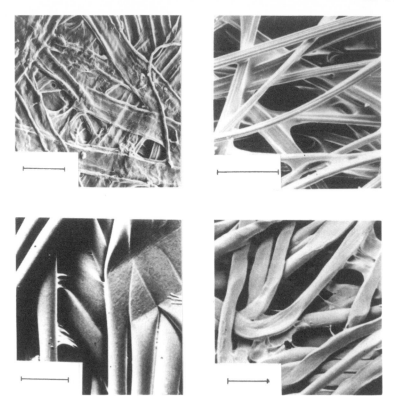

Fig. 6-35 Paper and nonwovens [25]. The marks correspond to 100 μm.
Top left: paper from pine wood   Top right: nonwoven from latex-bound rayon staple fibers.
Bottom left: spunlaid poly(ethylene)   Bottom right: heat-bound poly(propylene) fibers

Nonwovens (one word in industrial use!) are produced from fibers with far greater speeds than wovens and knits (Table 6-22). They resemble paper (Section 6.9) more than fabrics but differ from common papers by the arrangement of staple fibers and filaments, respectively (Fig. 6-35). Other non-woven sheet products from fibers are leathers (Section 6.9) whereas films and sheets are also two-dimensional but not made from fibers (Chapter 11).

Nonwovens are defined as flexible, porous, sheet-like entities from textile fibers that are bound thermally or chemically. Fibers may be rayon, polyamide, polyester, or olefin, either as staple fibers of at least 1 mm length or as filaments. Thicknesses of nonwovens vary between 25 μm and several centimeters and area weights between 10 g/m^2 and 1000 g/m^2. Typical applications are for hygiene (diapers, wipes, bandages, etc.), medical purposes (surgical gowns, caps, etc.), filters (water, coffee, gasoline, oil, etc.), geotextiles (barriers, soil stabilizers, frost protection, etc.), insulation, home textiles (pillows, upholstery padding, carpet backing, etc.), disposable clothing, etc.

Staple nonwovens are produced by a wet or dry process from either rayon or now mainly from polyester or poly(propylene). In the *wet process*, fibers of 5-40 mm length are suspended in a large volume of water and then placed at random on a sieve in a kind of filtration process. The necessary dilution ratio is proportional to the titer and inversely proportional to the square of the fiber length. Fibers are then bonded mechanically,

Table 6-23 Comparison of some properties of paper, nonwovens, and fabrics. *d* = thickness, *E* = tensile modulus, *σ* = bursting strength, *F* = tear strength, *f* = length-related breaking force, *ε* = elongation at break. PA = polyamide, PE = poly(ethylene).

Material	$\dfrac{d}{mm}$	$\dfrac{E}{MPa}$	$\dfrac{\sigma}{MPa}$	$\dfrac{F}{MPa}$	$\dfrac{f}{N\ cm^{-1}}$	$\dfrac{\varepsilon}{\%}$
Kraft paper	0.12	1000	0.18	1.3	49	4.3
PE nonwoven, spunlaid area-bonded filaments	0.20	860	1.1	4.0	79	29
PE nonwoven, spunlaid point-bonded filaments	0.15	147	0.39	-	19	19
Cotton fabric	0.26	46	0.60	13	57	12
Polyamide/rayon, nonwoven	0.26	8.6	0.25	5.3	12	72

thermally, or chemically, for example, by needles, pressure, glues, swelling of fiber surfaces by a solvent, etc.

The *dry process* uses longer staple fibers of 30-100 mm length or filaments that are placed on a sieve by an air stream or by carding (= brushing). The fibers are then bonded thermally or chemically.

For **spunlaid nonwovens**, fibers from polyolefins or polyester are spun and laid on a belt either mechanically by deflectors or by an air stream. The fibers are then bonded point-like or area-like with polyester binders. Point-bonded spunlaids behave like staple nonwovens whereas area-bonded spunlaids are more like paper (Table 6-23).

6.6   Textile Finishing

6.6.1   Overview

"Textile finishing" comprises all treatments that increase the perception and use properties of textiles, i.e., color, gloss, etc., on one hand and resistance to crease and shrinkage, non-iron, dirt repelling, etc., on the other. These treatments may take place on polymers before fiber spinning and then at every step of the textile chain (Fig. 6-20), i.e., at filaments, staple fibers, yarns, fabrics, or textiles. Which process is used at which part of the textile chain depends on the state of delivery and the properties of the materials, the desired effects, the economics, and the ability to react fast to the latest fashion trends.

Polymers for man-made fibers offer opportunities for such treatments during polymerization itself such as copolymerization with monomers that reduce pilling (Section 6.3.4), increase dyeability (Section 6.6.4), or reduce flammability (Section 3.5.6). Polyolefins are predominantly pigmented before fiber spinning because fibers, yarns, and textiles are difficult to dye.

Melt spinning is accompanied by the establishment of polymerization equilibria. The produced monomers, oligomers, and low-molecular weight degradation products accumulate at the fiber surface where they cause problems during fiber spinning, texturizing, and dyeing. They are therefore extracted.

Natural fibers cannot be used directly for textile or industrial purposes and must be worked up before yarn spinning by degreasing (wool), debasting (silk), etc.

6.6.2 Fiber and Fabric Treatment

Cellulose textiles (cotton, rayon) are **weighted** with mineral compounds (kaolin, talc, barium sulfate, magnesium salts) or urea in order to improve the bulk, hand, and area weight. These compounds are usually not resistant to washings. For the same reason, silk is **charged** with aqueous solutions of Na_2HPO_4 and $SnCl_4$ (p. 183).

Cellulosics with a softer hand are obtained by treatment of fibers and fabrics with **softening agents** such as waxes, oils and fats, hygroscopic compounds such as glycerol or poly(ethylene glycol)s, or reaction of the fiber with, for example, amides of N-hydroxymethyl fatty acids.

The flexural stiffness of fabrics can be improved by so-called **permanent starches** (**wash-fast starches**), for example, dispersions of poly(vinyl acetate) or poly(acrylic esters). Starches themselves as well as carboxymethyl celluloses and cellulose ethers, are easily washed out.

6.6.3 Easy Care

Fabrics and textiles from cellulose fibers have difficulties competing with those from synthetic fibers with respect to creasing and swelling by water. Cellulosic fabrics are therefore treated with various chemicals in order to improve the care properties.

Cellulosics crease because flexural bending causes crystalline regions to be shifted against amorphous regions. The mobility of cellulose chains in these regions is increased if cellulose is plasticized by water (p. 190), i.e., at high relative humidities or during washing. Crosslinking reduces the mobility of chains but it also decreases strength, especially that of cotton, and reduces moisture uptake and dyeability. Fabrics are therefore dyed *before* they are crosslinked. Crosslinking agents are aldehydes, urea-formaldehyde precondensates, and so-called reactant resins.

Crosslinking of rayon and cotton by aldehydes has been used for decades in order to improve the insufficient crease resistance in the dry state and the poor dimensional stability in the wet state (high stabilities are desired for wash-and-wear properties). The goal is the formation of formals as bridges between cellulose chains, for example,

(6-2)   cell–OH + HCHO + HO–cell →   cell–O–CH_2–O–cell + H_2O

However, formaldehyde also forms hemiacetals which reduce the formation of hydrogen bonds between adjacent hydroxy groups of a cellulose chain without adding anything to crosslinking. The loss of intramolecular hydrogen bonds reduces the rigidity of chains and is therefore probably responsible for the reduced wet strength of cotton. Indeed, such an effect is not shown by the tetrafunctional glyoxal, OHC–CHO.

Competition by synthetics also forced manufacturers of cotton fabrics to develop "no iron" types, also known as **durable press** or **permanent press**, by crosslinking cellulose chains. In principle, many crosslinking agents are available. However, such chemical compounds must not only be bifunctional or greater but must also be able to enter the amorphous regions of celluloses and be less hydrolyzable than the aldehydes used in the wash-and-wear finishing processes.

Commonly used are aqueous solutions of urea-formaldehyde precondensates (UF resins) with a formaldehyde/urea ratio of (1.3-1.8):1. These oligomers enter inter-micellar regions of cellulose where they harden at elevated temperatures by polycondensation (hence also known as resin formers or self-crosslinking agents). The increased crease resistance is thought to be caused mainly by the stiffening action of the resin "particles" and not by the crosslinking of cellulose chains by formaldehyde that is set free during the condensation reaction (see also Volume II, p. 484).

In contrast to fabrics treated with UF precondensates, those treated with so-called **reactant resins** can be boiled during washings. Such resins are mainly methylol derivatives of cyclic urea compounds (I-IV) which react mainly by intermolecular crosslinking of cellulose chains and less by self-condensation to resins:

1,3-dimethylolpropylene urea (I)

1,3-bis(methoxymethyl)-uron (II)

1,3-dimethylol-5-alkyl-triazone (III)

1,3-dimethylol ethylene urea (IV)

butadiene epoxide (V)

divinyl sulfone (VI)

However, all nitrogen-containing compounds I-IV react on washing with chlorine-containing water with formation of chloramine, NH_2Cl. Ironing such washed textiles releases hydrochloric acid which damages and discolors the fibers. These side reactions are not possible if nitrogen-free reactive compounds are applied, such as butadiene diepoxide (V), divinyl sulfone (VI), and others. Epoxy and vinyl groups add to the hydroxy groups of cellulose.

The constitution of the crosslinking agent does not matter much as far as the lowering of mechanical strength is concerned: crosslinking reduces the mobility of chains and therefore also the shear strength and strength at break. However, it does matter whether the fabric is crosslinked in the dry or in the wet state because different space and time averaged macroconformations are laid down in each state. After washing and drying of a dry-crosslinked fabric, chain segments will attain the same statistical distribution as they had before washing: the dry crease resistance will not change. However, the wet crease resistance of such a dry crosslinked fabric will differ. Conversely, a wet crosslinked fabric will have the same wet crease resistance but a different dry crease resistance.

6.6.4 Dyeing

Dyeing can take place at different stages of the textile chain (p. 180), depending on the origin of fibers, their chemical constitution, and the desired effect. Dyestuffs may be pigments (insoluble in the fiber) or dyes (soluble in or reacted with fibers) (Section 3.2).

Textile fibers and industrial fibers differ in the following aspects. In general, wear comfort requires textile fibers with considerable moisture uptake and good water retention; in technical fibers, these properties are usually not wanted. The hydrophobic polyolefins are therefore typical technical fibers which are, however, used for home textiles that do not require wear comfort such as carpets. Conversely, hydrophobicity of polymers may not necessarily mean very low moisture uptake and water regain because textile engineering may generate hydrophobic fibers that do take up fairly large proportions of water, for example, by capillary action. Textile fibers should also be extendable (fairly large elongations at break) and not too rigid (not too large tensile moduli) whereas industrial fibers usually do require large elastic moduli.

Natural industrial fibers (jute, sisal, etc.) are staple fibers (Table 6-24) whereas spun synthetic industrial fibers are obtained as filaments that are used as such or cut to staple fibers. Glass fibers are also milled. Fiber-like materials are also produced by dividing monoaxially stretched sheets in the stretch direction (Section 6.7.3).

6.7.2   Vegetable Fibers

Fibers from plants are commonly called **vegetable fibers** (UK: **plant fibers**), and more recently, **vegetal fibers**. Based on their origin, they are usually subdivided into **bast fibers** from the skin of stems (**stem fibers**) (examples: flax, jute), **leaf fibers** from leaves and fruits (example: sisal), and **seed hair fibers** (example: cotton).

The main chemical compound of all these fibers is cellulose (Table 6-15) but the compositions vary widely: cellulose from 93 % (cotton) to 64 % (flax, jute), polyoses (hemicelluloses, see Volume II, p. 420) from 0.2 % (jute) to 20 % (Manila hemp), lignins from 0 % (cotton) to 12 % (jute), and variable proportions of pectins (0.5-2 %), fats and waxes (0.2-1.5 %), and water-soluble compounds (1-5.3 %).

Cotton is the most important of these fibers (Fig. 6-1 and p. 188); it is used practically exclusively for textiles. All other vegetable fibers are now considered too stiff for comfort which is the reason that linen (from flax) has fallen out of favor for tropical suits. Vegetable fibers are still produced worldwide in considerable amounts: jute with ca. $4 \cdot 10^6$ t/a is the second most important natural fiber after cotton, followed by flax with almost $1 \cdot 10^6$ t/a (increasing), sisal (about constant), and hemp (strongly decreasing).

Bast Fibers

Bast fibers are obtained from stems of plants. Important bast fibers are

flax from *Linum usitatissimum*) for the production of linen;
hemp from *Cannabis sativa sativa* which contains only small amounts of psychoactive
    δ-tetrahydrocannabinol (hashish is from *Cannabis sativa indica* (makes poor fibers));
jute from two species of the genus *Corchoru* (*C. capsularis* and *C. olitorius*);
kenaf (*Hibiscus cannabinus*) delivers the **combo fiber** (**gambo fiber**);
ramie (**China grass, rhea**) from the nettle *Boehmeria nivea*;
rosella (**Java jute**) from the mallow *Hibiscus sabdariffa*;
sunn (**Madras hemp, Bombay hemp**) from *Crotalaria juncea L.* (pea family);
urena (**Congo jute, cadillo, aramina**) from the mallow *Urena labata*.

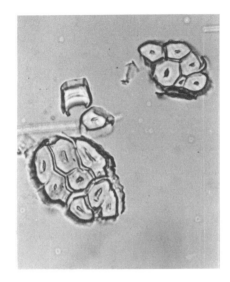

Fig. 6-36 Longitudinal view (left) of a flax fiber and cross-section of bundles of its ultimates (right) [14]. The ultimates are hollow fibers

Bast fibers are multicellular structures (Figs. 6-36 and 6-37) of considerable length, often ranging to several meters (see Table 6-24), which contrasts with the single-cell structure of cotton fibers. The fibers consist of bundles of interlocking structural elements called **ultimates** by botanists (short for **ultimate cells** or **ultimate fibers**). Ultimate fibers have lengths in the millimeter to centimeter range (they are the "fibers" of scientists). The ultimates form bundles (Fig. 6-37, left) that are held together be resins and other substances. In the bundles, ultimates are longitudinally disordered which allows some ultimates of a bundle to join up with other bundles. The whole fiber is therefore like a mesh in which interconnections between bundles are the links (see also the thickened portions in the flax fiber (Fig. 6-36, left)).

The ultimate fibers are hollow polygons (Fig. 6-36, right) with diameters of several micrometers that can take up considerable proportions of water (Table 6-24).

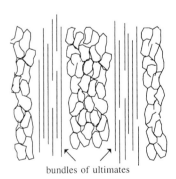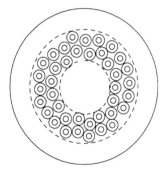

bundles of ultimates

Fig. 6-37 Longitudinal section (left) and cross-section (right) of multicellular fibers of a bast plant (schematic). The fibers consist of bundles (aggregates) of ultimates (shown left as vertical lines and right as circles) that are held together by various materials (resins, etc.) (dimensions not to scale). The cross-sections of ultimates are hollow polygons.

Table 6-24 Properties of important bast fibers at a relative humidity of 65 % [26]. 1 dtex = 1.11 den.

Property	Unit	Flax	Hemp	Jute	Kenaf	Ramie	Sunn
Length of fibers	cm	20-140	100-300	150-360	15-20	40-150+	
Length of ultimates	cm	0.44-7.7	0.5-5.5	0.07-0.7	0.15-1.1	2.5-25	0.2-1.4
Diameter of fibers	μm	40-620		30-140		60-900	
Diameter of ultimates	μm	5-76	10-51	15-25	12-36	16-126	8-61
Fineness	dtex	1.9-19.6	3.3-22	14-30	55	5-7	
Porosity of fibers	%	11		15		8	12
Water regain at 22°C	%		8	12		6	
Swelling in water	vol%	30		44		32	45
Textile modulus, initial	N/tex	19	18-22	13-26		15	
Strength at break	N/tex	0.46	0.57	0.38		0.58	
Elongation at break	%	3	1.6	1.2-1.7	2.7	4	

Bast fibers have generally higher textile moduli, greater strengths at break, and smaller elongations at break than cotton, which translates into less wear comfort (Table 6-24). Therefore, with the exception of ramie and flax, they are not used for textiles but for industrial applications such as twine, rope, carpet backing, canvas, geotextiles, etc.

Jute is the second most important natural fiber after cotton (Fig. 6-3). It is made into coarsely woven fabrics for textiles (Hessian cloth) and is now used as burlap (also from hemp or flax), for sacks (gunny bags (from Sanskrit: *goni* = sack)), etc. Since it is the least expensive of the strong bast fibers, it has found many other applications such as carpets and carpet backing, geotextiles, and reinforcing fiber for plastics.

Jute is a water-loving plant of the genus *Corchurus* that is grown in the tropics and subtropics where the days are long (Bangladesh, India, etc.). **Java Jute** (= **rosella**) and **Congo jute** (= **urena**) have similar properties to those of true jute but are botanically different; they are local crops and not traded world-wide. **Kenaf** is also a mallow like **rosella** and **urena**. It needs less water than jute and is now cultivated in China, Japan, South America, Mexico, the United States, and Southern European countries.

The word **flax** refers to both the flax plant and the unspun fibers of that plant. Flax fibers are obtained from *uprooted* plants which allows one to obtain the best yield of long fibers. The subsequent work-up by hand is still being practiced in agricultural societies but is increasingly being replaced by commercial processes in industrial countries. The traditional method places the flax plants for 3-5 weeks in the field where they were retted by the action of light, oxygen (oxidation of lignins), and bacteria (degradation of pectins). Field retting is preferred over retting in ponds, streams, and containers because it delivers the highest quality flax and the least pollution.

The stalks are then "dressed", i.e., the stalks are broken, the straw removed from the fiber by scutching (scraping with the edge of a knife), and then by pulling them over several beds of nails (hackles) (see also Volume II, p. 393). The "tow" coming from the densest hackles can be carded and spun like wool.

Commercial methods use retting in tanks, smashing stalks by fluted rollers, and scutching by rotating bladed wheels (Volume II, p. 393). The broken fibers plus stalk and shive are used for ropes and coarse fabrics. The remaining bundles of long fibers are separated by a combing machine to give individual fibers (line) which are spun to **linen**. The yield is 12-16 kg of line fiber per 100 kg of flax plants.

Ramie (Malay: rami), also called **Rhea** or **China grass**, is obtained from the 1-2.5 m high China grass. The stems consist of a thin exterior bark, the fiber containing cortex, a wooden layer, and the soft interior pith (medulla) which comprises ca. 2/3 of the stem diameter. In order to obtain the fibers, the stems are broken, the pith pushed out, and the fibers carefully isolated. The yield is ca. 2-4 wt% of the weight of the plant (80 wt% is water). Raw fibers contain 25-35 wt% xylans and arabans which are removed by cooking the fibers with alkali (degumming). The total yield of ramie fibers is 1-1.5 wt%.

For textile applications, ramie and flax are cut to staple fibers. Both flax and ramie have greater diameters, far smaller lumens, and fewer vacuoles than cotton (cf. Tables 6-24 and 6-25). The tensile modulus of flax fibers is therefore much greater than that of cotton fibers (60 versus 5.7 N/tex); it almost reaches the lattice modulus of cellulose I (84 N/tex, corresponding to 129 GPa) (for cellulose, see Volume II, p. 388 ff.). Flax and ramie are therefore stiff fibers with only small elongations at break (Table 6-24).

Hemp is grown in many countries but not in the United States since it may be used to produce marijuana from its dried flower clusters and leaves and hashish from the resin of the flower sprouts. The structure, work-up, and use of hemp is similar to that of flax. It now also serves as a reinforcing fiber in plastics for reinforced materials. The world-wide production of hemp has been declining for several years (Fig. 6-3).

Some other plants produce fibers with similar properties to those of hemp. These fibers are either stem fibers from another family or are leaf fibers. **Madras hemp (Bombay hemp)** is a stem fiber of the pea family that is generally known as **Sunn**. **Manila hemp** is a leaf fiber otherwise known as **abaca**. **Sisal hemp** is just another name for **sisal**.

Leaf Fibers

Leaf fibers are also called **hard fibers**. They are obtained from leaves in the widest botanical sense. For example, the stems of the banana-like plant *Musa textilis* are not woody trunks like those of flax or hemp but consist of layers of sheaths that are wrapped around each other. This system is called a leaf by botanists; hence, their fibers are classified as leaf fibers. The group of leaf fibers consist of

abaca (musa fibers, Manila hemp) from the banana-related *Musa textilis*;
banana from *Musa cavendishi* and *M.* sapientum;
esparto (alfa grass) from the grass *Stipa tenacissima*;
fique (Mauritius fiber) from the agave-like *Fuorcroya gigantea*;
henequen from the *Agave fourcroydes*;
phormium (New Zealand flax) from the Liliacae *Phormium tenax*;
sisal (sisal hemp) from the *Agave sisalana*.

Seed hair fibers

The group of seed hair fibers comprises

akon fiber (calotropis) from the seed hairs of *Asclepiadaceae* and *Apocynaceae*;
coir (coconut fiber) from the hairs of the nuts of *Cocos nucifera*;
cotton = seed hairs from the bolls of *Gossypium* (p. 188);
kapok (Bombax wool) from the fruits of the kapok tree *Ceiba pentadra*.

In addition, there are several other vegetable fibers that are only regionally important. Many vegetable fibers are known by different names: 340 names for 26 plants!

Table 6-25 Properties of some leaf fibers and seed hair fibers at a relative humidity of 65 % [26]. [1)]
Scoured (p. 197); mercerized cotton has a water regain of 8-12 %.

| Property | Unit | Leaf fibers | | | | Hair fibers | |
		Abaca	Banana	Henequen	Sisal	Coir	Cotton
Length of fibers	cm	60-200+	80-280		40-100+		1.5-7.5
Length of ultimates	cm	0.33-1.2	0.09-0.55		0.08-0.75	0.03-0.1	1.5-7.5
Diameter of fibers	μm	10-280	11-34		100-460	100-450	12-25
Diameter of ultimates	μm	6-46	18-30	8-33	7-47	12-24	12-25
Fineness	dtex	42-440	60-75	400-520	10-450		1-3.3
Porosity of fibers	%	17-21	35-53		17		
Water regain at 22°C	%	9.5	15		11	10	6 [1)]
Swelling in water	vol%	42			40		
Textile modulus, initial	N/tex			13	9-26		5.7
Strength at break	N/tex	0.64		0.39	0.40		0.35
Elongation at break	%	2-4.5		3.5-5	2-7	15-40	7

Properties of some leaf fibers and seed hair fibers are shown in Table 6-25. Abaca fibers are largely obtained by manual desheathing in the Philippines and by mechanized decortication in Central America and Indonesia. The world production of abaca is ca. 100 000 t/a whereas that of banana is only ca. 1000 t/a. The productions of sisal and henequen are each estimated to be in the 300 000 t/a range. All these leaf fibers are used for similar purposes to those of the stem fibers; sisal and henequen corner the cordage market (two-thirds of the world production).

Hair fibers are single cell fibers and not multicell fibers like stem and leaf fibers, i.e., hair fibers and ultimates (p. 217) are identical (Table 6-25). Cotton is the dominant textile fiber (p. 188). Coir comes in a brown variety from mature coconuts and a "white" variety from immature ones; 90 % are produced in Sri Lanka. The brown fiber is coarse and rigid and used in mattresses, car seats, brushes, geotextiles, and the like whereas the finer and more flexible variety serves in woven carpets, tufted floor mats, etc.

Besides these commercially important fibers, there are many others that are only locally important (for example, aloe) or have been used as substitute fibers in difficult times (broom, cattail (reed mace), hop, lupin, nettle, willow).

6.7.3 Split, Slit, and Fibrillated Fibers

Fibers and fiber-like products can not only be produced by fiber spinning but also by subdividing thin, endless products such as thin sheets (films) or thin bands that are up to 5 mm broad and up to 80 μm thick (Fig. 6-38). Such processes are attractive, not only because of lower investment and operational costs but also because they allow one to produce products with properties that cannot be obtained from fibers. Applications include yarns, ropes, bags, carpet backings, and geotextiles as well as other uses.

Early types were obtained from rayon, poly(styrene), poly(vinylidene chloride), or poly(vinyl chloride). They are now made almost exclusively from polyolefins (p. 162), especially from isotactic poly(propylene) since the ability to fibrillate increases in the order PES < PA < HDPE < PP, i.e., with increasing degree of crystallinity.

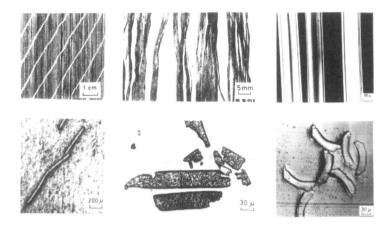

Fig. 6-38 Longitudinal view (top) and cross-sections (bottom) of slit fibers (left and right) and split fibers (center) from poly(propylene) [27]. The fibers were produced from films by needle rollers (left), by twisting slit film fibers (center), or by splitting fibers (right). 1 μ = 1 μm.

Slit film fibers are obtained by cutting extruded endless films with many parallel knives. The resulting fibers are then monoaxially stretched. Alternatively, one can stretch first and cut later. The resulting slit fibers are "homogeneous", i.e., they do not contain fibrils or capillaries as split fibers do.

Split film fibers result from the lengthwise division of bands or of endless films into a network of endless fibrils which may remain interconnected at certain points (Fig. 6-38, top center; Fig. 6-39) and may be curled or not. Such split film fibers can be obtained by three methods: uncontrolled mechanical, uncontrolled chemical-mechanical, and controlled mechanical.

In the *uncontrolled mechanical* method, groves of finite length and different depth are created by rotating cylindrical brushes. The thin portions of the groves break which results in a network of fibrils with very broad distributions of fibril sizes. For poly-(propylene), the strength of these split film fibers is about the same as that of the starting film whereas for polyamide and polyester it is about half as large.

In the *uncontrolled chemical-mechanical* method, the polymer is provided with an incompatible substance before film formation, for example, a salt or another polymer. The added material forms randomly distributed clusters that become oriented on film stretching. The film breaks at these inhomogeneities and forms a network of fibrils of irregular thickness.

The most important method is *controlled fibrillation* by rolls with teeth or needles or by closely packed knives at controlled distances with controlled slitting depths.

Fig. 6-39 Schematic representation of a split film fiber. Such fibers may contain capillaries and may be separated further into film fibers of finite length with more evenly distributed titers.

6.7.4   Carbon-chain Fibers

Poly(olefin) fibers (olefin fibers) (PO) of statistics often include not only fibers from poly(ethylene), $-\!\!\!\left(CH_2\!-\!CH_2\right)\!\!_n\!$-, and it-poly(propylene), $-\!\!\!\left(CH_2\!-\!CH(CH_3)\right)\!\!_n\!$-, but also those from poly(vinyl chloride), $-\!\!\!\left(CH_2\!-\!CH(Cl)\right)\!\!_n\!$-, and its copolymers, and even those from poly(styrene), $-\!\!\!\left(CH_2\!-\!CH(C_5H_6)\right)\!\!_n\!$-, all of which are fibers with carbon chains. It is therefore convenient to also treat in this Section the other carbon-chain fibers, i.e., halogenated ones and those from poly(vinyl alcohol), $-\!\!\!\left(CH_2\!-\!CH(OH)\right)\!\!_n\!$-. For all of these fibers, newer production statistics are either non-existent or are hard to find in the open literature.

True poly(olefin) fibers (PO) are produced in the greatest amounts of all carbon chain fibers because of the low cost of polymers and melt spinning; poly(olefin) fibers comprise 16 wt% of all manufactured fibers. China is now the largest producer. Since 2000, the production of poly(olefin) fibers has decreased in the United States.

90 wt% of poly(olefin) fibers are isotactic poly(propylene)s and 10 wt% are high and ultrahigh molecular weight poly(ethylene) (Fig. 6-2). 43 % of poly(olefin)s are used as filaments, 35 % as film fibers, and 22 % as staple fibers.

Poly(propylene) dominates because of its excellent fracture strength (Table 6-26). Because of its high crystallizability, it has a larger technical tensile modulus then poly-(ethylene) although its theoretical tensile modulus is only 1/8 of that of poly(ethylene) (Volume II, p. 545). However, gel spinning and ultradrawing of ultrahigh molecular weight poly(ethylene) (p. 167) delivers a highly oriented (> 95 %), highly crystalline (> 85 %) fiber with a density of 0.97 g/cm^3, a textile modulus of 90 N/tex (87 GPa), and a tensile strength at break of 2.7 N/tex (2.6 GPa) (Dyneema®, Spectra®).

All poly(olefin) fibers have good resistances against acids, alkali, and other chemicals. Because of this high inertness, they cannot by dyed but must be melt-pigmented.

Chlorofibers (**CLF**) contain either vinyl chloride units (VC units), $-CH_2\!-\!CHCl-$, or vinylidene chloride units (VDC units), $-CH_2\!-\!CCl_2-$. Fibers of the subclass **vinyon** (FTC) have at least 85 wt% (FTC) or 50 wt% (EU) VC units and those of the subclass **saran** at least 80 wt% (FTC) or 50 wt% (EU) VDC units (Table 6-1). Saran is now a free name in the United States but not necessarily in other countries.

PVC fibers are the oldest synthetic fibers (1934). Chlorofibers may be homopolymers of vinyl chloride, after-chlorinated poly(vinyl chloride)s, or copolymers of vinyl chloride with vinyl acetate, methyl methacrylate, or acrylonitrile. Copolymers of vinylidene chloride with acrylonitrile must have >65 wt% VDC and <35 wt% acrylonitrile. Chlorofibers have good resistances against chemicals and fire but low softening temperatures and dimensional stabilities. Their world production is estimated as 20 000 t/a (1987).

Fibers from **poly(tetrafluoroethylene)** (**PTFE**) are cherished because of their excellent resistance against chemicals, very good temperature resistance, low friction, and excellent triboelectric properties. The polymer was accidentally discovered by Roy Plunkett at Kinetic Chemicals (1938) which patented it in 1941; DuPont then took a license and marketed it as **Teflon** (now a free name in the United States). Because of the high melting temperature of 327°C, PTFE fibers cannot be spun from the melt. Fibers can also not be spun dry or wet from solution because suitable solvents are unknown. PTFE fibers are rather produced by dispersion spinning of PTFE particles in aqueous solutions of poly(vinyl alcohol) or of PTFE dispersions in kerosine.

Table 6-26 Average properties of olefin and vinyl filaments. HDPE = conventional high-density poly(ethylene), PP = it-poly(propylene), PS = atactic poly(styrene), PTFE = poly(tetrafluoroethylene), PVAL = poly(vinyl alcohol), PVC = poly(vinyl chloride). * Decomposition. High-modul fibers of ultrahigh HDPE-HD have values of $\rho = 0.97$ g/cm³; $E = 172$ GPa; $\sigma_B = 3.0$ GPa. a = amorphous.

Property	RF (%)	Physical unit	PP	HDPE	PTFE	PVC	PS	PVAL
Density	65	g/cm³	0.90	0.94	2.2	1.40	1.05	1.28
Textile modulus	65	N/tex	6.0	2.5	1.2	3.0		3.8
Strength at break	65	N/tex	0.62	0.27	0.13	0.25	0.48	0.66
	100	N/tex	0.62	0.27	0.13	0.24	0.50	0.50
Elongation at break	65	%	20	30	23	17	< 10	16
	100	%	20	30	23	16	< 10	21
Moisture uptake	65	%	0	0	0	0.1	< 0.1	4.2
Water regain	100	%	0	0		5		30
Melting temperature	0	°C	165	125	325	a	a	250*
Glass temperature	0	°C	– 14	– 80	– 52	68	100	81
Oxygen index (LOI-U) (p. 66)	0	%	17	18	95	32	18	22

In the 1930s, filaments from poly(vinyl alcohol) (PVAL) were produced in Germany for a short time by dry spinning of aqueous PVAL solutions. Considerable progress was then made in Japan in the 1940s by the treatment and acetalization of wet spun fibers (**vinylal (EU)**, **vinal (FTC)**). The resulting *crosslinked* fibers are resistant to water. Because of their hydrophilic character, they were originally used as a cotton substitute.

Wet-spun PVAL staple fibers are still produced in North Korea and China with estimated production capacities of 50 000 t/a and 150 000 t/a, respectively. A Japanese dry spinning process delivers PVAL filaments with considerably improved mechanical properties (Table 6-26), especially elongations at break (Japanese capacity 80 000 t/a, production 36 000 t/a (1993)). 90 % of the filaments are used as industrial fibers (fish nets, conveyor belts, etc.). Some are now also converted to short staple fibers and used as additives in papers and nonwovens.

6.7.5 Organic High-Performance Fibers

Synthetic fibers are commonly called **high-performance fibers, specialty fibers, high-functionality fibers**, or **high-temperature fibers** if they surpass the properties of common textile fibers with respect to moduli, strengths, and/or temperature resistance. Such fibers may be organic, inorganic (carbon, ceramic), or metallic.

Aramids

According to the FTC (and also to BISFA and EU), aramids are manufactured fibers "in which the fiber-forming substance is a long-chain synthetic polyamide in which at least 85 % of the amide linkages are attached directly to the aromatic rings." Commercial aramids comprise poly(*m*-phenylene isophthalamide) (Nomex®), poly(*p*-phenylene terephthalamide) (Kevlar®) and a meta-para copolyamide with ether units (Technora®) (see next page). The world production of aramid fibers is approximately 40 000 t/a.

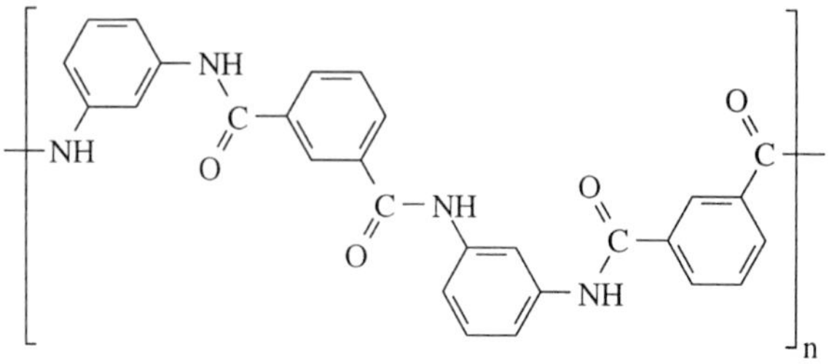

Nomex®

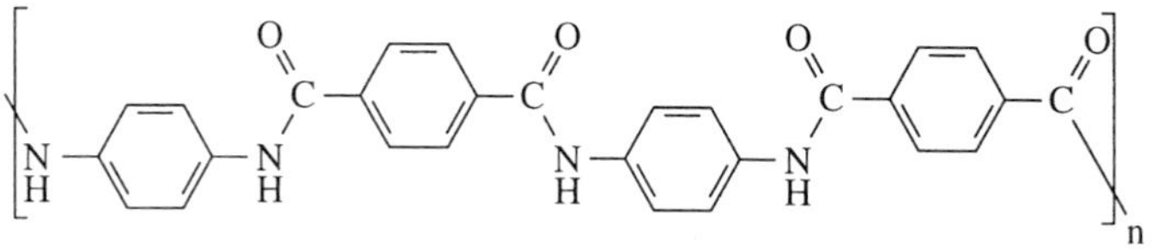

Kevlar®

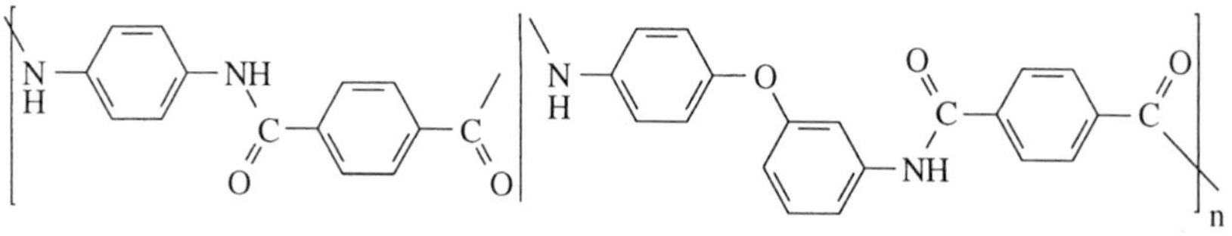

Technora®

Poly(*m*-phenylene isophthalamide) (**PMPI** or **PMIA**) is obtained from *m*-phenylene diamine and isophthaloyl chloride by solution polycondensation, interfacial polycondensation, or solution prepolycondensation followed by interfacial high polycondensation.

The solution polycondensation in dimethylacetamide uses NaOH as an HCl acceptor. After reaction, the solution is neutralized by Ca(OH)$_2$ and dry spun to fibers. The dope cannot be wet spun because of the high contents of CaCl$_2$ and water. CaCl$_2$ is subsequently washed out and the fiber heat-treated.

Interfacial polycondensation proceeds at the interface between, for example, tetrahydrofuran, and an aqueous alkali carbonate solution as an acid acceptor. The polymer is isolated, dissolved in *N*-methylpyrrolidone, and wet spun into an aqueous CaCl$_2$ solution; resulting fibers are preoriented and hot-drawn. Interfacial polycondensation delivers more oligomers than solution polycondensation. Hence, the resulting fibers (Nomex® (DuPont), Teijinconex® (Teijin), Fenilon (Russia), etc.) are less thermostable.

PMPI forms lyotropic liquid crystals. Fibers can thus be spun either form isotropic solutions at low polymer concentrations or from nematic solutions at higher ones (usually 20-23 wt%). Because of the greater orientation of chain segments, polymers from nematic solutions have greater tensile moduli and strengths and smaller fracture elongations than those from isotropic ones (Table 6-27).

Poly(*p*-phenylene terephthalamide) (**PPTA**) is obtained by the Schotten-Baumann reaction of *p*-phenylene diamine with terephthaloyl chloride in a 2:1 mixture of *N*-methylpyrrolidone and hexamethylphosphoric triamide (+ ca. 3 % LiCl) (DuPont) or *N*-methylpyrrolidone + CaCl$_2$ (Akzo). Akzo owned the polymerization patents and DuPont the patents for dope formation and fiber spinning which led to long patent disputes and then to cross-licensing (Kevlar® (DuPont), Twaron® (Teijin; formerly Arenka® (Akzo)).

Table 6-27 Properties of aramid fibers at room temperature. PBI = polybenzimidazole, PMPI = poly-(m-phenylene isophthalamide) (Nomex®), PPTA = poly(p-phenylene terephthalamide) (Kevlar®), PAH = polyamidehydrazide, PF = phenolic resin. I = isotropic, N = nematic; * decomposition.

Property	RH (%)	Physical unit	PBI I	PMPI I	PMPI N	PPTA 29	PPTA 49	PAH X-500	PF Kynol®
Density	65	g/cm^3	1.43		1.38	1.44	1.45	1.46	1.26
Tensile modulus	65	GPa	4.0	19	36	58	120	100	5.0
Fracture strength	65	GPa	0.33	0.35	0.70	2.8	2.8	2.1	0.28
Fracture elongation	65	%	29	6.7	4.7	4.0	2.5	3.5	40
Moisture uptake	65	%	15		8	7	4		10
Melting temperature	0	°C			380	460*			
Oxygen index (LOI-U)	0	%	42		32	29	29		32

Fibers are dry-jet wet spun from 20 % sulfuric acid solutions of PPTA. In this method, the dry jet exiting from the spinneret passes through an air gap before it enters the coagulation bath, whereas in conventional wet spinning the spinning die is located in the coagulation bath (Fig. 6-7). The very good tensile moduli and strengths can be increased by special spinning conditions (Kevlar 29 *versus* Kevlar 49).

PPTA fibers are difficult to dye with dyes because the polymer does not have sufficient accessible groups (there are regular hydrogen bonds between adjacent chains) and its liquid-crystalline structure is not "open". Dyeing with pigments generally disturbs crystal structures and decreases mechanical properties.

Kevlar®, Twaron® (Teijin), and Terlon (Russia) are temperature-resistant specialty fibers that are best known for their use in bullet-proof vests. However, their main application is as a reinforcing fiber for engineering plastics and to a smaller extent for electrical and thermal insulation. A disadvantage is the poor compression strength which is probably caused at least in part by the core-sheath structure of the fiber that arises from the spinning of sulfuric acid solutions. Compression of PPTA fibers leads to helical kink bands (Fig. 6-40). Compounds reinforced by PTTA thus do not fail because of a fracture of fiber or matrix but because of an elastic warping by shearing.

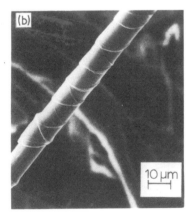

Fig. 6-40 Electron micrograph of a Kevlar 49 fiber in a polyamide matrix before (left) and after (right) a 3 % compression [28]. By kind permission of Chapman and Hall, New York.

Kevlar 49 thus has a very high ratio of tensile to shear moduli, ca. 70, which is much greater than that of glass fibers (2.0), wool (3.2), cotton (3.7), and nylon 6 (6.6). Only flax has a higher ratio of ca. 19, probably also from warping (see Fig. 6-36).

The polyamide-hydrazide X-500 contains only 50 mol% aromatic amide groups –NH(1,4-C$_6$H$_4$)CO– and is therefore not defined as an aramid. Nevertheless, its mechanical properties are similar to those of the Kevlar® grades (Table 6-27).

X-500

Imidazole and Azole Fibers

Aramids were followed by many other specialty fibers based on nitrogen-containing aromatic polymers (for the chemistry, see Volume II, Chapter 10).

Poly(benzimidazole) (PBI), short for poly(2,2'-*m*-phenylene-5,5'-bibenzimidazole) (Volume II, p. 509), was first developed as a flame-resistant textile fiber for space travel because of its high limiting oxygen index and very high glass temperatures of 430°C (unstabilized) and 500°C (annealed), respectively. At temperatures below 550°C, it does not melt or emit smoke or toxic fumes. It is mainly used for space and military applications but also as a replacement for asbestos (see properties, previous page).

polybenzimidazole (PBI)
= poly(2,2'-*m*-phenylene-5,5'-bibenzimidazole)

Industrially, **polybenzoxazole** (PBO) is called a "cis" compound and **polybenzthiazole** (PBTZ) a "trans" or "symmetric" compound. In the literature, PBTZ is usually called PBT; however, this is the official abbreviation of poly(1,4-butylene terephthalate).

`These compounds are synthesized by polycondensation of terephthalic acid and the hydrochlorides of 1,4-dithio-2,5-diamino-1,4-benzdithiol (for PBTZ) and 4,6-diamino-1,3-benzdiol (for PBO), respectively, in methanesulfonic acid, chlorosulfonic acid, or concentrated aqueous solutions of poly(phosphoric acid) (Volume II, p. 511). The polymers are directly spun from their lyotropic solutions. PBTZ is marketed as Zylon®.

On spinning, the preferential orientation of the molecule axes is maintained which leads to extraordinary values of mechanical properties in the fiber direction (Table 6-28). Annealing of PBTZ filaments leads to tensile moduli of ca. 330 GPa which corresponds to ca. 45 % of the theoretical value of 730 GPa (molecular orbital theory). PBTZ is more stable against oxidation and heat than PBO but not against compression.

PBO = poly[(1,2-*d*: 5,4-*d*)bisoxazole-2,6-diyl-1,4-phenylene]

PBTZ = poly[(1,2-*d*: 5,4-*d*)bisthiazole-2,6-diyl-1,4-phenylene]

Table 6-28 Tensile moduli E, compression moduli K, shear moduli G ("torsion moduli"), tensile strengths at break σ_B, compression strengths κ_B, and fracture elongations ε_B of PBO and PBTZ.

Polymer and form			E/GPa	K/GPa	G/GPa	σ_B/GPa	κ_B/GPa	ε_B/%
Poly(p-phenylene-2,5-benzobisthiazoldiyl), PBTZ								
Band			40			0.5		
Fiber,	untreated		<320		1.2	<4.2	< 0.4	<7.1
	annealed,	600°C				2.7		
Filament,	untreated		<170			2.3		<7.1
	annealed		<330			<4.2		<1.4
Film,	uniaxially oriented		270			2.0		0.88
	biaxially oriented		34			0.55		2.5
Poly(p-phenylene-4,6-benzobisoxazoldiyl), PBO								
Band			7.6			0.3		0.8
Fiber,	untreated		166	240	1	4.6	0.5	3.0
	annealed,	600°C	320			5.0		1.8
		650°C				3.4		1.3
		665°C	290			3.0		1.2

Polyetherimides and polyamide-imides

Many grades of many types of imide group-containing polymers are produced by many companies for use as plastics working materials, coatings, or fibers (examples: Kapton®, Kermel®, Torlon®, Ultem®) (Volume II, p. 498). These polymers may be either polyetherimides (PEI; examples: Kapton® and Ultem®) or polyamide-imides (PAI; examples: Kermel® and Torlon®):

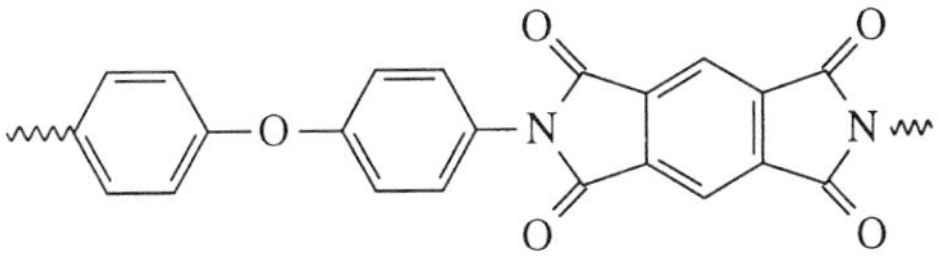

Kapton® SP1

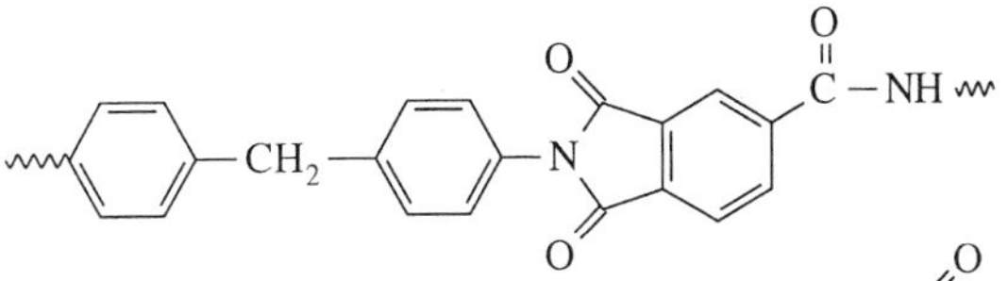

Kermel®

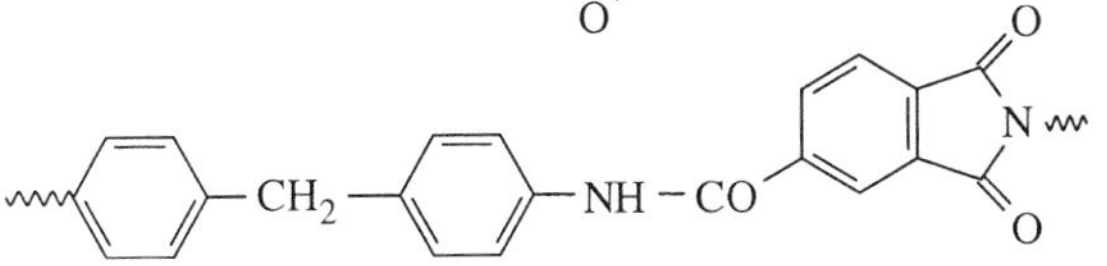

Torlon® 4000T

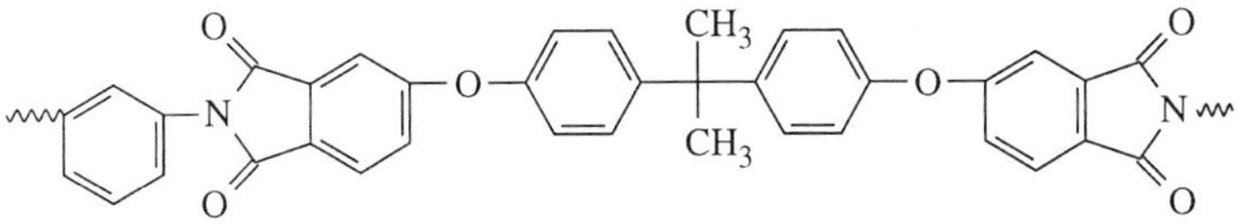

Ultem® 1000

These imide polymers are synthesized by reacting diamines such as 1,4-diaminobenzene (for Kermel®), 1,3-diaminobenzene (for Ultem®), 4,4'-diaminodiphenylether (for

Table 6-29 Properties of some fibers from polyetherimides PEI), polyamide-imides (PAI), aromatic polyesters, a polyetheretherketone (PEEK), and the novoloid Kynol® (PF). Data at 23°C.

Property	Physical unit	PEI		PAI		Polyesters		PEEK	PF
		Kapton®	Ultem®	Kermel®	Torlon® 2000	Econol®	Vectran®		Kynol®
Density	g/cm³	1.43	1.27	1.34	1.41	1.40	1.40	1.28	1.27
Tensile modulus	GPa		3	6.6	3.0	134	64	5.5	4.0
Flexural modulus	GPa	3.1	3.3		4.9				
Flexural strength	GPa	0.117	0.150		0.164				
Tensile strength	GPa	0.086	0.105	0.47	0.093	3.8	2.9	0.31	0.18
Fracture elongation	%	7.5	60	23	2.5	2.8	3.7	55	45
Moisture uptake	%	0.32	0.25	4		0.01	0	0.20	6
Oxygen index (LOI-U)	%	53		31	47		37	35	32

Kapton®), or 4,4'-diaminodiphenylmethane (for Torlon®) with an aromatic tricarboxylic acid anhydride (for Kermel® and Torlon®), a simple tetracarboxylic acid dianhydride (for Kapton®), or a more complex tetracarboxylic acid dianhydride (for Ultem®). In the first step, monomers are reacted in solution (for example, N-methylpyrrolidone for Kermel®) at elevated temperatures, which leads to an oligomeric polyamic acid precursor, for example, for Kapton®:

(6-3)

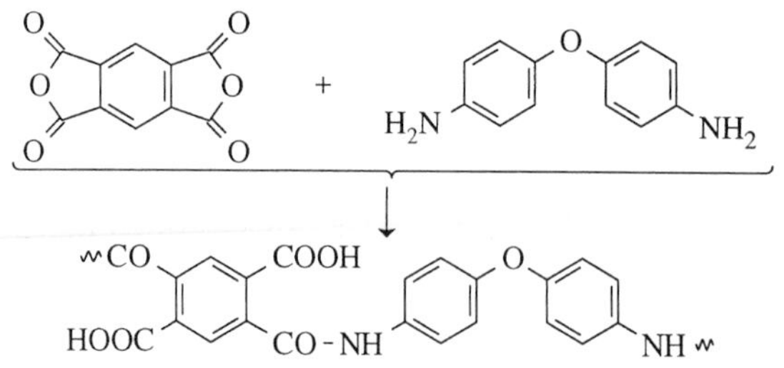

The resulting polyamic acid dope is spun to a precursor fiber. Heating the fiber results in intramolecular ring closure to the imide structure albeit with simultaneous intermolecular crosslinking to crosslinked fibers. Properties of polyetherimides and polyamide-imides are shown in Table 6-29.

Other Specialty Fibers

Because poly(p-hydroxybenzoate) is insoluble and unmeltable ($T_M > 550°C$), use has been made of its monomeric units in copolyesters for fiber spinning:

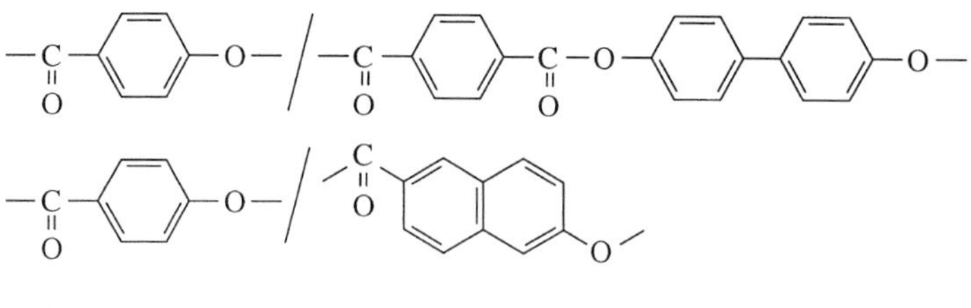

As expected, such copolyesters have high tensile moduli and tensile strengths but only small fracture elongations. Both fibers are apparently still under development.

Polyetheretherketone (PEEK), $\text{-[-O-}(p\text{-C}_6\text{H}_4)\text{-O-}(p\text{-C}_6\text{H}_4)\text{-CO-}(p\text{-C}_6\text{H}_4)\text{-]}_n$, is used as a thermoplastic ($T_M = 334°C$) but is also being tested as a fiber, especially for so-called fiber-fiber composites (FF) in which PEEK is hybridized with carbon fibers. Such FFs are advantageous because their manufacture proceeds without impregnation of the carbon fiber. The fibers also have better shape stabilities.

Phenol-formaldehyde polymers (PF) have long served as thermosets (Section 8.5.3) but can also be spun to fibers from their prepolymers (novolacs, see Volume II, p. 261). The as-spun fibers are then hardened (crosslinked) with formaldehyde. The resulting **novoloids** were developed as precursors for carbon fibers but are now marketed as heat resistant fibers (stable in air up to 150°C) with excellent flame retardancy and virtually no generation of gas or smoke on burning. They are used as a reinforcing fiber for plastics and rubber as well as for protective shields.

6.8   Inorganic Fibers

Inorganic fibers are fibers composed of inorganic elements and compounds, some of which are polymeric but many not. Most of these fibers are used for technical purposes, especially engineering applications. However, glass fibers are also used for textile applications albeit for home textiles and not for apparel and underwear.

Inorganic fibers are usually subdivided into **metallic fibers** and **mineral fibers**, a group that comprises all non-metallic inorganic fibers, both natural and synthetic. Mineral fibers find many applications, and it is therefore customary to subdivide them into textile glass fibers (Section 6.8.1), mineral fibers for insulation (Section 6.8.2), refractory fibers (Section 6.8.3), and carbon fibers (Section 6.8.4). Production statistics are hard to find but the annual world production of mineral insulating fibers was reported as $5.6 \cdot 10^6$ t/a in 1983 and that of textile glass fibers as $1.35 \cdot 10^6$ t/a in 1985.

6.8.1   Textile Glass Fibers

Glass fibers (US: **fiberglass**) are produced in many varieties (Table 6-30) which are distinguished by letter symbols according to their main applications. Textile glass is usually **E glass**, a borosilicate glass that was originally developed for the *e*lectrical industry (hence "E"). D glass has a better *d*ielectric strength. Acid-resistant glasses comprise types **A** (*a*lkali rich), **C** (*c*hemical/corrosion resistant), and **ECR** (E-CR; *e*lectrical, *c*orrosion *r*esistant). Type A can also be used for textiles purposes if it is treated with special adhesion promoters. Glasses may have good *s*trength (**S**) or good *t*ensile strength (**T**), or may fulfill mechanical *r*equirements (**R**). For properties of glass fibers, see Table 6-31.

E glass is a multipurpose borosilicate glass that is used not only by the electrical industry but also as textile fiber, reinforcing material for plastics (Chapter 9), insulation material (glass wool, glass bat, glass wadding), and in fiber optics (Section 15.1.5). Its polymeric chains have short side chains that prevent formation of band, sheet, and lattice structures (Volume II, p. 565); hence the relatively low glass temperature (355°C).

Table 6-30 Composition of technical silicate glasses in wt% [29, 30].

Component	A	C	D	E	ECR	R	S, T
SiO_2	70-72	60-65	73-74	52-56	58-63	60	60-65
Al_2O_3	0-2.5	2-6	0-0.3	12-16	10-13	25	20-25
CaO	5-10	14	0.3-0.6	16-25	21-23	9	0
MgO	1-4	1-3	0	0-5	2-4	6	10
B_2O_3	0-0.5	2-7	22-23	5-10	0.1	0	0-1.2
F	0	0	0	0-0.7	0.15	0	0
Na_2O	12-15	8-10	1-1.3	0-2	0-1.2	0	0-1.1
K_2O	0	0	1.5	< 1	0.4	0	0
Fe_2O_3	0	0.2	trace	0-0.8	0-0.4	0	0
TiO_2	0	0	0	< 1	2.1	0	0
ZnO	0	0	0	0	0-3.5	0	0

Textile glass fibers are produced as filaments or staple fibers. In the production of **filaments**, glass melts are drawn vertically through a bushing (similar to a spinneret in organic fiber spinning) that contains 200-4000 holes of 1-2.5 mm diameter in multiples of 200. The high drawing speeds of up to 80 m/s deliver highly drawn filaments: a hole diameter of 2 mm at the orifice of the bushing produces a 10 μm wide filament. The filaments are cooled and provided with sizes (p. 35). They are then processed directly to mats from continuous strands (definition: p. 205); chopped and then chemically bonded to tissues (also made from staple fibers) by a dry or wet process similar to the ones described on p. 210, bottom; or chopped to lengths of 3, 4.5, 6, or 12 mm or milled to 0.2 mm, both for reinforcing polymers. Staple fibers may also be assembled with a slight twist to **rovings** that are then processed to woven glass roving fabrics (US: roving cloth) or are chopped or milled.

Staple fibers are produced by either the *bushing blowing process* in which the exiting melt is fibrillated by pressured air or the *drum drawing process* in which solidifying fibers are broken while laid on a rotating drum. The fibers from either of the two processes are assembled in a finishing tube from which they are drawn as a **sliver** which is the analog to the roving, except for the absence of twist. Twisting converts the sliver to single glass staple fiber yarn which can be woven directly to a fabric or first converted to a folded yarn, a cabled yarn, or a multiple wound yarn and then woven to a fabric.

Tab. 6-31 Properties of some mineral fibers.

Property	RH (%)	Physical unit	Quartz	E Glass	S Glass	Asbestos (Chrysotile)	Al_2O_3	Al_2O_3 Whiskers
Diameter of filaments	65	μm	10	10	7	10	20	
Density	65	g/cm^3	2.2	2.59	2.48	2.55	3.1	3.96
Tensile modulus	65	GPa	69	72	86	80	345	2100
Tensile strength in composite	65	GPa	0.9	2.41	4.59	5.68	1.3	43
Elongation at break	65	%	1.3	3.5	2.8	11	0.8	
Melting temperature	0	°C	1650	1260		1520	2045	
Maximum use temperature	0	°C	900	600		1400		

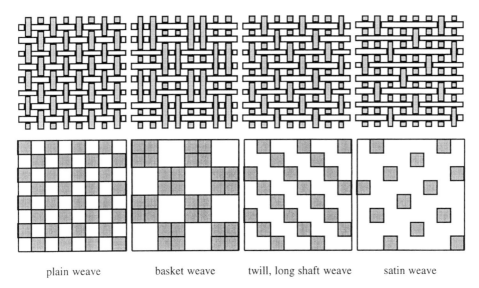

plain weave   basket weave   twill, long shaft weave   satin weave

Fig. 6-41 Glass fiber weaves with (top) arrangement of weft (white) and warp (gray) and (bottom) arrangement of bonding points. The terms plain weave, twill, and satin weave correspond to those of the organic textile fiber industry.

Woven **glass fiber fabrics** (Fig. 6-41) are produced in the same way and with the same types as the fabrics from organic textile fibers (Fig. 6-33). The types shown in Fig. 6-41 are called biaxial or 0/90 weaves because weft and warp cross each other at 90°. Triaxial weaves (0/45/90 or 0/60/–60) and tetraaxial weaves (0/45/90/–45) are also available.

There are many types of weaves available, some of which are shown in Fig. 6-42 together with the designations that are used in the fiberglass industry.

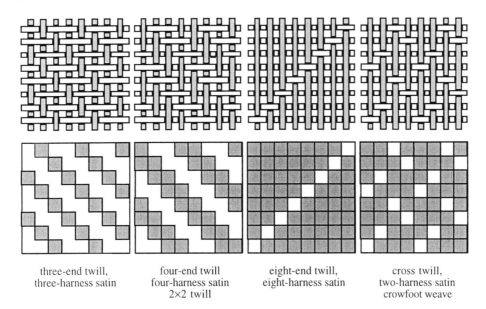

three-end twill,	four-end twill,	eight-end twill,	cross twill,
three-harness satin	four-harness satin	eight-harness satin	two-harness satin
	2×2 twill		crowfoot weave

Fig. 6-42 Some types of satin weaves.

6.8.2 Mineral Wool

Mineral fibers for insulation against heat, cold, fire, and/or noise are also known as mineral wool. They include three types: glass wool, rock wool, and slag wool. The main components of all three types of mineral wool are glass-like fibers (90-99.8 wt%) with 3-6 μm diameter that are held together by binders such as phenol-formaldehyde resins (less than 10 wt%), mineral oil (< 1 wt%), poly(methylsiloxanol) (< 0.5 wt%), and emulsifiers (< 0.2 wt%), usually poly(ethylene glycol)s.

Rock wools are prepared from the melts of a single rock-forming mineral such as basalt, rhyolite, or diabase by the Owens process or the rock wool process. In the *Owens process*, an air or steam jet is horizontally directed against the melt that exits vertically from the holes in the bushing. The resulting fibers are ca. 3-10 cm long.

Rock wool fibers of similar length are also produced by the *rock wool process*. In this process, the melt is directed to a horizontal rotor and then further redistributed by 2-3 other horizontal rotors. The centrifugal forces produce fine fibers but also some slug (i.e., lumps).

Glass wool is produced from mixtures of quartz sand, minerals (limestone, feldspar, heavy spar, dolomite, etc.), and industrial chemicals (Na_2CO_3, Na_2SO_4, $BaCO_3$, etc.) as well as recycled glass by the Owens process or the *rotary process*. In the latter process, a melt is distributed evenly from the inner walls of a rotating shaft into a rotating perforated basket. The exiting fibers are then drawn by hot combustion gases of a circular burner.

Slag wool is obtained from the residues (slag) of the smelting of metal ores. The melt of the residues is fed to a perforated centrifuge that rotates at an angle of ca. 45° and emits a stream of fibers.

All three processes deliver fibers with approximately Gaussian distributions of fiber diameters. The fibers form a porous mat in which they reside predominantly parallel to the surface of the mat. The thermal conductivity of the mat depends on the composition and the diameters of the fibers, the type and amount of the binder, and the bulk density. Thermal conductivities of mineral wools are 0.03-0.04 W m^{-1} K^{-1} at room temperature but may be 0.06-0.1 for glass wools at 200°C and 0.08-0.16 W m^{-1} K^{-1} for rock wools at 400°C. For comparison, thermal conductivities at room temperature are ca. 0.03 for polyurethane and poly(styrene) foams, 0.04 W m^{-1} K^{-1} for cork, ca. 0.1 W m^{-1} K^{-1} for bulk natural rubber, and ca. 0.6 W m^{-1} K^{-1} for solid low-density poly(ethylene).

6.8.3 Refractory Fibers

Refractory fibers are a diverse group of inorganic fibers that do not significantly deform or change chemically at temperatures above ca. 1000°C. Early types were natural minerals such as chrysotile asbestos, rock wool, or fibers produced from kaolin. Since ca. 1970, several other types have been developed based on boron (boron itself, boron nitride BN, boron carbide B_4C (Volume II, p. 554 ff.)). Since boron fibers are oxidized at moderate temperatures, newer fibers are based on silicon (silicon nitride Si_3N_4 or silicon carbide SiC (Volume III, p. 560)), or are advanced oxide fibers such as high-alumina products (SiO_2/Al_2O_3 = 0-0.3 wt/wt) and zirconia based fibers.

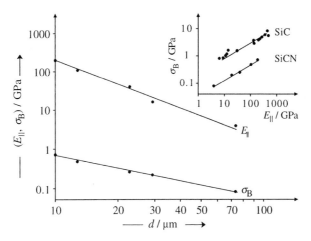

Fig. 6-43 Dependence of the logarithm of longitudinal tensile moduli $E_{\parallel}$ and strengths at break σ_B on the logarithm of the diameter d of $Si_xN_yC_z$ fibers [31]. Insert: lg-lg plot of fracture strengths as a function of longitudinal tensile moduli of $Si_xN_yC_z$ fibers [31] and SiC fibers [32]. The slopes are 1/2.

The mechanical properties of these fibers (Table 6-31) vary not only with their chemical composition but also with their preparation and especially with their diameters (Fig. 6-43). The logarithms of both tensile moduli and fracture strengths decrease linearly with increasing fiber diameter because thicker fibers have larger surfaces which can lead to more surface faults (for example, porosities) and, for certain manufacturing processes, also to variations of chemical composition from the exterior to the interior. The observed square-root dependence of fracture strengths on tensile moduli (Fig. 6-43, insert) agrees with the predictions of the Griffith theory for the effect of cracks and holes on mechanical properties (Volume III, p. 621).

The behavior of these refractory fibers is in stark contrast to that of solution-spun, drawn filaments from ultrahigh-molecular weight poly(ethylene) chains where the tensile strength at break is directly proportional to the tensile modulus (Fig. 6-44) for chains aligned in the fiber direction, just as predicted by theory (Volume III, p. 626).

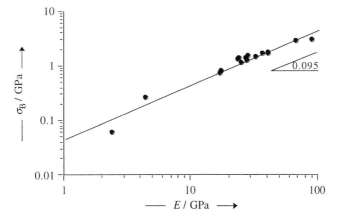

Fig. 6-44 Dependence of tensile strengths at break on tensile moduli for filaments from ultrahigh molecular weight (UHMW) poly(ethylene)s [13]. The slope of 0.095 is theoretical.

Table 6-32   Properties of selected refractory fibers [33]. am. = amorphous; α, β = crystal modifications of Al_2O_3 or SiC. A = Al_2O_3/x% Sil = α-Al_2O_3 + x% SiO_2 in mullite (3 $Al_2O_3\cdot2$ SiO_2 to 2 $Al_2O_3\cdot SiO_2$).

Property	Physical unit	B 100% Avco	SiC 100% Ube Tyranno	SiC 100% Avco SCS2	SiC 100% Tatcho SCW1	Al_2O_3 >99% DuPont Fiber FP	Al_2O_3/ 5% Sil ICI Saffil HA	Al_2O_3/ 20% Sil Denka Alcen
Morphology	-	am.	am.	β	β	α	α+A	α+A
Type	-	filament	yarn	filament	whisker	yarn	staple	yarn
Diameter (average)	μm	140	12	140	0.3	14	2.5	10
Density	g/cm^3	2.4	2.4	3.0	3.21	3.9	3.4	3.3
Melting temp.	°C	2200		2830		2040	1980	1890
Max. use temp.	°C	500	>1300	1600		1320	1600	1600
Tensile modulus	GPa	390	220	420	480	380	>300	170
Tensile strength (av.)	GPa	4.0	2.9	4.0	20.7	>1.4	1.5	1.7

The properties of refractory fibers of the same type can vary considerably depending on the fiber-forming process, the diameter of the fibers, and the product (staple fiber, filament, yarn, whisker) (Table 6-32). The highest moduli have been obtained for whiskers which are very thin single crystals that look like the very stiff hairs near the mouth of cats (hence the name). Whiskers from silicon carbides have tensile moduli between 380 and 700 GPa (Table 6-32) whereas those of aluminum oxide (sapphire whiskers) may reach 2100 GPa (Table 6-31). Whiskers are very expensive and are therefore used only in very special cases.

6.8.4   Carbon Fibers

Metallic and ceramic fibers have greater densities than fibers from organic polymers, which considerably reduces their performance since textile moduli and textile strengths are measured in stress per density. For example, steel with a density of 7.75 g cm^{-3} has a tensile modulus of 210 GPa and therefore a textile modulus of 210 GPa/(7.75 g cm^{-3}) = 27.1 MN/tex which is considerably lower than the textile modulus of ultradrawn fibers of ultrahigh-molecular weight high-density poly(ethylene) which is 172 GPa/(0.97 g cm^{-3}) = 177 MN/tex.

Carbon fibers (also called **graphite fibers** in the United States) have low densities and excellent properties. All carbon fibers are produced by carbonization of preformed organic fibers, first by Thomas Edison for filaments of incandescent lamps from cotton and rayon, then from poly(acrylonitrile) fibers, and finally from pitch (for manufacturing processes, see Volume II, p. 209 ff.). Depending on the process, carbon fibers from PAN or pitch can be obtained with high moduli or high strengths (Table 6-33). The world production was 6200 t/a in 1990.

All of these fibers serve as reinforcing materials for plastics. However, loops are not formed easily because filaments are fairly stiff. Carbon fibers are therefore difficult to weave from filaments. Instead, one first prepares wovens from poly(acrylonitrile) yarns that are later carbonized (CCPF = chemical conversion of a precursor fiber).

Table 6-33 Properties of carbon fibers from poly(acrylonitrile) (PAN) or pitch. HM = high modulus, UHM = ultrahigh modulus, HS = high strength, I = from isotropic liquid.

Property	Physical unit	PAN HS	PAN HM	PAN UHM	Pitch HM	Pitch I
Diameter of filaments	μm	7.5	7.5	8.5	10	
Density	g/cm^3	1.78	1.85	2.0	2.15	1.55
Tensile modulus	GPa	240	400	535	690	40
Tensile strength	GPa	3.75	2.45	1.85	2.24	0.9
Fracture elongation	%	1.55	0.65	0.35	0.3	2.3
Toughness	MPa	20	7.5	3		
Linear expansion coefficient	10^{-6} K^{-1}	–0.6	–1.0	–0.9		+3
Thermal conductivity	W m^{-1} K^{-1}	15-20	60-100			15

The ultimate carbon fibers are carbon nanotubes (**CNT**) that are now produced commercially in the ton per year range. Single wall nanotubes (**SWNT**) have walls of anellated unsaturated 6-membered carbon rings in graphene-like structures (L: *anellus* = small ring; in chemistry, usually as "annelated" instead of "anellated"); one of the ends or both of them are capped with a half-sphere of the same structure (Volume II, p. 208 ff.). Within the walls, carbon rings are joined in a zigzag configuration ($\sim$; α = 0°), in an armchair configuration ($\sim$; α = 30°), or in chiral configurations with helical arrangements of hexagons (0 < α < 30°) (see p. 603 for designations). The walls of multi-wall carbon nanotubes (**MWNT**) are two or more layers thick.

Carbon nanotubes have the highest tensile moduli and tensile strengths of all known materials (Table 6-34). Since they are excellent reinforcing fibers for plastics and could find many other applications, they are now (February 2008) offered by no less than 52 suppliers with an estimated total annual world production of ca. 1000 kg.

Table 6-34 Mechanical properties of carbon nanotubes.

Property	Physical unit	SWNT armchair theory	SWNT - experiment	SWNT zigzag theory	MWNT - experiment
Tensile modulus	GPa	940		940	800-900
Tensile strength	GPa	126	13-53	94.5	150
Elongation at break	%	23.1		16.5	

6.9 Paper

6.9.1 Introduction

Paper is a sheet-like porous product that is in general stiff, foldable, opaque, and printable. Papers based on cellulose or most synthetic polymers consist of felts of short fibers. So-called **plastic papers** are thin, extruded sheets of thermoplastics. Papers are porous but plastic papers are not.

The first papers were thin sheets that were produced in Egypt ca. 3000-4000 BC from the flattened medulla of the plant *Cyperus papyrus*. Since the supply of paper from Egypt ceased with the end of the Roman empire and the subtropical *Cyperus papyrus* does not grow in Europe, Western countries replaced paper for writing purposes by parchment from animal skins (vellum if from calves).

Modern papermaking by draining of water from aqueous supensions of fibers was invented in China ca. 200 B.C.-100 A.D. These papers were first made from hemp and later from the bark of mulberry trees or bamboo. Arabs learned the art of paper-making from Chinese prisoners they took in the battle of Talis, Turkestan (751 A.D.) and started paper-making in Baghdad. Europeans learned the art ca. 1100-1200 A.D. from the Arabs during the Reconquista in Spain. Initially, the main European raw material for paper was rags from old cloths. Mechanical wood pulp for papermaking was first produced in 1844/45, chemical wood pulp in 1874 (Volume II, p. 71 ff.).

China's production of paper was fostered by the invention of block printing from carved wooden blocks for the production of paper money (introduced in 1024) and newspapers and books (3 123 185 copies of the annals of the Yuan dynasty for the year 1328!). Printing with movable metal letters was invented ca. 1452 by Johannes Gutenberg (true name: Johannes Gensfleisch) in Germany.

The world consumption of paper and paperboard increases steadily: from $150 \cdot 10^6$ t/a in 1978 to $380 \cdot 10^6$ t/a in 1988. The latter figure corresponds to 80 kg per capita (US: 270 kg/a (2006)) and to $1.5 \cdot 10^9$ tons of harvested wood!

45 % of paper products serve for packaging, 39 % for writing and printing, and 16 % for diapers, toilet paper, and the like. Paper is also used for throw-away clothes, for example, in hospitals.

6.9.2 Standards

Internationally, the heaviness ("weight") of paper is quite logically expressed in mass per area: a 60-paper has an area-weight of 60 g/m^2. Writing and office papers have area-weights (called grammage) of 6-150 g/m^2 and cardboards those of more than 160 g/m^2. In the United States, the "weight" of cardboard is also an area-weight albeit expressed in pounds per square inch.

For printing papers, the United States uses a totally different system by specifying the "weight" W of paper as the total mass m of a ream of paper consisting of N pages of length L and width d. One ream of US letter size (8.5×11 inch) copy paper typically is a 20-lb paper (it actually weighs 5 lbs = 2.27 kg). A ream now has 500 sheets (formerly 480) and a printer's ream, 516 (Old French: *remme*, from Arabic: *rizmah* = bundle).

Paper sizes also differ in the United States and the rest of the world. In the ISO system, all paper sheets have the same ratio of height to width of $1:2^{1/2}$. The A series is for writing papers, the C series for envelopes of A series papers, and the B series for books (also A5 and A6).

Format A0 has an area of 1 square meter = 841 mm $\times$ 1189 mm. Format A1 equals one half of A0; with the format $1:2^{1/2}$, this translates to 594 mm $\times$ 841 mm. The letter size is A4 (210 mm $\times$ 297 mm) which fits into a C4 envelope (229 mm $\times$ 324 mm) or, when folded twice, into a C6 envelope (114 mm $\times$ 162 mm), etc.

6.9.3   Cellulose Papers

More than 99 % of all papers, cardboards, etc., are made from cellulose fibers. World-wide, these papers are obtained from wood pulp (66 wt%), recycled paper (25 wt%), other fibers such as linters (4 wt%), and fillers and pigments (5 wt%). The composition depends on the intended use: paper for high-gloss printing is rich in calcium carbonate; cardboard has a high percentage of recycled paper, etc.

Compositions of paper vary from country to country and from purpose to purpose. For example, German papers consist on average of 44 wt% recycled paper, 28 wt% chemical wood pulp, 16 wt% mechanical wood pulp, 5.4 wt% mineral fillers, 2.7 wt% starch, 1.9 wt% chemical products such as poly(acrylamide), 1.5 wt% aluminum sulfate, and 0.5 wt% resin glues and other additives. The production of 1 kg of paper required 4.7 kWh of energy and 250 kg of water but new production methods have reduced water usage to only 5 kg. Germany recycles 66 % of all paper and cardboard. Recycled paper and cardboard can only be used as additive but not for new paper because its fibers are too short. 20 % of consumed paper products cannot be recycled at all.

Wood pulp (mechanical pulp) is produced by mechanical grinding of debarked coni-fers (Volume II, p. 71). It is mainly used for cardboard and coarse papers which require adsorptivity, opacity, and voluminosity.

Chemical wood pulp is obtained from the bisulfite process or the sulfate process (kraft process). In the *bisulfite process*, non-coniferous wood is cooked with calcium (or Na, Mg, NH$_4$) hydrogen sulfite which converts insoluble lignins to soluble lignin sul-fonic acids and hydrolyzes so-called hemicelluloses to mono- and oligosaccharides (Volume II, p. 72). The remaining sulfite cellulose is defibrillated and bleached.

The *sulfate process* can process not only non-coniferous woods but also coniferous and highly resinous ones. The wood is cooked in an aqueous solution of NaOH, Na$_2$S, and Na$_2$CO$_3$. The resulting mixture is concentrated and provided with Na$_2$SO$_4$ (hence: sulfate process). Filtration separates the solution of alkali lignin (= black liquor) from the desired cellulose fibers. Sulfate cellulose leads to stronger (= more tear-resistant) papers and cardboard. The process is therefore also known as the *kraft process* (German: *Kraft* = strength, power, force) and the paper as kraft paper.

For paper making, 0.1-1 wt% aqueous suspensions of cellulose fibers, fillers, pig-ments, etc., are led with speeds of up to 60 km/h (≈ 38 mph) over a bronze sieve in an up to 100 m (≈ 109 yard) long paper making machine. Water and finer particles run off and the wet fleece is sent through roller presses and around cylindrical dryers in order to evaporate the remaining water and cause self-binding of cellulose fibers. For fine papers, the rough surface of the paper sheets is coated with starch or filled polymer latices.

All cellulose papers are opaque, stiff, barely stretchable, only a little dimensionally stable in the direction normal to the sheet, and not water resistant. Mechanical properties can be varied somewhat, depending on the cellulose source, the type of pulping process, the additives, etc. (Table 6-35). The raw material is inexpensive but paper machines are very costly. Cellulose is a renewable resource and paper can be relatively easily recycled but the manufacture of pulps and papers produce at lot of waste water and used air that have to be worked up at great cost. For these reasons, attempts have been made to develop papers from synthetic polymers (Sections 6.9.6 and 6.9.7) but none of them can beat the price of cellulose papers.

Table 6-35 Aspect ratio L/d, tensile moduli E, fracture strengths σ_B, and fracture elongations ε_B of some types of cellulose papers.

Paper type	L/d	E/MPa	σ_B/MPa	ε_B/%
Brown wrapping paper from mechanical wood pulp	176	380	17	9.5
Kraft paper from hard woods, bleached	120	470	20	6.5
Kraft paper from soft woods, bleached	45	350	15	10,0
Sulfite paper from soft woods, bleached	45	260	17	10.5
Paper with high proportion of recycled newspaper	35	260	17	14.0

6.9.4 Parchment Paper

Genuine **parchment** is collagen-based since it is obtained from untanned, oiled animal hides (Section 6.10). Parchment paper, on the other hand, is a cellulose paper from sulfite pulp or from linters. It is obtained by drawing cellulose paper for 5-20 s through a bath of 70-75 % cold sulfuric acid, followed by immediate washing with plenty of cold water, plasticizing with glycerol, and drying in calenders.

Parchment paper is predominantly used as packaging material for fatty foods, for example, sausages, cheeses, etc. Writable parchment papers are obtained by addition of animal or plant glues.

6.9.5 Vulcanized Fiber

Treating two or more layers of unglued paper from linters or soda pulp (from cooking wood with soda) at 50-70°C with a 70 wt% aqueous solution of $ZnCl_2$ or another agent (sulfuric acid, cuprammonium) under compression causes the sheets to gelatinize. For preparation of a parchment paper-like material, the web is washed to remove the gelatinizing agent, dried, and calendered to the desired thickness. For boards, gelatinized webs are wound up and ripened for days and weeks. The gelatinizing agent is then washed out slowly (8 days to 1 year!) and then compressed at 8-130°C.

The resulting vulcanized fiber has a low density but high tensile, impact, and flexural strengths; it is also shatterproof. Because of these properties, vulcanized fiber was widely used for suitcases. It is still the material of choice for spinning buckets (they last 30 years!), brake shoes, gears, washers, bushings, electrical insulation, and the like.

6.9.6 Synthetic Paper

Synthetic papers are fleeces of staple fibers from synthetic polymers that are obtained from polyolefins, polyamides, or polyesters by flash spinning. These staple fibers can be processed like cellulose pulp on regular paper-making machines and are therefore also called **synthetic wood pulp** (**SWP**). Fleeces from polyolefin fibers are thermally bonded without addition of chemical or physical binders. A well known example is Tyvek® (Du-Pont) from high-density poly(ethylene).

The resulting products comprise the whole range from fine writing papers and CD and DVD sleeves to heavy packaging papers, groundcloths, and house wrap. The material is difficult to tear but can be cut with scissors. Its surface is hydrophobic so that it cannot be penetrated by liquid water, yet it is porous so that capillary action allows writing with conventional, water-based inks. Printability of such synthetic papers is improved by surface coating similar to that of cellulose-based papers.

6.9.7   Plastic Papers

Plastic papers are extruded sheets of synthetic polymers with (at least) some surface porosity. Since conventional extruded sheets are not porous, such structures have to be produced by stretching, foaming, surface coating, or swelling. Porosity is required for printing and the desired low density.

The most important production method is drawing (see also Chapter 11) which has to be biaxial because sheets would otherwise fibrillate. The material for the sheets must have a large proportionality limit in the stress-strain curve (see Volume III, p. 520) since the plastic paper has to accommodate the stresses and pressures that it experiences on printing without permanent deformation. Printability and other properties are improved if the sheets are coated before or after biaxial stretching.

6.10   Leathers

Leathers are flat, porous materials composd of fibers and bonding additives. Natural leathers are prepared from animal hides and skins by chemical or physical crosslinking of their collagen fibers (Section 6.10.1). Artificial and synthetic leathers are working materials from synthetic polymers with properties that are similar to those of natural leathers (Section 6.10.2).

6.10.1   Natural Leathers

Hides and Skins

The surface organs of large animals (horses, cows, etc.) are known as hides whereas those of small animals (sheep, coats, etc.) are called skins. Both hides and skins consist of the thin, protective, nonvascular epidermis (epithelium), followed by the dermis with the *corium minor*, corium (cutis), and *subcutis* (Fig. 6-45). The division between epidermis and dermis is not sharp.

The corium minor contains all functional elements (blood vessels, sebaceous (fat) glands, erector muscles) as well as hairs with their perspiratory glands. These elements are embedded in a ground substance composed of mucopolysaccharides and globulins. The ground substance is reinforced by a network of collagen and elastin fibers. The leathery corium consists of bundles of collagen fibers whereas the subcutis is a loose connective tissue of muscle fibers, embedded fat, and blood vessels.

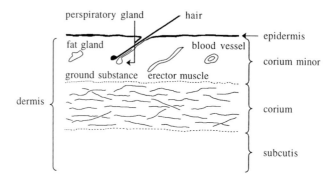

Fig. 6-45 Schematic representation of the skin.

Collagen Fibers

Collagen fibers (Volume I, p. 542) consist of laterally aggregated fibrils of 10-200 nm diameter (Fig. 6-46) that are composed of several subfibrils of 3.5 nm diameter. The subfibrils contain several protofibrils, called tropocollagen, which are triple helices consisting of two left-handed linear α_1 protein chains and a slightly different left-handed α_2 chain. The α_1 chains of calf skin consist of 1052 α-amino acid units of which 1011 are in 337 triplets of the type glycine-X-Y where X = proline, leucine, phenylalanine, or glutamic acid and Y mainly hydroxyproline or arginine. The remaining 41 α-amino acid units are in the chain ends as so-called telopeptides without triplet structure. Telopeptides are rich in lysine units that provide physical crosslinks between chains. Further covalent crosslinking between tropocollagens is caused by carbohydrates.

Protofibrils are aggregated in such a way that each amorphous region composed of polar sequences with predominantly positive charges is facing a crystalline apolar region with predominantly negative charges. Because of this alternating structure of subfibrils, fibrils show alternating dark (amorphous) and light (crystalline) bands after dyeing with uranyl salts.

Properties

In water, disordered protein chains in amorphous regions of collagen fibers become hydrated, which reduces the number of intermolecular hydrogen bonds and swells and extends the fiber. In acidic solutions, basic sidegroups in polar regions are neutralized; correspondingly acidic side groups are neutralized by added bases. In both cases, collagen fibers swell considerably. However, the small counterions of neutralizing agents cause an osmotic swelling pressure so that the fiber shortens.

Under the right conditions, one can destroy the original arrangement of fibers but still maintain the structure of the fibers themselves. Such swollen collagen fibers can be extruded to sausage casings or staple fibers. Collagen filaments are obtained by dissolving collagen fibers by enzymes and subsequent wet spinning.

Heating of water-swollen collagen to 40-60°C causes tropocollagen to dissociate into its α_1 and α_2 chains which then partially hydrolyze to give a solution of gelatin. Further heating produces bone glue.

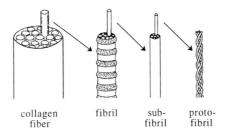

Fig. 6-46 Structure of collagen fibers (schematic).

Tanning

A considerable portion of hides and skins is converted to leathers. Slaughterhouses salt and dry hides and skins in order to prevent decomposition by bacteria before delivery to tanners which then wash and swell hides and skins. Fats and non-collagen proteins are removed in alkaline environments, either physically (emulsifying fats, extraction of water-soluble proteins) or chemically, often with proteolytic enzymes such as trypsin, papain, or lipase. Subsequent treatment of hides by calcium hydroxide/ sodium sulfide, followed by scraping, removes epidermis, corium minor, and subcutis.

During tanning, the loose web of remaining collagen fibrils is crosslinked by multifunctional agents. Crosslinking may be covalent by reaction with aldehydes, difluorodinitrodiphenylsulfones, or amino resins; with formation of ionic bonds by poly(phosphoric acid) or lignin sulfonic acid; with complex formation by chromium(III) salts; or with formation of hydrogen bonds from peptide and phenolic groups of vegetal tanning agents. The leathers are then provided with plasticizers and impregnating agents, such as emulsifying oils, silicon resins, acrylic polymers, or poly(isobutylene).

The variety of skins and hides and the multitude of tanning processes produces a great variety of leathers. However, more than 90 % of all natural leather are chrome leathers. Suede is a chrome leather with a soft, napped surface (from French: *gants de suède* = gloves of Sweden) (technically: a velour leather).

6.10.2   Synthetic Leathers

Leathers must be tear resistant, stretchable, and permeable to air and humidity, They should be easy to texture and dye so that they can follow the latest fashion. Since the raw material (hides and skins) is relatively scarce and the preparation of leathers very labor intensive, man has tried to imitate leather for decades.

Artificial leathers are flexible sheets from a variety of materials that look like leather. They may be made from scraps of true leathers that are bonded by polymers or may also consist of textile fabrics that were coated with a mixture of pyroxylin (= a cellulose nitrate), castor oil, pigments, and a solvent. Still other artificial leathers are fleeces of fibers provided with binders or even sheets of polymers such as polyurethanes, poly-(vinyl chloride)s, or polyamides with leather-like surfaces. Such artificial leathers lack the hand and the permeability to air and water vapor and are therefore used for handbags, easy chairs, and the like but not for clothing.

Modern **synthetic leathers** consist of two or more layers with pores of less than 1 μm diameter (**poromerics**). Such small pores allow the material to breathe (air, water vapor) but prevent the penetration by water droplets which have diameters of more than 1 μm.

Corfam® (DuPont), the first product of this type, consisted of three layers. The upper layer was a polyurethane foam that is permeable to water vapor. The center layer was a mixed fabric from 95 wt% poly(ethylene terephthalate) fibers and 5 wt% cotton. The lowest layer was composed of a porous fleece of poly(ethylene terephthalate) fibers that were bound by an elastomeric polyurethane binder. The Polish Polcorfam® and the Czech Barex® have the same composition (the latter is not identical with the Barex® of BP Amoco which is a poly(acrylonitrile-*co*-methyl methacrylate) barrier resin (see Section 11.2.2)).

The newer Clarino® is a two-layer poromeric with a support layer of polyurethane-bound polyamide fibers of 0.1-1 dtex and a surface layer of polyurethanes. It is more difficult to produce than Corfam® but has smoother and more glossy surfaces. Even finer fibrils of 0.01-0.001 dtex are used in the similar Sofrina®.

Quite different is the synthetic leather that is sold as Ultrasuede® in the United States, Alcantara® in Europe, and Escaine® in Japan, Australia, and South Africa. This Japanese invention is based on very fine poly(ethylene terephthalate) fibrils that are surface coated with poly(styrene). The fibrils are processed to a needle felt that is then impregnated by polyurethane. Subsequently, poly(styrene) is extracted by *N,N*-dimethylformamide. The result is a "porous" network of polyurethane in which the PES fibrils can move relatively freely, similarly to the collagen fibrils in the corium of natural leather. This synthetic suede has a far lower density than natural leather (Table 6-36); it is also crease resistant and washable.

Porvair® is used for the upper parts of shoes. This leather-like porous material consists of an upper polyester-urethane layer with very fine pores and a similar lower layer with somewhat bigger pores. Both layers are stabilized by reaction with carbodiimides.

The upper parts of shoes are also made from a porous "coagulated polyurethane" which is produced by soaking a fabric in a polyurethane solution and then in water.

Table 6-36 Typical properties of leathers. * Modulus at an elongation of 25 %.

Physical property	Physical unit	Natural leather	Clarino®	Alcantara®	Porvair®
Area weight	g/m^2	990	760	194	
Density	g/cm^3	0.68	0.49	0.14	
Tensile modulus	MPa	35	8		0.6*
Flexural modulus	MPa	2.6		1.1	
Tear strength	N/mm	64	40		
Fracture strength in-plane	MPa	27	5	3	1.5
Fracture elongation in-plane	%	54	120	80	350

Literature to Chapter 6

6.1.a INTRODUCTION: DICTIONARIES AND ENCYCLOPEDIAS
C.A.Farnfield, Ed., A Guide to Sources of Information in the Textile Industry, The Textile Institute, Manchester, 2nd ed. 1974
A.J.Hall, The Standard Handbook of Textiles, Wiley, New York 1975
M.Grayson, Ed., Encyclopedia of Textiles, Fibers, and Nonwoven Fabrics, Wiley-Interscience, New York 1984
J.I.Kroschwitz, Ed., Fibers and Textiles, A Compendium (= reprint of relevant chapters in H.F.Mark et al., Encyclopedia of Polymer Science and Engineering, 2nd ed.), Wiley, New York 1990
H.J.Koslowski, Dictionary of Man-Made Fibres, Internat.Business Press, Frankfurt/Main 1998

6.1.b INTRODUCTION: TEXTBOOKS and NEW FIBERS
H.F.Mark, S.M.Atlas, E.Cernia, Ed., Man-Made Fibers, Interscience, New York, 3 vols. 1968
J.G.Cook, Handbook of Textile Fibres, Merrow, Watford, UK 1968
M.E.Carter, Essential Fiber Chemistry, Dekker, New York 1971
P.Lennox-Kerr, The World Fibres Book, The Textile Trade Press, Manchester 1972
C.B.Chapman, Fibres, Butterworth, London 1974
R.W.Moncrieff, Man-Made Fibres, Halsted, New York, 6th ed. 1975
W.E.Morton, J.W.S.Hearle, Physical Properties of Textile Fibres, Heinemann, London, 2nd ed. 1975
F.Happey, Ed., Applied Fibre Science, Academic Press, New York 1978 ff. (3 volumes)
M.Lewin, S.B.Sello, Eds., Handbook of Fiber Science and Technology, Dekker, New York 1983
R.E.Fornes, R.D.Gilbert, Ed., Polymers in Fiber Science, VCH, Weinheim 1991
H.Brody, Ed., Synthetic Fibre Materials, Longman Higher Education, Harlow (Essex), UK 1994
T.Hongu, G.O. Phillips, New Fibres, Woodhead Publ., Cambridge, 2nd ed. 1997
M.Lewin, E.M.Pearce, Eds., Handbook of Fiber Chemistry, Dekker, New York, 2nd ed. 1998
K.K.Chwala, Fibrous Materials, Cambridge University Press, New York 1998
F.Fourné, Synthetic Fibers, Hanser, Munich 1999; Hanser Gardner, Cincinnati (OH) 1999
B.Wulfhort, T.Gries, Textile Technology–An Introduction, Hanser Gardner, Cincinnati (OH) 2005
M.Lewin, Ed., Handbook of Fiber Chemistry, CRC Press, Boca Raton (FL), 3rd ed. 2007

6.1.2 HISTORY AND PRODUCTION
C.H.Fisher, History of Natural Fibers, J.Macromol.Sci.-Chem. **A 15**/7 (1981) 1345; reprinted in R.B.Seymour, Ed., History of Polymer Science and Technology, Dekker, New York 1982
R.B.Seymour, R.S.Porter, Ed., History of Synthetic Fibers, Elsevier, Amsterdam 1992
S.Handley, Nylon: The Story of a Fashion Revolution, Johns Hopkins Univ.Press, Baltimore (MD) 2000

6.1.3 DESIGNATIONS OF FIBERS
J.E.McIntyre, P.N.Daniels, Eds., Textile Terms and Definitions, Textile Institute, Manchester, 10th ed. 1995
The International Bureau for the Standardisation of Man-Made Fibres, Terminology of Man-Made Fibres, BISFA, Brussels, 2000 Edition

6.2 MANUFACTURE OF FIBERS AND THREADS
C.Placek, Multicomponent Fibers, Noyes Development, Pearl River 1971
R.Jeffries, Bicomponent Fibers, Merrow, Watford, UK 1972
W.E.Morton, J.W.S.Hearle, Physical Properties of Textile Fibers, Wiley, New York, 2nd ed. 1975
A.Ziabicki, Fundamentals of Fibre Formation, Wiley, London 1976
S.L.Cannon, G.B.McKenna, W.O.Statton, Hard-Elastic Fibers (A Review of a Novel State for Crystalline Polymers), Macromol.Revs. **11** (1976) 209
Z.K.Walczak, Formation of Synthetic Fibers, Gordon and Breach, New York 1977
J.L.White, Fiber Structure and Properties, Wiley, New York 1979
J.S.Robinson, Ed., Spinning, Extruding & Processing of Fibers, Noyes, Park Ridge (NJ) 1980
J.L.White, Dynamics, Heat Transfer and Rheological Aspects of Melt Spinning: A Critical Review, Polym.Eng.Revs. **1** (1981) 297
A.Ziabicki, H.Kawai, Eds., High-Speed Fiber Spinning: Science and Engineering Aspects, Wiley-Interscience, New York 1985
D.R.Salem, Ed., Structure Formation in Polymeric Fibers, Hanser Gardner, Cincinnati (OH) 2001

Z.K.Walczak, Processes of Fiber Formation, Elsevier, Amsterdam 2003

6.3 TEXTILE FIBERS
W.E.Morton, J.W.S.Hearle, Physical Properties of Textile Fibres, Textile Institute, Manchester,
    3rd ed. 1993
B.P.Saville, Physical Testing of Textiles, CRC Press, Boca Raton (FL) 1999
C.M.Pastore, P.Kiekens, Surface Characteristics of Fibers and Textiles, Dekker, New York 2000
M.G.Northolt, P. den Decker, S.J.Picken, J.J.M.Balthussen, The Tensile Strength of Polymer
    Fibers, Adv.Polym.Sci. **178** (2005) 1

6.3.2a SILK TYPES: NATURAL SILK
R.V.Lewis, Spider Silk: the Unraveling of a Mystery, Acc.Chem.Res. **25** (1992) 392
D.Kaplan, W.Wade Adams, B.Farmer, C.Viney, Ed., Silk Polymers: Materials Science and
    Biotechnology (ACS Symp. Ser. **544**), American Chemical Society, Washington (DC) 1994
M.Okamoto, K.Kajiwara, Textile Fibers (Shingosen), in J.C.Salamone, Ed., Polymeric Materials
    Encyclopedia, CRC Press, Boca Raton (FL) 1996, p. 8299

6.3.2b SILK TYPES: POLYESTER
H.Szegö, Modified Polyethylene Terephthalate Fibers, Adv.Polym.Sci. **31** (1979) 89
J.Militký, J.Veníček, J.Kryštůfek, V.Hartych, Modified Polyester Fibres, Elsevier, Amsterdam 1991
D.Bunnschwiler, J.W.S.Hearle, Eds., Polyester: 50 Years of Achievement, Textile Institute,
    Manchester 1993

6.3.3a COTTON TYPES: COTTON
K.Ward, Jr., Chemistry and Chemical Technology of Cotton, Interscience, New York 1955
H.B.Brown, J.O.Ware, Cotton, McGraw-Hill, New York, 3rd ed. 1958
E.Lord, The Characteristics of Raw Cotton, Textile Institute, Manchester 1961
D.S.Hanby, Ed., The Cotton Fiber, American Cotton Handbook, Interscience, New York,
    3rd ed. 1965
L.Miles, Cotton, Wayland, Have UK 1980
R.M.Brown, Ed., Cellulose and Other Natural Polymer Systems, Plenum, New York 1982
A.D.Muir, N.D.Westcott, Eds., Flax: The genus Linum, CRC Press, Boca Raton (FL) 2003
P.J.Wakelyn, N.R.Bertoniere, A.D.French, D.P.Thibodeaux, B.A.Triplett, M.-A.Rousselle,
W.R.Goynes, Jr., J.V.Edwards, L.Hunter, D.D.McAlister, G.R.Gamble, Cotton Fiber Chemistry
    and Technology, CRC Press, Boca Raton (FL) 2006

6.3.3b COTTON TYPES: REGENERATED CELLULOSES
J.Dyer, G.C.Daul, Rayon Fibers, in M.Lewin, E.M.Pearce, Handbook of Fiber Chemistry, Dekker,
    New York, 2nd ed. (1998), Chapter 10, p. 725

6.3.4a WOOL TYPES: WOOL
C.Earland, Wool, Its Chemistry and Physics, Chapman and Hall, London, 2nd ed. 1963
W.von Bergen, Wool Handbook, Americam Wool Handbook Co., New York, 3rd ed. 1970
R.D.B.Fraser, T.P.MacRae, G.E.Rogers, Keratins, Their Composition, Structure and Biosynthesis,
    C.C.Thomas, Springfield (IL) 1972
R.Asquith, Chemistry of Natural Protein Fibres, Plenum, London 1977
J.A.Maclaren, B.Milligan, Wool Science, the Chemical Reactivity of the Wool Fibre, Science Press,
    Marrackville (N.S.W.), Australia 1981
J.D.Leeder, Wool, Nature's Wonder Fibre, Australasian Textiles, Ocean Grove, Australia 1984
R.Postle, G.A.Carnaby, S.de Jong, The Mechanics of Wool Structures, Wiley, New York 1988

6.3.4b WOOL TYPES: ACRYLICS
J.C.Masson, Acrylic Fiber Technology and Applications, Dekker, New York 1995

6.3.5 ELASTIC FIBERS
R.Meredith, Elastomeric Fibres, Merrow, Watford, UK 1971
G.H.R.Weiss, Natural Rubber Technol. **10** (1979) 80
K.H.Wolf, Textil Praxis **1981** (1981) 839 (in German)

6.4 YARNS
B.S.Goswami, J.G.Martindale, F.L.Scardino, Textile Yarns: Technology, Structure and Applications,
    Wiley-Interscience, New York 1977
J.S.Robinson, Ed., Manufacture of Yarns and Fabrics from Synthetic Fibers, Noyes Data Corp., Park
    Ridge (NJ) 1980

6.5 TEXTILES
A.J.Hall, The Standard Handbook of Textiles, Halsted, New York, 8th ed. 1975
D.S.Lyle, Modern Textiles, Wiley, New York 1976
D.S.Lyle, Performance of Textiles, Wiley, New York 1977
B.P.Corbman, Textiles. Fiber to Fabric, McGraw-Hill, New York, 6th ed. 1983
M.L.Joseph, Textile Science Introductory, Holt, Rinehart, and Winston, New York, 5th ed. 1986
S.Adanur, Ed., Wellington Sears Handbook of Industrial Textiles, Technomic, Lancaster (PA) 1995
M.Raheel, Ed., Modern Textile Characterization Methods, Dekker, New York 1996
B.Wulfhorst, T.Gries, Textile Technology–An Introduction, Hanser Gardner, Cincinnati (OH) 2005
S.J.Kadolph, Ed., Textiles, Pearson/Prentice Hall, Upper Saddle River (NJ), 10th ed. 2007
A.K.Sen, Coated Textiles, CRC Press, Boca Raton (FL), 2nd ed. 2007

6.5.2 NONWOVENS
R.A.A.Hentschel, Spunbonded Sheet Products, ChemTech **4** (1974) 32
M.T.Gilles, Nonwoven Materials. Recent Developments, Noyes, Park Ridge, NJ 1979
P.Schwartz, T.Rhodes, M.Mohammed, Fabric Forming Systems, Noyes, Park Ridge (NJ) 1982
H.A.Krässig, J.Lenz, H.F.Mark, Fiber Technology. From Film to Fiber, Dekker, New York 1984
J.Lunenschloss, W.Albrecht, Eds., Non-Woven Bonded Fabrics, Wiley, New York 1985
W.Albrecbt, H.Fuchs, W.Kittelmann, Eds., Nonwoven Fabrics, Wiley-VCH, Weinheim 2002

6.6 TEXTILE FINISHING
J.T.Marsh, Mercerising, Chapman and Hill, London 1941
H.Mark, N.S.Wooding, S.M.Atlas, Chemical Aftertreatment of Textiles, Wiley, New York 1971
I.D.Rattee, M.M.Breuer, The Physical Chemistry of Dye Adsorption, Academic Press, London 1974
E.R.Trotman, Dyeing and Chemical Technology of Textile Fibers, Griffen, London, 5th ed. 1975
R.H.Peters, The Physical Chemistry of Dyeing, Elsevier, Amsterdam 1975
C.S.Sodano, Water and Soil Repellants for Fabrics, Noyes Publ., Park Ridge, NJ 1979
R.W.Lee, Ed., Printing on Textiles by Direct and Transfer Techniques, Noyes Publ., Park Ridge (NJ)
    1981
J.E.Nettles, Handbook of Chemical Specialties, Textile Fiber Processing, Preparation and Bleaching,
    Wiley, New York 1984
M.Lewin, S.B.Sello, Ed., Handbook of Fiber Science and Technology, Vol. I, Chemical Processing
    of Fibers and Fabrics. Fundamentals and Preparation; Vol. II, Functional Finishes, Dekker, New
    York 1984 (in two parts, A and B)
K.V.Datye, A.A.Vaidya, Chemical Processing of Synthetic Fibers and Blends, Wiley, New York 1984
E.R.Trotman, Ed., Dyeing and Chemical Technology of Textile Fibres, Wiley, New York,
    6th ed. 1985
H.Zollinger, Color Chemistry. Syntheses, Properties and Applications of Organic Dyes and Pigments,
    VCH, Weinheim 1987
J.Wypych, Polymer Modified Textile Materials, Wiley, New York 1988
-, Textile Finishing Chemicals. An Industrial Guide, Noyes, Park Ridge (NJ) 1990
T.L.Vigo, Textile Processing and Properties: Preparation, Dyeing, Finishing and Performance,
    Elsevier Science, Amsterdam 1994
M.Raheel, Ed., Modern Textile Characterization Methods, Dekker, New York 1996
P.E.Slade, Handbook of Fiber Finish Technology, Dekker, New York 1998
H.K.Rouette, Encyclopedia of Textile Finishing, Springer, Berlin 2002
B.C.Goswani, R.D.Anandjiwala, D.M.Hall, Eds., Textile Sizing, Dekker, New York 2004
A.K.Sen, Coated Textiles. Principles and Applications, CRC Press, Boca Raton (FL), 2nd ed. 2008

6.7 TECHNICAL ORGANIC FIBERS
M.Lewin, J.Preston, Eds., Handbook of Fiber Science and Technology, Volume III, High Technology
    Fibers, Dekker, New York, Part A (1985), Part B (1989), Part C (1993)
A.R.Horrocks, S.C.Anand, Handbook of Technical Textiles, CRC Press, Boca Raton (FL) 2000

[19] W.S.Boston, in Kirk-Othmer, Encycl.Chem.Technol., Wiley, New York, 3rd ed., **24** (1984)
      612, Fig. 6; based on work by J.B.Speckman, J.Text.Inst. **18** (1927) T 431

[20] P.T.Naughton and 17 co-authors, in Kirk-Othmer, Encycl.Chem.Technol., Wiley-Interscience,
      4th ed., **25** (1998) 664, data of Table 6

[21] A.J.Ultee, in J.I.Kroschwitz, Concise Encyclopedia of Polymer Science and Engineering,
      Wiley, New York 1990, p. 382, Fig. 1

[22] B.C.Goswami, J.G.Martindale, F.L.Scardino, Textile Yarns: Technology, Structure and Appli-
      cations, Wiley, New York 1977; L.Rebenfeld, in Kirk-Othmer, Concise Encyclopedia of
      Chemical Technology, Wiley, New York, 4th ed. (1999), p. 828

[23] W.Albrecht, Kunststoffe **66** (1976) 660, Fig. 6

[24] J.W.S.Hearle, Fibers, 2. Structure, in Ullmann's Encyclopedia of Industrial Chemistry, Wiley-
      VCH, Weinheim, 6th ed. (2002), Vol. 13, p. 369, adapted from Fig. 23

[25] Courtesy of E.Treiber, private communication

[26] S.K.Batra, Other Long Vegetable Fibers, in M.Lewin, E.M.Pearce, Handbook of Fiber Chem-
      istry, Dekker, NewYork, 2nd ed. (1998), p. 505 ff., Tables 8.2-8.25

[27] H.Krässig, private communication 1976-05-07, see also Faserforschg.Textiltechn. **26** (1975)
      135

[28] S.J.Deteresa, S.R.Allen, R.J.Farris, R.S.Porter, J.Mater.Sci. **19** (1984) 57, Figs. 6a and 6c,
      courtesy of R.S.Porter

[29] R.Kleinholz, G.Heyn, R.Stolze, in R.Gächter, H.Müller, Eds., Plastics Additives Handbook,
      Hanser, Munich 1990, Chapter 10, Table 3

[30] S.Kessler, in J.Edenbaum, Ed., Plastics Additives and Modifiers Handbook, Van Nostrand
      Reinhold, New York 1992, Table 48-1

[31] B.G.Penn, F.E.Ledbetter III, J.E.Clemons, J.G.Daniels, J.Appl.Polym.Sci. **27** (1982) 3751,
      data of Table VI

[32] S.Yajima, Philos.Trans.Roy.Soc. (London) **A 294** (1980) 419

[33] E.Fitzer, R.Kleinholz, H.Tiesler, M.H.Stacey, R. De Bruyn, I.Lefever, M.Heine, Fibers.
      5. Synthetic Inorganic, in Ullmann's Encyclopedia of Industrial Chemistry, Wiley-VCH,
      Weinheim, 6th ed. (2002), Vol. 13, p. 563 ff., Table 15

7 Rubbers and Elastomers

7.1 Overview

7.1.1 Introduction

The word **elasticity** has very different specific meanings in automotive engineering, ecology, economics, electronics, and physics. In general, it indicates the condition or property of being flexible (bendable, pliable, persuadable, etc.) (G: *ela(s)tos* = beaten, from *elaunein* = to drive). In *physics*, elasticity denotes the ability of a deformed material to return to its original dimensions after the applied force has been removed. Physics also distinguishes between energy and entropy elasticity (Volume III, p. 517).

Steel is an iron-carbon alloy with less than 2 wt% carbon that was first manufactured in the 12th century. The ability of steel spheres to bounce back from steel plates goes hand-in-hand with the great hardness of steel. It was therefore very surprising that balls of a previously unknown soft material also behaved elastically. These balls were presented as curiosities to the Spanish court by Cristoforo Colombo (aka Christopher Columbus), an Italian in Spanish service, after his return to Spain from his famous voyage to the West Indies and Central America in 1492.

The soft material was obtained from a tree exudate (**gum**) and called *chauchuc* by the Spanish (obsolete; German: Kautschuk) after the Quechua: *cahuchu* = weeping tree; from *caa* = tree, *ochu* = tears). Only a few European people came to know this material, however, because there were no published reports and samples were rare. Later, some articles made from *cauchuc* were exported to Spain and Portugal but they did not generate much economic interest because of their inferior use properties.

Cauchuc was rediscovered by a French scientific expedition to the Amazon (see below). The report of the expedition mentioned this material which henceforth was referred to in England by its French name "caoutchouc." A few years later, its rubbing properties were noticed and caoutchouc now became known as **rubber** in English speaking countries (for the history, see Section 7.1.2), much later also as **India rubber** (because it was cultivated by British planters in Kerala, India), and finally as **natural rubber (NR)**. The later developed synthetic chemical compounds with soft elastic properties were then called **synthetic rubbers (SR)**. These chemical compounds are synthetic *elastic* poly*mers*, hence "**elastomers**." The elastic properties of both natural rubber and synthetic rubbers are much improved by chemical crosslinking of their polymer chains, a process that is generally known by its historic name, **vulcanization** (Section 7.2.4).

"Rubber" and "elastomer" are used by some as synonyms whereas "rubber" is for others only natural rubber and "elastomer" always a synthetic polymer with elastic properties. For other people, "rubber" is the uncrosslinked raw material and "elastomer" its crosslinked derivative. Or it is the other way around: "elastomer" is the uncrosslinked raw product and "rubber" its crosslinked counterpart.

This book adheres to the traditional designation of **rubber** as an *uncrosslinked*, natural or synthetic, entropy-elastic raw material (*vide* "natural rubber," "synthetic rubber," "Buna rubber," "rubber vulcanization," etc.). "Natural rubber" is one type of naturally produced rubbers (Section 7.3.1). "**Elastomer**" is defined in this book as a chemically or physically *lightly crosslinked* rubber with improved entropy-elastic properties. The crosslinks of *physically* crosslinked rubbers can be reversed at higher temperatures; these materials are therefore known as **thermoplastic elastomers** (Section 7.5). *Chemically* very strongly crosslinked natural rubber is no longer elastic and is known traditionally as **hard rubber**. The crosslinking of both hard rubber and *chemically* lightly crosslinked rubbers is irreversible; these materials are therefore "thermoset elastomers" (for thermosets, see Section 8.5).

7.1.2   Historical Development

Slight slitting of the trunk of rubber trees produces a milky fluid called **latex** (L: *latex* = liquid) from which natural rubber can be obtained by drying, coagulation, etc. (Section 7.3.1). Properties of the resulting elastic material were well known to the indigenous people of Central and South America who used it to prepare various useful goods; the oldest known artifacts are from 1600 B.C. For example, cloths and shoes were waterproofed by coating them with rubber latex. It was also known that drying such treated cloths and shoes over an open fire not only removed the water from the latex but improved the durability of the goods; to this day, some natural rubber is still traded as "smoked sheets." The Maya also knew of a kind of vulcanization. They treated balls of natural rubber with the sap of *Ipomoea alba* (moonflower vine (Morning Glory Family)). This process improved the elasticity of the rubber: balls of 9.5 cm diameter treated by the sap bounce back 2 meters high.

Columbus brought such balls to Europe as curiosities. In the 16th century, rubber shoes, cloaks, and bottles were exported to Spain and Portugal but these goods were not very useful because rubber shoes lost their elasticity in the cold European winters and rubber cloaks became sticky in the hot summer months. Lumps of natural rubber also lacked economic interest because they could not be processed in Europe. Melting resulted in degradation, and solvents for natural rubber were unknown. Rubber latex, on the other hand, could not be shipped to Europe because it was unstable and flocculated.

Natural rubber was almost forgotten until Charles-Marie de La Condamine brought back lumps of natural rubber from the Amazon. This gentleman was a member of a scientific expedition that was sent to South America by the Académie des Sciences, Paris, in 1735. The expedition ran into difficulties; Condamine traveled alone to Quito, did his measurements but did not have the money to return directly to Paris. He decided to travel down the Amazon where he learned from Indians how to harvest and use natural rubber. He was back in Paris in 1744.

The Academy dispatched two expeditions, one to Peru and the other to Lapland. The Peru expedition was instructed to determine the length of a degree of the meridian near the equator as the basis for the meter, which was later defined in 1793 as one ten-millionth of the length of the Earth's meridian at Paris along a quadrant (= distance from the equator to the North Pole). A prototype of the new unit "meter" was cast in brass in 1793 and in a platinum/iridium alloy in 1799.

In 1745, Condamine published his measurements and observations in the *Mémoirs de l'Academie des Sciences*. An English translation appeared in 1747 and a German one in 1748. This publication renewed the interest in caoutchouc which lost its good name after its rubbing properties were discovered accidentally in 1770 by Edward Nairne, a maker of scientific instruments, who sold cubic inch-sized pieces of these erasers for the extraordinary price of three shillings. In the same year, Joseph Priestley observed the same property (on Nairne's pieces?) and made it public. According to other sources, the rubbing properties of natural rubber were already known to Fernão de Magelhaes (aka Magellan, ca. 1480 to 1521 (died in battle)). These erasers are called "rubbers" in the United States but "rubber" is also slang for a condom.

In 1761, turpentine was discovered as a solvent for rubber by L.A.M.Hérrissant and P.J.Macquer. A 1791 patent by Samuel Pearl described a turpentine-based process for the manufacture of rubber-coated fabrics. In 1824, Charles Macintosh invented rain coats ("mackintoshes") made from a rubber sheet that was sandwiched between fabrics.

Unfortunately, articles manufactured from rubber solutions became sticky in hot weather. In 1824, Thomas Hancock observed that freshly cut pieces of natural rubber can be welded together again, i.e., that rubber has inherent tack (Section 7.2.2). The tacky lumps of rubber were difficult to process and Hancock therefore looked for a machining process to cut the big rubber lumps into small pieces. He built a **masticator** consisting of two counterrotating drums whose surfaces were studded with iron teeth that tore the rubber apart (G: *masthika*n = to grind the teeth). Masticators with corrugated rolls are still used for blending and formulating rubbers (Section 7.2.3).

However, mastication degrades rubber molecules, which makes rubber even more sticky. This stickiness was ascribed to the presence of liquids in rubber, and one therefore looked for "drying agents." According to some sources, Joseph Priestley found that adding carbon black to natural rubber reduced the inherent tackiness. After many experiments by many people, sulfur emerged as the best "drying agent" (F.L.Lüdersdorf (1832); N.Hayward (1834)). In 1839, the American Charles Goodyear observed that a fabric coated with a rubber/sulfur mixture became leathery after he accidently left it overnight in contact with a hot oven. After many trials, he found that this "hardening" of rubber by sulfur and heat was accelerated by addition of litharge (PbO) (1839) or white lead (magistery of lead; basic lead carbonate; $2\ PbCO_3 \cdot Pb(OH)_2$) (1841). Goodyear recognized the technological and economic importance of this process which he called "metallization" but he could not find investors and kept the invention secret before he applied for a patent three years later (US Patent 3633 (1844)).

In order to raise interest in his invention, Goodyear sent pieces of "metallic gum-elastic" to prospective investors. One of the pieces fell into the hands of the Englishman Thomas Hancock who was also interested in rubber. After analyzing the piece, Hancock applied for a patent (BP 9952 (1843)) which predated Goodyear's patent. Hancock called the process "vulcanization" since sulfur and fire are attributes of the Roman god *Volcanos* (*Vulcan*). He was probably also the first to produce solid rubber tires. Goodyear was awarded more than 200 patents (none of them for tires) but died deep in debt.

Goodyear's invention accelerated exports of **wild rubber**, first from Peru (ca. 25 t/a in 1822) and then increasingly from Brazil, reaching ca. 2700 t/a in 1860 (Fig. 7-1, statistics are incomplete). This wild rubber was harvested from several species of the genus *Hevea* and became known as **para rubber** after the state of Pará in Northern Brazil.

Wild rubber became a very profitable Brazilian monopoly after the Amazon became navigable in 1866. This, in turn triggered two reactions: collection of wild rubber in other countries and growth of rubber trees in plantations.

Additional wild rubber was collected not only in other Central and South American countries from trees but also from lianes in Africa, especially in what is now Ghana, Nigeria, and the Democratic Republic of Congo (Kinshasa), the latter known during 1888-1908 as Congo Free State, a personal property of the Belgian King Leopold II. In this "state", extremely brutal methods were used to collect rubber which ranged from forced labor, slavery, and cutting off right hands to burning down villages and killing all villagers (some speak of up to 10 million victims). Between 1880 and 1910 about half of the wild rubber came from Brazil and the other half from other countries (Fig. 7-1).

The British needed large amounts of rubber at moderate prices and therefore made several attempts to grow rubber trees. In a first attempt in 1873, J.Collins collected 2000 Brazilian seeds of which 12 germinated in England but all died.

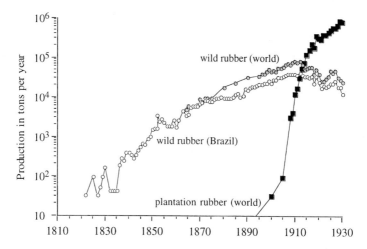

Fig. 7-1 Annual production of wild and plantation natural rubber 1822-1930 [1]. For production after 1930, see Fig. 7-4 on p. 256.

In 1876, a second attempt by Henry Wickham succeeded. From 70 000 seeds he shipped (and probably *not* smuggled!) to England, just ca. 2397 germinated. 1919 saplings were shipped to Ceylon (now: Sri Lanka) but 90 % died during the voyage. 100 of these plants were sent to Singapore; all died. Later that year, even plants (and seeds) were shipped to England from Brazil. From England, 100 saplings (which seem to have been crossbreeds) were sent to Ceylon and 22 further to Singapore. It is from these plants that most cultivars of *H. brasiliensis* are derived. Serious studies of cultivation and tapping in Singapore Botanical Gardens followed and commercial planting began in the Malay States in 1895; India (Kerala region), Indonesia, and Thailand followed.

The production of **plantation rubber** grew fast from 4 t/a (1900) and 145 t/a (1905) to 8200 t/a (1910) (Fig. 7-1) which caused the price of natural rubber to drop dramatically (Fig. 7-2). The proportion of wild rubber to the total annual natural rubber production decreased steadily and is now less than 1-2 %.

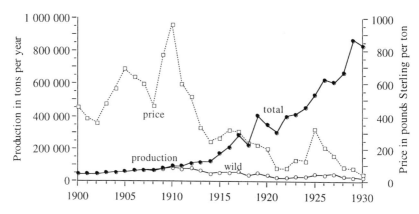

Fig. 7-2 Average annual price per ton of rubber in historical pound Sterling and annual production of wild natural rubber (O) and of all (wild and plantation) natural rubber (●) [1, 2]. For production after 1930, see Fig. 7-4 on p. 256.

Rubber was of both scientific and economic interest. It was known that natural rubber has the composition C_5H_8 (1824; M.Faraday) and is an isoprene derivative (1860; C.H.G.Williams). It was not clear whether it is a linear molecule (C.O.Weber (1900-1903), S.S.Pickles (1910), H.Staudinger (1920)) or cyclic (C.D.Harries (1911-1913)). One also knew that isoprene autopolymerized to a rubber in sealed bottles at room temperature in 8 years (W.A.Tilden, 1892) or 9 months (C.O.Weber, 1894), respectively.

Industrial syntheses of isoprene, $CH_2=C(CH_3)-CH=CH_2$, were not known, however, but one knew that rubber-like products can be obtained from the polymerization of "methyl isoprene" (= 2,3-dimethylbutadiene), $CH_2=C(CH_3)-C(CH_3)=CH_2$ (J.Kondakov, 1901). Since methyl isoprene can be obtained industrially from acetone, $(CH_3)_2CO$, via the pinacol, $(CH_3)_2C(OH)-C(OH)(CH_3)_2$, it was chosen as the monomer. The polymerization process was patented in 1910 (F.Hofmann) and automobile tires made from the resulting **methyl rubber** were tested successfully in Germany in 1912.

Two years later, World War I broke out. Germany was cut off from its supply of natural rubber and started industrial syntheses of methyl rubber (Table 7-1). The oxygen-initiated polymerization at 35-40°C delivered a rubber that was practically only good for hard rubber (hence type "H"). Polymerization at elevated temperatures was slower but yielded a soft rubber (German: *Weichgummi*, hence type "W"). Sodium was also tried as an initiator (known to polymerize isoprene (F.C.Matthews, E.H.Strange; C.D.Harries, O.Aschan (1910)) but there were technical difficulties (type "B"; from German: *beschleunigt* = accelerated). Polymerization with peroxides as initiators (A.Heinemann, 1910; A.Holt, 1911)) or ozone apparently never led to industrial products.

Methyl rubbers had inferior mechanical properties and were expensive, difficult to vulcanize, and fast ageing (no natural antioxidants). Hence, their production ceased after World War I. However, the price of natural rubber climbed very high because Dutch and British plantation owners formed a cartel. This led Henry Ford to found the rubber plantation Fordlândia in Brazil that employed 5000 workers (planned capacity: 300 000 t/a) but failed because the rubber trees were attacked by the fungus *Microcyclus ulei*.

The high rubber prices also renewed German interest in synthetic rubber. This time, there were promising new German sources of monomers either from acetylene (from calcium carbide) or from hydrocarbons (from the Bergius process for coal liquefaction). In 1926, the newly founded IG Farben syndicate chose butadiene as the monomer.

The forerunner of IG Farben was founded in 1904 by three German dyestuff companies (AGFA, BASF, Bayer); five other German companies joined in 1916. The syndicate was dissolved after WW I but refounded in 1925 as Interessen-Gemeinschaft Farbenindustrie AG (short: IG Farben ≈ *Dyestuff Syndicate*) by AGFA, Cassella, BASF, Bayer, Farbwerke Hoechst, and Chemische Fabrik Kalle; several smaller companies joined later. At the height of its existence, IG Farben consisted of 400 German and 500 foreign companies (all fully or partly owned; 190 000 employees). It was dissolved in 1945.

Table 7-1 World War I German industrial polymerizations of methyl isoprene [3].

Polymer	Reactor	Polymerization Initiator	$T/°C$	Months	Total production in tons
Methyl rubber H	200 L	O_2 from air, + added H	35 - 40	1.5 - 3	1800
Methyl rubber W	4000 L	O_2 from air	75 - 85	4 - 6	700
Methyl rubber B		0.5 % Na with CO_2	35 - 40	0.5 - 0.75	~ 10
Ozonide rubber		peroxides or ozonides		0.75 - 1	0

The first butadiene rubbers were called **Buna** because the polymerization of *bu*tadiene was initiated by sodium (D: *Na*trium) (S.Lebedev). The often explosive process was tamed by using regulators (ether, vinyl chloride) and continuous polymerization in extruders. The various types differed in their K number (a measure of molecular weight (Volume II, p. 14)) and were therefore known as **number Bunas** (Buna 32, etc.).

The year the process was ready for production, the bottom fell out of the market for natural rubber (from 4.55 RM/kg in 1925 to 0.35 RM/kg in 1932 (RM = Reichsmark, the then German currency). Buna production thus stopped but was resumed in 1937 because the German government wanted Germany to become self-sufficient.

The Soviet Union started to produce number Buna in 1933. It is unclear whether their Buna was independently developed on the basis of Lebedev's experiments or simply based on German patents: at the time, Germany patented only processes but not the composition of matter and the Soviets did not recognize patents at all. Soviet production of number Buna reached 77 000 t/a in 1939.

However, continuous polymerization in extruders delivered small throughputs, and discontinuous bulk polymerization was difficult to control because of the large heat of polymerization. Polymerization in aqueous emulsion produced an artificial latex but the rubber from this latex was inferior. Addition of reducing agents during emulsion polymerization (for the removal of the last traces of oxygen) and use of antioxidants finally resulted in a rubber that matched the number Buna from extrusion polymerization. Far better synthetic rubbers were then obtained by free-radical copolymerization of butadiene with styrene to **Buna S** (E.Tschunkur, W.Bock, DRP 570 980 (1929-07-21)) and with acrylonitrile to **Buna N** (E.Konrad, E.Tschunkur, H.Kleiner, DRP 658 172 (1930-04-26)). These **letter Bunas** completely replaced number Bunas (Section 7.3).

During the years number and letter Bunas were developed, another IG Farben activity turned out to be very important for the synthetic rubber industry. BASF, later a member company of IG Farben, was eager to develop the Bergius process for the conversion of brown coal (of which Germany had plenty) to gasoline and motor oils but had no experience in this business because Germany had only a very small petroleum industry (crude oil production: 10^3 barrels per day (Germany) *versus* $2.1 \cdot 10^6$ bbl/d (USA)). In 1925, BASF directors therefore visited Standard Oil of New Jersey (now: Exxon-Mobil) which was operating the biggest US oil refinery. BASF and Standard Oil subsequently became joint owners of the newly founded Gasolin AG in Germany.

In the next year, a Standard Oil executive visited Gasolin AG and also BASF (now IG Farben) where he learned about new catalysts for the Bergius process that not only turned brown coal into oil and further to gasoline and oil products such as monomers but could also convert low-quality oils and bitumens to high-quality gasoline. The "cracking" of crude oil by these catalysts soon replaced the previous simple distillation of petroleum which yielded only 1 L of gasoline from 4 L of oil.

In 1927, Standard Oil and IG Farben agreed to cross-licensing in the oil business (including Bergius process and butadiene rubber): each partner kept all the new patents it developed and also the licensing rights for these processes; the other partner got 50 % of the royalties. In 1930, Standard Oil bought the world-wide rights to the Bergius process (except for Germany). The same year, Standard Oil and IG Farben also formed the Joint American Study Company (JASCO). In 1933, a contract was signed with General Tire and Rubber Co. for the use of number Buna in tires and in 1938 with the US rubber industry about large-scale trials with Buna S. JASCO was reorganized in 1939: Standard Oil got the rights for all JASCO processes in the United States, Great Britain, and France and IG Farben the rights for all other countries. Standard Oil also bought back its stock from IG Farben.

In the United States, there was no commercial incentive to produce Buna S which was more expensive than natural rubber. In 1940, the United States government feared a delivery stop of natural rubber because of the second Japanese-Chinese War and established the Rubber Reserve Company which stockpiled natural rubber. In 1941, Standard Oil refused a government request to make the German Buna patents available to other US companies, was sued by the Justice Department for its cartel arrangement with IG Farben, and was called a traitor by Senator Henry Truman. At that time, the United States was not at war with Germany but had already confiscated German ships. Four days after the Japanese attack on Pearl Harbor on December 7, 1941, Germany declared war on the United States.

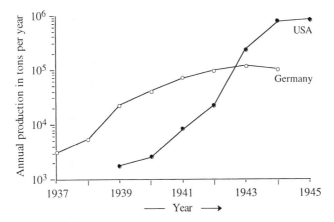

Fig. 7-3 Annual production of synthetic rubbers in Germany (number, Buna S, Buna N) and the United States (1939: GR-A and M; 1941: GR-A, M and S; 1942: GR-A, I, M, S)) before and during World War II [4]. US letter codes were unknown in 1939-1941; they were introduced in 1942.

Buna S was produced industrially in Germany since 1937; the production peaked in 1943 (Fig. 7-3). In the United States, Standard Oil (NJ) had trial productions of Buna S since 1939. There were also small productions of several specialty rubbers: sulfide rubber (Thiokol Co.: Thiokol® (1930)), chloroprene rubber (DuPont: Duprene® (1932) = Neoprene® (1938)), butadiene-methyl methacrylate rubber (B.F.Goodrich: Ameripol® (1940)), and isobutylene-isoprene rubber (butyl rubber; Standard Oil (NJ) (1942)).

But there was no synthetic general purpose rubber except Buna S. Standard Oil (NJ) was forced to give up its patent rights and, twelve days after Pearl Harbor, five US companies signed a patent-and-information sharing agreement (Standard Oil (NJ), Firestone, Goodrich, Goodyear, US Rubber). In 1942, these companies agreed to a "mutual recipe" for the production of butadiene-styrene rubber. The Baruch committee recommended in September 1942 to build 15 synthetic rubber plants under the umbrella of the government-owned Rubber Reserve Company. In 1945, US production of "government rubber" peaked at 834 000 t/a from which 731 000 t/a was styrene-butadiene rubber (Fig. 7-3).

Government rubbers (**GR**) were known by their 3-letter symbols: GR-A (butadiene-*a*crylonitrile), GR-I (*i*soprene-isobutylene), GR-M (= *m*onovinylacetylene → chloroprene), GR-P (*p*olysulfide → thiocol), GR-S (butadiene-*s*tyrene), GR-X (butadiene-styrene, *e*xperimental).

After World War II ended, Germany was forbidden to produce synthetic rubbers and the US production dropped (Fig. 7-4). In 1950, the economy started booming in Western countries: demand for automobiles increased and so did that for tires, especially for tire treads from so-called **cold rubber** that is produced by emulsion copolymerization of butadiene and styrene at 5°C. New polymerization initiators led to three new types of synthetic rubbers and the production of synthetic rubbers accelerated in 1960 (see Section 7.1.4). During 1979-1996, world synthetic rubber production stagnated but increased again since 1997 thanks to improving Asian economic conditions.

The production of natural rubber increased about linearly with time between 1950 and 1992 but accelerated after 1990* and more so after 2002 (Fig. 7-4). Natural rubber now (2006) comprises 40 % of the total rubber production, up from 30.7 % in 1976.

* The U.S.-Vietnamese war ended in 1975. The Vietnam-Cambodia war lasted from 1977 to 1980. China attacked Vietnam in 1979. Vietnamese-Cambodian fighting occurred in 1985.

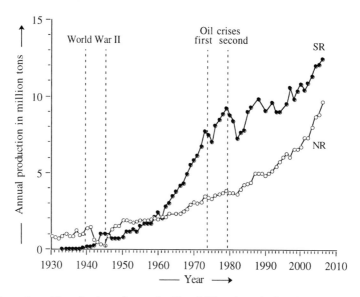

Fig. 7-4 Annual world production of natural rubber (NR) and synthetic rubbers (SR), including the former Soviet Union [1, 5a, 6]. Since 1951, the annual production of synthetic rubber also includes oil-extended rubbers (= general-purpose rubbers SBR, BR, IIR, and EPDM that contain mineral oil). Main producers of natural rubber are Thailand, Indonesia, Malaysia, Sri Lanka, Vietnam, and China.

7.1.3   Rubber Classification

Synthetic rubbers are subdivided into **diene** and **non-diene** rubbers according to the constitution of their monomer units or into **general purpose** and **specialty rubbers** with respect to their applications:

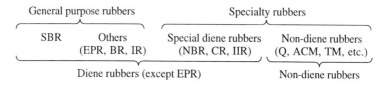

At present, ca. 30 main types and ca. 1800 subtypes of synthetic rubber are known. The various types are commonly distinguished by letters; unfortunately, abbreviations differ sometimes both nationally and internationally as well as between the various standardization agencies (ASTM, BS, DIN, ISO, IUPAC, etc.) (see Section 16.7).

The most commonly used ASTM system characterizes the various types by 1-5 capital letters of which the last letter indicates the chemical constitution of the polymer chain:

M   saturated carbon chains (M from "**m**ethylene")
O   carbon-**o**xygen chains
Q   silicone chains (Q from "**q**uartz")
R   unsaturated carbon chains (R from "**r**ubber")
T   carbon-sulfur chains (**t**hio rubbers)
U   rubbers with **u**rethane groups in chains

Monomeric units are characterized by preceding letters,

A	acrylic ester units	$-CH_2-CH(COOR)-$
B	butadiene units	$-CH_2-CH=CH-CH_2-$
C	chloroprene units	$-CH_2-CCl=CH-CH_2-$
E	ethylene units *or*	$-CH_2-CH_2-$
	ethylene oxide units	$-CH_2-CH_2-O-$
I	isoprene units *or*	$-CH_2-C(CH_3)=CH-CH_2-$
	isobutylene units	$-CH_2-C(CH_3)_2-$
N	acrylonitrile units	$-CH_2-CH(CN)-$
P	propylene units *or*	$-CH_2-CH(CH_3)-$
	2-methyl-5-vinyl **p**yridine units	$-CH_2-CH(C_5H_3(CH_3)N)-$
S	styrene units	$-CH_2-CH(C_6H_5)-$
T	carbon-sulfur chains (**t**hio rubbers)	various structures

and substituents by

F	**f**luorine	$-F$
G	allyl **g**lycidyl	$-CH_2-CH[CH_2OCH_2CH(ring-O)CH_2]$
M	**m**ethyl	$-CH_3$
P	**p**henyl	$-C_6H_5$
V	**v**inyl	$-CH=CH_2$

Additional prefixes symbolize modifications of polymer chains:

C	after-**c**hlorination
H	**h**ydrogenation
S	after-bromination (G: *bromos* = **s**tench)
X	introduced carboxyl groups

A *pre*fix Y– indicates a block (co)polymer. *Post*fixed letters show crosslinkabilities, for example, by **i**socyanates (–I) or **p**eroxides (–P). Examples (see also Table 16-12):

AU-I	synthetic rubber with chains containing polyester units (A) and urethane groups (U) that can be crosslinked by isocyanates;
CM	after-chlorinated (C) saturated poly(ethylene) chain (M);
IIR	unsaturated rubber (R) with isobutylene (I) and isoprene units (I);
PVMQ	silicone rubber(Q) with phenyl (P), vinyl (V), and methyl substituents (M);
SBR	unsaturated rubber (R) with styrene (S) and butadiene (B) units;
SIR	unsaturated rubber (R) with styrene (S) and isoprene units (I);
XNBR	unsaturated rubber (R) with acrylonitrile (N) and butadiene (B) units in which carboxylic groups were introduced by post-reaction (X);
YXSBR	unsaturated rubber (R) consisting of a triblock polymer (Y) with styrene (S) and butadiene blocks (B) and carboxylic groups (X).

The symbol for **t**hermo**p**lastic **e**lastomers is TPE. Letters postfixed to TPE indicate characteristic chain units:

TPE-A	segmented thermoplastic elastomers with ether and **a**mide units;
TPE-E	segmented thermoplastic elastomers with ether and **e**ster units;
TPE-O	thermoplastic **o**lefin elastomers;
TPE-S	thermoplastic elastomers with **s**tyrene-butadiene-styrene blocks;
TPE-U	thermoplastic **u**rethane elastomers.

7.1.4   Economic Importance

During the years 1996-2006, world production of natural rubber climbed faster than that of synthetic rubbers (Fig. 7-4). Asia is the dominant exporting region of natural rubber (96 %) and also the leading producer of synthetic rubbers (42.4 %), followed by the European Union (21.6 %) and North America (19.9 %) (Table 7-2). Three countries (Thailand, Indonesia, Malaysia) produced 70 % of all natural rubber.

Rubbers are mainly used for tires: in North America in 2006, ca. 75 wt% of all consumed natural rubber and ca. 52 wt% of all consumed synthetic rubbers, i.e., 60 wt% of all rubbers. About 20 wt% of all produced rubbers were used in cars and buildings and another 20 wt% for cable sheetings, boots, adhesives, etc.

In 2006, North America consumed 28.7 wt% natural rubber, 13.6 wt% thermoplastic elastomers, and 57.7 wt% vulcanizable synthetic rubbers (19.9 wt% solid SBR, 15.2 wt% BR, 8.1 wt% E/P, 2.2 wt% solid NBR, 1.5 wt% IR, 1.4 wt% CR, and 9.4 wt% other (SBR and NBR latices, specialty rubbers)).

Like all economic activities, synthetic rubber production is affected by political turmoil, economics of feedstocks (crude oil), emerging downstream products, and worldwide competition. The U.S. production of SBR (main use in car tires) is a prime example of these effects. The war effort in World War II launched the US synthetic rubber industry (Figs. 7-3 and 7-5) but demand dropped after the war ended. Since 1950, increasing prosperity led to generally affordable cars. The first oil crisis in 1973-74 (see Volume II, p. 49) caused the SBR production to decline sharply but the industry recovered fast until the second oil crisis hit in 1979. During these years, smaller cars and the longer lives of tires (from 16 000 to 64 000 miles) reduced the demand for SBR strongly. The United States also started to import more Japanese cars with Japanese tires. The new general purpose rubbers [butadiene rubbers (BR), isoprene rubbers (IR), and ethylene-propylene terpolymers (EPT (= EPDM, incl. EPM))] also cut into the SBR market.

Table 7-2 Production and consumption of natural rubber (NR) and synthetic rubbers (SR) in 2006 [5b], world population in 2006 [7], and major producers of natural rubber in 2006 [8]. "Synthetic rubber" also contains rubber dispersions and latices. "Production" and "consumption" refer to raw materials; they do not include the rubber content in imported and exported rubber goods.

Region	Production and consumption in 1000 tons per year						Population in millions	Consumption per capita in kg
	Natural rubber		Synthetic rubbers		Total rubber			
	Prod.	Cons.	Prod.	Cons.	Prod.	Cons.		
Asia and Oceania	9 316	5 962	5 295	5 730	14 811	11 692	4 035	2.9
European Union	0	1 280	2 709	2 508	2 709	3 788	496	7.6
North America	0	1 148	2 500	2 032	2 500	3 180	335	9.5
Latin America	203	520	665	805	868	1 325	568	2.3
Other Europe	0	177	1 301	913	1 301	1 090	234	4.7
Africa	421	118	67	101	488	219	934	0.23
Total	9 676	9 224	12 538	12 222	22 214	21 446	6 602	3.2

The six largest producers of natural rubber in 2006:

Thailand	2 968 000 t/a	Indonesia	2 515 000 t/a	Malaysia	1 288 000 t/a
India	853 000 t/a	Vietnam	540 000 t/a	China	533 000 t/a
Rest of the world: 979 000 t/a.					

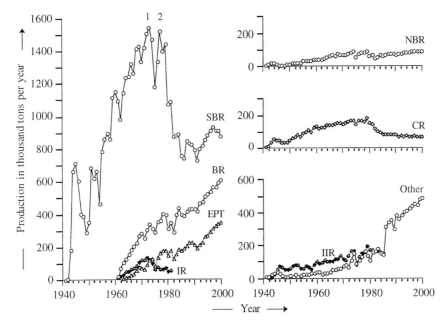

Fig. 7-5 Annual US production of synthetic rubbers [9]. Left: all-purpose rubbers (SBR, BR, EPT (= EPM, EPDM), IR); right, top and center: oil-resistant rubbers (NBR, CR); right, bottom: IIR (since 1983 in "Other"). "Other" includes SBR latices with more than 45 % styrene, NBR latex, fluoro rubbers, silicone rubbers, sulfide rubbers, acrylate rubbers, and exotics. 1, 2 = oil crises.

The US production of oil-resistant rubbers also experienced a cutback in 1979: a relatively small one for nitrile rubber (NBR) but a strong one for chloroprene rubber (CR) (Fig. 7-5). Thermoplastic elastomers (TPE) became strong players; in 2006, the US production reached 530 000 t/a and the world production ca. 1 100 000 t/a (Section 7.5.1). Acrylate and silicone rubbers remain niche products.

7.2 Raw Rubbers

Crude rubbers are raw rubbers before formulation and crosslinking. They are used only in very small amounts, for example, natural rubbers for shoes soles from crepe and a variety of synthetic rubbers in solution adhesives (Section 12.3). A small proportion of natural rubber is also isomerized and/or derivatized (Section 7.3.1).

Crude rubbers are usually delivered as bales. Some of the crude rubbers contain mineral oil (**oil-extended rubbers**). A prominent example is so-called **cold (SBR) rubber** which is delivered as crude rubber with 1 part mineral oil per 4 parts rubber.

Smaller amounts of crude rubbers are sold as sheets, particulates, powders, latices, or fluids. **Powdered rubbers** of CR, NBR, and SBR have particle diameters of ca. 1 mm (Section 7.3.8). Chloroprene, ethylene-propylene-diene, and styrene-butadiene rubbers are also offered as **particulate rubbers** with particle diameters of ≤ 10 mm. **Master batches** are partially formulated crude rubbers, especially carbon-black filled ones. Some rubbers are also offered as stabilized latices.

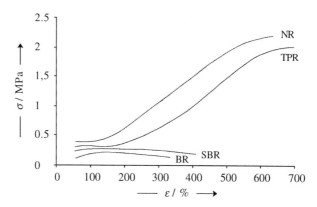

Fig. 7-6 Stress-strain curves of natural rubber (NR), a *trans*-poly(pentenamer) with 50 wt% carbon black (TPR), a styrene-butadiene copolymer (SBR), and a *cis*-butadiene rubber (BR) [10].

7.2.1   Green Strength

Raw rubbers must have sufficient strengths for direct uses as shoe soles (crepe soles) as well as for their subsequent processing to vulcanized rubbers. These **green strengths** are not officially defined. They may be maximum tensile strengths (upper yield strengths σ_Y *or* fracture strengths σ_B) or their difference $\sigma_B - \sigma_Y$. Good green strengths lead to dimensional stabilities and high tear resistances.

Good green strengths are probably caused by good crystallization on extension. Natural rubber NR has an excellent green strength (Fig. 7-6) that is higher than that of its synthetic counterpart IR although both are *cis*-1,4-poly(isoprene)s (see also Table 7-3). Styrene-butadiene rubbers SBR usually do not crystallize at all and ethylene-propylene-diene termonomer rubbers do so only if they contain more than 50 mol% of ethylene units. *Cis*-poly(butadiene) rubbers (BR) also have poor green strengths (Fig. 7-6).

The green strength of raw natural rubber decreases on mastication (Section 7.2.3) but increases if carbon black is worked in. Correspondingly, carbon black-filled *trans*-poly-(pentenamer) (TPR) also has a good green strength (Fig. 7-6). The green strength of general-purpose rubbers can be increased by added *trans*-poly(octenamer) (Table 7-3).

Table 7-3 Green strengths (as indicated by fracture strengths σ_B) and deformation hardnesses (Defo hardnesses) of carbon black-filled, unvulcanized rubbers that contained 0 wt% or 20 wt% *trans*-poly-(octenamer) (TOR) [11]. Formulation: 100 parts rubber, 50 parts carbon black N 550, 5 parts zinc oxide RS, and 1 part stearic acid.

Physical quantity	$T/°C$	TOR/wt%	NR	IR	SBR	BR	EPDM
σ_B/MPa	22	0	1.48	0.093	0.42	0.31	0.13
	22	20	2.95	1.00	1.00	0.94	1.44
Defo hardness	22	0	4 500	2 100	4 250	4 500	8 000
	22	20	19 800	18 000	16 500	9 700	15 000
	100	0	670	740	2 000	2 400	2 100
	100	20	470	430	1 080	1 390	1 000

The Defo hardness of rubbers is a measure of the deformation under load. Addition of *trans*-polyoctenamer (TOR) increases the Defo hardness of all rubbers at room temperature but decreases it at 100°C (Table 7-3) which allows a better working-in of rubber additives (Section 7.2.3). Stiff rubbers do not form a continuous band on rolling but disintegrate to crumbs.

7.2.2 Inherent Tack

Pieces of freshly cut, unvulcanized natural rubber are glued together irreversibly within seconds if they are pressed together. This **autohesion** of two identical specimens is called **(inherent) tack**. It plays an important role in the manufacture of rubber articles, welding of thermoplastics, film formation from polymer dispersions, etc.

The **pull strength** ("**bond strength**") increases first non-linearly with the square root of the contact time (range I), then linearly (ranges II and III), and finally becomes constant (range IV) (Fig. 7-7, top). Extrapolation of pull strength of range II to zero time delivers the initial pull strength σ_0.

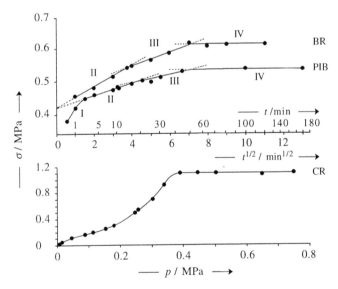

Fig. 7-7 Top: Pull strength σ as a function of the square root of time for a poly(butadiene) rubber (BR) with 51 % 1,2-, 20 % *cis*-1,4-, and 29 % *trans*-1,4-groups [12a]; $\overline{M}_n$ = 83 000 g/mol) and a poly(isobutylene) (IIR) ($\overline{M}_n$ = 1 300 000 g/mol) [13]. Bottom: Pull strength σ of a poly(chloroprene) rubber (CR) as a function of the contact pressure p after contact times of 5 minutes each [12b].

In the initial range I, the macroscopic and microscopic unevenness of the contacting surfaces is flattened by flow processes. The initial pull strength is strongly affected by the interaction between surface groups and therefore also by surface treatments. In this range, the pull strength increases with increasing pressure until it finally becomes constant (Figure 7-7, bottom).

Range II is controlled by movements of polymer segments which increase the number of contact points and hence the pull strength. Experiments with *cis*-1,4-poly(butadiene) have shown that diffusion coefficients are approximately $2 \cdot 10^{-12}$ cm^2/s and therefore in the same range as self-diffusion coefficients by tracer methods (see Volume III, p. 478). In this range, fracture surfaces are practically identical with the original surfaces, i.e., the specimen is fracturing adhesively and the fracture is a brittle fracture.

Range III is observed for entangled molecules. The diffusion rate of segments decreases but the pull strength increases since entanglements prevent segments from separating completely. With increasing time, "columns" are pulled out at some contact points. The columns break successively and the fracture becomes more and more cohesive.

Hence, inherent tack requires mobilities of chain segments: neither semicrystalline nor crosslinked rubbers exhibit autohesion. However, pressure-induced crystallizations do lead to increased tack which increases with increasing contact time. For this reason, natural rubber has more inherent tack than synthetic rubbers.

For many applications, inherent tack of synthetic rubbers is not large enough and is therefore improved by addition of **tackifiers**, usually alkylated phenol novolacs. An example is contact adhesives on the basis of novolac-filled poly(isobutylene)s. Tackifiers are also used to improve the adhesion of rubbers to tire cords. Examples are resorcinol-formaldehyde latices (RFL) for rayon, glass fiber, or polyamide cords; poly(butadiene-*co*-2-vinylpyridine) latices for polyamide cords; and water-dilutable tackifiers for polyester cords. For passenger cars, tire cords are mainly from rayon or polyester fibers, whereas steel is used for truck tires.

7.2.3 Mastication and Formulation

Rubber producers ship rubbers mainly as bales and, to a smaller extent, as latices. Rubber bales are masticated in special roll mills (see Fig. 4-33) with corrugated surfaces, so-called **masticators** (p. 251), in which the rubber is torn apart and kneaded. The combined action of heat and shearing during mastication degrades rubber molecules to smaller molecular weights. The resulting lower viscosity allows the mixing of different rubbers (= **blending**) as well as the easier and more homogeneous working-in of rubber additives (= **formulating**) since the hide is constantly ripped apart and glued together. Formulation includes the addition of crosslinking agents, accelerators, aging inhibitors, plasticizers, etc., in proportions between 0 wt% and 60 wt%.

Table 7-4 Rubber blends. ↑ Improvement, ↓ deterioration.

Parent	Add-in	Phases	Change of properties	
IIR	EPDM	2	↑ properties at 100°C	↓ permanent set
NR	BR	2	↑ tear propagation strength	↓ fracture propagation
NR	CR	2	↑ tear resistance	
NR	CR + EPDM	2	↑ ozone rsistance	
SBR	NR	2	↑ abrasion resistance	
SBR	BR	1	see Section 7.3.5	
SBR	EPDM	2	↑ paintability (car parts)	
SBR	CR	2	↑ freeze toughness	↓ discoloration by sun
EPDM	NBR	2	↑ oil, ozone, and freeze resistance	
IR	CR	2	↑ tensile strength	↓ fatigue

The desired rubber elasticity is achieved by crosslinking rubber molecules, usually at elevated temperatures (Section 7.2.4). The historic term **vulcanization** for the crosslinking is often used only for the crosslinking by sulfur and heat, the original crosslinking process for natural rubber. All other crosslinking processes of rubbers are called **curing**.

Rubber Blending

Rubbers are blended in order to improve some properties (Table 7-4). The resulting rubber blends may be one-phase or two-phase mixtures (Section 10.4.5).

Formulation

Thermosetting elastomers result from vulcanized mixtures of rubbers and rubber additives whose formulation, the by and large empirical composition and mixing, requires a lot of experience. The proportion of additives is usually not given as weight percent but as **phr** = *p*arts per *h*undred parts of *r*ubber. Rubber articles usually consist of ca. 50 wt% rubbers and 50 wt% mineral oil, fillers, vulcanization agents, and the like.

Typical formulations are shown in Table 7-5. "Rubber" in this table refers just to the rubber molecules themselves and not to the "rubbers" that are delivered to the rubber processing industry as **oil-extended rubbers (OER)**. The OERs contain up to 40 wt% mineral oil which was originally added as an extender of the expensive rubbers. However, the mineral oil also acts as a softener.

Table 7-5 Estimated content of rubber chemicals in various formulated rubbers [14]. Numbers do not contain the oil in oil-extended rubbers and fillers such as carbon black. Crosslinking by sulfur (S), peroxides (P), resins (R), or quinone dioxime (Q). By kind permission of Hanser Publ., Munich.

Additive	Parts rubber chemicals per 100 parts of rubber									
	NR	IR	SBR	BR	NBR	CR	EPT	EPT	IIR	IIR
	S	S	S	S	S	S	S	P	S	R,Q
Vulcanizing agents (see text)	1.2	1.3	1.7	1.7	1.9	0.8	3.5	8	2	4
Sulfur	1.8	1.7	1.6	1.6	1.0	0	1.3	0	2	0
Metal oxides (ZnO, MgO, PbO, etc.)	4	4	4	4	5	9	5	0	5	0
Fatty acids [a]	1.5	1.5	1.5	1.5	1.0	1.0	1.0	0	1.0	0
Stabilizers, aging inhibitors	2.5	3	3	3	3	2	1	1	1	1
Plasticizers (see text for definition)	8	8	8	8	20	8	50	30	8	8
Resins	0.5	1	1	1	3	1	0	0	0.5	0.5
Processing agents [b]	1.5	1.5	1	1	0.5	1	1	0.5	0.5	0.5
Waxes	1	1	0.8	0.8	1	0.8	0	0	0	0
Others [c]	1	1	1	1	1	0.5	0.5	2	0.5	0.5
Total	23.0	24.0	23.6	23.6	37.4	24.1	63.3	41.5	20.5	14.5

[a] Stearic acid, palmitic acid, and other lubricants;
[b] Metal soaps, fatty esters, polyglycols, factice (see Vol. II, p. 268), etc.;
[c] Mastication agents, activators for fillers, flame retardants, blowing agents, latex chemicals.

In the rubber industry, "plasticizer" denotes all agents that lower the viscosity of the rubbers, i.e., both physically acting agents (**extenders**) and chemically acting ones (**peptizers**). This use of the word "plasticizer" differs from that in the plastics industry where the word denotes an agent that lowers the glass temperature.

Fillers can be both **extenders** that reduce costs and **active fillers** that improve rubber properties such as elasticity, tear strength, etc., i.e., reinforce rubbers. The most important active fillers are carbon blacks which are offered in more than 100 types that differ in particle size and particle distribution, pore structure, type and content of chemical surface groups, hardness, etc. (Volume II, Section 6.1.9) and thus in their action as reinforcing agents.

Carbon blacks are used in proportions of 30-90 phr. Increasing proportions of carbon black increase moduli and fracture and tear strengths (all except for very high proportions of carbon black) as well as the hardness but decrease the fracture elongation (Table 7-6).

These actions are caused by free radicals on the surface of carbon black particles. Carbon blacks therefore act as additional crosslinking agents. The formation of radicals is increased by accelerators such as diphenylguanidine that are activated by zinc oxide which simultaneously serves as a filler.

Table 7-6 Effect of carbon black content on some properties of a styrene-butadiene rubber that was vulcanized at 154°C for 40 min. Composition: 100 phr SBR, 5 phr ZnO, 1 phr stearic acid, 5 phr ASTM oil 103, 0.9 phr accelerator 1, 0.1 phr accelerator 2, 1.4 phr sulfur, and varying proportions of a carbon black.

Properties	Physical unit	Proportion of carbon black in phr			
		20	40	60	80
300 % modulus	MPa	2.6	7.2	14.8	-
Fracture strength	MPa	17.6	24.8	24.7	21.0
Tear strength	kJ/m^2	22.6	45.7	58.1	45.5
Fracture elongation	%	690	620	460	290
Compression set (after 24 h)	%	24.2	21.2	22.3	20.6
Hardness (Shore A)	-	49	59	69	78

Peptizers are agents that promote the breakdown of rubber during mastication. These rubber peptizing agents (**RPA**) deliver radicals that accelerate the degradation of rubber molecules to give smaller molecular weights which decreases the viscosity and hence saves energy (in Table 7-5 included in "Others"). An example is RPA 6 = pentachlorothiophenol, C_6Cl_5SH. In natural rubber, the same effect can also be obtained by use of, for example, hydroxylamines which react with aldehyde groups of natural rubber. The reaction prevents a premature crosslinking of chains via aldehyde groups which in turn shortens the energy-intensive mastication.

More than 80 wt% of all rubbers are crosslinked by sulfur (Section 7.2.4) and less than 20 % by peroxides, crosslinking resins, or sulfur donors. "Vulcanizing agents" in Table 7-5 also contain vulcanization accelerators, retarders, and inhibitors.

Sulfur produces crosslinking bridges between polymer chains (Section 7.2.4), which shifts the property of the material from "viscous" to "elastic." Very high proportions of sulfur lead to hard rubber (Ebonite®, see p. 305).

Sulfur donors are thermally labile organic compounds that release sulfur at elevated temperatures, i.e., they prevent vulcanization during mastication which proceeds at lower temperatures. The resulting sulfur bridges $-S_i-$ may consist of i = 1-4 sulfur atoms as indicated to the right of the names of the donor compounds (see also Section 7.2.4):

OTOS [structure: O-morpholine N–C(=S)–S–N morpholine-O]   N-oxydiethylenedithiocarbamyl-N'-oxydiethylenesulfenamide   1

MBSS [structure: benzothiazole-S–S–N-morpholine-O]   2-morpholinodithiobenzothiazole   2

DPTT [structure: piperidine N–C(=S)–(S)₄–C(=S)–N piperidine]   dipentamethylenethiuramtetrasulfide   1, 2, (3, 4)

Vulcanization accelerators speed up the crosslinking by sulfur and therefore allow one to decrease sulfur contents and vulcanization times and temperatures. Lower sulfur contents reduce unwanted side reactions and over-vulcanizations. Shorter vulcanization times and temperatures save energy and are less damaging which increases the resistance to aging and permits the use of organic colorants instead of inorganic pigments.

The oldest accelerators are inorganic metal oxides such as ZnO, MgO, PbO, etc. They are still used as the sole accelerators for large parts. For smaller parts, they serve as activators for organic accelerators.

80 % of the more than 50 known organic vulcanization accelerators are thiazole compounds. Other accelerators include other sulfur compounds (thiurams, dithiocarbamates, sulfenamides, xanthogenates, thioureas, thiophosphates) as well as nitrogen compounds (amines, guanidines). Examples are

[structure: benzothiazole–SH]   2-mercaptobenzothiazole MBT

[structure: $(H_3C)_2N–C(=S)–S–S–C(=S)–N(CH_3)_2$]   tetramethylthiuramdisulfide

[structure: $((H_3C)_2N–C(=S)–S)_2 Zn$]   zinc dimethyldithiocarbamate

[structure: $(C_4H_9O–C(=S)–S)_2 Zn$]   zinc butylxanthogenate

[structure: $(C_6H_5NH)_2 C=NH$]   diphenylguanidine

[structure: hexamethylenetetramine]   hexamethylenetetramine

Vulcanization retarders prolong the time for the onset of vulcanization. This group comprises phthalic acid anhydride, N-nitrosodiphenylamine, and benzoic acid. Sulfenamides are **vulcanization inhibitors**; an example is N-cyclohexylthiophthalimide.

7.2.4 Vulcanization

Hot Vulcanization

Hot vulcanization with sulfur is by far the most important process for the crosslinking of rubbers. It is usually performed at 120-160°C in presses that are heated electrically or by steam. Vulcanization can also be caused by microwaves where it is faster for polar rubbers and slower for apolar ones. Vulcanization is sped up by carbon blacks but not by light-colored fillers.

The mechanism of *unaccelerated* sulfur vulcanization is not well known. The reaction is not accelerated by added free radicals but by organic acids and bases; it therefore seems to be an ionic process. The exact nature of this process is difficult to investigate, not only because of the high viscosity of the rubber mixture but also because sulfur itself undergoes some drastic changes at common vulcanization temperatures.

At 110°C, sulfur consists of ca. 93 % wt% cyclic S_8 molecules, 1 wt% linear poly(sulfur) chains, and 6 wt% π-sulfur (mixture of other cyclic rubber molecules: S_6, S_7, S_9, S_{10}, etc.) (Volume II, Fig. 12-13). The proportion of non-S_8 cyclics begins to increase to ca. 14 wt% at 140°C, passes through a maximum at 159°C and then becomes practically constant at ca. 10 wt%. At 140°C, the proportion of S_8 starts to drop while that of poly(sulfur) increases.

It is assumed that at vulcanization temperatures cyclooctasulfur either becomes polarized or even dissociates according to $S_8 \rightleftarrows S_m^{\oplus} + S_n^{\ominus}$. The sulfur cation then adds to the carbon-carbon double bond of diene rubbers, for example, to the isoprene unit I of natural rubber (Eq.(7-1)). The resulting cyclic intermediate II reacts with an isoprene unit I to form a carbon cation IV, then with S_8 to V, with I to VI, and then with I to VII:

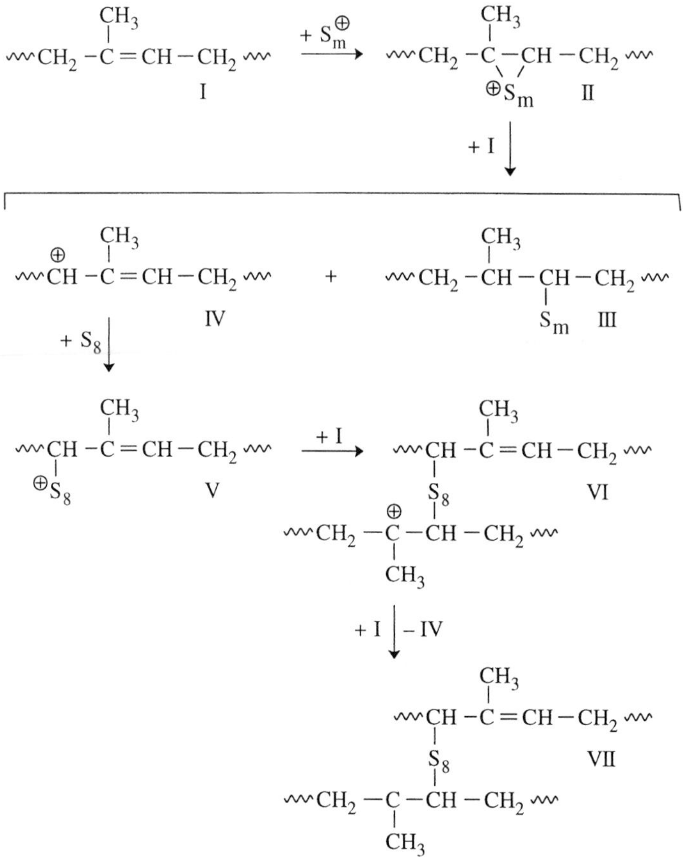

Vulcanized isoprene rubber contains 1 crosslinking bridge per 50 applied sulfur atoms. Besides polysulfide bridges $-S_i-$ with $i = 8$ (VII) or $i \leq 8$, there must therefore be other types of bridges, for example, vicinal double bridges VIII, cyclic sulfide IX, etc.:

$$
\begin{array}{ccc}
\underset{\displaystyle \underset{\displaystyle S_8}{|}}{\overset{\displaystyle \overset{\displaystyle CH_3}{|}}{\sim\!\!\sim\!\!CH-C=CH-CH_2\sim\!\!\sim}} &
\underset{\displaystyle \underset{\displaystyle S_m\ S_m}{|\ \ |}}{\overset{\displaystyle \overset{\displaystyle CH_3}{|}}{\sim\!\!\sim CH_2-C-CH-CH_2\sim\!\!\sim}} &
\overset{\displaystyle CH_3}{}
\end{array}
$$

(structures VII, VIII, IX)

Elastic properties are controlled by single bridges that are sufficiently far apart. Therefore, all structures VII-IX waste sulfur: the bridge of VII could be shorter, VIII behaves as a single bridge, and IX is an intramolecular group that does not contribute to elasticity at all. In hard rubber, 80-90 % of all the sulfur exists in intramolecular cyclic structures.

In industry, hot vulcanizations are never performed by sulfur alone but are always *accelerated* by compounds such as tetramethylthiuramdisulfide X, zinc dimethyldithiocarbamate XI, or 2,2'-dithiobisbenzthiazole XII as well as activators, for example, combinations of fatty acids and zinc oxide. These compounds lower vulcanization times and temperatures substantially.

(structures X, XI, XII)

According to model studies, polysulfides are formed if compounds X or XII react with cyclo-S_8, for example, $R_2N-C(S)-S-S_j-C(S)-NR_2$ with $j > 2$ from X. Zinc oxide is a chelating agent that produces $-S-Zn-S-$ groups similar to those in XI. Addition of accelerators strongly increases the addition of sulfur to the allyl position of diene monomeric units. Sulfur bridges probably also become shorter so that the same proportion of sulfur generates more crosslinks.

The vulcanization of polydienes seems to involve two different crosslinking reactions (Fig. 7-8). In general, one observes an induction period that helps to prevent **scorching**, i.e., a premature vulcanization. The induction period can be extended by adding retarding agents, usually fatty acids. It is also affected by vulcanization accelerators: amines need only a short time to cause scorching but sulfenamides require a longer time.

Scorching is unwanted because it impedes rapid extrusion and calendering of rubber mixes at elevated temperatures to preforms that are then molded and cured, for example, to tire treads. Good control of scorching is known in industry as "processing safety."

After onset of vulcanization, crosslinks should form fast (Fig. 7-8). The molar concentration of network junctions increases strongly and then passes through a maximum. The decrease is caused by chain degradation by shearing and/or oxidation which reduces the concentration of network junctions.

Finally, the formation of new junctions overwhelms the loss of network junctions by degradation and the molar concentration of junctions increases again. The relative proportion of crosslinking to degradation depends on the polydiene and on the vulcanization conditions. For example, poly(isoprene) chains are degraded by oxygen whereas styrene-butadiene rubbers are crosslinked. The subsequent increase of crosslinking densities is caused by the onset of long-term crosslinking.

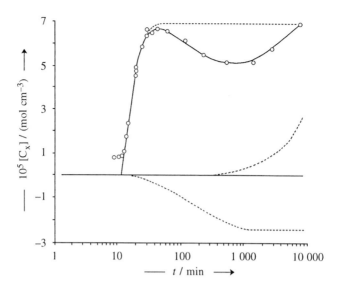

Fig. 7-8 Molar concentration of crosslinks, $[C_x]$, as a function of vulcanization time t during vulcanization of natural rubber with 2 phr sulfur, 5 phr zinc oxide, 1 phr 2-mercaptobenzthiazole, and 1 phr stearic acid at 140°C [15].

Cold Vulcanization

Thin articles from diene rubbers can be vulcanized at room temperature by disulfur dichloride or magnesium oxide. Preforms are treated either with S_2Cl_2 vapor or dipped for a few or several minutes into a solution of S_2Cl_2 in CS_2, benzene, or gasoline. Subsequent application of NH_3 vapors neutralizes the evolved HCl and removes surplus S_2Cl_2. The resulting cold vulcanizates have a poor aging resistance. They contain chlorine groups and monosulfidic sulfur bridges, for example,

$$(7\text{-}2) \quad 2\ \text{wwCH}_2-\text{CH}=\text{CH}-\text{CH}_2\text{ww} \quad \xrightarrow{+\,(1/2)\,S_2Cl_2} \quad \begin{array}{l} \text{wwCH}_2-\text{CH}-\text{CHCl}-\text{CH}_2\text{ww} \\ \qquad\qquad\quad | \\ \qquad\qquad\quad S \\ \qquad\qquad\quad | \\ \text{wwCH}_2-\overset{}{\text{C}}=\text{CH}-\text{CH}_2\text{ww} \end{array}$$

Crosslinking by Peroxides

Ethylene-propylene rubbers (EPR), ethylene-vinyl acetate rubbers (EVM),, ethylenemethyl acrylate rubbers (EAM), certain urethane rubbers, and high molecular-weight silicone rubbers (Q) are difficultly or not at all crosslinkable by sulfur. They are crosslinked by peroxides R_1–R_2 which dissociate to radicals $R_1{}^\bullet$ and $R_2{}^\bullet$ at elevated temperatures, for example dibenzoyl peroxide (to $R_1{}^\bullet = R_2{}^\bullet = C_6H_5COO^\bullet$), dicumyl peroxide (to $R_1{}^\bullet = C_6H_5C(CH_3)_2O^\bullet$ and $R_2{}^\bullet = CH_3{}^\bullet$), etc. The radicals abstract hydrogen atoms from the chains of hydrocarbon polymers which subsequently crosslink:

$$(7\text{-}3)$$

$$2\ \text{wwCH}_2-\text{CH}_2-\text{CH}_2\text{ww} \quad \xrightarrow[-\,2\,RH]{+\,2\,R^\bullet} \quad \begin{array}{l} \text{wwCH}_2-\text{CH}-\text{CH}_2\text{ww} \\ \qquad\qquad\ \bullet \\ \text{wwCH}_2-\text{CH}-\text{CH}_2\text{ww} \end{array} \quad \rangle\!\!\!\!\longrightarrow \quad \begin{array}{l} \text{wwCH}_2-\text{CH}-\text{CH}_2\text{ww} \\ \qquad\qquad\ | \\ \text{wwCH}_2-\text{CH}-\text{CH}_2\text{ww} \end{array}$$

Other Diene Vulcanizations

Rubber articles with improved thermal stability are sometimes produced by cross-linking diene rubbers with maleic imides, phenolic resins, or *p*-dinitrosobenzene. These reactions correspond to those of accelerated vulcanization by sulfur. The crosslinking reaction consists of an addition of the nitroso group to the diene chain with subsequent shift of the carbon-carbon double bond:

(7-4)

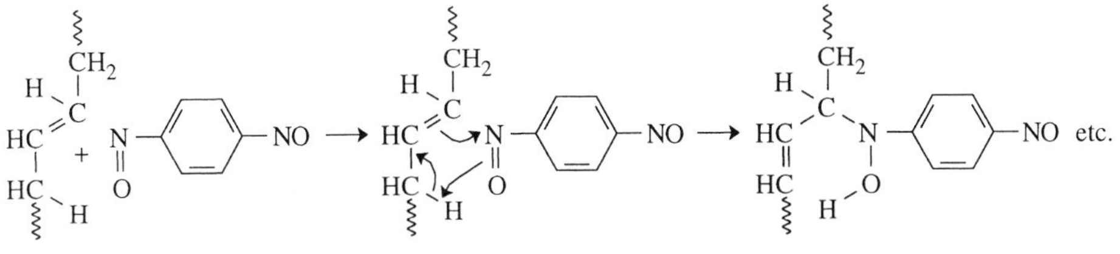

Dynamic Vulcanization

Blending of an olefin rubber such as EPM or EPDM with a thermoplastic polyolefin such as it-poly(propylene) in the melt produces a 2-phase system in which rubber particles of 1-2 μm diameter form domains in the continuous phase of the thermoplastic. The system is provided with a conventional sulfur-vulcanization system and subsequently masticated which results in the formation of polymer radicals and the crosslinking of the rubber domains. At not too large rubber contents, these blends behave as thermoplastic elastomers and can be processed like these (Section 7.5.3).

Crosslinking Density

Rubber elasticity is predominantly controlled by the crosslinking density, i.e., the amount of crosslinking bridges per volume of vulcanizate, which is a measure of the number N_X of monomeric units between crosslinks. Surgical gloves should be soft, pliable, and very stretchable ($N_X = 100$-150). Household rubber gloves are much more robust and stiff ($N_X = 50$-80). Much more crosslinked are inner tubes of tires ($N_X = 20$-30), tire treads ($N_X = 10$-20), and hard rubber ($N_X = 5$-10).

7.3   General Purpose Rubbers

General purpose rubbers comprise most diene rubbers and ethylene-propylene co-polymers and terpolymers except the corresponding thermoplastic elastomers. **Diene rubbers** are defined as rubbers that result from the polymerization of dienes and cyclo-alkenes, respectively. For the synthesis of diene rubbers, see Volume II, Chapter 6 ff.

All diene rubbers contain carbon-carbon double bonds, most often in the chains themselves and far less often in side groups; they can therefore be vulcanized (cross-linked) by sulfur. The carbon-carbon double bonds are isolated and not conjugated. The name "diene rubber" thus does not refer to the constitution of the polymers them-selves but to the constitution of the monomers from which they are derived.

7.3.1   Natural Rubber (NR)

Sources

Nature produces very many different types of elastic materials. The elastic compounds of *animalia* are all proteins whereas those of *plantae* are unsaturated hydrocarbon polymers. Elastic proteins comprise the abductin of molluscs, the resilin of arthropods, and the elastin of vertebrates (Volume II, p. 544 ff.).

The elastic materials of plants are based on isoprene units, $-CH_2-C(CH_3)=CH-CH_2-$, that are predominantly present in either cis or trans configurations:

Such isoprene rubbers are produced by thousands of different plants but only a few types are used by man. Economically dominating are cis rubbers; trans rubbers play only a minor role (see below). All rubbers are obtained from **rubber latices**, the milky exudates of rubber plants, which are aqueous dispersions of rubber particles.

Trees of *Castilla elastica and C. ulai* were very probably the primary source of latices of cis rubbers that were collected by MesoAmerican people. Guayule (pronounced 'why-You-lee') was the secondary source. African wild rubbers were not collected from trees but from vines (lianes). The rubber tree of living rooms is *Ficus elastica,* a plant that is not useful for rubber production because it dies after repeated cutting.

The Amazon wild rubbers of the 1820-1920s were mixtures from latices of 17 tropical species of the genus *Hevea* and so are probably still all presently sold Brazilian wild rubbers. All plantation rubber is obtained from the species *H. brasiliensis.* Plantation and wild *cis*-isoprene rubbers are the **natural rubbers** (**NR**) of commerce.

During World War II, other latex producing plants were investigated for their usefulness as a rubber source. The United States had a small production of rubber from the **Guayule** shrub (*Parthenium argentatum* Gray) whose roots hold 7-10 % rubber. The latex of this rubber also contains 25-30 wt% resins that have to be extracted by acetone. Guayule rubber does not have the allergy-producing proteins of Hevea rubber. Since 2003, small amounts of guayule rubber are therefore being produced again in Arizona.

In World War II, German scientists studied the extraction of rubber from the Russian variety of dandelions (*Taraxacum kok-saghyz* Rodin) that contains 6-10 % rubber in its roots (+ resins and inulin). Other candidates for rubber are the leaves of desert milk weed (*Asclepias erosa*: 2-12 %), goldenrod (*Solidago leavenworthii*: 7 %), and sunflower (ca. 1 %). Rubber is also found in the ca. 2160 species of the spurge family (*Euphorbia*).

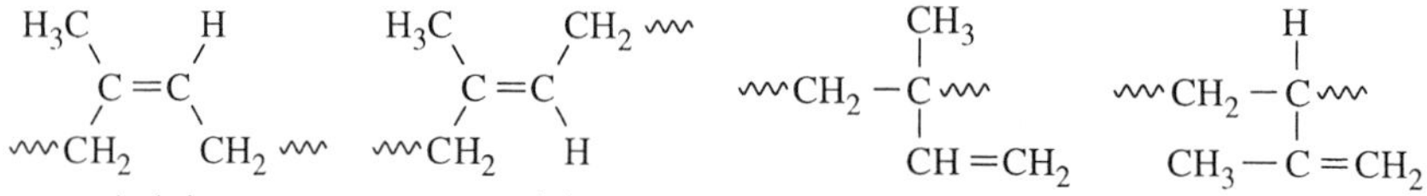

$$\underset{\textit{cis-}1,4}{\overset{\displaystyle H_3C \qquad\quad H}{\underset{\text{\tiny\(\sim\!\!\sim\!\!\sim\)}CH_2 \qquad CH_2\text{\tiny\(\sim\!\!\sim\!\!\sim\)}}{C=C}}} \qquad \underset{\textit{trans-}1,4}{\overset{\displaystyle H_3C \qquad CH_2\sim}{\underset{\sim CH_2 \qquad H}{C=C}}} \qquad \underset{1,2}{\sim\!\!\sim CH_2-\overset{\displaystyle CH_3}{\underset{\displaystyle CH=CH_2}{C}}\sim\!\!\sim} \qquad \underset{3,4}{\sim\!\!\sim CH_2-\overset{\displaystyle H}{\underset{\displaystyle CH_3-C=CH_2}{C}}\sim\!\!\sim}$$

Raw Natural Rubbers

Raw rubbers are obtained by tapping the trunks of rubber trees every other day. Each tapping delivers ca. 30 mL of milky latex which contains 30-45 wt% poly(isoprene) particles of 0.15-3 μm diameter. About 10 % of all latex is concentrated to 60 wt% by centrifugation or floatation and stabilized with ammonia. It may even be prevulcanized without precipitation. **Rubber latices** are used for dip molding and related processes.

Ca. 90 % of the produced latex is worked up directly in the plantations by coagulation, washing, drying, and shaping to either sheets or crepe. **Sheet rubber** is obtained by coagulation of the latex with formic or acetic acid and subsequent rolling of the rubber to layers of ca. 50 cm width and several millimeter thickness. The last roll embosses the sheet with a characteristic pattern. Sheets are cut to ca. 1 m length and either **air dried** or **smoke dried**, the latter in order to prevent attacks by microorganisms.

Crepe rubber is produced by coagulation of the latex with a solution of sodium sulfite in water. The coagulate runs through corrugated rolls and subsequently through smooth rolls. Unwashed rubber is sold as **brown crepe** and washed rubber as **pale crepe**. Like sheets, crepes are cut to ca. 1 m length, sorted according to quality, and sold as bales of up to 113.5 kg weight.

Commercial rubbers consist of 93-95 wt% poly(isoprene), 2.0-3.0 wt% proteins, 1.5-4.5 wt% acetone extractables, 0.2-0.5 wt% ash, and 0.3-1 wt% moisture. Sheet rubbers have a larger content of non-poly(isoprene)s than crepe rubbers. Most of these raw rubbers are now sold in standardized form (dimensions and weights of bales) and constant properties (content of dirt, rate of vulcanization) as **technically specified rubbers** (TSR). Examples are **TTR** (= Thai technical rubber), **SIR** (= standardized Indonesian rubber), **SMR** (standardized Malaysian rubber = Esemar), and **CNR** (Chinese natural rubber). **Technically classified** (natural) **rubbers** (TC-NR, TCR) are rubbers that are classified only qualitatively (but not numerically) with respect to their behavior on vulcanization.

Superior processing rubber (SP rubber) is produced in amounts of ca. 3000 t/a. It consists of vulcanized latex that is mixed with dilute regular latex, and then coagulated. SP rubber may contain up to 80 wt% crosslinked natural rubber.

The poly(isoprene)s of all of these rubbers consist of ca. 95 % *cis*-1,4 and ca. 3 % 3,4 units. Molecular weights are 10^5-10^7; molecular weight distributions are broad. It is not clear whether these broad distributions are natural or caused by the work-up.

Properties of Raw Rubbers

Raw natural rubbers are slightly yellow (crepes) to dark brown (para rubbers), amorphous materials. They become hard at ca. 0°C and lose all their strength above ca. 80°C. Short-time deformations lead to strong, reversible elongations. The stress-strain behavior (Fig. 7-9) follows the theory of entropy elasticity in both the compression range ($L/L_0 <$ 1) and the elongation range up to $L/L_0 \approx 5$, i.e., $\varepsilon \approx 400$ % (Volume III, Section 16.4). Energy elasticity contributes only a little.

The high stereoregularity of the chains leads to significant crystallization at elongations of more than ca. 500 % ($L/L_0 > 6$) (Fig. 7-9). As a result, natural rubbers possess both a high green strength and an excellent inherent tack. The crystallization increases with time so that long-time loads lead to irreversible deformations.

Vulcanized Natural Rubber

Sulfur vulcanization of natural rubbers to elastomers generates wide-meshed networks that preserve the macroconformations of segments between crosslinks. As a consequence, all those properties of raw rubber remain unchanged that do not depend on the load such as thermal properties and the compression modulus (Table 7-7). The tensile strength is strongly affected because it depends on the shortest crosslinks.

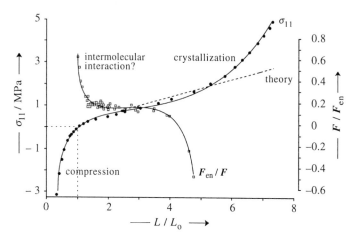

Fig. 7-9   Tensile strength σ_{11} [16] and energy-elastic proportion F_{en}/F [17] of the total force F as a function of strain ratio L/L_0 of natural rubbers.

The usual concentrations and particle sizes of carbon blacks (CB) as fillers also do not much change the macroconformations of segments between crosslinks and therefore do not affect the glass temperature. The density ρ and the thermal conductivity λ of CB-filled vulcanized natural rubbers (NR) is greater than that of unfilled ones because $\rho_{CB} > \rho_{NR}$ and $\lambda_{CB} > \lambda_{NR}$. The modulus of elasticity increases and the fracture elongation decreases because CB is a so-called **active filler**: its large specific surface leads to physical crosslinking by adsorption of rubber chains and its content of free radicals to chemical crosslinking by reaction with the carbon-carbon double bonds of diene segments.

Table 7-7 Properties of natural rubber (NR) at room temperature. CB = 50 wt% carbon black.

Property	Physical unit	Unvulcanized unfilled	Vulcanized unfilled	Vulcanized CB filled	Reclaimed CB filled
Density	g/cm³	0.91	0.97	1.12	1.19
Cubic expansion coefficient	10^5 K⁻¹	67	66	56	
Melting temperature	°C	36	40		
Glass temperature	°C	−72	−70	−70	
Heat capacity	J g⁻¹ K⁻¹	1.91	1.83	1.50	
Thermal conductivity	W m⁻¹ K⁻¹	0.13	0.15	0.28	
Modulus of elasticity	MPa		2	6	
Compression modulus	MPa	2000	2000	2200	
Tensile strength	MPa	14	32	28	9.0
Fracture elongation	%		780	500	500

7.3.2   Synthetic *cis*-1,4-Poly(isoprene) (IR)

cis-1,4-Poly(isoprene) with 98.5 % cis-1,4, 1% trans-1,4, and 0.5 % 3,4 units is obtained by polymerization of isoprene with Ziegler-Natta catalysts with Ti, Co, Ni, Nd alkyls in aromatic hydrocarbons; the lithium type (90 % cis-1,4) is no longer produced. IR is practically identical with NR but lacks the natural accelerators and antioxidants.

7.3.3 Balata and Guttapercha

Balata (from *Mimusops balata* in Brazil and Venezuela) and guttapercha (from *Palaquium gutta* and *P. oblongifolia* in Indonesia and Southern India) are *trans*-1,4-poly(isoprene)s with molecular weights of ca. 100 000. The hard materials (T_G = 38°C) become soft and pliable at ca. 60°C and melt at ca. 100°C [*trans*-1,4-modifications melt at 65°C (β; orthorhombic), 80°C (α; monoclinic), and 88°C (δ; monoclinic)].

Guttapercha (Malay: *getah* = sap; *percha* = strip of cloth) and balata (a Cariban word) can be vulcanized with sulfur. They are used for electrical cable sheathings, belt dressings, golf ball covers, dental fillings, and in the compounding of natural rubber.

7.3.4 Chicle

Chicle (Nahuatl: *tsictle*) is the coagulated latex of the plant *Achras sapota* of Southern Mexico, Guatemala, and Honduras which can be tapped once every three to four years to yield ca. 1.1 kg chicle. The purified commercial product consists of ca. 17 wt% rubber (= 1,4-*trans*-poly(isoprene)), 40 % resin (mainly triterpenes = molecules with six isoprene units), and sugars and starches. A typical chicle-based chewing gum consists of ca. 20 wt% chicle, 60 wt% saccharose, and glycerol, flavors, microcrystalline waxes, and diluents such as guttapercha or poly(vinyl acetate). Most chewing gums are now based on poly(vinyl acetate), poly(vinyl ether)s, poly(isobutylene), poly(ethylene), or SBR.

7.3.5 Butadiene Rubbers (BR)

Butadiene rubbers are offered in several types that differ in their content of cis, trans, 1,4, and 1,2 units (Table 7-8). The world nameplate capacity of BRs is ca. $9 \cdot 10^6$ t/a.

Table 7-8 Constitution, configuration, and thermal properties of butadiene rubbers (BR) and styrene-butadiene rubbers (SBR) as compared to *cis*-1,4-poly(isoprene) (IR). T_G = glass temperature, T_M = melting temperature, ΔH_M = Melting enthalpy per amount of repeating units.

Type	Composition of polymer chains in wt%					T_G in °C	T_M in °C	ΔH_M in kJ/mol
	Styrene units	cis-1,4	Diene units trans-1,4	1,2	3,4			
BR, stereo Co-type	0	98	1	1		−105	2	
BR, stereo Ni-type	0	97	2	1		−108	12	9.2
BR, stereo Ti-type	0	95	3	2				
BR, stereo Li-type	0	38	53	9		−95		
BR, emulsion type	0	10	69	21		−80		
BR, trans (La, Nd, Ni catalysts)	0	4	94	2		−83	80	4.2
BR, syndiotactic 1,2-	0	9	0	91		−5	156	
SBR, solution type A	19	30	42	9		−70		
SBR, solution type B	25	24	32	19		−47		
SBR, emulsion type (hot, 50°C)	23.5	18.3	65.4	16.3		−50		
SBR, emulsion type (cold, 5°C)	23.5	11.8	71.2	17.0		−50		
IR (Ti type)	0	98.5	1	0	0.5	28	−73	4.36

So-called **stereo-poly(butadiene)s** (= high-cis poly(butadiene)s) are obtained by polymerization of butadiene, $CH_2=CH–CH=CH_2$, with Ziegler-Natta catalysts based on Ti, Co, Ni, or Nd as transition metals (Volume II, Section 6.3.2). Neodymium produces polymers with the highest cis content (99 %) and practically no branching. Vulcanized Nd polymers have the best tensile strength and the lowest heat build-up (hysteresis) of the cis types. Ni catalysts produce some branching, and Co catalysts even more. Co types are used as modifiers for poly(styrene) and in acrylonitrile-butadiene-styrene polymers.

Lithium-type poly(butadiene)s (medium-cis poly(butadiene)s) are synthesized by direct butyl lithium initiated anionic polymerization of the C_4 cut from the naphtha cracking of crude oil; this cut contains 3-65 wt% butadiene. Addition of Lewis bases as randomizers leads to **vinyl butadiene rubbers** (**VBR**) with vinyl contents (= 1,2-units) that range from 90 % (using ethylene glycol methyl ether) to 20 % (using triethylamine). Glass temperatures of such VBRs cover the whole range of glass temperatures of other elastomers (Fig. 7-10). VBRs with 35-55 % vinyl groups have similar properties with respect to abrasion, elasticity, friction, and skid resistance to those of blends of SBR and *cis*-1,4-BR for tires. VBRs have therefore replaced in part SBR (see production of SBR, Fig. 7-5) which became too expensive because of the high price of styrene.

Since liquid rubbers are preferred for many applications, two types of **liquid BR rubbers** have been developed. Polymerization of butadiene with dianions as initiators and subsequent termination of chains by carbon dioxide delivers poly(butadiene)s with molecular weights of ca. 10 000 that can be crosslinked with polyisocyanates.

The proportion of 1,2-groups in liquid BRs can be varied: 10-20 % vinyl units are obtained from lithium alkyls in hydrocarbons, 30-70 % from sodium alkyls in hydrocarbons, and 90 % from sodium in hydrocarbons plus some tetrahydrofuran at low temperatures. The high initiator costs for liquid BRs can be reduced by employing chain transfer agents such as amines or 1,2-butadiene (see Volume I, Section 8.3.8).

Trans-BRs with >90 % trans units are obtained with La, Nd, or Ni catalysts. These BRs are similar to guttapercha (Section 7.3.3); they are used for the same purposes.

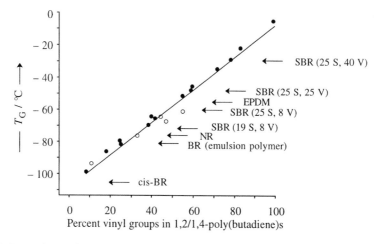

Fig. 7-10 Dependence of glass temperatures T_G on the content of vinyl groups in poly(butadiene)s [18] (data of [19] (O) and [20] (●)). For comparison, glass temperatures of other rubbers (BR, EPDM, NR, SBR) are indicated by arrows. Numbers in parentheses refer to the percentage of styrene units (S) and vinyl units (V) in conventional free-radically polymerized styrene-butadiene rubbers (SBR).

7.3.6   Styrene-butadiene Rubbers (SBR)

Types

The first styrene-butadiene rubbers were developed as **Buna S** in Germany before World War II and during World War II in the United States as **GRS** on the basis of the German patents (p. 254 ff.). They are now known as **SBR**.

The major workhorses are two emulsion-polymerized SBRs (**E-SBRs**): the "hot type" (polymerized at 50°C; originally at 41.5°C) with 23.5 wt % styrene units and 76.5 wt% butadiene units, and the "cold type" (polymerized at 5°C), respectively (see Table 7-8). The cold type has a lower cis-1,4 content than the hot type (11.8 % *versus* 18.3 %) which reduces the tendency to cyclize and thus undesirable increases of viscosity during kneading or roll-milling. Much more important is the fact that the weight-average critical chain length of primary chains before the onset of crosslinking ("gelation") is ca. 6.7 times greater for cold types than for hot types. Cold SBRs can thus be plasticized by low-cost petroleum oils without losing their mechanical properties.

The low stereoregularity of SBR prevents it from crystallizing under stress as natural rubber does. Vulcanized unfilled SBRs are therefore not self-reinforcing, i.e., they have only low tensile strengths of ca. 2.8-4.2 MPa. Loading SBRs with fine-particle carbon blacks increases the tensile strength to ca. 28 MPa, similar to that of natural rubber. Tensile strengths of vulcanized SBRs are also affected by the type of emulsifier that was used during polymerization. Fatty acid-emulsified SBRs have higher tensile properties than those emulsified with rosin acid; they also cure faster but have less tack. The high compression set of some E-SBRs can be reduced by compounding and blending. Vulcanized E-SBRs have properties similar to NRs but superior abrasion resistance (Table 7-9).

Table 7-9 Properties of carbon-black filled, vulcanized diene elastomers based on NR = natural rubber, IR = synthetic *cis*-1,4-poly(isoprene), BR = poly(butadiene) (Li type) (97 % cis), SBR = styrene-butadiene rubber (emulsion type with 40 % styrene), IIR = butyl rubber with ca. 3 % isoprene, EPDM = ethylene-propylene-non-conjugated diene rubber. Resistance: 6 = excellent, 5 = very good, 4 = good, 3 = sufficient, 2 = moderate, 1 = poor. a = amorphous.

Property	Phys. unit	NR	IR	BR	E-SBR	IIR	EPDM
Density	g/cm^3	0.93	0.93	0.94	0.94	0.93	0.86
Melting temperature	°C	36	2	1	a	–1.5	a
Glass temperature	°C	–72	–72	–95	–50	–66	–55
Maximum crystallization rate at	°C	–25		–55		–34	
Minimum service temperature	°C	–45	–45	–72	–40	–38	–35
Maximum service temperature	°C	100	100	90	140	150	150
300 % modulus	MPa	5.0	3.2	7.3	9.3	7.2	8.6
Tensile strength at break	MPa	32	26	14	29	22	13
Fracture elongation	%	780	620	510	650	620	320
Rebound elasticity	%	40	40	65	40	2	45
Hardness (Shore A)	-	50	55	60	60	55	65
Resistance to abrasion	-	3	3	6	5	3	4
acids, bases	-	4	4	4	4	5	6
oil	-	1	1	1	2	1	3
tear	-	5	5	2	4	4	4
weather	-	3	3	4	3	5	6
Impermeability to gases	-	2	2	3	3	6	3

Solution SBRs (S-SBR) are obtained by anionic copolymerization of styrene and butadiene with organo lithium compounds as initiators. The chemical structures of S-SBRs can be varied widely with respect to styrene content, proportion of butadiene structures (1,4-cis, 1,4-trans, 1,2-), sequence lengths, molecular weights, molecular weight distributions, and long-chain branching (see Table 7-8).

For example, S-SBRs with high content of 1,2 groups are known as **vinyl SBRs**. In turn, these variations lead to widely varying technological properties. For example, the narrower the molar mass distribution, the more carbon black and mineral can be taken up. The abrasion resistance of some S-SBRs surpass even that of E-SBR which is important for their use in tires.

Car Tires

The main application of SBRs is for tires of passenger cars and light trucks, especially their treads. Classic car tires consist of two parts: the tire itself and an air-filled, impermeable inner tube (usually from butyl rubber, Section 7.4.3). Tubeless tires contain an air-impermeable rubber layer on their inner side, usually a halobutyl rubber.

Car tires are built on a drum from many different parts. The load-bearing carcass consists of rubber that is reinforced with (usually) two plies of tire cord in the side walls and four plies in the tread base that supports the tread cap. Tire cords may consist of nylon (1947 ff.), polyester (1963 ff.), steel (1970 ff.), aramid (1980 ff.), or hybrids (1990 ff.); cotton (1900-1956), rayon (1938-1975), and fiber glass (1967-1985) are no longer used for tire cords. Side walls, tread bases, tread caps, and inner liners consist of at least four, and often many more, types of rubbers, rubber blends, and rubber formulations. For example, side walls must be highly elastic whereas tire treads should be abrasion resistant. The raw tire is then transferred to a heated mold and vulcanized.

Tires are thus very differently constructed from a great many materials. They are therefore not characterized according to their chemical composition or physical properties but according to their dimensions and certain use properties (ISO 4000). Tires are usually characterized by a quality code and an alphanumeric code.

The Uniform Tire Quality Grade (UTQG) consists of a number and two letters, for example, 250 A B, which means

Treadwear 250   Traction A   Temperature B

The value "250" indicates that the abrasion of the tire is 2.5 times less than that of a standard tire with a "100" rating. Traction and resistance against heating up are characterized by letters): AA (very high), A (high), B (medium), and C (low).

The next group indicates the manufacturer and certain tire properties, for example, for passenger cars and most light trucks (some light trucks use another system):

Michelin P 185/70 R 14 90S DOT B9 PA B55X 101

1st letter:	application (P = passenger car; LT = light truck, etc.);
1st number:	widest cross section of tire in mm (here "185 mm");
2nd number:	ratio of height of side wall to width in % (here "70"); a missing number is assumed to mean 82 %. A number higher than 200 indicates the tire diameter in mm (common for Japanese tires).
2nd letter:	type of tire (B = bias belt, D = diagonal, R = radial; a blank: crossply)

3rd number: diameter of wheel in inches (here "14" = 14 inches);

4th number: maximum load the tire can carry, e.g.,
(71 → 345 kg; 90 → 412 kg; 110 → 1060 kg);

3rd letter: letter code for maximally allowed speed in km/h during 10 min, e.g.,

      B = 50   G = 90   Q = 160   S = 180   U = 200

      H = 210   VR > 210   V = 240   Z > 240   Y = 300

last group: manufacturer, here using the code of the U.S. Department of Transportation (DOT) for the manufacturer and the plant (B9 = Michelin,
Lexington (SC)), another specification ("PA B55 X"), and the date of
manufacture (3 digits → before 2000; 4 digits → 2000 ff.), for example: 101 = *1*0th week of 199*1*, 1001 → *1*0th week of 200*1*.

There may also be other codes such as M+S ("mud and snow") if the tire behaves
better than average in such conditions (a free-for-all designation) or "snowflakes in a tri-
angle" (regulated). All European-made tires must also carry an "E" or "e" followed by a
numeric country code in a rectangle or circle and a number for the approved type.

Tires heat up to ca. 90°C for cruising speeds of 100-120 km/h (63-75 mph) and to
120-130°C for speeds of 180 km/h (113 mph). The critical value is the vulcanization
temperature of rubber, i.e., ca. 150°C.

The performance of tires depends on so many factors that generalizations about the
effect of polymer structure on properties are difficult. For example, abrasion and rolling
resistances of tires decrease with increasing glass temperature of butadiene-based elastomers whereas the traction on wet surfaces increases (Fig. 7-11).

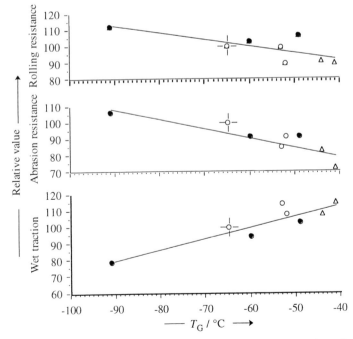

Fig. 7-11 Relative values of rolling resistance, abrasion, and traction on wet surfaces for oil-extended
SBR (O), vinyl SBR (Δ), and vinyl BR (●) [21]. Standards at –65°C are indicated by a cross.

Because of their oil resistance and versatility, nitrile rubbers have become the work-horses of the automotive industry for hoses, seals, etc., and also for other demanding applications such as conveyor belts. The oil resistance of these rubbers increases with in-creasing contents of acrylonitrile units that range between 18 % and 52 %, depending on the type and grade. The increase of oil resistance is usually accompanied by a decrease of rebound elasticity and low-temperature flexibility.

The oil resistance of **poly(chloroprene)s (CR)**, although good, is somewhat less than that of NBRs which, in combination with the enormous variety of nitrile rubbers, may be the reason for the decreasing importance of poly(chloroprene)s (see Fig. 7-3). Much better oil resistance is offered by peroxide-cured **polyurethane elastomers (AU)** that also have excellent tear strengths but a service temperature of only 75°C (Table 7-10).

NBRs and CRs are only moderately temperature resistant, which is the reason for the emergence of other specialty polymers. **Hydrogenated nitrile rubber (HNBR)** has a much improved heat stability compared to NBR (after 1000 hours: 155°C *versus* 125°C). HNBR is also resistant against chemicals and especially against swelling by oils which often allows it to replace the more expensive **fluoro rubbers (FKM)**, for example, for seals and hoses in oil production. Higher temperature resistances with excellent oil resi-stances are also shown by ethylene-acrylate rubbers, chlorosulfonated poly(ethylene), and silicone rubbers (Table 7-10).

Acrylate rubbers (ACM) are copolymers from ethyl, butyl, and/or octyl acrylates, $CH_2=CHCOOR$, with small proportions of a crosslinkable comonomer, such as allyl gly-cidyl ether. Polymers from ethyl acrylate have the best resistance to heat and oil and those from octyl acrylate the best resistance to low tempertures but a drastically reduced resistance against oil.

Ethylene-acrylate rubbers (EAM) from the terpolymerization of ethylene, methyl acrylate, and a carboxylic acid group carrying monomer such as methacrylic acid are crosslinked by amines. They have better tensile strengths and fracture elongations and smaller compression sets than ACM but are more easily swellable.

$-CH_2CH_2CH_2CH_2-/-CH_2CH-$ $\qquad\qquad\qquad\qquad\qquad\quad \underset{CN}{\mid}$	hydrogenated nitrile rubber (very low content of carbon- carbon double bonds)	HNBR
$-CH_2CH_2-/-CH_2CH-/-CH_2C(CH_3)-$ $\qquad\qquad\quad \underset{COOCH_3}{\mid} \qquad\quad \underset{COOH}{\mid}$	ethylene–acrylate rubber	EAM
$-CH_2CH_2-/-CH_2CH-/-CH_2CH-$ $\qquad\qquad\quad \underset{SO_2Cl}{\mid} \qquad \underset{Cl}{\mid}$	chlorosulfonated poly(ethylene)	CSM
$-CH_2CF_2-/-CF_2CF-/-Y-$ $\qquad\qquad\qquad \underset{CF_3}{\mid}$	fluoro rubber (example)	FKM
$\underset{CH_3}{\overset{CH_3}{\mid}}\quad \underset{CH_3}{\overset{CH_3}{\mid}}$ $-O-Si-/-O-Si-$ $\underset{CH_3}{\mid} \qquad \underset{CH_2CH_2CH=CH_2}{\mid}$	silicone rubber (example)	MVQ

$-O(CH_2)_2OOCNH-\langle\bigcirc\rangle-CH_2-\langle\bigcirc\rangle-NHCO-[O(CH_2)_2OOC(CH_2)_4CO)_n-$

polyurethane rubber (example)	AU

Table 7-10 Properties of reinforced oil- and/or temperature-resistant elastomers based on EAM = ethylene-acrylate rubber, AU = urethane-polyester rubber, CR = poly(chloroprene), CSM = perchlorosulfonated poly(ethylene), FKM = fluoro rubber, MVQ = silicone rubber with methyl and vinyl substituents, NBR = nitrile rubber (acrylonitrile-butadiene rubber), TM = sulfide rubber (Thiokol®).
Resistance: 6 = excellent, 5 = very good, 4 = good, 3 = sufficient, 2 = moderate, 1 = poor.

Property	Unit	CR	NBR	AU	TM	EAM	CSM	FKM	MVQ
Density	g/cm^3	1.25	1.00	1.25	1.35	1.10	1.25		1.25
Glass temperature	°C	–45	–34	–35	–50	–30	–25		–120
Minimum service temperature	°C	–25	–20	–22	–30	–15	–10	–10	–85
Maximum service temperature	°C	125	125	75	100	170	150	250	225
300 % modulus	MPa	4.3	3.5	13					2.2
Tensile strength	MPa	20	30	40	8	12	18	20	6
Fracture elongation	%	800	750	500	300	250	200	450	300
Hardness (Shore A)	-	50	45	70			90		50
Resistance to									
- abrasion	-	4	5	6		3	4	3	2
- acid, bases	-	5	3	2		2	5	6	2
- oil	-	5	6	6		6	5	6	6
- tear	-	5	4	6		3	4	4	3
- weather	-	5	5	5		5	5	6	6
Impermeability to gases	-	4	5	5		4	4	4	1

Fluoro rubbers (FKM) have the best heat resistance (Tables 7-10 and 7-11) but usually low resistance in the cold. Very many types are known, for example,

- random bipolymers from 75 % vinylidene fluoride, $CH_2=CF_2$, and 25 % hexafluoropropylene, $CF_2=CF(CF_3)$ that are crosslinked by amines;
- random terpolymers from vinylidene fluoride, hexafluoropropylene, and tetrafluoroethylene, $CF_2=CF_2$, for crosslinking with amines and bisphenol A;
- random quaterpolymers from vinylidene fluoride, hexafluoropropylene, tetrafluoroethylene, and halogen-containing monomers that are crosslinked by peroxides;
- *alternating* copolymers from tetrafluoroethylene and propylene, crosslinkable by peroxides;
- fluoro rubbers containing comonomeric units with nitrile groups in the side chains. Crosslinking here is *via* trimerization of –C≡N groups to 1,3,5-triazine rings.

Silicone rubbers (MVQ) (Volume II, p. 572) containing $-O-Si(CH_3)_2-$ dimethylsilicone units (M) and small proportions of silicone units with vinyl or allyl side groups (both V) (see p. 31) are cured at elevated temperatures by either peroxides or by compounds with Si–H groups that add to the carbon-carbon double bonds of V units.

These silicone elastomers have excellent cold resistances (Table 7-10). Their swelling behavior is substantially improved if they contain methyltrifluoropropyl siloxane groups, $-O-Si(CH_3)(CH_2CH_2CF_3)-$, as monomeric units (**FVMQ**); unfortunately, their compression set increases from 3 % (MVQ) to 30 % (FVMQ) (120°C).

Ethylene-vinyl acetate copolymers (EVM) with (40-60) % vinyl acetate units VAC from free-radical copolymerizations are elastomeric. Technical products with usually 40-45 % VAC units are amorphous and can be cured with peroxides. They are usually filled with carbon black because unfilled EVMs have low tensile strengths (Table 7-11).

At 120°C, EVM elastomers have very small compression sets similar to those of silicone elastomers. However, compression sets of EVMs increase strongly with decreasing temperature from 4 % at 120°C to 95 % at –20°C whereas those of MVQs increase only to ca. 10 %. EVMs are very resistant to ozone and the weather.

Phosphazene rubbers (PNF), $+P(OR)_2=N+_n$, are obtained from the alcoholysis of poly(phosphornitrile chloride), $+P(Cl_2)=N+_n$ with mixtures of fluorinated sodium alcoholates (Volume II, p. 582). They are cured by peroxides or sulfur.

Carboxynitroso rubbers (AFMU, CNR) are terpolymers from tetrafluoroethylene, trifluoronitrosomethane, CF_3NO, and ca. 1 mol% 4-nitrosoperfluorobutyric acid that are cured by diamines (Volume II, p. 288). Both elastomers have excellent flexibilities at very low temperatures; they are used in the arctic and in space.

$$-P=N-\ \underset{OCH_2CF_3}{|}\ /\ -P=N-\ \underset{OCH_2(CF_2)_4H}{|}\ /\ Y \qquad \text{phosphazene rubber} \qquad \text{PNF}$$

$$-CH_2CF_2-\ /\ -O-N-\ \underset{CF_3}{|}\ /\ -O-N-\ \underset{(CF_2)_3COOH}{|}\ \qquad \text{carboxynitroso rubber} \qquad \text{AFMU}$$

Table 7-11 Thermal properties and tensile strengths of elastomers [18b, 23, 24], arranged according to heat resistance. Data are typical values that vary with compounding, vulcanization/curing conditions, type and proportion of carbon black, etc. LST = lower service temperature, UST = upper service temperature. *) Medium content of acrylonitrile units. Crosslinked by peroxides [P] or sulfur [S].

Rubber	Crystal-lizability	T_G in °C	Cold and heat resistance retained at °C at elevated temperatures after					Tensile strength in MPa carbon black	
			LST	5 h	70 h	1000 h	UST	unfilled	filled
AU	–	–35	–22	170	100	70	75		
EU		–55	–35	170	100	70	75		
BR		–112	–72	170	110	75	90	3	20
TM		–50	–30	170	120	60	100	3	10
PNR	–	–45					100		
IR, NR	+	–72	–45	150	120	90	100	20	22
SBR	–	–50	–28	195	130	100	110	6	22
NBR *)	–	–34	–20	180	145	115	125	7	20
CR	+	–45	–25	180	130	110	125	20	22
ECO	–	–45	25	220	150	130	130		
EPDM [S]	–	–55	–35	200	170	130	135	7	18
EPDM [P]	–	–55	–35	220	180	140	150		
CM	–	–25	–12	180	160	140	150	10	
CSM	+	–25	–10	200	140	130	150	18	20
IIR	+	–66	–38	200	160	130	150	14	18
CO	–	–26	–10	240	170	140	150	6	15
H-NBR	+	–30	–18	230	180	150	160		
EVM	–	–30	–18	200	160	140	160	4	15
ACM	–	–22/–40	–10/–20	240	180	150	170	2	15
EAM		–40	–20	240			175		
PNF		–66	–42				175		7
CNR		–40					190	11	
FVMQ		–70	–45	>300	220	200	215		
MVQ	–	–120	–85	>300	275	180	225	3	
FKM	+	–18/–50	–10/–35	>300	280	220	250	12	15

7.4.2 Ozone and Oxygen Resistant Rubbers

In contrast to rubbers and elastomers with carbon-carbon double bonds, all elastomers with saturated polymer chains are fairly resistant to oxygen. Especially good resistances to O_2 and O_3 are shown by polymers with oxygen units in the main chains.

Ring-opening polymerization of α-epichlorohydrin leads to homopolymer (**CO**), with ethylene oxide to bipolymers (**ECO**), and with ethylene oxide and allylglycidyl ether to terpolymers (**ETER**). Homopolymers and bipolymers are usually cured with ethylene thiourea and the terpolymer by peroxides. Copolymerization of α-epichlorohydrin with non-conjugated dienes leads to rubbers that are vulcanizable with sulfur.

The polymers have only medium cold and heat resistances (Table 7-11) but excellent ozone resistances and very low degrees of swelling of only 5 % (CO) and 10 % (ECO) in ASTM oil no. 3 at 150°C. This is considerably lower than that of other highly ozone-resistant elastomers at this temperature, for example, 20 % for FKM and FVMQ, 50 % for MVQ, and 80 % for EVM and CSM.

	R	Name
	H	ethylene oxide
$H_2C\!-\!CHR$	CH_3	propylene oxide
$\diagdown\!\diagup$ O	CH_2Cl	α-epichlorohydrin
	$CH_2\!-\!O\!-\!CH_2\!-\!CH_2\!-\!CH\!=\!CH_2$	allylglycidyl ether

7.4.3 Butyl Rubbers

Natural rubber, synthetic *cis*-1,4-poly(isoprene), butadiene rubbers and butadiene-styrene rubbers are prone to oxidation because of their high contents of carbon-carbon double bonds. The desire to reduce the susceptibility to oxidation while maintaining the vulcanizability led to the development of butyl rubbers (Volume II, p. 275).

Poly(isobutylene)s (**PIB**), $+CH_2\!-\!C(CH_3)_2\!+_n$, are rubbery homopolymers of isobut-(yl)ene. Low-molecular weight PIBs are used as adhesives (Chapter 12) or viscosity improvers (Chapter 11) while high-molecular weight PIBs serve as additives for elastomers. Sealing films are usually copolymers of isobutene (= isobutylene) with 10 % styrene.

Butyl rubbers (**IIR**) are crosslinkable. They result from the Lewis acid-initiated, highly exothermic copolymerization of isobutene and 0.5-3 % isoprene in methyl chloride as a diluent at –90°C to –100°C and boiling liquid ethylene as a temperature regulator. Conventional sulfur-vulcanized IIRs are used mainly as inner tubes for car tires because of their very low permeability for gases. Their low rebound elasticity makes them excellent vibration absorbers. Heating tubes for tire vulcanizations are obtained from IIRs that have been cured with peroxides in the presence of phenolic resins because conventionally vulcanized IIRs have insufficient heat and aging resistances.

Halobutyls are butyl rubbers that have been after-chlorinated by elemental chlorine to **chlorobutyl rubbers** (**CIIR**) or after-brominated by elemental bromine to **bromobutyl rubbers**. Halobutyls cure faster than IIRs and can be covulcanized with natural rubber or styrene-butadiene rubbers. Their largest application is for innerliners, i.e., layers of halobutyl at the inner side of car tires that are firmly bound to the tire rubber by covulcanization. Halobutyls are also used for hoses, construction sealants, and the like. **Exxpro**® are brominated isobutene/α-methyl styrene copolymers for innerliners of tires.

7.4.4 Liquid Rubbers

The very high viscosity of rubbers makes classic rubber processing very labor and energy intensive: formulating and compounding must be done on heated roll mills and sulfur vulcanization in heated presses under pressure. Processing is much easier and less costly if liquids can be used. The first liquid rubbers were based on polysulfides. Later liquid rubbers comprised silicones, polyurethanes, polyesters, and polyethers and finally also polydienes.

The first polysulfides (1929) were *solid* **sulfide rubbers (TM)** from the polycondensation of 1,2-dichloroethane and sodium polysulfides, Na_2S_i (Thiocol®; Volume II, p. 431). Later, other chloro compounds were used, all leading to solid polysulfide rubbers of the type $+R-S_i+_n$ where R = organic unit and i = polysulfide rank which is i = 1-2 for solid rubbers that are cured with PbO, organic peroxides, or quinone dioxime.

Liquid polysulfide rubbers possess much higher polysulfide ranks of 3-4. They are cured by peroxides such as H_2O_2 or CaO_2, sodium perchromate, diisocyanates, epoxides, or phenol-formaldehyde resins. The main application is as a sealing compound.

Liquid silicone rubbers have much lower molecular weights than the peroxide-crosslinkable high-molecular weight types. The liquid types are dominated by the "cold" vulcanizing one-component types that cure at room temperature (**RTV** = *room temperature vulcanizing*). One of these types consists of branched poly(dimethylsiloxane)s with silanol endgroups that can be crosslinked with tetrabutyl titanate or methyltriacetoxysilane. In the presence of humidity, the latter crosslinking agent loses acetic acid and the resulting methyltrihydroxysilane reacts with the silanol groups of the polymer:

(7-5)

$$
3 \; \text{---Si-OH} + CH_3COO-\underset{\underset{OOCCH_3}{|}}{\overset{\overset{CH_3}{|}}{Si}}-OOCCH_3 \xrightarrow[-3\,CH_3COOH]{} \text{---Si-O-Si-O-Si---}
$$

Other types of RTV silicones cure by hydrosilation, which is the platinum compound catalyzed addition of $\gtrless$Si—H groups to vinyl or allyl groups of crosslinkable carbosilane polymers, for example:

(7-6)   ---Si—H + ---Si–CH=CH₂ ⟶ ---Si–CH₂–CH₂–Si---

Liquid polyurethane rubbers (AU, EU) are obtained from the reaction of diisocyanates such as diphenylmethane-4,4'-diisocyanate (MDI) or a mixture of tolylene-2,4 and tolylene-2,6 diisocyanates (TDI) with linear polyether (PEG) or polyester polyols to oligomeric prepolymers with isocyanate endgroups such as

$$OCN+(1,4\text{-}C_6H_4)NHCO-O(CH_2CH_2O)_n-CONH(1,4\text{-}C_6H_4)NCO$$

that are cured with basic polyamines and diamines such as methylene-*bis*-chloroaniline.

Table 7-12 Properties of vulcanized, regular or liquid, unfilled rubbers. Liquid SBRs have either OH or COOH end groups.

Physical property	Physical unit	Polyurethanes regular	Polyurethanes liquid	Butadiene-styrene copolymers regular	Butadiene-styrene copolymers liquid COOH	Butadiene-styrene copolymers liquid OH
Modulus of elasticity	MPa	13	0.1 - 35	61	63	85
Tensile strength	MPa	40	1 - 76	24	15	16
Ultimate elongation	%	500	10 - 1000	540	340	270
Hardness (Shore D)	-	20	4.5 - 85			
Rebound elasticity	%		10 - 64	45	41	50
Tire temperature after 25 min rolling	°C			72	97	123
Abrasion (after 1000 revolutions)	cm^3			0.21	1.21	0.43

Depending on the type and molar ratio of monomers, either thermoplastic or thermosetting elastomers are obtained. Before curing, prepolymers may be either liquid or solid. The latter usually become liquid at processing temperatures and are therefore also known as **cast elastomers**. The resulting elastomers have properties similar to those of regular polyurethanes (Table 7-12).

The simplest types of **liquid diene rubbers** are low-molecular weight degradation products of regular polydienes that are then crosslinked by peroxides via the remaining carbon-carbon double bonds. They cannot be vulcanized by sulfur.

Liquid diene rubbers with reactive end groups are of much greater interest. Such polymers are obtained by anionic polymerization of dienes with bifunctional initiators. The resulting macro dianions are subsequently reacted with carbon dioxide to ~COO$^\ominus$ end groups, with ethylene oxide to ~CH$_2$CH$_2$O$^\ominus$, or with ethylene sulfide to ~CH$_2$CH$_2$S$^\ominus$. The polymers are cured by multifunctional crosslinking agents, usually isocyanates, that need to be applied in high concentrations because of the low molecular weights of liquid diene rubbers. Many systems require exact stoichiometries.

Vulcanized liquid SBRs have far lower tensile strengths and fracture elongations than vulcanized regular SBRs (Table 7-12), which is probably caused by smaller numbers of crosslinks per primary molecule and increased contents of "free", i.e., uncrosslinked, chain ends. Tires from vulcanized liquid SBRs also show higher abrasion and get much warmer on rolling than those from conventional SBR. Liquid SBRs, although attractive for processing, are therefore not used for the manufacture of new car tires but for retreading, especially of medium and heavy-duty truck and aircraft tires.

Liquid natural rubber (LNR) results from the redox degradation of natural rubber by hydrazine and air at 60°C.

7.5 Thermoplastic Elastomers

7.5.1 Overview

Powder and liquid rubbers simplify formulation and compounding but still require costly and time-consuming vulcanization/curing procedures that lead to irreversible *chemical* crosslinks. Edge trims and other waste parts cannot be used again except after recycling, but this is expensive and does not result in virgin material (Section 7.7).

In this respect, *physically* crosslinked rubbers are much more advantageous. These "crosslinks" consist of "hard" domains in "soft" rubbery matrices where "hard" and "soft" does not refer to surface hardness but to the glass temperatures T_G of domains and matrices relative to the use temperature. "Hard domains" soften at temperatures $T > T_{G,hard}$ where these materials can be processed like thermoplastics. Cooling to $T < T_{G,hard}$ restores the elastic properties, hence the name **thermoplastic elastomer (TPE)**.

Thermoplastic elastomers have also been called elastoplastics, thermolastics, elastified plastics, flexible thermoplastics, low-modulus plastics, or plastomers. These materials are not rubbers (see definition, p. 249) and therefore neither "thermoplastic rubbers" nor "melt-processable rubbers." They should also not be called "rubber-plastic alloys" because such a designation would also include impact-modified thermoplastics which behave completely differently (Chapter 10).

Thermoplastic elastomers are never homopolymers but either block, segment, or graft polymers with usually two different types of monomeric units or blends of uncrosslinked and crosslinked polymers. According to chemical composition, one can distinguish between TPEs based on structural units of styrene/diene (**TPE-S**), olefin (**TPE-O, TPO**), ester (**TPE-E**), etheramide (**TPE-A**), or urethane (**TPE-U**). True TPE-Os are blends of crosslinked EPDM in poly(olefin)s but this designation is sometimes also used for blends of uncrosslinked EPDMs in poly(olefin)s, i.e., TPE-O/PO.

The special chemical structure of thermoplastic elastomers and the resulting domain character of the physical structure leads to a temperature dependence of moduli that differs characteristically from that of random copolymers. The property profile "thermoplastic elastomer" can be obtained from very different chemical and physical structures:

1. Amorphous thermoplastic (hard) domains in an amorphous elastomeric matrix.
   Example: triblock copolymers (styrene)$_m$-(butadiene)$_n$-(styrene)$_m$.
2. Amorphous thermoplastic (hard) domains in a plastified thermoplastic matrix.
   Example: flexible poly(vinyl chloride).
3. Crystalline thermoplastic (hard) domains in an amorphous elastomeric matrix.
   Example: polyetherester segmented copolymers.
4. Ion clusters and domains in an amorphous elastomeric matrix.
   Example: ionomers.

Application properties of thermoplastic elastomers range between those of chemically crosslinked elastomers and uncrosslinked thermoplastics. Like thermoplastics, TPEs creep and have large compression sets. Like elastomers, they are soft and have large fracture elongations. Because of the temperature sensitivity of their physical crosslinks, their use temperatures are usually lower than those of chemically crosslinked elastomers of similar chemical structure. TPEs are usually soluble, which is sometimes desirable and sometimes not. Thermoplastic elastomers can therefore not be used for car tires but find applications as shoe soles, toys, automotive parts, sports articles, tubes, cables, wires, adhesives, and medical products.

In 2006, 1 100 000 tons of thermoplastic elastomers were produced world-wide. The largest group were thermoplastic olefin-elastomers (60 %), followed by styrene-diene-styrene triblock TPE-S (19 %), and elastomeric polyurethanes TPE-U (19 %). Thermoplastic elastomers with ester-amide and ether-amide groups (total: 1.7 %) and other TPEs (0.3 %) are niche products.

Table 7-13 Properties of poly(styrene) and various styrene-butadiene polymers. x_S = mole fraction of styrene units, T_d = heat distortion temperature, ε_B = fracture elongation, K = impact strength (Charpy), τ_i = light transmission.

Polymer	x_S	$T_d/°C$	$\varepsilon_B/\%$	$K/(J\ m^{-1})$	$\tau_i/\%$
Poly(S)	100	97	7	21	90
Poly(B)-*block*-poly(S)	60	83	2	25	62
Poly(B)-*block*-poly(S)-*block*-poly(B)	60	78	2	24	51
Poly(B)-*block*-poly(S)-*block*-poly(B)-*block*-poly(S)	60	55	23	30	70
Poly(S)-*block*-poly(B)-*block*-poly(S)	60	67	70	43	71
Poly(B)-*blend*-poly(S)	60		2	5	0
Poly(B)-*graft/blend*-poly(S) (HIPS)	92	75	30	69	0
Poly(B-*stat*-S) (vulcanized SBR)	24		690		0

7.5.2   Thermoplastic Styrene-diene Elastomers (TPE-S, TPS)

Styrene-diene TPE-S are triblock polymers with "hard" ($T_G > T_{use}$) outer styrene blocks and "soft" ($T_G < T_{use}$) diene center blocks that may be butadiene (B), isoprene (I), or ethylene/butylene (EB) (sequential synthesis of S, B, and I blocks: see Volume I, p. 607). EB blocks are not produced by copolymerization of ethylene and 1-butene but by selective hydrogenation of butadiene blocks of styrene-butadiene-styrene triblock polymers, SBS. Since the anionic polymerization of the butadiene center block produces both 1,4-units, $-CH_2-CH=CH-CH_2-$, and 1,2-units, $-CH_2-CH(CH=CH_2)-$, hydrogenation of butadiene segments leads to SEBS triblock polymers with center blocks containing both ethylene units, $-CH_2-CH_2-$, and 1-butene units, $-CH_2-CH(C_2H_5)-$.

Thermoplastic elastomers are formed by triblock polymers, for example, $S_nB_mS_n$, with hard blocks S_n and soft blocks B_m and, less favorably, also by tetrablock polymers, for example, $S_nB_mS_nB_m$. Neither diblock polymers, e.g., B_mS_n, nor triblock polymers with "soft" exterior blocks, e.g., $B_mS_nB_m$, behave as thermoplastic elastomers (Table 7-13). Elastomeric properties are also not found for blends of butadiene rubbers with poly(styrene) that have the same overall composition of styrene and butadiene units as SBS block polymers and also not for graft polymers of styrene on poly(butadiene) that rather behave as high-impact strength poly(styrene)s (HIPS; Chapter 10).

Triblock polymers $S_n-B_m-S_n$ with styrene units as the minority components are thermoplastic elastomers for the following reasons:

1. Poly(styrene) molecules S_n and poly(butadiene) molecules B_m are thermodynamically immiscible; their melts will demix completely. Styrene blocks and butadiene blocks of diblock polymers S_n-B_m and triblock polymers $S_n-B_m-S_n$ would demix but they are coupled and therefore form domains of one type of block in the matrix of the other type. In the domains, blocks adopt the macroconformation of random coils.

2. If both blocks S_n and B_m of a diblock polymer S_n-B_m have the same volume requirement, blocks form layers (lamellae) in which layers of the same type face each other, i.e., $\cdots S_n-B_m\cdots B_m-S_n\cdots S_n-B_m\cdots B_m-S_n\cdots$ (Fig. 7-12, left).

Correspondingly, blocks of a diblock polymer S_n-B_m will cluster at the interface between two immiscible liquids 1 and 2 (solvents or melts) if S_n blocks are soluble in 1 and B_m blocks are soluble in 2 (Fig. 7-12, right).

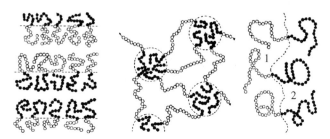

Fig. 7-12 Morphologies of some block polymers if blocks form random coils.
  Left: lamellae of diblock polymers formed if each block has the same volume requirement.
  Center: spherical domains of the minority endblocks of a triblock polymer in the continuous
           matrix of the majority center blocks.
  Right: diblock polymer at the interface of two continuous phases 1 and 2 (2 solvents or 2 melts).

3. At large B/S ratios ($m >> n$) in a diblock polymer $S_n\text{-}B_m$, minority blocks S_n would form spherical domains in a continuous matrix of B_m blocks because the surface of a sphere would provide the smallest contact area between S and B units.

4. A sufficiently high stress applied to such a diblock polymer would lead to a slip of molecule segments of those types of blocks that have sufficient mobility. In an $S_n B_m$ diblock polymer at room temperature, this would be the butadiene blocks because the matrix of the butadiene blocks is above its glass temperature ($T_G = -112°C$, Table 7-11) whereas the domains of styrene blocks are well below their glass temperature ($T_G = 100°C$ if domains are very large). On removal of the stress, the diblock polymer would be completely deformed, i.e., it behaves non-elastically.

5. However, the two styrene endblocks of a triblock molecule $S_n\text{-}B_m\text{-}S_n$ would be in two different spherical domains that are linked by the butadiene center block. These triblock polymers are therefore physically crosslinked (Fig. 7-12, center). Such physical crosslinks are absent in diblock polymers $S_n\text{-}B_m$ and less dominant in tetrablock polymers $S_n\text{-}B_m\text{-}S_n\text{-}B_m$ since one of the crosslinking blocks S_n here is a center block.

6. At room temperature, triblock polymers $S_n\text{-}B_m\text{-}S_n$ would therefore have good elastic properties if (a) styrene units are minority components, (b) styrene endblocks have sufficiently large molecular weights for entanglements which would provide styrene domains with high strengths and glass temperatures near that of homo-poly(styrene).

7. Styrene domains will "melt" reversibly at temperatures above their glass temperature. The triblock polymer can then be processed like a thermoplastic but will regain its elastic properties at room temperature. Hence, it will behave as a thermoplastic elastomer.

The ranges for spherical and lamellar morphologies can be calculated theoretically (see Volume III, p. 262 ff.). The simplest theory predicts for molecularly uniform diblock polymers in the strong segregation limit (sharp interfacial boundaries) that spherical domains of the minority component S should exist in a continuous matrix of B if the volume fraction ϕ_S is less than 0.21. Alternating lamellae should exist in the range $0.3 \leq \phi_S \leq 0.7$. In the ranges $0.21 \leq \phi_S \leq 0.3$ and $0.7 \leq \phi_S \leq 0.79$, one would expect cylinders since they can be viewed as elongated spheres or shrunk lamellae.

In real systems, these three simple morphologies (spheres, cylinders, lamellae) are complemented by more complicated ones such as bicontinuous phases or perforated layers. Ranges of simple morphologies are affected furthermore by the molecular weight distributions of the two types of blocks (Fig. 7-13).

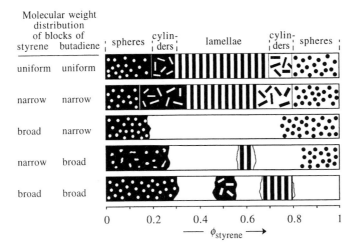

Fig. 7-13 Effect of volume fraction $\phi_{styrene}$ of styrene units on the morphology of poly(styrene)-*block*-poly(butadiene)s with narrow or broad molar mass distributions of styrene and butadiene blocks. White = domains of styrene units, black = domains of butadiene units.

Row 1:   simple theory for sharp domain borders and molecularly uniform blocks [25];
Rows 2-5: experimental findings for various widths of block length distributions [26].

Fig. 7-13 shows that the three predicted morphologies (spheres, cylinders, lamellae) are indeed observed experimentally in the correct sequence. However, experimental data show that blocks with narrow molecular weight distributions have broader ranges for cylinders and narrower ones for lamellae. Diblock polymers with broad molecular weight distributions of blocks have very broad ranges for spherical styrene domains.

Theory predicts that triblock polymers, $S_{m/2}B_nS_{m/2}$, behave morphologically like diblock polymers, S_mB_n, if the styrene segments of the former are half as long as the ones of the latter. Depending on the styrene/butadiene ratio and thus on morphology, very different stress-strain curves are observed (Fig. 7-14).

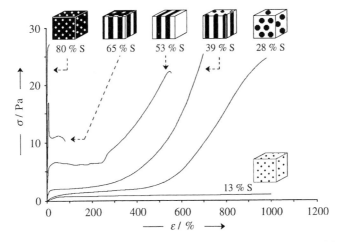

Fig. 7-14 Stress-strain diagrams of styrene-butadiene-styrene triblock copolymers with different percentages of styrene units S [27a].

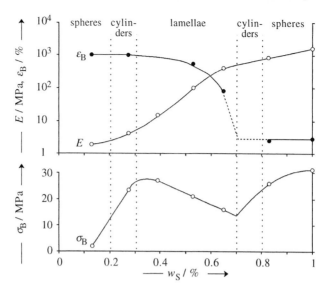

Fig. 7-15 Tensile moduli E, fracture strengths σ_B, and fracture elongations ε_B as functions of the weight fraction w_S of styrene units S in triblock copolymers S_n-B_m-S_n with butadiene units B [27b]. Dotted lines indicate ranges of morphologies for domains with sharp borders.

At w_S = 80 wt% styrene units, "soft" spherical butadiene domains exist in the "hard" styrene matrix: the triblock polymer acts as a rigid thermoplastic with high tensile strength and small fracture elongation (Fig. 7-15). In the opposite case (w_S = 13 wt%), the triblock polymer behaves as a filled elastomer with low tensile strength and very high fracture elongation. Styrene contents of 28-53 wt% lead to an intermediate behavior: elastomeric at small and medium elongations but strengthening at large ones. Property changes from one morphology to the other seem to be gradual from spheres to cylinders and *vice versa* but abrupt from lamellae to cylinders (Fig. 7-15).

Thus, thermoplastic diene elastomers from triblock polymers with "hard" spherical domains have properties similar to those of carbon-black filled classical diene elastomers, for example, SBR (Table 7-14). However, these tensile strengths and fracture elongations are observed only for the first stress-strain test since large elongations destroy the glassy styrene domains. Any type of elongation of triblock thermoplastic elastomers furthermore leads to **hysteresis** (Fig. 7-16), i.e., a difference in stress-strain curves from elongation and relaxation (G: *hysteros* = later, behind).

The upper service temperature of SBS thermoplastic elastomers with spherical styrenic domains is far lower than that of SBR elastomers (ca. 65°C versus 140°C) because of the low glass temperature of the domains which is caused by the low degree of polymerization of the styrene segments and the small size of the domains (ca. 15 nm) (see Volume III, p. 454). In contrast to vulcanized rubbers such as SBRs, TPE-S can therefore not be used near its processing temperature. This temperature must be higher than the hard segment dissolution temperature, i.e., the glass temperature T_G of the styrene spherical domains in the case of TPE-S and T_G or T_M in the case of other TPEs.

Viscosities of TPEs depend strongly on whether the T_G or T_M of the hard blocks are above or below the **order-disorder transformation temperature** (ODT), sometimes called the microphase separation transition temperature (MST). This temperature sepa-

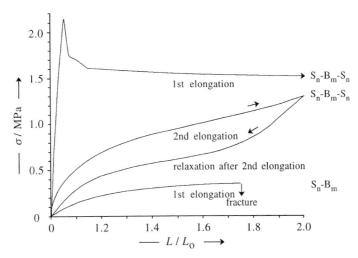

Fig. 7-16 Tensile strengths σ of an SBS triblock polymer with relatively high styrene content as a function of the draw ratio, L/L_0 for the first elongation (relaxation not shown) and the second elongation (with relaxation) as compared to the relaxation of a styrene-butadiene diblock polymer,

rates the strong **segregation limit** (where two relatively pure microphases, e.g., from S and B, are separated in equilibrium) from the weak segregation limit (where the microphases are intimately mixed). Melt viscosities of TPEs below the ODT are usually high at high shear rates (ca. 1000 Pa s) and even far higher at low shear rates (see Volume III, p. 507) but much smaller above the ODT because one-phase mixtures have considerably lower melt viscosities than two-phase mixtures. The temperature dependence of melt viscosity is also affected by shifts from one type of morphology to another. These shifts depend strongly on the molecular weight. They are easier for cylinder $\rightleftarrows$ lamella transformations but very difficult for the transformation sphere $\rightleftarrows$ cylinder because segments must diffuse through a hostile matrix.

7.5.3 Thermoplastic Olefin Elastomers (TPE-O, TPO)

TPE-Os have the highest production volume of all TPEs since their raw materials and polymerizations are fairly inexpensive. Most TPE-Os consist of uncrosslinked or crosslinked blends of isotactic poly(propylene)s and semicrystalline ethylene-propylene rubbers (EPM) or ethylene-diene-non-conjugated diene rubbers (EPDM).

EPMs and EPDMs for these TPE-Os differ from the conventionally crosslinkable grades because they must have especially long crystallizable segments of ethylene units, which leads to good physical crosslinking by crystallization. Such long sequences are obtained by pulsating feeding of monomers into the polymerization reactor. Conventional EPDMs, on the other hand, should have the best possible random distribution of segment lengths. Some commercial TPE-Os consist of interpenetrating networks. Table 7-14 compares properties of TPE-Os with those of other thermoplastic elastomers and conventionally vulcanized styrene-butadiene rubbers.

TPE-Os can be crosslinked by the special technique of **dynamic vulcanization** to socalled **thermoplastic vulcanizates (TPV)**, also known as **elastomeric alloy thermoplastic**

Table 7-14 Average properties of unfilled thermoplastic elastomers EVM and TPE compared to a carbon-black filled vulcanized SBR rubber (a random copolymer (Ran-CP)). EVM = ethylene vinyl acetate copolymer, TPE-E = polyetherester, TPE-O = ethylene-propylene copolymer, TPE-S = triblock copolymer styrene-butadiene-styrene, TPE-U = polyesterurethane.

Physical property	Physical unit	Random CP		Multiblock CP		Triblock CP	Ran-CP
		EVM	TPE-O	TPE-E	TPE-U	TPE-S	SBR
Density	g/cm^3	0.94	0.88	1.19	1.21	0.95	0.94
Upper service temperature	°C		<135	150	<80	<65	
Vicat temperature	°C	60	90		125		
Brittleness temperature	°C	<76	<−60	<−70	<−60	<−73	−50
100 % modulus	MPa		10	17	12		
200 % modulus	MPa		11		20		
300 % modulus	MPa				30	3.1	9.3
Tensile strength at break	MPa	14	15	30	39	21	29
Fracture elongation	%	850	650	650	600	800	650
Tear strength	kN/m	25	75	51	45	25	
Compression set (20 h)	%	100	85	34	90	50	
Rebound elasticity	%	50	35	50	30	65	40
Hardness (Shore A or D)		A80	A54	D55	D65	A75	A60

vulcanizates (EA-TPV). For this process, TPE-O is blended with incompatible thermoplastics such as poly(ethylene) (PE), it-poly(propylene) (PP), or polyamides PA 6 or 6.6, and sometimes also with poly(styrene) (PS) or styrene-acrylonitrile copolymers (SAN). Dynamic vulcanization can also be used for blends of ethylene-vinyl acetate rubbers with PP, PS, SAN, or PA 11, for blends of nitrile rubbers with PS, SAN, or PA 6.9, for blends of chloroprene rubbers with poly(ethylene terephthalate)s, and many more.

On mastication (usually with added crosslinking agents), polymer chains of TPE-Os are cleaved homolytically. The resulting polymer radicals then crosslink polymers to gel-like particles of ca. 1 μm diameter that touch each other. The resulting loose network of aggregates embeds the particles of the thermoplastics. On heating, aggregates are dissolved and the material can be processed like a conventional thermoplastic if the proportion of the thermoplastic is large enough.

7.5.4 Other Thermoplastic Elastomers

The property "thermoplastic elastomeric" ("elastoplastic") is also exhibited by a number of segmented polymers (= multiblock polymers) and graft polymers. In thermoplastic elastomers from segmented polymers, flexible ("soft") segments may consist of either aliphatic polyester or polyether sequences and rigid ("hard") segments with either urethane or aromatic polyester groups.

Elastoplastic graft polymers can be viewed as branched or degenerated star-like block polymers. Industrial types include graft polymers of isobutene on poly(ethylene), vinyl chloride on poly(ethylene-*co*-vinyl acetate), ethylene/propylene on poly(vinyl chloride), and styrene/acrylonitrile on acrylate rubbers.

Another type of thermoplastic elastomers results from the sulfonation of ENB-containing EPDM rubbers by acetanhydride/concentrated sulfuric acid which replaces some

=CH–CH$_3$ groups of ENB units by =C(SO$_3$H)–CH$_3$ (for ENB, see p. 278). Addition of zinc acetate leads to salts that form ion domains that act as crosslinks. The resulting polymers are thermoplastic elastomers (**CSM, CSR, CSPR**) since their ion clusters dissociate at processing temperatures but reform again on cooling to use temperature where the material again becomes a high-strength elastomer. Chemically, these polymers belong to the group of **ionomers** (Volume II, p. 231; Volume III, p. 267).

There are many other types of specialty elastomers of which only brominated poly-(isobutylene-*co*-(*p*-methyl styrene))s are mentioned here because of their use as inner liners for lighter-weight vehicle tires with increased inflation pressure retention.

7.6 Liquid-crystalline Elastomers

Liquid-crystalline elastomers are chemically crosslinked polymers that contain liquid-crystalline domains formed by either side chains or main-chain blocks. An example of the former is a siloxane polymer with various aromatic ester side chains. These side chains are coupled to the main chain by fairly short methylene chains. The polymer molecules are crosslinked via flexible bridges at the ends of some side chains. Examples are silicon bridges –O–Si[(CH$_3$)(CH$_2$)$_{11}$]– as shown below or ether structures containing bridges such as –(CH$_2$)$_{11}$–O–(1,4-C$_6$H$_4$)–O–(CH$_2$)$_{11}$O–.

Because of the orientation of their mesogens, liquid-crystalline elastomers have different stress-strain behaviors depending on whether mesurements are made in a longitudinal or transverse direction with respect to the mesogen axes (Fig. 7-17). These liquid-crystalline elastomers have considerably lower tensile moduli than conventional elastomers. Their 2% moduli (i.e., at $L/L_0 = 1.02$) are just 0.4 MPa in the longitudinal direction and only 0.013 MPa in the transverse one whereas 100% moduli of conventional elastomers are in the range of 4-10 MPa. These polymers are therefore probably not going to compete with conventional elastomers but they may be of interest because of their piezoelectric properties (Section 14.2.10).

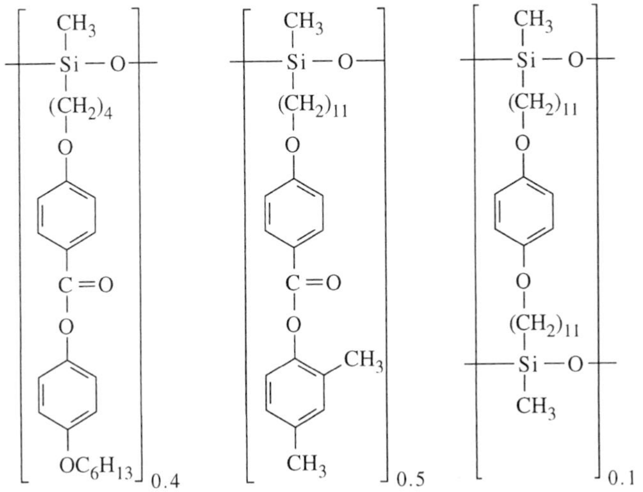

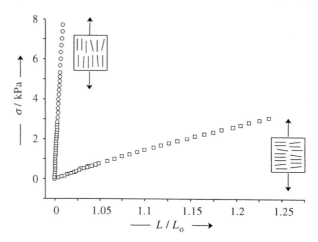

Fig. 7-17 Tensile strength σ as a function of the strain ratio L/L_0 for the smectic LC silicone elasto-mer of p. 293 [28]. Arrows indicate draw directions relative to the main axes of the mesogens.

7.7 Recycling and Reclaiming

Each year, civilization produces enormous amounts of discarded goods of which a small part is reused as such, a part is recycled, the major part is just abandoned, and a small part of the original materials of the goods is reclaimed. In 2006, North America consumed a total of $3.7 \cdot 10^6$ t of raw rubber ($2.0 \cdot 10^6$ t SR; $1.2 \cdot 10^6$ t NR; $0.5 \cdot 10^6$ t TPE) plus an unreported amount of rubber in imported goods, mainly in automobile tires.

The raw rubbers were provided with fillers, plasticizers, vulcanization agents, etc., in-cluding fabrics or steel wires for tires, and then processed to useful goods. Processing produces some scrap (trimmings, etc.) which the manufacturer can reuse from thermo-plastic elastomers but not from vulcanized rubbers. Neither the fate of scrap from the latter is reported nor that of discarded rubber articles, except of tires (see below). Like many other items, discarded rubber articles usually end up in landfills, at least in the United States. Many European countries have high-temperature incinerators.

In 2005, the United States produced ca. 299 million tires. In the same year, about 24 million old tires got their life extended by providing their running surfaces with new rubber that is bonded to the old carcass by vulcanization. This **retreading** is too costly and not very suitable for car tires because their treads consist of difficult to retread (and often unknown) rubber blends, for example, NR + SBR. Retreading is therefore used predominantly for heavy truck and airplanes tires whose treads are usually based only on NR. These retreaded tires are also less prone to failure because they are not used for long times at high speeds, i.e., at high tire temperatures.

In 2005, 259 million discarded tires with an average weight of 10.2 kg (22.5 lbs) each were utilized as **scrap tires** in the United States. This volume is equivalent to 87 % of the tire production of the same year, a marked contrast to the year 1990 where that figure was only 11 %. The stockpile of old tires decreased accordingly from 1000 million in 1990 to 188 million in 2005. Stockpiling creates potential health hazards: old tires are good breeding grounds for mosquitoes and produce carcinogens on open burning.

As reported for 2005 by the Rubber Manufacturers Association, scrap tires were traded for money (ca. 3 %; I) or used for energy (ca. 60 %; VII) or as a source of materials (ca. 37 %, II-VI):

	million scrap tires	
I. Exported scrap tires	7	
II. Baled scrap tires	2.7	
III. Cut, punched, or stamped sheets	6	
IV. Shredded pieces for road and landfill construction, septic tank leach fills, and other civil engineering projects	49	
V. Ground tire rubber (**GTR**), used	38	
a) directly as particles for athletic/recreational purposes		13
b) as molten material for molding or extrusion		12
c) as additive in asphalt		6
d) as such or devulcanized as an additive in tires		3
e) unknown		4
VI. Carbon source	1.3	
VII. Energy source	155	

Most scrap tires (155 million) were used as an *energy source* (VII) because of their high heat content wherefrom ca. 10 million directly as whole tires in dedicated facilities for the generation of electricity ("tires to energy", TTE) and 145 million after conversion to smaller pieces for the production of heat ("tire-derived fuel", TDF) for use in cement kilns (58 million), pulp and paper production (39 million), and in industrial (21 million) and utility (27 million) boilers. This use of scrap tires is known in industry as "recovery" (short for "energy recovery"), a misnomer since this energy is not "recovered" from any manufacturing process but *utilized* as the heat content of the material.

104 million scrap tires were used as *materials,* either directly (II), as sheets, shreds or particles (III-Va), as melts (Vb), in combination with other materials (Vc and Vd), or as a carbon source (VI).

About 2.7 million US scrap tires were baled (II), a process in which up to 100 tires are put on a rod and then are compressed to a bale. Such bales have been used as protective devices against landslides, etc., some successfully, some not.

The direct use of tire parts (III), shreds (IV), or particles (Va) is *not* considered recycling by the industry and the U.S. Environmental Protection Agency because **recycling** is defined as the use of GTR in new products such as mats, rubber-modified asphalt, and the like. Sheets for various purposes (III) are produced by cutting, punching, or stamping tire carcasses from bias-ply or fabric-bodied radial tires (III). Shredding delivers tire pieces of 5-30 cm diameter (IV) that are used for civil engineering projects such as road and landfill construction, septic tank leach fills, reduction of soil compaction, etc.

Cutting, shredding, or grinding are processes that consume enormous amounts of energy. Even in the brittle fracture of stones, only ca. 1.5 % of the applied energy is used to split chemical and physical bonds; the rest is wasted as heat, mainly from friction. For rubbers, the percentage of wasted heat is even greater because of their entropy elasticity. Grinding of tires thus requires heavy equipment such as hammer mills, guillotines, or rotary shear shredders which deliver chips of ca. 2.5-5 cm diameter. Steel wire fragments (if any) are removed by magnetic separators and organic fibers by cyclones.

The remains are ground further in mills to become **ground tire rubber (GTR)**. Very fine tire particles of ca. 30-100 mesh are obtained by pulverizing deeply frozen tires at liquid nitrogen temperature in a hammer mill. The high cost restricts this cryo process to expensive materials such as fluoro rubbers.

"Mesh" indicates the square root of the number of evenly distributed openings per square inch of a sieve. There are at least four different types of sieves. In the NIST system, a "5 mesh" laboratory sieve has 25 openings per square inch (645.16 mm²) which corresponds to apertures of holes of 4 mm. 30-100 mesh indicates apertures of 0.59-0.149 mm.

Coarse GTR particles are used directly for athletic/recreational purposes such as running tracks and playgrounds (p. 295, Va). GTRs are also mixed with urethanes or epoxies as binders and then molded or extruded to carpet underlay, flooring materials, etc. (Vb). Addition of GTR to asphalt (Vc) increases ductility, improves crack and skid resistance, and reduces street noise. For this purpose, GTR particles of 0.4-6.4 mm diameter are either mixed with aggregate and then blended with asphalt (dry process) or blended with asphalt (as a kind of master batch) and then added to the hot asphalt-aggregate mix (wet process). A major concern here is the higher density of GTR (1.15 *versus* 1 g/cm³ of asphalt) which may cause the insoluble GTR particles to settle in the "solid" asphalt. Disadvantageous is the air pollution from toxic emissions during the mixing of GTR and hot asphalt and a 50 % higher cost compared to conventional asphalt.

GTRs or devulcanized GTRs of relatively small particle size also serve as fillers in vulcanizable rubbers (Vd). One might expect that these GTR particles act as additional crosslinking agents and improve tensile strength. However, the tensile strength at break and the fracture elongation (Fig. 7-18) as well as the tear strength (not shown) decrease with increasing diameter of the particles and even more so after aging at 100°C (Fig. 7-18). Larger GTR particles retain the properties better than smaller ones. The reason for this behavior is the incompatibility of GTRs with their virgin rubbers.

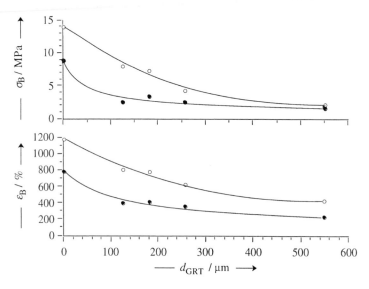

Fig. 7-18 Effect of particle diameter d_{GTR} of 30 phr cryo-ground tire rubber (GTR) from natural rubber vulcanizates on the tensile strength at break, σ_B, and the fracture elongation, ε_B, of virgin natural rubber vulcanizates filled with these GTR particles before (O) and after (●) 36 h aging of vulcanizates at 100°C [29]. All vulcanizates: 6 phr ZnO, 0.5 phr stearic acid, 0.5 phr MBT, 3.5 phr sulfur.

In principle, sulfur vulcanization should be a reversible process since bond energies of carbon-sulfur and sulfur-sulfur bonds are low, both absolute and with respect to that of the carbon-carbon bond. **Devulcanization** should therefore lead to a truly **reclaimed rubber** that should behave like and be compatible with its virgin counterpart.

Indeed, sulfur-vulcanized natural rubber can be completely decrosslinked at 200-220°C with diphenyldisulfide as devulcanization agent. Devulcanization of sulfur-vulcanized EPDM by the same agent for two hours at 275°C reduced the crosslinking density by 90 %. Several other chemical agents have been proposed but none seem to be in industrial use. There are also proposals for devulcanization by physical processes such as microwaves or ultrasound.

The amount of devulcanized GTR from scrap tires is relatively small (Vd on p. 295) and seems to be generated industrially mainly by two processes. The reclamator process heats fine GTR crumbs in oil to high temperatures under intensive stirring. The wet digester process cooks GTR with caustic and water. The less polluting dry digester process uses only steam.

Scrap tires serve also in electric furnaces as a carbon source for the making of high-carbon steel (VI on p. 295). In principle, GRTs can be pyrolyzed to carbon blacks and oils but this is apparently not done on an industrial scale.

Literature to Chapter 7

7.1.1.a GENERAL REVIEWS (textbooks, monographs)
G.S.Whitby, Ed., Synthetic Rubber, Wiley, New York 1954
C.Barlow, The Natural Rubber Industry, Oxford University Press, Kuala Lumpur 1978
C.M.Blow, C.Hepburn, Eds., Rubber Technology and Manufacture, Butterworth, London, 2nd ed. 1982
F.R.Eirich, Ed., Science and Technology of Rubber, Academic Press, New York 1983
D.C.Blackley, Synthetic Rubbers: Their Chemistry and Technology, Elsevier, London 1983
M.Morton, Ed., Rubber Technology, Van Nostrand Reinhold, New York, 3rd ed. 1987
J.E.Mark, B.Erman, F.R.Eirich, Science and Technology of Rubber, Academic Press, New York, 2nd ed. 1994
B.Erman, J.E.Mark, Structures and Properties of Rubberlike Networks, Oxford Univ.Press, New York 1997
A.Ciesielski, An Introduction to Rubber Technology, RAPRA, Shawbury, Shrewsbury, Shropshire, UK 2001
R.Kothandaraman, Rubber Materials, CRC Press, Boca Raton (FL) 2008

7.1.1.b GENERAL REVIEWS (handbooks, databanks, etc.)
K.F.Heinisch, Dictionary of Rubber, Elsevier, Amsterdam 1974
C.A.Harper, Ed., Handbook of Plastics and Elastomers, McGraw-Hill, New York 1975
E.R.Yescombe, Plasstics and Rubber: World Sources of Information, Appl.Sci. Publ., Barking, Essex, 2nd ed. 1976
J.A.Brydson, Rubber Materials and Their Compounds, Elsevier Appl.Sci., London 1988
W.Hofmann, Rubber Technology Handbook, Hanser, Munich 1989
R.F.Ohm, Ed., The Vanderbilt Rubber Handbook, Vanderbilt, Norwalk (CT), 13th ed. 1990
M.B.Ash, I.A.Ash, Eds., Handbook of Plastic Compounds, Elastomers, and Resins, VCH, Weinheim 1992 (trade names, manufacturers)
N.P.Cheremisinoff, Ed., Elastomer Technology Handbook, CRC Press, Boca Raton (FL) 1993
A.J.Tinker, K.P.Jones, Eds., Blends of Natural Rubber, Chapman and Hall, London 1998

A.S.Lodge, An Introduction to Elastomer Molecular Network Theory, Bannatek Press, Madison (WI) 1999
A.K.Bhowmick, H.L.Stephens, Eds., Handbook of Elastomers, Dekker, New York, 2nd ed. 2001
S.K.De, J.R.White, Eds., Rubber Technologist's Handbook, Rapra, Shawbury, Shrewsbury, Shropshire, UK 2001
C.F.Ruebensaal, Ed., Rubber Industry Statistical Report, International Institute of Synthetic Rubber Producers, New York (annually)
The International Plastics Selector, Inc., Desk-Top Data Bank, Elastomeric Materials, Cordura Publications, La Jolla (CA) (annually)
-, Elastomers Manual, International Institute of Synthetic Rubber Producers, New York (classes and names of rubbers and elastomers)

7.1.1.c GENERAL REVIEWS (testing), see also Chapter 5
R.P.Brown, Physical Testing of Rubbers, Appl.Sci.Publ., London 1979
W.C.Wake, M.J.R.Loadman, B.K.Tidds, Eds., Analysis of Rubber and Rubber-Like Polymers, Elsevier, New York, 3rd ed. 1983
M.J.R.Loadman, Analysis of Rubber-like Polymers, Kluwer, New York, 4th ed. 1989
R.P.Brown, T.Butler, Natural Ageing of Rubber, RAPRA, Shawbury, Shropshire, UK 2001

7.1.2 HISTORICAL DEVELOPMENT
H.Wolf, R.Wolf, Rubber: A Story of Glory and Greed, Covici, Friede, New York 1936
F.A.Howard, Buna Rubber: The Birth of an Industry, Van Nostrand, New York 1947
P.Schidrowitz, T.R.Dawson, Eds., History of the Rubber Industry, Heffer, Cambridge 1952
C.Heuck, Ein Beitrag zur Geschichte der Kautschuk-Synthese: Buna-Kautschuk IG (1926-1945), Chem.-Ztg. **94**/5 (1970) 147 (history of Buna by one of the original researchers)
H.Allen, The House of Goodyear (A Story of Rubber and Modern Business), Ayer, Manchester (NH) 1976 (first published in 1943)
P.E.Hurley, History of Natural Rubber, J.Macromol.Sci.-Chem. **A 15**/7 (1981) 1279; reprinted as Natural Rubber, in R.B.Seymour, Ed., History of Polymer Science and Technology, Dekker, New York 1982
M.Morton, History of Synthetic Rubber, J.Macromol.Sci.-Chem. **A 15**/7 (1981) 1289; reprinted as Synthetic Rubber, in R.B.Seymour, Ed., History of Polymer Science and Technology, Dekker, New York 1982
V.Herbert, A.Bisio, Synthetic Rubber: A Project That Had to Succeed. Greenwood, Westport (CT) 1985 (history of the U.S. synthetic rubber program 1940-1945)
P.J.T.Morris, The American Synthetic Rubber Program, University of Philadelphia Press, Philadelphia (PA) 1989
U.Giersch, Y.Kubisch, Gummi – die elastische Faszination, Nicolai, Berlin 1995. This German-language book contains a chapter about the history of wild rubber that is based on diaries and shipping manifests and tells quite a different story about the migration of seeds to Asia than the one that is usually told. An English translation of this chapter is available: M.J.R.Loadman, The Exploitation of Natural Rubber, Malaysian Rubber Producer's Research Association, Publication 1531; accessed as html version of the file www.cementex.com/pdf/history.pdf, 2007-12-16
J.Loadman, Tears of the Tree, The Story of Rubber–A Modern Marvel, Oxford University Press, New York 2005
International Rubber Study Group, World Rubber Statistics Historic Handbook 1900-1960, International Institute of Synthetic Rubber Producers, Wembley, UK 1996 (contains data back to 1822)

7.1.4 ECONOMIC IMPORTANCE
-, World Rubber Statistics Handbook **V** (1975-1995), International Institute of Synthetic Rubber Producers, Houston (TX)
-, Chem.Eng.News, June-July each year (various issues); reporting stopped in 2004
-, International Rubber Study Group (IRSG), newest data for the previous year can be found in the Rubber Statistical Bulletin; www.rubberstudy.com

7.2.3-7.2.4 MASTICATION, FORMULATION, VULCANIZATION (additives, s.a. Chapter 4)
G.Kraus, Ed., Reinforcement of Elastomers, Interscience, New York 1965
J.van Alphen, Rubber Chemicals, Reidel, Dordrecht 1973
G.Kraus, Reinforcement of Elastomers by Carbon Black, Adv.Polym.Sci. **8** (1971) 155

J.B.Donnet, Filler-Elastomer Interaction, Brit.Polym.J. **5** (1973) 213

M.Porter, Vulcanization of Rubber, in S.Oae, Ed., Organic Chemistry of Sulfur, Plenum, New York 1977

A.I.Medalia, Effect of Carbon Black on Dynamic Properties of Rubber Vulcanizates, Rubber Chem. Technol. **51** (1978) 437

Z.Rigbi, Reinforcement of Rubber by Carbon Black, Adv.Polym.Sci. **36** (1980) 21

A.Voet, Reinforcement of Elastomers by Fillers: Review of Period 1967-1976, Macromol.Revs. **15** (1980) 327

C.W.Evans, Practical Rubber Compounding and Processing, Appl.Sci.Publ., Barking, Essex 1981

E.M.Dannenberg, Filler Choices in the Rubber Industry, Rubber Chem.Technol. **55** (1982) 860

V.A.Shershnev, Vulcanization of Polydiene and Other Hydrocarbon Elastomers, Rubber Chem.Technol. **55** (1982) 537

A.Bhowmick, Ed., Rubber Products Manufacturing Technology, Dekker, New York1994

J.L.White, Rubber Processing. Technology, Materials and Principles, Hanser, Munich 1995 (without vulcanization)

F.W.Barlow, Rubber Compounding, Dekker, New York, 2nd Ed. 1995

W.Kleemann, K.Weber, Elastomer Processing. Formulas and Tables, Hanser, Munich 1998

P.A.Ciullo, N.Hewitt, The Rubber Formulary, W.Andrew, Norwich (NY) 1999

P.S.Johnson, Rubber Processing. An Introduction, Hanser, Munich 2001

J.S.Dick, Ed., Rubber Technology. Compounding and Testing for Performance, Hanser Gardner, Cincinnati (OH) 2001

B.Rodgers, Ed., Rubber Compounding, Chemistry and Applications, Dekker and CRC Press, Boca Raton (FL) 2005

N.Hewitt, Compounding Precipitated Silica in Elastomers, W.Andrew, Norwich (NY) 2007

V.C.Chandrasekaran, Essential Rubber Formulary: Formulas for Practitioners, W.Andrew, Norwich (NY) 2007

J.-M.Vergnaud, I.-D.Rosca, Rubber Curing nd Properties, CRC Press, Boca Raton (FL) 2008

7.3 GENERAL PURPOSE RUBBERS

W.Breuers, H.Luttrop, Buna: Herstellung, Prüfung, Eigenschaften, Berlin 1954 (Buna: Manufacture, Testing, Properties (historical account in German)

J.R.Scott, Ebonite, Its Nature, Properties and Compounding, MacLaren, London 1958

W.M.Saltman, Ed., The Stereo Rubbers, Wiley, New York 1977

C.W.Evans, Powdered and Particulate Rubber Technology, Appl.Sci.Publ., London 1978

W.D.Gunter, Butyl and Halogenated Butyl Rubbers, Dev.Rubber Technol. **2** (1981) 155

J.A.Brydson, Styrene-Butadiene-Rubber, Dev.Rubber Technol. **2** (1981) 21

L.Corbelli, Ethylene-Propylene Rubbers, Dev.Rubber Technol. **2** (1981) 87

M.J.Shuttleworth, A.A.Watson, Synthetic Polyisoprene Rubbers, Dev.Rubber Technol. **2** (1981) 233

A.D.Roberts, Ed., Natural Rubber Science and Technology, Oxford Univ.Press, Oxford 1988

7.4 SPECIALTY RUBBERS

R.G.Arnold, A.L.Barney, D.C.Thompson, Fluoroelastomers, Rubber Chem.Technol. **46** (1973) 619

E.L.Warrick, O.R.Pierce, K.E.Polmanteer, J.C.Saam, Silicone Elastomer Developments 1967-1977, Rubber Chem.Technol. **52** (1979) 437

J.C.Bament, J.G.Pillow, Developments with Polychloroprene, Dev.Rubber Technol. **2** (1981) 131

H.H.Bertram, Developments in Acrylonitrile-Butadiene Rubber (NBR) and Future Prospects, Dev. Rubber Technol. **2** (1981) 51

G.C.Sweet, Special Purpose Elastomers, Dev. Rubber Technol. **1** (1980) 45

R.J.Cush, H.W.Winnan, Silicone Rubbers, Dev. Rubber Technol. **2** (1981) 203

C.Hepburn, Polyurethane Elastomers, Elsevier, Amsterdam 1982

L.D.Albin, Current Trends in Fluoroelastomer Development, Rubber Chem.Technol. **55** (1982) 902

B.G.Willoughby, Liquid Rubbers, Dev. Block Copolymers, **2** (1985) 239

P.R.Dvornic, R.W.Lenz, High Temperature Siloxane Elastomers, Hüthig and Wepf, Basel 1990

A.L.Moore, Fluoroelastomers Handbook: The Definitive User's Guide, W.Andrew, Norwich (NY), 2006

R.C.Klingender, Ed., Handbook of Specialty Elastomers, CRC Press, Boca Raton (FL) 2008

M.Warner, E.M.Terentjev, Liquid Crystal Elastomers (International Series of Monographs in Physics 120), Oxford Univ.Press, New York, revised edition 2007

I.R.Clemitson, Castable Polyurethane Elastomers, CRC Press, Boca Raton (FL) 2008
R.C.Klingender, Ed., Handbook of Specialty Elastomers, CRC Press, Boca Raton (FL) 2008

7.5 THERMOPLASTIC ELASTOMERS
B.M.Walker, C.P.Rader, Eds., Handbook of Thermoplastic Elastomers, Van Nostrand-Reinhold, New
    York, 2nd ed. 1988
G.Holden, N.R.Legge, R.P.Quirk, H.Schroeder, Eds., Thermoplastic Elastomers, Hanser, Munich,
    2nd ed. 1996
G.Holden, Understanding Thermoplastic Elastomers, Hanser, Munich 2000
G.Holden, H.R.Kricheldorf, R.P.Quirk, Eds., Thermoplastic Elastomers, Hanser, Munich, 3rd ed.
    2004
S.Fakirov, Ed., Handbook of Condensation Thermoplastic Elastomers, Wiley-VCH, Weinheim 2005
J.G.Drobny, Handbook of Thermoplastic Elastomers, W.Andrew, Norwich (NY) 2007

7.7 RECYCLING AND RECLAIMING
C.P.Rader, Ed., Plastics, Rubber and Paper Recycling, Am.Chem.Soc., Washington (DC) 1995
K.Meardon, D.Russell, C.Clark, Scrap Tire Technology and Markets, W.Andrew, Norwich (NY)
    1993
R.H.Snyder, Scrap Tires: Disposal and Reuse, Society of Automotive Engineers, Inc., Wassendale
    (PA) 1998
W.E.Rudzinski, M.Y.Kariduraganavar, T.M.Aminabhavi, Effective Recycling of Scrap Rubber Tires.
    Alternative Solutions, Polymer News **26** (2001) 392
S.K.De, A.I.Isayev, K.Khait, Eds., Rubber Recycling, CRC Press, Boca Raton (FL) 2005

References to Chapter 7

[1]   International Rubber Study Group, World Rubber Statistics Historic Handbook, IRSG,
           Wembley, UK 1996 (contains data back to 1822)
[2]   Z.Frank, A.Musacchio, The International Natural Rubber Market, 1870-1930,
           EH Net*Encyclopedia*frank.international.rubber.market
[3]   C.Heuck, Ein Beitrag zur Geschichte der Kautschuk-Synthese: Buna-Kautschuk IG (1926-1945),
           Chem.-Ztg. **94**/5 (1970) 147 (history of Buna by one of the original IG Farben researchers;
           includes a large list of inventors and patents). Data reported by G.S.Whitby in G.S.Whitby,
           Ed., Synthetic Rubber, Wiley, New York 1954, sometimes deviate from those of Heuck.
[4]   R.F.Dunbrook, Historic Review, in G.S.Whitby, Ed., Synthetic Rubber, Wiley, New York
           1954, text and tables I, II, and IV
[5]   International Rubber Study Group (IRSG), Rubber Statistical Bulletin, (a) **56**/4 (2002), (b)
           November-December 2007; www.rubberstudy.com, accessed 2007-12-14
[6]   www.lgm.gov.my/nrstat/T1.htm, accessed 2007-12-12
[7]   www.wikipedia.org, accessed 2007-12-13
[8]   www.shfe.com.cn/528/4e14 (IRSG, 4th Shanghai Derivatives Market Forum, 2007-05-28)
           and single data from websites of Indian, Chinese, and Vietnamese rubber producers
[9]   Chem.Eng.News (May-July of each year). In 2004, C&EN stopped the publication of United
           States production data of synthetic rubbers.
[10]   F.Haas, D.Theisen, Kautschuk, Gummi, Kunststoffe **23** (1970) 502
[11]   G. Dall'Asta, Rubber Chem.Technol. **47** (1974) 511
[12]   L.Bothe, G.Rehage, Angew.Makromol.Chem. **100** (1981) 39, (a) data taken from Fig. 5,
           (b) Fig. 2
[13]   E.Bister, W.Borchard, G.Rehage, Kautschuk Gummi Kunststoffe **29** (1976) 527, Fig. 6
[14]   W.Hofmann, Kunststoffe **80** (1990) 847, data of Table 2
[15]   D.A.Smith, in W.M.Saltman, Ed., Stereorubbers, Wiley, New York 1977
[16]   L.R.G.Treloar, Trans.Faraday Soc. **40** (1944) 59, data taken from Fig. 5
[17]   M.C.Shen, D.A.McQuarrie, J.L.Jackson, J.Appl.Phys. **38** (1967) 791, Table 1 and Fig. 4
[18]   (a) H.-G.Elias, Neue polymere Werkstoffe 1969-1974 Hanser, Munich 1975; (b) H.-G.Elias,
           New Commercial Polymers 1969-1975, Gordon and Breach, New York 1977

[19] K.-H.Nordsiek, N.Sommer, Der Lichtbogen **33**/3 (No. 174) (1974) 8 (company journal of Che-
       mische Werke Hüls, Germany)

[20] A.E.Oberster, T.C.Bouton, J.K.Valaitis, Angew.Makromol.Chem. **29/30** (1973) 291

[21] S.Futamura, Angew.Makromol.Chem. **240** (1996) 137, data of Figs. 5-7

[22] A.K.Bhowmick, A.N.Gent, C.T.R.Pulford, Rubber Chem.Technol. **56** (1983) 226, Table II

[23] W.Hofmann, Kunststoffe **78** (1988) 132, Tables 1-3, selected data

[24] J.Falbe, M.Regitz, Eds., Römpp Chemie Lexikon, Thieme, Stuttgart, 9th ed. 1990, Vol. 2,
       p. 1106, selected data

[25] D.J.Meier, Theoretical Aspects of Block Copolymers, Fig. 5, in N.R.Legge, G.Holden,
       H.E.Schroeder, Eds., Thermoplastic Elastomers, Hanser, Munich 1987

[26] K.Gerberding, G.Heinz, W.Heckmann, Makromol.Chem., Rapid Comm. **1** (1980) 221 (data
       in text)

[27] G.Holden, E.T.Bishop, N.R.Legge, J.Polym.Sci. C **26** (1969) 37, (a) Fig. 5 (supplemented
       by drawings of morphologies with sharp borders of domains, (b) Table II

[28] E.Nishikawa, H.Finkelmann, H.R.Brand, Macromol.Rapid Comm. **18** (1997) 65, Fig. 4

[29] A.K.Naskar, P.K.Pramanik, R.Mukhopadhyay, S.K.De, A.K.Bhowmick, Rubber Chem.
       Technol. **73** (2000) 902, Table IV

Table 8-1 Classification of polymeric materials according to the architecture of their molecules and the mobility of their segments. T = use temperature, T_{trans} = highest transformation temperature, i.e., the melting temperature T_M of (semi)crystalline polymers, the transition temperature T_{NI} (nematic $\rightleftarrows$ isotropic liquid) of liquid-crystalline polymers, or the glass temperature T_G of amorphous polymers.

Molecular architecture ↓ Segmental mobility →	$T < T_{trans}$ small	$T > T_{trans}$ large
Linear or slightly branched	thermoplastic	liquid
Physically (reversibly) crosslinked	*no special name*	thermoplastic elastomer
Chemically (irreversibly) crosslinked	thermoset (plastic)	(thermoset) elastomer

Polymers are traditionally classified according to their processing and end use properties, i.e., whether the same material can be processed repeatedly or not (thermoplastic *versus* thermoset) and whether the material is highly elastic or not (elastomer *versus* plastic). These properties depend on molecular architectures on one hand and on segmental mobilities on the other. Since molecules may be uncrosslinked, physically crosslinked, or chemically crosslinked, three different types of polymer behavior are observed for polymers with large segmental mobilities: liquids, thermoplastic elastomers, and elastomers (which are really elastomeric *thermosets* as far as their chemical structure and processability is concerned). Similarly, one has to subdivide the class of thermosets into thermosets proper and "reversible thermosets" which are thermoplastics that are physically crosslinked by crystal bridges, ion clusters, or domain structures.

As far as these properties are concerned, it really does not matter whether the material is three-, two,- or one-dimensional, i.e., a working material, sheet/film, or fiber. Properties of three-dimensional materials (bulk materials) can therefore be conveniently treated in a chapter on polymers with elastic properties (Chapter 7) and in a chapter on polymers with plastic properties (this Chapter).

However, working materials (bulk materials) on one hand and sheet/films and fibers on the other differ with respect to the effect of surface properties of the latter. For this reason, coatings/adhesives (Chapter 12) and packaging materials (Chapter 13) are not included in this Chapter which treats only basic plastics. Filled/reinforced plastics are discussed in Chapter 9 and polymer blends in Chapter 10.

8.1.2 Historical Development

The first "plastics" were used by prehistoric artists who searched for easily processable, long-lived materials with optimal artistic effects. Egg white and blood proteins were used as binders for pigments as early as 15 000 B.C. in the cave paintings of Altamira (Spain). Later paints were based on gelatin (the degradation product of the protein collagen) or on polymers that result from the crosslinking of unsaturated natural oils. The latter are still being used in today's oil paints.

The first real plastic materials were shaped natural products. In the medieval ages, cow horns were split, flattened by steaming and pressing, and then used as windows for lanterns or as dyeable marquetries in wood. However, these sheets from the protein keratin rolled back to their original shape after some time.

Products based on casein are much more processable (first known report by the Bavarian monk Wolfgang Seidel (1492-1562)). This protein from skim milk was treated with warm lye and shaped while still warm. The desired shape was then frozen-in by dipping the material into cold water. The same process was rediscovered and patented in 1885 by the American Emery Edward Childs. Better products were obtained by chemical crosslinking of casein with formaldehyde, patented in 1895 by Germans Wilhelm Krische and Adolf Spitteler. The resulting **Galalith**® (G: *gala* = milk, *lithos* = stone) was used originally to manufacture white chalk boards for German schools. It is still produced for haberdashery since it is very easy to dye in brilliant colors.

Another early thermoset was **ebonite** (G: *ebenos* = ebony tree (*Diospyros ebenum*, a tree with dark-colored heartwood); from Egyptian *hebni*), a hard, black material from the vulcanization (cross-linking) of natural rubber with 30-50 wt% sulfur under the action of basic lead carbonate (magistery of lead, 2 $PbCO_3 \cdot Pb(OH)_2$, one of the many "white leads". Ebonite was invented in 1851 by the American Nelson Goodyear, the brother of Charles Goodyear (discoverer of the vulcanization of natural rubber to an elastomer (with PbO in 1839, and with basic lead carbonate in 1841; see p. 251)).

At first, ebonite had no commercial use. However, it became important for the electrical industry which was being developed during the last decades of the 19th century and urgently needed electrical insulators that were easier to process than wood and minerals. But natural rubber was at that time a fairly scarce and expensive raw material (p. 251 ff.) and so one searched for less expensive substitutes. In 1899, the Englishman Arthur Smith took out a patent for the use of phenol-formaldehyde resins as electrical insulators; these resins were known from the 1872 work of the German Adolf von Baeyer. The breakthrough came with the 1906 heat-and-pressure patent of the Belgian born, naturalized American Leo H. Baekeland. In 1909, the resulting **Bakelite**® was recognized as affording easily processable, excellent electrical insulating materials (see Chapter 14).

The years after the mid 1800s saw not only the discovery of semisynthetic and fully synthetic thermosetting materials but also that of thermoplastics. The first semisynthetic thermoplastic materials were based on cotton which contains ca. 93 % cellulosic fibers (p. 189). In the mid 1800s, many attempts were made to improve cotton fibers and fabrics. **Mercerization** = treatment of cotton with soda lye (aqueous NaOH solution) by the Englishman John Mercer led in 1844 to cotton with improved textile properties (p. 190). The Frenchman L.Figuier demonstrated in 1846 that cellulose paper becomes more stable mechanically by treating it with sulfuric acid; in 1853, the Englishman W.E.Gaine got a patent for the manufacture of **parchment paper** (p. 238) by the same process. Another English patent by Thomas Taylor described the production of **vulcanized fiber** by treating stacks of paper sheets with zinc chloride under pressure (p. 238). All of these procedures were physical transformations.

The first chemical transformation of cellulose was discovered in 1846 by Christian Friedrich Schönbein, a chemistry professor in Basel, Switzerland, who accidentally treated cotton with a mixture of nitric acid and sulfuric acid. The resulting cellulose nitrate (commonly called "nitrocellulose" although it is not a nitro compound but an ester) soon found use as so-called gun cotton and, in diethyl ether/ethanol solutions, as **collodion** for waterproofing woven fabrics (1856 English patent to Alexander Perkins). The same inventor in 1862 found a method for the preparation of plastic masses from cellulose nitrate in a mixture of camphor, castor oil, and dyestuffs. This process as well as a similar one by Daniel Spill for cellulose nitrate in alcoholic camphor solutions was not commercially successful. The breakthrough came after the American John Wesley Hyatt patented in 1869 the coating of surfaces of billiard balls made from ivory or from cloth reinforced, ivory dust filled shellac with collodion and in 1870 for the manufacture of a horn-like material from the solvent-free mixture of cellulose nitrate and camphor which became known in 1872 as **celluloid**®.

Other thermoplastics were discovered earlier but commercialized much later (Volume II, Table 1-1; for details see the relevant chapters in Volume II). Styrene was apparently discovered in 1786 by an English chemist called Neuman and its conversion to a solid material in 1839 by the German pharmacist Eduard Simon. The polymeric nature of this "metastyrene" was recognized in the 1920s by the German Hermann Staudinger (Nobel prize 1953). I.G. Farben started to produce **poly(styrene)** (PS) in 1930.

Vinyl chloride (VCM) was synthesized in 1838 by the French professor Henri Victor Regnault who also observed the formation of resins. In 1912, targeted polymerizations of VCM were performed by the German industrial chemist Fritz Klatte and the Russian chemist I.I.Ostromislensky. In 1931, I.G. Farben began the industrial production of **poly(vinyl chloride)** (PVC).

In 1901, the German Otto Röhm submitted his Ph.D. thesis about "Polymerization products of acrylic acids." He later started his own company, Röhm & Haas Co, and, in 1911, again began to research acrylic compounds which led to methyl methacrylate in 1928, and to the marketing of **poly(methyl methacrylate)** (PMMA) in 1933.

While trying to compress ethylene by very high pressures, R.O.Gibson of Imperial Chemical Industries (ICI), England, observed in 1933 the formation of traces of lightly branched poly(ethylene). ICI started the commercial production of this **low-density poly(ethylene)** (LDPE) in 1939.

Less branched **high-density poly(ethylene)s** (HDPE) were first obtained in 1953 by low pressure processes using transition metal catalysts (Karl Ziegler (Nobel prize 1963)), chromium oxide catalysts (J.P.Hogan, R.L.Banks, Phillips Petroleum), or molybdenum oxide catalysts (E.Field, M.Feller, Standard Oil of Indiana). Production started in 1955 in several countries.

Ziegler catalysts were used by the Italians Guilio Natta (Nobel prize 1963), P.Pino, and G.Mazzanti for the stereospecific polymerization of 1-olefins such as propylene to **isotactic poly(propylene)** (PP) (patents to Montecatini). This crystallizable polymer was also obtained by three other research groups using supported metal oxides (J.P.Hogan, R.L.Banks, Phillips Petroleum; A.Zletz, Standard Oil of Indiana) or organometal catalysts W.N.Baxter, N.G.Merkling, I.M.Robinson, G.S.Stamatoff, DuPont). The production of this polymer started in 1957 simultaneously in Germany, Italy, and the United States. After a patent process of 22 years (1958-1980), the court determined that Phillips Petroleum was the owner of the composition-of-matter patent to which the owners of process patents had to pay royalties.

Oligomeric formaldehydes were described as early as 1839 by Justus von Liebig but it took extensive scientific investigations on polymeric formaldehydes in the 1920s and 1930s and 14 years of development by DuPont de Nemours before **poly-(formaldehyde)** (poly(oxymethylene), POM) reached the market in 1959.

The recognition of the macromolecular structure of plastic, elastomeric, and fibrous materials by Hermann Staudinger in the mid 1920s and the subsequent scientific investi-

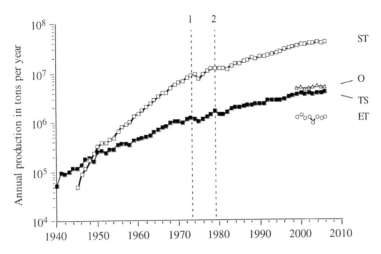

Fig. 8-1 Logarithm of annual North American production of plastics: standard thermoplastics (ST), engineering thermoplastics (ET), thermosets (TS), and various others (O), exclusive of fibers, elastomers, some coatings, etc. [1]. 1 = first oil crisis (see Volume II, p. 49), 2 = second oil crisis. Data do not include poly(methyl methacrylate) (ca. 300 000 t/a (?)) and several small-volume plastics.

ST: HDPE, LDPE, LLDPE, PP, PS, ABS and other styrenics, PVC, aliphatic PA (exclusive of fibers, and thermoplastic polyesters (PET, PTT, PBT, PCT). Newer data also include Canadian or Canadian/Mexican productions (North American Free Trade Association (NAFTA)).

TS: EP, PF, UF, MF. Newer data for PF, UF, and MF include Canadian productions.

ET: US and Canadian productions of POM, granular fluoropolymers, PI, PAI, PEI, PC, modified PPO, PPS, PSU, LCP.

O: PUR (TDI, MDI, polyols), UP, and various others (not identified).

Note that the originator of these data, the American Chemistry Council, groups plastics differently from common practice: ABS, aliphatic polyamides (nylons), and thermoplastic polyesters are included in the data for standard polymers and not in those for engineering polymers. The group of engineering plastics also includes high-performance plastics such as granular fluoropolymers, various polyimides (PI, PAI, PEI), poly(1,4-phenylene sulfide) (PPS), polysulfones (PSU), and liquid-crystalline polymers (LCP). Thermosetting monomers for polyurethanes PUR (TDI, MDI, polyols) and unsaturated polyesters (UP) are included in "others" but not in "thermosets."

gation of polymerization methods in the 1930s and 1940s by both academe and industry promoted the development of commercially useful polymers (Fig. 1-1). In the 1930s and 1940s, syntheses of petrochemicals were developed based on new catalysts (for the beginnings of petrochemistry, see p. 254 and Volume II, p. 45 ff.). These petrochemicals differed not only from the older coal-based chemicals but were also available in large amounts at lower cost (see Volume II). Petrochemistry thus replaced coal chemistry, first in the United States, the motherland of petrochemistry, and a few years later also in Europe. As a result, production of standard thermoplastics based on petrochemicals saw a rapid growth. In the United States, production of standard thermoplastics surpassed that of the older thermosets in 1950 (Fig. 8-1).

The production of standard thermoplastics decreased markedly in the year after the first oil crisis (1973) and stayed flat for three years after the second one (1979) (see Fig. 8-1). The market price of a barrel of Arabian light oil jumped from US$ 2.18 (1973-01-01) to US$ 11.65 (1974-01-01) during the first oil crisis and from US$ 13.39 (1979-01-01) to US$ 34.00 (1981-10-01) during the second oil crisis (see also Volume II, p. 49). After these market turbulences, the US growth rate of standard plastics became smaller.

8.1.3   Types and Classifications of Plastics

Types of Plastics

Standard plastics have fairly simple chemical structures because they are based on simple petrochemical monomers such as $CH_2=CH_2$ for the various poly(ethylene)s (LDPE, HDPE, LLDPE, etc.), $CH_2=CH(CH_3)$ for isotactic poly(propylene), $CH_2=CHCl$ for poly(vinyl chloride), and $CH_2=CH(C_6H_5)$ for atactic poly(styrene) (PS). They were soon produced in such large amounts that they also became known as **volume plastics**, **commodity plastics**, or **bulk thermoplastics** (Section 8.2). Standard plastics are mainly used for low-tech applications such as packaging, household goods, agriculture, and the like (see Section 8.1.4).

The success of standard polymers led to the development of plastics with improved mechanical and thermal properties that could be used for engineering purposes. These **engineering thermoplastics (technical plastics, technoplastics)** (Section 8.3) include various copolymers of styrene with acrylonitrile (SAN) or styrene/acrylonitrile plus butadiene rubbers (ABS); aliphatic polyamides known as nylons (PA 6, 6.6, etc.); and "thermoplastic polyesters" with aromatic polyesters from terephthalic acid and ethylene glycol (PET), trimethylene glycol (PTT), 1,4-butylene glycol (PBT), or cyclohexane-1,4-dimethylol (PCT, PCDT). Nylons and thermoplastic polyesters for thermoplastics have the same chemical structures as their counterparts for textile fibers albeit higher molecular weights (see also Volume II, p. 352).

Engineering plastics combine *several* improved properties such as larger tensile moduli, greater impact strengths, smaller cold flows, and the like. They are often defined as those thermoplastics that maintain their dimensional stabilities and mechanical properties both above 100°C and below 0°C.

Even better properties are exhibited by **high-performance thermoplastics** (Section 8.4) albeit often only for a single property such as heat stability or chemical inertness. High-performance plastics comprise polymers with widely different chemical structures such as poly(1,4-phenylene sulfide) PPS, polysulfones and polyethersulfones (PSU; many different structures), polyimides (PI), polyamide-imines (PAI), polyetherimines (PEI), and polyetheretherketones (PEEK; many different compositions). Poly(tetrafluoroethylene) (Teflon; PTFE), poly(vinylidene fluoride) (PVDF), and other fluoroplastics are often considered a special class as are liquid-crystalline polymers.

Functional plastics are plastic materials that serve a single, non-mechanical purpose. An example is poly(ethylene-*co*-vinyl alcohol) with a large proportion of vinyl groups that serves as a barrier polymer in packaging. Other functional polymers are used as resists (Section 14.2.11), in optoelectronics (Section 15.3), as piezoelectric materials, etc. "Functional plastics" should not be confused with "*functionalized polymers*" which are polymers that carry functional chemical groups with a desired chemical reactivity that are either already present in the monomer or have been introduced into the polymer by post-polymerization reactions ("function polymers").

Thermosets are polymers that result from chemical crosslinking reactions of thermosetting materials during shaping whereby the materials are "set." Since shaping requires a chemical reaction, thermosetting resins are sometimes also called reaction polymers or reactive polymers. Thermosetting materials may be oligomers as in the so-called B stages of phenolic or amino resins or polymers such as unsaturated polyesters. Examples are alkyd resins and the condensation products of formaldehyde with phenol (PF), urea (UF), or melamine (MF). Epoxy resins (EP), polyurethanes (PUR), and allylic resins (various symbols starting with DA) are sometimes considered thermosets and sometimes technoplastics.

No sharp dividing lines exist between the four main types of thermoplastics: standard plastics, engineering plastics, high-performance plastics, and functional plastics. A certain type of plastic can sometimes belong to one group or to another, depending on the particular property one is interested in. It must also be emphasized that "plastic" and "polymer" are not synonyms because the former usually contain adjuvants that may considerably change the properties of a polymer. Hence, there are no generally accepted criteria for the categorization of a plastic according to application.

An example is the common categorization of plastics according to their tensile moduli and tensile strengths (Fig. 8-2). Poly(ethylene)s and poly(1-olefin)s clearly belong to the group of standard plastics but no clear dividing lines exist between standard and engineering plastics on one hand and engineering plastics and high-performance plastics on the other. Liquid-crystalline polymers are a distinct class by themselves.

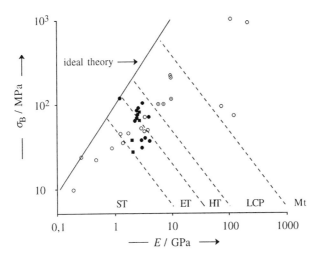

Fig. 8-2 Tensile strengths at break σ_B of plastics (for metals: upper yield strength σ_Y) as a function of tensile moduli E of metals and unfilled, unplasticized plastics. o Crystalline materials, ⊘ glasses of liquid-crystalline polymers, ● amorphous, ■ crosslinked. The solid line gives the ideal theoretical function $\sigma_B = 0.095\,E$ (Volume III, pp. 574, 627); broken lines indicate empirical ranges for standard polymers (ST), engineering thermoplastics (ET), high-performance thermoplastics (HT), liquid-crystalline polymers polymere (LCP), and metals (Mt). Thermosets (■) behave similarly to engineering thermoplastics or high-performance plastics as far as tensile strengths and moduli are concerned.

Even liquid-crystalline polymers cannot compete with metals as far as tensile moduli and tensile strengths are concerned (Fig. 8-2). However, such a comparison is not relevant. Although all materials are sold by weight, all of them are used according to their volume requirements as far as their mechanical properties are concerned. One therefore has to compare not tensile moduli and tensile strengths *per se* but the corresponding "specific" properties, i.e., the tensile modulus per density and the tensile strength per density.

For example, steel has a tensile modulus of E = 210 GPa and the ultradrawn high-molecular weight, high-density poly(ethylene) Spectra 1000® one of "only" 172 GPa (for comparison: ca. 0.15 GPa for low-density poly(ethylene) and ca. 1.5 GPa for high-density poly(ethylene)). However, steel has a density of ρ = 7.75 g cm^{-3} whereas Spectra 1000® has one of only ca. 1 g cm^{-3}. This particular poly(ethylene) therefore has a much greater "specific" tensile modulus E/ρ than steel, 177 *versus* 27 GPa cm^3 g^{-1}.

With few exceptions, plastics are not polymers *per se*, i.e., chemical substances, but *systems* of polymers and adjuvants (Chapter 3). Very many different types of adjuvants are used for both processing and use (Table 8-2) although not necessarily all of them in a specific polymer grade (see below). It is estimated that, on average, 50 % of the cost of a plastic comes from the polymer and another 50 % from adjuvants.

Properties of plastics depend not only on the chemical and physical structure of polymers and the proportion and properties of adjuvants but also on the processing and testing conditions. It is therefore important that testing conditions are standardized (see next Section). All of these effects can be summarized in the **4P rule**:

(8-1)   Property = **p**olymer + **p**olymer adjuvants + **p**rocessing

Table 8-2 Adjuvants (exclusive of fillers and dyestuffs) used in some important plastics (though not necessarily all of them simultaneously). MM = PMMA = poly(methyl methacrylate). Polymer blends also usually contain compatibilizers (p. 44 and Section 10.4.6).

Type of adjuvant	PE	PP	PS	ABS	PVC	MM	PET	PC	PA	POM	UP	EP	PUR
Processing aids													
Lubricants	+	+						+					
Parting agents		+				+	+					+	+
Plasticating agents					+								
Slip/antiblocking agents	+	+			+	+							
Nucleation agents	+	+											
Antifoaming agents											+	+	+
Antishrinking agents											+		
Use additives													
Stabilizers													
Antioxidants	+	+	+	+	+	+			+	+	+	+	+
UV absorbers	+	+	+	+	+	+	+	+	+	+	+	+	+
Heat stabilizers	+	+	+	+	+	+			+	+	+	+	+
Flame retardants	+	+	+	+	+	+	+	+	+	+	+	+	+
Biocides	+				+								+
Modifiers of bulk properties													
Fibers											+		
Coupling agents	+	+	+	+	+	+	+	+	+	+	+	+	+
Wetting agents	+	+						+				+	+
Plasticizers	+	+	+	+	+	+			+	+	+	+	+
Thickeners											+		
Impact modifiers	+	+			+			+	+		+	+	+
Blowing agents	+	+	+	+	+	+	+	+	+	+	+		
Crosslinking agents	+										+	+	
Modifiers of surface properties													
Antistatic agents	+	+	+	+	+	+	+	+	+	+	+	+	+
Antifogging agents	+	+											
Antiblocking (slip) agents	+		+	+									

The four classes of plastics are usually further subdivided according to their specific kind of processing or their special application. Each type of *processing* requires specific rheological and/or chemical properties of polymers that depend on molecular and super-molecular properties such as constitution, molecular weight, molecular weight distribution, configuration, chain flexibility, entanglement, etc. One thus distinguishes between plastics for extrusion, injection molding, blow forming, and the like.

Plastics are furthermore subdivided according to their special *applications*: for coatings, packaging, insulating materials, tubes and pipes, household articles, floor coverings, insulating materials, thickeners, etc.

In general, one specific type of polymer may satisfy several different requirements, depending on its **grade** which is usually characterized by the manufacturer's trade name for this line of products and letters or numbers or combinations thereof for the specific grade. There are ca. 25 000 trade names for plastics, elastomers, and fibers. Plastics are offered in ca. 500 types with a total of ca. 13 000 grades. Some of these grades carry trade names that differ from those of other grades of the same basic type of polymer. Conversely, the same trade name is sometimes used for very different polymers. An example is Bexloy® of the DuPont de Nemours company which is used for an amorphous polyamide (Bexloy C), a high-compact polyester (Bexloy J), a thermoplastic elastomer (Bexloy V), and an ionomer (Bexloy W).

The polymers for these plastics are usually neither identified by their Registry numbers (CAS numbers) nor by their IUPAC names or their trivial names but by abbreviations or acronyms. For example, the polymer with the chemical constitution (structure) $-OCH_2CH_2O-CO(1,4-C_6H_4)CO-$ of its repeating unit has the IUPAC name poly(oxyethyleneoxyterephthaloyl) and is generally called poly(ethylene terephthalate) in polymer science (usually written unsystematically as polyethylene terephthalate). In the plastics industry, it is known as *the* "thermoplastic polyester" (sometimes as "saturated polyester") in order to distinguish it from "unsaturated polyester" with the repeating unit $-OCH_2CH_2O-COCH=CHCO-$. Poly(ethylene terephthalate) is known by four acronyms: PET or PETP for plastics, PES for fibers, and PETE (or "1") for PET (PETP).

ISO Classification of Thermoplastics

Properties of individual plastics and/or their base polymers can be obtained from company literature. Collections of plastics/polymer properties in books, encyclopedias, and the like usually contain "averaged" data that are based on published properties which are obtained using different testing conditions. These conditions not only vary from country to country but are sometimes specific for a manufacturing company and sometimes not disclosed at all (see also Chapter 5).

For these reasons, ISO introduced an international classification systems for thermoplastics. In this system, a thermoplastic grade is characterized by a number of qualitative or quantitative data blocks similar to the international standards for rubber tires (p. 276). The system should allow processors to switch between grades of various plastics manufacturers and compounders, respectively. A typical code consists of a general description, the testing standard, and up to five data blocks, for example:

Molding material DIN 7744-PC, XF, 55-045, GF 30
(for an explanation of the numbers and letters, see next page)

The first code indicates the intended application, here a molding material. It is followed by the standard which is used for testing, here the German standard DIN 7744.

The subsequent data blocks are separated from the first block by a hyphen and from each other by commas. The *first data block* indicates that the molding material is based on a polycarbonate (PC). In the example, the type of "PC" is not specified further (for example, based on bisphenol A, bisphenol I, etc. (Volume II, p. 343)). For other polymers, the first data block may contain other informations about the base polymer such as EVAC-30 for a copolymer of ethylene with 30 % vinyl acetate, PVC-P for a plasticized poly(vinyl chloride), PS-E for a poly(styrene) from emulsion polymerization, etc.

The *second data block* with up to four letters delivers qualitative informations. The first letter indicates the intended application, for example, B = blow molding, G = general purpose, P = paste, and Y = textile yarn. An "X" (as in the example) indicates that the application is not specified.

The second, third, and fourth letters of this data block refer to important adjuvants and/or additional information. Examples are: A = grade contains antioxidants, D = grade is a dry blend, L = material contains a light stabilizer, and S = slip additive (= lubricant). The "F" in the example indicates the presence of a flame-retarding agent.

The *third data block* provides quantitative information about up to three indicators (Table 8-3). These data are not comprehensive; for example, they are not identical with some of the 50 standards of the CAMPUS® system (p. 133). They also vary from polymer to polymer, depending on the chemical structure of the polymer and physical properties that are especially important for that particular polymer. Indicators are used for the following properties:

chemical structure:	content of monomeric units such as vinyl acetate units (VAC) or acrylonitrile units (AN),
	degre of isotacticity (IT) (p. 14),
	density (D) as a measure of the degree of branching, etc.;
molecular weight:	intrinsic viscosity (IV),
	Fikentscher K value (Volume II, p. 14), etc.;
bulk density:	BD (p. 77);
rheological data:	melt volume index (MVI) (p. 86), etc.;
thermal data:	Vicat temperature (VT) (p. 135),
	torsional stiffness temperature (TST), etc.;
mechanical data:	tensile modulus = modulus of elasticity (E) (p. 138),
	tensile strength at 100 % elongation (TS) (p. 128 ff.),
	Shore hardness (SH) (p. 151),
	notched impact strength (ISN) (p. 145), etc.

Some of these indicators do not give the actual values but only the range in which the actual value can be found. The ranges differ from indicator to indicator; they are also coded. The codes can be found in published tables.

For example, polycarbonates should be characterized by three indicators (IV, MVI, E) but only two of these were used in the example (IV, MVI). The actual intrinsic viscosity of the specimen is $[\eta] = 56$ mL/g which places it into the range 50-60 mL/g. This range is coded as "55."

Similarly, the melt volume index of the specimen is MVI = 5.5 g/(10 min) which places it into a cell range that is coded as "045."

Table 8-3 Indicators used for specific polymers in the third ISO data block. * Thermoplastic copolymer with ≤ 25 wt% butadiene (*not* SBR).

Symbol	Polymer or copolymer	Indicators					
PE	poly(ethylene)	D		MVI			
PP	poly(propylene)	IT		MVI			
PS	poly(styrene)			MVI	VT		
SAN	poly(styrene-*co*-acrylonitrile)			MVI	VT		
SB	poly(styrene-*co*-butadiene) *			MVI	VT		ISN
ABS	poly(acrylonitrile-*co*–butadiene-*co*-styrene)			MVI	VT		ISN
ASA	poly(acrylonitrile-*co*-styrene-*co*-acrylic ester)			MVI	VT		ISN
EVAC	poly(ethylene-*co*-vinyl acetate)	VAC		MVI			
PVC-U	poly(vinyl chloride), unplasticized	VT	FK			E	ISN
PVC-P	poly(vinyl chloride), plasticized				TST	E, TS, SH	
PA	polyamide, aliphatic (= nylon)	IV				E	
PC	polycarbonate	IV	MVI			E	
PMMA	poly(methyl methacrylate)	IV			VT		
PET	poly(ethylene terephthalate)	IV					

The *fourth data block* informs one about the type and proportion of fillers and reinforcing agents. The first letter indicates the type, for example, C = carbon or G = glass, and the second letter the shape, for example, S = spherical or F = fibrous. The letters are followed by coded numbers that give the range of content of the filler/reinforcing agent in weight percent, for example, "15" for the range 12.5-17.5 wt%. The PC of the above example thus contains ca. 30 wt% glass fibers.

The *fifth data block* is reserved for individual data that are exchanged between sellers and buyers, for example, additional standards, restrictions, and the like.

A similar system is used by ASTM. An example is

Molding Material ASTM D-4000 PI000G42360

which is a molding material from a polyimide (PI). Since no individual property tables had been specified for polyimides ("000"), cell table G of standard ASTM D-4000 was used. This cell table specifies five indicators: "4" stands for the value of tensile strength (actual: 85 MPa), "2" for the value of flexural modulus (actual: 3500 MPa), "3" for the value of Izod impact strength (actual: 50 J/m^2), and "6" for the value of the heat distortion temperature (actual: 300°C). The fifth indicator was not specified ("0").

8.1.4 Prices and Materials Competition

Plastics have properties such as light weight and extensibility that sets them apart from traditional materials such as wood, glass, metals, and concrete. They do compete with these materials for other properties such as stiffness. These properties are markedly improved if polymers are filled with particulates or are reinforced with fibers (Chapter 9).

But with the exception of a few properties, materials properties *per se* are not the determining factor for engineering applications. For example, a desired stiffness of the walls of an article may be achieved by using a material with a sufficiently large tensile modulus but it may also be obtained from a material with a smaller tensile modulus by

Table 8-4 Average price of prime grades of naturally colored, unfilled thermosetting raw materials and thermoplastics (TP) [2]. E = extrusion grade, GP = general purpose, I = injection molding. * 1.4 t/a.

Price in US dollars per kilogram for indicated annual purchases in tons/year (t/a) on 2008-03-27			
Thermosetting 91-230 t/a	Standard TP 910-2300 t/a	Engineering TP 140-230 t/a	High-performance TP 91 t/a
UF 1.60	PVC 1.74	PET 2.05	PSU 8.62
PF (GP) 1.65	HDPE (I) 1.94	PBT 2.09	PTFE 15.78
MF 2.09	ABS (I) 2.11	POM 2.57	PVDF 16.20
EP (GP) 2.20	PS 2.12	SAN 2.68	PEI 17.60
PUR 2.49	PP (E) 2.13	PA 6 3.76	LCP 23.38
UP (GP) 4.07	LDPE (I) 2.22	PA 6.6 3.89	ECTFE 29.61
VE 4.71	PMMA 2.51	PC (I) 3.89	PEEK 99*

making the walls thicker or by reinforcing the walls with ribs. The designer thus has to balance material properties and costs of materials and processing.

Table 8-4 compares the prices of some important polymers. The lowest prices are asked for the oldest plastics materials, i.e., the raw materials for the thermosetting materials from formaldehyde with phenol (PF), melamine (MF), or urea (UF). Later developed thermosetting materials are offered at higher prices: epoxy resins (EP), raw materials (polyols, diisocyanates) for polyurethanes (PUR), unsaturated polyesters (UP), and so-called vinyl (ester) resins (VE), the latter being macromonomers from the reaction of an epoxy resin with (meth)acrylic acid (see Volume II, p. 291). The prices of the latter two types (UP, VE) exceed those of engineering plastics. Prices may also vary markedly with the grade. The price asked for a general purpose EP was 2.20 $/kg but that of a semiconductor grade, 6.53 $/kg.

With respect to price, standard thermoplastics compete directly with thermosetting materials (Table 8-4) because they are bulk polymers based on oil and natural gas: poly(vinyl chloride) (PVC), poly(ethylene)s (high and low density: HDPE and LDPE), isotactic poly(propylene) (PP), poly(methyl methacrylate) (PMMA), atactic poly(styrene) (PS), and the copolymer/blend from acrylonitrile, butadiene, and styrene (ABS).

Even some engineering plastics are in this price range which may be one of the reasons why the American Plastics Council counts thermoplastic polyesters (PET, PBT) as standard thermoplastics (Fig. 8-1); the other reasons may be the high production volume and the predominant use of PET for plastics bottles. Other polymers in this group comprise the two poly(oxymethylene) (POM), styrene-acrylonitrile copolymers (SAN), the very many types of aliphatic polyamides (nylons) (PA 6, 6.6, etc.), and many types of polycarbonates (PC) with bisphenol A polycarbonate as the main product.

Much greater prices are commanded by high-performance plastics such as the many types of polysulfones (PSU), the various fluorinated polymers [poly(tetrafluoroethylene) (PTFE), poly(vinylidene fluoride) (PVDF), poly(ethylene-*co*-chlorotrifluoroethylene) (ECTFE)], polyetherimide (PEI), liquid-crystalline polymer (LCP) (probably the Vectra® type), and polyetheretherketone (PEEK).

These materials costs have to be related to the costs of formulation and processing which vary widely and are not easy to access. However, one can compare some mechanical properties with cost per weight and especially with cost per volume (Table 8-5) since materials in articles are utilized per volume and not per weight.

Table 8-5 Average densities ρ, tensile moduli E, upper yield strengths σ_Y, heat distortion temperatures (type A) T_{HDT}, and prices per mass (P_M) [2, 3], as well as prices per volume (P_V) and volume
prices per modulus (P_V/E) or yield strength (P_V/σ_Y) of some plastics, metals, and other materials.

Material	$\dfrac{\rho}{g\ cm^{-3}}$	$\dfrac{E}{GPa}$	$\dfrac{\sigma_Y}{MPa}$	$\dfrac{T_{HDT}}{°C}$	$\dfrac{P_M}{\$\ kg^{-1}}$	$\dfrac{P_V}{\$\ L^{-1}}$	$\dfrac{P_V/(\$/L)}{E\ /\ GPa}$	$\dfrac{P_V/(\$/L)}{\sigma_Y\ /\ GPa}$
Standard thermoplastics								
Poly(ethylene), HD	0.96	1.1	32	49	1.94	1.85	1.68	57.8
Poly(propylene), it	0.90	1.2	36	56	2.13	1.91	1.59	53.1
Poly(styrene), at	1.05	3.2	46	73	2.12	2.22	0.69	48.3
Poly(vinyl chloride)	1.35	2.9	60	68	1.74	3.18	1.09	53.0
Engineering thermoplastics								
ABS	1.07	2.9	56	93	2.11	2.26	0.78	40.4
PET	1.21	2.9	81	85	2.05	2.48	0.86	30.6
Poly(oxymethylene)	1.42	3.7	72	124	2.57	3.65	0.99	50.7
Polyamide 6.6	1.10	2.9	65	75	3.89	4.27	1.47	65.7
Polycarbonate A	1.20	2.9	79	135	3.89	4.66	1.61	59.0
High-performance thermoplastics								
Polysulfone	1.24	26	73	203	8.62	10.69	0.41	14.6
Polyimide	1.40	5.0	72	243	12.37	17.32	3.46	24.1
LCP	1.40	9.6	144	180	23.35	32.69	3.40	24.8
Thermosetting resins								
Phenol-formaldehyde	1.36	8.6	50	121	1.65	2.25	0.26	45.0
Urea-formaldehyde	1.56	10.0	43		1.60	2.50	0.25	58.1
Epoxy resin	1.20	3.6	72	110	2.20	2.64	0.73	36.7
Metals								
Steel	7.86	200	430	-	0.30	2.36	0.012	5.49
Aluminum, primary	2.67	73	93	-	1.76	4.17	0.057	44.8
Copper	8.77	122	72	-	3.33	29.21	0.024	406
Miscellaneous								
Wood (oak)	0.5	11	70	-		1.68	0.153	24
Cement	2.3	30	5.5	-	0.078	0.18	0.006	32.7
Glass (E glass)	2.54	72	125	-				

As a simple example, assume that the material in a part has a volume of 1 cm^3 and that the
material needs to have an upper yield strength of 72 MPa. This strength can be obtained from copper
with a density of ρ = 8.77 g/cm^3 or from POM with ρ = 1.2 g/cm^3. Based on the prices P_M in Table
8-5, the cost of 1 cm^3 would be 29.2 cts for copper but only 3.1 cts for POM.

Tensile moduli of standard and engineering plastics are smaller than those of traditional engineering materials such as wood, concrete, or metals (Table 8-5). These plastics
can therefore not be used for load-bearing properties. On the other hand, the upper
yield strength of plastics is substantially greater than that of unreinforced concrete and
about the same as that of wood, aluminum, and copper but lower than that of steel.

The load-bearing properties of thermoplastics can be markedly improved by reinforcing them with fillers (Chapter 10), especially with fibers (Chapter 9). Examples are
glass fiber-reinforced unsaturated polyesters for boat hulls, even for smaller war ships
(replacement of steel) and carbon fiber-reinforced resins for aeroplane sections (replacement of aluminum).

8.1.5   Economic Importance

Because of low cost and some special properties, the production of thermosetting materials is still increasing, both in the world (Table 8-6) and in the United States (Fig. 8-1). However, their growth rate and their proportion of annual plastics production are much smaller than those of standard thermoplastics which are based on less expensive raw materials, can be produced in large amounts per time unit (continuous versus discontinuous), and can be processed faster in greater volumes.

The largest producing regions of plastics in 2005 were Asia (36.5 %), Western Europe (25 %), and North America (24 %) (Table 8-6). With respect to individual countries, China is probably the leader (percentage unknown), followed by the United States (percentage unknown), Germany (8 %), and Japan (6.5 %). Benelux is also a large producer (5 %); the plastics production of South Korea is also probably large.

Table 8-6   World production and consumption per capita of plastics raw materials (= polymers for thermoplastics; monomers for thermosetting materials) [4]. For 2000 ff., world regions are defined as follows: North America = NAFTA (United States (US), Canada (CA), Mexico (MX)); Western Europe = European Union (25 countries) plus Norway and Switzerland; Asia = all countries from Pakistan to Japan, including Indonesia, Taiwan, South Korea, and China; Eastern Europe = all other European countries including Russia and other former members of the Soviet Union; Central and South America = all countries of the Western Hemisphere except North America; Africa and Middle East = all African countries plus countries from Egypt to Iran; Oceania = Australia, New Zealand, Philippines and other Pacific countries except Indonesia. Reference [1] reported for North America (US+CA+MX (part)) only $49.8 \cdot 10^6$ t/a (2005) instead of $55.2 \cdot 10^6$ t/a.

Region	Production in million tons per year					
	1920	1940	1960	1980	2000	2005
North America		0.1	3.0	17.4	45.4	55.2
Western Europe			2.3	19.7	46.2	57.5
Asia			0.74	10.1		84.0
Eastern Europe			0.54	6.9		8.0
Central and South America			0.07	2.4		10.4
Africa and Middle East			0.05	1.7		13.8
Oceania			~ 0	0.8		1.1
Total World	*0.052*	*0.36*	*6.65*	*59.0*	*172*	*230*
wherefrom thermosets		0.25	2.53		9.0	
wherefrom thermoplastics		0.11	4.12		163	
World population (billions)	1.81	2.25	3.04	4.43	6.08	6.45
Consumption in kg per capita	0.03	0.16	2.2	13.3	28.3	35.7

In 2005, the average world plastics consumption climbed to 35.7 kg per capita and year. But annual consumptions per capita varied widely: those of Germany (128 kg), United States (104 kg), and Japan (81 kg) contrast with those of sub-Saharan Africa (1.2 kg), Bolivia (1.1 kg), and India (0.4 kg).

The ca. 16-fold increase in plastics consumption between 1960 and 2005 is caused less by population growth ("only" 2.1-fold) but mainly by increased living standards combined with a throw-away mentality. The replacement of older materials by plastics seems to be far less important, as one can see from the following.

Until ca. 1960, most countries were practically closed economic regions. Production was approximately identical with consumption and storekeeping was usually negligible. However, the last decades saw the establishment of a global market so that the consumption of *unprocessed materials* is given by production + import − export. The true consumption of *processed* materials is unknown since no statistics exists for the material content of imported and exported goods.

An example is the United States where lumber is imported only a little and exported practically not at all. The production and consumption of lumber therefore follows the population growth (Fig. 1-2). Raw steel is imported in about the same amount as it is exported as scrap metal, and again, the consumption of raw steel follows approximately the population growth. Hence, neither lumber nor steel seem to have been replaced in sizable proportion by plastics.

Quite different is the US situation for aluminum, which lost some ground to plastics in the packaging sector and saw a slight decrease in production (Fig. 1-2). Zinc is different: in 1986, the United States produced 203 000 t but imported 664 000 t! It can be concluded therefore that plastics did not replace many other materials but rather opened opportunities for hitherto impossible applications.

Comprehensive statistics of *world productions* of individual *synthetic polymers* do not seem to exist. Scattered older data for 2000 (Table 8-7) indicate a total world production of 220 million tons which includes plastic working materials (72.5 %), fibers (13.2 %), coatings (9.1 %), and elastomers (5.2 %).

Table 8-7 Annual world production (in million tons) of synthetic polymers in 2000. "Plastics" includes working materials, packaging materials, expanded polymers, etc. "Coatings" also comprises adhesives and "rubbers" include dispersions and latices (see Table 7-2).

Raw materials	Plastics	Fibers	Elastomers	Coatings
Standard polymer				
Poly(ethylene)	52			
Poly(propylene)	29	5		
Poly(vinyl chloride), hard and plasticized	26			
Poly(styrene), included expanded poly(styrene)	13			
Technical polymers				
Saturated Polyesters	10.5	20		
Polyamides	2	4		
Polycarbonates	1.7			
Poly(methyl methacrylate)	1.8			
Polyurethanes	8.4		0.6	2
Other, including blends	5.5			
High-performance polymers				
Fluorinated polymers	0.08			
Sulfur-containing polymers	0.06			
Other	0.06			
Unspecified thermoplastics	2			18
Thermosets	7.7			2.3
Rubbers			10.9	
Total	*159.8*	*29*	*11.5*	*20*

Table 8-8   Percentage of production of raw materials for important plastics in North America (NAFTA = Canada, Mexico, United States) in 2006 [1] and in Japan in 2007 [5]. [1] Includes LLDPE, [2] oPS = other styrene-based polymers including ABS.

Region	LDPE [1]	PP	HDPE	PVC	PET	PS	PF	UF/MF	oPS [2]	PA	Other
North America	18.4	16.1	15.5	13.1	7.3	5.5	4.2	3.0	2.7	1.1	13.1
Japan	13.2	21.8	8.0	15.2	4.9	7.6	2.1	1.8	4.7	1.9	18.8

In North America and in Japan, the largest group of all produced plastics raw materials is poly(olefin)s (PP, HDPE, LDPE, LLDPE): 50.0 % in North America and 43 % in Japan (Table 8-8). The next big group comprises poly(vinyl chloride) (13.1 % *versus* 15.2 %) and styrene-based polymers (PS and other) (8.2 % *versus* 12.3 %).

However, polymers with the same basic monomeric unit such as $-CH_2-CH_2-$ for the poly(ethylene)s are not necessarily homopolymers as the next Sections show. For example, so-called linear low-density poly(ethylene)s always contain a few percentages of 1-olefin units. The various representatives of a basic chemical type may also differ in the degree of branching (if any), the molecular weight, and the molecular weight distribution. Some grades of a certain polymer type may also be polymer blends although that is rarely mentioned in company literature.

Polymer producers thus offer very many grades of a basic type, for example, for aliphatic polyamides (nylons) 93 grades by LNP Corp., 62 grades by Rilsan Corp., 43 grades by the Fiberfil Division of Dart Industries, and 35 grades by Emser Werke, Switzerland. Polycarbonates are sold in 24 grades each by Mobay and Teijin, etc.

In Western nations in 2000, ca. 33 % of all plastics were used on average for packaging (containers, bottles, sheets, etc.), ca. 22 % for building and construction, ca. 8 % for short-lived household goods and furniture, 7 % for electrical/electronic purposes, 7 % for transportation, and ca. 23 % for all other purposes (Table 8-9). These percentages do not differ much for highly industrialized countries. Polymers for packaging comprise mainly poly(1-olefin)s (> 80 %) and PET (ca. 10 %) whereas PVC dominates for building and construction (ca. 50 %). Because of their low density, engineering plastics are gaining ground in transportation: in 2003, no less than 16 000 new polymer grades were introduced to the automotive industry.

Table 8-9   Relative consumption of plastics per sector in Europe and in selected countries (United Kingdom (UK), United States (US), and Germany).

Type of consumption Year →	Europe 1990	Europe 2000	UK 2000	US 2000	Germany 2003
Packaging	23.3	37.3	36.0	28.0	33.0
Building/construction	27.8	18.9	21.7	24.0	23.5
Household/furniture	11.1	21.3	7.1		9.5
Electrical/electronic	16.7	7.3	8.1	4.5	7.5
Transportation	7.8	7.2	7.3	5.5	9.0
Agriculture		2.6	7.0		2.5
Medical			1.9		1.0
Other or not specified	13.3	5.4	10.9	38.0	14.0

8.2   Standard Thermoplastics

All manufacturing processes for bulk products require inexpensive raw materials and low-cost energy. Materials and energy are both provided for standard thermoplastics by crude oil and natural gas (Volume II, p. 45 ff.). The cracking of crude oil furthermore directly delivers polymerizable monomers such as ethylene and 1-olefins. Standard thermoplastics are therefore usually based on monomers of the type $CH_2=CHR$ where R is H, CH_3, etc.

The direct production of these monomers from the primary raw materials crude oil and natural gas fosters vertical integration. Monomers are produced in very large plants with capacities of often more than 100 000 tons per year, which need a considerable capital outlay for their construction (Table 8-10). As a result, research is mainly concerned with the development of new processes and less with inventing new polymers.

8.2.1   Poly(ethylene)s

Overview

The designations poly(1-olefin), **poly(α-olefin)**, or **olefin polymer** usually refer to polymers of ethylene (IUPAC: ethene), $CH_2=CH_2$, and/or aliphatic 1-alkyl ethylenes, $CH_2=CHR$, with $R = C_iH_{2i+1}$. Despite the name, poly(1-olefin)s do not contain unsaturated olefin groups. These polymers are rather polyalkanes with the idealized constitution $+CH_2-CHR+_n$ where R is a linear or branched alkane group. However, poly(1-olefin) sometimes also refers to polymers from *all* 1-substituted ethenes.

Poly(ethylene)s are produced in the greatest volume of all synthetic polymers because the monomer can be obtained from crude oil or natural gas and, to a lesser extent, also from ethanol (Volume II, p. 99). The "poly(ethylene)s" of industry are not macromolecular substances with the ideal constitution $+CH_2-CH_2+_n$ but rather homopolymers of ethene or copolymers of ethene and small proportions of 1-alkyl ethylenes, $CH_2=CHR$, with $R = CH_3$, C_4H_9, C_6H_{13}, and/or C_8H_{17}. All industrial ethylene homopolymers are more or less branched. The practically unbranched poly(methylene), $+CH_2+_n$, from the polymerization of diazomethane, CH_2N_2, has the same constitution as the truly linear poly(ethylene), $+CH_2-CH_2+_n$, but is not an industrial product.

Table 8-10 Economical characteristics for thermosets and standard, engineering, and high-performance plastics. *) Based on Tables 8-4 and 8-5.

	Thermosets	Standard thermo-plastics	Engineering plastics	High performance plastics
Research and development costs	low	fairly low	medium	high
Vertical integration with raw materials	little	large	medium	little
Investment cost of plants	low	high	medium	low
Emphasis during development	process	process	process/product	product
Technical know-how of customers	low	low	large	medium
Price in US$ per kg (April 2008) *)	1.6-2.3	1.8-2.0	2.5-5.5	15-100

Table 8-11 Classification and syntheses of poly(ethylene)s.

Designation			Density	High-pressure syntheses		Low-pressure syntheses		
ASTM	ISO	ASTM	in g/cm^3	Tube	Tank	Gas	Solution	Slurry
IV			> 0.96					
III	PE-HD	HDPE	0.94 – 0.96	-	-	+	+	+
II	-	LLDPE	0.925 – 0.94	-	-	+	+	+
I	PE-LD	LDPE	0.90 – 0.925	+	+	-	-	-
-	-	mLLDPE	0.886 – 0.935	-	-	+	+	+
-	-	VLDPE	0.863 – 0.885	-	-	+	+	+

The various industrial poly(ethylene)s differ in their chemical structure, especially in the presence or absence of comonomeric units, the type and extent of branching, and hence the degree of crystallinity (Table 8-11). Poly(ethylene)s are usually categorized according to their density as high density (HDPE, PE-HD), medium density (MDPE), linear low density (LLDPE, PE-LLD), low density (LDPE, PE-LD), and very low density poly(ethylene)s (VLDPE). Correlations between densities and designations vary somewhat according to source and regulating body; for example, densities of MDPE are also listed as 0.926-0.940 g/cm^3, those of LLDPE as 0.915-0.925 g/cm^3, and those of VLDPE as less than 0.915 g/cm^3 (see also Table 8-12, below). Linear low-density poly-(ethylene)s from the polymerization of ethylene with metallocene catalysts are often considered a separate class of these polymers (metallocene-poly(ethylene), mLLDPE).

The greater the density, the larger in general the degree of crystallinity; density variations caused by comonomeric units (if any) play a minor role. The degree of crystallinity, in turn, is predominantly controlled by the type and degree of branching. These branches are caused either by branching reactions in the polymerization process as in LDPEs or by smaller or larger proportions of comonomeric units as in MDPEs, LLDPEs, and VLDPEs. Industrially used comonomers comprise 1-butene, 1-hexene, 1-octene, and 4-methyl-1-pentene. They are used in proportions of 1-2 mol% for MDPEs, 2.5-3.5 mol% for LLDPEs, and > 4 mol% and up to 24 mol% for VLDPEs.

Polymer constitutions and molar mass distributions are furthermore affected by the type of polymerization process (gas phase, solution, suspension) and the type of reactor (see also Volume II, p. 178 ff.). The resulting polymer constitutions are schematically shown in Fig. 8-3.

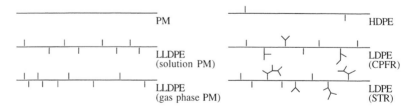

Fig. 8-3 Schematic representation of the constitution of various poly(ethylene)s where I represents short alkyl groups and $\curlywedge$, $\curlyvee$, etc., branched alkyl groups. PM = poly(methylene) from the polymerization of diazomethane, HDPE = high-density poly(ethylene), LDPE from the polymerization in continuous plug-flow reactors (CPFR) or stirred tank reactors (STR), LLDPE linear low-density poly(ethylene) from the polymerization (PM) in solution or in the gas phase. The latter polymers are called "linear" because their branched structure is *not* caused by a branching reaction of ethylene itself.

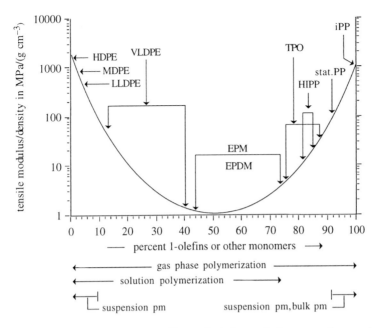

Fig. 8-4 Schematic represenation of the stiffness of ethylene, ethylene–propylene, and propylene based polymers [6]. Stiffness is defined here as the ratio of tensile modulus to density. HDPE is usually a homopolymer. MDPE, LLDPE, and VLDPE are copolymers with increasing proportions of higher 1-olefins (see text), EPM is an ethylene-propylene copolymer and EPDM an ethylene-propylene-diene monomer copolymer (see p. 278). Thermoplastic-elastomeric poly(olefin)s (TPO) are usually blends containing ethylene-propylene copolymers (see p. 291). High-impact poly(propylene)s (HIPP) are blends of isotactic poly(propylene)s (iPP) with EPDM.

The greater the degree of intrinsic or extrinsic branching, the lower the density and also the degree of crystallinity. This, in turn, lowers the stiffness of the polymers (Fig. 8-4): HDPE, MDPE and LLDPE are thermoplastics whereas VLDPE , EPM, and EPDM are elastomers and TPOs (= TPE-Os) are thermoplastic elastomers. With increasing content of 1-olefins, the stiffness of ethylene-*co*-1-olefin polymers thus passes through a minimum before it increases again with higher contents of 1-olefin units.

Ethylene Polymers
 Low-density poly(ethylene)s (LDPE (ASTM), PE-LD (ISO)) and **medium-density poly(ethylene)s (MDPE, PE-MD)** are obtained by the free-radical polymerization of ethylene at pressures of up to 280 MPa (Volume II, p. 218 ff.). The reaction conditions cause chain transfer reactions that lead to short alkyl branches $-(CH_2)_jH$ with $j = 2$, 4-6, a few unsaturated sidegroups such as $-CH=CH_2$, and some long chain branching (see also Fig. 8-3). The extent of branching is controlled by the type of initiator (oxygen, peroxides), polymerization medium (solution, gas phase), and the distribution of residence times in the reactors. The branching leads to less crystallization and thus to lower polymer densities. Molecular weight distributions are very broad for polymers from polymerizations in continuous stirred tank reactors ($\overline{M}_w/\overline{M}_n = 12$-16) but narrower from those in continuous plug flow reactors (6-8 from multiple feeds and 3-4 from single feeds).

High-density poly(ethylene)s (HDPE (ASTM), **PE-HD** (ISO) and also some **low-density** (LDPE) and **very low-density poly(ethylene)s (VLDPE)** are polymers from the polymerization of ethylene by transition metal catalysts at low pressures (Volume II, p. 222 ff.). HDPEs are far less branched than LDPEs which leads to higher densities and greater extents of crystallinity (see Table 8-12).

Linear low-density poly(ethylene)s (LLDPE (ASTM), PE-LLD (ISO)) are copoly-mers of ethene with 5-12 % 1-butene, 1-hexene, or 1-octene. Copolymers with 10-20 % 1-octene are marketed as "linear poly(ethylene)s" because their branches (in the topo-logical sense) do not result from branching reactions of growing poly(ethylene) chains as in the free-radical polymerization of ethylene but rather are inherent to the structure of their comonomers.

Metallocene-poly(ethylene)s (mLLDPE, MPE-LLD) result from the polymerization of ethene with special transition metal catalysts that are activated by methylaluminoxanes (Volume I, p. 289 ff.). The transition metal catalysts are metallocenes, examples of which are Cp_2ZrCl_2 and $iPr[1\text{-}Flu;Cp]ZrCl_2$ where Cp = cyclopentadienyl, Flu = fluor-enyl, and iPr = isopropyl. These polymers are also called **single-site catalyzed poly-(ethylene)s, poly(olefin) plastomers,** or **homogeneous ethylene copolymers,** depending on their structure, properties, and the marketing prowess of the manufacturing company.

mLLDPEs have relatively narrow molecular weight distributions and are therefore more difficult to process than poly(ethylene)s with broad distributions where the low-molecular weight fractions act as a plasticizer and a lubricant. However, their low density has opened up new applications for poly(ethylene)s.

Table 8-12 Designations and properties of some unfilled injection-molding grades of poly(ethylene)s of the Neste company. Data at 23°C unless noted otherwise. nf = no fracture. All grades have: relative permittivity $\varepsilon_r = 2.3$, resistivity $\rho = 10^{15}$ Ω cm, electrical resistance $R = 10^{14}$ Ω, water absorption w = 0.01 % (saturation), moisture uptake w = 0.01 % (50 % relative humidity).

Physical property		Physical unit	LD NCPE 1515	LLD NCPE 8030	LLD NCPE 8644	LD NCPE 3416	HD NCPE 7003	HD NCPE 7007
Density		g/cm³	0.915	0.919	0.935	0.958	0.958	0.964
Heat distortion temperature B		°C	41	46			84	80
Heat distortion temperature A		°C			39	58	50	54
Vicat temperature A		°C		75	86	110	130	130
Vicat temperature B		°C			48	65	75	85	80
Tensile modulus	(1 mm/min)	MPa	150	830	650	1350	1350	1600
Yield strength	(50 mm/min)	MPa	8	12	17	30	30	31
Elongation at yield	(50 mm/min)	%		12	12	8.1	8.2	7.1
Fracture elongation	(50 mm/min)		> 50	> 50	> 50	> 50	> 50	> 50
Impact strength	(23°C)	kJ/m²	nf	nf	nf	nf	nf	nf
Impact strength	(-30°C)	kJ/m²	nf	nf	nf	nf	nf	nf
Notched impact str.	(23°C)	kJ/m²	nf	nf	6.5	nf	5.2	5.6
Notched impact str.	(-30°C)	kJ/m²	nf	11	5.9	5	5.4	4.6
Melt flow volume	(190°C, 2.16 kg)	mL/10 min	18	36	9.5	0.24	3.6	8.3
Density of melt	(225°C)	g/mL		0.787	0.794		0.814	0.819
Thermal conductivity	(melt)	W/(m K)		0.287	0.303		0.326	0.353
Specific heat capacity	(melt)	J/(kg K)		3070	3150		3350	3420

Average molecular weights of the polymers of all these poly(ethylene) families are usually lower than ca. $3 \cdot 10^5$. Poly(ethylene)s with $3 \cdot 10^5 < M_r < 4 \cdot 10^5$ are known as "high molecular weight" (HMW-HDPE), those with $4 \cdot 10^5 < M_r < 1,5 \cdot 10^6$ as "very high molecular weight," and those with $M_r > 3 \cdot 10^6$ as "ultrahigh molecular weight" (**UHMWPE**).

Table 8-12 contains data for 6 grades from a total of 17 poly(ethylene)s of the Neste company. The selected grades are those with the lowest and the highest densities of each of the three major types of poly(ethylene)s (LDPE, HDPE, LLDPE). Comparison of the designations of the poly(ethylene) grades in Table 8-12 with the ASTM designations in Table 8-11 shows that designations are usually according to the polymerization process and not according to density (LDPE 3416 has the same density as HDPE 7003).

All poly(ethylene)s are semicrystalline. Their degrees of crystallinity, heat distortion and Vicat temperatures, tensile moduli, and yield strengths increase with decreasing degree of branching as measured by the density of the grades (Table 8-12) whereas impact strengths decrease. Hardnesses, opacities, and resistances against chemicals also increase whereas the resistance against stress cracking (Volume II, p. 617) decreases.

Thermodynamic melting temperatures are usually given as ca. 115°C for LDPEs and LLDPEs, ca. 135°C for HDPEs, and 144°C for ideal crystallites of unbranched poly-(ethylene)s. Glass temperatures are controversial: reported data for poly(ethylene)s range from –125°C to +2°C according to the Polymer Handbook.

Low-density poly(ethylene)s are easy to process to packaging films and refuse bags (up to 65 % of the total market), squeeze bottles, extruded pipes, toys, and various coatings. High-density poly(ethylene)s are much more rigid and are therefore used for small and very large containers (200 gallons) (up to 40 % of the market), household goods, grocery bags, wire and cable coatings, and the like.

Ethene Copolymers

Free-radical copolymerization of ethylene and *vinyl acetate* delivers copolymers (**E/VAC, EVA**) that are used as films if the content of VAC units is 1-40 % and as elastomers, hot melt adhesives, and PVC modifiers if it is more than 40 %. Ethylene-vinyl acetate copolymers are also transesterified (not saponified!) to pseudo-copolymers with ethylene and vinyl alcohol units (EVAL®, **E/VAL, EVOH**). Such polymers with high VAL contents are used as gas tight barrier polymers (Section 13.2.2) and those with low VAL contents for fluidized bed coatings (p. 107).

Free-radical copolymerization of ethylene with up to 20 % acrylic acid delivers tough packaging films. Copolymerization of ethylene with up to 15 % methacrylic acid results in copolymers that are partially neutralized with Na^+ or Zn^{2+} to so-called **ionomers** for extrusion coatings (Volume II, p. 231). Fluorinated ionomers serve as membranes.

8.2.2 Poly(1-alkylene)s

Poly(propylene)s

Tactic poly(propylene)s (**PP**), $+CH_2-CH(CH_3)+_n$, result from the polymerization of propylene with transition metal catalysts to isotactic PPs (**PP, iPP**, it-PP) or with metallocene catalysts to either isotactic PPs (**mPP**) or syndiotactic ones (**sPP**) (Volume II, p. 232). These poly(propylene)s are semicrystalline polymers of high stiffness (Fig. 8-4).

Table 8-13 Six of the 93 Vestolene® grades (= propylene polymers) of the Evonik company (former-ly: Degussa, originally Hüls AG). At 23°C, all grades have the same thermal expansion coefficient of $\alpha = 14 \cdot 10^{-5}$ K^{-1}, relative permittivity of $\varepsilon_r = 2.3$, resistivity of $\rho = 10^{15}$ Ω cm, electrical resistance of $R = 10^{13}$ Ω, dielectric loss factor of tan $\delta = 5 \cdot 10^{-4}$ (at 50 Hz), tracking resistance (600 drops (p. 588)), and water absorption $w = 0.01$ % (saturation). SCoPM = statistical copolymer, nb = no break.

Physical property		Physical unit	Homopolymers 2000	5000	7000	Block-CoPM 9500	5300	SCoPM 9421
Density		g/cm^3	0.903	0.902	0.901	0.901	0.898	0.901
Isotacticity index		-	95	95	95	95	85	85
Intrinsic viscosity		mL/g	150	220	310	450	190	450
Volume flow index (230°C, 2.16 kg)		mL/10 min	55	15	3.2	0.5	18	0.5
Heat distortion temperature A (1.82)		°C	60	55	55	55	50	45
Heat distortion temperature B (0.45)		°C	105	100	100	90	80	75
Vicat temperature (B)		°C	105	100	100	90	70	65
Yield stress	(50 mm/min)	MPa	38	37	35	32	25	26
Elongation at yield	(50 mm/min)	%	8	8	8	10	12	12
Fracture elongation	(50 mm/min)	%	25	> 50	> 50	> 50	> 50	> 50
Tensile modulus	(1 mm/min)	MPa	1800	1600	1500	1300	800	700
Creep modulus	(1 h)	MPa			1150	1200		700
Creep modulus	(1000 h)	MPa			500	400		300
Impact strength	(Izod, 23°C)	kJ/m^2	50	65	130	nb	nb	nb
Impact strength	(Izod, –30°C)	kJ/m^2	12	15	17	40	15	30
Notched impact strength	(23°C)	kJ/m^2	2	2.5	5	20	3.5	25
Notched impact strength	(–30°C)	kJ/m^2	1.5	1.6	2	3	1.5	2.5

So-called atactic poly(propylene) (**APP**) is a highly branched, non-crystallizable polymer that serves as an adhesive, replacement for bitumen, and the like. Originally, it was an unwanted by-product of isotactic poly(propylene)s from polymerizations with early Ziegler catalysts but it is now produced in dedicated plants.

The poly(propylene)s of industry are not necessarily isotactic homopolymers but may also be random copolymers of propylene with small proportions of ethylene or 1-butene or block copolymers of the type poly(propylene)-*block*-poly(ethylene-*co*-propylene). So-called impact poly(propylene)s (**IPP**) and high-impact poly(propylene)s (**HIPP**) of commerce are polymer blends of it-PP with 15-25 wt% of EPDM rubbers. EPDM contents of 60-65 wt% lead to thermoplastic olefin elastomers (TPO). Elasto-meric poly(propylene)s (**ePP**) are stereoblock copolymers of propene.

Isotactic poly(propylene)s are semicrystalline polymers that combine good mechani-cal properties and thermal properties (Table 8-13) with chemical resistance and moder-ate costs. They are used for a wide variety of goods from packaging films and fibers to containers, automotive parts, appliances, and artificial turf. The intended use depends on the desired stiffness of the poly(propylene) grade which is controlled by the molecular weight. The higher the molecular weight, the lower the melt flow index (p. 85), an im-portant indicator of processability. Applications of PP are shown in Table 8-14.

Syndiotactic poly(propylene) (**sPP**) is extraordinarily transparent and highly impact resistant but not very rigid. It melts at lower temperatures than iPP (160°C *versus* 184°C) and is used for special articles such as containers for cosmetics and for films and spun-bonded sheet products.

Table 8-14 Effect of melt flow index MFI on the choice of processing methods and the intended end-use application of poly(propylene) [7].

MFI/(dg min^{-1})	Processing method and/or shape	End-use application
>200	melt-blown fibers	diapers, food containers
35-70	thin-wall injection molding	food containers
30-40	spun-bonded fibers	apparel, diapers
15-35	continuous filaments	upholstery fabrics, carpet face
10-35	blow molding	bottles
10-30	staple fibers	carpet backing
10-20	injection molding	appliances, outdoor furniture
6-12	cast/blown films	adhesive tape
2-5	bioriented PP films	food packaging
2-4	slit tape	woven fabrics, geotextiles
2	compression molding	beverage caps
1-5	monofilaments	produce bags, cordage, twine
1	thermoforming sheet	food containers

Other Poly(1-olefin)s

Isotactic **poly(1-butene)** (**PB**), $+CH_2-CH(C_2H_5)+_n$ is produced by higher generation Ziegler catalysts. It has a glass temperature of $-20°C$, a heat distortion temperature (at 1.82 MPa) of 57°C, a Vicat temperature B of 110°C (greater than iPP), and a thermodynamic melting temperature of 140°C. Because of its continuous service temperature of 100°C, it is used mainly for hot water pipes and large containers.

Poly(4-methyl-1-pentene) (**PMP**), $+CH_2-CH(CH_2CH(CH_3)_2)+_n$ is an isotactic polymer that has the lowest density of all thermoplastics ($\rho = 0.83$ g/cm^3) and is glass clear despite being highly crystalline. An excellent continuous service temperature of 120°C ($T_M = 242°C$, $T_G = 29°C$) allows its use for sterilizable medical instruments, containers for ready-to-eat meals, and the like.

8.2.3 Styrene Polymers

Poly(styrene)s

Styrenics is the common commercial term for polymers, copolymers, and graft polymers of styrene (vinyl benzene, phenyl ethylene), $CH_2=CH(C_6H_5)$. The name of the homopolymer is poly(styrene) (**PS**) but this name is sometimes used for all styrenics.

Poly(styrene) itself is obtained by free-radical polymerization of styrene in bulk, suspension, emulsion, or solution. Molding masses of this atactic polymer (**aPS**) have molecular weights of ca. $(2-3)\cdot10^5$. Syndiotactic poly(styrene) (**sPS**) results from the polymerization of styrene with special transition metal catalysts.

All commercial, atactic poly(styrene)s are amorphous, linear polymers that do not crystallize on drawing. So-called **crystal poly(styrene)s** result from the thermal polymerization of neat styrene. These polymers are atactic, amorphous, and crystal clear (hence the name) because of the absence of residues from catalysts, emulsifiers, etc., and the large refractive index ($n_D = 1.59$); they are not crystalline.

Easy flow grades may have lower molecular weights or may contain up to 4.5 wt% white mineral oil. Resilience is improved by adding small proportions of SBS triblock

Table 8-15 Properties of several grades of BASF poly(styrene)s. Note that grades of all industrial polymers are usually not pure polymers but may have been modified and/or may contain adjuvants.
144 C = very easy flow;   143 E = easy flow;   165 H = high molecular weight;
158 K = heat resistant;   168 N = heat resistant, high molecular weight.

Physical property	Test condition	Physical unit	Grade				
			144 C	143 E	165 H	158 K	168 N
Intrinsic viscosity	not disclosed	mL/g	74	96	119	96	119
Heat distortion temperature B	0.45 MPa	°C	80	82	84	98	98
Heat distortion temperature A	1.82 MPa	°C	70	72	76	86	86
Vicat temperature A	10 N	°C	88	88	92	106	106
Vicat temperature B	50 N	°C	84	84	89	101	101
Tensile modulus	1 mm/min	MPa	3150	3200	3150	3200	3250
Creep modulus	1000 h	MPa		2300	2830	2700	2850
Tensile strength	5 mm/min	MPa	46	50	56	50	63
Fracture elongation	5 mm/min	%	1.5	2	2	2	3
Impact strength	−30 to +23°C	kJ/m^2	6	9	11	10	13
Notched impact strength	−30 to +23°C	kJ/m^2	2	2	2	2	2

polymers. Heat distortion temperatures and creep moduli increase with increasing molecular weight (grades 144 C → 143 E → 165 H in Table 8-15) because these properties are affected by increasing entanglement of polymer chains and thus by the resistance against flow under load. The high heat grades 158 K and 168 N are modified (not disclosed) which leads not only to greater heat distortion temperatures but also to lower dielectric loss factors, tan δ.

Standard poly(styrene)s are easy to dye and process. They are mainly used for food packaging, for household appliances such as refrigerators, and in electrical/electronic parts. **Expandable poly(styrene)s** (Section 10.5) are either uncrosslinked (**EPS**) or crosslinked (**XPS**). They serve for thermal insulation, mainly in construction. Their damming properties can be considerably increased by incorporated infrared absorbing compounds.

Commercial syndiotactic poly(styrene) (**SPS, sPS, st-PS**) is obtained from the polymerization of styrene with a Ti metallocene catalyst plus methaluminoxane or borate. The polymer has the same glass temperature as atactic poly(styrene) (100°C) but a heat distortion temperature A of 96°C (PS = 70-86°C), and by virtue of its crystallinity, a melting temperature of 270°C (see also Volume II, p. 272). However, neat SPS is brittle and is therefore offered only as grades that are reinforced with 30 wt% glass fibers.

About 40 wt% of styrene is used for homopolymers and ca. 60 wt% for very many different types of copolymers and impact-modified grades (for syntheses, see Volume II, p. 273). The copolymers are engineering thermoplastics (see Section 8.3.1).

Substituted Poly(styrene)s

Polymers from free radical-polymerized substituted styrenes are niche thermoplastics. The most widely used of these seems to be **poly(*p*-methyl styrene)** (**PMS**) which has a greater hardness and a higher heat distortion temperature than poly(styrene) itself. Also marketed are bipolymers of vinyl toluene (= mixture of 33 wt% para and 65 wt% meta isomers), bipolymers of *p*-methyl styrene and acrylonitrile, and terpolymers of *p*-methyl styrene, acrylonitrile, and butadiene.

8.2.4 Vinyls

"Vinyl" is the common name of the group $CH_2=CH-$ (IUPAC name: ethenyl) and "vinyl monomer" the name of a monomer of the type $CH_2=CHR$ where R is a substituent that is neither an alkyl group, a nitrile group $-CN$, nor an acid or ester group $-COO-$. Hence, "vinyl polymers" are the polymers of vinyl monomers. However, in the world of the plastics industry, **vinyls** include only the homopolymers and copolymers of vinyl chloride monomer, $CH_2=CHCl$ (**VCM, VC**).

90 % of these vinyls are homopolymers (**PVC**), either rigid (= unplasticized) (**PVC-U**) or plasticized (**PVC-P**). Other vinyls are copolymers of vinyl chloride with 3-20 % vinyl acetate, 3-10 % propylene, or 87 % vinylidene chloride, $CH_2=CCl_2$. Vinyls may also be graft polymers of vinyl chloride on butyl acrylate polymers or ethylene-vinyl acetate copolymers. Several grades of vinyls are also marketed with impact modifiers and other adjuvants (Table 8-16).

All PVCs are obtained by free-radical polymerization of vinyl chloride (VCM) in bulk (gas phase), emulsion, suspension, or microsuspension. VCM can lead to liver angiosarcoma, a rare type of cancer if one is routinely exposed to it for many years. Newer polymerization and work-up methods result in residual monomer contents of PVC of less than 1 ppm which is considered an insignificant risk for health.

Table 8-16 Six of the 163 grades of rigid PVC of the Solvay company. All grades contain lubricants and flame retardants. Also added: aging inhibitors (A), reinforcing fillers (F), heat stabilizers (H), impact modifiers (I), and/or UV absorbers (U). Grades are delivered as pellets (Pe) or powder (Po) for extrusion (EX), extrusion blow molding (EB), or injection molding (IM). nf = no fracture.

Physical property	Physical	Grades of Benvic® PVC-U					
PVC designation →	unit	IR	PEB	EB	EB	ER	S-ER
PVC grade →		346	918	950	928	999	820
Additives →		A	-	I	I	AFHIU	AFHIU
Delivered as →		Pe	Po	Pe	Pe	Pe	Pe
Processing →		IM	EB	EB	EB	EX	EX
Application →		electro	food	food	food	tubes	profiles
Density	g/cm^3	1.32	1.41	1.31	1.39	1.23	1.51
Heat distortion temperature B	°C		68				73
Vicat temperature B	°C	78	77	75	76	96	82
Thermal expansion coeff. ‖ (23-80°C)	K^{-1}	$7 \cdot 10^5$	$7 \cdot 10^5$	$7 \cdot 10^5$	$7 \cdot 10^5$	$8 \cdot 10^5$	$3 \cdot 10^5$
Tensile modulus (1 mm/min)	MPa	3300	3400	2300	2800	2700	3300
Yield stress (50 mm/min)	MPa	49	58	46	53	45	46
Elongation at yield (50 mm/min)	%		3.5	4	4	2.5	3.7
Fracture elongation (50 mm/min)	%	12	15			28	35
Impact strength (Izod. 23°C)	kJ/m^2		nf				nf
Impact strength (Izod. -30°C)	kJ/m^2		75				228
Notched impact strength (23°C)	kJ/m^2		3.5				16
Notched impact strength (-30°C)	kJ/m^2		3.1				5.5
Resistivity	Ω cm		$>10^{15}$				$>10^{15}$
Electrical resistance	Ω		10^{13}				$>10^{15}$
Water absorption (23°C, saturation)	%	0.3	0.3	0.3	0.3	0.3	0.2
Moisture uptake (23°C, 50 % RH)	%	0.03	0.03	0.03	0.03	0.03	0.04

All PVCs must be stabilized by appropriate adjuvants for processing (temperature) and use (light, oxygen of air) against discoloration (from the formation of HCl and conjugated double bonds) and the accompanying change of mechanical properties by crosslinking (see Section 3.6.3).

About 2/3 of PVCs are used as rigid or hard PVC (**PVC-U, rigid PVC**) and ca. 1/3 as plasticized polymer (**PVC-P**). However, rigid PVCs may contain up to 10 wt% plasticizer. Some previously used plasticizers were present in proportions that were toxic for small children and are therefore no longer allowed (see Section 3.4.3). Rigid PVCs are mostly suspension-polymerized polymers; their main application is for tubes and pipes. Plasticized PVC serves mainly as packaging film or cable sheathing and injection-molded shoe soles, bumpers, and the like.

Poly(vinylidene chloride) (PVDC) is not produced as a homopolymer but always as a copolymer of vinylidene chloride (VDCM), $CH_2=CCl_2$, with 15-20 % vinyl chloride or 13 % vinyl chloride and 2 % acrylonitrile (Saran®). PVDC has the lowest gas permeabilities of all plastics (Table 11-1) and is therefore a valuable packaging material.

8.2.5 Production Statistics

Comprehensive global production statistics are not available in the open literature but it can be surmised that the total world production of standard plastics will continue to increase (see Fig.1-2), especially in China and in the Arab countries.

The situation is different for both individual standard thermoplastics and individual countries where either no data are accessible at all or data are inconsistent because an individual thermoplastic may be counted as a member of a certain group for a number of years and then as a member of another group. Production volumes of some polymers are furthermore often not included at all in figures of "total thermoplastics" production, for example, poly(methyl methacrylate) in the US production data of standard or engineering thermoplastics. In addition, the designation of a thermoplastic as a "standard thermoplastic" may change with time.

For North America (= members of NAFTA = North American Free Trade Association (US, Canada, Mexico)), consistent data are available only for the years 2000-2006 (Table 8-17). These data include the common standard types of isotactic poly(propylene) (PP), low-density poly(ethylene) (LDPE), high-density poly(ethylene) (HDPE), linear low-density poly(ethylene) (LLDPE), poly(vinyl chloride) (PVC) (only the polymer itself or all PVC types (PVC-U + PVC-P), i.e., including plasticizer if any) and (atactic) poly(styrene) (PS).

Probably because of production volumes, these statistics for standard thermoplastics also comprise the blend/copolymer ABS (apparently also including other styrenics such as SAN and SB), thermoplastic polyesters PET (exclusive of PES fibers), and aliphatic polyamides PA ("nylons"; exclusive of fibers).

From 2000 to 2006, NAFTA countries increased production of thermoplastic polyesters (18 %), linear low-density poly(ethylene)s (17 %), it-poly(propylene)s (16 %), and high-density poly(ethylene)s (10 %). Production remained constant for poly(vinyl chloride)s (+3 %), other styrenics (oSty; +1 %), and nylons (–1%) but decreased for low-density poly(ethylene)s (–7 %), poly(styrene) (–8 %), and ABS (–11 %).

Table 8-17 Production of standard thermoplastics in 2000-2006 [1]. Data for the United States and Canada. Data for ABS, PET, and PA also contain Mexican productions. oSty = other styrenics.

Year	Annual North American production in thousands of tons									
	PP	HDPE	LLDPE	LDPE	PVC	PS	oSty	ABS	PET	PA
2000	7145	7288	5058	3826	6557	3107	813	662	3191	582
2001	7234	6939	4663	3494	6473	2776	746	553	3132	517
2002	7698	7250	5073	3650	6945	3028	786	552	3290	578
2003	8020	7132	5056	3543	6675	2902	780	573	3444	581
2004	8423	7966	5645	3766	7257	3065	848	619	3751	608
2005	8156	7334	5399	3560	6928	2857	849	565	3518	568
2006	8308	8011	5935	3565	6773	2846	825	587	3764	577
2006/2000	1.16	1.10	1.17	0.93	1.03	0.92	1.01	0.89	1.18	0.99

8.3 Engineering Thermoplastics

Engineering thermoplastics are plastics that can be used for load bearing and other engineered articles. They must combine easy processability with stiffness and good yield, also good fracture, compression, shear, and impact strengths. These properties should be maintained at low and high temperatures and when exposed to the elements, chemicals, and mechanical stresses. Of course, there is no sharp dividing line between standard and engineering plastics with respect to properties but engineering plastics are usually produced in smaller volumes and therefore command higher prices (Table 8-4).

Engineering plastics are usually never used as pure polymers but always as compounds with various adjuvants and, in general, also as filled and fiber-reinforced grades. Fiber reinforcement increases density, tensile modulus, tensile strengths and, for semicrystalline polymers, also heat distortion temperature, but decreases shrinkage and tensile elongation (see also Chapter 9). Especially important is the significant increase of tensile moduli to values greater than 3.4-4 GPa (Table 8-18) because this leads to such additional rigidity that one can use thinner object walls.

Improvements vary widely with the type of polymer for a reinforcement by 30 wt% glass fibers, the industry standard (Table 8-18). However, it is not possible to correlate these changes with the chemical and/or the physical structure of the matrix, with changes in the physical structure of the matrix (if any) in the presence of fibers, and/or with interactions fiber/matrix, in part because fundamental investigations are lacking and in part because the industrial habit of comparing physical properties at a constant *weight* fraction of fillers is scientifically unsound. Instead, comparisons should be made on a per volume basis (see Chapter 9).

In general, reinforcement of plastics by 30 wt% of short glass fibers leads to 3-8 times greater tensile moduli and 2-4 times greater fracture strengths, while fracture elongations decrease relatively little for thermosets but dramatically for thermoplastics (Table 8-18). Heat distortion temperatures improve markedly for semicrystalline thermoplastics but very little for amorphous ones. Continuous service temperatures vary only a little for amorphous thermoplastics but markedly for thermosets and crystalline thermoplastics.

Table 8-18 Guide values for the effect of reinforcement of plastics with 30 wt% of short glass fibers on some properties. ρ = Density, Δ = shrinkage, σ_F = tensile strength at fracture, E = tensile modulus, ε_F = fracture elongation, T_{HD} = heat distortion temperature (1.82 MPa), T_{CS} = continuous service temperature. U = unfilled, GF = filled with glass fibers.

Plastic	ρ/(g cm⁻³)		Δ/%		E/GPa		σ_F/MPa		ε_F/%		T_{HD}/°C		T_{CS}/°C	
	U	GF	U	GF	U	GF	U	GF	U	GF	U	GF	U	GF
Amorphous thermoplastics														
PVC	1.4	1.6	0.3		2.9	13.6	50	129	60	4	74	74		
SAN	1.1	1.4	0.4	0.1	3.6	12.2	72	129	4	2	93	104	96	104
ABS	1.1	1.3	0.4		2.9	7.2	50	115	20	3	110	110	110	110
PC	1.2	1.5	0.6	0.2	2.9	10.8	79	158	125	2	145	150	135	135
Semicrystalline thermoplastics														
PP	0.9	1.2	1.1	0.5	0.7	5.7	35	72	500	2	60	150	115	160
PET	1.3	1.5	1.7	0.5	2.9	10.8	57	129	300	5	71	235		188
PA 6	1.1	1.7	1.1	0.4	2.9	14.3	65	230	29	10	65	260	150	204
Thermosets														
PF	1.3	1.8	1.1	0.5	7.2	18	54	72	0.6	0.2			150	230
UP	1.2	1.8	0.6	0.1	3.6	18	86	143	2.6	0.2		230	150	204
EP	1.2	1.8	0.2	0.5	3.6	26	72	430	4.4	3.0			88	260

In order to show the effect of chemical structure on the use properties of polymers, only data for unfilled plastics are shown in the other tables of Section 8.3. However, most grades of these polymers are either filled or reinforced (Chapter 9) or blended/grafted with elastomers (Chapter 10).

8.3.1 Styrene Copolymers

Styrene delivers a relatively brittle and not very impact-resistant homopolymer (Table 8-15) but can be converted to engineering polymers by free-radical copolymerization and/or graft copolymerization. Styrene forms alternating copolymers with maleic anhydride (**SMA**) or *N*-phenyl maleimide (**SMI**). Random bipolymers of styrene with acrylonitrile contain either 24 % or 60-70 % acrylonitrile (both **SAN**).

All of these bipolymers have higher glass and heat distortion temperatures than poly-(styrene) and also improved mechanical properties as shown for SAN in Tables 8-19 and 8-22. They are used mainly for household appliances and furniture. SAN with 60-70 wt% acrylonitrile units are barrier resins for gases (Chapter 11).

Pseudo-statistical copolymers of ethylene and 15-77 wt% styrene are produced with titanium half-sandwich catalysts. These polymers are marketed as ethylene-styrene "interpolymers" (**ESI**) (*all* copolymers were formerly called "interpolymers"!). ESIs with up to 50 wt% styrene are semicrystalline and elastic but those with greater contents of styrene units are amorphous and either rubbery or glassy.

Polymers of ca. 75 wt% styrene blocks and ca. 25 wt% butadiene blocks are thermoplastics (**SB**) that are produced by anionic polymerization. Triblock polymers of styrene with butadiene or isoprene center blocks (**SBS**, **SIS**) are thermoplastic elastomers (**TPE-S**) as are their hydrogenation products styrene-ethylene/butanediyl-styrene (**SEBS**, from SBS) and styrene-ethylene/propylene-styrene (**SEPS**, from SIS) (see p. 286).

Table 8-19 Properties of some styrene copolymer grades of the BASF company.

Physical property		Physical unit	SAN Luran® 368 R	ASA Luran® S 776S	SB 2711	SB 466 I	SB-*b* Styrolux® 637 D	ABS Terluran® 2877
Density		g/cm³	1.08	1.07	1.05	1.05	1.02	1.07
Heat distortion temperature B		°C	101	101	77	90	78	104
Heat distortion temperature A		°C	98	96	67	82	68	99
Vicat temperature B		°C	101	93	79	92	68	101
Tensile modulus	(1 mm/min)	MPa	3800	2200	1300	2250	2000	3000
Creep modulus	(1000 h)	MPa	2800	1170				1870
Yield stress	(50 mm/min)	MPa		47	15	31	36	60
Tensile strength	(5 mm/min)	MPa	75	47				
Elongation at yield	(50 mm/min)	%	5	3.3	1.4	1.7	2.2	3
Fracture elongation	(50 mm/min)	%		30	>50	40	35	15
Impact strength	(23°C)	kJ/m²	19		94	72		60
Impact strength	(–30°C)	kJ/m²	19		54	47		30
Notched impact str.	(23°C)	kJ/m²	2		8	10		5
Notched impact str.	(–30°C)	kJ/m²	2		6	6		3

The most important styrene copolymers are known as impact poly(styrene)s (**IPS**) or high-impact poly(styrene)s (**HIPS**). They are obtained by free-radical copolymerization of styrene/acrylonitrile mixtures in the presence of various rubbers (Table 8-20), which leads to a simultaneous graft copolymerization of the former on the latter. Graft copolymerizations of styrene + acrylonitrile on *cis*-butadiene rubbers (BR) or on acrylonitrile-butadiene rubbers (NBR) results in **ABS** polymers, on chlorinated poly(ethylene) rubbers in **ACS**, on EPDM rubbers in **AES**, and on acrylate rubbers in **ASA**.

Table 8-20 Graft copolymers of styrene.

Abbreviation	Comonomers	Grafting on	After-treatment
ABS	styrene + acrylonitrile	butadiene rubbers	
ACS	styrene + acrylonitrile	chlorinated poly(ethylene)	blending with SAN
AES	styrene + acrylonitrile	EPDM rubbers	
ASA	styrene + acrylonitrile	acrylate rubbers	
MBS	styrene + methyl methacrylate	butadiene rubbers	
SMA	styrene + maleic anhydride	rubbers	

General features of these processes are discussed in Section 10.4.7 but details are usually trade secrets and the chemical and physical processes during polymer manufacturing and the resulting chemical and physical structures are often not known or not revealed. All polymers of Table 8-20 are complex, multiphase polymers in which copolymerization generated or purposely added graft and/or block copolymers act as compatibilizers. The physical structure of **ABS** polymers is that of a "hard" poly(styrene-*co*-acrylonitrile) matrix with interdispersed soft, crosslinked elastomer particles whose compatibility with the matrix is assured by the styrene/acrylonitrile chains grafted upon them. The strength and rigidity of the ABS polymers decrease with increasing rubber content but increase with increasing molecular weight of the styrene/acrylonitrile matrix.

Table 8-21 Mass fractions w of styrene (S), acrylonitrile (AN), and butadiene units (BU) in commercial styrenics with densities ρ, tensile moduli E, yield strengths σ_Y, fracture elongations ε_B, and notched impact strengths $\gamma_{i,N}$. PS = Standard poly(styrene), SAN = styrene-acrylonitrile bipolymer, IPS = impact-modified poly(styrene)s, ABS = acrylonitrile/butadiene/styrene polymers.

Type	$\dfrac{w_S}{\%}$	$\dfrac{w_{BU}}{\%}$	$\dfrac{w_{AN}}{\%}$	$\dfrac{\rho}{\mathrm{g\ cm^{-3}}}$	$\dfrac{E}{\mathrm{GPa}}$	$\dfrac{\sigma_Y}{\mathrm{MPa}}$	$\dfrac{\varepsilon_B}{\%}$	$\dfrac{\gamma_{i,N}}{\mathrm{kJ\ m^{-2}}}$
PS	100.0	0.0	0.0	1.05	4.1	37	0.9	14
IPS, medium impact	96.6	3.4	0.0	1.05	3.4	27	1.4	32
IPS, high impact	94.9	5.1	0.0	1.05	2.4	22	3.5	69
IPS, super-high impact	85.5	14.5	0.0	1.02	1.8	14	17.0	240
SAN	74.7	0.0	25.3	1.08	3.9	61	1.6	17
ABS, medium, type 1	78.9	4.9	16.2	1.05	2.6	24	40.0	70
ABS, medium, type 2	57.4	13.3	29.3	1.05	3.0	49	9.4	150
ABS, standard	53.9	19.3	26.8	1.04	2.7	39	2.9	360
ABS, superhigh impact	49.9	26.7	23.4	1.04	1.9	29	2.2	370

The dispersed rubber particles act on impact as "shock absorbers." However, the increase of impact strength is accompanied by a decrease of tensile modulus and an increase of fracture elongation (Table 8-21). These polymers are also not clear like poly(styrene) itself but rather turbid because the styrenic matrix and the graft copolymer domains have different refractive indices and the diameters of the dispersed domains exceed one-half of the wavelength of incident light.

Styrene-butadiene diblock polymers (**SB**), on the other hand, are clear thermoplastics and neither thermoplastic elastomers like their triblock cousins SBS nor rubbers like the random styrene-butadiene copolymers SBR. Like ABS, SB polymers have rubbery domains of poly(butadiene) blocks in a continuous thermoplastic matrix (here: poly(styrene) blocks) with a different refractive index. SB is clear because the rubbery domains are smaller than one-half of the wavelength of incident light; these polymers are thus called **glass-clear high-impact poly(styrene)s** (glass-clear **HIPS**) (see Table 8-19).

ABS polymers have good heat distortion temperatures, excellent resistance against chemicals, and outstanding surface characteristics. However, their weatherability is not very good because the double bonds of the butadiene units can be fairly easily oxidized. Weatherability can be improved if the unsaturated butadiene units are replaced by saturated units, for example, after-chlorinated poly(ethylene) (Volume II, p. 228).

All impact grades are good electrical insulators with relative permittivities of $\varepsilon_r = 2.3$, volume resistivities of $\rho \approx 10^{17}\ \Omega$ cm, electrical resistances of $R \approx 10^{14}\ \Omega$, electric strengths of $S \approx 135$ kV/mm, and $\tan \delta \approx 0.7$ (0.5 for grades 158 K and 168 N).

8.3.2 Polycarbonates

Polycarbonates have the general constitution $+\!\mathrm{R\!-\!O\!-\!CO\!-\!O}\!\!\displaystyle{\frac{}{}}_n$ where R is an aliphatic, cycloaliphatic, or aromatic residue. The simplest aliphatic representative with R = CH_2CH_2 is a biodegradable, clear elastomer that is not used commercially. Poly(propylene carbonate) from propylene oxide and CO_2 serves as a binder resin in semiconducting devices.

Table 8-22 Guide values for physical properties of unfilled engineering thermoplastics. H = Homopolymer, MS = medium impact strength, a = amorphous, nc = usually not crystalline.

Physical property	Physical unit	SAN	ABS MS-2	PMMA H	PC H	PPC	POM H	PA 6	PET H
Density	g/cm^3	1.08	1.05	1.19	1.21	1.2	1.42	1.13	1.21
Water absorption (24 h)	%	0.2		0.25	0.15	0.16	0.22	1.6	0.08
Linear expansion coefficient	10^{-5} K^{-1}	6.6	8.3	8.0	7.0	5.0	9.0	9.0	7.2
Shrinkage	%			0.4	0.6	0.8	2.8	1.0	0.6
Melting temperature	°C	a	a	a	nc		175	215	265
Glass temperature	°C	120	99	105	150		-82	50	70
Heat distortion temperature A	°C	103	98	90	132	152	124	65	85
Tensile modulus	GPa	3.5	3.0	3.0	2.3		2.8	2.7	2.9
Flexural modulus	GPa	3.7	2.4		2.2	2.9	2.9	2.8	2.4
Upper yield strength	MPa	77			55	67			81
Upper flexural strength	MPa		74						
Tensile strength at fracture	MPa	75	49	65	65	73	69	81	73
Flexural strength	MPa	90			95	97			118
Elongation at upper yield	MPa				6.0				
Fracture elongation	%	1.6	9.4	6	110	120	<70	100	300
Notched impact strength	J/m	28	300	23	870	530	80	60	53
Hardness (Rockwell R)	-		83	100	75	115	122	119	
Hardness (Rockwell M)	-	80			70	82	94		106
Relative permittivity (1 MHz)	-			3.3	2.9	3.0	3.7	3.4	3.4
Dielectric loss factor ($\cdot 10^4$)	-			400	100	240	50	200	200

The polycarbonates of trade are either fully or predominantly aromatic. Their homopolymers result from the polycondensation of diphenols HO–Ar–OH with derivatives of carbonic acid, H_2CO_3, such as diphenyl carbonate, $(C_6H_5O)_2CO$, or phosgene, $COCl_2$. Common polycarbonates are based on bisphenol A, HO(p-C_6H_4)–Z–(p-C_6H_4)OH with Z = >C(CH$_3$)$_2$ (**BPA**). Other homopolycarbonates may contain units such as Z = >CH$_2$ (**BPF**), Z = >C=CCl$_2$ (**BPC**), and many others (see Volume II, p. 343).

Copolymers play a very minor role (Volume II, p. 346). **Polyphthalate carbonates (PPC)** are copolymers with bisphenol A units, carbonic acid units, and terephthalic acid units (for properties, see Table 8-22). Poly(ester-co-carbonate)s with less than 10 wt% dodecanoic units serve for thin-walled articles and poly(ether-co-carbonate)s with up to 20 wt% poly(ethylene glycol) units for dialysis membranes.

Polycarbonates are transparent, glossy, amorphous engineering thermoplastics with excellent dimensional stabilities and impact strengths, moderate heat stabilities, and low water absorptions (PC in Table 8-22). They can be easily processed and are used for CDs, prescription lenses, and other injection-molded articles (for molecular engineering, see Volume II, p. 346). Baby bottles from PCs are now suspect because they contain residual BPA, a suspected carcinogen.

8.3.3 Thermoplastic Polyesters

In the plastics industry, "polyester" refers to polymers with ester groups –COO– in the main chain, except carbonic ester groups, –OCOO–. The industry furthermore distinguishes thermoplastic polyesters (Fig. 8-5) from unsaturated polyesters (Section 8.5.3).

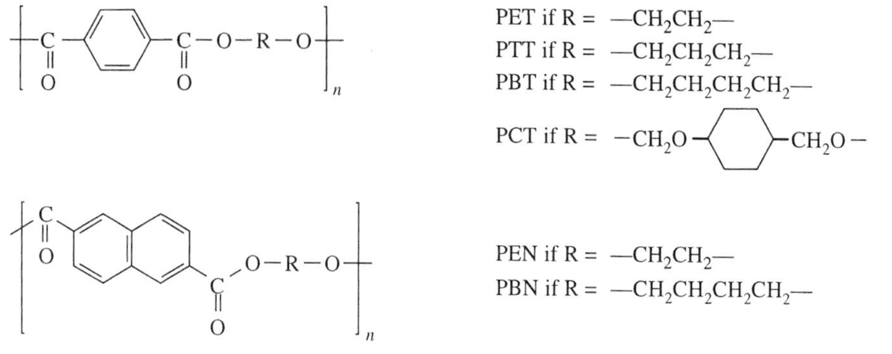

PET if R = —CH₂CH₂—

PTT if R = —CH₂CH₂CH₂—

PBT if R = —CH₂CH₂CH₂CH₂—

PCT if R = —CH₂O—⟨C₆H₁₀⟩—CH₂O—

PEN if R = —CH₂CH₂—

PBN if R = —CH₂CH₂CH₂CH₂—

Fig. 8-5 Chemical structure of thermoplastic polyesters. The cyclohexane-1,4-dimethylol units of PCT (PCDT) are present in the ratio 30:70 (cis/trans).

The most important thermoplastic polyester is **poly(ethylene terephthalate) (PET)** that was originally obtained by polycondensation of dimethyl terephthalate with ethylene glycol (Volume II, p. 350 ff.) and developed as a synthetic fiber with molecular weights of 15 000-20 000 (PES; Chapter 6). The as-spun fibers (p. 164) are drawn and textured (p. 205), which converts them into textile fibers. Newer syntheses use terephthalic acid instead of dimethyl terephthalate (see Volume II, p. 350 ff.).

Fiber grades of poly(ethylene terephthalate)s could not be injection molded because of their slow crystallization. The development of improved methods for the polycondensation of terephthalic acid with ethylene glycol led much later to low crystallinity bottle grades with molecular weights of 24 000-36 000, technical fibers with molecular weights of 40 000-50 000, and, with the discovery of nucleating agents (Section 3.3.1), to high-molecular weight engineering grade PETs with high crystallinities (see Table 8-23). These nucleated PETs are injection molded into ca. 140°C warm molds. Cycle times can be reduced by using PETs that contain ca. 5 mol% non-symmetrical diols as comonomers, for example, 3-methylpentane-2,4-diol.

High stiffnesses are obtained by reinforcement with glass fibers. However, this may lead to warping which can be reduced by addition of particulate fillers such as mica. Particulate fillers may lead to warping, too, which can be off-set by addition of acrylic-type rubbers.

Conventional enginering PETs are not suitable for blow molding to bottles since their melts are not viscous enough. Blow molding grades therefore have much higher molecular weights, which leads to higher concentrations of chain entanglements and therefore also to higher stabilities of the parisons.

Bottle-grade PETs should lead to glass-clear bottles, which means that polymers should be totally devoid of nucleating agents. Their ability to crystallize is reduced further by incorporating flexibilizing monomeric units from comonomers such as iso-phthalic acid HOOC(i-C₆H₄)COOH, neopentylglycol HOCH₂C(CH₃)₂CH₂OH, cyclohexane-1,4-dimethylol HOCH₂(1,4-C₆H₁₀)CH₂OH, or 4,4'-di(hydroxyethoxy)diphenylmethane CH₂[(p-C₆H₄)OCH₂CH₂OH]₂. Barrier properties are improved by blending PET or by coating bottle walls with other polymers.

Poly(1,4-butylene terephthalate) (PBT) (Fig. 8-5) contains the 1,4-butylene group which is more flexible than the ethylene group of PET. For this reason, PBT crystallizes

Table 8-23 Degree of crystallinity w_c, heat distortion temperature T_{HDT}, appearance, and application of poly(ethylene terephthalate)s.

Physical state	$w_c/\%$	$T_{HDT}/°C$	Appearance	Application
amorphous	0 - 5	< 67	clear	blister packages
amorphous, oriented	5 - 20	< 73	clear	bottles
crystalline	25 - 35	< 127	opaque	food containers
crystalline, oriented	35 - 45	< 160	clear	films, hot-fill containers

faster than PET and can therefore be injection molded with short cycle times at mold temperatures of only 50-80°C. Mold temperatures of PBT (and also of PET) can be further reduced by adding plasticizers and lubricating agents, respectively, such as ethylene-acrylic ester copolymers, butyl benzyl phthalate, or aromatic sulfone amides. A large proportion of PBT is used in blends with, for example, PET or PC, especially for interior and exterior applications in automobiles.

Poly(trimethylene terephthalate) (PTT) (Fig. 8-5) is a fairly new engineering plastic that became available only after a commercially viable syntheses for 1,3-trimethylene glycol was developed (see Volume II, p. 355). PTT crystallizes faster than PBT but slower than PET. The main application of this engineering plastic is as fiber for home textiles. It is not called PPT because "P" is the symbol for "propylene" $-CH_2CH(CH_3)-$.

Poly(ethylene 2,6-naphthalate) (PEN) (Fig. 8-5) is a liquid-crystalline engineering polymer that is presently used mainly for photographic films of the Adventix® photo system. Amorphous PEN (= **APEN**) can be injection molded.

Fairly new as plastics are the aromatic polyesters **poly(1,4-butylene 2,6-naphthalate) (PBN)** and **poly(1,4-cyclohexylene terephthalate) (PCT)** (see Fig. 8-5).

A special group of thermoplastic polyesters are the **polyarylates** which are mostly co-polyesters of diphenols and mixtures of aromatic acids such as terephthalic acid, isophthalic acid, and *p*-hydroxybenzoic acid (Volume II, p. 360; see also Section 8.4.8). These amber-colored polymers have good mechanical properties and high thermal stabilities of up to 240°C (unreinforced) and 343°C (reinforced); they are used for components of engines, power tools, and the like.

8.3.4 Aliphatic Nylon Plastics

Nylon plastics comprise the second largest group of engineering plastics. The industrial term "nylon plastics" refers to semicrystalline or amorphous polyamides that are used as plastics. Some of these "nylon plastics" are indeed "nylons" within the meaning of the Federal Trade Commission (FTC) definition of nylon *fibers* (= more than 85 % of the amide linkages –NH–CO– attached directly to aliphatic groups (see Table 16-8)) but some are not, since their amide linkages may be to aromatic groups (Section 8.4.1).

Overall, the group of **nylon plastics** consists rather not only of aliphatic polyamides but also of some semi-aromatic polyamides (or semi-aliphatic ones, depending on the view) of the AABB type which are not "nylons" according to the FTC definition. Completely aromatic polyamides (**aramids**) of the AB and AABB type are not used as plastics but exclusively as fibers (p. 223) and to a smaller extent also for coatings.

Table 8-24 Guide values for physical properties of common types of unfilled semicrystalline aliphatic nylon plastics at 23°C. RH = relative humidity. For codes, see text below.

Physical property	RH	Physical unit	6	6.6	6.9	6.12	11	12	6-3.T
Density	50	g/cm^3	1.13	1.14	1.09	1.07	1.04	1.02	1.12
Water absorption (24 h)	50	%	1.6	1.2	0.5	0.5	0.3	0.25	0.25
(equil.)	50	%	2.7	2.5	1.8	1.4	0.8	0.7	
(equil.)	100	%	9.5	8.5	4.5	3.0	1.9	1.5	7.5
Linear expansion coeff.	50	$10^{-5}\,K^{-1}$	10	10			15	11	6
Shrinkage	50	%	1.0	1.0	1.7	1.2	1.2		0.5
Melting temperature	50	°C	215	265	205	217	194	179	-
Heat distortion temperature B	50	°C	185	235	150	165	150	150	141
Heat distortion temperature A	50	°C	65	75	55	82	55	55	127
Glass temperature	50	°C	50	50			47	40	150
Tensile modulus	50	GPa	2.70		2.90		1.27	1.24	3.00
Flexural modulus	0	GPa	2.80	2.80	2.90	2.00	1.20	1.10	2.60
	50	GPa	0.97	1.20			0.64	0.60	
Upper yield strength	50	MPa	81	83	55	55	55	55	85
Fracture strength	0	MPa	81	83	59	61	55	55	60
	50	MPa	45	58			46	48	
Fracture elongation, average	50	%	100	60	125	150	200	200	70
Notched impact str. (Izod)	0	J/m	60	59	58	58	54	95	65
	50	J/m	215	113			110	120	
Hardness, Rockwell R	50	-	119	121	111	114	108	107	
Relative permittivity	50	-	3.4	3.5	6.9	3.5	3.1	3.1	3.5
Dielectr. loss factor $(\cdot 10^4)$	0	-	200					200	300

Commercial semicrystalline engineering plastics of the nylon type are mainly aliphatic polyamides of the AB type with repeating units $-NH(CH_2)_qCO-$ (PA 6, 11, and 12) or of the AABB type with repeating units $-NH(CH_2)_rNH-CO(CH_2)_sCO-$ (PA 4.2, 4.6, 6.6, 6.9, 6.10, and 6.12) (see p. 184-185). These aliphatic polyamides are less rigid than polycarbonates and thermoplastic polyesters (see Tables 8-22 and 8-24). Because of their hydrophilic amide groups, they absorb moisture and thus need to be "conditioned" before processing. The absorbed water acts as a plasticizer: it decreases tensile strengths and increases notched impact strengths.

The group of aliphatic polyamides also includes the polymer **PA MXD 6** from *m*-xylylene diamine (MXD) and adipic acid (6). MXD 6 is called "aliphatic" because its amine groups are bound to CH_2 groups and not directly to the 1,3-phenylene group. Since MXD 6 is a high-performance polymer, it will be discussed in Section 8.4.1.

The glass-clear, amorphous polyamide **PA 6-3.T** (or **PA TMDT**) is a copolymer from a 50:50 mixture of 2,2,4- and 2,4,4-*tri*methyl-1,6-hexamethylene *d*iamine (TMD or 6-3 because of the 6-carbon hexamethylene unit with 3 methyl substituents) and terephthalic acid (T). Its amorphous character is caused by irregular incorporations of its diamine units into the polymer chain. This polymer has a far higher glass temperature (150°C) than the common aliphatic AB and AABB type nylon plastics (40-50°C, Table 8-24) which is lowered only a little to 135°C by an equilibrium water uptake of 7.5 wt%.

Nylon plastics are mainly used in the electrical/electronic and automotive industries, usually reinforced by fillers (for improved rigidity and thermal stability) or elastomers (for improved impact resistance) (see also Table 8-18).

8.3.5 Polyacetals

Alkylene diethers R'O–CHR–OR" are known as acetals, and industrial polymers with the chain structure ~CHR–O–CHR–O–CHR–O~ are therefore referred to as **polyacetals**, **acetal polymers**, or **acetal plastics**. The only industrially important acetal polymers are those with oxymethylene groups –O–CH$_2$– as the main repeating unit; these polymers are therefore chemically known as **poly(oxymethylene)s (POM)**.

Industrial poly(oxymethylene)s comprise two types. **Poly(formaldehyde)s** are obtained by polymerization of formaldehyde, HCHO, and subsequent capping of the half-acetal endgroups with acetanhydride to ester endgroups. **Poly(trioxane)s** result from the ring-opening copolymerization of s-trioxane, the symmetric trimer C$_3$H$_6$O$_3$ of HCHO, with small proportions of dimethylformal or cyclic ethers such as dioxolane. The two types are known industrially as **polyacetal homopolymer** (= poly(formaldehyde)) and **polyacetal copolymer** (= poly(trioxane)). Although the polyacetal homopolymer is obtained by polymerization of formaldehyde, it is not called a polyaldehyde because this name is reserved for polymers from higher aldehydes such as acetaldehyde, CH$_3$CHO.

The homopolymer has a more regular chain structure than the copolymer and therefore a higher crystallinity (70 % (homo) *versus* 55 % (co)) and higher melting temperature and a greater Vicat B temperatures (Table 8-25). Because of the higher crystallinities, homopolymers have higher tensile and flexural strengths, hardnesses, and fatigue resistances than copolymers. Because of the randomly incorporated comonomeric units, polyacetal copolymers unzip less than homopolymers and are therefore more stable against heat, alkali, and hot water.

Poly(oxymethylene) chains are very compact (Volume III, p. 447). Polyacetals therefore take up very little moisture (less then 0.7 wt%) and have excellent dimensional stabilities, high strengths, and tensile moduli. Their metal-like properties allow them to replace press-molded zinc (Table 8-25). However, most polyacetals are used for automotive parts, especially for fuel systems, since they are not prone to stress corrosion by aliphatics, aromatics, halogenated hydrocarbons, and detergents. Polyacetals for outdoor applications need to be stabilized by UV stabilizers or carbon black. Limiting oxygen indices (LOI) are ca. 15 %.

Polyacetals compete with nylons 6 and 6.6. These nylons are better with respect to abrasion and impact but polyacetals are less prone to fatigue and creep.

Table 8-25 Comparison of some properties of acetal polymers and zinc.

Physical property	Physical unit	Acetal homo-PM	Acetal Co-PM	Zinc	Zinc alloy 8
Density	g/cm^3	1.42	1.41	7.13	6.37
Melting temperature	°C	175	165	419.5	≈380
Vicat temperature B	°C	173	163		
Heat distortion temperature A	°C	124	120		
Tensile modulus	MPa	2800	3200		
Upper yield strength	MPa	70	72		207
Tensile strength, molded	MPa	69	71	35	220
press molded	MPa			145	
Fracture elongation	%	25-70	25-70		1-2

Table 8-27 Average thermal and mechanical properties of unfilled semicrystalline (sc) or amorphous (a), partly aromatic (6.T/6.I/6.6, 6.T/6, 6-3.T) or "aliphatic" polyamides (MXD 6) at 23°C as compared to poly(hexamethylene diamine) (PA 6.6).

Numbers and letters in polymer designations: "6" refers to 1,6-hexamethylene diamine in PA 6.6, 6.T, and 6.I; to adipic acid in PA 6.6 and MXD 6, and to ε-caprolactam in 6.T/6. "T" symbolizes terephthalic acid in 6.T, 6.T/6, and 6-3.T, "I" refers to isophthalic acid in 6.I, and MXD to *m*-xylylene diamine in MXD 6. "6-3" is a 50/50 mixture of 2,2,4- and 2,4,4-trimethyl-1,6-hexamethylene diamine (also called TMD). RH = relative humidity, nb = no break, nc = not crystalline.

Property		Physical unit	PA 6.6 (sc)	6.T/6.I/6.6 (sc)	6.T/6 (sc)	MXD 6 (sc)	6-3.T (a)
Density		g/cm^3	1.14	1.17	1.18	1.21	1.12
Moisture uptake	50 % RH, 24 h	%	1.2			0.31	0.25
	50 % RH, equil.	%	2.5	1.5	1.8	1.9	3.5
	100 % RH, equil.	%	8.5	5.8	7.5	3.0	7.5
Melting temperature (DSC), dry		°C	265	310	298	243	nc
Heat distortion temperature B (0.45 MPa)		°C	235				140
Vicat temperature B		°C	200	130	120		146
Heat distortion temperature A (1.8 MPa)		°C	75	120		93	120
Glass temperature (DSC, dry)		°C	50	127	113	85	140
	(DSC, 50 % RH)	°C	15		40	52	
	(DSC, 100 % RH)	°C	–32			15	
Linear thermal expansion coefficient		10^{-5} K^{-1}	10			7.2	6.0
Shrinkage	(50 % RH)	%	1.0				0.5
Specific heat capacity		J/(K g)	1.7				1.45
Thermal conductivity (20°C)		W/(m K)	0.23	0.24		0.38	0.21
Tensile modulus	dry	MPa	3200	3200	3200	4400	2800
	50 % RH	MPa	1600		3600		3000
Flexural modulus	dry	MPa	2830	3650	3500	4500	2660
	50 % RH	MPa	1200				
Upper yield strength	dry	MPa	83	105	100		90
	50 % RH	MPa	59		110		75
Fracture strength	dry	MPa	83	110	120	99	60
	50 % RH	MPa	77		159	75	
Flexural strength	dry	MPa	117	310		160	135
Elongation at yield	dry	%	5	3	4.5		8
	50 % RH	%	25				
Elongation at fracture	dry	%	60		> 10	2	70
	50 % RH	%	> 300			> 10	150
Impact strength, Izod	dry	J/m	59			20	nb
	50 % RH	J/m	107				
Notched impact str., Izod	dry	J/m	59	110	35		360
	50 % RH	J/m	113	100	40	20	13
Rockwell hardness	dry	-	R 121			M107	M93
	50 % RH	-	R 108				
Relative permittivity	dry, 1 kHz	-	3.8		4.0		3.5
	1 MHz	-	3.5			3.9	3.1
	50 % RH, 1 kHz	-	6.5				3.9
	1 MHz	-	4.1				3.4
Dissipation factor	dry, 1 MHz	-	0.03				0.02
	50 % RH, 1 MHz	-	0.08				0.03
Volume resistivity	dry	Ω cm	10^{15}			10^{16}	>10^{15}
	50 % RH	Ω cm	10^{13}				>10^{14}

8.4.2 Polyimides

"Polyimide" (**PI**) is the umbrella term for two types of polymers with imide groups in the main chain: thermosetting and thermoplastic. In the "thermoset type", diamines such as 4,4'-diaminodiphenyl ether, *p,p'*-diaminodiphenylmethane, or 1,4-phenylenediamine are reacted with tetracarboxy dianhydrides, trimellitic anhydride, or more complex dianhydrides. In the first step, a polyamic acid is formed (Volume II, p. 498 ff.). The subsequent intramolecular ring-closure is accompanied by intermolecular condensation, i.e., thermosetting.

Obviously, such thermosetting polyimides can be produced from very different types of monomers so that there are very many grades of many types of polyimides from very many companies. The following chemical structures are just representatives of the three major types of **polyimides (PI**, example: Kapton® (chemically a poly*ether*imide!)), **polyamide-imides (PAI**; example: Torlon®), **polyetherimides** (example: Ultem®), and **polyesterimides** (not shown). The acronym **PEI** is used for both poly*ester*imides (ISO, ASTM) and poly*ether*imides (DIN). These polymers are used as molding compounds as well as films and lacquers. Some physical properties are listed in Table 8-28.

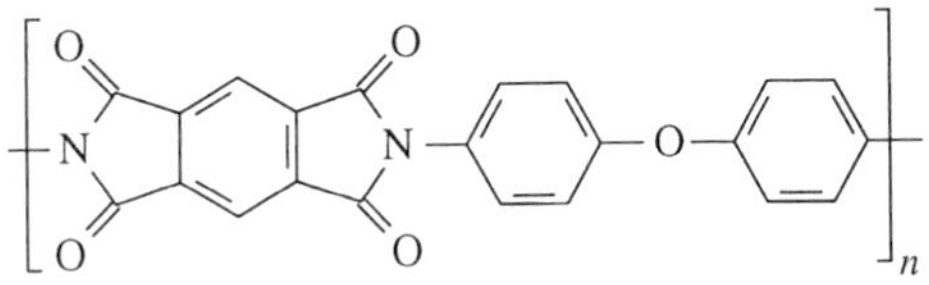

PI
Kapton®

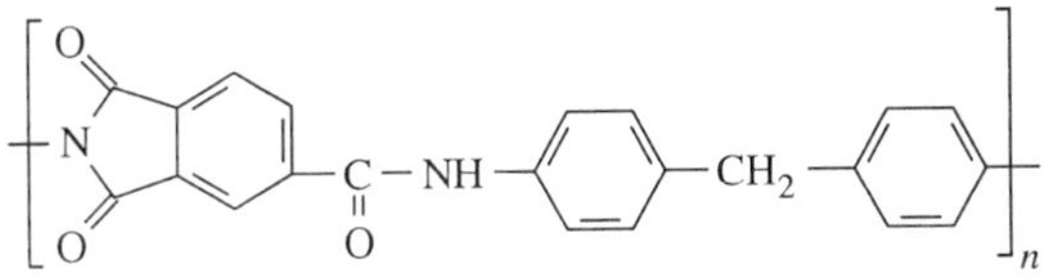

PAI
Torlon®

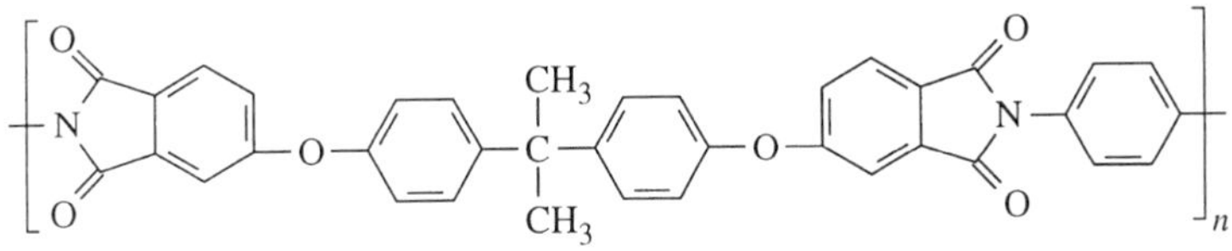

PEI
Ultem®

Intermolecular ring closure is avoided in the synthesis of **thermoplastic polyimides**. In these syntheses, bismaleimides with preformed imide groups are reacted with diamines, disulfides, aldoximes, and the like to **polybismaleimide (BMI)** with the following schematic structure where Ar, Ar' = aromatic groups and Z = NH, S, etc.:

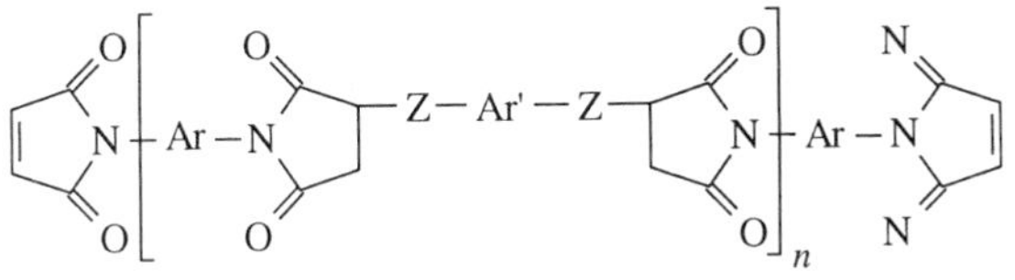

These polymers are cured (crosslinked) by peroxides or azo compounds. The cured polymers are tough materials with low thermal expansion coefficients, outstanding temperature resistance, low flammability, good dielectric properties, and high radiation resistance. They are used in the aerospace and automotive industries.

Table 8-28 Typical physical properties of some unfilled high-performance plastics.

Physical property	Physical unit	PI Kapton	PEI Ultem 1000	PAI Torlon	PEEK Victrex	PES Victrex 200P	PAS Radel R 5000	PSF Udel P1700	PPS Ryton
For the chemical structure, see page:		341	341	341	343	343	343	343	342
Density	g/cm^3	1.43	1.27	1.38	1.30	1.37	1.29	1.24	1.35
Water absorption, 24 h	%	0.03	0.25	0.28	0.14	0.43		0.22	0.02
Thermal expansion coeff.	$10^{-5}\,K^{-1}$	5.0	6.1	4.0	4.7	5.5	5.6	5.6	4.9
Shrinkage	%		0.6	1.4	1.1			0.7	
Melting temperature	°C	-	-	-	334	-	-		285
Glass temperature	°C	235	220		144	230	220	188	185
Heat distortion temp. A	°C	135	200	274	156	215	204	174	135
Tensile modulus	GPa	1.3	3.3		4.0	2.4	2.14	2.25	3.8
Flexural modulus	GPa	3.4	3.3	4.6	3.9	2.6	2.28	2.69	3.8
Upper yield strength	MPa				91	85	72	70	
Flexural strength	MPa		145			130	86	106	96
Tensile strength at break	MPa	120	110	95	100	83		76	66
Elongation at break	%	10	60	3	150	60	60	75	1.6
Notched impact strength	J/m	37	153	135	48	90	640	70	16
Hardness (Rockwell R)	-				126	88		69	120
Relative permittivity (1 MHz)	-	-	3.4	3.2	3.9	3.3	3.5	3.4	3.2
Dielectric loss factor ($\cdot 10^4$)	-	-	20	20	6	30	22	20	7

8.4.3 Poly(*p*-phenylene sulfide)

Poly(*p*-phenylene sulfide) (**PPS**), $\text{--S-(1,4-C}_6\text{H}_4\text{--)}_n$, results from the polycondensation of disodium sulfide Na_2S with 1,4-dichlorobenzene in *N*-methyl-2-pyrrolidone as solvent at 260°C. Polymers are slighly yellowish (from $FeCl_3$ produced by reactor walls) and have molecular weights of 15 000-20 000. These low-molecular weight polymers can be employed for coatings but their melt viscosities are too low for injection molding and extrusion. Coatings are cured (= oxidized) with hot air which leads to brown, insoluble products. Higher molecular weights are obtained if alkali carboxylates are employed as modifiers which makes curing unnecessary.

PPS is flame retardant and stable in air up to 500°C; melting temperatures of un-crosslinked polymers are up to 315°C. Polymer grades are usually filled with minerals (for electrical switches, automotive pumps, etc.) or reinforced by glass fibers (for example, for wing profiles of newer Airbus jet aircraft). Unfilled PPS is converted to fibers, films, and sheets and as coatings for pumps, valves, and cooking utensils.

8.4.4 Polyarylene Ether Sulfones

Aromatic polysulfones with ether groups in the main chain are often simply called **polysulfones** (**PSU**). These polymers are either homopolymers or alternating or random copolymers and are known by many names and abbreviations (see below; other abbreviations include PPSF and PSO). The slightly yellow, transparent polymers are often stabilized against degradation by ultraviolet light. Melt viscosities are high but shear thinning is fairly small; injection-molded parts are therefore practically isotropic.

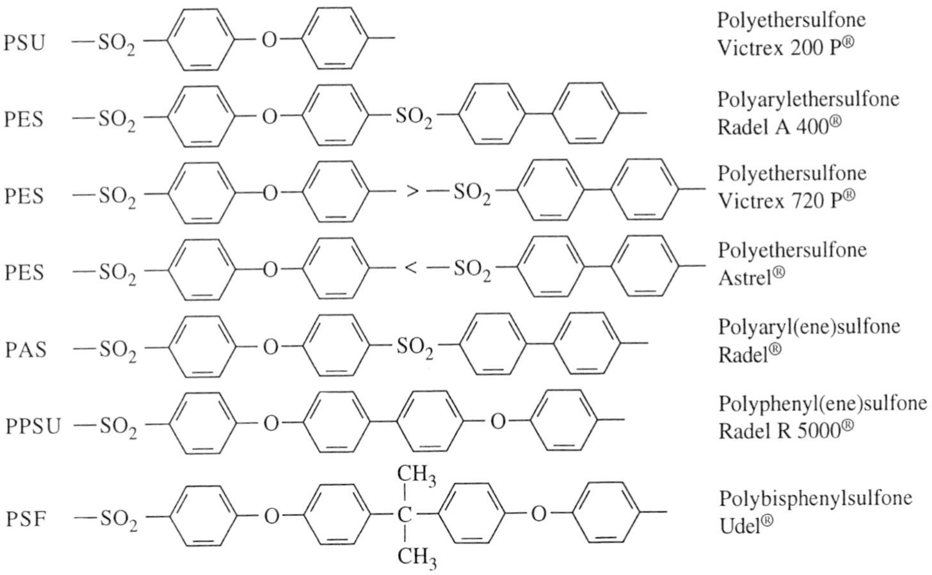

PSU —SO$_2$ [structure] Polyethersulfone
Victrex 200 P®

PES —SO$_2$ [structure] Polyarylethersulfone
Radel A 400®

PES —SO$_2$ [structure] Polyethersulfone
Victrex 720 P®

PES —SO$_2$ [structure] Polyethersulfone
Astrel®

PAS —SO$_2$ [structure] Polyaryl(ene)sulfone
Radel®

PPSU —SO$_2$ [structure] Polyphenyl(ene)sulfone
Radel R 5000®

PSF —SO$_2$ [structure] Polybisphenylsulfone
Udel®

In these polymers, arylene sulfone groups are responsible for the excellent heat stability whereas ether groups provide chain flexibility and good processability (for properties, see Table 8-28). Some grades of polysulfones are blends with ABS or SAN and other grades are reinforced with 30 wt% glass fibers or carbon fibers.

8.4.5 Polyarylenetherketones

Polyaryl(en)etherketones (PAEK) are commonly known as **polyetherketones (PEK)**. Similarly to polyaryl(en)ethersulfones, they are produced with many different types of repeating units. In contrast to symbols for PSUs, symbols for PEKs are much more logical: ether units are symbolized by E and ketone units by K. The number and sequence of these two letters characterizes the polymer. The repeating unit of PEKK thus has one ether unit followed by two ketone units whereas PEEK has two ether units followed by one ketone unit. Ether and ketone groups are always separated by 1,4-phenylene units.

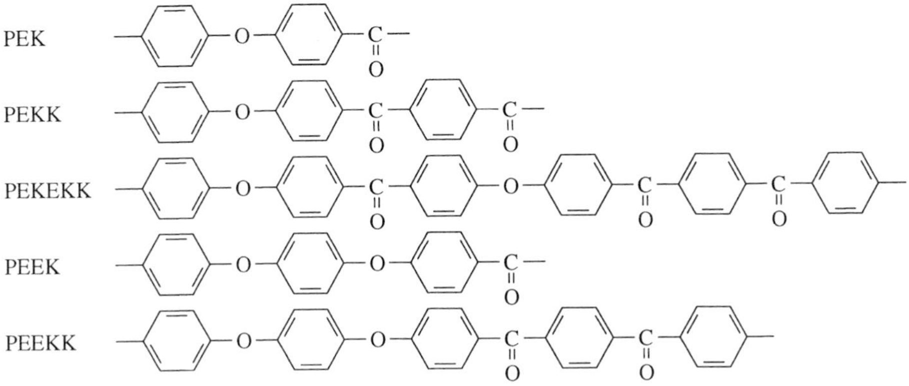

PEK [structure]

PEKK [structure]

PEKEKK [structure]

PEEK [structure]

PEEKK [structure]

Table 8-29 Properties of some polyaryletherketones.

Property	Physical unit	PEK Kadel®	PEEK Victrex®	PEEKK Hostatek®	PEKEKK Ultrapek®
Density, 100 % amorphous	g/cm^3	1.272	1.264		
commercial grade	g/cm^3		1.320	1.30	1.32
100 % crystalline	g/cm^3	1.430	1.401		
Melting temperature (DSC)	°C	364	334	363	377
Heat distortion temperature (ISO A)	°C		140	103	170
Glass temperature (DSC)	°C	153	143	167	175
Continuous service temperature	°C		250	220	260
Linear thermal expansion coefficient	K^{-1}		$4.7 \cdot 10^{-6}$		$4.2 \cdot 10^{-6}$
Thermal conductivity (20°C)	W/(m K)		0.25		0.24
Tensile modulus	MPa	3190	3650	4000	4700
Tensile strength at yield	MPa		92	100	
Tensile strength at fracture	MPa	104	92	90	118
Elongation at yield	%		4.9	5.5	
Elongation at fracture	%		50	28	13
Impact strength (Izod, 3.1 mm)	J/m		no break		
(Charpy)	kJ/m^2		no break	no break	
Notched impact strength (Izod, 3.1 mm)	J/m	59	83		80
(Charpy)	kJ/m^2		8.2	8	10
Hardness (Rockwell)	-		M99		
Relative permittivity (50 Hz)	1		3.2		3.4
Resistivity (volume resistance)	Ω cm		$5 \cdot 10^{16}$		$>10^{16}$
Dissipation factor (50 Hz)	1		0.003		0.002
Water absorption (24 h)	%		0.5		0.2

All polyetherketones are semicrystalline polymers with melting temperatures between 330°C and 380°C and glass temperatures between 130°C and 180°C (exmples: Table 8-29). They are rigid (3.2 ≤ E/GPa ≤ 4.7), strong (90 ≤ σ_B/MPa ≤ 120), and impact resistant (Izod and Charpy tests: no fracture). PEEK is a matrix resin in fiber-reinforced composites for aerospace applications.

8.4.6   Syndiotactic Poly(styrene)

Syndiotactic poly(styrene) (**sPS**) (Volume II, p. 272) is obtained by polymerization of styrene with metallocene catalysts (Volume I, p. 290). The semicrystalline polymer has the same glass temperature as free radical-polymerized atactic poly(styrene) (T_G = 100°C). Its high melting temperature (T_M ≈ 270°C) is combined with approximately equal densities of crystalline and amorphous regions, which provides SPS with good shape retention. All industrial grades are reinforced by glass fibers.

8.4.7   Fluoroplastics

Fluorinated thermoplastics (fluoroplastics) have tensile moduli and tensile strengths that are typical for standard plastics (Table 8-30). Their notched impact strengths and fairly high heat distortion temperatures would put them into the class of engineering plastics. However, it is their surface properties (low wettability, small friction coefficients, non-stick characteristics) that elevate them to the class of high-performance plastics. Fluoroplastics also have excellent weatherability and flame retardancy and are physiologically benign at conventional use conditions.

Fluoroplastics may be homopolymers such as

PTFE	poly(tetrafluoroethylene)	$-[CF_2-CF_2]_n-$
PCTFE	poly(chlorotrifluoroethylene)	$-[CCIF-CF_2]_n-$
PVDF	poly(vinylidene fluoride)	$-[CH_2-CF_2]_n-$
PVF	poly(vinyl fluoride)	$-[CH_2-CHF]_n-$

alternating copolymers such as

| ETFE | poly(ethylene-*alt*-tetrafluoroethylene) | $-[CH_2-CH_2-CF_2-CF_2]_n-$ |
| CM-1 | poly(vinylidene fluoride-*alt*-hexafluoropropylene) | $-[CH_2-CF_2-CF_2-CF(CF_3)]_n-$ |

or random copolymers such as

CTFE	poly(chlorotrifluoroethylene-*co*-tetrafluoroethylene)	$-[CCIF-CF_2-/-CF_2-CF_2]_n-$
ECTFE	poly(ethylene-*co*-chlorotrifluoroethylene)	$-[CH_2-CH_2-/-CCIF-CF_2]_n-$
FEP	poly(tetrafluoroethylene-*co*-hexafluoropropylene)	$-[CF_2-CF_2-/-CF_2-CF(CF_3)]_n-$
PFA	a copolymer from $CF_2=CF_2$ and a small proportion of perfluorovinyl methyl ether	$-[CF_2-CF_2-/-CF_2-CF(OCF_3)]_n-$

The parent compound of all fluoroplastics is poly(tetrafluoroethylene) (**PTFE**). The original trade name **Teflon** for this polymer has now become a generic name for most fluorinated plastics; a "teflon" of the DuPont company may therefore be a PTFE, a FEP, or a PFA. Unfortunately, abbreviations may also vary; for example, poly(chlorotrifluoroethylene) is known as PCTFE or just as CTFE although the latter is the symbol for poly(chlorotrifluoroethylene-*co*-tetrafluoroethylene).

Table 8-30 Typical physical properties of unfilled fluoroplastics. nb = no break.

Property	Physical unit	PTFE	FEP	PFA	ETFE	CTFE	ECTFE	CM-1	PVDF
Density	g/cm^3	2.2	2.15	2.15	1.70	2.10	1.68	1.88	1.76
Moisture uptake (24 h)	%	0	0.4	0.03	0.1	0	0.01		0.03
Thermal expansion coeff.	10^{-5} K^{-1}	10	9	12	7	7	8	4	9
Shrinkage	%	4.5	4.5			1.2			3.0
Melting temperature	°C	327	275	305	270	220	240	327	175
Glass temperature	°C				110	40	−76		
Heat distortion temp. A	°C	49	51	48	74	66	78	220	90
Continuous service temp.	°C	260	205	260	150	175	150		150
Brittleness temperature	°C	−200	−100	−200	−100	−40	−100		−60
Tensile modulus	GPa	0.42	0.36	0.70	0.85	2.20	6.70	3.90	0.86
Flexural modulus	GPa	0.68	0.69	0.68	1.40	1.28	1.70	4.70	1.43
Flexural strength	MPa	nb	18	15	28	58	50		55
Compressive strength	MPa		21		49				
Tensile strength	MPa	25	21	28	45	43	56	39	45
Fracture elongation	%	350	300	300	200	250	200	2	150
Notched impact strength	J/m	160	nb	nb	nb	150	nb	21	200
Hardness, Rockwell R	-	45							
Shore B	-		56		75				
Relative Permittivity (1 MHz)	-	2.1	2.1	2.1	2.6	2.5	2.5	2.3	8
Dielectr. loss factor ($\cdot 10^4$)	-	2	7	20	5	130	10	20	1000

Because of the extremely high melt viscosity above its high melting temperature of 327°C (Table 8-30), PTFE cannot be processed by extrusion or injection molding but only by machining or sintering. The polymer flows somewhat under pressure; because of this, bearings and seals from teflon contain fillers.

With the exception of CM-1, all other fluorinated thermoplastics have lower melting temperatures than PTFE (Table 8-30). In general, upper continuous service temperatures are lower and brittleness temperatures are higher than those of PTFE.

8.4.8   LCP Glasses

LCP glasses are liquid-crystalline polymers (LCP) below their glass temperatures. These polymers contain rodlike mesogenic units in their main chains or in their side chains (Volume III, p. 240 ff.) and may be either thermotropic (liquid crystallinity caused by temperature change) or lyotropic (liquid crystallinity above a certain critical concentration). LCP thermoplastics are glasses of thermotropic main-chain polymers. Thermotropic side-chain LCPs serve as optical storage elements and lyotropic side-chain LCPs as fibers (see Section 6.7.5).

Thermotropic main-chain LCPs contain rod-like mesogenic units (shown in Fig. 8-6 as rectangular boxes) which are practically oriented parallel in domains of LCPs. The preferential directions of mesogenic units differ from domain to domain. Rigid segments of these mesogenic units can be flexibilized by incorporating rigidity breakers (aka flexible spacers) which may be side-chain carrying units, kinked monomeric units, or flexible chain segments (top to bottom in Fig. 8-6).

During processing of liquid-crystalline states (usually nematic), flow causes the longitudinal axes of mesogens to orient themselves parallel to each other. This preferential orientation is maintained if the melt is cooled below the glass temperature. It leads to a considerable increase of tensile moduli and fracture strengths in the longitudinal mesogen direction (∥) but a decrease in the transverse direction (⊥) relative to the isotropic state (Table 8-31). Because such effects are usually obtained by reinforcement of plastics with oriented glass fibers, LCP glasses are often called **self-reinforcing plastics**.

The anisotropy caused by the self-reinforcing effect improves the strength of well-designed supporting parts in the desired directions (compare wood (Volume II, p. 67)). In supporting parts from isotropic engineering plastics, macroscopic anisotropies can only be obtained through part design, for example, by incorporated additional stiffening ribs.

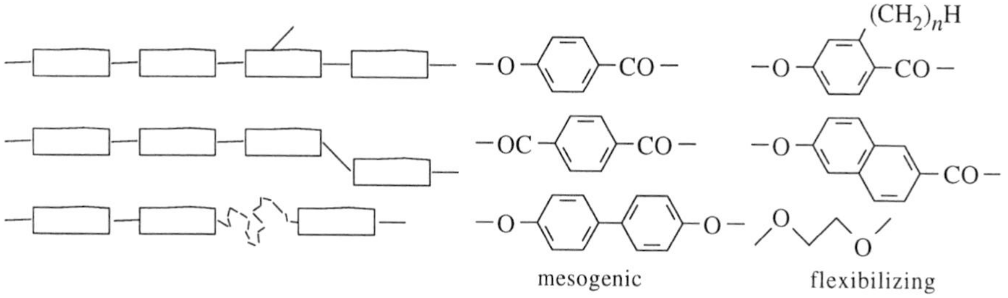

Fig. 8-6 Left: flexibilizing of chains with rigid mesogenic units by rigidity-breaking flexible spacers (schematic). Center: rigid mesogenic units. Right: rigidity breakers with (top to bottom) frustrated crystallization, non-linear chain units, flexible chain units (see schematic effect to the left).

Table 8-31 Tensile moduli E and fracture strengths σ_B of low-density poly(ethylene), poly(hexamethylene adipamide) (PA 6.6), and poly(ethylene terephthalate) (PET) as isotropic molding compounds and in the longitudinal direction of their fibers as well as properties of thermotropic LCP glasses (TT) and lyotropic LCP fibers (LT), respectively, in longitudinal and transverse directions as compared to isotropic states. PPBT = 30 wt% poly(p-phenylene benzbisthiazole) in poly(2,5-benzimidazole); HX 2000 = amorphous LC copolyester with 30 wt% glass fibers; for X7G, Vectra, Kevlar: see text. [1] With 30 wt% glass fibers, * flexural moduli.

Polymers		E/GPa			σ_B/MPa		
		longitudinal	transverse	isotropic	longitudinal	transverse	isotropic
-	PE-LD			0.15			23
-	PA 6.6	13		2.5	1000		74
-	PET, PES	19		0.13	1400		54
TT	X7G	54	1.4	2.2	151	10	63
TT	Vectra®	11	2.6	5.0	144	54	97
TT	HX 2000 [1]	17*	4.2*		159	66	
LT	Kevlar®	138	7		2800		
LT	PPBT	120	17	62	1500	680	700

Most industrially produced thermotropic main-chain LCPs are copolymers based on p-hydroxybenzoic acid. The homopolymer, poly(p-hydroxybenzoic acid) (**PHB**) has excellent mechanical properties (Ekonol P-3000®; Table 8-32) but a melting temperature of at least 550°C and such a high melt viscosity that it cannot be conventionally processed. It was not offered as such but as a mixture with powders of poly(tetrafluoroethylene) (Ekonol T-4000®), aluminum, or bronze or as an aqueous dispersion with PTFE. Processing of these metal-like compounds was by sintering, flame spraying, or forging.

p-Hydroxybenzoic acid has therefore been used as a stiffening comonomer. The first self-reinforcing plastic was a copolyester X7G with 60 mol% p-hydroxybenzoyl units (I), 20 mol% terephthaloyl units (II), and 20 mol% ethylene glycol units (III). This composition was chosen because it leads to minima in melt viscosities and maxima in many mechanical properties (Fig. 8-7).

X7G is not a random copolymer but consists of a phase from a polymer with blocks of I and blocks of II + III and another, amorphous phase of unknown composition. The polymer is no longer produced because similar mechanical properties can be obtained from far less expensive glass fiber-reinforced saturated polyesters.

Rodrun® is a copolyester with the same units I, II, and III as X7G but with an unknown composition and a different block length of terephthaloyl glycol and p-hydroxybenzoic acid units. Its properties differ considerably from those of X7G (Table 8-32).

Much better processable than poly(p-hydroxybenzoate) (I) itself or X7G (I+II+III) are various other copolymers containing monomeric units I. Examples are the series of

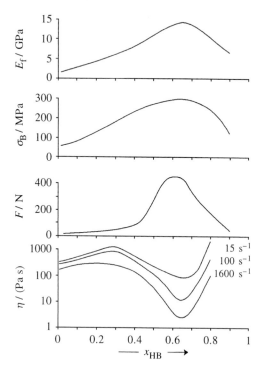

Fig. 8-7 Flexural moduli E_f, fracture strengths σ_B, impact strengths F, and melt viscosities η of co-polyesters of terephthalic acid, ethylene glycol, and p-hydroxybenzoic acid (HB) as a function of the mole fraction x_{HB} in their block copolymers [8].

Vectra® polymers with different compositions, for example, Vectra® 950 with 67 mol% I and 33 mol% V; Xydar 300®, also with 67 mol% I; Ekkcel I-2000® with units I, II, and IV; and Ekkcel C-1000® with units I, VI, and VIII (Table 8-32). Even better process-abilities are exhibited by copolymers from II+VI+VIII (U-Polymer®, Ardel®, Arylef®) but these polymers have smaller moduli and strengths (example in Table 8-32).

PHB as well as the copolymers I+II+VI and I+V+VII are self-lubricating. This prop-erty is further improved by the addition of graphite, boron nitride, molybdenum disul-fide, or poly(tetrafluoroethylene). Such polymers are used for bearings, gears, electrical and electronic parts, cooking utensils, and under-the-hood automotive parts.

8.5   Thermosets

8.5.1   Overview

Thermosets result from simultaneous shaping and irreversible chemical crosslinking (curing, hardening) of **prepolymers (reaction polymers, thermosetting compounds)** by polyaddition, polycondensation, or chain polymerization, usually at elevated tempera-tures. Prepolymers may be liquid or solid. Powdery prepolymers must have glass tem-peratures of at least 40-50°C so that they do not sinter or stick together while stored.

Table 8-32 Typical physical properties of unfilled LCP glasses.

Property	Physical unit	X7G	Rodrun 300	Xydar A 950	Ekonol P-3000	Ekkcel I-2000	Ekkcel C-1000	Ardel 1000
Density	g/cm³	1.39		1.35	1.45	1.40	1.35	1.21
Water absorption (24 h)	%				0.4	0.025	0.040	
Thermal expansion coeff.	$10^{-5}\,K^{-1}$	0		0.1	1.5	2.9	5.2	6.1
Shrinkage	%							
Melting temperature	°C				>550	413	370	
Heat distortion temp. A	°C	145	170	355		293	300	175
Tensile modulus	GPa			9.7	7.25	2.55	1.33	2.0
Compressive modulus	GPa					3.5	2.1	
Flexural modulus	GPa	6.0	10.0	11.0	0.51	4.9	3.2	2.0
Flexural strength	MPa	115	145	131		120	106	81
Fracture strength	MPa		220	116		99	70	70
Elongation at break	%	100	4.5	4.9		8.0	8.0	50
Notched impact strength	J/m	18	40	127		55	22	210
Hardness (Rockwell M)	-					88	124	
Relative permittivity (1 MHz)	-			3.9		3.16	3.68	3.0
Dielectric loss factor ($\cdot 10^4$)	-			390	264	30		150

Phenolic resins, amino resins, furan resins, and certain polyimides harden by polycondensation (Section 8.5.3), urethanes and epoxides by polyaddition (Section 8.5.4), and unsaturated polyesters, allyl compounds, *endo*-dicyclopentadiene, and some so-called vinyl esters and acrylic esters by chain polymerization (Section 8.5.5).

All of these compounds have high concentrations of reactive groups per monomer or prepolymer molecule, respectively, which subsequently leads to high crosslinking densities of the cured polymers. Consequently, segment lengths between crosslinking sites are short, which leads to low shrinkage and low segment mobilities and therefore to good creep resistances and high heat distortion temperatures (Table 8-33).

Table 8-33 Typical physical properties of unfilled thermosets. DAP = diallyl phthalate, UP = unsaturated polyester, PF = phenol-formaldehyde, MF = melamine-formaldehyde, UF= urea-formaldehyde, EP = epoxy, PUR = polyurethane.

Property	Physical unit	DAP	UP	PF	MF	UF	EP	PUR
Density	g/cm³	1.27	1.3	1.25	1.48		1.2	1.05
Moisture uptake (24 h)	%	0.2	0.4	0.15			0.13	0.2
Thermal expansion coefficient	$10^{-5}\,K^{-1}$	11	8	8			5	
Shrinkage	%	1.0	0.6	1.1	0.7	1.3	0.5	1.0
Heat distortion temperature A	°C	155	130	121	148		170	91
Tensile modulus	GPa	2.2	3.4	2.8			2.5	
Flexural modulus	GPa	2.1						4.4
Tensile strength at break	MPa	28	70	65			70	
Fracture elongation	%		2	1.8			6	
Notched impact strength	J/m	17	16	16			35	21
Hardness (Rockwell M)	-	98	90	126			95	
Relative permittivity (1 MHz)	-		3.4	3.5	4.7		3.7	3.5
Dielectric loss factor ($\cdot 10^4$)	-	500	200	200			400	30

Thermosetting materials have the disadvantage of relatively long processing times and the irreversibility of the thermosetting reaction. These disadvantages are compounded by the fairly high cost of some thermosetting materials (p. 314) and also by the impossibility of recycling most thermosets to their monomers. Energetic recycling is possible but requires high combustion temperatures (Section 8.6).

For these reasons, market shares of thermosetting materials have shrunk considerably from 69 % in 1940 to 38 % in 1960 and just 5 % in 2005 (p. 316) although their market volume increased ca. 3.5 times between 1960 and 2005. More than 50 wt% of all thermosetting materials are presently consumed by the electrical and electronic industries and approximately 14 wt% each for household goods and sanitary articles.

8.5.2   Curing

The **curing** of thermosetting "prepolymeric" compounds consists of branching and subsequently **crosslinking** reactions that lead to the **hardening** of matter. During these reactions, two processes overlap: gel formation and glassy solidification (Fig. 8-8). At temperatures above a temperature T_{Gg}, liquid prepolymers are converted to branched molecules of greater molecular weight with increasing numbers of functional groups per molecule. At a certain critical conversion of the groups (the **gel point**), many branched molecules interconnect to an "infinitely" large molecule that is swollen by the remaining lower molecular weight molecules (prepolymers and their reaction products). With increasing time, more and more of the remaining soluble molecules are incorporated into the network and the whole system solidifies to a glass (for details, see Volume I, p. 491). At sufficiently low temperatures, no gel is formed and the whole systems solidifies directly to a glass. The temperature T_{Gg} can therefore be identified as that temperature at which the time for the gel formation equals the time for the glass formation.

This behavior is general but cannot always be observed. For example, in the vulcanization of rubbers, glasses are not formed because most rubbers and elastomers have very low glass temperatures. Conversely, glass temperatures of cured phenolic resins are higher than their decomposition temperatures.

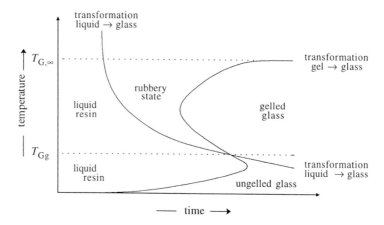

Fig. 8-8 Processes during thermosetting reactions at various curing temperatures [9].

Ideally, thermosetting should proceed without shrinkage. But contractions do occur during curing because in these polymerization processes some intermolecular distances of monomer and prepolymer molecules, respectively, are replaced by shorter intra-molecular covalent bonds (Volume I, p. 187). Obviously, one needs to reduce the concentration of reacting groups immediately before crosslinking sets in (i.e., in the so-called **B stage**). This can be achieved by (a) using thermosetting resins with greater molecular weight, (b) adding fillers to the resins, and/or (c) applying after-pressure.

Crosslinking is a statistical process which leaves some functional groups unreacted. At the moment crosslinking sets in, these groups do not find reaction partners because crosslinking drastically reduces segment mobilities. However, they may diffuse slowly and react much later. As a result, raw strengths of thermosets immediately after shaping are considerably lower than those after long times.

Crosslinks are formed at random which leads to heterogenous networks and less than optimal physical properties of thermosets. Free chain ends act not only as "plasticizers" but also react more readily with oxygen, etc., than with intramolecular groups. For this reason, thermosetting materials have been developed that can only be crosslinked via their endgroups, for example, acetylene end groups of oligophenylenes. Such resins include so-called macromonomers which are oligomers with reactive endgroups.

8.5.3 Thermosets by Polycondensation

Resins for thermosetting by polycondensation comprise certain imides (Section 8.4.2) and the reaction products of formaldehyde (**F**) with phenols (mainly phenol itself) to **phenolic resins (PF)** or with amines to **amino resins** such as **urea resins (UF)** or **melamine resins (MF)**. The use of other aldehydes or other amino compounds is rare.

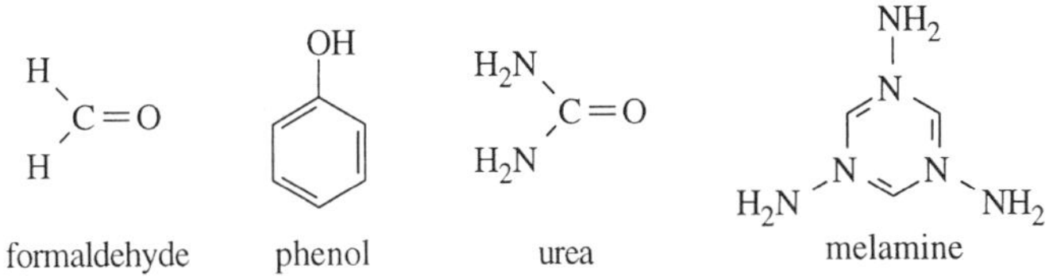

formaldehyde   phenol   urea   melamine

Phenolic Resins

Phenolic resins result from the reaction of phenol and formaldehyde to methylol phenols such as I and further mainly to methylene phenols such as II and also to open-chain formals such as III (for details and the mechanism, see Volume II, p. 261 ff.):

(8-2)

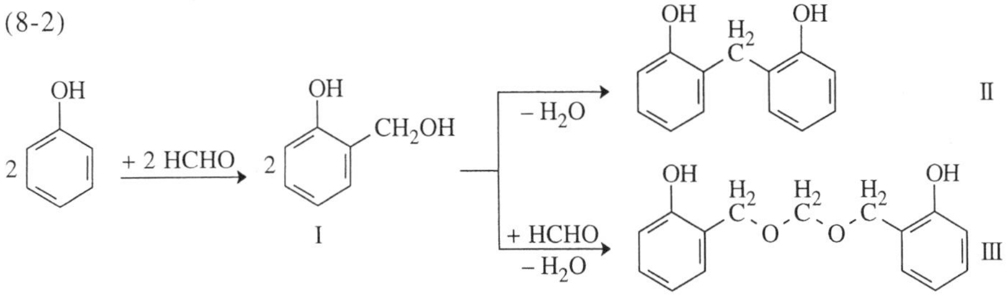

Each phenolic unit has three reactive positions (2 ortho, 1 para) that finally lead to crosslinking. Acid-catalyzed reactions lead to **novolacs** (= new (shel)lacs); L: *novus* = new; Italian: *lacca* = lacquer) while base catalysis results first in **resols** (A state), then in **resitols** (B stage), and finally in **resits** (C stage).

In stage A, resols are meltable and still soluble in certain solvents. In stage B, the pre-polymeric resitols still soften on heating and swell in certain solvents but are no longer soluble. In stage C, resits are insoluble and no longer meltable.

Curing produces water, (Eq.(8-2)), which is volatile at the curing temperatures but becomes partially trapped in the polymer at higher degrees of reaction where the melt is very viscous. Shaping takes place when the polymer enters the C stage, i.e., at the beginning of crosslinking. Since crosslinking requires only three reacting groups per molecule, proportions of split-off low molecular byproducts (mainly H_2O) are small and only a few voids are produced. The proportion of voids can be reduced further by adding absorbing materials, for example, wood meal in the case of water.

Phenolic resins are produced in many variations, for example, as so-called plastified (or elastified) resins that are either etherified with alcohols or esterified with fatty acids (see Volume I, p. 261 ff.). The world production of phenolic resins is ca. $3 \cdot 10^6$ t/a.

Amino Resins

These resins are obtained from the condensation of NH group-containing components such as urea or melamine with carbonyl-group containing components such as formaldehyde R' = H) and nucleophilic (H-acidic, OH-acidic, NH-acidic, etc.) components in a kind of Mannich reaction:

$$(8\text{-}3) \quad \text{\textasciitilde} Z-H \;+\; \underset{\underset{R'}{|}}{\overset{\overset{R}{|}}{C}}=O \;+\; H-N\diagup_{\diagdown} \;\longrightarrow\; \text{\textasciitilde} Z-\underset{\underset{R'}{|}}{\overset{\overset{R}{|}}{C}}-N\diagup_{\diagdown} \;+\; H_2O$$

nucleophile   carbonyl comp.   NH compound   amino resin

The systems crosslink because all NH compounds have functionalities greater than 2 ($f = 4$ for urea, $f = 6$ for melamine).

Worldwide, approximately $11 \cdot 10^6$ t/a are produced of which ca. 86 % are urea resins and 14 % melamine resins. Urea resins are water-soluble and mainly used as glue for wood products. Amino resins and phenolic resins are used as molding compounds, lacquers, adhesives, binders for molding sands, fibers (PF), expanded plastics (UF), and many other applications (see also Volume II, p. 261 ff. (PF) and p. 483 ff. (UF, MF)).

8.5.4   Thermosets by Polyaddition

Epoxy resins or isocyanates plus diols or polyols as polyurethane precursors crosslink with hydroxyl group-containing compounds by polyaddition (not an "addition polymerization", see p. 10)). Prereactions to B stages are not necessary since the starting materials are already highly viscous and no low-molecular weight compounds are released Thus, polymerization and shaping occur in a single stage.

Epoxy Resins

Epoxy resins (epoxies, EP) are oligomers that contain oxirane groups (Volume II, p. 322). Most epoxy resins are glycidyl ethers of bisphenol A from the reaction of bisphenol A with epichlorohydrin which leads to epoxide groups I, bisphenol A units II, and glycidyl groups III. Epoxies with improved weatherability are based on cycloaliphatic epoxides such as vinyl cyclohexanediepoxide (IV) or dicyclopentadiene diepoxide (V).

Conventional aliphatic epoxies have the constitution I–[II–III]$_q$–II–I. They are liquids if $0.1 \leq q \leq 0.6$ and solids if $2 \leq q \leq 25$. All commercial epoxies are formulated with plasticizers, fillers, pigments, and the like.

Epoxies are cured (crosslinked) *warm* with multifunctional carboxylic acids such as VI or carboxylic anhydrides such as VII which deliver ester structures (from oxirane + HO–) or ether structures (from oxirane). They may also be cured *cold* with polyfunctional amines such as diethylene triamine VIII or isophorone diamine IX to give β-hydroxypropyl amine groups, >N–CH$_2$–CHOH–CH$_2$–.

Cycloaliphatic epoxides are also cured with *p*-vinyl phenol, CH$_2$=CH(1,4-C$_6$H$_4$OH), or poly(azelaic anhydride), HO[OC(CH$_2$)$_7$CO]$_m$O[OC(CH$_2$)$_7$CO]$_n$OH. During warm curing, *p*-vinyl phenol polymerizes thermally without added initiator to poly(*p*-vinyl phenol) which is the true crosslinking agent.

The many possible types of epoxy resins and curing agents as well as their ability to take up large proportions of many different adjuvants allow taylor-made applications of epoxy resins for many different uses such as two-component adhesives, lacquers, structural elements, large containers, electrical/electronic parts, and the like. Also advantageous are the small shrinkage on curing, the drying in air without heating, and the high continuous service temperatures (Table 8-33).

Disadvantageous are the required exact dosage of resin + curing agent, the long curing times (aftercures may take days), the high viscosities during hardening (problems with degassing and impregnation of fabrics), and the difficult demolding. For less sophisticated applications such as glass fiber-reinforced engineering plastics, epoxy resins compete with the less expensive unsaturated polyesters (Section 8.5.5).

More than 50 % of epoxy resins are used for coatings. Other uses comprise adhesives, printed wiring boards, flooring and paving, binders in mortars and concrete, bonding agents for fiberglass laminates, and many others.

Polyurethanes

These polymers (**PUR**), with the characteristic urethane group –NH–CO–O– in the main chain, result practically exclusively from the polyaddition of isocyanate groups –N=C=O of diisocyanates and triisocyanates to hydroxyl groups HO– of diols and poly-ols (see Volume II, p. 489 ff.).

More than 95 % of all PURs are based on diphenylmethane diisocyanate (= methane-diphenyl-4,4'-diisocyanate (MDI)) and toluene diisocyanate (TDI) (industrially also known as tolylene diisocyanate or toluylene diisocyanate), which is either a 80/20 or a 65/35 mixture of the 2,4 and 2,6 isomers. Other isocyanate compounds comprise hydro-genated MDI (H$_{12}$MDI), *p*-phenylene diisocyanate (PPDI), 1,5-naphthalene diisocyanate (NDI), and 2,4-methylene diphenylisocyanate (MDPI).

TDI 80:20 or TDI 65:35 (2,4:2,6)   diphenylmethane-4,4'-diisocyanate (MDI)

H$_{12}$MDI   PPDI   NDI   MDPI

Functionalities f of the low-molecular weight hydroxy compounds used range from f = 2 of 1,4-butanediol and f = 3 of glycerol to f = 6 of sorbitol and f = 8 of saccharose. The vast majority (90 %) of the hydroxy compounds used are so-called **polyether-poly-ols** HO–Z$_n$–H where Z$_n$ may be $+CH_2CH(CH_3)O+_n$ or $+(CH_2)_mO+_n$ with m = 2 or 4. So-called modified polyether-polyols consist of polyether-polyols that are grafted upon polymer particles from acrylonitrile-styrene copolymers (= polymer-polyols), polyurea-polyol dispersions, or polyurethane dispersions.

The world-wide production of polyurethanes is ca. $8.6 \cdot 10^6$ t/a (2000). About 70 % are used for foams (expanded plastics; Section 10.5) and 30 % for elastomers, lacquer systems, and plastics.

Reaction Injection Molding

A special processing method for very fast-polymerizing chemicals is **reaction injec-tion molding** (**RIM**) in which these compounds are mixed with catalysts and then in-jected immediately into a mold where they are simultaneously polymerized and shaped. Polymerizing chemicals must be liquid at processing temperatures of 100°C or less and must have very low viscosities of less than 1 Pa s for mixing and 0.01-0.1 Pa s for injec-tion. Suitable compounds comprise certain monomers for polyurethanes, *endo*-dicyclo-pentadiene, and SiO$_2$-filled methyl methacrylate. ε-Caprolactam is no longer polymer-ized by the RIM process.

The polymerizing material must gel within seconds to shape-stabilized articles, which is achieved by either crosslinking reactions or by fast crystallizations. Cycle times should be ca. 30 s for mass-produced articles and not more than 3 min for special parts. RIM processes require less capital investment, provide faster production, use less energy, and deliver better products than conventional processing methods.

The most important RIM processes comprise the production of PUR elastomers and flexible expanded polyurethanes. Raw materials are usually TDI and MDI, polymer polyols, and conventional propylene-based polyols that are capped with ethylene oxide. Chain extenders are conventional ethylene glycol types. Use of amine-type chain extenders leads to hybrids of polyurethanes and polyureas.

8.5.5   Thermosets by Chain Polymerization

Thermosets are also obtained by free-radical crosslinking polymerization of compounds with functionalities of three or more. Such compounds comprise allyl com-monomers, unsaturated polyesters, and so-called vinyl esters.

Allyl Polymers

Crosslinking requires functionalities of $f \geq 3$ which is provided by diallyl compounds ($f = 4$) such as *diallyl diglycol carbonate* (DADC), *diallyl phthalate* (DAP), *diallyl iso-phthalate* (DAIP), triallyl compounds ($f = 6$) such as *triallyl cyanurate* (TAC; 2,4,6-tris-(allyloxy)-*s*-triazine), and many other allyl compounds (see Volume II, p. 298):

$$CH_2{=}CHCH_2OCOOCH_2CH_2$$
$$CH_2{=}CHCH_2OCOOCH_2CH_2$$

DADC

$$CH_2{=}CHCH_2O{-}N{=}N{-}OCH_2CH{=}CH_2$$
$$N$$
$$CH_2{=}CHCH_2O \quad N \quad OCH_2CH{=}CH_2$$

TAC

$$CH_2{=}CHCH_2OOC$$
$$CH_2{=}CHCH_2OOC$$

DAP

These monomers are low-viscosity substances which need to be prepolymerized by free radicals to a B stage, often to less than 25 % with respect to allyl groups. In the subsequent C stage, prepolymers are simultaneously crosslinked and shaped.

Polymers from DAP and DAIP are excellent insulating materials. Allyl monomers such as TAC are also used as crosslinking comonomers in glass fiber mats that have been pre-impregnated with unsaturated polyester resins. Polymers of DADC have about the same light transmission as poly(methyl methacrylate) but a 30-40 % greater scratch resistance. Their main use is in sun-glasses where they compete with polycarbonates.

Unsaturated Polyesters

Industrial unsaturated polyesters result from the polycondensation of glycols with carboxylic acid anhydrides or dicarboxylic acids, especially ethylene glycol and maleic anhydride:

$$(8\text{-}4) \qquad HOCH_2CH_2OH \; + \; \text{[maleic anhydride]} \; \xrightarrow{-H_2O} \; \left[OCH_2CH_2O-C{\overset{O}{\diagdown}}{\underset{H}{\overset{}{C}}}{=}{\underset{\overset{}{C}{\diagdown}O}{\overset{H}{C}}} \right]$$

During polymerization, a large proportion of cis double bonds of maleic anhydride molecules are isomerized to trans double bonds, delivering fumaric acid units. In this reaction, glycols have to be used in excess since up to 15 mol% of the glycols add to the C–C double bonds where they form ether bonds.

The resulting molecules $+CO\text{–}CH{=}CH\text{–}CO\text{–}OCH_2CH_2O+_n$ act as crosslinking agents, mainly for styrene but also for methyl methacrylate and other monomers. Industry considers these monomers as the crosslinking agents, probably because it is they that are added. Industrially, "unsaturated polyester" (**UP**) also refers to the crosslinked polymer and not to the true unsaturated polyester molecules.

Since unsaturated polyester molecules contain many carbon-carbon double bonds, only a relatively small proportion of them are needed for crosslinking reactions. The remaining carbon-carbon double bonds can later react with oxygen and other reactive gases, which causes the less than optimal weatherability of unsaturated polyesters. Crosslinking also generates many free chain ends that either add nothing to the mechanical properties of the cured polymers or affect them adversely.

Unsaturated polyester resins are usually reinforced by glass fibers; these compounds serve for many purposes, from opaque structural elements to boat hulls. Applications are made easier by the use of sheet molding compounds (**SMC**) which are glass fiber mats that are pre-impregnated with formulated mixtures of, for example, 24 wt% styrenated unsaturated polyester, 41 wt% $CaCO_3$, 30 wt% of 5-11.2 cm long glass fibers, plus initiators, lubricants, etc., and ca. 3 w% added thermoplastics that smooth the surface. Bulk molding compounds (**BMC**) contain short fibers instead of fiber mats.

SMCs and BMCs are delivered as resins between two poly(ethylene) sheets. For their use, one of the peel plies is peeled off. The resin layer is pressed against the mold by rollers, which also welds the edges of the new resin layer to the edges of the previously applied resin layers. The subsequent curing of the resin layers by heating causes a part of the styrene monomer to evaporate, which "foams" the resin somewhat and compensates for the shrinkage caused by the crosslinking polymerization.

On curing, unfilled unsaturated polyester resins shrink by 6-8 % and BMCs by ca. 0.2-0.4 %, which leads to rough surfaces. Smoother surfaces are obtained if unsaturated polyester resins are mixed with syrupy solutions of up to 25 wt% thermoplastics such as poly(styrene), poly(methyl methacrylate), or cellulose acetobutyrate. Unsaturated polyesters and added thermoplastics demix to some extent, which causes the thermoplastic to diffuse to the surface where its presence in combination with the "foaming" leads to a smoother surface of these **low-profile resins**.

Vinyl Ester Resins

Surface roughness can also be reduced or avoided by use of so-called vinyl ester resins (= **vinyl resins, modified vinyl resins**). Some of these resins contain vinyl groups $CH_2{=}CH–$ but they are neither vinyl esters, $CH_2{=}CH–O–OC–R'$, nor oligomeric poly-

(vinyl ester)s, $\text{+CH}_2\text{-CH(OOCR')-}_n$. They are rather so-called macromonomers of the type $[CH_2=C(CH_3)CO\text{-}OCH_2CH_2O]_2Z$ that contain two methacryl (or acryl) endgroups in which the $-OCH_2CH_2O-$ may also be replaced by $-OCH_2CH(OH)CH_2-$. Z may be one of the following groups:

where Z' is $-CO-NH-Z''-NH-CO-$.

Vinyl ester resins are sold as solutions in styrene (S) or vinyl toluene (VT). They are hardened like unsaturated polyester resins, i.e., by free-radical polymerization of the solvents S or VT with the vinyl ester molecules acting as crosslinking agents. Vinyl ester resins lead to better mechanical properties than the thermosets from unsaturated polyester resins albeit at higher costs. They are used as matrix resins for composites and for tanks, pipes, pumps, etc.

8.6 Disposal and Recycling

8.6.1 Introduction

Like most materials, plastics are used to produce goods or parts of goods, either short-lived ones like packaging materials, medium-lived ones like automotive parts, or long-lived ones like construction materials for houses. The manufacture of these goods produces scrap and waste: in the polymer industry from polymer syntheses, in the plastics industry from processing, and from the consumer after the useful lifetime of the goods or after a change of fashion.

Industrial scrap and waste is usually produced in fairly large amounts of the same polymer or even the same grade. Thermoplastic and rubber wastes can usually be easily cleaned and then reused, either directly or in mixtures with virgin polymers. Depending on the economy, they may also be incinerated for energy production. However, because of their crosslinked state, scrap from thermosets and elastomers can only be worked up mechanically and then used as fillers.

Much more difficult is the disposal and recycling of plastic refuse generated by *households* and *trades* because this refuse is usually soiled and consists of many different types of plastics, all in small amounts. Although plastics constitutes only a fairly small proportion of city refuse (US: ca. 7 wt% ≈ 8.5 vol%), it is very visible, especially if it is thrown away carelessly.

Plastics in city refuse can be disposed of by biological degradation, mechanical recycling, feedstock recycling, energy recovery, and burying in landfills. The proportion of these types varies from country to country for historical, economical, psychological, and legal reasons. In the United States, only ca. 7 wt% of the annually generated plastics waste is materially recycled (2006) but 21 wt% of aluminum, 22 wt% of glass, 22 wt% of cardboard, and 52 wt% of paper and paperboard. About 10 wt% of the US refuse is burned (Section 8.6.5) and 80 wt% ends up in landfills.

The situation is quite different in Switzerland which recycles about 91 wt% of glass (1998) and much of the steel of tin cans but no aluminum because of the very small amount of the latter. The Swiss burn 77 wt% of their remaining refuse (75 % of which for energy production) and bury only 23 wt% in landfills.

According to the (European) Association of Plastics Manufacturers, in 2002 Western Europe produced ca. $50 \cdot 10^6$ t of new plastics and $20 \cdot 10^6$ t of plastics waste of which 38 wt% ($7.6 \cdot 10^6$ t) was recycled: 62 % for energy recovery, 30 wt% by mechanical recycling, and 3.3 % by feedstock recycling, while 4.7 % was exported (mainly to Asia).

8.6.2 Biological Degradation

Discarded bottles and packaging films are a visual nuisance, and therefore it has often been demanded that those packaging materials should be naturally degradable by light, water, oxygen, or microorganisms. However, many polymers used for packaging purposes (poly(olefin)s, poly(styrene)s, poly(vinyl chloride)) have carbon-carbon chains and lack chromophores and therefore do not degrade "naturally" in reasonable times.

It has been proposed to use starch particles as a filler for polymers in packaging. Such particles degrade relatively fast under the action of light, water, and oxygen, which then leads to a breakup of plastics into small particles. Such an action would remove an eyesore but not the plastic itself which would ultimately end up in rivers and finally in the oceans. Indeed, it has been reported that the Sargasso Sea harbors a very large field of small plastics particles that are bobbing in the water and harm maritime life.

Faster natural degradations are possible if the polymers contain easily degradable groups either in the monomeric units themself (polyesters, cellulose, etc.) or as comonomeric units or are provided with sensitizers. In any case, degradations in open air are still very slow, and in the depths of landfills even slower: after 25 years, one can still read the headlines of buried newspapers. Fast degradations *are* possible, however, e.g., by tropical microorganisms that ate US uniforms during jungle warfare in World War II.

Natural degradation of many polymers to their oligomers may be harmful because some oligomers might be toxic or carcinogenic. It is also not economical because it destroys the value of plastics as materials, feedstocks, or energy sources.

8.6.3 Mechanical Recycling

In principle, mechanical recycling should be the most economical and environmentally prudent method for the removal of plastics scrap and waste. In practice, several problems arise for post-consumer plastics because the discarded items consist of different types of polymers that must be sorted according to type before grinding, pelletizing, and melting since various types of thermoplastics are generally incompatible and their

PP

PE-HD

Fig. 8-9 Codes for recyclable plastics. Left: US number code, here with additional letter code (example: it-poly(propylene)). Center and right: European codes with letters, numbers, and/or bar codes.

mixtures have generally inadequate properties (Chapter 10). Addition of compatibilizers is generally very expensive and faces the additional problem that the composition of polymer mixtures may change with locality and time.

Waste plastic must therefore be sorted before further processing. Since most plastics look alike, plastic containers are now provided with number, letter, or bar codes for easy sorting by unskilled labor (Fig. 8-9). Bar codes are impractical since containers must be oriented by hand before they can be read by machines. Most practical are simple number codes that may or may not be supplemented by letter codes:

1 = PETE	= poly(ethylene terephthalate)	
2 = HDPE	= high-density poly(ethylene)	
3 = V	= poly(vinyl chloride)	
4 = LDPE	= low-density poly(ethylene)	
5 = PP	= isotactic poly(propylene)	
6 = PS	= atactic poly(styrene)	
7	= all other plastics, including multilayer materials.	

The major cost of mechanical recycling is the curbside collection (including shipping), followed by hand sorting and work-up (washing, regrinding, pelletizing). In general, prices of mechanically recycled plastics (R) are ca. 50 % of those of virgin material (V) (Table 8-34); those with natural colors costing ca. 20 % more than mixed ones and pellets costi ca. 30 % more than regrinds. Exceptions are the lost cost ratio, R/V = 0.36, for regrind poly(propylene) (low demand) and the high cost ratio, R/V = 0.98, for clear poly(ethylene terephthalate). In the United States, about 35 % of the recycled PET is exported to China. Recycled PET is also used for nonwoven fabrics and pillow stuffings.

Table 8-34 Comparison of average prices in US-$ per kilogram for virgin commodity plastics (for purchases of ca. 1000 tons per year) and post-consumer (P) or industrial (I) recycled plastics [10].

Type	Virgin plastics (V) Type	$/kg	Recycled plastics (R) Type	$/kg	R/V $/$
PVC	dispersion polymer, GP	1.78	I: clear	0.88	0.49
HDPE	general purpose	2.00	P: natural color, regrind	1.10	0.55
			P: mixed colors, regrind	0.93	0.47
			P: mixed colors, pellets	1.26	0.63
LDPE	clear film	2.25	P: clear film or pellets	1.17	0.52
PP	general purpose	2.27	I: regrind	0.81	0.36
			I: pellets	1.04	0.46
PS	crystal, general purpose	2.11	P	0.99	0.47
			I	1.19	0.56
PET	bottle grade	2.03	I: clear	1.98	0.98
			I: mixed colors	1.48	0.73

Table 8-35 Dependence of properties of three styrenic plastics on the number of processing cycles [11]. ABS = graft copolymer of styrene/acrylonitrile on butadiene rubber, HIPS = high-impact poly-(styrene) from the polymerization of a styrenic solution of butadiene rubber; SAN = random bipoly-mer with 76 mol% styrene and 24 mol% acrylonitrile units.

Property	Physical unit	Plastic	Number of processing cycles					
			1	2	3	4	5	6
Melt flow index	10 g/min	HIPS	3.8	3.9	3.9	4.0	4.2	4.4
		ABS	27	31	34	39	41	48
		SAN	8.5	9.8	10.5	11.2	11.7	12.2
Tensile strength	MPa	HIPS	36.8	37.8	38.4	37.9	38.4	38.3
		ABS	37.2	42.2	44.8	45.1	50.1	50.7
		SAN	69.8	69.4	68.4	67.0	68.3	69.4
Elongation at break	%	HIPS	10.3	7.9	8.0	7.6	7.1	7.2
		ABS	24.0	10.4	8.6	8.9	8.0	7.7
		SAN	6.4	6.3	5.2	5.1	5.4	5.8
Flexural strength	MPa	HIPS	52.2	51.8	51.7	51.2	51.6	51.5
		ABS	45.8	48.2	48.8	49.5	50.4	49.8
		SAN	71.0	73.2	70.6	69.9	70.7	70.1
Impact strength	kJ/m^2	HIPS	17.0	15.0	14.0	15.0	13.0	12.0
		ABS	49.0	33.0	18.0	15.0	10.0	6.1
		SAN	9.2	9.2	9.1	8.8	8.9	8.7
Notched impact str.	kJ/m^2	HIPS	5.4	5.1	3.3	4.0	2.5	2.1
		ABS	25.0	13.0	10.5	7.6	4.1	2.3

Recycled thermoplastics may be used as such or mixed with virgin material of the same type. However, recycled plastics cannot go through an infinite number of use/reuse steps since each processing step causes some degradation as indicated by the increase in melt flow indices with increasing number of processing steps (Table 8-35) (melt flow indices are related to inverse molecular weights (p. 85)).

Tensile strengths of amorphous polymers are practically unaffected by the molecular weight above a certain critical molecular weight (p. 143). Indeed, tensile strengths of SAN and HIPS remain constant after 6 processing cycles (Table 8-35). Tensile strengths of ABS do increase with increasing number of processing cycles, which indicates morphological changes and/or increasing crosslinking. Indeed, fracture elongations and impact strengths of ABS decrease dramatically with increasing number of processing cycles whereas those of HIPS and especially SAN do so far less.

Another problem of recycled plastics can be the presence of other unwanted materials, for example, small proportions of poly(lactide) (PLA) in recycled poly(ethylene terephthalate) (PETE). Both polymers are used for bottles and both look similar. A small proportion of PLA may easily end up in recycled PETE, especially since optical sorting by infrared is only 97.5 % accurate. However, as little as 1 wt% PLA in a PETE recycling stream causes problems with drying since PLAs have much lower melting temperatures (170°C (D,L form); 188°C (L form)) than PETE (255°C).

In the United States, standard thermoplastics are relatively little recycled, engineering plastics to a diminishing extent, and thermosets not at all. However, thermosets may be ground up very finely and then added in proportions of 15-30 wt% to thermosetting

Table 8-36 Material recycling of standard (commodity) plastics in municipal solid waste (MSW) in the United States [12], amended by data of [13,14].

Plastic	Post-consumer recyclates in 1000 tons per year								
	1988	1993	1995	1997	1999	2001	2003	2005	2006
PET	162	204	360	360	470	470	410	540	620
HDPE		218	370	420	490	430	470	520	580
LDPE/LLDPE		54	90	100	130	150	150	190	280
PP		100	130	120	120	10	10	10	20
PS	39	18	20	10	10				10
PVC	34	5.1	4.5	4.1	4.1				
PA	27								
Other	1	36	25.5	95.9	125.9	330	350	390	530
Total recyclate	263	635.1	1 000	1 110	1 350	1 390	1 390	1 650	2040
Plastics in MSW			18 990	21 460	24 170	25 380	26 650	28 910	29 940
US production of thermoplastics			27 573	31 525	32 199	36 524	38 706	39 736	41 191
Percentage of recyclates of thermoplastics									
of municipal solid waste			5.3	5.2	5.6	5.5	5.2	5.7	6.9
of annual US production			3.6	3.5	4.2	3.8	3.6	4.2	5.0

resins without adverse effects on the properties of thermosets. "Recycling" of standard thermoplastics is often a euphemism as far as plastics from municipal waste are concerned since many of these recyclates are "downcycled" to products of lesser quality.

An exception is beverage bottles because they are usually clear, not soiled much (if at all) and made from easy to process plastics such as poly(ethylene terephthalate) and high-density poly(ethylene). The US recycling of PET, HDPE, and also LDPE/LLDPE has therefore climbed steadily (Table 8-36) whereas the recognized recycling of poly-(propylene), poly(styrene), poly(vinyl chloride), and polyamides is very small. There is also a high proportion of unspecified "other." In general, the percentage of US recycled plastics has remained constant at 3.5-5 % of the US *annually domestically produced* plastics during 1995-2006, i.e., without imported plastics and plastics goods.

8.6.4   Feedstock Recycling

In principle, some polymers may be recycled to their feedstocks. Polyamides and polyesters can be hydrolyzed at elevated temperatures to their monomers but it is questionable whether such processes and the subsequent work-ups by fractional distillation, etc., are economical.

Another possiblity is brute force "recycling" of plastics or mixtures of plastics to feedstocks at temperatures of 700-800°C and subsequent work-up by distillation. The type and proportion of the resulting low-molecular weight compounds varies not only with the constitution of the feed polymers (Table 8-37) and the cracking temperature but also with the composition of the feed if unsorted plastics waste is used. The proportion of valuable aromatic compounds is usually ca. 20 %. It can be boosted to ca. 40 % if one avoids the formation of carbon black by blowing in water vapor. Alternatively, one can use a low-temperature pyrolysis of waste plastics in poly(ethylene) wax at 400°C which leads to low-boiling oils with high olefin contents plus waxes and carbon blacks.

Table 8-37 Products from the fluidized bed pyrolysis of poly(ethylene) (PE), poly(ethylene terephthalate) (PET), a polyurethane (PUR), an ethylene/propylene/non-conjugated diene terpolymer (EPDM), a styrene-butadiene rubber (SBR), and unspecified tire treads at the indicated pyrolysis temperatures [15]. * Formulated elastomers.

Pyrolysis products	Weight percent of pyrolysis products from					
	PE 740°C	PET 768°C	PUR 760°C	EPDM* 750°C	SBR* 740°C	Tire treads 700°C
Hydrogen	0.75	0.30	0.23	1.92	0.8	0.42
Methane	23.6	3.8	3.5	22.75	10.2	6.06
Ethane	6.7	0.24	0.76	1.08	1.2	2.34
Ethene	19.8	1.5	5.4	8.67	2.6	1.65
Propane	0.08	0.01	0.01	0.02	0.1	0.43
Propene	5.5	0.08	1.7	0.36	0.7	1.53
Other aliphatic hydrocarbons	2.16	0.07	1.4	0.77	8.4	3.03
Benzene	19.1	18.3	5.0	7.38	4.2	2.42
Toluene	3.9	2.5	1.8	2.33	3.8	2.65
Other aromatics	4.76	6.12	4.36	7.96	13.98	19.11
Nitriles	0	0	3.33			
Ketones, acetals	0	1.55	9.5			
Other organics	11.3	11.7	34.1			
Carbon black, inorganic fillers	1.75	7.1	0.5	41.85	50.7	40.0
Carbon dioxide	0	17.2	14.4	0.75	0.1	1.74
Carbon monoxide	0	27.4	8.7	1.25	0.2	1.48
Water	0	2.1	5.0	1.27	0.4	5.11
H_2S, thiophene				0.15	0.3	0.27
Tire cord	0	0	0	0	0	11.30
Total	99.4	99.97	99.69	98.51	97.68	99.32

8.6.5 Energy Recovery

Plastics wastes are also incinerated to generate energy (called "energy recovery" or "energy recycling"). Plastics contain high energy contents between ca. 18 MJ/kg for poly(vinyl chloride) and ca. 87 MJ/kg for ABS polymers. These energy contents are highly desirable for the incineration of municipal wastes which otherwise would need the addition of heating oil for the complete combustion of wastes.

The incineration of plastics, both separately or as part of municipal wastes, has led to ecological concerns with respect to production of HCl from poly(vinyl chloride), generation of dioxins, air pollution in general, and global warming by the CO_2 produced. The type and proportion of waste gases depends of course on the incineration temperature which is generally higher in Europe than the United States.

In general, air pollution from the burning of plastics is far less than that from coal, brown coal, and wood. Furthermore, the CO_2 production by the burning of plastics is similar to that of oil (from which plastics are produced).

Incineration of plastics waste contributes relatively little to the concentration of HCl in the air. In Germany, incineration in 1978 produced a total of 131 500 tons of HCl of which 77.6 % came from power stations (coal, brown coal, oil), 17.5 % from wastes, and 4.9 % from all other sources. Because wastes contain at most 50 wt% PVC, not more than 8.8 % of the produced HCl can stem from PVC. Furthermore, the amounts of HCl emitted are much smaller than the amounts produced because 10-15 % of the generated HCl is bound by the produced ash and most of the remaining HCl is removed by washing the waste gases.

A potentially more serious problem is the generation of dioxins, which are cyclic compounds that contain two oxygen atoms in one doubly unsaturated ring. Examples:

1,4-dioxin   1,2-dioxin   dibenzo-*p*-dioxin   2,3,7,8-tetrachlorodibenzo-*p*-dioxin (TCDD)

Chlorinated dioxins, especially TCDD ("agent orange"), are toxic: the smallest deadliest dose of $LD_{50} = 0.6$ µg/(kg body weight) was found for male guinea pigs. In humans, TCDD may cause chloracne on long exposure but there does not seem to be a correlation between symptoms and intensity of exposure (Seveso accident where workers were exposed to massive doses of dioxins). Epidemiologic investigations did not provide clear indications of cancerogeneous, teratogeneous, or gene-toxic actions of TCDD for the usual concentrations of TCDD equivalents in human bodies.

Dioxins are generated by the burning of all organic compounds, not only plastics. Their generation can be prevented by incineration temperatures of ca. 1200°C. For economic reasons, however, most incinerators operate at ca. 800°C, however.

8.6.6 Landfills and Ecological Balances

With respect to investment and operation, the deposition of wastes in landfills is the most economic method if sufficient land is available. But landfills, even ones lined with clay or poly(ethylene) sheets, are not unproblematic because they may lead to the pollution of ground water by substances that are extracted from wastes. Polymers themselves are not very problematic in this respect because they are usually not water soluble and very difficult to decompose naturally, if at all (Section 8.6.2).

However, plastics may be problematic because they usually contain soluble and/or degradable adjuvants. Since the type and concentration of these adjuvants are regulated for food containers and extractions are very slow, no major problems can be expected.

There are not many comprehensive studies of the effect of various materials on the ecology. An example is the packaging of milk which can be or could be bought in stores in glass bottles, poly(ethylene)-coated cardboard containers, or bottles or tubes made from poly(ethylene) (Table 8-38). Glass bottles are usually reused several times but all other containers are used only once and then discarded. As expected, the shipping of milk in light-weight poly(ethylene) or cardboard containers is less costly and less energy consuming than the shipping of milk in glass containers, which is the reason why supermarkets practically no longer sell milk in glass bottles (fine food and health food stores still do at a premium).

However, glass bottles can be reused after washing 20 or 40 times before they break. The total energy consumption for the manufacture, circulation, and disposal (exclusive of recycling?) of containers shows that glass bottles have a clear advantage over poly(ethylene) or cardboard containers but flexible poly(ethylene) tubes do even better. These tubes are also least ecologically damaging (air, water, soil). Unfortunately, such

Table 8-38 Ecological balances for 1000 milk containers of 1 liter each. Swiss data [16].

	Glass bottle	Glass bottle	PE tube	PE bottle	Paper, cardboard
Weight of 1000 containers in kg	484	484	7.0	22.0	25.0
Number of uses of a container	40	20	1	1	1
Effective weight in kg of 1000 containers per use	12.1	24.2	7.0	22.0	25.0
Energy consumption in MJ for the manufacture, circulation, and disposal of 1000 containers	650	730	400	1510	1770
Ecological damage (critical volumes) of					
air (10^6 m^3)	5.4	7.5	5.9	21	31
water (m^3) (including bottle washes)	32	33	3.0	17	150
soil (if landfill) (dm^3)	6.8	5.2	2.3	9.8	8.3
soil (average Swiss disposal) (dm^3)	3.3	11	0.35	3.0	1.3

tubes are not very consumer friendly since milk cannot be stored in tubes once they are opened unless the tubes are resealable.

Glass bottles are also advantageous compared to poly(ethylene) or cardboard containers with respect to air pollution and soil pollution in landfills. However, their production, use, and recycling pollutes more water than that of poly(ethylene) containers and also more soil if their circulation rate is low. Paperboard containers are always the worst except for the average Swiss disposal which includes incineration. Since these data refer to the customary Swiss one-liter containers (ca. one quart), one can only speculate how the numbers would change for milk which is sold in one-gallon containers in the United States.

Literature to Chapter 8

ENCYCLOPEDIAS AND HANDBOOKS
H.-G.Elias, New Commercial Polymers 1969-1975, Gordon and Breach, New York 1977
H.-G.Elias, F.Vohwinkel, New Commercial Polymers 2, Gordon and Breach, New York 1986
F.T.Traceski, Specifications and Standards for Plastics and Composites, ASM International, Materials Park (OH) 1990
M.J.Berius, Plastics Engineering Handbook, Chapman and Hall, London, 5th ed. 1991
H.Domininghaus, Plastics for Engineers, Hanser, Munich 1992
Rosato's Plastics Encyclopedic Dictionary, Hanser, Munich 1993
J.F.Carley, Ed., Whittington's Dictionary of Plastics, Technomic, Lancaster (PA), 3rd ed. 1993
H.Saechtling, Kunststoff-Taschenbuch, Hanser, Munich, 24th ed. 1989; -, International Plastics Handbook, Hanser, Munich 1995
D.P.Bashford, Thermoplastics Directory and Databook, Chapman and Hall, London 1997
J.A.Brydson, Plastics Materials, Butterworth-Heinemann, Oxford, 7th ed. 1999
M.Chanda, S.K.Roy, Plastics Technology Handbook, Dekker, New York, 3rd ed. 1998
O.Olabisi, Ed., Handbook of Thermoplastics, Dekker, New York 1997
W.V.Titow, Technological Dictionary of Plastics Materials, Elsevier (Pergamon), Amsterdam 1999
E.S.Wilks, Ed., Industrial Polymers Handbook,Wiley-VCH, Weinheim 2001 (4 Vols) (= relevant chapters from Ullmann's Encyclopedia of Industrial Chemistry, 5th edition

C.A.Harper, E.M.Petrie, Plastics Materials and Processes. A Concise Encyclopedia, Wiley, Hoboken (NJ) 2003

M.Chanda, S.K.Roy, Plastics Technology Handbook, CRC Press, Boca Raton (FL), 4th ed., 2007

DATA BANKS

W.J.Roff, J.R.Scott, Handbook of Common Polymers, Butterworths, London 1971

-, Parat-Index of Polymer Trade Names, Fachinformationszentrum Chemie, Berlin 1989

J.C.Bittence, Ed., Engineering Plastics and Composites, ASM International, Materials Park (OH) 1990 (trade names, manufacturers, and application of engineering plastics)

E.W.Flick, Industrial Synthetic Resins Handbook, Noyes, Park Ridge (NJ), 2nd ed. 1991 (descriptions of ca. 3000 plastics)

M.Ash, I.Ash, Handbook of Plastic Compounds, Elastomers and Resins, VCH, New York 1991 (ca. 15 000 commercial products)

The International Plastics Selector, Commercial Names and Sources for Plastics and Adhesives, Business Publishing, Englewood (CO) 1992 (2 vols., 14 600 types and grades)

C.A.Harper, Modern Plastics Handbook, McGrawHill, New York 2000

Plastics Databank CAMPUS (= Computer Aided Material Preselection by Uniform Standards) www.campusplastics.com

-, Modern Plastics Encyclopedia, McGraw-Hill, New York (annually in October)

-, Encyclopédie Française des Matières Plastiques, Les Publicateurs Techniques Association, Paris (annually)

TEXTBOOKS

G.R.Moore, D.E.Kline, Properties and Processing of Polymers for Engineers, Prentice-Hall, Englewood Cliffs (NJ) 1984

A.W.Birley, R.J.Heath, M.J.Scott, Plastics Materials, Blackie, Glasgow, 2nd ed. 1988

L.Mascia, Thermoplastics: Materials Engineering, Elsevier, London 1982, 2nd ed. 1989

W.Michaeli, H.Greif, H.Kaufmann, F.-J.Vossebürger, Plastics Technology, Hanser, Munich 1994

H.-G.Elias, An Introduction to Plastics, Wiley-VCH, Weinheim, 2nd ed. 2003

T.A.Osswald, G.Menges, Materials Science of Polymers for Engineers, Hanser, Munich, 2nd ed. 2003

ANALYSIS, TEST METHODS, SPECIAL PROPERTIES

J.Urbanski, W.Czewinski, K.Janicka, F.Majewska, H.Zowall, Handbook of Analysis of Synthetic Polymers and Plastics, Wiley, New York 1977

A.Krause, A.Lange, M.Ezrin, Plastics Analysis Guide, Hanser, Munich 1983

T.R.Crompton, The Analysis of Plastics, Pergamon, Oxford 1984

D.Braun, Simple Methods for the Identification of Plastics, Hanser, Munich, 2nd ed. 1986

G.Kämpf, Characterization of Plastics by Physical Methods, Hanser, Munich 1987

Yu.V.Moiseev, G.E.Zaikov, Chemical Resistance of Polymers in Aggressive Media, Plenum, New York 1987

J.I.Kroschwitz, Ed., Polymer Characterization and Analysis, Wiley, New York 1990 (reprints of articles in H.F.Mark et al., Eds., Encyclopedia of Polymer Science and Engineering

D.O.Hummel, Atlas of Polymer and Plastics Analysis, Vol. 1, Polymers, 2nd ed. 1979; Vol. 2, Plastics, Fibres, Rubbers, Resins; Starting and Auxiliary Materials, Degradation Products, 2nd ed. in two parts, 1984 and 1988; Hanser, Munich and VCH, Weinheim

F.Scholl, Atlas of Polymer and Plastics Analysis, Vol. 3, Additives and Processing Aids. Spectra and Methods of Identification, Hanser, Munich and VCH, Weinheim

M.Ezrin, Plastics Failure Guide. Cause and Prevention, Hanser Gardner, Cincinnati (OH) 1996

D.C.Hylton, Understanding Plastics Testing, Hanser Gardner, Cincinnati (OH) 2004

W.Grellmann, S.Seidler, Plastics Testing, Hanser, Munich 2007

BIBLIOGRAPHIES

E.R.Yescombe, Sources of Information on the Rubber, Plastics and Allied Industries, Pergamon, Oxford 1968; Plastics and Rubbers: World Sources of Information, Appl.Sci.Publ., Barking, Essex, UK, 2nd ed. 1976

G.J.Patterson, Plastics Book List, Technomic Publ., Westport (CT) 1975

O.A.Battista, The Polymer Index, McGraw Hill, New York 1976

S.M.Kaback, Literature of Polymers, Encycl.Polym.Sci.Technol. **8** (1968) 273

J.T.Lee, Literature of Polymers, Encycl.Polym.Sci.Eng, 2nd ed., **9** (1987) 62
R.T.Adkins, Ed., Information Sources in Polymers and Plastics, K.G.Saur, New York 1989

HISTORICAL DEVELOPMENT
F.M.McMillan, The Chain Straighteners: Fruitful Innovation. The Discovery of Linear and Stereo-
    regular Polymers, MacMillan, London 1981
H.R.Sailors, J.P.Hogan, A History of Polyolefins, Polym. News **7** (1981) 152
R.B.Seymour, T.Cheng, Eds., History of Polyolefins, Reidel, Dordrecht 1986
D.B.Sicilia, A Most Invented Invention, Amer. Heritage of Invention and Technology **6**/1 (1990) 45
S.T.I.Mossman, P.J.T.Morris, Eds., The Development of Plastics, R.Soc.Chem., Cambridge 1994
S.Fenichel, Plastic: The Making of a Synthetic Century, Harper Business, New York 1996
H.Martin, Polymers, Patents, Profits. A Classic Case Sudy for Patent Infighting, Wiley-VCH,
    Weinheim 2007 (history of Ziegler poly(ethylene))

8.2.1 POLY(1-OLEFIN)S, General
L.F.Albright, Processes for Major Addition-Type Plastics and Their Monomers, Krieger,
    Melbourne (FL), 2nd ed. 1985 (ethylene, propylene, vinyl chloride, styrene)
K.Soga, T.Shiono, Ziegler-Natta Catalysts for Olefin Polymerizations, Progr.Polym.Sci. **22** (1997)
    1503
L.A.M.Utracki, Polyolefin Alloys and Blends, Macromol.Symp. **188** (1997) 335
I.Scheirs, W.Kaminsky, Eds., Metallocene-Based Polyolefins, Wiley, New York 1999 (2 vols.)
C.Vasile, R.B.Seymour, Eds., Handbook of Polyolefins, Dekker, New York, 2nd ed. 2000
J.L.White, D.D.Choi, Polyolefins. Processing, Structure Development, and Properties, Hanser,
    Munich 2005

8.2.1.a POLY(ETHYLENE)S
P.Ehrlich, G.A.Mortimer, Fundamentals of the Free-Radical Polymerization of Ethylene,
    Adv.Polym.Sci. **7** (1970) 386
S.Cesca, The Chemistry of Unsaturated Ethylene-Propylene Based Terpolymers, Macromol.Rev.
    10 (1975) 1
A.J.Peacock, Handbook of Polyethylene. Structures, Properties, and Applications, Dekker,
    New York 2000

8.2.2.a POLY(PROPYLENE)S
T.O.J.Kresser, Polypropylene, Reinhold, New York 1960
H.P.Frank, Polypropylene, Gordon and Breach, New York 1968
E.G.Hancock, Ed., Propylene and Its Industrial Derivatives, Halsted, New York 1973
S. van der Ven, Polypropylene and Other Polyolefins. Polymerization and Characterization,
    Elsevier, Amsterdam 1990
D.B.Sicilia, A Most Invented Invention, Amer.Heritage of Invention and Technology **6**/1 (1990) 45
J.Karger-Kocsis, Ed., Polypropylene, Chapman and Hall, London 1995 (Vol. I: Structure and
    Morphology, Vol. II: Copolymers and Blends, Vol. III: Composites)
E.P.Moore, Jr., Ed., Polypropylene Handbook. Polymerization, Characterization, Properties,
    Processing, Applications, Hanser, Munich 1996
E.P.Moore, Jr., The Rebirth of Polypropylene: Supported Catalysts, Hanser, Munich 1999
H.G.Karian, Handbook of Polypropylene and Polypropylene Composites, Dekker, New York 1999
J.Karger-Kocsis, Polypropylene, Kluwer Academic, Dordrecht (NL) 1999
N.Pasquini, Polypropylene Handbook, Hanser, Munich, 2nd ed. 2005

8.2.2.b POLY(1-BUTENE)
I.D.Rubin, Poly(1-butene), Gordon and Breach, New York 1968
B.A.Krentsel, Y.V.Kissin, V.I.Kleiner, L.L.Stoskaya, Eds., Polymers and Copolymers of Higher
    α-Olefins, Hanser, Munich 1997

8.2.2.c POLY(4-METHYL-1-PENTENE)
K.J.Clark, R.P.Palmer, Transparent Polymers from 4-Methylpentene-1, Soc.Chem.Ind., Mono-
    graph No.20, London 1966, p.82

8.2.3 STYRENE POLYMERS

R.H.Boundy, R.F.Boyer, Styrene, Reinhold, New York 1952

C.H.Basdekis, ABS Plastics, Reinhold, New York 1964

C.A.Brighton, G.Pritchard, G.A.Skinner, Styrene Polymers: Technology and Environmental
Aspects, Appl.Sci.Publ., London 1979

R.Po', N.Cardi, Synthesis of Syndiotactic Polystyrene: Reaction Mechanism and Catalysis,
Progr.Polym.Sci. **21** (1996) 47

J.Scheirs, D.Priddy, Eds., Modern Styrenic Polymers, Wiley, Hoboken (NJ) 2003

8.2.4 VINYLS

M.Kaufman, The History of PVC - The Chemistry and Industrial Production of Polyvinylchloride,
MacLaren, London 1969

J.V.Koleske, L.H.Wartman, Poly(vinylchloride), Gordon and Breach, New York 1969

W.S.Penn, PVC Technology, MacLaren, London, 3rd ed. 1972

H.A.Sarvetnick, Ed., Plastisols and Organosols, Van Nostrand Reinhold, New York 1972

R.A.Wessling, Polyvinylidene Chloride, Gordon and Breach, New York 1975

R.H.Burgess, Ed., Manufacture and Processing of PVC, Hanser, Munich 1981

G.Butters, Ed., Particulate Nature of PVC, Elsevier, New York 1982

W.V.Titow, Ed., PVC Technology, Elsevier Appl.Sci., New York, 4th ed. 1984

E.D.Owen, Ed., Degradation and Stabilization of PVC, Elsevier Appl.Sci., New York 1984

M.K.Naqvi, Structure and Stability of Polyvinyl Chloride, J.Macromol.Sci.-Rev.Macromol.
Chem.Phys. **C 25** (1985) 119

J.Wypych, Polyvinyl Chloride Stabilization, Elsevier, Amsterdam 1986

L.I.Nass, C.A.Heiberger, Ed., Encyclopedia of PVC, Dekker, New York, 2nd ed. 1988-1992
(4 vols.)

C.E.Wilkes, J.W.Summers, C.A.Daniels, Eds., PVC Handbook, Hanser, Munich 2005

8.3 ENGINEERING THERMOPLASTICS, General

Z.D.Jastrzebski, The Nature and Properties of Engineering Plastics, Wiley, New York 1976

R.Burns, Polyester Molding Compounds, Dekker, New York 1982

J.M.Margolis, Ed., Engineering Thermoplastics, Dekker, New York 1985

R.B.Seymour, Polymers for Engineering Applications, ASM International, Metals Park (OH) 1987

E.W.Flick, Engineering Resins. An Industrial Guide, Noyes, Park Ridge, NJ, 1988 (2500 industrial
engineering plastics)

L.Bottenbruch, Ed., Engineering Thermoplastics (Polycarbonates-Polyacetals-Polyesters-
Cellulose Esters), in G.W.Becker, D.Braun, Eds., Kunststoff-Handbuch **3/1** (1992), Hanser,
Munich 1996

G.W.Becker, D.Braun, Eds., Techn. Thermoplaste, Techn. Polymer-Blends, Kunststoff-Handbuch
3/2 (1993), Hanser, Munich

W.A.Woishnis, Ed., Engineering Plastics and Composites, ASM International, Materials Park
(OH), 2nd ed. 1993

8.3.2 POLYCARBONATES

D.G.LeGrand, J.T.Bendler, Handbook of Polycarbonate Science and Technology, Dekker, New York
1999

8.3.3 THERMOPLASTIC POLYESTERS

S.Fakirov, Ed., Handbook of Thermoplastic Polyesters, Wiley, Hoboken (NJ), 2 vols. 2002

J.Scheirs, T.E.Long, Eds., Modern Polyesters, Wiley, Hoboken (NJ) 2003

8.3.4 ALIPHATIC NYLON PLASTICS

M.I.Kohan, Nylon Plastics Handbook, Hanser, Munich 1995

8.3.6 POLY(METHYL METHACRYLATE)

M.B.Horn, Acrylic Resins, Reinhold, New York 1960

8.4 HIGH PERFORMANCE THERMOPLASTICS

R.B.Seymour, G.S.Kirshenbaum, Eds., High Performance Polymers: Their Origin and Their
Development, Elsevier, New York 1985

S.Béland, High Performance Thermoplastic Resins and Their Composites, Noyes, Park Ridge (NJ) 1990
E.Baer, A.Moet, Ed., High Performance Polymers, Hanser, Munich 1991 (composites, fibers)
J.I.Kroschwitz, Ed., High Performance Polymers and Composites, Wiley, New York 1991

8.4.2 POLYIMIDES
M.I.Bessonov, M.M.Koton, V.V.Kudryavtsev, L.A.Laius, Polyimides, Plenum, New York 1987
M.K.Ghosh, K.L.Mittal, Polyimides, Dekker, New York 1996

8.4.7 FLUOROPLASTICS
L.A.Wall, Ed., Fluoropolymers, Wiley, New York 1972
R.E.Banks, Ed., Preparation, Properties and Industrial Applications of Organofluoro Compounds, Wiley, New York 1982
D.P.Carlson, W.Schmiegel, Fluoropolymers, Organic, Ullmann's Encyclopedia of Industrial Chemistry **A 11** (1986) 393
A.E.Feiring, J.F.Imbalzano, D.L.Kerbow, Developments in Commercial Fluoroplastics, Elsevier Science, Amsterdam, 1992
R.E.Banks, B.E.Smart, J.C.Tatlow, Eds., Organofluorine Chemistry, Plenum, New York 1995
J.Scheirs, Ed., Modern Fluoropolymers, Wiley, New York 1997
J.G.Drobny, Technology of Fluoropolymers, CRC Press, Boca Raton (FL) 2000
S.Ebnesajjad, P.Khaladkar, Fluoropolymer Applications in Chemical Processing Industries, W.Andrew, Norwich (NY) 2005
S.Ebnesajjad, Fluoroplastics, Vol. I, Non-Melt Processible Fluoroplastics, W.Andrew, Norwich (NY) 2001; Vol. 2, Melt Processible Fluoroplastics, W.Andrew, Norwich (NY) 2002

8.4.8 LCP GLASSES
A.Ciferri, I.M.Ward, Eds., Ultra-High Modulus Polymers, Appl.Sci.Publ., London 1979
A.E.Zachariades, R.S.Porter, Eds., High Modulus Polymers, Dekker, New York 1987
G.W.Gray, Ed., Thermotropic Liquid Crystals, Wiley, Chichester 1987
A.Donald, A.Windle, S.Hanna, Liquid Crystalline Polymers, Cambridge Univ.Press, Cambridge, 2nd. ed. 2006

8.5 THERMOSETS
A.Whelan, J.A.Brydson, Eds., Developments with Thermosetting Plastics, Halsted, New York 1975
L.H.Sperling, Interpenetrating Polymer Networks and Related Materials, Plenum, New York 1980
J.M.Margolis, Ed., Advanced Thermoset Composites, Van Nostrand Reinhold, New York 1986
S.G.Entelis, V.V.Evreinov, A.I.Kuzaev, Reactive Oligomers, VSP, Zeist, Netherlands 1989
R.E.Wright, Molded Thermosets, Hanser, Munich 1991
W.F.Gum, W.Riese, H.Ulrich, Reaction Polymers (Polyurethanes, Epoxies, Unsaturated Polyesters, Phenolics, Special Monomers, and Additives), Hanser, Munich 1992
S.-C.Lin, E.M.Pearce, High Performance Thermosets–Chemistry, Properties, Applications, Hanser, Munich 1993
S.H.Goodman, Handbook of Thermoset Plastics, Noyes Publ., Westwood (NJ), 2nd ed. 1999
J.-P.Pascault, H.Sautereau, J.Verdu, R.J.J.Williams, Eds., Thermosetting Polymers, Dekker, New York 2002
M.Biron, Thermosets and Composites, Elsevier, Amsterdam 2004

8.5.3.a PHENOLIC RESINS
T.S.Carswell, Phenoplasts, Interscience, New York 1947
K.Hultzsch, Chemie der Phenolharze, Springer, Berlin 1950
R.W.Martin, The Chemistry of Phenolic Resins, Wiley, New York 1956
N.J.L.Megson, Phenolic Resin Chemistry, Butterworths, London 1958
D.F.Gould, Phenolic Resins, Reinhold, New York 1959
A.A.K.Whitehouse, E.G.K.Pritchett, G.Barnet, Phenolic Resins, Iliffe, London 1967
G.W.Becker, D.Braun, W.Woebcken, Eds., Duroplaste (= Kunststoff-Handbuch, vol. 10), Hanser, Munich 1988
A.Knop, L.A.Pilato, Phenolic Resins, Springer, Berlin 1985
A.Gardziella, L.A.Pilato, A.Knop, Phenolic Resins, Springer, Berlin 2000

8.5.3b AMINO RESINS
B.Meyer, Urea-Formaldehyde Resins, Addison-Wesley, Reading (MA) 1979

8.5.4a EPOXY RESINS
J.I.DiStasio, Ed., Epoxy Resin Technology, Noyes, Park Ridge (NJ) 1982

8.5.4b POLYURETHANES
G.Woods, The ICI Polyurethanes Book, Wiley, New York 1987
D.Randall, S.Lee, Eds., The Polyurethanes Book, Wiley, Hoboken (NJ), 2003

8.5.4.c RIM PROCESS
E.Martuscelli, C.Marchetta, Eds., New Polymeric Materials: Reactive Processes and Physical
    Properties, VNU Science Press, Utrecht 1987

8.5.5 UNSATURATED POLYESTERS
P.F.Bruins, Unsaturated Polyester Technology, Gordon and Breach, New York 1975
P.Penczek, P.Czub, J.Pielichowski, Unsaturated Polyester Resins: Chemistry and Technology,
    Adv.Polym.Sci. **184** (2005) 1

8.6 DISPOSAL AND RECYCLING
J.E.Guillet, Ed., Polymers and Ecological Problems, Plenum, London 1973
J.Leidner, Plastics Waste: Recovery of Economic Value, Dekker, New York 1981
U.S. Environmental Protection Agency (T.R.Curlee, S.Das), Plastic Wastes. Management,
    Control, Recycling, and Disposal, Noyes, Park Ridge (NJ), 1991
W.Hoyle, D.R.Karsa, Eds., Chemical Aspects of Plastics Recycling, R.Soc.Chem., Cambridge, UK,
    1997

8.6.2 BIOLOGICAL DEGRADATION
S.J.Huang, Ed., Biodegradable Polymers, Hanser, Munich 1989
E.A.Dawes, Ed., Novel Biodegradable Microbial Polymers, Kluwer, Dordrecht 1990
M.Vert, J.Feijen, A.Albertsson, G.Scott, E.Chiellini, Biodegradable Polymers and Plastics,
    CRC Press, Boca Raton (FL) 1992
G.Griffin, Chemistry and Technology of Biodegradable Polymers, Blackie, New York 1993
A.-C.Albertsson, S.Huang, Eds., Degradable Polymers, Recycling, and Plastics Waste Management,
    Dekker, New York 1995
G.S.Moore, S.M.Saunders, Advances in Biodegradable Polymers, Rapra, Shropshire, UK, 1998
E.S.Stevens, Green Plastics. An Introduction to the New Science of Biodegradable Plastics, Princeton
    Univ. Press, Princeton (NJ) 2001
A.L.Andrady, Ed., Plastics and the Environment, Wiley, Hoboken (NJ) 2003
A.Azapagic, A.Emsley, I.Hamerton, Eds., Polymers, the Environment, and Sustainable Develop-
    ment, Wiley, Chichester, UK 2003
P.Smith, Ed., Biodegradable Polymers for Industrial Applications, CRC Press, Boca Raton (FL) 2005

8.6.3 MECHANICAL RECYCLING
R.B.Seymour, J.M.Sosa, Plastics from Plastics, ChemTech **7** (1977) 507
R.J.Ehrig, Ed., Plastics Recycling - Products and Processes, Hanser, Munich 1991
B.A.Hegberg et al., Mixed Plastics Recycling Technology, Noyes Data, Park Ridge (NJ) 1992
J.Brandrup, M.Bittner, W.Michaeli, G.Menges, Eds., Recycling and Recovery of Plastics, Hanser,
    Munich 1996
J.Scheirs, Polymer Recycling - Science, Technology and Applications, Wiley, New York 1998

8.6.4 FEEDSTOCK RECYCLING
J.Scheirs, W.Kaminsky, Eds., Feedstock Recycling and Pyrolysis of Waste Plastics: Converting
    Waste Plastics into Diesel and Other Fuels, Wiley, Chichester 2006

8.6.5 ENERGY RECOVERY
C.F.Cullins, M.M. Hirschler, The Combustion of Organic Polymers, Oxford Univ.Press, Oxford
    1981
R.M.Aseeva, G.E.Zaikov, Combustion of Polymer Materials, Hanser, Munich 1986

References to Chapter 8

[1]   Data for 1999-2006: American Plastics Industry Producers' Statistical Group, as compiled
        by Veris Consulting, LLC. Data for 1940-1998: data from very many sources collected by
        the author since 1957

[2]   -, Plastics News (March 31, 2008) p. 29

[3]   www.census.gov/compendia/statab/tables; www.virtualsteel.com; www.hardwoodint.com;
        www.lme.co.uk; all accessed on 2008-03-31

[4]   2005 data: Plastics Europe. WG Market Research & Statistics. www.plasticseurope.org,
        accessed 2008-03-28. Earlier data: scattered data from very many different sources.

[5]   The Japan Plastics Industry Federation, Production of Plastics Materials 2007, Japan
        www/jpif.gr.jp/english/statistics/monthly/2007/2007_production_materials_e.htm,
        accessed 2008-03-31

[6]   L.L.Böhm, H.-F.Enderle, M.Fleissner, F.Kloos, Angew.Makromol.Chem. **244** (1997) 93,
        Fig. 13

[7]   R.E.King III, U.Stadler, Angew.Makromol.Chem. **261/262** (1998) 189, Table 1

[8]   W.J.Jackson, Jr., H.F.Kuhfuss, J.Polym.Sci.-Polym.Chem.Ed. **14** (1976) 2043, Fig. 1

[9]   J.B.Enns, J.K.Gillham, J.Appl.Polym.Sci. **28** (1983) 2567, Fig. 1

[10]   Plastics News (May 26, 2008), p. 34

[11]   M.Heneczkowski, Kunststoffe **83**/6 (1993) 473, Tables 2 and 3

[12]   www.epa.gov/epaoswer/non-hw/muncpl/msw99.html#links, accessed 2008-06-08

[13]   www.plasticsrecycling.com/statistics, accessed 2002-01-05

[14]   S.E.Selke, in C.A.Harper, Ed., Modern Plastics Handbook, McGraw-Hill, New York 2000,
        Chapter 12

[15]   W.Kaminsky, Makromol.Chem., Macromol.Symp. **48/49** (1991) 381, Tables 2 and 3

[16]   Bundesamt für Umweltschutz (= Swiss Environmental Protection Agency), Oekobilanzen von
        Packstoffen ("Ecological balances of packaging materials"), Schriftenreihe Umweltschutz,
        No. 24, Berne 1984

9 Polymer Composites

9.1 Polymer Systems

9.1.1 Introduction

Polymer systems consist of two or more components of which at least one is a polymer (L: *componere* = to put together; from *com* = together and *ponere* = to put, to carry). The other component(s) may be polymeric or non-polymeric and may exist as solid(s), liquid(s), or gas(es). As solid materials, these other components may be present as powders, platelets, fibers, fabrics, or mats.

Polymer systems may be homogeneous (entities of components have submicroscopic dimensions) or heterogeneous (at least one of the components is present as entities with greater than microscopic dimensions) (Table 9-1). Polymer systems can be subdivided into very many categories depending on the emphasis that is put on the chemical make-up, the physical structures and states of components, the geometric dimensions of and interactions between components, and the like.

The two main groups of polymer systems are composites and blends. However, the use of these terms in polymer science and engineering differs somewhat from their general meanings. In common language, a **system** is a group of different entities that make up a collective entity (G: *sunistanai* = to bring together, from *sun* = together and *histanai* = to cause to stand). A **composite** is made up of distinguishable components (L: *compositus*, past participle of *componere*; see above) whereas a **blend** is a mixture (see Chapter 10) in which the constituent components are indistinguishable from one another. For example, a "blended whiskey" refers to a homogeneous liquid that results from the mixing of two or more brands of whiskies. A "dry blend" is a macroscopic mixture of a polymer powder with powdered adjuvants, especially fillers.

In a **polymer composite**, one of components is a polymer that forms a continuous **matrix** (L: *matrix* = womb, from *mater* = mother). All other component(s) are particulates with *distinct shapes* (spheres, spheroids, platelets, fibers, etc.) that are distributed or arranged in the matrix. The particulates may be solids of any size, from nano-sized particles to decimeter long fibers or fiber mats. However, particle-filled polymers are often not considered "composites" and the term "polymer composite" is then applied exclusively to polymers filled with fibers or fibrous materials such as fiber mats. Obviously, there is no sharp dividing line between "particle" and "fiber" or "platelet" because this would require a defined minimum value of the aspect ratio (see p. 162 for the sizes of fibers and p. 36 for geometric properties). For the purpose of this chapter, both particle-filled and fiber-filled polymers are therefore considered "composites."

Properties of composites are controlled by both polymeric matrices and added components as well as by any interactions between the matrix and the components. In some composites, the contribution of the embedded components to the properties of the resulting composite is just proportional to the volume fraction of the embedded component. Since these embedded components are less costly than the polymer matrix, they act as **extenders**. Other embedded components contribute more than proportionally to the properties of the composite and are therefore considered **reinforcing agents**. Such reinforcements are especially prominent for fibrous fillers, which is the reason why "polymer composite" is often used exclusively for fiber-filled composites.

Especially strong reinforcements are obtained if very strong and stiff fibers are aligned in the polymer matrix. The polymeric matrix then acts as a glue or binder. Since the load is carried predominantly by the fibers and not by the polymer matrix, it is not correct to call such a system a "reinforced polymer." These bonded-fiber materials are therefore often distinguished as **advanced polymer composites**.

Furthermore, "reinforcement" may mean many things since not all polymer properties are improved. For example, a "reinforced polymer" may have a greater strength than the unreinforced parent polymer but not necessarily other improved properties. "Reinforcement" is thus used for the improvement of many different properties such as the various strengths (tensile, flexural, impact, etc.) as well as for such properties as abrasion.

Very different materials are used as fillers and/or reinforcing agents (Section 3.3, p. 30 ff.) but not all are suitable for all polymers (Table 9-1). These added materials may be organic or inorganic; particulates, fibers, or platelets; and compact or porous. Most popular are glass fibers (US: fiberglass) which constitute in the United States ca. 93 wt% of all fillers + reinforcing agents for plastics, mostly for unsaturated polyester resins. Other thermosets and thermoplastics are less often provided with fillers; examples are epoxides, phenolic resins, amino resins, polyamides, poly(propylene), saturated polyester resins, styrene polymers, polycarbonates, and a few other thermoplastics. In the United States, most reinforced plastics are used in transportation (55 wt%), followed by construction materials (20 wt%), corrosion resistant working materials (15 wt%), and electrical components (10 wt%).

Table 9-1 Fillers for thermoplastics, thermosets, and elastomers.

Filler	Used with	wt%	Improved property
Inorganic			
Chalk	PE, PVC, PPS, PB, UP	<33	price, gloss
K_2TiO_3	PA	40	dimensional stability
Heavy spar	PVC, PUR	<25	density
Talc	PUR, UP, PVC, EP, PE, PS, PP	<30	white pigment, impact strength, plasticizer uptake
Mica	PUR, UP, PP	<25	dimensional stability, stiffness, hardness
Kaolin	UP, vinyl polymers	<60	demolding
Glass spheres	thermoplastics, thermosets	<40	moduli, shrinkage, compression strength, surface properties
Glass fibers	thermoplastics, thermosets	<40	tensile and impact strength
SiO_2, pyrogenic	thermoplastics, thermosets, SI rubbers	<30	tensile strength, viscosity increase
SiO_2, precipitated	diene rubbers		
Quartz	PE, PMMA, EP	<45	thermal stability, fracture
Sand	EP, UP, PF	<60	shrinkage
Al, Zn, Cu, Ni	PA, POM, PP	<100	thermal and electrical conductivity
MgO	UP	<70	stiffness, hardness
ZnO	PP, PUR, UP, EP	<70	UV stability, thermal conductivity
Organic			
Carbon blacks	PVC, HDPE, PUR, PI, elastomers	<60	UV stability, pigmentation, hardness
Graphite	EP, MF, PB, PI, PPS, UP, PTFE	<50	rigidity, creep
Wood flour	PF, MF, UF, UP, PP	<5	shrinkage, impact strength
Starch	PVAL, PE	<7	biological degradation

9.1.2 Historical Development

Nature uses various types of reinforcements for different purposes. Examples:

- Shells of lobsters and other crustacea consist of the polysaccharide chitin with embedded calcium carbonate.
- In bones, the long axes of platelets of hydroxy apatite crystals ($\phi = 0.4$) are parallel to the long axes of collagen fibers ($\phi = 0.5$) that are plasticized with water ($\phi = 0.1$).
- Cartilage consists of a network of collagen fibers (20 wt%) that is bound to a gel of highly branched glycosaminoglycans (5 wt%) in water (75 wt%).
- Wood consists of a network of lignins that is reinforced by cellulose fibers, plasticized by water, and "foamed" by air. Properties of wood are mainly controlled by the arrangement of cellulose fibers: their axial orientation is responsible for the large flexural strength of bows from the wood of the osage-orange tree whereas their random orientation in the wood of *Ficus sycomorus* of northeastern Africa and adjacent Asia leads to excellent impact strengths (see Volume II, p. 64 ff.)..

Nature creates these composites by simultaneous generation of the components, for example, by a growth of inorganic crystals in the polymerizing organic matrix. Synthetic composites and blends, on the other hand, are usually produced by mixing of two (or more) independently generated components. Examples are the mixing of two polymers, the compounding of polymers with fillers, the impregnation of fabrics by resins, or the simultaneous spinning of two polymers to bicomponent fibers. Composites can also be generated by polymerization of a monomer in a polymer of a different constitution. An example is the formation of rubber-reinforced poly(styrene)s.

The first known man-made composites were used in Babylon in ca. 4000 BC where clay for pottery was filled with crushed stones for reducing shrinkage during drying and fracture during firing. In ca. 3000 BC, long-fiber reinforced boat hulls were made from bundles of papyrus stalks in a matrix of bitumen in both Mesopotamia and Egypt. Early polymeric composites comprise straw-reinforced bricks from clay (Mesopotamia), with natural resins impregnated linen fabrics for the wrapping of mummies (Egypt), bows from animal tendons, wood, silk, and adhesives (Mongolia), and whetstones from sand and shellac (India). Just like their modern successors, ancient composites had greater elastic moduli, improved fracture strengths, decreased shrinkages, etc. (Table 9-2).

Table 9-2 Increase ↑ or decrease ↓ of properties of amorphous (A) and crystalline (C) plastics by addition of extenders or reinforcing agents. [↑] indicates a weak increase. Properties may also increase *or* decrease (↓↑); [a] brittle plastics become more ductile but ductile ones more brittle.

Property	Extenders		Reinforcing agents	
	A	C	A	C
Flexural modulus	↑	↑	↑	↑↑
Tensile modulus	(↑)	(↑)	↑	↑
Fracture strength	↓	↓	↑	↑
Brittleness	↑	↑	↑↓[a]	↑↓[a]
Melt viscosity	↓	↓↑	↓	↓↑
Heat distortion	–	(↑↓)	–	↑
Shrinkage	↓	↓	↓	↓

9.2   Mixture Rules

9.2.1   Moduli

Properties of polymer systems change with the proportions of their components. In many cases, these changes can be described by simple or modified mixing rules that are known by different names in different disciplines (Table 9-3). These rules assume that properties are affected only by the proportions and intrinsic properties of the components and not by any interactions, interlayers, or quantum effects (Section 9.2.3).

For example, an applied mechanical stress is taken up by the matrix and is then transferred to the additive, for example, fibers. In the simplest case, one assumes that (a) both matrix and fibers deform only elastically and not plastically, (b) matrix and fibers have the same Poisson's ratios, and (c) mechanical stresses do not cause a slip between matrix and fibers. Matrix and fiber should have good physical contacts but neither physical bonds (adsorption) nor chemical bonds are assumed to be present.

These conditions are fulfilled for tensile moduli since these quantities are determined for very small elongations $\Delta L/L_0$ below the proportionality limit. As a result, tensile moduli of a glass sphere-filled epoxy resin are independent of the presence or absence of coupling agents (coupling finishes) (initial slope in Fig. 9-1).

In contrast, fracture strengths are obtained at larger elongations where stresses are no longer proportional to elongations even in unfilled polymers (see pp. 138, 140). In filled polymers, there may be also slippage between the surfaces of the added particles and the matrix, and there may be also deformation of particle shapes. Fracture strengths therefore also depend on interactions between the particles and the matrix.

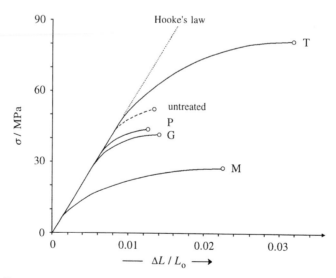

Fig. 9-1   Tensile stress σ as a function of the elongation $\Delta L/L_0$ of a cured epoxy resin that is filled with 30 vol% glass spheres [1]. Glass spheres were untreated or treated with various chemicals.

Glass spheres were treated with phenyltriethoxysilane (P), γ-methacryloxypropyl trimethoxysilane (G), or a mixture of various methylchlorosilanes (M) and then mixed with the filled epoxy resin.

Alternatively, glass spheres were first pretreated with triethylenetetramine (T) and then coated with a ca. 0.1 μm thick layer of the epoxy resin. The layer was cured and the coated spheres were mixed with the matrix which was then cured. O indicates fracture.

Table 9-3 Names of mixing rules. * After a mathematical manipulation.

Discipline	Exponent in Eq.(9-5)		
	n = 1	(n = 0) *	n = – 1
Mathematics	arithmetic average	geometric average	harmonic average
Chemical engineering	mixing rule	logarithmic mixing rule	inverse mixing rule
Mechanics	Voigt model	-	Reuss model
Electrical engineering	parallel	-	series
Materials science	upper limit	-	lower limit

Simple mixing rules are obtained from the following physical models. In one limiting case, a test specimen consists of a mixture of infinitely long, rigid fibers F with tensile modulus E_F and tensile strength σ_F in a matrix with modulus E_M and strength σ_M. The specimen is drawn to a value of $\varepsilon = (L - L_0)/L_0 = \Delta L/L_0$ in the direction of the completely oriented long axes of the fibers where L_0 and L are the original and the final lengths of the specimen, respectively. The force $F = \sigma A = \varepsilon E A$ for this drawing attacks the total cross-sectional area of the specimen and is distributed to the two components F and M according to their cross-sectional areas A_F and A_M, respectively:

$$(9\text{-}1) \qquad F = F_M + F_F = \sigma_M A_M + \sigma_F A_F = \varepsilon E_M A_M + \varepsilon E_F A_F$$

Because cross-sectional areas are perpendicular to the fiber lengths, one has $A_M/A = \phi_M$ and $A_F/A = \phi_F$. With $\phi_M + \phi_F \equiv 1$ and $F_i/A = E_i\varepsilon$ (for i = F, M), one obtains the simple **rule of mixtures (rule of mixing)** which predicts a linear increase of the longitudinal tensile modulus $E = E_\parallel$ with increasing volume fraction ϕ_F of the fiber (Fig. 9-2):

$$(9\text{-}2) \qquad E = E_M\phi_M + E_F\phi_F = E_M + (E_F - E_M)\phi_F = E_\parallel$$

In the other limiting case (long axes of rigid fibers are perpendicular to the draw direction), fractional moduli or stresses are not additive but fractional elongations are:

$$(9\text{-}3) \qquad \varepsilon = \varepsilon_M\phi_M + \varepsilon_F\phi_F = (\sigma_M/E_M)\phi_M + (\sigma_F/E_F)\phi_F = \sigma/E_\perp$$

Since in this case $\sigma = \sigma_M = \sigma_F$, one obtains for transverse moduli $E_\perp$ the **inverse rule of mixtures (inverse rule of mixing)** (Fig. 9-2):

$$(9\text{-}4) \qquad 1/E_\perp = (\phi_M/E_M) + (\phi_F/E_F)$$

Eqs.(9-2) and (9-4) can be written with exponents n = +1 and n = –1, respectively:

$$(9\text{-}5) \qquad E^n = (E_M)^n\phi_M + (E_F)^n\phi_F$$

After some mathematical manipulations, one obtains from Eq.(9-5) with n $\rightarrow$ 0 the so-called **logarithmic mixing rule (logarithmic mixing law)**:

$$(9\text{-}6) \qquad \lg E = (\lg E_M)\phi_M + (\lg E_F)\phi_F$$

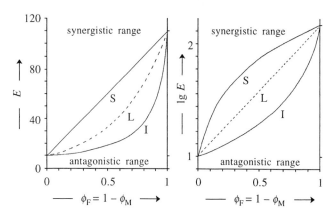

Fig. 9-2 Dependence of the tensile modulus E of composites or blends on the volume fraction ϕ_F of its component F (fiber, filler, etc.) if the simple (S), logarithmic (L), or inverse (I) rules of mixing apply. ϕ_M = volume fraction of the matrix M. Calculations for $E_M = 10$ and $E_F = 110$ (arbitrary physical units, for example, megapascals). Left: conventional plot; right: logarithmic plot.
At $\phi_F = 0$, $E = E_M$ and at $\phi_F = 1$, $E = E_F$.

Fig. 9-2 shows the dependence of moduli E of composites or blends of F and M on the volume fraction ϕ_F of their component F. In principle, E may be any physical property and F and M any type of component. According to the assumptions inherent in deriving the mixing laws, properties E can only have values in the range between the limits that are given by the simple and the inverse rule of mixtures.

Sometimes, effective property values E_{eff} are observed that are either above the limits given by the simple rule of mixtures (**synergistic effects**) or below the limits imposed by the inverse law of mixtures (**antagonistic effects**). For example, if the component F is a rod and the rules of mixing do not apply, then one would observe a synergism for $E_{eff} > E_{\parallel}$ and an antagonism for $E_{eff} < E_{\perp}$ (see also Section 10.4.6).

Some authors define synergism and antagonism differently by postulating that true synergism is present only if the function $E = f(\phi_F)$ has a maximum, i.e., if $E > E_F$ at some higher volume fractions $\phi_F < 1$. Correspondingly, antagonism is present only if the function $E = f(\phi_F)$ has a minimum, i.e., that some values of E are smaller than E_M.

9.2.2 Densities

Mechanical properties of simple mixtures, as defined in Section 9.2.1, depend on volume fractions. However, true volume fractions are not easy to determine and physical properties are therefore usually presented as functions of mass fractions of components.

Volume fractions ϕ and mass fractions w of components M and F can be inter-converted if volumes are additive:

$$(9\text{-}7) \qquad \phi_F = \frac{w_F}{w_F + w_M(\rho_F / \rho_M)}$$

For non-additive volumes, the true volume fraction of F can be calculated from $\phi_F = w_F(\rho/\rho_F)$ where ρ = density of the specimen composed of M and F and w_F and ρ_F are mass fraction and density of F, respectively.

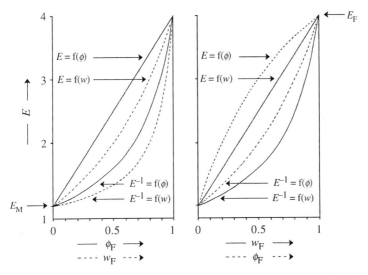

Fig. 9-3 Illustration of the effect of choosing the wrong concentration parameter for an ideal binary system with $E_M = 1$, $E_F = 4$, and $\rho_F/\rho_M = 2$. Left: E values of mixtures depend correctly on volume fractions ϕ_F (——) or erroneously on mass fractions w_F (----). Right: "true" dependence of E on w_F (——) and "wrong" dependence of E on ϕ_p (----) (see text).

The presentation of properties E of mixtures as a function of mass fractions instead of volume fractions may lead to erroneous conclusions. If then the density of F is greater than that of M, the true mixing rule degenerates to the logarithmic mixing rule whereas the inverse mixing rule apparently indicates an antagonistic effect (Fig. 9-3, left). For the physically unlikely situation of a true dependence of properties on mass fractions, one would see apparent synergistic effects for the rule of mixtures and apparent logarithmic dependencies for the inverse rule of mixing (Fig. 9-3, right).

The assumed additivity of volumes always has to be checked. A calculated density

$$(9\text{-}8) \qquad \rho_{calc} = \frac{1}{(w_F/\rho_F) + (w_M/\rho_M)}$$

that is greater than the experimental density ρ_{exp} is caused by internal voids from improper compounding. The inverse case, $\rho_{exp} > \rho_{calc}$, indicates a densification of the matrix M which is usually much easier to compress than the fiber or filler F, for example, by a filler/fiber-induced crystallization (see next Section).

Other factors have to be considered for hybrid composites consisting of one matrix M and two types A and B of fibers of which a part is assembled in bundles (cables). The composite contains Y_A cables of A with N_A fibers A and Y_B cables of B with N_B fibers B. Single fibers have cross-sectional areas A_A and A_B, respectively, and densities ρ_A and ρ_B. The number fractions of cables are $y_A = Y_A/(Y_A + Y_B) = 1/(1 + (Y_B/Y_A))$ for type A and $y_B = Y_B/(Y_A + Y_B) = 1/(1 + (Y_A/Y_B))$ for type B. The generalized fraction of A is

$$(9\text{-}9) \qquad f_A = \frac{1}{1 + \dfrac{Y_B}{Y_A} \cdot \dfrac{N_B}{N_A} \cdot \dfrac{A_B}{A_A} \cdot \dfrac{\rho_B}{\rho_A}}$$

where f_A = mass fraction w_A for any Y, N, A, and ρ. The fraction f_A becomes the volume fraction ϕ_A if $\rho_B/\rho_A = 1$, the number fraction x_A if $\rho_B/\rho_A = A_B/A_A = 1$, and a fraction Y_A if $\rho_B/\rho_A = A_B/A_A = N_B/N_A = 1$ since $Y_A + Y_B \equiv 1$.

9.2.3 Interlayers

A composite of matrix M and fiber/filler F may contain more than the two "phases" of continuous M and dispersed F. For example, compounding resins with fillers can trap air; these voids mostly disappear after the first application of pressure.

Filler particles are usually hydrophilic and adsorb water. Such water layers on surfaces are mostly 3 molecules deep, i.e., ca. 0.5 nm.

The physical structure of the matrix near particles (fibers, particulates) may also differ from that in the interior of the matrix. Such differences can be caused by conformational changes of matrix molecules, their adsorption or transcrystallization on particle surfaces, or even chemical reactions of matrix molecules with reactive surface groups of the fillers/fibers. These interlayers are often called "interphases" although they are usually not thick enough to fulfill the thermodynamic requirements for a "phase."

These interlayers consist of polymer molecules that are bound so firmly to the surface of the fiber/filler particles that they cannot be extracted (Fig. 9-4). The proportion of these is calculated from the ratio of the mass of residue *minus* the mass of pure filler divided by the mass of the original polymer. The calculated bound polymer fraction depends on the concentration and the specific surface of the filler.

It is more expedient to consider volumes instead of masses and report ratios V_b/V_F of volumes V_b of bound polymers and volumes V_F of fillers, which requires a conversion of experimentally observed masses to volumes using densities. The thickness a_b of the bound polymer layer is then calculated from $V_b/V_F = [(d_F + 2\,a_b)^3 - d_F^3]/d_F^3$ where d_F is the experimental diameter of the filler particles.

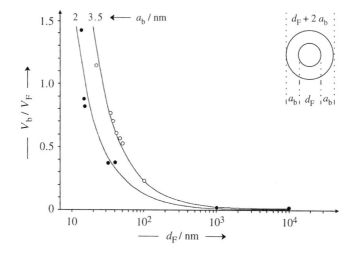

Fig. 9-4 Volume ratio V_b/V_F of bound polymer and filler as a function of the diameter d_F of filler particles. ● Poly(ethylene) at pyrogeneous silica [2a]; O SBR rubber at carbon black [3]. Solid lines: calculated for thicknesses a_b of interlayers of 2.0 nm and 3.5 nm, respectively.

Fig. 9-5 Transcrystalline layers of polyamide 6.6 on fibers of poly(*p*-phenyleneterephthalamide) (Kevlar 49®) [4]. The layers have widths of 5-23 μm; the PA 6.6 matrix is filled with spherulites.

Such layers were found to be ca. 2 nm thick for poly(ethylene)s at finely divided pyrogenic silicon dioxide particles (Aerosil®) and 3.5 nm thick for styrene-butadiene rubbers at carbon black particles (Fig. 9-4). Dynamic mechanical measurements also indicated similarly sized "glassy" interlayers of SBR on carbon black particles. The physical or chemical reasons for the presence of such layers are usually unknown.

Fillers and especially fibers may also act as nucleation agents for crystallizable thermoplastics where they cause **trans-crystallization** that leads to **epitaxial growth** (Fig. 9-5). Such 5-23 μm thick layers are known for many organic fibers; glass fibers seem to be less effective in inducing trans-crystallization of matrix molecules.

Fillers and fibers may not only exert physical effects but also chemical ones, either directly or indirectly. Direct effects occur in carbon black-filled diene rubbers since the surfaces of carbon black particles contain free radicals. These radicals initiate a **grafting** of diene units onto the carbon black particles, which leads not only to a chemically bound "interlayer" of diene units but also to a crosslinking of the rubber molecules.

Another example of grafting is that of epoxy units on the silanol groups of glass fibers. However, the situation is quite different for epoxy resins that are filled with carbon fibers. Curing these resins with amines leads to a hard epoxy matrix and soft layers near the fiber surfaces which are up to 500 nm thick. It was suggested that these layers are caused by a purely physical effect, i.e., by a loss of random conformations of epoxy molecules that is caused by the barrier action of the fiber surface. However, the great thickness of this layer seems rather to point to a chemical effect. Carbon fibers, especially surface-treated ones, contain polar surface groups which strongly bind water molecules that can react with epoxy groups. The resulting OH groups do not react with the amine. As a consequence, crosslinkable epoxy groups are lost and one gets low-molecular weight products and/or wide-mesh networks near the fiber surface. This view is supported by the fact that the soft interlayer disappears if carbon fibers are deactivated by coating them with cured phenolic resins.

Interlayers and epitaxial layers act not only as third "components" besides the matrix molecules and filler/fiber materials but may also promote the formation of physical networks of filler/fiber particles. If spheres and rods are present in the same volume fractions, then rods will have a greater probability to contact each other and form a physical network. This probability is further advanced if both types of bodies have interlayers with the same thickness (Fig. 9-6).

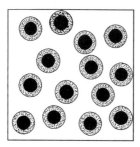 

Fig. 9-6 Effect of particle shape on the formation of physical networks from spheres and rods (black) of the same volume per sphere or rod. Interlayers have the same thickness (dotted). Both types of entities are present with the same volume fraction of $\phi_F = 0.112$. At this concentration, randomly oriented rods have a far higher probability to contact each other than the spheres.

On drawing, shearing, or impact, physical contacts between matrix molecules and fibers/fillers are loosened, which increases ductibility but reduces strength. Fibers and also some fillers are therefore treated with agents that improve adhesion and often also chemical bonding between fibers and matrix. These agents are known by many names, depending on industry, for example, adhesion promoters (coatings), primers (coatings), finishes (reinforcing fibers), sizing agents (glass fibers), or coupling agents (fillers).

For glass fibers, coupling agents are usually silanes (p. 41), most often aminopropyl-alkoxysilanes, which are physically adsorbed on glass surfaces according to infrared measurements. In the presence of primary or secondary amino groups (but not in their absence), they also react chemically with silanol groups on the glass surface. For example, vinylalkoxysilanes, $CH_2=CHSi(OR)_3$, are probably first hydrolyzed to vinyltri-hydroxysilane II before they react with silanol groups I of the glass surface:

$$(9\text{-}10) \quad \underset{\overset{|}{OH}}{\overset{\xi}{\underset{|}{\sim\!\!\sim Si}}} - O\sim\!\!\sim \ +\ \underset{\overset{|}{OH}}{HO - \overset{\overset{OH}{|}}{Si} - CH = CH_2} \longrightarrow \underset{O - Si(OH)_2CH = CH_2}{\overset{\xi}{\underset{|}{\sim\!\!\sim Si}} - O\sim\!\!\sim} \quad + H_2O$$

$$\hspace{4cm} \text{I} \hspace{3.5cm} \text{II}$$

The various silane coupling agents change the stress-strain behavior of filled polymers quite differently (p. 374). Some lead to increased extensibilities but reduced fracture strengths (M in Fig. 9-1); others increase both extensibilities and fracture strengths (T) and still others have little effect (G, P). The action of these coupling agents may interfere with those of some sizes (p. 35) with which glass fibers are treated in order to improve processability and prevent fiber fracture.

9.3   Tensile Moduli

9.3.1. Long-Fiber Reinforcement

Tensile moduli are obtained from the initial slopes of stress-strain curves. Because stresses and elongations are small, neither the nature of the interface between fiber/filler and matrix nor the presence of interlayers should matter.

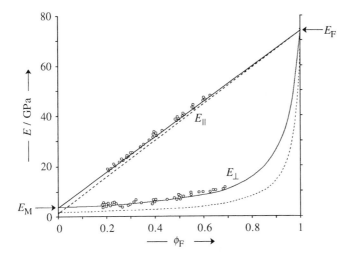

Fig. 9-7 Dependence of the longitudinal (∥) and transverse (⊥) tensile moduli E of cured, fiber-filled unsaturated polyester resins on the volume fraction of long, unidirectionally oriented glass fibers [5]. o Experimental values immediately after preparation of the fiber-filled polymers, —— calculated from the simple or inverse rule of mixtures, - - - experimental values after 10^4 h (ca. 13.7 months) at room temperature (for clarity, experimental data points are not shown).

In the simplest case, fiber-filled polymers are drawn in the direction of the longitudinal axes of infinitely long, non-aggregated, rigid fibers. Tensile moduli of freshly prepared specimens obey exactly the simple (∥) or inverse (⊥) laws of mixing (Fig. 9-7). After long times, however, longitudinal moduli $E_∥$ are somewhat smaller, especially at small fiber contents, which is caused by creeping of the matrix. The time effect is stronger for transverse moduli where it is *percentage-wise* more pronounced for small fiber contents than for large ones. This change is probably caused by a strong bending of the embedded fibers which is easier for isolated fibers (small fiber content) than for more densely packed ones (high fiber content).

For fiber directions other than $\vartheta = 0°$ or $90°$ to the draw direction, moduli E_ϑ of the composites can be calculated from the longitudinal modulus E_0, the transverse modulus E_{90}, the shear modulus G, and the Poisson ratio v (transverse elongation along long-fiber axes and longitudinal elongation transverse to long-fiber axes):

$$(9\text{-}11) \qquad \frac{1}{E_\vartheta} = \frac{\cos^4 \vartheta}{E_0} + \left(\frac{1}{G} - \frac{2v}{E_{90}}\right)(\sin^2 \vartheta)(\cos^2 \vartheta) + \frac{1}{E_{90}}\sin^4 \vartheta$$

The Poisson ratio can be set to zero because elongations perpendicular to the fiber long axes are hampered by the rigidity of fibers. Tensile moduli are therefore controlled only by the moduli E_0, E_{90}, and G and especially the orientation angle of the fibers. The resulting functions $E_\vartheta = f(\vartheta)$ have similar shapes for different glass fiber contents in different thermoplastics (Fig. 9-8).

In a compound, fibers may be oriented at more than one angle ϑ_i to the strain direction, which can be expressed by a weighted orientation factor. For equal deformation of fibers and matrix (Voigt model, see Volume III, p. 546 ff.), one obtains the **Krenchel orientation factor**, f_O, as

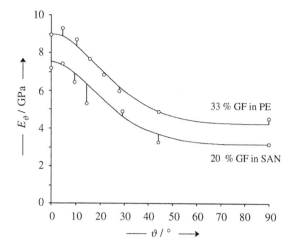

Fig. 9-8   Tensile moduli E_ϑ as a function of the angle ϑ of the long axes of fibers relative to the draw direction for glass fiber (GF)-reinforced composites of poly(ethylene) and styrene-acrylonitrile copolymers [6]. Lines: calculated from Eq.(9-11) and $v \to 0$.

(9-12) $$f_O = \sum_i \phi_i \cos^4 \vartheta_i$$

where ϕ_i = volume fraction of fibers with long axes at an angle ϑ_i to the applied load (continuous distributions of fiber axes require integration). For example, $f_O = 3/8$ if fibers are oriented in the four directions $\vartheta_1 = 0°$, $\vartheta_2 = \pi/2$, $\vartheta_3 = \pi/4$, and $\vartheta_4 = -\pi/4$ in the same proportions, $\phi_1 = \phi_2 = \phi_3 = \phi_4 = 1/4$. Some orientations and orientation factors are shown in Fig. 9-9.

Of course, the largest rigidities are always obtained for the long axes of fibers in the draw direction ($f_O = 1$) but such anisotropic composites have also very poor transverse properties (Fig. 9-7). These composites are only useful in applications where stresses are completely unidirectional. Since this is not true for most applications one needs to select fiber directions that are optimal for the intended purpose (Fig. 9-9). For example, the three-dimensional random orientations of fiber long axes are not optimal for composites that are randomly stressed in all directions since the orientation factor is only 1/6.

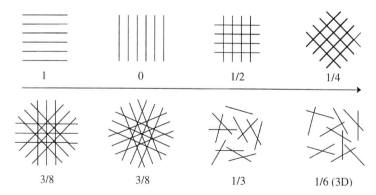

Fig. 9-9   Factors f_O for the orientation of long fibers with respect to the strain direction $\to$ in two-dimensional cases and in one three-dimensional case with random orientations (3D).

One would expect that stiffening is larger, the greater the tensile modulus of long fibers. However, it was found experimentally that the rigidities of fibers are not fully exploited if the ratio E_F/E_M of moduli exceeds circa 50-100. For example, experimental longitudinal tensile moduli of epoxy resins ($E_M = 3.2$ GPa with $\phi_F = 60$ %) matched the theoretical ones for glass fibers, poly(phenylene terephthalate) fibers (PPTA), and carbon fibers with $E_F/E_M \leq 80$ but reached only 90 % of the theoretical value for boron fibers ($E_F/E_M = 121$) and 53 % for HM (high-modulus) carbon fibers ($E_F/E_M = 216$). Ratios of the transverse moduli of these composites were either smaller than theoretical ones (PPTA: 0.71; HM carbon fibers: 0.87) or larger (S glass: 2.11; boron fiber: 2.41).

For technical purposes, moduli *per se* are not important since most materials are not used per mass but per volume, a measure of which is the "specific" modulus = modulus divided by density. Because of their lower densities, organic fibers with small fiber moduli are therefore often better reinforcement agents than boron or steel fibers with larger fiber moduli. In transportation, such lightweight composites also save energy.

Fiber-reinforced composites should also resist solvents such as gasoline or toluene (in fuels) and chlorinated hydrocarbons (cleansing agents). For these reasons, high-performance resins are usually thermosets that are reinforced by high-modulus fibers. One would prefer high-performance thermoplastics because they are more easy to process.

The placing of correctly orientated long fibers in plastics is a difficult and time-consuming process. The work-intensive hand lay-up process (p. 107) thus uses fiber mats instead of fibers, and the cost-intensive filament winding process (p. 109) employs fiber rovings. Resins for the matrices of these composites should not be very viscous in order to allow easy flow between the fibers. For these reasons, long fibers are practically used only for thermosetting resins.

9.3.2   Reinforcement by Short Fibers and Particulates

In contrast to long-fiber filled plastics, short-fiber filled ones can be processed with conventional processing equipment. However, these machines must be made from specially hardened steels since glass fibers and certain hard particulates can be very abrasive.

Like long fibers, efficiencies of short glass fibers depend on their axial orientations and, in addition, also on their axial ratios. A complete or at least controlled orientation of short fibers would be very desirable since it would allow a comparison of experimental data with micromechanical models. Alas, such experiments are extraordinarily difficult; no good data seem to exist.

Conventional processing of short-fiber reinforced plastics by extrusion, injection molding, calendering, etc., leads to a certain, usually desirable, orientation of fiber axes. Such orientations range from two-dimensional random orientations of fibers in molded materials to almost unidirectional ones in injection-molded ones.

The influence of axial ratios and orientations of short fibers on properties E can be described by an efficiency factor f for the property E_F of the fiber in the simple rule of mixtures. This factor is the product of the orientation factor f_O (Eq.(9-12)) and a factor f_L for the axial ratio of the fibers:

$$(9\text{-}13) \qquad E = E_M\phi_M + fE_F\phi_F = E_M + (fE_F - E_M)\phi_F \quad ; \quad f = f_O f_L$$

Cox Equation

The efficiency factor f of Eq.(9-13) has been calculated theoretically by Cox for circular fibers of length L_F and cross-sectional area A_F that are subjected to a load parallel to the fiber axis. It is assumed that the matrix is stressed homogeneously, both fiber and matrix behave elastically (no shearing!), no load is transferred by the ends of the fibers, and Hooke's law applies. The effective modulus is then

$$(9\text{-}14) \qquad E_{Cox} = E_M \phi_M + \left(1 - \frac{\tanh b[(L_F/2)]}{b[L_F/2]}\right) E_F \phi_F \; ; \quad b = (Q E_F^{-1} A_F^{-1})^{1/2}$$

with $b = (Q E_F^{-1} A_F^{-1})^{1/2}$ and $Q = 2\pi\, G_M / lg\, (d/r_F)$ where d = distance between centers of fibers perpendicular to their long axes, G_M = shear modulus of the matrix, and $r_F = d_F/2$ = radius of the fibers. The volume fraction of hexagonally packed fibers is obtained from d and r_F as $\phi_F = 3^{-1/2}\, 2\pi\, (r_F/d)^2$.

A very lengthy calculation delivers the longitudinal modulus as

$$(9\text{-}15) \qquad E_{\parallel,Cox} = E_M \phi_M + f_\parallel E_F \phi_F \qquad ; \qquad f_\parallel = 1 - [(\tanh Z)/Z)$$

$$Z = \left(\frac{2\, G_M}{E_F \ln(2 \cdot 3^{-1/2}\, \pi/\phi_F)^{1/2}}\right)^{1/2} \frac{L_F}{d_F}$$

The reduced longitudinal modulus $E_\parallel/E_{\parallel,\infty}$ depends strongly on the axial ratio L_F/d_F where $E_\parallel$ = longitudinal modulus of a short fiber and $E_{\parallel,\infty}$ = longitudinal modulus of an infinitely long fiber (corresponds to that from the simple rule of mixing, Fig. 9-10). High efficiencies of $E_\parallel/E_{\parallel,\infty} > 0.90$ are therefore obtained only if short fibers have axial ratios of more than 100 and are also truly parallel to each other in the draw direction. Of course, fiber contents should not be very high.

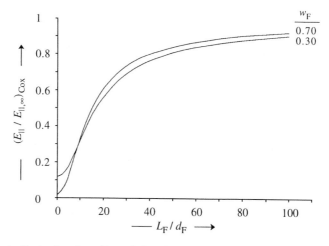

Fig. 9-10 Longitudinal reduced tensile moduli as functions of the axial ratio L_F/d_F of glass fibers (E_F = 72 GPa) in it-poly(propylene) (E_M = 1.6 GPa; G_M = 0.56 GPa) as calculated from Eq.(9-15). Volume fractions ϕ_F and ϕ_M were obtained from densities ρ_F = 2.54 g/cm^3 and ρ_M = 0.90 g/cm^3 by assuming additivity of volumes (see Eq.(9-7)).

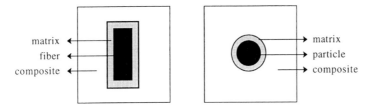

Fig. 9-11 Kerner model for the properties of particle-filled composites. The fiber (left) or spherical particle (right) is surrounded by a layer of the matrix. This "coated" entity is embedded in the macroscopically homogeneous, isotropic composite.

Halpin-Tsai Equation

The Cox approach applies to the reinforcement of plastics by a series of parallel fibers in a continuous matrix. Another approach by Kerner embeds the particle (fiber, sphere, etc.) in the pure matrix which in turn is surrounded by a macroscopically homogeneous, isotropic matter with the properties of the composite (Fig. 9-11). Elaborate matrix calculations for stresses and strains in various directions lead to the very complex **Kerner equation** for shear and bulk moduli.

The Kerner equation can be simplified, for example, by assuming equal Poisson's ratios of matrices and particles ($\mu_M = \mu_F$). The resulting simplified Kerner equation is known as the **Halpin-Tsai equation** and is usually written as

$$(9\text{-}16) \qquad \frac{E}{E_M} = \frac{1 + K_1 K_2 \phi_F}{1 - K_2 \phi_F} \quad ; \quad K_2 = \frac{(E_F / E_M) - 1}{(E_F / E_M) + K_1}$$

It becomes more transparent after rearrangement with $\phi_M + \phi_F \equiv 1$ which gives

$$(9\text{-}17) \qquad E = \frac{E_F + K_1 [E_M \phi_M + E_F \phi_F]}{K_1 + E_F \left[\dfrac{\phi_M}{E_M} + \dfrac{\phi_F}{E_F} \right]} \quad \text{or} \quad K_1 = E_F \frac{1 - E \left[\dfrac{\phi_M}{E_M} + \dfrac{\phi_F}{E_F} \right]}{E - [E_M \phi_M + E_F \phi_F]}$$

Eq.(9-17) delivers the rule of mixtures (= upper limit) for $K_1 \to \infty$ and the inverse rule of mixtures (= lower limit) for $K_1 = 0$ but cannot describe synergistic or antagonistic effects. Curves for the range $\infty > K_1 > 0$ do not comprise the predictions of the logarithmic mixing law (compare Fig. 9-2) but show a slight S shape for medium values of K_1 for plots of $\lg E = f(\phi_F)$ (Fig. 9-12). It is identical with the Takayanagi Eq.(10-24).

The adaptable parameter K_1 of Eq.(9-17) is a strengthening factor for the moduli of particles. Because of its dependence on the modulus of the compound, it is an empirical factor that is affected by many quantities such as the shape and the packaging of the filler particles and the direction of the load relative to the orientation of anisotropic particles. For example, $K_1 = 2\, L_F/d_F$ for longitudinal and transverse moduli if E is the tensile modulus but $K_1 = 3^{1/2} \lg (L_F/d_F)$ if E is the shear modulus of plastics that have been unidirectionally reinforced with short fibers or platelets.

The adaptable quantity K_2 indicates the use of fiber volumes. It becomes $K_2 = 0$ for $K_1 = \infty$ (upper limit, stress parallel to the fiber axes) but $K_2 = 1 - (E_M/E_F)$ for $K_1 = 0$ (lower limit, stress perpendicular to the fiber axes).

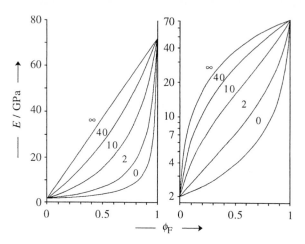

Fig. 9-12 Normal (left) and semi-logarithmic (right) plots of the modulus E as a function of the volume fraction ϕ_F, assuming Eq.(9-17) with $E_F = 72$ GPa and $E_M = 2$ GPa. Numbers indicate K_1.

Lewis-Nielsen Equation

Experimental tensile moduli of a polyamide fiber-reinforced rubber follow well the Halpin-Tsai equation if the fiber content is not very large ($\phi_F = 0.35$ in Fig. 9-13). The Cox equation can also represent the data reasonably well in the industrially interesting range $10 \leq L_F/d_F \leq 39$ but predicts too low values of $E_\parallel$ for L_F/d_F values below ca. 15 and too high ones for values above ca. 25. The Lewis-Nielsen equation (see below) deviates from experimental data for length ratios greater than ca. 4.

The effect of spheroids (spheres, etc.) on the properties of composites is usually also described by the Kerner equation or the Halpin-Tsai equation, Eq.(9-16), respectively, assuming $K_1 = (7 - 5\,\mu_M)/(8 - 10\,\mu_M)$.

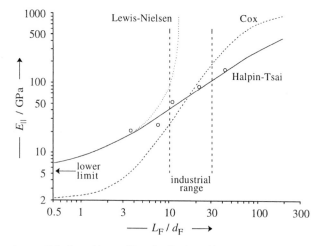

Fig. 9-13 Dependence of the logarithms of longitudinal tensile moduli $E_\parallel$ on the logarithms of axial ratios L_F/d_F of fibers for a rubber that was unidirectionally reinforced with $\phi_F = 0.35$ polyamide fibers [7]. Curves were calculated using equations of Halpin-Tsai (Eq.(9-17)), Cox (Eq.(9-15)), and Lewis-Nielsen (Eq.(9-18)). The arrow indicates the lower limit (= modulus of polyamide). $E_F/E_M = 973$.

For an epoxy resin with $\mu_M = 0.4$ (and thus $K_1 \approx 0.406$) that is filled with glass spheres, the resulting equation describes well the dependence of tensile moduli E on volume fractions ϕ_F of the filler if $\phi_F \leq 0.45$ (Fig. 9-14). Deviations occur at $\phi_F = 0.5$, regardless of the choice of Poisson's ratio (0.4 or 0.5).

Such deviations may be caused by a "fiber effect", i.e., the linear association of spheres. Indeed, a linear line-up is mathematically the most stable organization for 3-56 spheres (exact mathematical proof of this "sausage effect" is still pending).

Alternatively, the deviation from the Halpin-Tsai equation at higher volume fractions of spheres may be caused by a more highly ordered interlayer around the spheres with a correspondingly greater modulus. At higher concentrations of spheres, these interlayers would contact each other and produce a "network" (see Fig. 9-6).

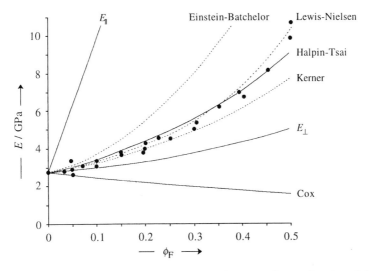

Fig. 9-14 Tensile moduli E of a glass sphere filled, cured epoxy resin as a function of the volume fraction ϕ_F of glass spheres. ● Experimental data [8]. Lines are calculated using the Cox Eq.(9-15), Einstein-Batchelor Eq.(9-19), Halpin-Tsai Eq.(9-17), a simplified Kerner equation, and the Lewis-Nielsen Eq.(9-18). Mixture rules deliver the upper limit $E_\parallel$ (longitudinal moduli) and the lower limit $E_\perp$. In the range $0 < \phi_F < 0.2$, data can be approximated by a straight line (see also Fig. 9-15).

The Cox, Kerner, and Halpin-Tsai equations all assume that filler particles may be present up to volume fractions of $\phi_F = 1$. However, particles generally pack less efficiently (p. 77) and one therefore has to consider the maximally possible packing fraction $\phi_{F,max}$ of the particular filler shape. The **Lewis-Nielsen equation** thus introduces a new correction factor K_3 into the denominator of the expression for E/E_M (compare left expressions of Eqs.(9-16) and (9-18)):

$$(9\text{-}18) \qquad \frac{E}{E_M} = \frac{1 + K_1 K_2 \phi_F}{1 - K_2 K_3 \phi_F} \quad ; \quad K_2 = \frac{(E_F/E_M) - 1}{(E_F/E_M) + K_1} \quad ; \quad K_3 = 1 + \left(\frac{1 - \phi_{F,max}}{\phi_{F,max}^2} \right) \phi_F$$

Values of $\phi_{F,max}$ can be obtained from Table 4-1 if the shape and arrangement of particles is known. For unknown shapes and arrangements, they may be obtained from the sedimentation volume of fillers.

Table 9-4 Einstein coefficients k_E as a function of the shape and orientation of rigid spheres, cubes, and fibers and of the stress direction. L = length, d = diameter, ϕ_{max} = maximum packing density of spheres in sphere-like aggregates.

Dispersed entities	Orientation of entities	Stress direction and slippage	k_E
Spheres	any	none; no slip	2.50
Spheres	any	none; slip	1.00
Spheres, aggregated	any		$2.50/\phi_{max}$
Cubes	random		≈ 3.1
Fibers	uniaxial	tension parallel to fiber axes	$2\,L/d$
Fibers	uniaxial	tension perpendicular to fiber axes	1.50
Fibers	uniaxial	compression	1.0
Fibers	uniaxial	shear longitudinal/transverse	2.0
Fibers	uniaxial	shear transverse/transverse	1.50
Fibers	random, three-dimensional	shear	$\approx 0.5\,L/d$

The constant K_2 is calculated from the material constants E_F and E_M, which leaves K_1 as the only quantity that needs a theoretical underpinning. For rigid spheres in a matrix with Poisson's ratio $\mu_M = 1/2$, Lewis-Nielsen maintain that $K_1 = k_E - 1$ is related to the Einstein coefficient k_E in the dependence of viscosities η on the volume fraction ϕ_2 of the solute, $\eta = \eta_1(1 + k_E\phi_2 + ...)$ if wall effects are absent (see Volume III, p. 406 ff.).

Other k_E values apply for other shapes and orientations of particles and for different stress directions (Table 9-4). Furthermore, a correctional factor f_E has to be introduced, $K_1 = f_E(k_E - 1)$, if Poisson's ratio μ_M of the matrix differs from 1/2. For these correction factors, only estimates are known: $f_E = 0.90$ for $\mu_M = 0.40$, 0.867 for $\mu_M = 0.35$, 0.84 for $\mu_M = 0.30$, and 0.80 for $\mu_M = 0.2$.

The Lewis-Nielsen approach describes tensile moduli of glass sphere-filled, cured epoxy resins slightly better than the Halpin-Tsai equation (Fig. 9-14). For large filler contents, it often delivers too high tensile moduli since it is very sensitive to small variations of filler contents in these concentration ranges (cf. Fig. 9-13 for fibers). The description of the concentration dependence of moduli by an **Einstein-Batchelor** type of equation (see Volume III, p. 406) results in too high moduli (Fig. 9-14):

$$(9\text{-}19) \qquad E = E_M(1 + 2.5\ \phi_F + 6.2\ \phi_F^2 + ...)$$

Many filler particles do not exist in geometrically simple shapes such as spheres, spheroids, or rods but in irregular shapes (see p. 37). The shape of such particles is not only usually unknown but it may also change during processing. Brittle particles may break during injection molding and spheroidal particles may aggregate to fiberlike entities. Non-spheroidal particles may also become oriented in the strain direction. Furthermore, their effect on properties of composites is also affected by their size distribution, which contrasts with the effect of spheres where only volume requirements are important.

The dependence of the moduli of particulate-filled composites on filler contents is usually determined empirically. For filler contents of less than $\phi_F \approx 0.3$, it mainly follows the generalized rule of mixtures, Eq.(9-13), with efficiency factors smaller than unity as shown in Fig. 9-15. This range is industrially important because standard filler contents are usually 30 wt%.

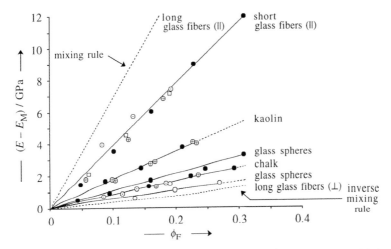

Fig. 9-15 Dependence of the difference of moduli E of composites and moduli E_M of their matrices on the volume fraction of various fillers [9a]. ○ Poly(ethylene) (PE), ⊘ poly(propylene) (PP), ⊥ poly(oxymethylene) (POM), ● polyamide 6 (PA 6), ⊕ poly(butylene terephthalate) (PBT).

Efficiency factors are sometimes determined only by the filler and not by the matrix; i.e., the contribution of the matrix to the slope $(fE_F - E_M)$ of the function $E = f(\phi_F)$ is negligible; chalk in poly(propylene) and polyamide 6 is an example (Fig. 9-15). The contribution of the matrix, however, is not negligible for glass spheres in poly(ethylene) or polyamide 6 (Fig. 9-15).

Efficiency factors are also independent of the matrix for thermoplastics (PE, PP, PBT, PA 6) filled with short glass fibers. However, its value of $f = 0.54 \pm 0.06$ is far greater than expected for $f = f_O f_L$, Eq.(9-13), since a random orientation of these fibers would lead to $f_O = 1/6$ (Fig. 9-9) and the length factor should not exceed $f_L = 1/2$ if these fibers are broken just once during processing. However, short glass fibers are sold with lengths of ca. 2 mm and diameters of 10-20 μm (axial ratios 100-200) but are broken more than once during injection molding so that the processed composite contains fibers of just 0.2 mm average length, i.e., axial ratios of only 10-20. Larger than theoretically expected efficiency factors are also found for glass spheres and kaolin, a platelet-like filler with an axial ratio of $L_F/d_F \approx 5\text{-}100$.

None of the proposed equations can describe the experimental data correctly (Table 9-5), which may be due to theoretical simplifications and/or unconsidered facts (inter-layers, actual filler sizes, orientation of anisotropic fillers, etc.).

Table 9-6 Experimental and calculated tensile moduli of composites of a polyamide 6 ($E_M = 3.0$ GPa) with glass fibers ($E_F = 72$ GPa; $\phi_F = 0.193$; various axial ratios L_F/d_F) [9b].

L_F/d_F	Upper limit	Halpin-Tsai	Lewis-Nielsen	$E_\parallel$/GPa Experi-mental	Lewis-Nielsen	Cox	Lower limit
16.67	16.32	11.64	11.60	9.6	9.98	9.29	3.68
10.00	16.32	10.07	9.96	8.5	7.07	7.43	3.68
7.69	16.32	9.24	9.09	8.2	6.24	6.01	3.68

9.4   Strengths

9.4.1   Overview

Strength indicates the stress (= force per area) that a body can endure without failure. For tensile experiments, the limiting strength is the fracture strength of brittle materials and the upper yield strength of ductile ones. Other important limiting strengths are flexural strength, shear strength, and compressive strength. For anisotropic materials, these strengths cannot be interconverted. Examples are shown in Table 9-6.

Table 9-9 Strengths σ of fiber-reinforced epoxy composites [10]. σ_F = longitudinal tensile strength of fiber, σ_M = tensile strength of cured epoxy resin (= 60 MPa).

Fiber	σ_F/MPa tension	σ/MPa of composites				
		tension		compression		shear
		‖	⊥	‖	⊥	in plane
S glass	4900	1620	40	690	140	80
Carbon fiber, high strength	3800	1240	41	1240	170	80
E glass	3500	1020	40	620	140	70
Boron fiber	3000	1240	70	3310	280	90
PPTA (Kevlar®)	2800	1240	30	280	140	60
Carbon fiber, high modulus	2500	760	28	690	170	70

For both rigid and soft matrices, tensile strengths at fracture increase with increasing fiber content whereas maximum elongations frequently (but not always) decrease (Fig. 9-16). The upper yield of soft polymers disappears at higher fiber contents and the composite fails by brittle fracture.

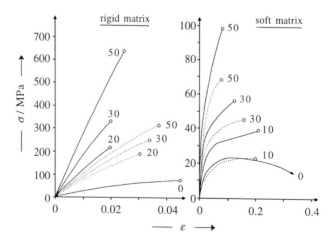

Fig. 9-16 Tensile stress of fiber reinforced plastics as a function of strain. Stress in fiber direction (——) or for two-dimensional random distributions of fibers (- - -). O Fracture. Numbers indicate fiber content in volume percent.
  Left: poly(*p*-phenylene terephthalamide) fibers (Kevlar 49®) in rigid-strong ("hard-strong") poly-(methyl methacrylate) [11a]; right: poly(ethylene terephthalate) fibers in soft-ductile ("soft-tough") low-density poly(ethylene) [11b]. For classification of plastics, see Volume III, p. 531.

In tensile stress-strain experiments, composites from rigid plastics usually fracture at particle-polymer interfaces; fibers rarely break themselves. This behavior occurs because shear forces transfer local shear stresses to particle-matrix interfaces where they are distributed over the larger surface area of the particles. The distribution is easier, the larger the specific surface area of the particle, the fewer defects present on the particle surface, and the more elastic the interlayers (if any) between particles and matrix (see last paragraphs of Section 9.4.2).

Compact rods (fibers) always have greater specific surfaces than compact spheres of the same material with the same volumes; rods are therefore more efficient than spheres. For the same reason, rough surfaces are more effective than smooth ones, for example, carbon blacks with specific surfaces up to 1000 m^2/g (p. 34) *versus* synthetic chalks with up to 40 m^2/g and natural chalks with up to 15 m^2 (p. 31). However, only a part of the interior surfaces of particulates is accessible by polymer molecules. In addition, active fillers such as carbon blacks contain free radicals that can react with matrix molecules. Efficiencies of various fillers are therefore difficult to compare (Fig. 9-17, left).

The fracture strength of carbon black-filled elastomers decreases steeply with increasing surface average of diameters (p. 38) of filler particles (Fig. 9-17, right). Since thicknesses of interlayers are independent of particle diameters (Fig. 9-4), interlayers of larger filler particles contain fewer polymer molecules per mass of filler particle than those of smaller filler particles. Hence, the smaller the number of polymer molecules in interlayers per mass of filler particle, the less the interlayers are able to absorb and distribute stresses.

The transfer of stresses depends also on the proportion of defects in interlayers and on the surface of particles. Such defects with sizes of more than 10 µm are caused by the compounding of polymers with fillers and the subsequent processing of compounds.

Smaller defects develop if compounds contract: long intermolecular distances are replaced by short intramolecular ones if thermosetting materials are cured, and interseg-mental distances are shortened if thermoplastics crystallize. The contracting matrices detach from filler surfaces so that filler/matrix contact areas are reduced.

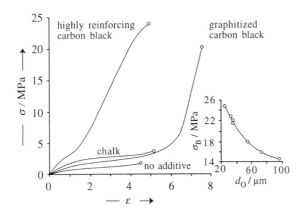

Fig. 9-17 Left: effect of fillers on stress-strain curves of a vulcanized rubber [12]. Highly reinforcing carbon blacks have large specific surfaces and graphitized carbon blacks relatively small ones. Chalk particles are large.

Right: dependence of fracture strength of a carbon black-reinforced elastomer on the surface area average d_0 of the diameters of carbon black particles.

These effects may also arise on cooling of amorphous thermoplastics since fillers have smaller thermal expansion coefficients than polymeric matrices: polymers contract more than fillers. The interplay of all of these effects and the action of mechanical stresses and environmental effects generates macroscopic defects, for example, by the growth and interconnection of microcracks.

The greater the stress, the more important become all of these effects. The nature of the matrix/filler interface is therefore much more important for stresses than for moduli which are determined for very small stresses.

9.4.2 Fiber-reinforced Polymers

Tensile Strength: Long Fibers

The *upper bound* of strength of long fiber-reinforced polymers is observed if stresses attack longitudinally along the long axes of fibers, and both fibers F and matrix M are deformed isometrically ($\varepsilon_M = \varepsilon_F$) until the component with the smallest fracture elongation breaks (usually the fiber) at the fracture elongation $\varepsilon_{F,B}$.

The failure of a non-yielding composite is described by Hooke's law, $\sigma_B = E\varepsilon_B$. The upper bound of the modulus E of the composite is given by the simple rule of mixtures, $E = E_M\phi_M + E_F{}^*\phi_F$, where $\phi_M \equiv 1 - \phi_F$ and $E_F{}^*$ = effective tensile modulus of fibers. For $\phi_F \to 0$, the fracture elongation ε_B of the composite equals the fracture elongation of the matrix ($\varepsilon_B = \varepsilon_{B,M}$). At $\phi_F \neq 0$, it equals the fracture elongation of that component that elongates the least of all of the components, usually the fiber ($\varepsilon_B = \varepsilon_{B,F}$) if the matrix adheres completely to the fiber.

The fracture strength σ_B of the composite is therefore given by

(9-20) $\sigma_B = \sigma_{B,M} + (\sigma_{B,F}{}^* - \sigma_{B,M})\phi_F$

where $\sigma_{B,F*}$ = effective fracture strength of the fiber. Similarly for the lower bound:

(9-21) $1/\sigma_B = (1/\sigma_{B,M}) - [(1/\sigma_{B,M}) - (1/\sigma_{B,F}{}^*)]\phi_F$

The tensile strength of a long fiber-reinforced poly(methyl methacrylate), a rigid-brittle ("hard-brittle") polymer, indeed increases linearly with increasing content of 2-dimensionally randomly oriented fibers (Fig. 9-18). These data deliver an effective tensile strength of the fiber of $\sigma_{B,F*}$ = 587 MPa, which is lower than both the strength $\sigma_{B,F}$ = 2760 MPa of the fiber itself and the value of (1/3)×2760 MPa = 920 MPa after correcting for the 2-dimensional random orientation of the fibers (1/3; p. 382).

The shapes of the functions $\sigma_B = f(\phi_F)$ for unidirectionally oriented fibers (longitudinal or transverse) suggest that different fracture mechanisms operate at large and small fiber concentrations ϕ_F (Fig. 9-18). For example, transverse tensile strengths $\sigma_{B,\perp}$ of a reinforced poly(methyl methacrylate) (PMMA) passed through a minimum at an $\phi_F \approx$ 0.1 of the PPTA fibers (Fig. 9-18). Another example is a PMMA with PA 6.6 fibers (not shown) which had at ϕ_F = 0.2 a longitudinal tensile strength of $\sigma_{B,\parallel}$ = 23.5 MPa that is considerably lower than the $\sigma_{B,M}$ = 73.1 MPa of the matrix itself. At the same volume fraction ϕ_F, a bend exists in the $\sigma_{B,\parallel} = f(\phi_F)$ function of composites of PMMA with unidirectional PPTA fibers (Fig. 9-18).

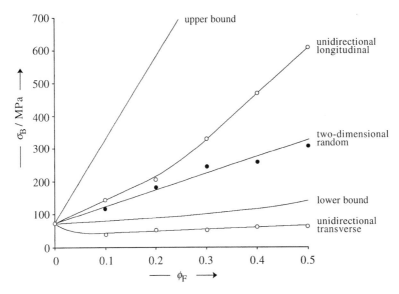

Fig. 9-18 Tensile strength σ_B of a fiber reinforced poly(methyl methacrylate) (PMMA) as a function of the volume fraction ϕ_F of long ($L_F/d_F = 822$) poly(p-phenylene terephthalamide) fibers (Kevlar® 49; PPTA) [11]. Legends: upper bound, lower bound (rule of mixtures); three fiber directions (experiment).

PMMA:	E_M = 2.63 GPa	$\sigma_{B,M}$ = 73.1 MPa	$\varepsilon_{B,M}$ = 0.046
PPTA:	E_F = 131 GPa	$\sigma_{B,F}$ = 2760 MPa	$\varepsilon_{B,F}$ = 0.02

These phenomena are quite common for rigid-brittle polymers that are reinforced with fibers of not too high moduli; examples are PA 6.6 fibers in polycarbonate PC but not in the softer PE or PA 12. A possible reason for these phenomena is the formation of voids on the surface of fibers. Electron micrographs show that at low fiber contents, planar cracks spread through the matrices and debonded fibers are pulled out of either surface of the crack. In the borderline case, the resin does not adhere to the fiber at all and the "interlayer" is just a layer of voids. The fiber-matrix contact area is reduced, which corresponds to a smaller effective fiber volume.

Voids near fiber surfaces reduce fiber/matrix contact areas, which corresponds to a smaller effective fiber volume. In the Halpin-Tsai equation for moduli of composites, this effective fiber volume can be expressed by $K_2\phi_F$ (see Eq.(9-16)). If voids prevent any reinforcing effect of the fibers, the effective fiber modulus becomes zero and K_2 of Eq.(9-16) reduces to $K_2 = -1/K_1$. The Halpin-Tsai equation then reduces to

$$(9\text{-}22) \qquad E = E_M\left(\frac{1-\phi_F}{1+(\phi_F/K_1)}\right)$$

Insertion of Hooke's laws, $\sigma_{B,M} = E_M\sigma_{B,M}$ (matrix), and $\sigma_B = E\sigma_B$ (composite) delivers

$$(9\text{-}23) \qquad \frac{\sigma_B}{\sigma_{B,M}} = \frac{\varepsilon_B}{\varepsilon_{B,M}}\left(\frac{1-\phi_F}{1+(\phi_F/K_1)}\right)$$

For these conditions, composite and matrix expand to the same content, $\varepsilon_B = \varepsilon_{B,M}$, and Eq.(9-23) becomes

(9-24) $\dfrac{\sigma_B}{\sigma_{B,M}} = \left(\dfrac{1 - \phi_F}{1 + (\phi_F / K_1)} \right)$

This equation describes correctly that the transverse tensile strength is lower than that of the matrix itself for composites that are unidirectionally reinforced with long fibers. However, the prediction is only semiquantitative and fails to reproduce the shape of the function $\sigma_B = f(\phi_F)$. The behavior of such composites must therefore also be affected by other factors.

Tensile Strength: Short Fibers

In short fiber-reinforced composites, stresses are not transferred from the matrix to the fiber ends. Instead, short fibers act as **stress concentrators** for the following reason. The tensile stress applied to the fiber increases over a length $L_{crit}/2$ from zero at the fiber ends to a maximum value σ_{max} at the mid-section of the fiber (Fig. 9-19). At the fiber ends, the matrix with its lower modulus is deformed more strongly than the fiber with its higher modulus, which leads to shear stresses τ. The components of the composite are thus subjected to average tensile stresses that differ from those of the components themselves.

In general, one assumes that the average tensile stress of the matrix is practically identical with the tensile stress of the matrix itself, $\overline{\sigma}_M = \sigma_M$. However, the average tensile stress of the fiber in the composite is smaller than the tensile stress of the fiber itself, $\overline{\sigma}_F < \sigma_F$. This quantity $\overline{\sigma}_F$ can be calculated as follows.

The stress distribution along the fiber can be approximated by a trapezoid (UK: trapezium) (Fig. 9-19). Since the dotted areas above and below $\overline{\sigma}$ must equal each other, the average tensile strength of the fiber must be

(9-25) $\overline{\sigma}_F L_F = \sigma_F[L_F - (L_{crit}/2)]$

and, with the simple rule of mixtures, $\sigma = \Sigma_i \, \sigma_i \phi_i$,

(9-26) $\sigma = \sigma_M \phi_M + \overline{\sigma}_F \phi_F = \sigma_F[1 - \{L_{crit}/(2\,L_F)\}]\phi_F$

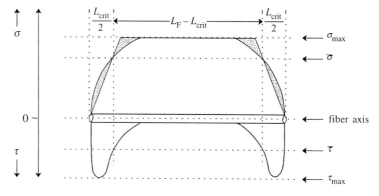

Fig. 9-19 Distribution of tensile stresses σ (top) and shear stresses τ (bottom) along a short fiber F of length L_F and critical length L_{crit}. See text for definitions of average and maximum tensile stresses and shear stresses.

The **critical fiber length** L_{crit} is defined as the length of those fibers that can be pulled from the matrix without any breakage. This quantity can be calculated from the equilibrium of forces, i.e., the tensile force $F_\sigma = \sigma_F Q_F$ along the fiber and the shear force $F_\tau = \tau_{F/M} O_E$ that is distributed across the surface $O_E = 2 \pi r_F (L_{crit}/2)$ of the two fiber ends.

From the equilibrium condition $F_\sigma = F_\tau$ and thus $\sigma_F(\pi r_F^2) = \tau_{F/M}(2 \pi r_F (L_{crit}/2))$, it follows for the critical fiber length:

$$(9\text{-}27) \qquad L_{crit} = \sigma_F r_F / \tau_{F/M} = (\sigma_F d_F)/(2 \tau_{F/M})$$

In a similar manner, one can calculate the equilibrium of forces for fibers with lengths that are larger than their critical lengths. The average tensile strength of such fibers is $\overline{\sigma}_F = 2 \overline{\tau}_{F/M} (L_F / d_F)$. It follows that the tensile strength of the composite increases with the axial ratio L_F/d_F of the fibers:

$$(9\text{-}28) \qquad \sigma = \sigma_M \phi_M + \overline{\sigma}_F \phi_F = \sigma_M \phi_M + 2 \overline{\tau}_{F/M} (L_F / d_F) \phi_F$$

A linear dependence of σ_B on the axial ratio L_F/d_F of fibers is indeed found for glass-fiber filled polyamide 6 and poly(butylene terephthalate) for axial ratios L_F/d_F smaller than ca. 20 (Fig. 9-20). Eq.(9-28) furthermore demands $\sigma_B = \sigma_{B,M}\phi_M$ for $L_F/d_F = 0$ which is obeyed by PA 6 and PBT but not by it-PP where σ_B seems to pass through a minimum.

Stress concentrations at interfaces of fibers in rigid-brittle matrices can be reduced by thin elastic fiber coatings, for example, thin layers of flexibilized epoxy resins on the surface of glass fibers in matrices of conventional EP resins. Since such "rubber buffers" are less stiff than the matrix; they deform locally, reduce stress concentrations, and thus increase fracture strengths relative to untreated fibers. For the same reason, polyamide as matrix leads to larger increases of tensile strengths after glass fiber reinforcement than the less ductile cured epoxy resin (see Section 9.4.3).

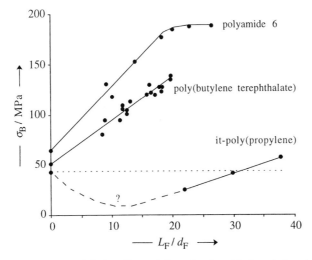

Fig. 9-20 Fracture strengths σ_B of glass fiber-reinforced polyamide 6, poly(butylene terephthalate), and it-poly(propylene) as a function of the axial ratio L_F/d_F of fibers [9c]. The proportions of glass fibers were 25 wt% in PA 6 and 30 wt% in PBT and PP.

"Rubber buffers" also reduce the concentration of voids that act as "nucleating enti-ties" for the onset of brittle fracture. By acting as spacers, they likewise reduce the for-mation of fiber bundles, which in turn increases the tensile strength of the composites.

Real fibers have a distribution of strengths (Fig. 5-8); they do not break simultane-ously as ideal fibers would. The load is thus distributed to fewer and fewer fibers, espe-cially if the matrix does not adhere to the fiber: the effects of early fiber breaks are inconsequential for ultimate failure. Fiber bundles in matrices with no adhesion are thus weaker than composites with the same concentration of separated fibers.

Creep Strengths

Tensile data of plastics are determined by short-time experiments, with drawing speeds of 1 mm/min for tensile moduli, 5 mm/min for fracture strengths and elongations, and 50 mm/min for yield strengths and elongations (ISO 527). For engineering purposes, one is also interested in the behavior of plastics at long times. All thermoplas-tics creep; at constant load, logarithms of tensile strengths decrease with the logarithm of time, $\lg \sigma = \lg \sigma_0 - K \lg t$, where K = empirical constant and σ_0 = strength at time zero (Fig. 9-21). After a certain "critical" time, tensile strengths decrease faster.

The decrease of tensile strengths is caused by flow processes that lead to intramolec-ular conformational changes at short times (regime I) and altered phase morphologies (packing of chains, etc.) at long ones (regime II). The onset of regime II is earlier, the higher the testing temperature. For example, tensile strengths of the LDPE of Fig. 9-21 (heat distortion temperature 45°C at a load of 0.45 MPa) decreased very little in ten years at 20°C but very strongly after one day at 60°C.

Heat-resistant plastics should have heat distortion temperatures of 180°C at 0.45 MPa, tensile strengths of at least 45 MPa, and flexural moduli of at least 2.2 GPa. In air at 115°C, at least 50 % of these values should be retained after 100 000 hours = 11.5 years.

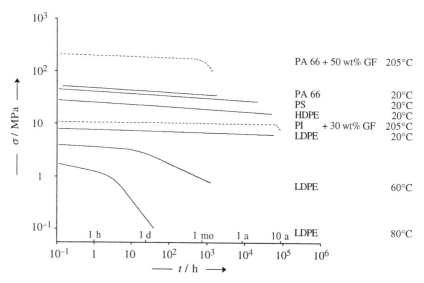

Fig. 9-21 Time dependence of tensile strengths σ of unfilled (————) and glass fiber (GF) reinforced (- - - -) plastics at various temperatures.

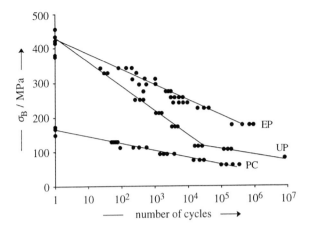

Fig. 9-22 Dependence of fracture strength σ_B on the logarithm of the number of cycles, N_{cycl}, during periodic loading of a cured epoxy resin EP that was reinforced by long glass fibers in a $0°/90°$ arrangement, a cured unsaturated polyester resin UP in a mat of glass fibers, and a polycarbonate PC with randomly distributed short glass fibers [13].

For **dynamic stresses**, one observes a linear dependence of fracture strengths σ_B on the logarithm of the number of cycles N_{cycl}, $\sigma_B = \sigma_0 - K \lg N_{cycl}$ (Fig. 9-22). In some cases, a strong decrease is followed by a weaker one at a greater number of cycles (UP in Fig. 9-22). Non-impregnated fiber bundles show the same fatigue as their composites with plastics: neither the matrix itself nor the fiber-matrix interface contributes significantly to the observed fatigue.

9.4.3 Strengths of Particulate-filled Polymers

Tensile Strengths

Tensile strengths of composites of plastics with randomly filled fibers are generally greater than those of the matrices themselves (Figs. 9-20, 9-23). The tensile strengths of particulate-filled composites with the same filler concentration of $w_F \approx 0.3$, on the other hand, are either equal to or lower than that of the matrix (Fig. 9-23). In both cases, tensile strengths of filled semicrystalline matrices tend to be, on average, lower than those of amorphous ones. The lowering of tensile strengths may be caused by void formation at the filler-matrix interface due to the rough surface of filler particles.

For the same type of matrix, efficiencies of fillers increase with their anisotropy, i.e., spheroids < platelets < fibers, for example, chalk < talc < glass fibers.

The tensile yield stress σ_Y of particle-filled rigid-ductile isotactic poly(propylene) decreases with both volume fraction ϕ_F and diameter d_F of the chalk particles; it increases with increasing d_F for pyrogenic silica (Fig. 9-24). This behavior can be explained by assuming zero adhesion between matrix molecules and spherical filler particles where the total load is carried by the matrix.

The yield stress $\sigma_Y = \phi_M \sigma_{Y,M}$ of the composite depends only on the yield stress $\sigma_{Y,M}$ and on the available *cross-sectional* fraction $\phi_M = 1 - \phi_F$ of the matrix. The matrix loses its continuity, however, if the filler fraction ϕ_F approaches $\phi_{F,max}$, the cross-sectional

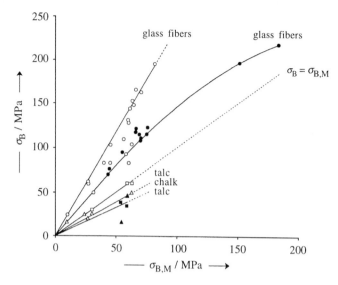

Fig. 9-23 Tensile stress σ_B of composites of amorphous (O,□,Δ) and semicrystalline (●,■,▲) polymers with $w_F \approx 0.3$ of short glass fibers (O,●), talc platelets (□,■), or spheroidal chalk particles
(Δ,▲) as fillers. Solid lines are empirical guides for the eye. - - - Tensile strength of the composite
identical to that of the matrix. Data of [14,15].

fraction for the maximum packing of circles. $\varphi_F = 1$ is never fulfilled since $\varphi_F \leq \varphi_{F,max}$.
In order to fulfil the conditions $\phi_F = \varphi_F = 0$ and $\phi_F = \varphi_F = 1$, the term $1 - \phi_F$ needs to be
replaced, e.g., by a hyperbolic function $Q = (1 - \phi_F)/(1 + K\phi_F)$ (see Eq.(9-29)). The
parameter K can be calculated from the maximum spatial packing of spheres and planar
circles, i.e., $\phi_{F,max}$ and $\varphi_{F,max}$, respectively (Table 9-7).

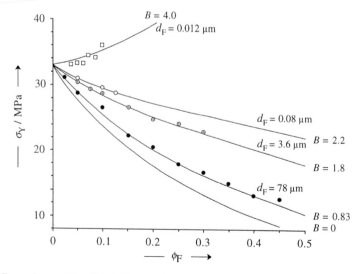

Fig. 9-24 Dependence of tensile yield strength, σ_Y, on the volume fraction ϕ_F of composites of an
isotactic poly(propylene) filled with pyrogenic silica (□) or chalk (O, ⊕, ●) of various particle diameters d_F [16]. Lines were calculated with Eq.(9-29), $K = 5/2$ and the indicated values of the empirical
parameter B.

Table 9-7 Maximum spatial packing of spheres and maximum planar packing of circles.

Spatial packing	$\varphi_{F,max}$	Planar packing	$\phi_{F,max}$	K
Hexagonal close	0.740	Hexagonal close	0.907	2.427
Face centered cubic	0.524	Square array	0.785	2.317
Random loose	0.601	Random	0.82	2.024

However, such a function would be universal, which is not the case (Fig. 9-24). The equation thus has to be amended by an exponential term, exp $(B\phi_F)$, where the adjustable quantity B may contain contributions of the structure and size of interlayers and the structure of the filler:

$$(9\text{-}29) \qquad \sigma_Y = \sigma_{Y,M} \frac{1-\phi_F}{1+K\,\phi_F} \exp(B\phi_F) \quad ; \quad K = \frac{\varphi_{F,max}-\phi_{F,max}}{(1-\varphi_{F,max})\,\phi_{F,max}}$$

Flexural Strengths

Flexural properties are not intrinsic properties of polymers, composites, and blends, since flexibilities are composite quantities that depend also on the geometry of the testing specimen and the number of supports (p. 141). These quantities control the relative proportion of the various forces to which the specimen is subjected.

In bending tests (p. 141), the top surface of the specimen is in tension and the bottom surface in compression; in addition, there is always some transverse shearing. Because stresses are ideally directly proportional to moduli, $\sigma = E\varepsilon$, and flexural and tensile moduli are interrelated (Fig. 5-5), dependences of flexural strengths σ_{FS} on ultimate tensile strengths σ_B can be expected. Experimentally, a ratio $\sigma_{FS}/\sigma_B \approx 1.45$–1.6 is found for filler contents of $w_F = 0.3$ (Fig. 9-25).

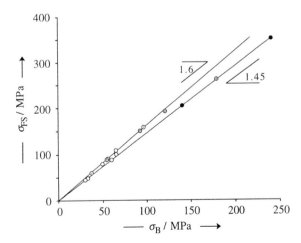

Fig. 9-25 Flexural strength σ_{FS} as a function of tensile strength σ_B [17]. Filler content: 30 wt%.
   ○   Unfilled plastics (from left to right: PE, PP, PS, PBT, PC, PMMA, PSU);
   ◎   mineral-filled plastics (PP + talcum, PA + kaolin);
   ⊕   glass fiber-reinforced plastics (two PPs, PBT, PA);
   ●   carbon fiber-reinforced plastics (PBT, PA).

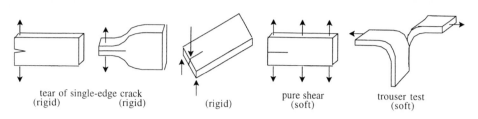

tear of single-edge crack
(rigid)   (rigid)   (rigid)   pure shear   trouser test
                                                                    (soft)   (soft)

Fig. 9-26 Some notched specimens for testing of tear strengths of rigid and soft specimens.

9.4.4   Tear Strengths

Tear strengths Z can be measured in many different ways (Fig. 9-26). Despite the name, a tear strength is not a "strength" (= pressure = force per area = energy per volume) but a force per length = energy per area. The tear strength is one-half of the so-called **tear energy**, the energy required to produce two new surfaces on tearing.

Tear strengths of poly(ethylene) composites are independent of the chemical nature of highly active fillers, such as fumed silica and carbon black (Fig. 9-27). The action of these fillers becomes noticeable for volume fractions ϕ_F exceeding ca. 1 %. Beyond this concentration, the tear strength of the composites of low-molecular weight, low-density poly(ethylene) ($M_r \approx 30\ 000$) *decreases* with increasing filler concentration and so does that of a low-molecular weight, high-density poly(ethylene). Electron micrographs indicate that this decrease is probably caused by a weaker bonding between interlayers and matrices.

Reduced tear strengths decrease less – and even increase for a certain range of filler concentrations – for higher molar mass poly(ethylene)s because of increased entanglements of coiled polymer molecules at the matrix-interlayer interface. This shift to higher reduced tear strengths is also aided by crosslinking of the matrices.

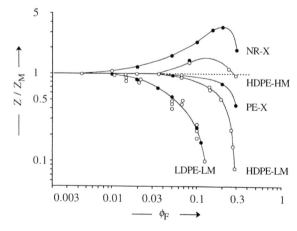

Fig. 9-27 Ratios Z/Z_M of tear strengths Z of composites and Z_M of matrices (natural rubber (NR) or poly(ethylene)s (PE)) as a function of the volume fraction ϕ_F of fillers carbon black (●) or fumed silica (○) (data of [2b, 18, 19]). HD = high density, LD = low density, HM = high molecular weight, LM = low molecular weight, X = crosslinked.

The effect of filler fraction on the tear strength Z_{TS} of filled, crosslinked, low-density poly(ethylene)s can be described by the equation

$$(9\text{-}30) \qquad Z = Z_M(1 - K_M \phi_F)$$

where Z_M is the tear strength of the matrix. The adjustable constant K_M measures the detour the crack has to make if it hits a filler particle. K_M is thus related to the diameter of spheroids and the axial ratio of rods. It assumes, for example, a value of $K_M = 7.7$ for the LDPE-HM of Fig. 9-27.

Tear strength *increases* with increasing filler fraction for crosslinked polymers and then passes through a maximum, provided the network chains are mobile enough. The effect is greater for chemically crosslinked, elastomeric NR-X with its high mobility of network chains than for the physically crosslinked (= entangled), semicrystalline HDPE-HM ($T_G \ll T$ of "amorphous" regions). The chemically crosslinked PE-X exhibits no maximum at all because of the short lengths of network chains.

Z/Z_M values decrease dramatically at filler concentrations $\phi_F > 0.3$, probably because of maximum packing of filler particles. For example, silica particles have diameters of $d_F \approx 16$ nm and a bound poly(ethylene) layer of $l_I \approx 2$ nm; their effective diameter is thus $d_{eff} = (16 + (2 \times 2))$ nm. The effective volume of such particles is 1.95 times higher than their volume which boosts their effective volume fraction to 0.585 % from 0.30 %. Since randomly loose packed spheres can occupy no more than 60.1 % of the total volume, a regular matrix no longer exists between such "coated" filler particles. The particles just touch each other with their interlayers and the composite becomes brittle.

9.4.5 Impact Strengths

Fundamental differences exist between the impact strengths of unnotched and notched specimens. With increasing filler content, more and more voids are formed around filler particles in *unnotched* specimens and the unnotched impact strength decreases. In *notched* specimens, the notch is the largest "void"; thus, the effect of additional voids created by the addition of filler particles is relatively minor. The more filler particles present, the greater the number of obstacles a crack has to circumvent and the larger is the notched impact strength.

Larger anisotropies of filler particles increase impact strengths for two reasons. They lead to larger detours, especially if the long axes are transverse to the stress direction. A larger anisotropy at the same filler content also means fewer chain ends acting as stress concentrators and thus fewer cracks.

Good adhesion does not necessarily mean improved impact strength, however. If the adhesion is so good that matrix and fibers are not separated upon impact, cracks travel farther into the matrix and the impact strength is low. Poor adhesion, on the other hand, causes fibers to separate from the matrix on impact; the crack will be deflected and the impact energy absorbed by the matrix.

The resistance to impact certainly depends on the strength of the matrix. One can thus relate the logarithm of the ratio of notched impact strengths of composites and matrices, $\eta_N / \eta_{N,M}$, to the logarithm of the corresponding ratio of tensile strengths, $\sigma_B / \sigma_{B,M}$, all at constant volume fraction of the filler (Fig. 9-28). The resulting straight lines depend on

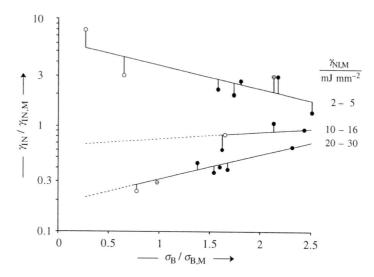

Fig. 9-28 Logarithm of relative Charpy notched impact strength as a function of relative tensile strength of composites of various thermoplastics with $\phi_F = 0.3$ of glass fibers (●), talc (⊕), and chalk (O) at 23°C. Data of [14]. Numbers indicate the notched impact strengths of the matrix. Data are reported as energy per area which is related to crack propagation (p. 145, bottom).

the notched impact strength of the matrices: the slope is negative at small values of $\gamma_{N,M}$ but becomes positive at large ones. For a given value of $\gamma_{NI,M}$, lines seem to be independent of the chemical nature and anisotropy of the filler as well as the chemical and physical structure of the matrix.

9.5   Fracture Elongation

Drawing of unfilled polymers deforms the equilibrium shapes of polymer coils and leads to flow in the draw direction. Both effects are reduced if filler particles are present since filler surfaces act as barriers for polymer coils. Near the filler surface, coils can adopt fewer macroconformations and the matrix becomes more rigid. Furthermore, flow is hampered by filler particles, which act as obstacles. As a result, fracture elongations ε_B of composites decrease strongly upon the addition of 10 vol% of filler (Table 9-8). ε_B decreases further or may increase for $\phi_F \geq 0.2$.

Table 9-8 Variation of fracture elongation with volume fraction of randomly oriented fibers [11].

Polymer	Fiber	$10^2 \, \varepsilon_B$ for $\phi_F =$					
		0	0.1	0.2	0.3	0.4	0.5
Poly(ethylene)	glass	10	3.1	1.9	1.9	1.7	
Poly(methyl methacrylate)	glass	4.6	1.6	0.90	1.0	0.80	
Poly(methyl methacrylate)	Kevlar 49	4.6	2.7	3.0	3.0	3.9	3.6

Table 9-9 Tensile moduli E_M, tensile strengths, $\sigma_{B,M}$, elongations at break, $\varepsilon_{B,M}$, and strength products $\sigma_B\varepsilon_B$ of unfilled thermoplastic polymers and ratio of fracture elongations, $\varepsilon_B/\varepsilon_{B,M}$, of composites with random orientation of anisotropic fillers and matrix polymers [15]. In all cases, the weight fraction of fillers is $w_F = 0.3$, except * where it is $w_F = 0.4$.

Polymer	Unfilled resin				$\varepsilon_B/\varepsilon_{B,M}$ of composite with			
	$\dfrac{E_M}{MPa}$	$\dfrac{\sigma_{B,M}}{MPa}$	$\dfrac{\varepsilon_{B,M}}{\%}$	$\sigma_B\varepsilon_B$	chalk	talc	glass fiber	asbestos
Axial ratio of filler $\rightarrow$					1-5	9	20	1000
PP, it	1600	31	620	192	0.29*	0.017	0.008	0.011*
PA 6	1200	64	220	141	0.14		0.016	0.014
PA 12	1200	60	270	162			0.022	
PET	2800	66	160	106			0.025	
LDPE	210	10	500	50	0.44*	0.08	0.089	
POM	2700	63	45	28		0.067	0.13	0.04
PVC-S	2700	60	8	4.8	1.0	0.75*	0.38	0.15
PS	3800	55	4	2.2	0.5	0.4*	0.5	
SAN	3600	70	5	3.5			0.6	

The relative fracture elongation $\varepsilon_B/\varepsilon_{B,M}$ of composites with ϕ_F = constant also decreases with increasing axial ratios of fillers (Table 9-9), probably because of the greater detour a polymer molecule must take on elongation of the matrix. In general, fracture elongations are reduced more, the greater the product $\sigma_{B,M}\varepsilon_{B,M}$ of the matrix.

9.6 Thermal Properties

9.6.1 Expansion Coefficients

Addition of inorganic fillers (small thermal expansion) to thermoplastics (large thermal expansion) reduces the proportion of effects of the latter and, according to the simple rule of mixing, reduces the *shrinkage* of plastics on processing. A small contribution to smaller expansions of filled polymers comes from reduced macroconformations of polymer molecules near filler surfaces. The shrinkage can even be eliminated if fillers with negative thermal expansion coefficients are used, for example, carbon fibers or the mineral eucryptite, $LiAl(SO_4)_2$.

Linear thermal expansions of fiber-filled polymers are inversely proportional to the square root of the volume fraction of fibers, $(L - L_M)/L_M = \Delta L/L_M = K\phi_F^{-1/2}$ (Fig. 9-29). This dependence indicates that the thermal expansion of molecule coils is hampered more, the shorter the distance between filler particles.

The smaller the thermal expansion, the less the polymer will shrink after processing, often by 50 %. Shrinkages may even be greater for filled crystallizing polymers than for unfilled ones because filler particles may constitute obstacles to crystallization, although some of them may act as nucleating agents. Another contribution to smaller shrinkages comes from the diminished presence of voids and other defects. These reduced postshrinkages lead to parts with greater dimensional stabilities.

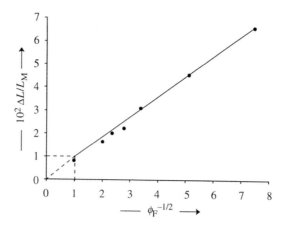

Fig. 9-29 Relative thermal expansion, $\Delta L/L_M$, of a glass fiber-reinforced poly(ethylene) as a function of the inverse square root of the volume fraction ϕ_F of the fibers [20].

Thermal volume expansions $\Delta V/V_0$ of composites containing spheroidal fillers are proportional to cubic thermal expansion coefficients $\beta = (\Delta V/V_0)/\Delta T$. Since a 3-dimensional expansion can be expressed by the compression coefficient, $K = p(-\Delta V/V_0)$, and the pressure p is the total stress, σ, acting on both the filler and the matrix, it follows that $\sigma = K\beta\Delta T$ for the composite and $\sigma_M = K_M\beta_M\Delta T$ for the matrix, and thus

(9-31) $\beta/\beta_M = (\sigma/\sigma_M)(K_M/K)$

The decrease of thermal expansion by addition of fillers thus equals approximately the increase of compression moduli if the same stress acts on both matrix and filler. This is not true for fibrous and platelet-like fillers where the filler and the matrix are neither subjected to the same stress nor are experiencing the same elongation.

9.6.2   Transformation Temperatures

Glass temperatures T_G of matrices of amorphous thermoplastics with filler contents of 30 wt% or less equal approximately the glass temperatures $T_{G,M}$ of the matrices since conventional measurements of T_G are made by static methods and changes of physical structures of such matrices near filler particles can be neglected because of the small volume percentage of the fillers. Glass temperatures of filled elastomers, on the other hand, are generally higher than those of unfilled ones because filler contents are usually greater than 30 wt%. Melting temperatures T_M of conventionally filled semicrystalline thermoplastics also do not deviate much, if any, from those of their unfilled polymers.

The situation is quite different for heat distortion temperatures HDT because they are determined under load (Table 5-2). These methods are in effect creep experiments, and HDTs therefore decrease with increasing load (see Tables 8-12 ff.).

Heat distortion temperatures (at 1.82 MPa) of filled amorphous polymers are little greater than those of unfilled matrices if HDTs are smaller than ca. 150°C but somewhat greater for HDTs of more than ca. 150°C (Table 9-10, Fig. 9-30).

Table 9-10 Heat distortion temperatures (HDT-A; 1.80 MPa) of unfilled polymers (matrices) and their composites with 30 wt% glass spheres (GS), chalk (CH), talc (TA), asbestos fibers (AF), short glass fibers (GF), carbon fibers (CF), or wollastonite fibers (WF) [14, 15, 17]. ● Amorphous polymer, O semicrystalline polymer. Other filler contents: * 50 wt%, ** 40 wt%, *** 20 wt%.

Polymer	HDT/°C unfilled	GS	CH	TA	AF	GF	CF	WF
● PS	86				91	93		
● SAN	90					100		
● PC	143					145		266*
● PSU	174		177			177		
O HDPE	50				77**	108		
O PP	62		73**	95**	96**	120		85**
O PA 6	65		60		193	208		
O PBT	67	70***				205		130*
O ETFE	74					210	240	
O PA 6.6	80	74				226	257	215
O POM	125	101	102***			161		

At a constant load of 1.82 MPa, HDTs of fiber-filled crystalline polymers are always greater than those of unfilled matrices (Table 9-10, Fig. 9-30). They also depend strongly on the type and shape of the filler (Table 9-10) as well as on the chemical structure of the matrix. Spheroidal fillers such as glass spheres and chalk increase HDTs only a little, if any, but platelets and fibers do so strongly.

For 30 wt% glass fibers as fillers, two groups of matrices can be distinguished. For composites from high density poly(ethylene), isotactic poly(propylene), poly(oxymethylene), and poly(1,4-phenylene sulfide), HDTs are ca. 60°C higher than those of the matrix (Fig. 9-30) whereas those of other matrices are increased on average by 130°C (actually between 115°C (PA 11) and 160°C (PEEK)).

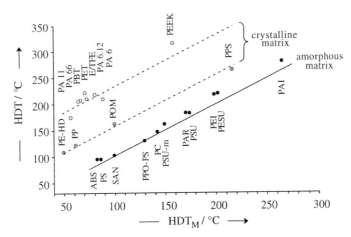

Fig. 9-30 Heat distortion temperatures HDT (load: 1.82 MPa) of amorphous (●) and semicrystalline (O,⊕) polymers containing 30 wt% short glass fibers as a function of the heat distortion temperature HDT$_M$ of the matrix polymers.
Solid line: amorphous matrices (HDT = HDT$_M$); broken lines: range of semicrystalline matrices with increases of 60°C (lower broken line) and 130°C (upper broken line).

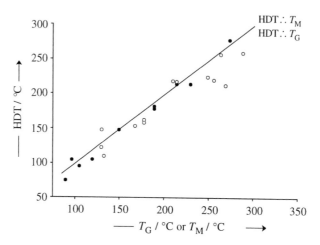

Fig. 9-31 Heat distortion temperatures HDT (load: 1.82 MPa) of amorphous (●) and semicrystalline (O) polymers containing 30 wt% short glass fibers as a function of the glass temperature T_G (amorphous polymers) or the melting temperature T_M (semicrystalline polymers) of the matrix. The solid line corresponds to melting (HDT ∴ T_M) or glass transformation (HDT ∴ T_G), respectively.

Heat distortion temperatures HDT (load: 1.82 MPa) of amorphous polymers that are randomly filled with 30 wt% short glass fibers are practically identical to the glass temperatures of the matrices (Fig. 9-31). In contrast, HDTs of composites from semicrystalline polymers are often lower than the melting temperatures of the matrices.

The effects of particle shapes of fillers and polymer structures may be due to the effects of fairly thick transcrystalline interlayers (p. 379). A 30 µm thick glass fiber (ρ_M = 2.59 g/cm^3) with a 15 µm thick transcrystalline layer of a polar polymer would have an effective diameter of 60 µm, which would boost the effective volume fraction of fibers from ϕ_F = 0.16 for "naked" fibers (= 30 wt%) to $\phi_{F,eff}$ = 0.60 if the matrix and transcrystalline layers have the same density of 1.1 g/cm^3.

At such a large volume fraction, the composite would consist practically only of contacting interlayers (see maximum possible filler fraction, p. 77). The resulting physically crosslinked network would resist the applied load, creep would be reduced, and HDTs increased. Glass fibers in apolar matrices would have smaller polymer adsorptions and thinner interlayers (see also p. 400 ff.).

On spheres and spheroids, transcrystallizations would be radial and not transverse, i.e., with less order and over shorter distances (cf. growth of spherulites; Volume III, p. 222). Interlayers would be thinner and contacts between "coated" spheroids less probable (Fig. 9-6). As a result, matrices retain their identities and HDTs do not change.

9.6.3 Heat Capacities and Thermal Conductivities

Addition of fillers to polymers reduces energy demands for processing and cycle times for injection molding and deep drawing. These effects are caused by the large difference in specific heat capacities c_p of polymers and fillers. At 127°C, $c_p/(\text{J g}^{-1}\text{ K}^{-1})$ of polymers range from 1.66 for rigid poly(vinyl chloride) to 2.52 for polyamide 6, whereas mineral fillers have values of only 0.86-0.92.

After processing, articles must be cooled, which takes a long time since thermal conductivities $\lambda/(W\ m^{-1}\ K^{-1})$ of polymers range from 0.12 for it-poly(propylene) to 0.53 for poly(ethylene), leading to long cycle times. Mineral fillers possess ca. 20 times greater thermal conductivities ($CaCO_3$ up to 3 W m^{-1} K^{-1}) so that mineral-filled plastics cool much faster. As a rule of thumb, cycle times are reduced by the same percentage as the weight percentage of fillers.

9.7 Rheological Properties

Rheological properties of filled polymers are usually characterized by melt flow indices MFI or melt volume indices MVI which are proportional to the inverse apparent viscosities (p. 85). MVIs of filled amorphous polymers are always smaller than those of their unfilled parent polymers, i.e., their apparent viscosities are greater, as one would expect for dispersions. However, MVIs of filled semicrystalline polymers are sometimes greater than those of unfilled ones as shown in Table 9-11 for poly(propylene).

Table 9-11 Melt volume indices $MVI_{190/5}$ (190°C, load 5 kg) in grams of extruded material per ten minutes for some polyolefins with various fillers [14, 17]. (Numbers) indicate axial ratios.

Polymer	MVI after filling with 30 wt% (* 40 wt%) of						
	Unfilled	Glass spheres	Chalk	Talc	Mica	Glass fibers	Asbestos
	-	(1)	(1-5)	(9)	(40)	(30)	(3000)
LDPE	1.7	2.6	0.2*	1.1	1.2	1.6	2.3
HDPE	22	-	18	-	-	6.0	0.7
PP	1.0	-	2.5*	3.0*	-	6.0	4.0*

9.8 Macrocomposites

Composites are usually subdivided according to the major dimension L of their particles in macrocomposites ($L > 1$ mm), composites (1 µm $< L <$ 300 µm), and nanocomposites ($L <$ 100 nm). Examples of macrocomposites are polymer concrete, fiberboard, plywood, and polymers reinforced with long fibers, mats, fabrics, or rovings.

Polymer concrete is composed of polymers (usually polyester or poly(methyl methacrylate)), aggregates (sand, ground limestone, etc.), and other additives (dyes, etc.); it does *not* contain cement (Vol. II, p. 570). Polymer-concrete hybrids include polymer-cement hybrids (cement is partially replaced by polymers) and polymer-impregnated concrete (conventional concrete saturated with a monomer that is polymerized *in situ*). Polymer concretes have higher moduli and higher tensile and compressive strengths than ordinary concretes. They are not vulnerable to damage by freeze-thaw cycles.

Fiberboards usually consist of bonded wood particles. They are subdivided into those with low density (particle boards, chipboards) from wood chips, sawmill shavings or sawdust, medium density, and high density (hardboards). The particles of low and medium density fiberboards and also some hardboards are glued together by urea-formaldehyde resins (interior applications) and polyurethanes or phenolic resins (exterior applications). Some hardboards are "glued" together by compression.

Long fiber-filled polymer macrocomposites range in structure from polymers filled with "long" fibers of 6–12 mm length to fiber strands and mats filled with polymers. The filler content can often reach 65 wt%. Upon application of tensile stress, local stress concentrations are transferred by shear forces onto the polymer-fiber interface and distributed over the much greater fiber surface area. The fibers therefore must bond well to the polymer (use of coupling agents) and must have a certain length in order to avoid slipping (see also p. 380 ff. (moduli) and p. 392 ff. (tensile strengths)).

This group of macrocomposites also includes prepregs (fiber mats soaked with unsaturated polyester resins) and the composites generated by filament winding (wound filaments saturated with epoxies or unsaturated polyesters), all of which are subsequently polymerized (p. 108). Sheet molding compounds (SMCs) are now available for both thermosetting resins and thermoplastics.

Thermal conductivities, tensile moduli, and flexural, compression, tensile, and impact strengths of two-dimensionally reinforced thermosets increase with increasing weight percentage of reinforcing agents, whereas thermal expansion coefficients decrease (Table 9-12). Rovings give the best results, followed by woven fabrics and mats.

Table 9-12 Properties of cured prepregs from unsaturated polyester resins [21].

Property wt% of reinforcement →	Phys. unit	Mats		Woven fabrics		Rovings
		20-30	40-50	40-50	50-60	70-80
Density	g/cm^3	1.3–1.5	1.5–1.75	1.5–1.75	1.6–1.85	1.9–2.1
Thermal expansion coeff. ($\cdot 10^6$)	K^{-1}	40–30	24–20	22–18	18–16	14–12
Thermal conductivity ($\cdot 10^{-2}$)	W/(K m)	14–19	22–31	19–25	25–31	37–41
Tensile or flexural modulus	GPa	5–7	9–10	10–14	14–18	21–26
Shear modulus	GPa	2–3	4–5	2–3	3–4	–
Elongation at break	%	2	2	2	2	2
Flexural strength	MPa	115–145	180–220	220–260	260–300	400–500
Compression strength	MPa	110–135	165–200	150–180	180–200	–
Tensile strength	MPa	65–90	130–170	200–240	240–275	700–800
Impact strength	kJ/m^2	25–55	75–105	130–16–	160–190	

Historical Notes to Chapter 9

COX EQUATION
H.L.Cox, Brit.J.Appl.Phys. **3** (1952) 72

KERNER EQUATION
E.H.Kerner, The Elastic and Thermo-elastic Properties of Composite Media, Proc.Phys.Soc. [London]
   B 69 (1956) 808

HALPIN-TSAI EQUATION
J.E.Ashton, J.C.Halpin, P.H.Petit, Primer on Composite Analysis, Technomic Publ. Co., Stamford
   (CT) 1969, Chapter 5
S.W.Tsai, N.J.Pagano, Invariant Properties of Composite Materials, in S.W.Tsai, N.J.Pagano, Eds.,
   Composite Materials Workshop, Technomic Publ. Co, Stamford (CT), p. 233
J.C.Halpin, J.L.Kardos, The Halpin-Tsai Equations: A Review, Polym.Eng.Sci. **16** (1976) 344

NIELSEN EQUATION
L.E.Nielsen, J.Appl.Phys. **41** (1970) 4626
L.E.Nielsen, Predicting the Properties of Mixtures, Dekker, New York 1978

Literature to Chapter 9

R.L.McCullough, Concepts of Fiber-Resin Composites, Dekker, New York 1971
M.O.W.Richardson, Ed., Polymer Engineering Composites, Appl.Sci.Publ., London 1977
Yu.S.Lipatov, Physical Chemistry of Filled Polymers, Khimiya, Moscow 1977 (in Russian);
   English translation: RAPRA, Shawbury, Shrewsbury, England 1979
V.K.Tewary, Mechanics of Fibre Composites, Wiley, Chichester, Sussex 1978
R.G.Weatherhead, FRP Technology, Appl.Sci.Publ., Barking, Essex 1980
D.Hull, An Introduction to Composite Materials, Cambridge University Press, Cambridge 1981
N.L.Hancox, Ed., Fibre Composite Hybrid Materials, Hanser International, Munich 1981
R.P.Sheldon, Composite Polymeric Materials, Appl.Sci.Publ., London 1982
M.J.Folkes, Short Fiber Reinforced Thermoplastics, Wiley, New York 1982
E.K.Sichel, Ed., Carbon Black-Polymer Composites, Dekker, New York 1982
M.Grayson, Ed., Encyclopedia of Composite Materials and Components (= reprint of articles in Kirk-
   Othmer, Encyclopedia of Chemical Technology, Wiley, New York, 3rd ed. 1983)
M.Holmes, D.J.Just, Structural Properties of Glass Reinforced Plastics, Appl.Sci.Publ., London 1983
P.C.Powell, Engineering with Polymers, Chapman and Hall, London 1983
H.Ishida, A Review of Recent Progress in the Studies of Molecular and Microstructure of Coupling
   Agents and Their Functions in Composites, Coatings and Adhesive Joints, Polym.Composites
   5 (1984) 101
D.H.Kaelble, Computer-Aided Design of Polymers and Composites, Dekker, New York 1985
H.Ishida, G.Kumar, Eds., Molecular Characterization of Composite Interfaces, Plenum,
   New York 1985
D.W.Clegg, A.A.Collyer, Eds., Mechanical Properties of Reinforced Plastics, Elsevier Appl.Sci.,
   London 1986
A.A.Berlin, S.A.Volfson, N.S.Enikolopian, S.S.Negmatov, Principles of Polymer Composites,
   Springer, New York 1986
S.K.Bhattacharya, Ed., Metal-Filled Polymers, Dekker, New York 1986
J.M.Margolis, Ed., Advanced Thermoset Composites, Van Nostrand Reinhold, New York 1986
R.B.Seymour, R.D.Deanin, Eds., History of Polymeric Composites, Dekker, New York 1987
P.Calvert, S.Mann, Synthetic and Biological Composites Formed by In-Situ Precipitation,
   J. Mater.Sci. **23** (1988) 3801
S.M.Lee, Ed., International Encyclopedia of Composites, VCH, Weinheim 1989 (6 vols.)

R.Yosomiya, K.Morimoto, A.Nakajima, Y.Ikada, T.Suzuki, Adhesion and Bonding in Composites, Dekker, New York 1989

F.R.Jones, Ed., Interfacial Phenomena in Composite Materials '89', Butterworths, London 1989

H.Ishida, Ed., Interfaces in Polymer, Ceramic, and Metal Matrix Composites, Elsevier Sci.Publ., New York 1989

S.M.Lee, Ed., International Encyclopedia of Composites, VCH, Weinheim 1990-1991 (6 Vols.)

F.T.Traceski, Specifications and Standards for Plastics and Composites, ASM International, Materials Park (OH) 1990

P.K.Mallick, S.Newman, Ed., Composite Materials Technology, Hanser, Munich 1990

B.D.Agarwal, L.J.Broutman, Analysis and Performance of Fiber Composites, Wiley, New York, 2nd ed. 1990

E.P.Plueddeman, Silane Coupling Agents, Plenum, New York, 2nd ed. 1991

E.Baer, A.Moet, Eds., High Performance Polymers, Hanser, Munich 1991 (includes composites)

M.T.Bryk, Degradation of Filled Polymers, Ellis Horwood, New York 1991

T.L.Vigo, B.J.Kinzig, Eds., Composite Applications. The Role of Matrix, Fiber, and Interface, VCH, Weinheim 1992

J.Kroschwitz, Ed., High Performance Polymers and Composites, Wiley, New York 1992
(= reprints of articles in the 2nd edition of Encyclopedia of Polymer Science and Engineering)

S.M.Lee, Ed., Handbook of Composites. Reinforcements, VCH, Weinheim 1993

H.-H.Kausch, Ed., Advanced Thermoplastic Composites, Hanser, Munich 1993

F.R.Jones, Ed., Handbook of Polymer-Fibre Composites, Longman Scientific, Essex 1994

L.A.Pilato, M.J.Michno, Advanced Composite Materials, Springer, Berlin 1995

R.E.Shalin, Ed., Polymer Matrix Composites, Chapman & Hall, London 1995

V.P.Privalko, V.V.Novikov, The Science of Heterogeneous Polymers, Wiley, New York 1995

L.E.Nielsen, R.F.Landel, Mechanical Properties of Polymers and Composites, Dekker, New York 1996

L.H.Sperling, Polymeric Multicomponent Materials: An Introduction, Wiley, New York 1997

D.J.Lohse, T.P.Russell, L.H.Sperling, Eds., Interfacial Aspects of Multicomponent Polymer Materials, Plenum, New York 1997

E.Fitzer, L.M.Manocha, Carbon Reinforcements and Carbon/Carbon Composites, Springer, Heidelberg 1998

R.F.Jones, Ed., Guide to Short Fiber Reinforced Plastics, Hanser, Munich 1998

P.R.Hornsby, Rheology, Compounding and Processing of Filled Thermoplastics, Adv.Polym.Sci. **139** (1999) 155

R.S.Davé, A.Loos, Processing of Composites, Hanser Gardner, Cincinnati (OH) 2000

R.Talreja, J.-A.E.Månson, Polymer Matrix Composites, Elsevier Science, Amsterdam 2001

T.J.Pinnavaia, G.W.Beall, Eds., Polymer-Clay Nanocomposites, Wiley, New York 2001

J.-F.Gerard, I.Meisel, C.S.Kniep, S.Spiegel, K.Grieve, Fillers and Filled Polymers, Wiley, New York 2001

G.Akovali, Ed., Handbook of Composite Fabrication, RAPRA, Shawbury, Shrewsbury, Shropshire, UK 2002

L.Tong, et al., 3D Fiber Reinforced Polymer Compoites, Elsevier, Amsterdam 2003

R.M.Rowell, Ed., Handbook of Wood Chemistry and Wood Composites, CRC Press, Boca Raton (FL) 2005

A.K.Mohanty, M.Misra, L.T.Drzal, Eds., Natural Fibers, Biopolymers, and Biocomposites, CRC Press, Boca Raton (FL) 2005 (development and application of biopolymer-based composites)

Y.-Q.Mai, Z.-Z.Yu, Eds., Polymer Nanocomposites, CRC Press, Boca Raton (FL) 2006

J.C.Gerdeen. R.A.L.Rorrer, H.W.Lord, Engineering Design with Polymers and Composites, CRC Press, Boca Raton (FL) 2006 (includes strength-to-weight and cost-to-strength ratios)

R.F.Gibson, Principles of Composite Material Mechanics, CRC Press, Boca Raton (FL), 2nd ed. 2007

P.K.Mallick, Fiber-Reinforced Composites. Materials, Manufacturing, and Design, CRC Press, Boca Raton (FL), 3rd ed. 2008

References to Chapter 9

[1] A.S.Kenyon, J.Colloid Interface Sci. **27** (1968) 761 (supplemented), A.S.Kenyon, H.J.Duffy, J.Polym.Eng. **7** (1970) 1

[2] K.Kendall, F.R.Sherliker, Brit.Polym.J. **12** (1980) 85, (a) Fig. 3, (b) Fig. 1

[3] G.Kraus, Reinforcement of Elastomers, Wiley, London 1965, p. 140; see also [2], Fig. 2

[4] I.Mildner, A.K.Bledzki, Angew.Makromol.Chem. **272** (1999) 27, Figs. 3b and 3e

[5] H.Brintrup, PhD Thesis, RWTH Aachen 1975; H.Brintrup, H.Derek, B.Meffert, in G.Menges, H.Potente, G.Wiegand, Eds., Kunststofftechn.Koll.IKV Aachen [Ber.] **9** (1978) 89

[6] J.K.Lees, Polym.Eng.Sci. **8** (1968) 186, Table 1, Figs. 2 and 3

[7] J.C.Halpin, J.Composite Materials **3** (1969) 732, Fig. 1; J.C.Halpin, J.L.Kardos, Polym.Engng.Sci. **16** (1976) 344, Fig. 4a

[8] J.C.Smith, J.Res.Natl.Bur.Standards **80A** (1976) 45

[9] G.W.Ehrenstein, R.Wurmb, Angew.Makromol.Chem. **60/61** (1977) 157, (a) Figs. 23 and 26, Table 8, (b) Table 4, (c) taken from Fig. 3

[10] C.Zweban, H.T.Hahn, R.B.Pipes, Composite Design Guide, Volume I, Mechanical Properties, Section 1.7, (1980), p. 65; quoted by J.Davis in Kirk-Othmer, Encyclopedia of Chemical Technology, Wiley, New York, 3rd ed. (1984), Suppl.Vol., p. 260

[11] B.F.Blumentritt, B.T.Vu, S.L.Cooper, Polym.Eng.Sci. **14** (1974) 633, data of Tables 2 and 3, Figs. 1 and 2; Polym.Eng.Sci. **15** (1975) 428, Table 2, Fig. 3 (a) and 2 (b)

[12] G.Kraus, Adv.Polym.Sci. **8** (1971) 155, Fig. 24; Angew.Makromol.Chem. **60/61** (1977) 215, Fig. 12

[13] J.F.Mandell, D.D.Huang, F.J.McGarry, MIT, Dept.Mater.Sci.Eng., Res.Rep. R 80-4 (November 1980), Fig. 1

[14] A.W.Bosshard, H.P.Schlumpf, in R.Gächter, H.Müller, Eds., Plastics Additives Handbook, Hanser, Munich, 2nd ed. 1984, data of various tables

[15] -, Modern Plastics Encyclopedia 1984-1985, various tables

[16] B.Pukánszky, B.Turcsányi, F.Tüdös, in H.Ishida, Ed., Interfaces in Polymer, Ceramic, and Metal Matrix Composites, Elsevier, New York 1988, p. 467, Fig. 1; see also: B.Turcsányi, B.Pukánszky, F.Tüdös, J.Mater.Sci.Lett. **7** (1988) 160

[17] H.Schlumpf, in R.Gächter, H.Müller, Eds., Plastics Additives Handbook, Hanser, Munich, 3rd ed. 1990, different tables

[18] K.Kendall, F.R.Sherliker, Brit.Polym.J. **12** (1980) 111, Figs. 2 and 4

[19] After data of G.Kraus, see also [3].

[20] J.K.Lees, Polym.Eng.Sci. **8**/7 (1968) 195, Table I

[21] S.Atmeyer, K.-H.Seemann, G.Zieschank, G.Spur, W.Brockmann, P.Berns, K.Wiebusch, Ullmann's Encyklopädie der technischen Chemie, Verlag Chemie, Weinheim, 4th ed., **15** (1978) 281, data of Table 2

10   Polymer Mixtures

10.1   Overview

10.1.1   Definitions

The dictionary defines a mixture as "any composition of two or more substances that are not chemically bound to each other" (American Heritage Dictionary). In common language, the word "mixture" is used as an umbrella term for specific types of mixtures such as composite, blend, compound, amalgam, etc., that have different meanings in different fields. "**Compound**" indicates generally an unspecified combination of two parts but "(chemical) compound" is used in chemistry as a synonym for "chemical substance" (a substance composed of many like molecules (p. 7)) whereas the "compound" of technology (p. 371) designates a carefully prepared mixture of a polymer (p. 7) and adjuvants (p. 20). A "**composite**" in polymer technology is a mixture of a polymeric substance with another chemical substance that is in a distinct physical shape (sphere, fiber, platelet, cube, etc.).

In polymer science and technology, **mixture** is then left as a general term for a material composed of a polymer and a second component (another solid or liquid polymer, a solvent, or a gas) that is *not* in a physically distinct shape. These mixtures are then subdivided into various subtypes according to their components and/or their homogeneity or heterogeneity.

Mixtures of *two* (rarely more than two) molecularly or microscopically "*dispersed*" *solid polymers* are usually called **polymer blends** (for a more selective definition, see Section 10.4). The term "blend" is also used for the mixture of a polymer powder with a powdery filler, i.e., a **dry blend** (p. 22). "Polymer blend" refers in general to those mixtures of polymers in which none of the polymeric components is present in a distinct geometric shape. Neither bicomponent fibers (p. 168) nor polymeric fiber-reinforced polymers (p. 380 ff.) are therefore considered polymer blends.

In principle, the two polymers of a polymer blend may be both thermoplastics, a thermoplastic and a rubber, or two rubbers. Their mixture may be homogeneous or heterogeneous. Mixtures of two thermoplastics are sometimes also distinguished as **polymer alloys**. However, "polymer alloy" may also mean a blend of two miscible or "compatible" polymers (see Section 10.1.2), a blend of two polymers with a distinct feature (for example, only one glass temperature), or a blend with a synergistic behavior of certain mechanical properties, etc. The polymers of such blends may be both amorphous or both semicrystalline or one may be amorphous and the other one semicrystalline. An "alloy" of two polymers may therefore be quite different from an alloy of two metals which always contains crystallites of both components.

Industry uses homogeneous and heterogeneous blends of thermoplastics. Whether a blend consists of one phase or multiple phases depends on the thermodynamic parameters of the components and mixtures as well as often on the (kinetically controlled!) preparation of the blend. If both polymeric components are amorphous and the thermodynamic parameters are of the right magnitudes, one-phase mixtures may exist for the whole concentration range (Section 10.2.2). Mixtures of amorphous and semicrystalline polymers may show miscibility only (if any) for the melt and the amorphous phase.

Table 10-1 Effect of polymer morphology on the properties of heterogeneous polymer blends.

Common name	Continuous phase	Discontinuous phase	Improved property
Elastomer blend	soft	soft	abrasion
Polymer-filled elastomer	soft	hard	tensile modulus
High-impact thermoplastic	hard	soft	impact strength
Thermoplastic alloy	hard	hard	impact strength, melt viscosity

Crystalline polymers cocrystallize only if they have similar unit cells (Volume III, p. 202); i.e., if the two polymers can form an isomorphous system. Homogeneous polymer blends arise if the two components form polymer complexes, for example, by acid-base interactions, hydrogen bonds, or steric fits of their monomeric units.

Properties of homogeneous blends can be often estimated by interpolating the properties of their components. Heterogeneous blends, on the other hand, often show jumps in their property-composition diagrams (see Table 10-1). For example, the two phases may be both hard, both soft, or one soft and the other one hard where "hard" and "soft" refer to the position of the highest transformation temperature (melting temperature, glass temperature) relative to the observation temperature and not to surface hardness.

Mixtures of a *solid polymer* and a polymeric or non-polymeric *liquid* (Section 10.3) are not called "blends" although the two components are indistinguishable from each other in such a mixture (p. 371). Depending on the chemical nature of the two components, such polymer-containing mixtures are known as **oil-extended rubbers, plasticized polymers**, etc.

In principle, this group also comprises mixtures of a thermoplastic and a rubber since commercial rubbers are (very viscous) liquids because they are often not crystalline at all and have very low glass temperatures (p. 273). Such mixtures consist of a thermoplastic matrix in which rubber particles are dispersed (Section 10.4.7). They have improved mechanical properties, especially impact strengths, and are therefore also known as **rubber-toughened plastics** or **impact-modified plastics**. Usually, they are not true physical dispersions of rubber particles in a thermoplastic matrix but partially grafted and/or crosslinked materials.

Mixtures of two rubbers are called **rubber blends** and, after crosslinking, **elastomer blends** (see p. 249 for the definitions of "rubber" and "elastomer").

The combination of a polymer and a gas leads in general to a "dispersion" of the gas in the polymer. Depending on the components and the resulting physical structure, these materials are known by many different names such as **expanded plastics, expanded rubbers, plastic foams, cellular plastics, sponge polymers**, etc. (see Section 10.5).

Many marketed polymer grades are not polymers (+ adjuvants!) *per se* but polymer blends although that is not reflected in the name of the polymer grade. Examples are some grades of LDPEs which are blends of low-density poly(ethylene)s LDPE and linear low-density poly(ethylene)s LLDPE (Vol. II, p. 216). Other examples are blends of isotactic poly(propylene) (iPP) and ethylene-propylene-diene terpolymers (EPDM) such as impact copolymers of propylene (**ICP**) (iPP with 15-20 wt% EPDM) and thermoplastic poly(olefin) elastomers (TPO) (iPP + 60-65 wt% EPDM) (Vol. II, p. 235). Poly(vinylidene chloride)s of commerce are always copolymers (Vol. II, p. 285).

10.1.2 State of Matter

States of matter of polymer mixtures are often described by the "solubility", "miscibility", or "compatibility" of their components, which leads to "homogeneous," "heterogeneous," "compatible," or "incompatible," etc., "mixtures," "solutions," "blends," etc., without further explanation or definition of the meaning of these words.

Miscibility refers to the ability of different matter *in the same physical state* to mix with a random distribution of their smallest components. The smallest components are molecules of liquids or particles of powders. Such mixtures are "homogeneous" with respect to their constituting entities, even for powders, provided one of the components does not agglomerize to larger particles, which renders the mixture "heterogeneous."

However, the terms "homogeneous" and "heterogeneous" are also used with respect to the formation of phases. Thermodynamically, a **phase** is a physically identical region which is separated from other phases (with other physically identical regions) by a sharp boundary (Volume III, p. 314 ff.). The stable coexistence of two or more phases is called **phase equilibrium**. An example is the phase equilibrium of a mixture of 0.5 wt% poly(ethylene) PE in xylene X that at 98°C forms two phases: phase A is a dilute solution L of PE in X and phase B a dispersion of PE "particles" in a solution L' of PE in X (Volume III, Fig. 10-15, left). At this point of the liquidus curve, the *system* PE/X is heterogeneous because it consists of two phases that are in equilibrium. Phase A is macroscopically homogeneous (= physically identical) down to the level of molecules and phase B is macroscopically identical down to the level of PE particles which are in equilibrium with the poly(ethylene) molecules that are dissolved in the solution L'. In this terminology, a mixture of two different powders Q and R is always heterogeneous because Q and R are not in equilibrium although the mixture may appear to be homogeneous if inspected with the naked eye.

Solubility is a thermodynamic term that describes the statistical distribution of two (or more) components down to the molecular level in a single thermodynamic phase. By definition, one of the components of the solution is always a condensed matter. Hence, there are solutions of gases, liquids, or solids in liquids or solids but not solutions of gases in gases.

Partial solubilities and **partial miscibilities**, respectively, are exhibited by systems that form one-phase regions only in certain concentration ranges but not in other ones. An example is certain plasticized polymers in which liquids are only partially dissolved.

Compatibility is an operational term for the behavior of certain polymer-polymer mixtures that are not in thermodynamic equilibrium but behave as if they are in time intervals that are necessary to determine certain physical properties. Compatible polymers may be blended to heterogeneous microblends that do not separate into macroscopic phases at experimental conditions.

Properties of mixtures depend strongly on the thermodynamic miscibility or immiscibility of components or their compatibility or incompatibility for the duration of the targeted time span. For example, blends of two incompatible plastics are mostly undesired because they may lead to bad tensile properties, but some of these blends lead to better processabilities and improved impact strengths. Exudation of partially miscible plasticizers is generally undesirable but a compatibility of polymers may improve impact strengths. Miscibilities can be predicted if the thermodynamic parameters are known.

10.2   Miscibility

10.2.1   Introduction

Two components A and B are miscible if two thermodynamic conditions are obeyed: (1) the Gibbs energy of mixing must be negative or zero,

$$(10\text{-}1) \qquad \Delta G_{mix} = \Delta H_{mix} - T \Delta S_{mix} \le 0$$

and (2) the second derivative of the Gibbs energy of mixing and the first derivative of the chemical potential $\Delta \mu_A$ of the component A with respect to the volume fraction ϕ_B of the component B must be positive or zero (see also Volume III, Chapter 10):

$$(10\text{-}2) \qquad (\partial^2 \Delta G_{mix}/\partial \phi_B{}^2)_{T,p} = \partial \Delta \mu_A/\partial \phi_B \ge 0$$

A polymer is usually not a single chemical substance but a mixture of polymer molecules with the same constitution but different molecular weights, Even "practically molecularly homogeneous" ("monodisperse") polymers have fairly broad molecular weight distributions. Furthermore, molecules of a homopolymer are usually not homogeneous with respect to the constitution of their molecules (branches, head-to-head junctions, etc.) and also not with respect to their configurations (non-ideal tacticity). Molecules of copolymers also usually differ with respect to their chemical composition.

Enthalpies and entropies of mixing, ΔH_{mix} and ΔS_{mix}, respectively, are mainly affected by three thermodynamic effects:

- combinatorial entropy of mixing (so-called "configurational" mixing entropy),
- intermolecular interactions (enthalpy of mixing), and
- free volume effects (affect both enthalpy and entropy).

Miscibility may thus be caused by a negative enthalpy of mixing, a positive entropy of mixing, or a combination of both.

By definition, an **ideal solution** is characterized by a zero enthalpy of mixing, an entropy of mixing that is controlled exclusively by combinatorial effects, and an unchanged free volume. The first two conditions mean zero interaction energies E between units which are either low-molecular weight molecules (such as plasticizer molecules) or monomeric units of polymers.

An **ideal mixture** of a polymer with monomeric units 1 and another polymer with monomeric units 2 is characterized by interaction energies of $E_{11} = E_{22} = E_{12} = E_{21}$. As a result, the change of entropy on mixing is not affected by contributions that depend on the environment of units: neither translational entropies nor internal rotational entropies or vibrational entropies are changed upon mixing. However, the two types of units can be arranged in many different "configurations." A mixture is therefore more disordered than its pure components: the combinatorial entropy of mixing is always positive which, in turn, leads to a negative contribution to the Gibbs energy of mixing since, by definition, mixing enthalpies of ideal mixtures are zero ($\Delta H_{mix} \equiv 0$). Positive combinatorial entropies thus favor miscibilities.

For equal interactions 1-1, 1-2, and 2-2, the degree of disorder of a mixture of units 1 and 2 is largest if both types of units have low molecular weights. Mixtures are less disordered if monomeric units 2 are dissolved in solvents 1 and even more ordered if monomeric units 1 are blended with monomeric units 2. Disorder thus decreases in the sequence solvent-solvent > solvent-polymer > polymer-polymer. The coupling of monomeric units in polymer chains automatically decreases disorder because coupling reduces the number of possible spatial arrangements of monomeric units. A large degree of polymerization allows therefore only a few contacts between units which, in turn, leads to a drastically lower combinatorial entropy: the combinatorial entropy of a mixture of two polymers with infinitely large degrees of polymerization is zero.

Therefore, the following conditions apply for miscibilities. In mixtures of *two low-molecular weight liquids*, the combinatorial entropy is so large and the entropy term $-T\Delta S_{mix}$ is therefore so negative that the enthalpy of mixing ΔH_{mix} can adopt fairly positive values without the Gibbs energy of mixing, ΔG_{mix}, becoming positive. A positive ΔG_{mix} indicates that the contribution of repulsive forces is greater than that of attractive ones. In these cases, positive enthalpies of mixing can no longer be compensated by negative entropy terms: the two liquids are not miscible.

In mixtures of a *polymer* and a *low-molecular weight liquid*, combinatorial entropies are lower than those of mixtures of two solvents. For miscibilities, mixing enthalpies are thus restricted to far lower positive values compared to the mixing of two solvents. The range for repulsive forces is much smaller, and, as a result, there are far fewer true solutions of polymers in solvents than mixtures of two solvents. Many mixtures of polymers and plasticizers are therefore thermodynamically unstable (Section 10.3).

The range of miscibilities is further reduced for mixtures of *two polymers* because combinatorial entropies are much smaller. Negative Gibbs energies of mixing of two polymers can be obtained only if the following conditions apply:

1. A small combinatorial entropy allows for a small positive enthalpy of mixing. However, the theoretical treatment of the solubility parameter (Section 11.2) shows that dispersion forces as well as interactions between permanent and/or induced dipoles always lead to positive enthalpies of mixing. The contribution of these forces to the enthalpy of mixing must therefore be minimized if one wants to obtain true polymer mixtures (i.e., homogeneous polymer blends) with a negative Gibbs energy of mixing. However, this is possible only for polymers with very similar chemical constitutions, i.e., for pairs of polymers with so-called non-specific interactions.

2. A negative Gibbs energy of mixing is also caused by a negative enthalpy of mixing which requires strong attractive forces between components. Such "specific interactions" lead to exothermic heats of mixing which is observed for pairs of polymers with complementary groups, resulting in, for example

- formation of polysalts from polyacids and polybases,
- generation of hydrogen bonds between two types of polymers,
- development of stereocomplexes between differently configured monomeric units, or
- cocrystallization of segments of chemically different polymers.

10.2.2 Parameters

Two components mix if changes of entropy, enthalpy, and/or volume lead to negative Gibbs energies of mixing. Contributions by changes of entropies and enthalpies can be calculated with the **Flory-Huggins lattice theory** (Volume III, p. 308), a statistical thermodynamics theory that places monomeric units on a lattice. The change of entropy is calculated from the mutual arrangement of the monomeric units and the change of enthalpy from the interactions that occur. Compressibilities of mixtures are neglected.

The theory calculates the **molar combinatorial entropy of mixing**, $\Delta S_{mix,m}$, as

(10-3) $\Delta S_{mix,m} = - R[(\phi_1/X_1) \ln \phi_1 + (\phi_2/X_2) \ln \phi_2] \equiv \Delta S_m$

where R = molar gas constant, X_i = degree of polymerization of component i, and ϕ_i = volume fraction of component i (i = 1, 2). A solvent molecule 1 with a degree of polymerization of X_1 = 1 occupies only one lattice point whereas a polymer molecule 2 with $X_2 \gg 1$ monomeric units takes up X_2 lattice points.

The mixing of two polymers 1 and 2 drastically reduces the number of possible arrangements of monomeric units. According to Eq.(10-3), a solution of a polymer with X_2 = 4.46 in a solvent with X_1 = 1 has a combinatorial molar entropy of mixing of $\Delta S_{mix,m}/(\text{J K}^{-1} \text{ mol}^{-1})$ = +2.159 at a volume fraction of ϕ_1 = 0.79. Molar entropies of mixing decrease rapidly if component 1 is also a polymer, leading to values of +0.720 for X_1 = 14.2, +0.679 for X_1 = 22.6, and +0.616 for X_1 = 293. They approach a limiting value of +0.6133 for infinitely large degrees of polymerization.

However, these relatively small differences in mixing entropies lead to very large differences in the mixing/demixing behaviors as shown by the phase diagrams for mixtures of a poly(isobutylene) with X_2 = 4.46 with various poly(dimethylsiloxane)s with degrees of polymerization of X_1 = 14.2, 22.6, or 293 (Fig. 10-1).

In contrast to entropies of polymer-solvent systems, molar entropies of mixing, $\Delta S_{mix,m}$, of polymer blends are only slightly positive. The resulting entropy term $-T\Delta S_{mix,m}$ becomes slightly negative and can no longer compensate the usually positive enthalpy term $\Delta H_{mix,m}$: the molar Gibbs energy of mixing becomes positive (Eq.(10-1)) and the two polymers do not mix which is the commonest case.

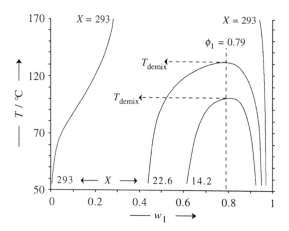

Fig. 10-1 Demixing temperatures T as a function of the mass fraction w_1 of poly(dimethylsiloxane)s with degrees of polymerization of X_1 = 14.2, 22.6, and 293 in mixtures with a poly(isobutylene) with X_2 = 4.46 [1]. The systems have upper critical demixing temperatures, for example, T_{demix} = 100°C for X_1 = 14.2 and 132°C for X_1 = 22.6 (this temperature is not visible for X_1 = 293 since it is outside the range of the y axis). They form one phase above the curves but two phases below them.

The mixing of two components 1 and 2 can be treated as a quasichemical reaction of their monomeric or chain units, i.e., pairs of units 1-1 react with pairs of units 2-2 to form pairs 1-2. These pairs have interaction energies of ε_{11}, ε_{22}, and ε_{12}, respectively. The change of interaction energies per contact is therefore $\Delta\varepsilon = \varepsilon_{12} - (1/2)(\varepsilon_{11} + \varepsilon_{22})$.

The energy of mixing refers to the total system and hence is controlled by the number N_g of the lattice sites, the number z of the nearest neighbors of a pair 1-2, and the probabilities ϕ_1 and ϕ_2 that the next lattice site is occupied by a unit 1 or 2:

$$(10\text{-}4) \qquad \Delta H = N_g z \phi_1 \phi_2 \Delta\varepsilon$$

The product of the two unknown parameters z and $\Delta\varepsilon$ is divided by the thermal energy $k_B T$ to yield a new parameter χ, the **Flory-Huggins interaction parameter**:

$$(10\text{-}5) \qquad \chi = z\Delta\varepsilon/(k_B T) = B/T \quad ; \quad B = z\Delta\varepsilon/k_B$$

An exclusively enthalpic Flory-Huggins parameter should therefore be inversely proportional to the temperature. However, experiments show an additional entropic term:

$$(10\text{-}6) \qquad \chi = A + (B/T) \quad ; \quad A, B = \text{system-specific constants}$$

These additional entropic contributions are much more pronounced for polymeric units than for solvent molecules and should therefore also depend on the volume fraction ϕ_2 of the polymeric units. Very often, it is found that $\chi = \chi_0 + k\phi_2 + f(\phi_2)$ where $f(\phi_2)$ may be $k'\phi_2^2$ or $k'' \ln \phi_2$ (k', k'' = constants). In contrast to the original assumptions of the theory, the Flory-Huggins parameter is therefore not a purely enthalpic quantity and also not a measure of the enthalpy of mixing but a measure of an *energy* of mixing that contains the enthalpy of mixing and the non-combinatorial proportions of the entropy of mixing. This non-combinatorial part depends on the volume of the system whereas the combinatorial part is controlled by the number of units.

For exclusively enthalpic interactions, the molar enthalpy of mixing is given by

$$(10\text{-}7) \qquad \Delta H_{\text{mix,m}} = \Delta H/n_g = RT \, \chi_0 \phi_1 \phi_2$$

where $n_g = N_g/N_A$ = amount of units, N_g = number of lattice sites, N_A = Avogadro constant, $R = k_B N_A$ = molar gas constant, and k_B = Boltzmann constant.

Comparison of Eqs.(10-3), (10-7), and (10-1) shows that two polymers are thermodynamically miscible only if the interaction parameter χ_0 is negative (presence of specific interactions) or zero or is slightly positive (non-specific interactions). If two polymers with the same degrees of polymerization of $X_1 = X_2 = 1000$ are present in equal volume fractions, $\phi_1 = \phi_2 = 1/2$, Eq.(10-3) predicts a molar entropy of mixing of $\Delta S_{\text{mix,m}} = +5{,}76 \cdot 10^{-3}$ J K^{-1} mol^{-1} and, for 200°C, an entropy term of $-T\Delta S_{\text{mix,m}} = -2.73$ J/mol. Since true miscibilities require $\Delta G_{\text{mix,m}} \leq 0$ according to Eq.(10-1), interaction parameters should be $\chi_0 \leq 0.002\ 77$ according to Eq.(10-7).

Only very few polymer mixtures have such low interaction parameters, for example, the pair poly(2,6-dimethyl-1,4-phenyleneoxide) + poly(styrene) (PPE/PS) with $\chi_0 = -0.1$ at 200°C (the blend PPE/PS was at one time known industrially as PPO; the present modified PPO® is a blend of PPE and rubber-modified PS (see Volume II, p. 327).

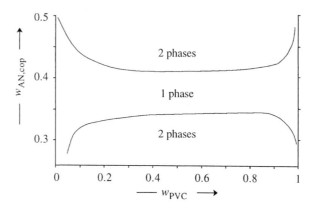

Fig. 10-2 Miscibility of a poly(vinyl chloride) with nitrile rubbers containing various mass fractions $w_{AN,cop}$ of acrylonitrile units [2]. Molecular weights are always 100 000; $T = 25°C$.

Small differences in chemical constitution can change interaction parameters marked-ly: at the same temperature, blends of PPE with poly(o-fluorostyrene) have a value of χ = +0.013, with poly(o-chlorostyrene) +0.03, and with poly(p-chlorostyrene) +0.045. Very small variations in copolymer composition may also change miscibility to immis-cibility (Fig. 10.2).

For polymer solutions in low-molecular weight solvents, solubility predictions are commonly made with the help of the more easily accessible solubility parameters δ in-stead of interaction parameters, especially in the paint industry (see also Section 12.2). Solubility parameters δ are defined as the square roots of cohesion energy densities (Volume III, p. 299), for example, $\delta_1 = [z\,(\varepsilon_{11}/2)/V_{1,mol}]^{1/2}$ where $V_{1,mol}$ = molar vol-ume of component 1.

The interaction energy ε_{12} of the pair 1-2 is given by the square root of the product of homo interaction energies, $\varepsilon_{12} = (\varepsilon_{11}\varepsilon_{22})^{1/2}$. Since $\Delta\varepsilon = \varepsilon_{12} - (1/2)(\varepsilon_{11} + \varepsilon_{22})$ (p. 419, top) and solubility parameters and interaction parameters are related (see above), one obtains for the enthalpy of mixing

(10-8) $\Delta H_{mix} = V\phi_1\phi_2(\delta_1 - \delta_2)^2$

The molar enthalpy of mixing is $\Delta H_{mix,m} = \Delta H_{mix}/n$ and the specific enthalpy of mixing, $\Delta h_{mix} = \Delta H_{mix}/m$, where n = total amount of the system, m = total mass, V = total volume, ϕ_i = volume fraction, and δ_i = solubility parameters of components i.

Hence, the concept of solubility parameters is based exclusively on dispersion forces between 1-1, 1-2, and 2-2. Because of the square of the difference of solubility parame-ters in Eq.(10-8), enthalpies of mixing are always positive for the sole presence of dis-persion forces.

However, stronger attraction forces (dipole-dipole interactions, hydrogen bonds) may also be present so that the enthalpy of mixing becomes negative. For example, cis-1,4-poly(butadiene) (BR) and styrene-butadiene rubber (SBR) form 1-phase blends with a specific enthalpy of mixing of $\Delta h_{mix} = -2.1$ J/g. Blends of BR and poly(styrene) lead also to 1-phase blends with $\Delta h_{mix} = -1.2$ J/g whereas blends of SBR and natural rubber form two phases with $\Delta h_{mix} = +1.2$ J/g.

Blends of chemically similar polymers may not only show fairly large entropies of mixing but also large mixing enthalpies. An example is the specific mixing enthalpy of poly(methyl acrylate) and poly(methyl methacrylate (+10.5 J/g). Predictions of miscibilities based on chemical structures ("like dissolves like") are therefore not very reliable.

10.2.3   Phase Diagrams of Fluid Systems

Homogeneous (1-phase) fluid systems of two components (polymer solutions and homogeneous melts of blends) demix at certain temperatures and concentrations and form two phases. This section discusses systems that remain fluid at all temperatures.

Reduced molar Gibbs energies of mixing are calculated from the second law of thermodynamics (Eq.(10-1)) and the lattice model (Eqs.(10-3) and (10-7)) as

$$(10\text{-}9) \qquad \Delta G_{mix,m}/RT = (\phi_1/X_1)\ln\phi_1 + (\phi_2/X_2)\ln\phi_2 + \chi_0\phi_1\phi_2 \; ; \; \text{(Volume III, p. 312)}$$

Mixtures of low-molecular weight substances such as solvents with $X_1 = X_2 = 1$ may therefore have interaction parameters up to $\chi_0 = +2.77$ without demixing.

Polymer solutions ($X_2 \gg 1; X_1 = 1$) and polymer blends ($X_2 \gg 1; X_1 \gg 1$) demix at far lower interaction parameters. Furthermore, the function $\Delta G_{mix,m}/RT = f(\phi_2)$ becomes asymmetric if $X_2 \neq X_1$. In addition, this function not only has a minimum (if $\Delta G_{mix,m}$ is negative) or a maximum (positive $\Delta G_{mix,m}$) but both minima and maxima (Fig. 10-3)

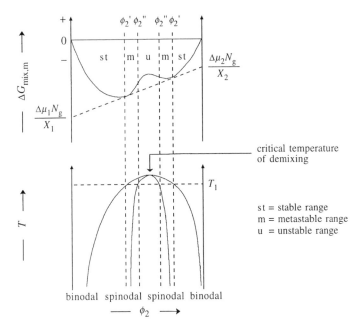

Fig. 10-3 Demixing temperatures T (bottom) and molar Gibbs energies of mixing, ΔG_m, for the temperature T_1 (top) as functions of the volume fraction ϕ_2 of the component 2 for a partially miscible fluid system.

The extrapolation of tangents to the $\Delta G_{mix,m} = f(\phi_2)$ curve delivers $\Delta \mu_1 N_g/X_1$ for $\phi_2 \to 0$ and $\Delta \mu_2 N_g/X_2$ for $\phi_2 \to 1$. In a 2-phase system, chemical potentials $\Delta \mu_i$ of each component i = 1, 2 must be the same in each phase ' and ", i.e., $\Delta \mu_1' = \Delta \mu_1"$ and $\Delta \mu_2' = \Delta \mu_2"$. These equalities exist only if two points on the curves have a common tangent. The contact points of this tangent at the $\Delta G_{mix,m} = f(\phi_2)$ curve thus determine the compositions ϕ_2' and $\phi_2"$ of the two phases ' and " (Fig. 10-3).

Systems with compositions $\phi_2 < \phi_2'$ and $\phi_2 > \phi_2"$ are stable one-phase mixtures that do not demix. The border between the one-phase and two-phase regions is called **bi-nodal** (L: *bi* = two; *nodus* = knot). The points of inflection of the $\Delta G_{mix,m} = f(\phi_2)$ curve separate the two-phase region into two metastable ranges and one unstable range (Fig. 10-3, top); the border between these regions is called **spinodal** (L: *spina* = spine).

In the metastable two-phase ranges, systems are stable if the two phases differ only slightly in their compositions but they will phase separate into a dispersion of one phase in the other if the composition of the two phases differs more strongly. In the unstable two-phase range, mixtures phase separate into two continuous phases similar to inter-penetrating networks.

The constant B of Eq.(10-6) is positive for endothermic mixtures: the interaction parameter χ_0 decreases with increasing temperature. *Above* a so-called **upper critical solution temperature (UCST)**, all compositions form only one phase (Figs. 10-3 and 10-4, I). There are also systems with **lower critical solution temperatures (LCST)** *below* which systems are completely miscible (Fig. 10-4, II). Note that *upper* and *lower* do not refer to the *absolute* positions of these temperatures but to their *relative* positions with respect to the two-phase regions (compare Figs. 10-4, III and 10-4, IV). In principle, all systems should have both a UCST and an LCST but in most cases only one is observable because of interference by melting or boiling temperatures, onset of decompositions, etc.

Most mixed solvents and most polymer solutions have Type I phase diagrams (Fig. 10-4, left); they demix on cooling. The interaction parameters χ of these systems decrease with increasing temperature so that B in Eq.(10-6) is positive. The enthalpy of mixing is always positive (endothermic mixture) if χ does not depend on the volume fraction ϕ_i. The miscibility increases with increasing temperature.

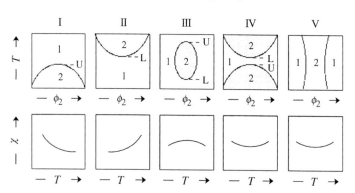

Fig. 10-4 Top: types of idealized ($X_1 = X_2$, $\chi \neq f(\phi_2)$) phase diagrams, $T = f(\phi_2)$, for polymer solutions or polymer blends with one-phase (1) and two-phase (2) ranges and upper (U) and lower (L) critical solution temperatures. Type III is known as a critical miscibility gap and type V as an hourglass diagram. Because of $X_1 \neq X_2$ and/or $\chi = f(\phi_2)$, real systems are asymmetric.
Bottom: temperature dependence of interaction parameters.

Type II systems, on the other hand, demix on heating: $d\chi/dT$ is positive, the enthalpy of mixing is negative, and the mixture is exothermic. Here, miscibilities increase with decreasing temperatures.

Type III systems have closed miscibility gaps with UCST > LCST because interaction parameters pass through maxima with increasing temperature. Such phase diagrams are observed for aqueous solutions of poly(vinyl alcohol) or methyl cellulose where they are caused by desolvations that increase with increasing temperature.

The LCST of type IV systems with UCST < LCST can be observed for polymer solutions if the solvents are above their boiling temperatures and under pressure. The mixing of the dense polymer with the highly expanded solvent vapor causes a contraction that leads to negative entropies of mixing and thus to lower critical solution temperatures. With increasing temperature, interaction parameters pass through minima.

The concentration dependence of demixing temperatures at both sides of extremal points becomes more pronounced with increasing molecular weight of polymers. Finally, phase diagrams degenerate to the "tree trunk" diagrams of Type V (Fig. 10-4). Such phase diagrams are observed not only for high-molecular weight polymers in low-molecular weight solvents above boiling temperatures but also for more than 90 % of all polymer-polymer mixtures. Here, temperature changes do not lead to true miscibilities for the indicated range of compositions although such polymer blends may behave as "compatible" for certain mechanical tests.

10.2.4   Phase Diagrams with Solid Phases

Section 10.2.3 discussed polymer systems that remain fluid for the whole ranges of concentrations and temperatures regardless of the existence of phase separations, if any, i.e., solutions and melts. However, the cooling of liquids (L) may lead to solids, i.e., crystalline (C) or amorphous ones (G) (for liquid crystals, see Volume III, Chapter 8). Phase transformations L → C occur for *one-component* liquids at the melting temperature T_M of crystallites and transformations L → G at the glass temperature T_G of amorphous solids.

On cooling below transformation temperatures, *two-component* liquids will exhibit different phase diagrams (Fig. 10-5, I-V). On cooling a homogenous liquid L_{A+B} with components A and B, isotropic amorphous glasses G_{A+B} are formed with the same composition as L_{A+B}. Glass temperatures T_G of these glasses vary monotonously with the composition from $T_{G,A}$ to $T_{G,B}$ (Fig. 10-5, I) if the glass temperatures $T_{G,A}$ and. $T_{G,B}$, respectively, of the components are greater than the upper critical solution temperature UCST. The shape of the resulting curve $T_G = f(\phi_A)$ depends on the interactions A/B, the expansion coefficients, and/or the specific heat capacities. Even if a UCST is reached upon further cooling, such isotropic glasses would demix only very slowly because of the very low diffusion coefficients of polymer molecules in glasses (Volume III, p. 479).

A different pattern is obtained if component A does not crystallize but component B does (Fig. 10-5, II). On cooling solutions L with small concentrations of A (= high concentrations of B), liquids L will phase separate first into a mixture 2 of crystallized B in a liquid L' that has a different composition than the original liquid L. Above a critical ϕ_A, the original liquid L forms a glass G from both components A and B.

A pattern of $T = f(\phi_A)$ similar to that of I is obtained for a system with two crystallizable components A and B that form a homogeneous liquid L_{A+B} and cocrystallize on cooling. Here, each composition leads to a single transformation temperature, i.e., the melting temperatures T_M of the cocrystallized solids, that varies with the original volume fraction ϕ_A (Fig. 10-5, IV).

Two crystallizable components may also form a eutectic (Fig. 10-5, III) (G: *eutektos*, from *eu* = well, *tektos* = melted). Below the eutectic point (= minimum of the $T_M = f(\phi_A)$ curve), a mixture of C_A and C_B exists. Below the liquidus curve and above the eutectic temperature, one observes a mixture of a liquid and C_B for $\phi_A < \phi_{A,eu}$ and a mixture of a liquid and C_A for $\phi_A > \phi_{A,eu}$.

Such phase diagrams become much more complicated if two crystallizable components A and B mix in the melt but do not cocrystallize (Fig. 10-5, V). Here, one also observes two regions below the melting temperature T_M, a region 2 with L + C_B at small ϕ_A and a region 1 with L + C_A at large ϕ_A. Regions 2 and 1 are separated in a certain medium concentration range by a region 6 that consists of L + C_A + C_B. At even lower temperatures, the liquids become gel-like solids G and the sequence of regions 2 → 6 → 1 for $\phi_A > 0$ is replaced by a sequence 4 → 7 → 3 where 4 = G + C_B, 7 = G + C_A + C_B, and 3 = G + C_A.

Cooling the melt of mixtures of two or more polymers will never lead to *true* phase equilibria because the cooling proces will "freeze" non-equilibrium states, and equilibrium states are rarely available because of the very slow diffusion of polymer segments which are in the range of 10^{-14}-10^{-18} cm^2 s^{-1} (see also Volume III, p. 479).

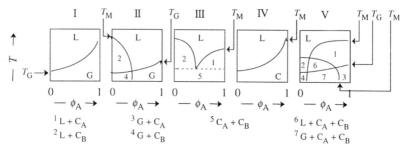

Fig. 10-5 Phase diagrams $T = f(\phi_A)$ for transformations of two-component, single phase liquids L into various solids (glasses G, crystalline solids C_A and/or C_B, various mixed phases 1-7). T_M = melting temperature, T_G = glass temperature.

I. Solidification of a one-component liquid into a glass. II. Phase separation for a liquid into a crystallizable solid B and a non-crystallizing polymer A. III. Formation of a eutectic mixture from two crystallizable polymers. IV. Formation of isomorphous crystals. V. Phase diagram for two semicrystalline polymers.

10.3 Plasticized Polymers

10.3.1 Introduction

Plasticizers (p. 42) are added to thermoplastics or elastomers in order to improve deformabilities, sometimes processing (see also p. 102) and often also expandabilities of foamable polymers. This technological definition of a plasticizer does not always conform with that of science, according to which a plasticizer improves segmental mobilities of polymer chains and lowers the glass temperature.

Plasticizers are usually low-molecular weight liquids (Section 3.4.3); solid plasticizers such as camphor are no longer used. Oligomeric and polymeric plasticizers are relatively rare; they constitute less than 5 wt% of all consumed plasticizers. For example, low-molecular weight polyesters serve as true plasticizers for polymers whereas their higher molecular weight grades are used in polymer blends.

Polymers are not only subject to targeted plasticizings but also to unintended ones by water, water vapor, or carbon dioxide. Polymers with hydrophilic groups such as amide groups of polyamides absorb several percent H_2O from the humidity of air. They are therefore **conditioned** by subjecting them to controlled conditions, usually 24 h at 23°C and 50 % relative humidity. However, equilibria are established only after long times (p. 79; see also Figs. 3-3, 4-2, and 4-9). Such conditioned polyamides have much higher impact strengths than unconditioned ones.

Because of its many hydroxyl groups, cellulose also takes up moisture. Cotton is plasticized by water which reduces the glass temperature of cotton to values below 20°C (p. 190). Wet cotton fabrics crease but can also be ironed easily. Cellulose papers are always plasticized by water. The water is desorbed at temperatures above 110°C whereupon paper becomes brittle. Conversely, starch can be plasticized with 5-14 wt% water and then injection molded at temperatures of ca. 170°C. Plasticized starch has fracture elongations of ca. 25 % and higher fracture strengths than poly(styrene).

10.3.2 Molecular Plasticizer Efficiencies

The efficiency of plasticizers is defined as the magnitude of change of a property per proportion of plasticizer. In the molecular picture, plasticizers should increase the mobility of chain segments, which in turn leads to several technological effects such as lowered glass temperatures, increased fracture elongations, improved impact strengths, etc. However, only glass temperatures depend exclusively on segmental mobilities. All other technological data are also affected by other parameters (see below). It follows that the determination of plasticizer efficiencies by changes of glass temperatures, tensile moduli, fracture strengths, yield strengths, fracture elongations, hardnesses, etc., with plasticizer concentration does not necessarily deliver coherent results.

Plasticizers are efficient only if they form thermodynamically stable mixtures with the polymer; insoluble plasticizers are not effective. However, a plasticizer should not be a thermodynamically good solvent for the polymer because this would lead to strong interactions between polymer and plasticizer. On one hand, such interactions are desired because they dissolve helical structures and/or crystalline regions of polymers, which in turn enhances the mobility of chain segments. Polar plasticizers also increase the ratio of gauche and trans microconformations in polar chains, which reduces the average of potential barriers (for the gauche effect, see Volume III, pp. 45, 199).

On the other hand, too strong interactions between polymer and plasticizer will lead to solvations. The sheath of solvent molecules acts as a physical substituent: it increases the rotational barrier and makes the chains more rigid. The larger the molecules of solvents of similar constitution, the thicker is the solvent sheath and the less effective is the plasticizer as measured by the decrease of glass temperatures (Fig. 10-6).

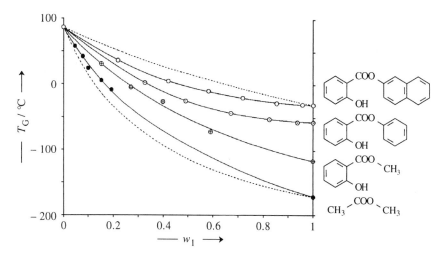

Fig. 10-6 Glass temperature T_G of a poly(styrene) (T_G = 85.5°C) as a function of the mass fraction w_1 of added plasticizer (experimental data of [3]). Solid lines: Couchman Eq.(10-10) with k = 2.5 (naphthyl salicylate), 2.7 (phenyl salicylate), 1.3 (methyl salicylate), and 1.5 (methyl acetate). Broken lines: Fox Eq.(10-11) for β-naphthyl salicylate (T_G = –29.5) and methyl acetate (T_G = –170°C).

The dependence of glass temperature T_G of plasticized polymers 2 on the weight fraction $w_1 = 1 - w_2$ of the plasticizer 1 can be described by the **Couchman equation** which was derived originally for entropic effects in copolymers (see Volume III, p. 464):

$$(10\text{-}10)\qquad \ln T_G = \frac{w_1 \ln T_{G,1} + k w_2 \ln T_{G,2}}{w_1 + k w_2} \quad ; \quad k = \Delta c_{p,1}/\Delta c_{p,2}$$

In the original theory, k is the ratio of specific heat capacities of species 1 and 2; for plasticized polymers, it may be treated as an adoptable constant.

The logarithms of temperature may be replaced by the temperatures themselves if the ratio $T_{G,1}/T_{G,2}$ does not differ much from unity. If furthermore $k = T_{G,2}/T_{G,1}$, the Couchman equation converts to the often used **Fox equation**

$$(10\text{-}11)\qquad 1/T_G = (w_1/T_{G,1}) + (w_2/T_{G,2})$$

High molecular efficiencies are provided by plasticizers with low glass temperatures, i.e., those with small interactions between plasticizer molecules. Plasticizers with strong interactions between their molecules will form a kind of network that polymer segments must deform, which requires more energy. Weak interactions between plasticizer molecules generally lead to small plasticizer viscosities.

10.3.3 Technological Plasticizer Efficiencies

Molecularly, good plasticizer efficiencies should be obtained by small plasticizer molecules with only weak interactions between them. However, such molecules also have high vapor pressures and therefore also great volatilities (example: spherical camphor molecules). They would also sweat out easily. Technologically, one therefore seeks a compromise between a large molecular plasticizer efficiency (requiring only small proportions of expensive plasticizers) and the technological usefulness (little exudation of plasticizers (p. 23)).

For polar polymers such as poly(vinyl chloride), one prefers somewhat polar plasticizers such as dialkyl phthalates that are able to dissolve crystalline regions at elevated temperatures. After processing, on cooling to lower temperatures, the newly formed crystallites form a kind of physical network that provides the resulting physical gel with additional strength.

For polymers with high processing and application temperatures, plasticizers must be thermally stable. An interesting case is the use of plasticizers for polymers with such high melt viscosities that they must be processed at very high temperatures during which the polymers degrade. Such polymers are provided with crosslinkable plasticizers of low viscosity that reduce the melt viscosity of the mixture and hence the processing temperature. A subsequent crosslinking of the plasticizer provides the plasticized polymers with high glass temperatures.

Technological plasticizer efficiencies do not refer to the lowering of glass temperatures *per se* but to the desired increases of extensibilities and impact strengths; moduli, tensile and flexural strengths as well as hardnesses are usually reduced if greater concentrations of plasticizers are present.

Opposite effects are sometimes found at small plasticizer contents (usually smaller than 10 wt%): with increasing plasticizer concentration, tensile strengths first increase (anti-plasticization) and then decrease (plasticization) (Fig. 10-7). Such an effect cannot be caused by a stiffening of polymer segments by solvating plasticizer molecules since a

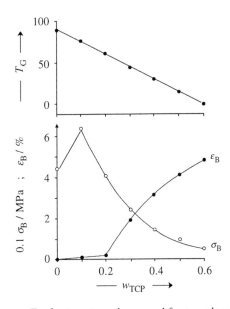

Fig. 10-7 Glass temperatures T_G, fracture strengths σ_B, and fracture elongations ε_B of a plasticized poly(vinyl chloride) as a function of the weight fraction w_{TCP} of the plasticizer tricresyl phosphate (TCP) showing anti-plasticization at $w_{TCP} < 0.1$ and plasticization at $w_{TCP} > 0.1$ [4]. The linear extrapolation of T_G to $w_{TCP} \rightarrow 1$ leads to the glass temperature of TCP, $T_G = -64.5°C$ (not shown).

stiffening of molecule segments should lead to an increase of glass temperatures, which is not found (Fig. 10-6). A healing of voids is much more likely since one observes that densities of this plasticized, non-crystallizing polymer pass through a maximum with increasing plasticizer concentration.

According to DSC measurements, **antiplasticizations** of semicrystalline polymers are caused by crystallizations that are more likely at greater plasticizer concentrations. The increased crystallinity leads to greater fracture strengths and maybe also to larger tensile moduli. At even larger plasticizer contents, dilution effects and true plasticizing take over. It seems that antiplasticizations arise primarily from morphological changes and not from specific interactions.

There may also be so-called "specific" or (better) "selective" plasticizations. Sulfonation of diene groups of EPDM polymers from ethylene, propylene, and small proportions of a non-conjugated diene results in ionomers (Volume II, p. 231). The ionic clusters and domains (Volume III, p. 269) of these polymers can be locally (i.e., selectively) plasticized by zinc stearate or glycerol. Glycerol acts similarly in sulfonated poly(styrene)s. The local plasticization reduces the physical crosslinking by ion clusters and domains and lowers melt viscosities: 5 wt% glycerol in sulfonated poly(styrene) lowers the melt viscosity by a factor of 1000 compared to 40 wt% of the plasticizer dioctyl phthalate for the same effect. Molecularly, these effects are not plasticizations but dissociations of physical crosslinks.

10.4   Polymer Blends

10.4.1   Introduction

Polymer blends are defined in this book as those mixtures of two or more polymers in which the polymers are not present in a specific shape. This group of "polymer blends" thus does not comprise bicomponent fibers (p. 168) and also not polymers that are reinforced with polymer fibers (see Chapter 9).

The polymeric components of polymer blends may be uncrosslinked or crosslinked and below or above their glass or melt temperatures, i.e., they may be thermoplastics (common), thermosets (uncommon), or rubbers or elastomers (common). Polymer blends may therefore consist of two rubbers or elastomers (Section 10.4.5), two thermoplastics (Section 10.4.6), or a thermoplastic and a rubber or elastomer. Polymer blends with three different polymeric components are not uncommon for rubber-reinforced thermoplastics but rare for rubber blends and blends of thermoplastics.

Blends are obtained by very different processes (Section 10.4.2) that deliver one-phase or multiphase blends (Section 10.4.6) with different morphologies (Section 10.4.4). The main components of blends containing plastics are always uncrosslinked. Melts of such blends can therefore be processed by extrusion or high-duty mixing.

Polymer blends are mostly prepared in order to improve certain properties of standard thermoplastics or elastomers in an economic way (see Table 10.2). However, especially impact strengths of thermoplastics are sometimes improved so drastically by blending with other polymers that one can tolerate an increased price per weight because the cost per property value is much lower (see Table 8-5).

Table 10-2 Properties of commercial polymer blends and their components. E = tensile modulus, σ_B = tensile strength, F_B = impact strength, HDT = heat distortion temperature. "Relative cost" refers to the component with the lowest cost per weight. ABS = acrylonitrile-butadiene-styrene polymer, PC = bisphenol A polycarbonate, IPS = impact poly(styrene), *MMA = methacrylate-based polymeric impact modifier, PPE = poly(oxy-2,6-dimethyl-1,4-phenylene), PS = poly(styrene), PVC = poly(vinyl chloride), SBS = poly(styrene-*block*-butadiene-*block*-styrene).

Components	$\dfrac{E}{\text{GPa}}$	$\dfrac{\sigma_B}{\text{MPa}}$	$\dfrac{F_B}{\text{J m}^{-1}}$	$\dfrac{\text{HDT}}{°\text{C}}$	Relative costs
ABS	2.06	35	320	86	1.00
PC	2.41	66	800	132	1.53
ABS + PC	2.55	43	530	105	1.19
PVC	2.76	48	<530	66	1.00
*PMMA	3.10	69	25	85	2.08
PVC + *MMA	2.34	45	800	75	2.44
PS	2.06	33	64	88	1.00
PPE	2.55	72	85	192	4.10
PPO® = PS + PPE	2.41	66	160	129	1.31
IPS		24	85	87	
IPS + 5 wt% SBS		21	128	87	
IPS + 10 wt% SBS		19	208	87	
IPS + 20 wt% SBS		17	320	86	

Polymer blends may have properties similar to the properties of one of the components, have weighted averages of properties, or show synergism (Table 10-2). For example, poly(oxy-2,6-dimethyl-1,4-phenylene) (PPE) has 1.2 times greater tensile modulus, 2.2 times greater tensile strength and 1.3 times greater impact strength than poly(styrene) but costs 4.1 times as much (Table 10-2). Blends of PPE + PS (i.e., PPO®) have similar tensile moduli and only 10 % smaller tensile strengths than PPE itself but 2.5 times greater impact strength, and all that at a price per weight which is only ca. 32 % of that of PPE and only 1.3 times greater than that of PS. The blends have also greater heat distortion temperatures than PS. Since the HDT of the blend is considerably lower than that of PPE, blends can be processed at lower temperatures which saves energy.

Production statistics of polymer blends are scarce and not comprehensive. For example, the 1998 world consumption was 320 000 t/a of PPE/PS plus PPE/PA. Blends of polycarbonate with either ABS or ASA amounted also to 320 000 t/a whereas the consumption of blends of PBR with PET or PC was only 60 000 t/a. It is estimated that ca. 10 % of all commercial grades of thermoplastics are really blends. For rubbers, this proportion is about 75 %.

10.4.2 Manufacture of Polymer Blends

Polymer blends are obtained *physically* by mechanical mixing of melts, latices, or solutions of two separately produced polymers or *chemically* by *in situ* polymerization of a monomer in the presence of a preformed polymer. All processes have advantages and disadvantages with respect to process control, properties of blends, and economics.

Melt Mixing

Blends can be prepared by mixing bales of rubbers or pellets or powders of thermoplastics using heated rolls (p. 112), kneaders, or extruders (p. 115). Because of the high melt viscosities, good mixing can only be achieved at higher temperatures and with strong shear fields. These conditions lead to homolytic degradation. The resulting polymer radicals cause grafting and/or crosslinking.

Shear degradation can be surpressed by use of low-molecular weight polymers but is often desired since it may cause the formation of graft polymers that act as compatibilizers (p. 44). Multiphase mixtures are always produced by thermodynamically immiscible components but can also be generated from some thermodynamically miscible ones.

Domain sizes in multiphase systems are controlled by mixing conditions and melt rheologies. Stirring and kneading provides the system with energy that is consumed in part by the formation of new surfaces by subdividing macrophases into microphases. The energy introduced is also used for flow processes or is released as heat.

The rate of heat absorption by the system is not unlimited because domain sizes finally become constant. They cannot be reduced further because high melt viscosities lead to very small diffusion coefficients of polymer molecules: the melt does not demix despite a thermodynamic immiscibility.

Extruders and intensive mixers cause laminar mixing which is more intensive if both components have the same viscosity. If not, the low-viscosity component will be stressed more intensely than the high-viscosity one. If the low-viscosity component is also the major component, a blend will be formed in which domains of the high-viscosity minority component swim in a "sea" of the low-viscosity majority one.

Which component goes into which phase seems to be controlled in a first approximation by a factor Q that is given by the viscosities η and the volume fractions ϕ of components 1 and 2:

$$(10\text{-}12) \quad Q = \frac{\eta_1}{\eta_2} \cdot \frac{\phi_2}{\phi_1}$$

Continuous phases are formed from component 1 if $Q < 1$ and from 2 if $Q > 1$.

Many empirical equations have been proposed for the dependence of melt viscosities η on the volume fractions ϕ_i or weight fractions w_i of components i = 1, 2. Examples for homogeneous melts are

$$(10\text{-}13) \quad \ln \eta = \phi_1 \eta_1 + \phi_2 \eta_2 + K_1 \phi_1 \phi_2 \qquad \text{or}$$
$$(10\text{-}14) \quad \eta^a = \phi_1 \eta_1^a + \phi_2 \eta_2^a$$

and for heterogeneous blends

$$(10\text{-}15) \quad \ln \eta = w_1 \ln \eta_1 + w_2 \ln \eta_2 \qquad \text{or}$$
$$(10\text{-}16) \quad \ln \eta = \phi_1 \ln \eta_1 + \phi_2 \ln \eta_2 + K_2 \phi_1 \phi_2 \qquad \text{or}$$
$$(10\text{-}17) \quad 1/\eta = K_3[(w_1/\eta_1) + (w_2/\eta_2)]$$

where K_1, K_2, K_3, and a are adoptable constants. Usually, a low melt viscosity is desirable for fast and low energy consuming processing.

Latex Mixing

Latices (= aqueous dispersions) of polymers are mixed at lower temperatures and smaller shear rates than melts. The resulting intimate mixture of latex *particles* is preserved on coagulation. However, homogenization of *molecules* can be achieved only by melting, which causes the same problems as the mixing of melts.

Latex mixing does not produce polymer radicals and therefore neither grafting nor crosslinking. Hence, microphases are neither chemically nor physically interconnected. In contrast to polymer blends of *cis*-1,4-poly(butadiene) and poly(styrene) from melt blending or *in-situ* polymerizations (see below), polymer blends from latex mixing therefore have only small notched impact strengths.

Graft copolymers as *compatibilizers* for blends from latex mixing are obtained in the formation of **ABS in emulsion**. In this process, one lets styrene diffuse into dispersed butadiene-containing rubbers. The subsequent free-radical polymerization leads to grafted dispersions that are mixed with a separately produced dispersion of styrene/acrylonitrile copolymers (SAN). The resulting blend is precipitated. This process is more expensive and less effective than *in-situ* polymerization and is practically no longer used.

Solution Mixing

In this method, a polymer 2 is dissolved in a solvent 1 (solution 2+1) and then combined with a separately prepared solution of a polymer 3 in the same solvent 1 (solution 3+1). The resulting mixture 3+2+1 is homogeneous in certain ranges of concentrations and temperatures that depend on the interaction parameters χ_{12}, χ_{13}, and χ_{23}. According to theoretically calculated spinodals, solubilities at high polymer concentrations are controlled by the polymer-polymer interaction parameter χ_{23} and those at small polymer concentrations by the differences $\chi_{12} - \chi_{23}$ and $\chi_{13} - \chi_{23}$, respectively.

The homogeneity of the polymer mixture cannot be maintained by fast freezing the solution or by evaporating or subliming the solvent. Instead, polymers 2 and 3 demix; their domain sizes become especially large on evaporation.

Concentrating solutions of thermodynamically miscible polymers 2 and 3 in a solvent 1 leads to either one-phase or two-phase systems, depending on the interaction parameters χ_{ij}. One-phase systems are obtained (if no two-phase ranges are encountered) by passing through the whole range of temperatures and polymer concentrations. Two-phase ranges are formed if differences between χ_{12} and χ_{13} are large and a certain total polymer concentration is exceeded. At still higher total polymer concentrations, one-phase regions are favored again. However, the domains are now so large and the diffusion coefficients so small that neither homogeneous concentrated solutions nor one-phase solids can be formed on further concentrating: for kinetic reasons, the system remains two-phase despite thermodynamic miscibility.

For the same components, the three physical methods (mixing of melts, latices, or solutions) deliver different blend morphologies. An example is the mixing of atactic poly(methyl methacrylate) (PMMA) with bisphenol A polycarbonate (PC). PMMA is amorphous and PC is also usually amorphous although it is crystallizable:

– Melt mixing of PMMA and PC at 220°C leads to a polymer blend that shows the two glass temperatures T_G of PMMA and PC, respectively, as well as the melting temperature T_M of PC: the blend must consist of at least 3 phases.

– Mixing of PMMA and PC solutions in tetrahydrofuran (THF) and subsequent boiling off or slow evaporation of THF delivers films with two T_Gs but no T_M of PC.

– Precipitation of PMMA-PC from their THF solution by heptane results in a blend with only one glass temperature.

In-situ Polymerization

Polymer blends can also be produced by polymerization of a monomer A in the presence of a polymer B. The technologically most important case is the polymerization of the monomer of a thermoplastic in a graftable and crosslinkable diene rubber, which may lead to phase reversals (Fig. 10-8). Another important case is the polymerization of a crosslinkable monomer in an already crosslinked polymer, which leads to "interpenetrating networks" (IPN) that are usually not molecularly interpenetrating networks but more or less pronounced microphases of the two network-forming polymers.

In-situ polymerization of a monomer 1 in a thermodynamically miscible polymer 3 does not necessarily lead to one-phase polymer blends since 1 is both a solvent for the pre-existing 3 as well as a solvent for the newly formed polymer 2. Two phases are formed if the difference between interaction parameters, $\chi_{12} - \chi_{13}$, is large. For kinetic reasons, such 2-phase regions may be preserved on further polymerization although the polymers are now thermodynamically miscible (see "solution mixing," above).

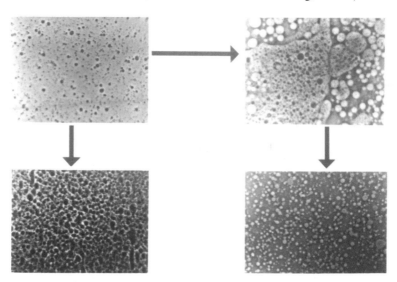

Fig. 10-8 Phase reversal during the free-radical polymerization of styrene in stirred styrenic poly-(butadiene) solutions [5]. In positive phase contrast microscopy (Section 10.4.3), phases with the greater refractive indices appear black (PS) and those with the smaller refractive indices show up white (BR and BR-*graft*-PS).

From top left to bottom left: no phase reversal in unstirred solutions. Poly(styrene) particles PS formed at small monomer conversions grow to large particles in the poly(styrene)-grafted continuous network of poly(butadiene) BR.

From top left to top right and then to bottom right: phase reversal in stirred styrene-BR systems at small monomer conversions (top right) followed by the formation of PS-grafted BR particles in a continuous PS matrix (bottom right).

Courtesy of Dr. Henno Keskkula, The Dow Chemical Company, Midland (Michigan).

All three types of *in-situ* polymerizations to immiscible blends often show phase reversals. An example is the free-radical polymerization of a 5-10 wt% styrenic solution of a poly(butadiene) to form high-impact poly(styrene) (**HIPS**). In the first stage, poly(styrene) is formed in the presence of the unchanged poly(butadiene). Since poly(styrene) is not miscible with poly(butadiene), small poly(styrene) particles are already formed at small monomer conversions. It then depends on whether the system is stirred or not.

At higher styrene conversions in unstirred systems, styrene is grafted onto poly(butadiene) chains that finally become crosslinked. The poly(styrene) particles from the early polymerization stages are then embedded in the continuous network of grafted poly(butadiene) chains (Fig. 10-8).

In stirred systems, phases are reversed if the styrene conversion is greater than 9-12 %, i.e., if the concentration of poly(styrene) becomes comparable to that of the originally present poly(butadiene). After that, grafted and crosslinked poly(butadiene) particles are formed in a continuous poly(styrene) matrix. The smaller the initial poly(butadiene) concentration, the earlier the phase reversal happens. Such "rubber-modified thermoplastics" are very important industrial polymers (see Section 10.4.7).

Whether or when phase reversals occur in unstirred *in-situ* polymerizations also depends on the degree of grafting. An example is the polymerization of unstirred styrenic solutions of poly(chloroprene) (CR) which either leads to phase reversals or not, depending on the initially present type of CR.

The physical structure of blends of thermoplastic polymers poly(S) with their elastomeric graft polymers poly(B-*graft*-S) is controlled by the ratio of the molecular weight M_{gS} of the styrene branches S of poly(B-*graft*-S) to the molecular weight M_{PS} of poly(S) (Fig. 10-9).

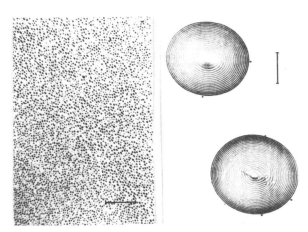

Fig. 10-9 Physical structures of blends that were prepared by mixing poly(styrene)s with a poly(butadiene)-*graft*-poly(styrene) that had been freed from poly(styrene) [6]. The mass average molecular weight of the grafted poly(styrene) chains was $\overline{M}_w$ = 60 000 g/mol. The bar is 2 μm long.

Left: micelles of the graft polymer in the matrix of a poly(styrene) with a molecular weight $\overline{M}_w$ = 5 000 g/mol, which is smaller than the average molecular weight of styrene branches.

Right: onion-like microphases of the graft polymer in the matrix of a poly(styrene) with a molecular weight of $\overline{M}_w$ = 330 000 g/mol, which is greater than the average molecular weight of styrene branches. Black: butadiene chains, white: styrene chains.

Courtesy of Professor Dietrich Braun, German Plastics Institute (DKI), Darmstadt.

Poly(*B-graft*-S) forms spherical associates in the matrix of poly(styrene) if $M_{PS} < M_{gS}$ (Fig. 10-9, left) but lamellae of onion-type entities if $M_{PS} > M_{gS}$ (Fig. 10-9, right). This phenomenon is caused by the different macroconformations of tethered poly(styrene) branches in the onions and "free" poly(styrene) molecules and the incompatibility of polymers resulting therefrom (see Volume III, p. 290).

10.4.3   Analysis of Polymer Blends

Most properties of polymer blends are controlled by the phase properties of the specimen, and one should therefore know the compositions, dimensions, and morphologies of phases, the composition and structure of interlayers (if any), and the thermodynamic state of the systems. For these analyses, many methods are available but some of them are laborious or not clear-cut.

The *appearance* of a specimen may or may not indicate multiphase phase systems. Opaque specimens always consist of multiple phases but the clarity of a specimen is not proof of its homogeneity. Opacity (Section 15.2.2) is caused by light scattering and can only be shown by multiphase systems if phase dimensions are larger than one-half of the wavelength of incident light, phases differ sufficiently in their refractive indices, *and* the dispersed phases are not arranged in such a way that the scattered light interferes destructively.

For example, multiphase blends of polymers with different refractive indices are not opaque if large phases are arranged in layers, for example, in lamellae. Transparent blends are also formed if the dispersed particles have a core-shell structure and this structure has the same (average) refractive index as the surrounding matrix (see Section 15.2.1); the diameter of the particles is not relevant.

For example, one can make transparent, impact-resistant poly(styrene)s from styrene and butadiene monomers although poly(styrene) and poly(butadiene) have quite different refractive indices (1.59 *versus* 1.52) and the dimensions of the resulting dispersed elastomer particles in the poly(styrene) matrix are similar to the wavelength of the incident light. Multiphase blends may thus be transparent or not.

A clear specimen consists of multiple phases if it becomes turbid when temperatures are changed. However, continued clarity does not indicate homogeneity since (a) multiphase systems may be clear and (b) demixing may be very slow because of very small diffusion coefficients.

On the other hand, a turbidity that increases slowly with time may indeed be caused by a slow demixing but may also be produced by the uptake of moisture. For example, blends of poly(styrene) and poly(vinyl methyl ether) are thermodynamically miscible at temperatures up to 112°C. At $T < 112$°C, these blends are transparent at low relative humidities but opaque at higher ones.

Phases with dimensions greater than the wavelength of incident light can be detected by optical microscopy and those with smaller dimensions by electron microscopy, sometimes directly but often only after "dyeing" of one of the components. For example, diene units of poly(olefin)-poly(diene) blends react with osmium tetroxide to form deeply colored esters of osmic acid that appear black under the electron microscope; this technique was used in the electron micrographs of Figs. 10-8, 10-29, and 10-31.

(10-18)

This reaction causes specimens to swell, which leads to shifts in the dimensions of phases that have to be considered for quantitative analyses. RuO_4 is an even stronger oxidizing agent than OsO_4. It is used to "dye" aromatic units, ether and amine groups, and tertiary carbon atoms.

Amorphous polymers form random coils with unperturbed dimensions both above and below their glass temperatures (Volume III, p. 92 ff.). *Coil dimensions* in these states can be determined by small-angle X-ray or neutron scattering (Volume III, Chapter 5). In thermodynamically miscible systems, major components act as "solvents" for minority components and, because miscibilities are usually caused by specific interactions, as thermodynamically good solvents. Coil dimensions are greater in good solvents than in bad ones.

The comparison of coil dimensions of amorphous components in their pure state and in their blends therefore shows heterogeneity (same coil dimensions) or homogeneity (larger dimensions in blends). For example, coils of poly(ε-caprolacton) (PCL) in poly-(vinyl chloride) matrices (PVC) are 35-50 % larger than in pure PCL of the same molecular weight. Hence, PCL and PVC form a thermodynamically miscible blend.

The same phenomenon manifests itself macroscopically in the volumes and *densities*, respectively. Volumes are additive in immiscible systems of components A and B with masses m_A and m_B, i.e., $V = V_A + V_B$. The density ρ of an immiscible system with the total mass $m = m_A + m_B$ is calculated from the densities, $\rho_A = m_A/V_A$ and $\rho_B = m_B/V_B$ and the weight fractions, $w_A = m_A/m$ and $w_B = m_B/m$, of components as

(10-19) $$\rho = \frac{m_A + m_B}{V_A + V_B} = \frac{\rho_B}{w_B + w_A(\rho_B/\rho_A)} \neq w_A\rho_A + w_B\rho_B$$

Densities of immiscible systems do not vary linearly with the mass fraction (expression to the right of the inequality sign) but follow the inverse rule of mixtures:

(10-20) $1/\rho = (w_A/\rho_A) + (w_B/\rho_B)$

For additivity of volumes, i.e., absence of interactions, plots of densities ρ as a function of mass fractions w lead to non-linear, downwardly bent curves for both $\rho_B/\rho_A > 1$ and $\rho_B/\rho_A < 1$ (Fig. 10-10). In systems with interactions of components, total volumes V are not sums of volumes V_A and V_B but contain an additional term, $V = V_A + V_B + \Delta V$. All data points above the additivity curve indicate a contraction (negative ΔV) and therefore attractive forces between A and B. Hence, density measurements allow one to estimate interactions but are problematic for solid systems because voids may be present.

The existence of two or more physical *transformation temperatures* always proves the presence of multiple phases (two glass temperatures, one glass temperature and one melting temperature, etc.). Conversely, the presence of only one transformation temperature is not proof of a single-phase system. Whether or not several transformation tem-

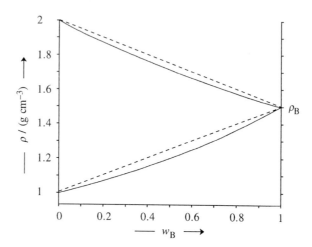

Fig. 10-10 Density of mixtures as a function of the weight fraction of component B.

Solid curves: Dependence of density ρ on the weight fraction w_B of component B ($\rho_B = 1.5$ g/cm^3) in mixtures with components A ($\rho_A = 2.0$ g/cm^3 (top curve) or 1.0 g/cm^3 (bottom curve)) if volumes of A and B are additive. The curves obey the inverse rule of mixing.

Broken straight lines: additivity of densities (no physical meaning). Systems with specific attraction forces between A and B show contraction, and densities are then above the solid lines although densities do not vary linearly with w_B except in very special cases.

perature are observable depends on the size of phases (at least 15 nm in diameter), the relative position of transformation temperatures, and the measuring conditions.

In order to recognize two transformation temperatures, glass temperatures must be at least 30 K apart in differential scanning calorimetry experiments (DSC) and at least 20 K apart in dynamic-mechanical analysis (DMA). The range of a glass temperature T_G should also not be much broader than the sum of the ranges of the glass temperatures of the two components. A broader T_G range indicates partial miscibility and the presence of additional microphases. The minor phase must also be present in a sufficiently high proportion, 15 % for DSC and 10 % for DMA. For example, two glass temperatures were found by DMA in blends of poly(styrene) and poly(oxy-2,6-dimethyl-1,4-phenylene) but only one by the less sensitive DSC.

According to the loss peaks from dynamic-mechanical measurements, atactic poly-(styrene) has a glass temperature at 100°C and emulsion polymerized, amorphous poly-(butadiene) (E-BR) one at –80°C (Fig. 10-11). The graft polymer of styrene on E-BR (E-BR-g-S) shows only one loss peak; hence, it is probably a one-phase system. However, high-impact poly(styrene) (HIPS) from the polymerization of a styrenic solution of E-BR exhibits two loss peaks and consists of two phases.

Fluorescence phenomena are very sensitive to phase separations since excimer fluorescences arises from interactions of groups that are 0.3-0.6 nm apart and radiationless transfers from groups at distances of ca. 2 nm. For example, blends of poly(styrene) and poly(vinyl methyl ether) are thermodynamically miscible over the whole concentration range. At a certain mixing ratio, the fluorescence intensity is constant and independent of temperature up to 120°C ($w_{PS} = 0.01$) and up to 110°C ($w_{PS} = 0.3$), respectively, because phenyl/phenyl contacts are randomly distributed and their concentration in a homogeneous mixture is constant (Fig. 10-12). At temperatures greater than these tem-

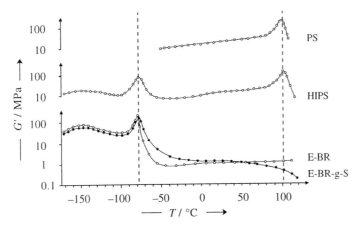

Fig. 10-11 Temperature dependence of the mechanical loss moduli G' of a free-radically polymerized poly(styrene) (PS), a poly(butadiene) rubber (E-BR) by emulsion polymerization, the high-impact poly(styrene) (HIPS) from the polymerization of the styrenic solution of E-BR, and the graft polymer of styrene on E-BR ((E-BR-g-S) [7]. Broken lines indicate dynamic glass temperatures.

peratures, phase separation occurs which leads to many more phenyl-phenyl contacts and the fluorescence intensity increases drastically.

Mechanical and rheological measurements are far less sensitive and more difficult to interpret. In a first approximation, tensile strengths of blends of compatible polymers are weighted averages of the strengths of both components, whereas such averages pass through minima if the polymers are incompatible (see also Table 10-2).

The composition of phases in multiphase blends can be determined by spectroscopic methods, especially by nuclear magnetic resonance. Thermodynamic parameters of blends can be obtained by many methods, for example, by determination of phase equilibria, inverse gas chromatography, neutron scattering, heats of mixing, absorption of solvent vapor, and many more.

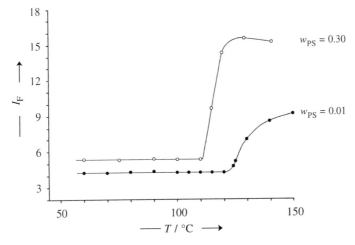

Fig. 10-12 Temperature dependence of fluorescence intensities I_F of blends of poly(vinyl methyl ether) with different weight fractions w_{PS} of poly(styrene) [8].

10.4.4   Phase Morphology

Whether a polymer system is one-phase or multiphase depends on thermodynamic functions as well as the kinetic conditions during preparation. The structure of multiphase systems is determined by the weight ratio, solubility and viscosity of the two components, as well as the shape and packing of the dispersed component.

The weight ratio often determines which component forms the dispersed phase and which the continuous one, provided both components were sufficiently mobile during the preparation of the blend. Dispersed components are often present as spheroidal particles. Since equal-sized spheres can occupy no more than 74 vol% of the space (hexagonal close packing, see p. 77), a component with less than 25 vol% can never form a continuous phase. An exception to this rule exists if one component is immobilized by crosslinking during the preparation of the blend.

Both components are often present in such proportions that either one can form the continuous phase. In melt mixing, a component will self-associate if it is much more polar than the other component. The more polar component (usually the one with the greater solubiliy parameter) will then form the dispersed phase. An exception to this rule exists if one component is much more viscous than the other. In this case, the high-viscosity phase is enveloped by the less viscous one (see also Eq.(10-12)).

In solution mixing, the least soluble component will form the dispersed phase. An example is the benzene solution of a blend of *cis*-1,4-poly(isoprene) (IR) and poly-(methyl methacrylate) (PMMA). On the addition of methanol, IR forms the dispersed phase because methanol is a stronger precipitant for IR than for PMMA. On the addition of petroleum ether, it is just the opposite.

10.4.5   Rubber Blends

Most types of rubber do not have all desired properties and therefore are blended with other rubbers (Table 7-4). The miscibility of rubbers can often be estimated by examining the difference between their solubility parameters δ, often measured in non-SI Hildebrand units H and reported without units (1 H = 1 $(cal/cm^3)^{1/2}$; see Table 11-2). Blends of two rubbers usually form two phases if the difference between the solubility parameters is greater than ca. 0.8 H and only one phase if $H \leq 0.8$ (Fig. 10-13).

Both one-phase and two-phase blends are used commercially. Two-phase blends are especially common in tire treads because they are much more abrasion resistant than single elastomers. Common blends from styrene-butadiene rubbers (SBR) and natural rubber (NR) have a positive specific enthalpy of mixing of Δh = +1.2 J/g and, because of the negligible entropy term (Section 10.2.2), also a positive Gibbs energy of mixing. The also used blends of SBR with *cis*-1,4-poly(butadiene) (BR) have a negative specific enthalpy of mixing of –2.1 J/g and form one-phase blends.

One-phase rubber blends are advantageous over two-phase blends in formulating since vulcanization agents (sulfur, accelerators, peptizers, etc. (p. 264)) may be distributed unevenly in the two types of phases of two-phase blends. On vulcanization, one phase may then be overvulcanized and the other phase undervulcanized which leads to useless products.

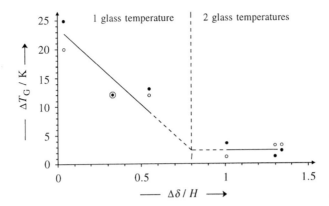

Fig. 10-13 Difference $\Delta T_G = T_{G,i} - T_{G,i,\text{ blend}}$ between the glass temperatures $T_{G,i}$ of pure components I ($\bullet$) and II ($\circ$) and the glass temperatures $T_{G,i,\text{ blend}}$ that were observed for these components in rubber blends as a function of the difference $\Delta\delta$ between the solubility parameters of the two components. Points on top of each other at a fixed value of $\Delta\delta$ belong to one pair I/II. Solubility data are in conventional Hildebrand units, $1\ H = 1\ (\text{cal/cm}^3)^{1/2} = 2.046\ (\text{J/cm}^3)^{1/2}$.

10.4.6 Blends of Thermoplastics

Mixing Rules

Properties of blends of two thermoplastics depend not only on the proportions and properties of the two components and the number of types of phases but for multiphase systems also on the interaction between the phases at the interfaces, the presence of interlayers, and action of compatibilizers, if any.

Blends of *completely immiscible* thermoplastic components without interactions at phase surfaces, no interlayers, and no compatibilizers are simple dispersions of the minority component in the continuous matrix. Their mechanical properties can be described by the simple rule of mixing (Section 9.2). For anisotropic minor components in isotropic matrices, orientation of the dispersed particles relative to the direction of force decides whether properties follow the simple or the inverse rule of mixtures.

For *completely miscible* blends, interactions between components must be considered, in the simplest case by an empirical parameter I_{AB}. The property E of the blend is then given by the fractions f_A and f_B and the properties E_A and E_B of the components as

$$(10\text{-}21) \qquad E = E_A f_A + E_B f_B + I_{AB} f_A f_B$$

The nature of the fractions f is left open (volume fractions, weight fractions, surface fractions). Most authors assume that weight fractions have to be used for mechanical properties and volume fractions for thermodynamic properties.

Eq.(10-21) is a quasi-thermodynamic equation in which the terms $E_A f_A$ and $E_B f_B$ describe the contributions of components A and B ("entropic factors") and the term $I_{AB} f_A f_B$ an interaction ("enthalpic factor") (see the last term of Eq.(10-9)). I_{AB} is therefore often split into the two properties E_A and E_B and a cross term E_{AB}, similar to the treatment of thermodynamic interaction energies, $\Delta\varepsilon = \varepsilon_{12} - (1/2)(\varepsilon_{11} + \varepsilon_{22})$ (see Section 10.2.2). The cross term is identical with the value of E at $f_A = f_B = 1/2$:

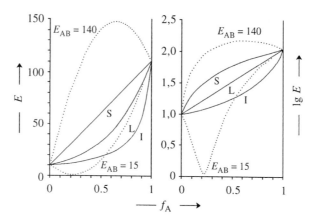

Fig. 10-14   Broken curves: Dependence of properties E on the fraction f_A of the component A in mixtures of A with $E_A = 110$ and B with $E_B = 10$ and the cross terms $E_{AB} = 140$ ($I_{AB} = 320$) or $E_{AB} = 15$ ($I_{AB} = -180$), respectively. Left: normal representation; right: logarithmic E axis.
   Solid curves: simple (S), logarithmic (L), and inverse (I) rules of mixing for immiscible systems. Curves for S correspond to $I_{AB} = 0$ in Eq.(10-22).

(10-22) $I_{AB} = 4 E_{AB} - 2 E_A - 2 E_B$

   The parameter I_{AB} equals zero for $E_{AB} = E_A = E_B$, which leads to the simple rule of mixing. Values of $I_{AB} > 0$ lead to "synergistic" effects: at any fraction f_A, properties E are larger than predicted by the simple rule of mixing (Fig. 10-14). However, the maximum of the curve is not at $f_A = 1/2$ but at a value of f_A that is controlled by E_{AB}.

   "Antagonistic" effects occur if $I_{AB} > 0$. Depending on the value of I_{AB}, values of E are lower than those predicted by the inverse rule of mixing for small f_A and between the values for the logarithmic and inverse rules of mixing at greater values of f_A. The latter values are often interpreted as "non-antagonistic." It is questionable whether such effects are really purely energetic. If not, then it is linguistically incorrect to call values of $E_{AB} > 1$ syn*erg*istic effects (G: *erg* = energy!).

   According to literature, Eq.(10-21) is able to describe *all* properties of miscible blends (thermodynamic, mechanical, rheological, electrical, optical, etc.). However, these statements are usually based on data from fairly narrow concentration ranges and/or from blends of components with fairly similar properties (see below). For totally immiscible components, one has to use the equations that are presented in Chapter 9.

   Similar empirical, semi-empirical, or theoretical equations have been presented for many special properties (mechanical, rheological, thermal, etc.) or for all properties, using different types of specific weights (volume, weight), direct properties or their inverse values or logarithms, additional orientation factors, etc. Whether or not such equations are physically meaningful depends on the properties and very often also on the measuring conditions. In very many cases, they are just useful for interpolations (see below).

Two Amorphous Polymers

   Commercially useful blends of two amorphous polymers are relatively rare; an example is the 50/50 wt/wt blend of atactic poly(styrene) (PS) and atactic poly(vinyl methyl ether). However, some polymers are crystallizable but do not crystallize during conven-

Table 10-3 Some commercially used blends of two amorphous thermoplastics. M = thermodynamically miscible, C = mechanically compatible in the industrially used range of compositions.

Polymer A	B	Miscibility	Improvement	Applications
PVC	chlorinated PE	M	reduction of plasticizer exudation	cable insulation, swimming pool linings, shoes
PVC	MeSAN	M	processing, heat distortion temp.	house sidings
PVC	PMMA	C		
PPE	PS	M	impact strength, processing temperature, cost	electronic casings

tional processing or in a mixture with other polymers. Examples are blends of poly(styrene) (PS) and poly(oxy-2,6-dimethyl-1,4-phenylene) (PPE) and blends of atactic poly(vinyl chloride) (PVC), and poly(ε-caprolactone) (PCL). Because of difficult crystallizations, published melting temperatures of these polymers vary widely: 262-307°C (PPE), 212-310°C (PVC), and 55-64°C (PCL). Blends of such quasi-amorphous polymers with other quasi-amorphous or amorphous polymers are usually amorphous (no melting temperatures) but may crystallize on annealing.

Blends of two amorphous polymers are often completely miscible. They are immiscible if interaction parameters of their components are unfavorable. In the latter case, polymers of such blends can be made compatible by addition of compatibilizers or by mechanical (mixing processes) or chemical (polymerization) *in situ* formation of graft polymers that act as compatibilizers. Commercially, both miscible and compatible blends are used (Table 10-3).

Miscible (1-phase) polymer blends have negative or very slightly positive interaction parameters χ (negative energy of mixing). They may demix on lowering the temperature (UCST) and/or increasing it (LCST) (see Section 10.2.3) and are only partially miscible above LCST and below UCST, i.e., in certain concentration ranges. Examples include:

UCST	oligomers	poly(isobutylene)/poly(dimethylsiloxane), poly(ethylene oxide)/poly(propylene oxide)
LCST	specific interactions between components	poly(vinyl methyl ether)/poly(styrene)
UCST + LCST	chem. similar polymers	poly(styrene)/poly(*o*-chlorostyrene)

So far, neither a UCST nor an LCST has been found for the blend of PS + PPE. In principle, this blend is therefore thermodynamically miscible. However, it is also difficult to homogenize by melt mixing and therefore only compatible.

Blends of miscible amorphous polymers such as PCL and PVC show only one glass temperature, which varies monotonously with the composition of the blend (Fig. 10-15). This case can be described by both the Couchman equation, Eq.(10-10), with $T_G = f(w_A)$, and the Kelley-Bueche equation, Eq.(10-23), with $T_G = f(\phi_A)$,

$$(10\text{-}23) \quad T_G = \frac{K\phi_A T_{G,A} + \phi_B T_{G,B}}{K\phi_A + \phi_B}$$

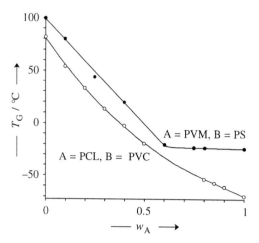

Fig. 10-15 Dependence of the glass temperature T_G on the weight fraction w_A of the component A in polymer blends of poly(ε-caprolactone) (PCL) and poly(vinyl chloride) (PVC) as well as poly(vinyl methyl ether) (PVM) and poly(styrene) (PS). Experimental values of $\alpha_{L,B} = 5.5 \cdot 10^{-4}$ K^{-1}, $\alpha_{G,B} = 2.4 \cdot 10^{-4}$ K^{-1}, and $\alpha_{L,A} = 6.2 \cdot 10^{-4}$ K^{-1}).

if one sets the constant $k = \Delta c_{p,A}/\Delta c_{p,B}$ of the Couchman equation equal to 1.5 and the constant $K = (\alpha_L - \alpha_G)_B/\alpha_{L,A}$ of the Kelley-Bueche equation equal to 0.5 where $\alpha_i =$ cubic thermal expansion coefficient of A or B in the glassy (G) or liquid (L) state.

In contrast, the glass temperature of the system PVM-PS decreases practically linearly with increasing weight fraction $w_A = w_{PVM}$ in the range $0 \le w_A \le 0.6$ and then becomes constant, indicating a changed phase morphology. The system forms one-phase blends if prepared from solutions in toluene or xylene but 2-phase blends if from solutions in chloroform or trichloroethene. This behavior is probably caused by differences in interaction parameters and passing through miscibility gaps during removal of chlorinated hydrocarbons. On heating above 125°C, blends form two phases.

Mechanical properties of blends of two amorphous polymers depend strongly on their preparation. Blends of bisphenol A polycarbonate (PC) and poly(methyl methacrylate) (PMMA) by solution mixing and subsequent precipitation show additivity of volumes (approximately equal to additivity of densities because of small density differences) (Fig. 10-16), indicating a non-miscible system.

Neither solution mixing nor *in-situ* polymerization of methyl methacrylate solutions of PC is expected to lead to graft polymers. Consequently, tensile moduli of blends decrease linearly with increasing content of PC. However, mixing of PC and PMMA powders and subsequent melting showed a minimum/maximum behavior of the function $E = f(w_{PC})$, probably because the blends were not homogeneous and shearing during stirring may have formed free radicals and graft polymers.

In contrast to the system PC-PMMA, densities of the solution-blended system PS-PPE are not additive but pass through a maximum with increasing weight fraction of PPE (Fig. 10-17). This contraction of volumes points toward specific interactions between PS and PPE, i.e., negative thermodynamic interaction parameters χ. The tensile moduli of these blends are also larger than those from the simple rule of mixing. This variation of E with w_{PPE} cannot be described by Eq.(10-22) with a constant parameter I_{AB}, probably because of a concentration dependence of I_{AB}.

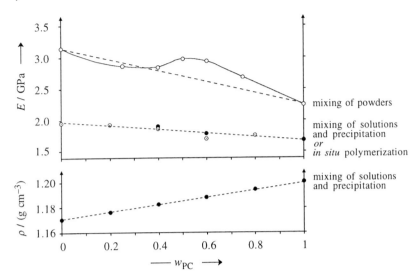

Fig. 10-16 Dependence of tensile moduli E and densities ρ of blends of a bisphenol A polycarbonate (PC) and a poly(methyl methacrylate) (PMMA) on the weight fraction w_{PC} of the polycarbonate [9,10].

The solid line is empirical. Broken lines correspond to the simple rule of mixtures, which implies approximate additivity of volumes of polymers with similar densities.

Preparation of specimen:
○ mixing of powders and subsequent melting;
○ mixing of solutions and subsequent precipitation by a precipitant;
● *in-situ* polymerization of methyl methacrylate solutions of the polycarbonate.

Polymers for powder mixing differed from those for solution mixing and *in-situ* polymerization.

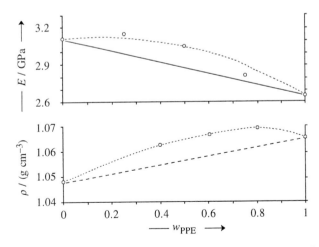

Fig. 10-17 Dependence of tensile moduli E and densities ρ of blends of poly(oxy-2,6-dimethyl-1,4-phenylene) (PPE) and poly(styrene) (PS) from the mixing of toluene solutions of the polymers and subsequent precipitation by methanol [11].

Tensile moduli: the broken line for $E = f(w_{PPE})$ corresponds to Eq.(10-22) with $I_{AB} = 0.6$ and the solid line to the simple rule of mixtures.

Densities: both lines are empirical.

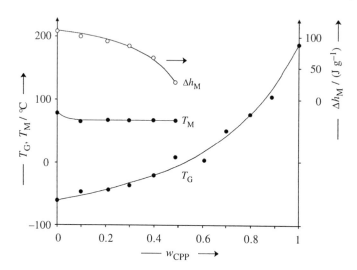

Fig. 10-18 Glass temperatures T_G, melting temperatures T_M, and specific heats of melting Δh_M of blends of chlorinated isotactic poly(propylene) (CPP) and poly(ε-caprolactone) (PCL) as functions of the weight fraction w_{CPP} of CPP [12]. The solid curve for T_G was calculated with the Couchman Eq.(10-10) and $k = 2.3$. Other curves are empirical.

An Amorphous and a Crystalline Polymer

Polymer blends of an amorphous and a crystallizable polymer may or may not have crystalline phases. For example, crystallizable poly(ε-caprolactone) (PCL; $T_M = 64°C$; $T_G = -60°C$) crystallizes in its blends with the non-crystallizable, chlorinated, isotactic poly(propylene) (CPP) in the range 100-50 wt% PCL (Fig. 10-18). The melting temperatures of PCL domains in CPP are only slightly smaller than the melting temperature of PCL itself whereas the specific heat of melting decreases with increasing content of CPP. The glass temperature of the blends increases continuously with increasing content of CPP. The two polymers thus form a partially miscible blend.

Densities of extruded blends of the usually non-crystallizing bisphenol A polycarbonate (PC) and the well-crystallizing isotactic poly(propylene) (PP) are lower then calculated for additivity of volumes. The volume of the blend is therefore greater than the sum of the volumes of its components which points to a decrease of crystallinity of PP regions. This effect cannot come from repulsion forces because volumes are additive in blends of immiscible polymers. For unknown reasons, tensile moduli seem to show a maximum at large PP contents (Fig. 10-19).

Extruded blends of PC with either a non-crystallizing atactic poly(styrene) (PS) or a well-crystallizing high-density poly(ethylene) (HDPE) have moduli that are smaller than predicted by the inverse rule of mixtures (Fig. 10-20). This effect is caused by morphological changes. According to electron micrographs, spheroidal poly(ethylene) domains exist in a continuous PC matrix at small HDPE concentrations. The HDPE domains become rodlike at greater HDPE contents. At still higher HDPE concentrations, phase inversion occurs and the blend now contains rodlike PC domains in a continuous HDPE matrix. Such morphological changes are not considered by any of the reported empirical or theoretical equations for the dependence of properties on composition.

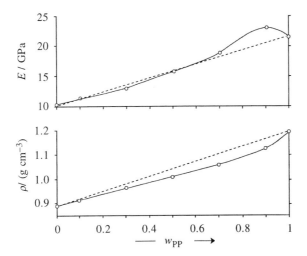

Fig. 10-19 Tensile moduli E and densities ρ of extruded blends of granulates of bisphenol A poly-carbonate (PC) and it-poly(propylene) (PP) [13]. Broken lines indicate the simple rule of mixtures for moduli and an approximate additivity of volumes for polymers with slightly different densities.

The presence of rodlike domains in the PC-HDPE composite of Fig. 10-20 is caused by the shear gradient during extrusion. At small HDPE contents, phase separation of the immiscible components leads to domains of HDPE in the matrix of PC. These domains are spherical because of interfacial tension. At larger HDPE contents, domains become larger but shear forces can now somewhat overcome interfacial forces and spheres are now deformed to rods that are partially oriented in the shear direction. At still greater HDPE concentrations, phase inversion occurs, most likely via an intermediate lamellar stage.

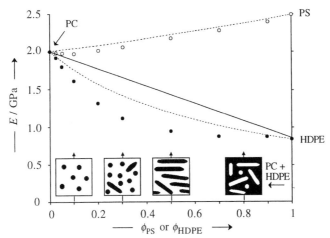

Fig. 10-20 Tensile moduli E of extruded blends from powders of a bisphenol A polycarbonate (PC) with either atactic poly(styrene) (PS) or high-density poly(ethylene) (HDPE) as a function of the volume fraction ϕ_{PS} of PS or ϕ_{HDPE} of HDPE [14]. Solid line: simple rule of mixtures (for clarity, not shown for PS); broken lines: inverse rule of mixtures. Insert: schematic representation of the phase morphology of PC/HDPE at various PC contents; black: PE, white: PC.

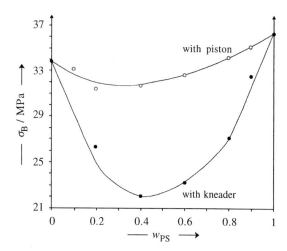

Fig. 10-21 Dependence of fracture strength σ_B of blends of poly(styrene) (PS) and high-density poly-(ethylene) on the weight fraction w_{PS} of poly(styrene) [15]. Test bars were prepared by injection molding machines with a piston or a kneader. By kind permission of Academic Press, New York.

Hence, phase morphologies of blends can be greatly affected by processing conditions (Fig. 10-21). Injection molding machines with pistons or kneaders, single-screw and double-screw extruders, etc., all mix with different intensities so that the resulting phases of blends differ in size, shape, and orientation (if any). Machines also shear differently, which leads to different proportions and types of graft polymers that act as compatibilizers (see below). Fracture strengths of such blends differ considerably for injection-molded blends of poly(styrene) and high-density poly(ethylene) (Fig. 10-21).

The effect of phase morphology on mechanical properties is especially pronounced for blends of amorphous polymers and glassy liquid-crystalline polymers (LCP) (Fig. 10-22) since the macromolecules of the former are present as unperturbed coils (isotropic phase) whereas those of the latter exist in liquid-crystalline domains (LC) with preferential directions of segment axes (anisotropic phase).

The statistical-thermodynamic theory predicts that coils are expelled from the (usually nematic) LC domains if specific interactions are absent. Small proportions of LCP should therefore lead to LC domains in an amorphous matrix. On drawing, LCP segments will orient themselves in the draw direction and the tensile moduli in the draw direction obey the simple rule of mixing (Fig. 10-22), similarly to the drawing of fiber-reinforced polymers parallel to the fiber direction (Fig. 9-7).

On compression molding, shear flow leads to spherical LCP domains. The domains are smaller the greater the shear rates in simple shear flow. Tensile moduli of compression molded plastics parts are therefore much lower than those prepared by drawing. They are below the values predicted by the inverse rule of mixing and seem to exhibit a shallow minimum (Fig. 10-22).

Commercial blends of amorphous and semicrystalline polymers are used for many purposes; Table 10-4 lists some examples. Most of these blends are thermodynamically immiscible or only partially miscible. Their phases must therefore be mutually anchored by compatibilizers that are either formed *in situ* during polymerization or processing or are added purposely.

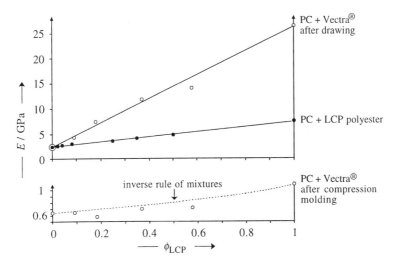

Fig. 10-22 Tensile moduli E of blends of a bisphenol A polycarbonate (PC) and liquid-crystalline polymers (LCP) as functions of the volume fraction ϕ_{LCP} of the LCP.
(O) Blend of PC and Vectra® RD 500 (from 52 mol% 4-hydroxybenzoic acid, 28 mol% 6-hydroxy-naphthalene-2-carboxylic acid, 10 mol% terephthalic acid, and 10 mol% hydroquinone);
(●) Blend of PC and a polyester from a t-C_4H_9 substituted hydroquinone and a terephthalic acid derivative, HOOC–$C_6H_3(CH_2CH_2C_6H_5)$–COOH.

Models

The many possible effects make one realize why it is not yet possible to describe *all* mechanical properties of *all* blends from amorphous and crystalline polymers for the *whole range* of compositions with *one theoretical model*. Instead, two approaches are often used. In one of these, one simply uses a polynomial with variables raised to integral powers, for example, $E = E_A(1 + 2.5\ \phi_A + 6.2\ \phi_A^2 + ...)$, to describe properties E, in analogy to the Einstein-Batchelor equation, Eq.(9-19).

The treatment of Takayanagi assumes that some portions of heterogeneous blends are under uniform stress and other portions are under uniform extension. Blends may be amorphous/amorphous, amorphous/crystalline, or crystalline/crystalline.

Table 10-4 Commercial blends of an amorphous and a crystalline thermoplastic. M = thermodynamically miscible, C = mechanically compatible in the technically used range of compositions. Improvements refer to both individual components, except where noted.

Polymers A	B	Misci-bility	Improvement	Application
PC	PE	C	flow properties (PC only)	automotive parts
PC	PET or PBT	C	processing, chemical resistance	bumpers, casings for office equipments, tubes and pipes
PET	PMMA	C	less deformation, less shrinkage (PET only)	electrical/electronics
PVDF	PMMA	M	clarity (PVDF only), resistance to chemicals and UV light (PMMA only)	sheets for outdoor applications
PA	PO-CoPM	C	impact strength	

The mathematical treatment leads to an equation with one adoptable parameter α:

$$(10\text{-}24) \quad \frac{E}{E_A} = \frac{\phi_A E_A + (\alpha + \phi_B)E_B}{(1 + \alpha\phi_B)E_A + \alpha\phi_A E_B} \quad ; \quad E = \frac{E_B + (1/\alpha)(\phi_A E_A + \phi_B E_B)}{(1/\alpha) + E_B[(\phi_A/E_A) + (\phi_B/E_B)]}$$

The Takayanagi equation, Eq.(10-24), left, is identical with the Halpin-Tsai Eq.(9-17) if one sets $E_A = E_M$, $E_B = E_F$, $1/\alpha = K_1$, and $\phi_A + \phi_B \equiv 1$. It uses only one adjustable parameter α and can also describe the S-shaped curves that are often observed for plots of $\lg E = f(\phi_B)$ (Fig. 10-23). Eq.(10-24) becomes the simple rule of mixing for $\alpha = 0$ (uniform extension), $E = \phi_A E_A + \phi_B E_B$, and the inverse rule of mixing, $1/E = (\phi_A/E_A) + (\phi_B/E_B)$, for $\alpha = \infty$ (uniform stress). Eq.(10-24) applies only for the absence of flow, i.e., for moduli but not for large deformations.

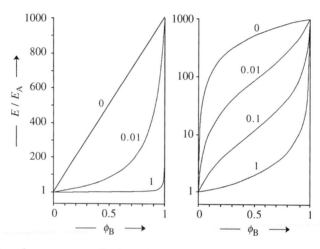

Fig. 10-23 Effect of parameter α on the dependence of reduced moduli E/E_A on volume fractions ϕ_B of components B according to the Takayanagi model, Eq. (10-24). $E_B = 10^3 E_A$ with $\alpha = 0$, 0.01, 0.1 or 1 as $E/E_A = f(\phi_B)$ (left) and $\lg E/E_A = f(\phi_B)$ (right).

Compatibilizers

Coil-forming polymer molecules at phase interfaces in heterogeneous systems try to move into the other phase because this would increase the combinatorial entropy of the system. However, the larger the interaction parameter χ, the stronger is the repulsion and the less probable such a movement. Interlayers between phases are therefore very thin (<1 nm) and far smaller than diameters of polymer coils (>10 nm). For this reason, polymer phases cannot be mutually anchored by their constituent polymer molecules alone.

However, such anchorings can be provided by compatibilizer molecules. Compatibilizers are block or graft polymers that improve the adherence of one phase to the other (p. 44). They are either deliberately added to heterogeneous blends (Section 3.4.4) or are formed in situ if a monomer is polymerized in the presence of a polymer with different monomeric units. Compatibilizer molecules contain two types of long segments with different chemical constitution of which one type can reside in one type of the phases of a blend and the other type in the other type (Fig. 10-24).

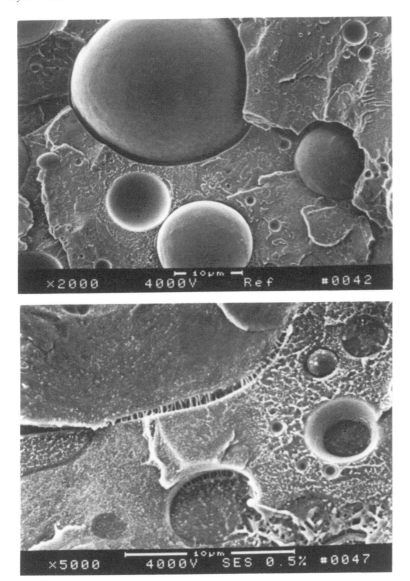

Fig. 10-24 Scanning electron micrographs of a 30/70 wt/wt blend of a poly(ethylene) (PE) and a poly(styrene) (PS) [16]. PE and PS were blended for 10 min at 150°C in a Brabender kneader.

Top: PE forms spherical domains in the continuous matrix of PS. On cooling the blend, spheres shrink because of PE crystallization and mostly have no direct contact with the matrix

Bottom: addition of 0.5 wt% of a styrene-*block*-ethylene-*block*-styrene copolymer to the blend causes triblock polymer molecules to assemble in the phase surfaces and form "threads" between phases, i.e., to act as a compatibilizer.

Compatibilizer molecules contain long, chemically different segments A_{seg} and B_{seg} such as blocks of diblock polymers, segments of segmented polymers, or long side chains of comb-like polymers (Fig. 10-25). Segments A_{seg} will reside in A phases and segments B_{seg} in B phases, if A_{seg} is thermodynamically miscible or can cocrystallize with A components of blends (ditto for B_{seg} in B components).

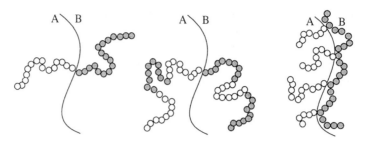

Fig. 10-25 Compatibilizer molecules of diblock polymers (left), segmented bipolymers (center), and graft polymers (right) at the interfaces (solid lines) of phases A and B. Segments of units O are miscible or cocrystallizable with phase A and segments of units ◑ with phase B of the blend.

Thermodynamic miscibility is always achievable if segments A_{seg} of the compatibilizer molecules and molecules A of the blends have the same chemical constitution *and* are present in the same macroconformation; however, this is not true for the pairing of random coil-forming molecules A with tethered chains A_{seg} of compatibilizer molecules. Interestingly, A and A_{seg} do not need to be chemically identical; it suffices that they are thermodynamically miscible. In the following, only compatibilization in amorphous systems will be discussed since cocrystallizations are very rare.

Diblock polymers (Fig. 10-25, left) are the most efficient of all types of compatibilizers because in principle each of their blocks can find an accommodating phase. In contrast, steric reasons prevent some segments of a segmented bipolymer (Fig. 10-25, center) to enter the appropriate phase. Segmented bipolymers are therefore less efficient than diblock polymers. Graft polymers behave similarly to segmented bipolymers (Fig. 10-25, right). In each case, efficient compatibilizer molecules reside at interphases.

Efficiencies of compatibilizers thus decrease with increasing number of segments per molecule

$$\text{diblock} > \text{triblock} > \text{multiblock} > \text{random bipolymer}$$

and also with increasing crowding at segment-segment bonds:

$$\text{diblock} > \text{3-star} > \text{comb polymer (graft polymer)}$$

Segments should also have the same molecular weights and be sufficiently large so that they can entangle with the molecules of the host phase.

The required mass m_C of compatibilizer C per volume V_{blend} can be calculated as

$$(10\text{-}25) \qquad \frac{m_C}{V_{blend}} = \frac{3\,\phi_A M_C}{A_C R_A N_A} \quad ; \quad N_A = \text{Avogadro constant}$$

from the volume fraction ϕ_A of blend component A_m with domain radii of R_A that are spherically dispersed in the matrix of components B_n if the compatibilizer segment with the molar mass M_C occupies an area A_c on the surface of the sphere of A_m.

For example, for a volume fraction $\phi_A = 0.1$ of the blend component A_m and a radius $R_A = 1$ μm of the spherical domain one needs a concentration of the compatibilizer of $2 \cdot 10^{-3}$ g/cm^{-3} (≈ 0.2 %) if the molar mass of the segment is $M_C = 1 \cdot 10^5$ g/mol and its contact area, $A_C = (5$ nm$)^2$. The greater the molar mass of the compatibilizer, the more compatibilizer is needed.

In equilibrium, a certain percentage of compatibilizer molecules reside at the interface and the other percentages in the interiors of either the dispersed phase or the continuous phase. In each of these two phases, "alien" blocks of compatibilizer molecules tend to micellize above a certain critical concentration, similar to that of amphiphiles in water. At concentrations of the compatibilizer molecules C in the dispersed phase above the critical micelle concentration, micelles are formed which reduces the proportion of C molecules that is available for the interface between the phases. The same phenomenon is present in the continuous phase. For these reasons, efficiences of compatibilizers are less than proportional to their concentrations above a certain compatibilizer concentration. This effect is more pronounced for compatibilizing diblock polymers than for triblock ones because the former have lower critical micelle concentrations.

The efficiency of compatibilizers is also diminished if the high viscosity of the matrix reduces the ability of compatibilizer molecules to diffuse to the interface of the phases. For these kinetic reasons, compatibilizer molecules should have large diffusion coefficients, i.e., low molecular weights (not high ones) and be highly branched (graft and star polymers) and not linear (block polymers). Kinetic demands are therefore just the opposite of thermodynamic ones and a compromise between the two demands must be sought in each particular case.

Two Crystallizable Polymers

Blends of two crystallizable polymers are relatively rare. Examples are the compatible blend of it-poly(propylene) and poly(1-butene) for films and sheets and the blend of poly(butylene terephthalate) and poly(ethylene terephthalate) for gasoline pipes and electrical/electronic parts.

Two crystallizable polymers rarely form isomorphous blends. An example is the blend of poly(vinyl fluoride) and poly(vinylidene fluoride) (PVF + PVDF) (Fig. 10-26) which is isomorphous just like the corresponding copolymer of vinyl fluoride and vinylidene fluoride (P(VDF + VF)).

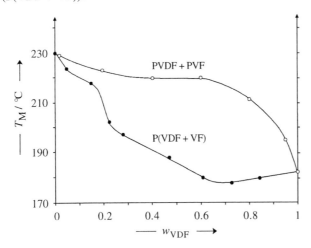

Fig. 10-26 Melting temperatures of copolymers P(VDF + VF) from vinylidene fluoride (VDF) and vinyl fluoride (VF) and of blends of the two homopolymeres PVDF and PVF as a function of the weight fraction w_{VDF} of VDF units [17].

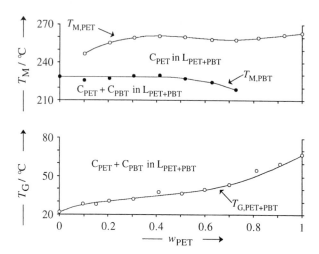

Fig. 10-27 Melting temperatures T_M (top) of poly(ethylene terephthalate) (PET) and poly(butylene terephthalate) (PBT) crystallites in PET/PBT blends and glass temperatures T_G (bottom) of blends as a function of the weight fraction w_{PET} of PET in the PET-PBT blends [18].

Most solid blends of crystallizable polymers are not isomorphous. Molten blends of poly(ethylene terephthalate) (PET) and poly(butylene terephthalate) (PBT) are homogeneous. On cooling, PET crystallites appear at weight fractions of $w_{PET} > 0.09$. The melting temperatures of the crystallites first increase with increasing PET concentrations, then decrease somewhat, and finally increase again at higher PET weight fractions .

Below these melting temperatures, PET crystallites C_{PET} reside in a liquid $L_{PET+PBT}$ whose composition differs from that of the original one. On further cooling, PBT crystallites appear if $w_{PET} < 0.73$ (Fig. 10-27). The melting temperature of these PBT crystallites is constant and identical with that of pure PBT if $w_{PET} < 0.5$. Below the melting temperatures $T_{M,PBT}$, the system consists of crystallites $C_{PET} + C_{PBT}$ in a liquid $L_{PET+PBT}$. The crystallites obviously form a kind of physical network so that the blend appears as "solid." The liquid $L_{PET+PBT}$ solidifies below the glass temperature $T_{G,PET+PBT}$.

Literature calls such blends "miscible." In fact, they contain PET and PBT crystallites as well as an isotropic glass $G_{PET+PBT}$. Depending on the cooling conditions, crystallites are either dispersed or united into a kind of physical network.

Molecular Polymer Composites

So-called molecular polymer composites are blends of "chemically similar" polymers of which the dispersed polymer contains rodlike segments and therefore has liquid-crystalline character whereas the matrix consists of a polymer with coil-like molecules. An example is the blend of *trans*-poly(phenylene bisbenzothiazole) (PBTZ) and a "chemically similar" polyimide (ABPBI).

Ideally, "rodlike" PBTZ molecules are molecularly dispersed in the matrix of ABPBI, which should lead to a kind of molecular fiber reinforcement and good adherence between fibers and matrix. In reality, only segments of PBTZ molecules are rodlike but not the whole molecule. For mixtures of amorphous polymers with such liquid-crystalline polymers, lattice theories of athermal solutions (absence of specific interactions) predict demixing of the system (Volume III, p. 314 ff.). Blends of PBTZ and ABPBI thus do not form molecular solutions but bundles or domains of PBTZ in ABPBI.

Above a certain critical polymer concentration c_{crit} and below a transformation temperature T_{nem}, dissolved lyotropic polymers such as PBTZ convert from isotropic solutions to (usually nematic) mesophases with liquid-crystalline structures.

From ternary systems of an LC polymer (PBTZ), a coil-forming polymer (ABPBI), and a solvent (methane sulfonic acid) one obtains after solvent removal similarly above c_{crit} and below T_{nem}, respectively, phase-separated (heterogeneous) molecular composites that consist of liquid-crystalline PBTZ domains of several micrometers diameter in an ABPBI matrix.

At concentrations below c_{crit} and above T_{nem}, respectively, one gets similarly "molecularly homogeneous" composites consisting of molecular bundles of PBTZ molecules of several nanometers diameter in an ABPBI matrix that at equal fracture strengths have greater tensile moduli than their heterogeneous counterparts (Fig. 10-28).

The fracture strengths $\sigma_{B,\parallel}$ of these molecular composites are directly proportional to their tensile moduli $E_\parallel$, $E_\parallel = K\sigma_{B,\parallel}$, where K is a constant with a maximum theoretical value of 0.095 according to semi-theoretical calculations (Volume III, p. 627). Fibers and films of homogeneous PBTZ-ABPBI composites have values of $K \approx 0.01$ (≈ 11 % of theory) and those of heterogeneous ones, $K \approx 0.03$ (≈ 32 % of theory). Ultradrawn poly-(ethylene)s follow the theoretical relationship, $E_\parallel = 0.095\ \sigma_{B,\parallel}$.

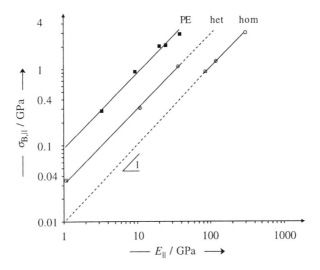

Fig. 10-28 Fracture strengths $\sigma_{B,\parallel}$ as a function of tensile moduli $E_\parallel$ of molecularly homogeneous (hom) and heterogeneous (het) composites from PBTZ and ABPBI (see text) [19], as compared to those of ultradrawn specimens of an ultra-high molecular weight poly(ethylene) (PE) [20].
■ ultradrawn PE, O ABPBI fibers, ● PBTZ fibers,
⊕ fibers from 30 wt% PBTZ in ABPBI, ⊞ films from 30 wt% PBTZ in ABPBI.

10.4.7   Rubber-modified Thermoplastics

Addition of rubbers or elastomers to thermoplastics delivers various types of blends, depending on the components and processing conditions. For example, mechanical mixing of thermoplastic-elastomeric poly(styrene-*block*-(ethylene-*co*-butylene)-*block*-styrene) and thermoplastic isotactic poly(propylene) results in a thermoplastic elastomer with improved properties. On mechanical mixing, mixtures of a chemically crosslinked rubber and a thermoplastic with or without added crosslinking agents deliver semi-interpenetrating networks (dynamic vulcanization; Section 7.5.3).

The most important type of such blends is dispersions of 5-20 % of rubber particles in a thermoplastic (thermosets are rare). Morphologies of these blends depend strongly on manufacturing conditions (Fig. 10-29). Addition of rubber increases fracture elongations and notched impact strengths and lowers tensile moduli and fracture strengths: the plastic toughens but the glass temperature does not change. Depending on the components, the manufacturing conditions, the manufacturer, and the country such blends are known as **rubber-modified**, **rubber-reinforced**, **rubber-toughened** or **toughened plastics** or, according to the most important improvement of properties, as **impact-modified plastics** or simply as **impact** or **high-impact plastics**. Deliberately added rubbers are therefore also called impact modifiers or tougheners, depending on the plastic (p. 44).

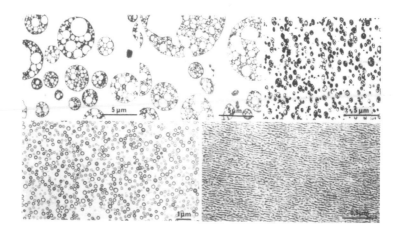

Fig. 10-29 Morphologies of rubber-modified poly(styrene)s (PS) [21].
   Top left: convential high-impact poly(styrene) (HIPS);
   Top center: HIPS with decreased stress cracking (Volume III, p. 614);
   Top right: HIPS with high surface gloss;
   Bottom left: capsule morphology of a HIPS with increased transparency;
   Bottom right: finely dispersed rubber phase in a glass-clear HIPS.
By kind permission of Hanser-Verlag, Munich.

The conversion of relatively inexpensive but not very tough standard thermoplastics to impact-resistant materials is technologically and commercially very important. Impact modifying is most often used for thermoplastics but is not restricted to this class of plastics alone. It can also be used for thermosets. An example is the modification of epoxy resins by telechelic oligobutadienes with carboxy endgroups.

Rubber-reinforced Styrenics

Rubber-modified poly(styrene)s (IPS, HIPS, ABS) are prepared by *in-situ* polymerization of styrene in a styrene solution of an unsaturated rubber. All other rubber-modified thermoplastics are obtained by melt mixing of thermoplastics and rubbers.

These two types of manufacturing differ because of the differences in the syntheses and chemical behavior of the components of polymer blends. Styrene is a good solvent for various unsaturated rubbers and can be polymerized easily by free-radical polymerization. Monomers for all other rubber-modifiable thermoplastics are non-solvents for rubbers, and their polymers must be obtained by special free-radical polymerizations (for example, PVC), polycondensation (for example, PC), or by coordination polymerization (for example, PE or PP). In addition, free-radical polymerization of a monomer such as styrene in the presence of diene rubbers leads to graft polymers that act as compatibilizers for the rubber domains in the poly(styrene) matrix. Too little grafting by *in-situ* polymerization or melt mixing renders the components of the blend incompatible: the blend disintegrates to crumbs. Too much grafting leads to a brittle blend.

Rubber-modified poly(styrene)s from the polymerization of styrene solutions (S) of rubbers (R) such as poly(butadiene) rubbers (BR; both medium cis from lithium butyl-initiated polymerizations and high cis from Ziegler-Natta polymerizations), styrene-butadiene rubbers (SBR), or unsaturated ethylene-propylene rubbers (EPDM) are known as **impact poly(styrene)s (IPS)**, **high-impact poly(styrene)s (HIPS)**, or **toughened poly(styrene)s (TPS)** (see also p. 433). These blends are usually obtained by bulk polymerization of styrene in the presence of rubbers; the polymerization is mostly thermally initiated (Volume I, p. 314). In the less-used bulk–suspension process, the polymerization is started as a bulk polymerization with added peroxide initiators and then continued as a suspension polymerization in water. The advantage of the latter process is the easy removal of process heat. In principle, direct suspension polymerizations are possible but they are not commercial because the dispersed organic phase is not mixed thoroughly enough for phase inversion (Fig. 10-8) and the desired domain size (Fig. 10-29).

The synthesis of **acrylonitrile-butadiene-styrene** plastics (**ABS**) is mainly by emulsion polymerization. In this process, an initial aqueous dispersion of poly(butadiene) with the desired particle size is free-radically grafted by a mixture of styrene and acrylonitrile monomers. Alternatively, the graft copolymer is prepared separately and then mixed with a styrene-acrylonitrile polymer (SAN) to achieve the desired composition.

ABS is also synthesized by suspension or bulk polymerization. In suspension polymerization, a solution of poly(butadiene) in a mixture of styrene (S) and acrylonitrile (AN) is polymerized discontinuously by an oil-soluble free-radical initiator to 30-35 % conversion of S + AN. The solution is then dispersed in water and subsequently polymerized at steadily increasing temperatures, using free-radical initiators with different decomposition rates.

Bulk polymerizations to ABS polymers are especially important in the United States. They proceed similarly to that of HIPS, i.e., by polymerization of a mixture of styrene and acrylonitrile in the presence of a butadiene rubber (BR) at temperatures of 100-170°C, either thermally or by added free-radical initiators. Replacement of BR by SBR with less than 40 wt% styrenic units leads to products with good surface gloss.

All of these polymerizations lead to multiphase systems (Section 10.4.3) in which the rubber forms the dispersed phase which in turn may contain poly(styrene) domains (in

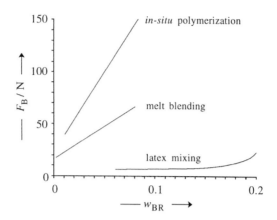

Fig. 10-30 Notched impact strengths F_B of rubber-reinforced poly(styrene) as function of the weight fraction w_{BR} of the butadiene rubber [22]. The blends were prepared by latex mixing or melt blending of poly(styrene) and *cis*-poly(butadiene) or by *in-situ* polymerization of styrene in a styrenic solution of the rubber. By kind permission of Plenum Press, New York.

HIPS) or SAN domains (in ABS). Impact strengths of the various ABS, HIPS, etc., polymers may thus very widely, depending on the proportion, size, size distribution, and structure of the dispersed phases (see Deformation Processes below).

Other Rubber-reinforced Thermoplastics

In-situ polymerization of monomers in the presence of rubbers delivers blends with greater notched impact strengths than those obtained from mixing of melts or latices (Fig. 10-30). This process leads to the highest notched impact strengths but is used only for rubber-reinforced styrenics (IPS, HIPS, ABS). For reasons listed below, all other rubber-modified thermoplastics are obtained by melt mixing of rubbers and thermoplastics:

- Free radicals cannot graft vinyl chloride onto nitrile rubbers (NBR) because of unfavorable copolymerization parameters Q and e (Volume I, p. 410).

- Grafting of monomers onto saturated rubbers such as chlorinated poly(ethylene) is difficult if the rubbers do not contain chemical groups that are easily transferable. Other examples are blends of poly(butadiene)s and butyl rubber and blends of poly(methyl methacrylate) and acrylic rubbers.

- Melt mixing is the only possible process if thermoplastics can be synthesized only by Ziegler-Natta or similar polymrizations since these processes do not involve chain transfer processes.

In principle, melt mixing can produce polymer radicals and therefore also grafting but the proportion of graft polymers, if any, is usually too small to anchor dispersed rubber domains in the continuous matrix of the thermoplastic. For this reason, compatibilizing block or graft polymers (p. 44) are usually added to melt-mixed rubber-modified thermoplastics.

Commerce uses many different rubber-modified thermoplastics but the scientific and technical literature does not reveal for most of these whether they are miscible, immiscible, or compatible. Some examples are listed in Table 10-5.

Table 10-5 Some commercially used rubber-modified thermoplastics from melt mixing.
Miscibility: C = mechanically compatible, I = immiscible, M = miscible.
Improved properties: C = chemical resistance, F = flame retardance, H = heat resistance, I = impact resistance, PL = plasticization, PR = processing.

Polymers A	B	Miscibility	Improvement of C	F	H	I	PL	PR	Application
PE	ethylene copolym.	I	+			+			films, sheets
PP	EPDM	I				+		+	wire and cable insulation, tubes, bumpers, sealings
PVC	NBR	M					+	+	wire and cable insulation, parts in contact with foodstuffs
PVC	ACR	I				+			interiors of mass transportation systems, casings of devices
PVC	ABS	C		+	+	+		+	same as PVC-ACR
PVC	CPE	I				+	+		wire and cable insultation, automobile interiors, shoes, pond liners, tubes
PC	ABS	C			+	+		+	casings, automotive parts
PPE	ABS	M			+	+		+	casings

Deformation Processes

On drawing, thermoplastics and their rubber-modified counterparts are essentially deformed by the same mechanism. However, rubber-modified thermoplastics behave quite differently than non-modified ones. For example, atactic poly(styrene) is a rigid-brittle thermoplastic whose tensile stress increases almost linearly with increasing strain (PS(t) in Fig. 5-3). The polymer develops crazes at a tensile stress of ca. 35 MPa but has no yield point and fractures at 0.9 % elongation with a fracture strength of ca. 55 MPa.

In contrast, high-impact poly(styrene) (HIPS) has an upper yield point at ca. 2 % elongation and a shallow lower yield region, followed by strain hardening until it ruptures at ca. 20 % elongation (Fig. 10-31, right). The tensile modulus and the fracture strength of HIPS are both smaller than that of PS but the notched impact strength is much larger. The upper yield stress of ABS polymers is even greater than that of HIPS, especially if the disperse rubber particles are small (Fig. 10-31, left).

The strong increase of fracture elongation cannot come from the dispersed rubber phase because rubber domains are surrounded by the rigid poly(styrene) matrix. It is caused by processes within the matrix, i.e., either by shear deformation, by craze formation, or a combination thereof. The first two processes can be distinguished by measuring the volume of the specimen: the volume remains constant for shear deformations but increases for craze formation because the latter produces voids (Volume III, p. 580). Volume changes can be measured by creep experiments which show that rubber-modified poly(styrene) deforms practically completely by crazing, whereas rubber-modified poly(propylene) deforms predominantly by shear flow. These behaviors are not influenced by the magnitude of the applied stress but they contrast with that of rubber-modified poly(oxymethylene) where a small increase of applied stress leads to a large increase of crazing. In the latter polymer, crazing and shear deformation are obviously not independent of each other.

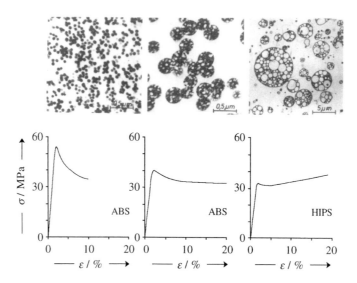

Fig. 10-31 Effect of composition and morphology on stress-strain curves of HIPS (right) and ABS polymers with small (center) and large (left) rubber particles [23a]. The upper yield strength increases in the order HIPS < ABS (small rubber particles) < ABS (large rubber particles).

Deformation Mechanisms

Rubber-modified poly(styrene)s and poly(propylene)s behave differently because the glass temperature of the poly(styrene) matrix (100°C) is greater than the temperature during the creep experiments (30°C) whereas that of the poly(propylene) matrix is lower (−10°C). On deformation, PP molecules begin to flow, which is easier the smaller the obstacles. Therefore, impact strengths *de*crease with *in*creasing size of rubber domains.

Shear flow is not possible for the poly(styrene) matrix, which therefore can deform only by multiple crazing. The bigger the dispersed rubber particles, the more they can absorb the energy: impact strengths *in*crease with *in*creasing size of the rubber domains.

The following discussion is concerned with the latter case, the rubber modification of thermoplastics that have glass temperatures above the measuring temperature, i.e., deformations that are dominated by crazing. For example, the notched impact strength of poly(styrene) is small and independent of temperature in the range −100°C to +52°C (Fig. 10-32). Modification of poly(styrene) with *cis*-1,4-poly(butadiene) causes the notched impact strength to start to increase at temperatures above −80°C, i.e., above the glass temperature of this rubber (−90°C). At the same temperature, a white zone becomes visible near the notch, which indicates the formation of crazes. This **stress whitening** grows with increasing temperature. At +12°C, it comprises the whole specimen and the notched impact strength starts to increase dramatically with increasing temperature.

This behavior is caused by the uneven deformation of thermoplastic matrix and rubber particles. The largest deformation is near the equator of rubber particles, which causes stress peaks that are removed by formation of crazes in the matrix. The crazes grow equatorially to the rubber particles (i.e., normal to the stress direction) until they are stopped by an obstacle (Fig. 10-33) which may be another rubber particle or a shear band or because the stress concentration at the tip of the craze becomes too small.

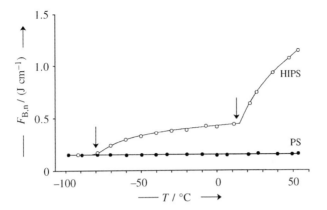

Fig. 10-32 Temperature dependence of the notched impact strength $F_{B,n}$ of poly(styrene) (PS) and rubber-modified poly(styrene) (HIPS) [24]. Arrows indicate the onset of craze formation (stress whitening) at –80°C and the completion of stress whitening at +12°C.
By kind permission of Elsevier Applied Science, London.

The high concentration of rubber particles produces many crazes and thus an even distribution of stresses. In contrast, stress peaks in unmodified rigid-brittle thermoplastics are concentrated on only a few weak spots that are already present at small deformations. Unmodified rigid-ductile thermoplastics fare better because stress peaks can be smoothened by necking (telescope effect) (Volume III, Fig. 16-2, top, and p. 576).

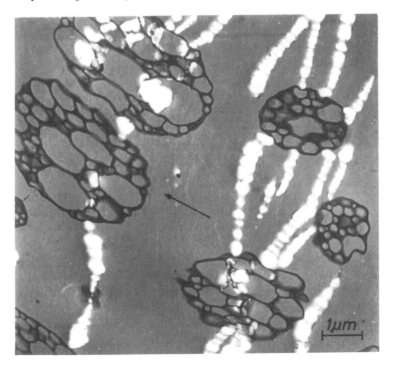

Fig. 10-33 Development of crazes at the equator of rubber domains of a rubber-modified poly(styrene) perpendicular to the direction of stress (arrow) [23b].

In rubber-modified thermoplastics, telescope effects cannot be the primary reason for the observed reinforcement. In such plastics, greater tensile and impact strengths are observed even in the absence of strain softening (compare HIPS in Fig. 10-31).

The effectiveness of rubber particles in rubber-modified thermoplastics depends on the size and morphology of their domains (see ABS in Fig. 10-31). Very small rubber domains are not effective because they cannot initiate craze formations and rather promote shear flow. Very large rubber domains do initiate the formation of crazes but create only a few of them and are therefore relatively ineffective.

Rubber domains therefore have optimal sizes for the best notched impact strengths. These optimal sizes are ca. 1 μm for rubber-modified, amorphous poly(styrene) and ca. 0.1 μm for slightly crystalline (ca. 5 %) poly(vinyl chloride). The effectiveness also depends on the morphology of the rubber domains: compact domains are not very effective whereas domains with occluded thermoplastic particles are most effective.

10.5   Porous Polymers

10.5.1   Overview

Porous polymers can be considered "blends" of gases with thermoplastics, thermosets, or elastomers. Pores can be open or closed and can be produced by many different processes. The most important group of these processes involves chemical or physical foaming. Porous materials can be used three-dimensionally as working materials, two-dimensionally as sheets, and "one-dimensionally" as fibers.

Foamed plastics are also known as **plastic foams, cellular plastics**®, **sponge plastics**, or **expanded polymers**. They are subdivided according to

- their density as **dense plastic foams** with densities of 0.4-0.6 g/cm^3 and **light plastic foams** with densities of 0.01-0.10 g/cm^3;
- their glass temperatures T_G relative to the application temperature T as **hard** or **rigid plastic foams** ($T_G \gg T$) and **soft** or **flexible plastic foams** ($T_G \ll T$);
- their cell structure as **open-celled plastic foams** with interconnected air cells that can take up liquids and **closed-cell plastic foams** where each air cell is completely surrounded by a polymeric wall;
- their skin as **structural foams** or **integral skin foams** if a low-density core of the item is surrounded by a dense skin;
- and as **syntactic foams** if the enclosed gases are not in direct contact with the polymeric matrix but are enclosed in hollow filler particles such as glass spheres or ceramic or plastic particles.

Elastomeric polymeric foams are subdivided according to their cell structure and the use of blowing agents (foaming agents) into

- **foamed rubbers**, or **cellular rubbers** with closed cell structures that were prepared without blowing agents;
- **expanded rubbers** (ISO) with closed cell structures that were obtained with the help of blowing agents; and
- **sponge rubbers** (ISO) with open cells.

Foamed polymers have low heat conductivities since they are blends of polymers with low heat conductivities $\lambda/(W\ m^{-1}\ K^{-1})$ between 0.13 (*cis*-1,4-poly(isoprene)) and 0.44 (high-density poly(ethylene)) and gases with even lower ones between 0.0073 (CCl_3F) and 0.026 (air). They are therefore used as thermal insulating materials. Since they also possess low densities and because their polymeric matrix behaves elastically, they also serve as shockproof packaging materials. Structural foams exhibit high strengths at good values of stiffness per mass and, depending on the polymer, also good mechanical and thermal properties which makes them useful materials for bumpers.

Production and consumption statistics are not very comprehensive and often unreliable: in 1990, the same author claimed a US consumption of 2 million tons of foamed plastics in one paper and 6 million tons per year in another one.

There are also large differences in the type of consumption. The proportions of foamed plastics as insulating materials were 65 % in Western Europe, 41 % in the United States, and 15 % in Japan, for drinking cups 1 % in Western Europe and 33 % in the US, and for packaging purposes 32 % in Western Europe, 20 % in the US, and 79 % in Japan (about half of it for the packaging of fish). Percentages of other uses were 2% in Western Europe and 6 % each in the US and Japan.

10.5.2 Manufacturing

Porous plastics are manufactured from polymers either chemically by decomposition of gas-producing foaming agents or by physically combining polymers and preformed gases such as air, water vapor, or haloalkanes, either directly into the wanted shapes or, more often, to blocks of foamed plastics. The blocks are then cut to the desired shape and size, for example, by saws, electrically heated wires, or even hot water jets.

Physical Processes

Extensive **drawing** of plastic films produces foamed films with long elongated pores.

Sintering of powdery thermoplastics at temperatures just below their glass temperature (amorphous polymers) or melting temperature (semi-crystalline polymers) produces materials with open pores that can serve as filters. The process is used in special cases for poly(tetrafluoroethylene), poly(ethylene), and poly(styrene).

Pouring of polymer solutions onto heated surfaces causes solvents to evaporate which leads to many small channels in polymers.

Extraction methods remove particulates from polymeric matrices, for example, sodium chloride crystals from NaCl-filled epoxy resins, poly(ethylene), and poly(vinyl chloride). For the manufacture of viscose sponges, sodium sulfate crystals are mixed into viscose, a concentrated solution of cellulose xanthogenate in aqueous sodium lye (sodium hydroxide) (Volume II, p. 399) which causes cellulose xanthogenate to precipitate. Heating the resulting mass causes the xanthogenate to decompose to cellulose and the sponge structure to stabilize. Sodium sulfate is then extracted with water.

The so-called **mechanical foaming** of polymers involves an intensive stirring or beating of polymer latices or prepolymers in mixtures with surface active materials. The resulting foam is subsequently stabilized by chemical crosslinking. Examples are soft foams from urea resins or plasticized poly(vinyl chloride) (Table 10-6).

Table 10-6 Production of important hard (H) and soft (S) polymer foams from thermoplastics (TP), thermosets (TS), and rubbers (R). MF = melamine-formaldehyde resin, PF = phenolic resin, UF = urea-formaldehyde resin.

| Polymer | Type | Production of foams | | |
		mechanically	physically	chemically
PF, MF, or UF resins	TS	H	H	H
Polyurethanes	TS	-	H, S	H, S
Poly(styrene)	TP	-	H	-
Poly(vinyl chloride)	TP	S	H, S	H, S
Poly(ethylene)	TP	-	-	H, S
Silicone	R	-	-	S
Natural rubber	R	-	H, S	H, S
Natural rubber latex	R	S	-	S

So-called **physical foaming** consists of the expansion of volatile liquids or pressurized gases in a polymeric matrix. A wide variety of polymers, gases, and methods are used in this most important group of foaming processes.

For example, PVC plastisols can be caused to extend by compressed nitrogen, carbon dioxide, or air to integral skin foams. Molded parts from expanded poly(styrene), poly-(ethylene), or poly(propylene) are manufactured by blowing foamed polymer spheres of 1-8 mm diameter into porous aluminum molds (autoclave process). The spheres are then sintered together by hot water vapor at 1.2 bar for PS and at 3 bar for PP to foamed articles with densities between 12 and 300 kg/m^3.

Foamed polymer particles can also be extruded. Extrusion requires fewer processing steps than the autoclave process but more energy.

Physical foaming is also used for expanded poly(styrene), known as Styrofoam® in the United States and Styropor® in Europe. In the Styropor process, poly(styrene) containing ca. 5 wt% pentane is expanded 40-80 fold by 105°C hot water vapor. Temporary storage for 3-48 h at room temperature lets ca. 40 % pentane diffuse out and air diffuse in which avoids formation of vacua. The prefoamed PS is filled into molds and further expanded by heating with water vapor to 130°C which causes the particle surfaces to melt together.

Integral skin foams can be obtained by injection molding of a "blend" of a thermoplastic and a volatile liquid, either by directly adding a few percent of the liquid to the granulate or by pumping the liquid into the polymer melt under high pressure.

Physical Blowing Agents

In order to be volatile, physical blowing agents must be liquids with low boiling temperatures and small enthalpies of vaporization. Such compounds are low-molecular weight hydrocarbons and fluorinated hydrocarbons (Table 10-7). A fraction of these compounds escapes during foam formation and use of foamed articles and another fraction becomes "permanently" entrapped.

Some of these compounds are very harmful environmentally and therefore their use is regulated nationally and internationally. For simplification, they are also usually not addressed by their chemical names but by their letter and number codes (Table 10-7).

Table 10-7 Systematic, common, and trivial designationss of haloalkanes. Codes consist of a combination of letters and numbers. ( ) incorrect designation.
   Letter codes: CFC = *chlorofluorocarbons* (no hydrogen or bromine), HCFC = *hydrochlorofluorocarbons*, HFC = *hydrofluorocarbons*, PFC = *perfluorocarbons*. HCFCs will be phased out by 2010.
   CFC number codes: 1st number = number of C atoms minus 1; 2nd number = number of H atoms plus 1; 3rd number = number of F atoms; a = asymmetric, e = equal (symmetric).

Composition	Freon	CFC	HCFC	HFC	PFC	R	Halon	Other
CCl_4	10	10					104	
CCl_3F	11	11				11		
CCl_2F_2	12	12				12	122	
$CClF_3$		13						
CF_4	14	(14)			14		14	
$CHCl_2F$			21					
$CHClF_2$			22			22		
CH_2ClF	31							
CHF_3				23				
CH_2F_2				32				
CH_3F				41				
CH_2BrCl							1011	
$CBrClF_2$	12B1						1211	(BCF)
$CBrF_3$	13B1						1301	
$CCl_3–CCl_3$		110						
$CCl_2F–CCl_2F$		112						
$CCl_2F–CClF_2$		113						
$CCl_3–CF_3$		113a						
$CClF_2–CClF_2$		114				114	242	
$CClF_2–CF_3$		115						
$CF_3–CHCl_2$				123				
$CHClF–CF_3$				124				
$CHF_2–CF_3$				125				
$CHF_2–CHF_2$				134				
$CF_3–CH_2F$				134a				Suva-134a
$CCl_2F–CH_3$			141b					
$CClF_2–CH_3$			142b					
$CH_2Cl–CH_2Cl$	150							
$CHCl_2–CH_3$	150a							
$CHF_2–CH_3$	152a							
$CF_3–CHF–CF_3$				227ea				FE-227, FM-200
$CHF_2–CHF–CHF_2$				245a				
$CF_3–CF_2–CF_2–CF_3$						610		PFB, CEA-410

   Under sunlight, some of the haloalkanes, especially CCl_3F and CCl_2F_2, form chlorine radicals that react with ozone molecules of the ozone layer of the upper atmosphere which protects the earth from ultraviolet light. The depletion of this layer ("ozone hole") is presumed to add to global warming. Other volatile liquids are less problematic in this respect but are prone to produce smog by photochemical reactions (Table 10-8).
   Physical blowing agents must also have very many other properties. They should not be toxic or flammable. They should also dissolve well in monomers if polymerization and foam formation proceed simultaneously, for example, fiber or mineral powder containing foamed polyurethanes from isocyanates and, for example, polyols, by reinforced reaction injection molding (RRIM), since good solubility leads to foams with low densities.

Table 10-8 Physical blowing agents. T_{bp} = boiling temperature at normal pressure, ΔH_v = vaporization enthalpy, λ = thermal conductivity at 30°C (CO_2: 25°C), F = flammability, ODP = ozone depletion potential, GWP = global warming potential, SP = smog potential. * Sublimation.

Blowing agent		T_{bp} °C	ΔH_v kJ mol^{-1}	λ W m^{-1} K^{-1}	F	ODP	GWP	SP
Air	N_2, O_2, etc.			0.0259		0		
Carbon dioxide	CO_2	−78.5		0.0168	−		0.00025	
i-Butane	C_4H_{10}	−11.7		0.0161	+			
Pentane	C_5H_{12}	36.1	25.8	0.0135	+		0.001	500
i-Pentane	C_5H_{12}	27.9	23.7		+		0.001	500
Methyl chloride	CH_3Cl	−23.8		0.0105				
Methylene chloride	CH_2Cl_2	40.0	28.0	0.0063				
Trichloroethylene	$CHCl{=}CCl_2$	86.9						
CFC-11	CCl_3F	23.8	24.9	0.0073	−	1	0.4	<0.1
CFC-12	CCl_2F_2	−29.8		0.0058	−	0.96	3.1	<0.1
CFC-112	$CCl_2F{-}CCl_2F$	47.6	27.5					
CFC-113	$C_2Cl_3F_3$			0.0063				
CFC-114	$CClF_2{-}CClF_2$	3.6		0.0054	−	1	<1.5	
HCFC-21	$CHCl_2F$	8.9	24.0	<0.0073				
HCFC-22	$CHClF_2$	−40.8		0.011*	−	0.049	0.036	0.6
HCFC-123	$CF_3{-}CHCl_2$	23.8		0.010*	−	0.016	0.019	5
HFC-134a	$CF_3{-}CH_2F$	−26.5		0.014	−	0	0.27	0.6
HFC-141b	$CCl_2F{-}CH_3$	32.0		0.010*	+	0.079	0.092	1
HFC-142b	$CClF_2{-}CH_3$	−10.0		0.0094*	+	0.056	0.37	0.5
HFC-152a	$CHF_2{-}CH_3$	−24.7		0.0094*	+	0	0.029	5
HFC 245a	$CHF_2{-}CHF{-}H_3F_5$							

However, physical blowing agents should not dissolve well in the polymer itself since all dissolved gases are potential plasticizers. Some of the haloalkanes of Table 10-8 reduce the compression strength of foamed plastics significantly.

During the foaming process, physical blowing gases become trapped in closed polymer cells. They should not diffuse through cell walls (or only very slowly) and escape because this would reduce the thermal insulation power of expanded plastics considerably: the thermal conductivity $\lambda/(\text{W m}^{-1}\,\text{K}^{-1})$ of HCFC-22 is only 0.011 whereas that of natural rubber is 0.15 and that of LDPE is 0.40.

Furthermore, a diffusing out of blowing gases that is faster than the diffusing in of air would lead in cells to pressures that are lower than the atmospheric pressure and hence to a shrinkage of the expanded polymer. Shrinking can be reduced or prevented by greater gross densities (definition on p. 77), use of integral skin foams, and partial use of air as the blowing gas.

Chemical Blowing Agents

Polymeric foams can also be produced by gases that are produced during the polymerization processes themselves or by the decomposition of **blowing agents (foaming agents, expanding agents)** that are deliberately added to polymers.

Foaming occurs during the formation of polyamides from polyisocyanates and polycarboxylic acids (Volume II, p. 491) or polycarbodiimides from diisocyanates (Volume II, p. 506). Neither process is commercial.

Table 10-9 Names, abbreviations of names, and constitution of commercial chemical blowing agents.

Abbr.	Name	Structure
ADC	azodicarbonamide (= 1,1-azobisformamide)	$H_2NOC-N=N-CONH_2$
ADCDIP	azodicarbonamide diisopropyl ester	$(CH_3)_2CHOOC-N=N-COOCH(CH_3)_2$
AZDN	*N,N*-azobisisobutyronitrile (= AIBN)	$(CH_3)_2C(CN)-N=N-C(CN)(CH_3)_2$
BDSH	benzene-1,3-disulfonylhydrazide	$H_2N-NH-SO_2-(i\text{-}C_6H_4)-SO_2-NH-NH_2$
BSH	benzenesulfonylhydrazide	$C_6H_5-SO_2-NH-NH_2$
DAB	diazoaminobenzene	$C_6H_5-NH-N=N-C_6H_5$
DFSDSH	diphenylsulfone-3,3'-disulfonylhydrazide	$[H_2N-NH-SO_2-(i\text{-}C_6H_5)]_2SO_2$
DNPT	*N,N*-dinitrosopentamethylenetetramine	see I below
IPT	isophthalic acid bis(carbonic acid ethylester anhydride)	$(i\text{-}C_6H_4)(COOCOOC_2H_5)_2$
NTA	*N,N'*-dimethyl-*N,N'*-dinitrosoterephthalamide	$[CH_3-N(NO)-CO]_2(p\text{-}C_6H_4)$
OBSH	4,4'-oxybis(benzenesulfonyl hydrazide)	$[H_2N-NH-SO_2-(p\text{-}C_6H_4)]_2O$
PTZ	5-phenyltetrazole (5-PT)	see II below
TSH	toluene-4-sulfonylhydrazide	$CH_3-(p\text{-}C_6H_4)-SO_2-NH-NH_2$
TSSC	toluene-4-sulfonyl-semicarbazide	$CH_3-(p\text{-}C_6H_4)-SO_2-NH-NH-CONH_2$

Chemical blowing agents (CBA) are added to polymers in proportions of 5-15 wt% in compression molding, 1-5 wt% in the manufacturing of parts from PVC plastisols, 0.3 wt% in the extrusion of profiles, 0.2-0.8 wt% in the injection molding of integral skin foams, and 0.1 wt% for preventing sink marks (sunken spots) during the injection molding of molded parts.

Commercially, many different blowing agents are used (Table 10-9). Most of these agents have fairly complex chemical structures and are therefore known in the trade only by the abbreviations or acronyms of their names. About 90 wt% of all polymeric foams by chemical blowing are obtained by using 1,1-azobisformamide (ABFA), also known as azodicarbonamide (ADC). About 4 wt% are hydrazine derivatives, especially OBSH, and also BDSH, BSH, DFSDSH, and TSSC.

Chemical blowing agents are used mainly for foamed rigid and plasticized poly(vinyl chloride)s (ca. 2/3 of total CBA consumption) and crosslinked poly(olefin) foams (ca. 1/5). CBAs are also used for PS, ABS, modified PPO®, PC, PET, and various PAs.

ABFA has a fairly high decomposition temperature (Table 10-10) which can be lowered by use of activators, i.e., chemical catalysts. These activators are alkaline metal compounds, mostly zinc derivatives. The decomposition is inhibited by dicarboxylic and tricarboxylic acids or their anhydrides. Use of these inhibitors enhances the sharpness of profiles because it allows the polymer melt to fill all the nooks and crannies before pores are formed.

ABFA (IV) decomposes mainly to nitrogen, carbon monoxide, and urea (V), which decomposes further into ammonia and isocyanic acid (VI) which is in equilibrium with

$$(10\text{-}26) \quad H_2N-\underset{\underset{O}{\|}}{C}-N=N-\underset{\underset{O}{\|}}{C}-NH_2 \xrightarrow{-N_2,\ -CO} H_2N-\underset{\underset{O}{\|}}{C}-NH_2 \xrightarrow{-NH_3} HNCO$$

IV, V, VI

$$+ IV \Big\downarrow -N_2,\ -2\ HNCO \qquad\qquad +2\ HNCO \Big\downarrow$$

$$H_2N-\underset{\underset{O}{\|}}{C}-\underset{H}{N}-\underset{H}{N}-\underset{\underset{O}{\|}}{C}-NH_2 \qquad VII$$

$$3\ H-N=C=O \rightleftarrows 3\ HO-C\equiv N \quad IX$$

$$\Big\downarrow -NH_3$$

VIII — HN—NH / $O=C$—N(H)—$C=O$

X, XI

its tautomeric form, cyanic acid (IX) (Eq.(10-26)). Isocyanic acid (VI) trimerizes to iso-cyanuric acid (X) and cyanic acid (IX) to cyanuric acid (XI). Isocyanic acid also poly-merizes to cyamelide, $+O-C(=NH)+_n$. Dimerization of ABFA (IV) yields nitrogen, iso-cyanic acid (VI), and hydrazodicarbonamide (VII), which splits off ammonia and forms urazole (1,2,4-triazolidine-3,5-dione; VIII). The decomposition of ABFA in air thus leads to 32 wt% of gases (N_2, CO, NH_3), 41 wt% solids, and 27 wt% sublimates.

NTA has a low decomposition temperature and is therefore used for temperature-sen-sitive polymers such as PVC plastisols. However, it needs to be stabilized by mineral oil.

The physical structure of the foamed plastics obtained by the use of CBAs depends strongly on the particle size of the agents. Each CBA particle produces a great volume of gas which forms a gas cell around the decomposing particle. If the viscosity of the melt is adjusted correctly, cell walls remain intact and the polymer is transformed into a closed-cell expanded material. Low polymer viscosities cause cell walls to burst, which leads to open-pore structures.

Table 10-10 Average decomposition temperatures T_{dec} of chemical blowing agents in air, produced gases, yields v_{gas} of gases, and uses of chemical blowing agents.

Code	$\dfrac{T_{dec}}{°C}$	Produced gases	$\dfrac{v_{gas}}{cm^3\ g^{-1}}$	Used for
ADC (= ABFA)	205-215	N_2, CO, NH_3, CO_2	190-240	ABS, PA, PE, PP, PS, PVC, TPE
ADCDIP	100	N_2, CO, CO_2		PE, PP, PVC, PS, ABS, PA, PUR
AZDN	90-120	N_2	110-130	PVC
BDSH	115-130		120	
BSH	95-110		130	PE, PVC, EP, PF, elastomers
DAB	95-110	N_2	115	
DFSDSH	130-155		110	PVC, PE
DNPT (DPT)	60-180	N_2, NO, NH_3	195	BR, CR, HIPS, NR, SBR, PVC-P
IPT	190	CO_2	75	PC
NTA	90-105		265	PVC, PUR, SI
OBSH	130-165	N_2, H_2O	125	CR, EVAC, LDPE, NBR, NR, PUR
PTZ (5-PT)	240-250	N_2	200	ABS, LCP, PA, PBT, PC, TPS
TSH	105-110	N_2, H_2O	120	CR, EPDM, EVAC, NR, NBR, SBR
TSSC	210-270	N_2, CO_2, NH_3	140	ABS, HDPE, HIPS, PA, PP, PVC-U

Large CBA particles produce too much gas too fast. Cell walls become too thin; they break which results in open-pore expanded polymers. The ratio of open/closed pores can be widely regulated by adjusting polymer viscosities and proportions, particle sizes and particle size distributions of CBAs; and concentrations of activators.

10.5.3 Properties

Mechanical Properties

Expanded polymers are characterized by their **overall density** (= **gross density**), sometimes also called **volumetric weight**, which is the mass of the polymer (the mass of gas is negligible) per volume of the body. For integral skin foams with dense outer regions and less dense inner ones, one also determines the **core density** of the interior.

The **strength** of an expanded polymer is usually expressed as compressive strength for a defined reduction of the height of a test specimen, for example, 40 %. Also used are indentation hardness (p. 150 ff.). Strengths are usually proportional to the compression modulus of the polymer and the third power of the thickness of the specimen. They usually do not depend directly on the volumetric weight since they are not only controlled by the proportion of polymer per volume of foam but also by the shape of the pores and the thickness and distribution of the *lamellae* (= the walls of the pores) and *knots* (= the intersection of pore walls) of the foam, i.e., by the foaming process itself.

Mechanical properties of expanded polymers depend primarily on the glass temperature of the polymers themselves and secondarily on the density of the foams. Thermosets such as crosslinked phenolic, urea, and epoxy resins, polyurethanes and poly(isocyanurate)s as well as rigid thermoplastics such as poly(styrene) and unplastisized poly(vinyl chloride) lead to rigid ("hard") expanded polymers. Flexible ("soft") expanded polymers result from soft thermoplastics such as low-density poly(ethylene) and plasticized poly(vinyl chloride) and from crosslinked rubbers (natural, chloroprene, styrene-butadiene, silicone).

The density ρ_E of expanded polymers is usually taken as a measure of the polymer content of expanded plastics. The lower the density, the higher is the gas content and the lower the mechanical properties such as moduli, strengths, etc. In the case of the simple rule of mixtures, Eq.(9-2), one would expect that the mechanical properties E_E of expanded plastics would depend directly on the properties E_P of the polymer and the ratio ρ_E/ρ_P of densities of expanded plastics (ρ_E) to those of their polymers (ρ_P):

$$(10\text{-}27) \qquad E_E = E_P\phi_P + E_G\phi_G \approx E_P(\rho_E/\rho_P) = E_P(\rho_E/\rho_P)^{2z} \quad ; \quad z \equiv 1/2$$

since mechanical properties of gases are practically zero ($E_G \approx 0$) and the volume fraction ρ_P of polymers can be expressed by $\phi_P = V_P/V_E = (m_P/\rho_P)/(m_E/\rho_E) \approx \rho_E/\rho_P$ because the mass of gas in the expanded polymer is practically zero, i.e., $m_E = m_P + m_G \approx m_P$. The resulting expression $E_E = E_P(\rho_E/\rho_P)$ is written $E_P(\rho_E/\rho_P)^{2z}$ with $z = 1/2$ in order to accommodate any experimentally found deviation from $E_E = E_P(\rho_E/\rho_P)$ (see below).

In *stable* liquid-gas foams, however, surface tension causes the borders of cells to always form from three lamellae of the liquid at angles of 120° [first rule of J.A.F. Plateau; the second rule says that a knot is always formed by four borders with angles of

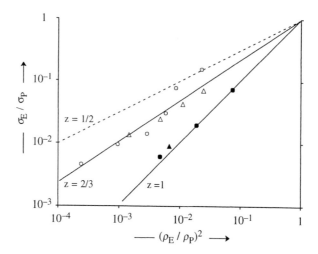

Fig. 10-34 Ratios σ_E/σ_P of fracture strengths σ_E of expanded polymers and σ_P of parent polymers as a function of the square of density ratios, $(\rho_E/\rho_P)^2$. O extruded poly(styrene) sheets, O extruded poly(styrene) boards, $\triangle$ extruded poly(ethylene) boards, $\bullet$ epoxides, $\blacktriangle$ phenolic resins.
 Broken line: simple rule of mixtures ($z = 1/2$), solid lines: empirical correlations for exponents of $z = 2/3$ (Plateau borders) and $z = 1$, respectively.

ca. 109° 28' 16", the tetrahedral angle $\tau = \arccos (1 - (1/3)]$. Liquid-gas foams that do not follow **Plateau's rules** are unstable and reorganize until the rules are fulfilled.

 If Plateau's borders are infinitely thin and deliver the main contribution to mechanical properties such as moduli or tensile strengths, then the exponent z should be 2/3:

$$(10\text{-}28) \qquad E_E/E_P = (\rho_E/\rho_P)^{4/3} = (\rho_E/\rho_P)^{2z} \quad ; \quad z = 2/3$$

 However, experimental values of z for fracture strengths of expanded polymers are sometimes neither $z = 1/2$ (Eq.(10-27)) nor $z = 2/3$ (Eq.(10-28)) but $z = 1$ (Fig. 10-34).

 In general, relative fracture strengths σ_E/σ_P of extruded *sheets* of expanded poly(styrene) seem to follow Eq.(10-27) with $z = 1/2$ (Fig. 10-34 (top curve)) as do also flexural moduli of high density integral skin foams with thick skins (not shown). However, fracture strengths of extruded *boards* of both expanded poly(styrene) and expanded poly(ethylene) are given by $(\rho_E/\rho_P)^{4/3}$, i.e., a value of $z = 2/3$. Foamed crosslinked epoxy and phenolic resins showed dependences $E_E/E_P = (\rho_E/\rho_P)$, i.e., $z = 1$. Similar dependences of property ratios on powers of density ratios also seem to exist for compression and shear moduli as well as for creep and fatigue properties.

 The deformation of expanded rigid thermoplastics such as Styrofoam® first increases rapidly with time but becomes constant after several days (Fig. 10-35). Integral skin foams with 1-2 mm thick skins are considerably more rigid and show no time-dependent deformation if loads are not too large. Their advantageous strength/density ratios make them suitable for engineering purposes, i.e., for structures. Such structural foams are used for computer casings, many sports articles, bumpers, and airplane parts.

 The shape of the force–deformation curves determines whether upholstery is comfortable or not (Fig. 10-36). Both types show initially an almost linear increase of deformation with increasing force, but bad upholstery has an S-shaped deformation-

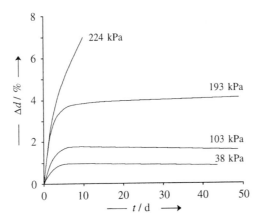

Fig. 10-35 Time dependence of the deformation Δd of an expanded poly(styrene) ($\rho = 0.048$ g/cm^3) caused by various static loads [25].

force behavior in which further slight increases of force cause deformations to increase steeply until they finally become practically constant. For good upholstery, the initial response is similar to that of bad upholstery but the deformation then increases almost linearly with the applied force. Bad upholstery also wears out after repeated uses: the threshold of the onset of the steep increase of deformation with small increases of force has shifted to small forces.

Thermal Conductivity

Many foamed plastics are used as thermal insulating materials. Their low thermal conductivity is not caused by the polymers but by their high proportion of gases.

At room temperature, thermal conductivities $\lambda/(W\ m^{-1}\ K^{-1})$ of amorphous polymers vary between 0.13 and 0.19. Examples are natural rubber (0.134), poly(styrene) (0.142), polyurethanes (0.147), and poly(dimethylsiloxane) (0.163). Thermal conductivities of gases are considerably lower: air has 0.026, carbon dioxide 0.017, and CFC 11 only 0.0073.

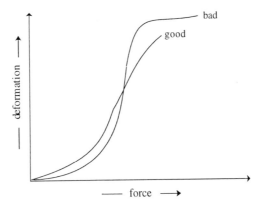

Fig. 10-36 Schematic representation of the effect of increasing force on the resulting deformation of good and bad upholstery from a soft foamed plastic.

From the simple rule of mixtures one obtains with the volume fraction $\phi_P = m_P/(\rho_P V)$ $= 1 - \phi_G$ of the polymer and the density $\rho_E = (m_P + m_G)/V \approx m_P/V = V_P\rho_P/V$ of the expanded polymer:

(10-29) $\lambda_E = \lambda_P\phi_P + \lambda_G\phi_G \approx \lambda_G + (\lambda_P - \lambda_G)(\rho_P/\rho_E)$

Foaming polyurethane ($\rho_P/(cm^3 \, g^{-1})$) = 1.05) with fluorotrichloromethane (CFC 11) ($\lambda_G/(W \, m^{-1} \, K^{-1})$) = 0.0073 (Table 10-8) to a closed cell material with a density of $\rho_E/(cm^3 \, g^{-1})$ = 0.016 should thus deliver a foamed plastic with a thermal conductivity of $\lambda_E/(W \, m^{-1} \, K^{-1})$ = 0.0081. In fact, the actual thermal conductivity is about twice as large (Table 10-11).

Foaming polyurethanes with carbon dioxide or pentane delivers expanded polymers with greater thermal conductivities than those with CFC 11 because the former have greater thermal conductivities than the latter. In general, expanded polymers have lower densities than other thermal insulating materials (Table 10-11).

The thermal conductivity of air ($\lambda/(W \, m^{-1} \, K^{-1})$) = 0.0259) is considerably higher than that of blowing gases such as $CFCl_3$ (0.0073), pentane (0.0135), and CO_2 0.0168). Thermal conductivities will thus increase if blowing gases diffuse out of expanded polymers and are replaced by air. An example is the thermal conductivity, $\lambda/(W \, m^{-1} \, K^{-1})$, of a $CFCl_3$ containing hard (rigid) polyurethane foam which increased from 0.017 to 0.0182 after 6 weeks and to 0.0216 after 12 months.

During this time, the thermal conductivity of an integral skin foam of the same polyurethane remained constant. The same phenomenon is also observed if hard polyurethane foams are formed with CO_2: the thermal conductivity increased within 12 months from 0.023 to 0.032 W m^{-1} K^{-1}. In contrast, the thermal conductivity for the corresponding integral skin foam remained constant for the same time period of time (for the permeation of gases through polymers, see Section 13.2).

Table 10-11 Densities ρ_P of polymers and ρ_{EP} of expanded polymers and other materials and thermal conductivities λ_G of gases and λ_{EP} of expanded polymers and other insulating materials.

Material		Gas	$\dfrac{\rho_P}{g \, cm^{-3}}$	$\dfrac{\rho_{EP}}{g \, cm^{-3}}$	$\dfrac{\lambda_G}{W \, m^{-1} K^{-1}}$	$\dfrac{\lambda_{EP}}{W \, m^{-1} K^{-1}}$
Polyurethane	rigid	$CFCl_3$	1.05	0.016	0.0073	0.017
Polyurethane	rigid	CO_2	1.05	0.016	0.0168	0.033
Polyurethane	flexible	CO_2	1.05	0.016	0.0168	0.044
Poly(styrene)	rigid	C_5H_{12}	1.05	0.016	0.0135	0.036
Poly(styrene)	flexible	C_5H_{12}	1.05	0.032	0.0135	0.033
Poly(ethylene)	flexible	air	0.93	0.035		0.058
Paper-kapok-paper		air		0.016	0.0259	0.035
Cork	low density	air		0.086	0.0259	0.036
Cork	high density	air		0.22	0.0259	0.049
Glass wool		air	2.60	0.12	0.0259	0.041
Hair felt		air		0.19	0.0259	0.037
Pine		air, H_2O		0.50		0.112

Literature to Chapter 10

10.2 MISCIBILITY
G.Olabisi, L.M.Robeson, M.T.Shaw, Polymer-Polymer Miscibility, Academic Press, New York 1979
K.Solc, Ed., Polymer Compatibility and Incompatibility: Principles and Practices, Harwood
    Acad.Publ., New York 1982
Y.S.Lipatov, A.E.Nesterov, Thermodynamics of Polymer Blends, Technomic, Lancaster (PA) 1997
T.Araki, Q.Tran-Cong, M.Shibayama, Structure and Properties of Multiphase Polymeric Materials,
    Dekker, New York 1998

10.3 PLASTICIZED POLYMERS
K.Thinius, Chemie, Physik und Technologie der Weichmacher, VEB Deutscher Verlag für Grundstoff-
    industrie, Leipzig 1962
I.Mellan, Industrial Plasticisers, Pergamon, Oxford 1963
P.D.Ritchie, Ed., Plasticisers, Stabilisers, and Fillers, Butterworth, London, 8th ed. 1980 (3 vols.)
J.K.Sears, J.R.Darby, The Technology of Plasticizers, Wiley, Chichester 1980

10.4a POLYMER BLENDS
D.R.Paul, S.Newman, Eds., Polymer Blends, Academic Press, New York 1978 (2 vols.)
E.Martuscelli, R.Palumbo, M.Kryszewski, Eds., Polymer Blends, Plenum, New York 1980
I.C.Sanchez, Bulk and Interface Thermodynamics of Polymer Alloys, Ann.Rev.Mater.Sci. **13**
    (1983) 387
D.J.Walsh, J.S.Higgins, A.Maconnachie, Ed., Polymer Blends and Mixtures, Nijhoff, Dordrecht
    1985 (NATO ASI Ser. E, No. 89)
D.J.Walsh, S.Rostami, The Miscibility of High Polymers: The Role of Specific Interactions,
    Adv.Polym.Sci. **70** (1985) 119
L.A.Utracki, A.P.Plochocki, Industrial Polymer Blends and Alloys, Hanser, Munich 1985
L.A.Utracki, Polymer Alloys and Blends, Hanser, Munich 1989
M.M.Coleman, J.F.Graf, P.C.Painter, Specific Interactions and the Miscibility of Polymer Blends,
    Technomic, Lancaster (PA) 1991
L.A.Utracki, Ed., Two-Phase Polymer Systems, Hanser, Munich 1992
M.J.Folkes, P.S.Hope, Polymer Blends and Alloys, Blackie Academic, London 1993
L.A.Utracki, Ed., Encyclopedic Dictionary of Commercial Polymer Blends, ChemTec Publ.,
    Toronto 1994
V.P.Privalko, V.V.Novikov, The Science of Heterogeneous Polymers, Wiley, New York 1995
L.H.Sperling, Polymeric Multicomponent Materials, Wiley-VCH, Weinheim 1997
L.A.Utracki, Commercial Polymer Blends, Chapman and Hall, New York 1998
G.O.Shonaike, G.P.Simon, Eds., Polymer Blends and Alloys, Dekker, New York 1999
D.R.Paul, C.B.Bucknall, Eds., Polymer Blends. Formulation and Performance, Wiley, New York
    2000 (2 vols.)
S.Mazuumdar, Composites Manufacturing, CRC Press, Boca Raton (FL) 2002
G.Simon, Ed., Polymer Characterization Techniques and Their Application to Blends, ACS Books,
    Washington (DC) 2003
R.M.Rowell, Ed., Handbook of Wood Chemistry and Wood Composites, CRC Press, Boca Raton
    (FL) 2005
C.Harrats, G.Groenickx, S.Thomas, Eds., Micro- and Nanostructured Multiphase Polymer Blend
    Systems. Phase Morphology and Interfaces, CRC Press, Boca Raton (FL) 2006
L.M.Robeson, Polymer Blends, Hanser, Munich 2007
C.Harrats, Multiphase Polymer-Based Materials. An Atlas of Phase Morphology at the Nano and
    Micro Scale, CRC Press, Boca Raton (FL) 2009

10.4.b RUBBER BLENDS
P.J.Corish, Fundamental Studies of Rubber Blends, Rubber Chem.Technol. **40** (1967) 324
C.B.Bucknall, Toughened Plastics, Appl.Sci.Publ., London 1977
A.J.Tinker, K.P.Jones, Ed., Blends of Natural Rubber, Chapman and Hall, London 1998

10.4.c RUBBER-MODIFIED THERMOPLASTICS
J.R.Dunn, Blends of Elastomers and Thermoplastics - A Review, Rubber Chem.Technol. **49**
    (1976) 978

A.A.Collyer, Rubber Toughened Engineering Plastics, Chapman and Hall, London 1994
C.B.Arends, Ed., Polymer Toughening, Dekker, New York 1996

10.4.d INTERPENETRATING NETWORKS
L.H.Sperling, Interpenetrating Polymer Networks and Related Materials, Plenum, New York 1980

10.4.e COMPATIBILIZERS
J.T.Lutz, Jr., D.L.Dunkelberger, Impact Modifiers for PVC, Wiley, New York 1992
S.Datta, D.J.Lohse, Polymeric Compatibilisers, Hanser, Munich 1996
I.W.Hamley, The Physics of Block Copolymers, Oxford Univ. Press, New York 1999

10.5 POROUS POLYMERS
T.H.Ferrigno, Rigid Plastic Foams, Reinhold, New York 1963
E.A.Meinecke, R.C.Clark, The Mechanical Properties of Polymeric Foams, Technomic, Westport
    (CT) 1972
K.C.Frisch, J.H.Saunders, Eds., Plastic Foams, Dekker, New York, 2 Vols. (1972, 1973)
-, Foams (Desk Top Data Bank), Internat.Plastics Selector, San Diego 1978
N.C.Hilyard, Ed., Mechanics of Cellular Plastics, Hanser International, Munich 1982
G.Woods, Flexible Polyurethane Foams: Chemistry and Technology, Appl.Sci.Publ., London 1982
S.Semerdijev, Introduction to Structural Foam, Soc.Plastics Industry, Brookfield Center CT) 1982
F.A.Shutov, Foamed Plastics, Cellular Structure and Properties, Adv.Polym.Sci. **51** (1983) 155
Y.L.Meltzer, Expanded Plastics and Related Products, Noyes, Park Ridge (NJ) 1983
B.C.Wendle, Structural Foam, Dekker, New York 1985
F.A.Shutov, Integral/Structural Polymer Foams, Springer, Berlin 1986
M.F.Ashby, L.J.Gibson, Cellular Solids. Structure and Properties, Pergamon Press, Oxford 1988
D.Klempner, K.C.Frisch, Eds., Handbook of Polymeric Foams and Foam Technology, Hanser,
    Munich 1991
A.H.Landrock, Handbook of Plastic Foams, Noyes Publ., Park Ridge (NJ) 1995
J.L.Throne, Thermoplastic Foam Extrusion, Hanser, Munich 2004
D.Klempner, V.Sendijarevic, Eds., Handbook of Polymeric Foams and Foam Technology, Hanser,
    Munich, 2nd ed. 2004
S.-T.Lee, N.S.Ramesh, C.B.Park, Polymeric Foams. Science and Technology, CRC Press, Boca
    Raton (FL) 2007
K.Ashida, Polyurethane and Related Foams. Chemistry and Technology, CRC Press, Boca Raton
    (FL) 2008
S.-T.Lee, D.Scholz, Ed., Polymeric Foams. Technology and Development in Regulation, Process,
    and Products. CRC Press, Boca Raton (FL) 2008

References to Chapter 10

[1]   G.Allen, G.Gee, J.P.Nicholson, Polymer **2** (1961) 8, taken from Fig. 2
[2]   R.Casper, L.Morbitzer, Angew.Makromol.Chem. **58-59** (1977) 1, Fig. 4
[3]   E.Jenckel, R.Heusch, Kolloid-Z. **130** (1953) 89, Table 1
[4]   R.S.Spencer, R.F.Boyer, J.Appl.Phys. **16** (1945) 594 (T_G); G.J. van Amerongen, in
        R.Houwink, Ed., Chemie und Technologie der Kunststoffe, Akadem.Verlagsges.Geest und
        Portig, Leipzig, Vol. I (1954), Fig. 195 (σ_B, ε_B)
[5]   G.H.Molau, H.Keskkula, J.Polym.Sci. [A-1] **4** (1966) 1695, Figs. 1, 2, 4, 6
[6]   D.Braun, M.Fischer, G.P.Hellmann, Macromol.Symp. **83** (1994) 77, Fig. 3
[7]   H.Willersinn, Makromol.Chem. **101** (1967) 296, Figs. 3 and 9
[8]   R.Qian, X.Gu, Angew.Makromol.Chem. **141** (1986) 1, Figs. 4 A and 4 B
[9]   H.Antychowicz, Z.Wielgosz, Polimery **24** (1979) 193
[10]   H.Polańska, T.Koomoto, T.Kawai, Polimery **25** (1980) 365, 396
[11]   L.W.Kleiner, F.E.Karasz, W.J.MacKnight, Polym.Eng.Sci. **19/7** (1979) 519, Figs. 1, 3

[12]   D.Allard, R.E.Prud'homme, J.Appl.Polym.Sci. **19** (1981) 1245; M.Aubin, D.Bussières,
       D.Duchesne, R.E.Prud'homme, in K.Solc, Ed., Polymer Compatibility and Incompatibil-
       ity, Harwood Academic Publ., Chur, Switzerland 1982, Table III
[13]   Z.Dobkowski, Polimery **25** (1980) 110
[14]   T.Kunori, P.H.Geil, J.Macromol.Sci.-Phys. **18** (1980) 99, Figs. 3 and 19
[15]   L.D.Han, Rheology in Polymer Processing, Academic Press, New York (1976), Fig. 1.11
[16]   T.Imai, D.J.Meier, Michigan Molecular Institute, personal communication
[17]   G.Natta, G.Allegra, I.W.Bassi, D.Sianesi, G.Caporiccio, E.Torti, J.Polym.Sci. **A 3**
       (1965) 4631, Fig. 3
[18]   A.Escala, E.Balizer, R.S.Stein, ACS Polym.Prepr. **19**/1 (1978) 152, Figs. 1 and 2
[19]   PBTZ/APPBI data of S.J.Krause, T.Haddock, G.E.Price, P.G.Lenhert, J.F.O'Brien,
       T.E.Helminiak, W.W.Adams, J.Polym.Sci. B-Polym.Phys. **24** (1986) 1991, Table I
[20]   P.Smith, P.J.Lemstra, Makromol.Chem. **180** (1979) 2983, Table 1;
       -, J.Polym.Sci..-Polym.Phys.Ed. **19** (1981) 1007, Table 1, Fig. 1
[21]   H.Jenne, Kunststoffe **74** (1984) 551, Fig. 2
[22]   J.A.Mason, L.H.Sperling, Polymer Blends and Composites, Plenum, New York (1976),
       Fig. 3.15
[23]   J.Stabenow, F.Haaf, Angew.Makromol.Chem. **29/30** (1973) 1, (a) Fig. 17, (b) Fig. 3
[24]   C.B.Bucknall, Toughened Plastics, Elsevier Appl.Sci., London (1977), Fig. 10
[25]   W.B.Brown, Plast.Progr. **1959** (1960) 149; quoted by K.W.Suh, R.E.Skochdopole,
       Kirk-Othmer, Encyclopedia of Chemical Technology, Wiley, New York, 3rd ed.,
       11 (1980) 82, Fig. 7

11 Polymer Solutions and Gels

11.1 Overview

Polymer solutions are industrially prepared and used for two different reasons. On one hand, they are employed as aids for the processing of polymers to useful materials, for example, in solution spinning to fibers (Chapter 6), preparation of polymer blends (Chapter 10), and manufacture of paints, coatings, and adhesives (Chapter 12). The choice of a solvent for these purposes is dictated primarily by the constitution of the polymers, i.e., their solubility parameters (Section 11.2), and secondarily by technological, economical, and ecological considerations.

On the other hand, polymer solutions are technologically used because of their special properties. In these cases, it is the solvent that dictates the choice of polymers. The most important solvent for these purposes is water, which is used not only as a solvent for water-soluble foodstuffs but also as a carrier for pharmaceuticals, in the production of paper, as a solvent for productivity enhancers in oil fields, and for many other uses.

The most important water-soluble polymer is starch (Volume II, p. 376) with a world production of ca. $55 \cdot 10^6$ t/a (2005). The ratio of food/non-food applications of starch is estimated as 55/45. About 21 % of the non-food applications are in paper-making, 6 % for cardboards, 6 % as an aid in chemical processes, and 20 % for other technical uses. Starch also serves as a base material for glucose syrup and for various derivatives as well as a chemical feedstock for the production of ethanol and lactic acid.

Water-soluble polymers include other natural materials such as plant gums as well as semi-synthetic polymers based on plant gums, starch, or cellulose (Volume II, p. 414) and synthetic polymers such as poly(vinyl alcohol), poly(acrylamide), poly(ethylene glycol), etc. The total global consumption of water-soluble polymers for technical applications is presently estimated as ca. 9 million tons per year; most of which are natural products or derivatives therefrom. Up-to-date statistics for the various technological uses of water as a solvent for polymers do not seem to exist, but older data (1981) do indicate the relative importance of water-soluble polymers in various industries (Table 11-1).

Table 11-1 World consumption (without former Eastern block countries) of industrially used water-soluble polymers in 1981, exclusive of industrial uses of starch and starch derivatives. Numbers also do not include the consumption of water-soluble polymers as foods and food additives.

Application	World consumption in tons per year		
	Based on natural products	Fully synthetic products	Total
Paper production	4 300 000	600 000	4 900 000
Textile industry	400 000	100 000	500 000
Oil production	250 000	30 000	280 000
Flocculation processes	50 000	150 000	200 000
Detergents	60 000	20 000	80 000
Paints and lacquers	≈ 0	70 000	70 000
Cosmetics	18 000	10 000	28 000
Pharmaceuticals	7 000	5 000	12 000
Total	*5 085 000*	*985 000*	*6 070 000*

The second most important group of polymer solvents is comprised of hydrocarbon oils. Polymers for these solvents are mainly poly(isobutylene)s, methacrylate copolymers, ethylene-propylene copolymers, and hydrogenated ethylene-styrene copolymers. The US consumption of polymers for such solutions is estimated as ca. 50 000 t/a and the world consumption as ca. 500 000 t/a.

These polymers are added to mineral oils because of their thickening action, i.e., because of the strong increase of viscosity with increasing polymer concentration. The variation of solution viscosity with polymer concentration is controlled by the polymer constitution, polymer-solvent interactions (Section 11.3), and especially by the self-association of polymer molecules, which may even lead to swollen physical networks (gels; Section 11.6). In some applications, use is made of the temperature dependence of the viscosities of polymer solutions and, in others, of the polymer degradation by shearing.

11.2   Solubility of Polymers

Whether a polymer dissolves in a solvent or not is controlled enthalpically by the interactions polymer-solvent, polymer-polymer, and solvent-solvent and entropically by the number and types of arrangements of molecules and molecule segments relative to the polymer and the solvent. At present, neither enthalpies and entropies of mixing nor Gibbs mixing energies can be calculated from individual properties of polymers and solvents. However, solubilities can be estimated semi-empirically by the so-called **solubility parameters**.

The concept of the solubility parameter treats the mixing of polymer and solvent molecules as a quasichemical reaction in which a pair of polymer segments, PP, reacts with a pair of solvent molecules, SS, to form two pairs of solvated polymer segments, PS:

(11-1)   PP + SS $\rightleftarrows$ 2 PS

Each pair possesses a cohesion energy E_i and per volume correspondingly a **cohesion energy density**, E_i/V_i. The square root of the pairwise cohesion energy density is defined as the **solubility parameter**, $\delta_i = (E_i/V_i)^{1/2}$.

Probabilities of forming solvated polymer segments PS from PP and SS pairs depend on the volume fractions of components P and S as well as on the net change in the squared solubility parameters. Hence, the energy of mixing E_{mix} per total volume V is

(11-2) $E_{mix}/V = \phi_S \phi_P (\delta_S^2 + \delta_P^2 - 2\,\delta_{PS}^2)$

if each solvent molecule and each polymer segment occupies the same volume.

The energies for separations of pairs, i.e., for the "reactions" SS $\rightarrow$ 2 S, PP $\rightarrow$ 2 P, and PS $\rightarrow$ P + S, are assumed to be the vaporization energies of these reactions. This assumption reveals the approximative character of this approach because the polymers to be dissolved are usually not liquids but solids, i.e., either semicrystalline substances or frozen-in liquids (= amorphous substances). The approach thus neglects the energies that are required for the conversion of solids into liquids.

The approach also assumes that the interaction energy between solvent molecules S and polymer segments P can be identified with the geometric mean of the interaction energies of pairs of solvent molecules SS and polymer segments PP, respectively, i.e., $\delta_{PS}^2 = (\delta_S^2 \delta_P^2)^{1/2}$. The energy term for PS can thus be replaced by those for PP and SS:

$$(11\text{-}3) \qquad E_{mix}/V = \phi_S \phi_P[\delta_S^2 + \delta_P^2 - 2\,(\delta_S^2 \delta_P^2)^{1/2}] = \phi_S \phi_P(\delta_S - \delta_P)^2$$

According to this approach, the solubility of a polymer in a solvent is controlled only by the difference of solubility parameters of the solvent and polymer. Solubility parameters δ_S are calculated from evaporation energies of solvents and solubility parameters δ_P of polymers from intrinsic viscosities $[\eta]$ of uncrosslinked polymers or the degree of swelling of gels of crosslinked ones in various solvents (Volume III, p. 299). The solubility parameters of the solvent with the largest $[\eta]$ or the smallest volume fraction ϕ_2 of polymers in gels are taken as the solubility parameter of the polymer. Alternatively, solubility parameters can be calculated by adding empirical increments that have been assigned to each chemical group. The approach can be refined by introducing "three-dimensional" solubility parameters that split solubility parameters δ into three parts coming from dispersion forces (δ_d), polar forces (δ_p), and hydrogen bonds (δ_h) (Table 11-2).

Two liquid substances mix if their Gibbs mixing energy is zero or negative, $\Delta G_{mix} = \Delta H_{mix} - T\Delta S_{mix} \le 0$. The entropy term $T\Delta S_{mix}$ is always positive since mixing causes the entropy to increase. Hence, Gibbs mixing energies remain negative as long as $T\Delta S_{mix}$ is greater than ΔH_{mix}.

Mixing enthalpies ΔH_{mix} reach their lowest value of zero if interactions S-S and P–P are identical. In this case, one also has $\delta_P - \delta_S = 0$ since solubility parameters are purely enthalpic quantities. The more unequal the interactions P-P and S-S, the more positive ΔH_{mix} becomes until it finally can no longer be surpassed by the term $-T\Delta S_{mix}$. Polymer and solvent will therefore no longer mix above a certain value of $|\delta_P - \delta_S|$.

Solubility ranges are ca. $(\delta_P \pm 1)\,(cal/cm^3)^{1/2}$ for apolar polymers in apolar solvents, up to ca. $(\delta_P \pm 2)\,(cal/cm^3)^{1/2}$ for polar polymers in moderately polar solvents, and up to $(\delta_P \pm 3)\,(cal/cm^3)^{1/2}$ for polar polymers in polar solvents. The greater ranges for polar polymers are caused because the use of $\delta_{PS}^2 = (\delta_S^2 \delta_P^2)^{1/2}$ in Eq.(11-3) is restricted to dispersion forces. For this reason, deviations from a "singular" value of δ are expected for polar polymers and especially for those that can form hydrogen bonds. For this reason, the concept of solubility parameters has been extended by introducing "three-dimensional" solubility parameters as the sum of the squares of the contributions of dispersion forces, polar interactions, and hydrogen bonds:

$$(11\text{-}4) \qquad \delta^2 = \delta_d^2 + \delta_p^2 + \delta_h^2$$

The split of the "polar" forces into dipole-dipole interactions and hydrogen bonds is questionable since both types of interactions involve interactions between Lewis bases and Lewis acids. Such interactions are always exothermic whereas interactions by dispersion forces are always endothermic.

Predictions of solubilities by solubility parameters are satisfactory for not too polar, amorphous polymers in not too polar solvents. These conditions usually apply to the polymer-solvent systems that are used in the lacquer and paint industries.

Table 11-2 Solubility parameters δ, so-called three-dimensional solubility parameter δ_d, δ_p and δ_h of solvents and polymers, and experimentally observed solubilities (+) and insolubilities (-) of poly(iso-butylene) (PIB), poly(vinyl acetate) (PVAC), and poly(hexamethylene adipamide) (PA 6.6) (numbers indicate solubility parameters of these polymers). Values of solubility parameters are given in the predominantly used Hildebrand units, $1\ H = 1\ (cal/cm^3)^{1/2}$, in the literature mostly without physical units. These units are related to SI units by $1\ (cal/cm^3)^{1/2} = 2.046\ (J/cm^3)^{1/2}$. na = not applicable.

Compounds	δ	δ_d	δ_p	δ_h	PIB 7.9	PVAC 11.3	PA 6.6 13.6
Apolar solvents							
Heptane	7.40	7.40	0.00	0.00	+	–	–
Cyclohexane	8.18	8.18	0.00	0.00	+	–	–
Carbon tetrachloride	8.65	8.85	0.00	0.00	+	–	–
Carbon disulfide	9.97	9.97	0.00	0.00	+	–	–
Lewis acids							
Chloroform	9.33	8.75	1.65	2.80	+	+	–
Dichloromethane	9.90	8.91	3.10	3.00	+	+	–
t-Butanol	10.84	7.70	2.80	7.10	–	–	–
Methanol	14.49	7.42	6.00	10.90	–	–	–
Water	23.43	7.00	8.00	20.88	–	–	–
Lewis bases							
Diethyl ether	7.61	7.05	1.40	2.50	–	–	–
Benzene	9.06	8.99	0.50	1.00	+	+	–
Tetrahydrofuran	9.51	8.22	3.25	3.50	+	+	–
Acetone	9.69	7.58	5.70	2.00	–	+	–
Pyridine	10.60	9.25	4.30	2.90	–	+	–
m-Cresol	11.50	8.82	3.00	6.10	–	–	+
Dimethylformamide	12.14	8.50	6.70	5.50	–	+	–
Polymers							
Poly(propylene oxide)	7.50	7.00	2.50	1.00	na	na	na
Poly(ethylene)	8.10	8.10	0.00	0.00	na	na	na
Poly(styrene)	9.33	8.60	3.00	2.00	na	na	na
Poly(vinyl chloride)	11.00	9.40	4.50	3.50	na	na	na
Poly(vinyl acetate)	11.29	9.30	5.00	4.00	na	na	na

In such cases, solubility parameters can also serve to predict whether a polymer dissolves in a mixed solvent. Such mixed solvents may even be two non-solvents where one has a larger solubility parameter than the polymer and the other one a smaller one. An example is cellulose nitrate ($\delta = 10.6$ H) which is insoluble in diethyl ether ($\delta = 7.4$ H) or ethanol ($\delta = 12.7$ H) but dissolves in the mixture of these two liquids. Conversely, a mixture of two polymer-dissolving liquids may be a non-solvent for that polymer. An example is poly(acrylonitrile) ($\delta = 12.8$ H) which is soluble in dimethylformamide ($\delta = 12.1$ H) and in malononitrile ($\delta = 15.1$) but not in the mixture of these liquids.

The concept of the solubility parameter usually fails for strongly polar polymers and also for highly crystalline ones. The reason for this behavior can be found in the basic assumptions: the use of dispersion forces as the basis for the elimination of interactions P-S and that of evaporation energies as the measure of separation energies. However, dipole-dipole interactions are much more important for polar polymers than dispersion forces, and melting energies are much more significant for crystalline polymers than evaporation energies.

11.3   Viscosity of Polymer Solutions

11.3.1   Introduction

Dilute solutions of polymers are often highly viscous, which is especially appreciated in the food industry. Technologically, viscosity effects are used in tertiary oil recovery (Section 11.4.1), for the reduction of drag (Section 11.4.2), and for improving the so-called VI index (Section 11.4.3). The desired viscosity effects for these applications depend on the molecular weight and the molecular weight distribution of the polymer molecules and always on the shape of polymer molecules, the polymer-solvent interaction (Sections 11.3.2-11.3.5), polymer associations (Section 11.3.6), electrolyte effects (Section 11.3.7), and the influence of shearing (Section 11.3.8). Some of these effects are so strong that highly viscous gels are formed (Section 11.6). Conversely, viscosities of polymer dispersions are very low (Section 11.5).

11.3.2   Spheres in Solution

Macromolecular *substances* (= polymers) adopt spherical shapes relatively easily in their dispersions. Macro*molecules* (polymer molecules), on the other hand, rarely exist as true spheres. The time-averaged shape of a coil-forming macromolecule is that of a sphere but the momentary shape is kidney-like (Volume III, p. 84).

The only known example of "hard," practically unsolvated (unswollen) spheres from single macromolecules is that from dendritic poly(α,ε-lysine)s with monomeric units $-NH-CH(CH_2CH_2CH_2CH_2N<)-CO-$ in DMF + 1 % LiCl (see Fig. 11-1). The protein molecule apoferritin is an example of a solvated spherical molecule (Volume I, p. 635), and the protein hemoglobulin an example of a solvated spherelike associate consisting of four protein molecules (Volume I, p. 535). Solvated spherical associates of very many molecules are formed by high-molecular weight globulins (Fig. 11-1).

The viscosity η of a highly diluted solution of unsolvated compact spheres in a solvent of viscosity η_1 is a function of the volume fraction ϕ_2 of the spheres, where $K_1 = 5/2$ is the theoretically calculated **Einstein value** (Volume III, p. 407):

$$(11-5) \qquad \eta = \eta_1(1 + K_1\phi_2) = \eta_1(1 + (5/2)\phi_2) \qquad \text{(in the limit } \phi_2 \to 0)$$

At larger volume fractions ϕ_2, interactions have to be considered and η becomes

$$(11-6) \qquad \eta = \eta_1(1 + (5/2)\phi_2 + K_2\phi_2^2 + K_3\phi_2^3 + ...)$$

where the volume fraction ϕ_2 of the solute can be expressed by $\phi_2 = (V_H N_A/M_2)c$ since $\phi_2 = V_2/V = N_2 V_H/V$, $V_H = m_2/\rho_2$, $m_2 = M_2/N_A$, and $c \equiv c_2 = (N_2 M_2)/(N_A V)$ where $V_H =$ hydrodyamic volume of spheres, $N_A =$ Avogadro constant, $M_2 =$ molar mass of a sphere, $m_2 =$ mass of spheres, $N_2 =$ number of spheres, $\rho_2 =$ density of spheres, and $c =$ mass concentration of spheres, giving

$$(11-7) \qquad (\eta/\eta_1) - 1 = (5/2)(V_H N_A/M_2)c + K_2(V_H N_A/M_2)^2 c^2 + K_3(V_H N_A/M_2)^3 c^3 + ...$$

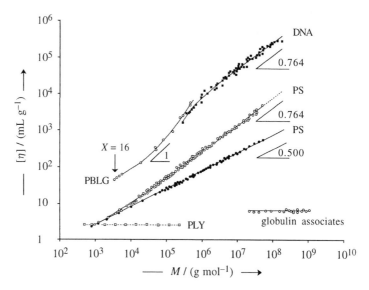

Fig. 11-1 Logarithm of intrinsic viscosities $[\eta]$ of various polymers as a function of the logarithm of the molar mass M of practically molecularly uniform ("monodisperse") polymers.
———— Empirical, - - - - theory for spheres with $\rho_2 = 1$ g/mL. X = degree of polymerization.
DNA = Wormlike double helices of deoxyribonucleic acids in dilute aqueous NaCl solutions at 20°C [1]. For clarity, values of $[\eta]$ have been multiplied by a factor 10.
PBLG = Wormlike single helices of poly(γ-benzyl-L-glutamate)s in the helicogenic solvent N,N-dimethylformamide DMF at 25°C [2]. Values of $[\eta]$ were multiplied by a factor 10.
PS = Coil molecules of flexible polymer chains of poly(styrene)s in the thermodynamically good solvent benzene at 25°C (O) and in the theta solvent cyclohexane at 35.4°C (●) [3].
PLY = Dendritic poly(α,ε-lysine)s with variable branching lengths between branching points and t-butyloxycarbonyl endgroups in DMF + 1 % LiCl at 25°C [4].
Globulin associates in water at 20°C [5].

The ratio η/η_1 is known as **relative viscosity** η_r (in the literature usually as η_{rel}). The term $(\eta/\eta_1) - 1$ is the **relative viscosity increment** (IUPAC) but literature calls it **specific viscosity**, η_{sp}, although it is not a specific quantity (see Appendix 16.4).

Division of the relative viscosity increment, $(\eta/\eta_1) - 1$, by the mass concentration c of the solute delivers $\eta_{red} \equiv \eta_{sp}/c$ which traditionally (and by IUPAC) is called **reduced viscosity** or **viscosity number** although it is neither a viscosity nor a number but a specific volume, i.e., the hydrodynamic volume occupied by the unit mass of the solute. Literature also uses a **logarithmic viscosity number**, $\eta_{ln} \equiv (\ln \eta_{rel})/c_2$, which is not a number but a specific volume without any physical meaning. The same quantity is also called **inherent viscosity**, $\eta_{inh} \equiv \eta_{ln}$, but it is not a viscosity either.

The limiting value of the reduced viscosity is the **limiting viscosity number**, $[\eta]$, commonly called **intrinsic viscosity**, which is obtained from Eq.(11-7) with the definition $[\eta] \equiv (5 N_A V_H)/(2 M_2)$ so that

(11-8) $\eta_{red} \equiv \eta_{sp}/c = [\eta] + k_H[\eta]^2 c + k''[\eta]^3 c^2 + ... \; ; \quad k_H = (2/5)^2 K_2$

where k_H = **Huggins constant**. Intrinsic viscosities can also be obtained as limiting values of the function $\eta_{ln} = f(c)$. Modern scientific literature reports values of $[\eta]$ in mL/g whereas older literature usually give $[\eta]$ in 100 mL/g or 1000 mL/g.

Molecules that form hard spheres are not only "round" but also have a homogeneous interior that does not contain solvent molecules. Because of $M/(N_A V_H) \equiv \rho_2$, such spherical molecules should have intrinsic viscosities of $[\eta] = 5/(2\,\rho_2)$ that do not depend on molecular weights (see PLY in Fig. 11-1).

Associates of spherical globulin molecules behave similarly but consist not only of globulin molecules but also of solvent molecules so that their intrinsic viscosity $[\eta]$ is greater than $5/(2\,\rho_2)$. The degree of swelling of the associates increases slightly with the "molecular weight" of the associates and so do the intrinsic viscosities (Fig. 11-1).

11.3.3 Polymer Coils in Dilute Solutions

Flexible chain molecules exist in dilute solutions as loose random coils with the momentary shape of a kidney (Volume II, p. 84). On time average, this shape can be replaced by that of a hydrodynamically equivalent sphere, which allows one to describe the concentration dependence of reduced viscosities of random coils by the same type of equation as Eq.(11-8) for hard spheres. However, constants k_H and k'' now have different meanings and values.

For hard, unsolvated spheres, both theory and experiments on glass spheres deliver $K_2 = 6.2$ (Eq.(11-6)) and thus a Huggins constant of $k_H = 0.992$. As a consequence, dilute solutions of hard spheres have low relative viscosities, for example, $\eta/\eta_1 \approx 1.026$ at $\phi_2 = 0.01$ and $\eta/\eta_1 = 1.312$ at $\phi_2 = 0.1$: spheres are not good thickeners.

Solutions of coil molecules have much greater viscosities. Flexible chain molecules form statistical coils that are swollen by the solvent and have very low segment concentrations (Volume III, p. 102). The flow of such molecules creates large friction coefficients and hence large viscosities even in dilute solutions (this Section). These effects are enhanced if polymer molecules self-associate (Section 11.3.6).

The third term of the right side of Eq.(11-8) can often be neglected at not too large concentrations c_2. In this concentration range, the resulting linear dependence of $\eta_{sp}/c = f(c)$ leads to a function $\eta/\eta_1 = f(c)$ that is convex against the c axis (not shown) and a function $\lg (\eta/\eta_1) = f(c)$ that is concave against the c axis (Fig. 11-2, left).

Strong polymer-solvent interactions swell polymer coils, which leads to high solution viscosities. For example, the poly(styrene) of Fig. 11-2 has an $[\eta] = 133$ mL/g in the good solvent benzene but only an $[\eta] = 55$ mL/g in the bad solvent decalin. In the dilute solution range, viscosities of polymer solutions in good solvents (such as PS in benzene) are always greater than those in bad solvents (such as PS in decalin). However, moderately good solvents such as methylethylketone for PS ($[\eta] = 65.5$ mL/g) may lead at higher concentrations to lower viscosities than those shown by bad ones although they produce higher viscosities in the low concentration range (see M in Fig. 11-2, left).

Coil densities vary with molecular weight M_2 of the polymers and so do the intrinsic viscosities of coil molecules (Fig. 11-1). This molecular weight dependence can be described empirically by the **Kuhn-Mark-Houwink-Sakurada equation**:

(11-9) $[\eta] = K_v M_2^\alpha$

The constants K_v and α depend on the polymer, the solvent, and the temperature, and the constant K_v also on the type and width of the molecular weight distribution.

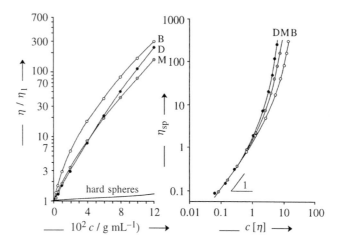

Fig. 11-2 Viscosity parameters of a poly(styrene) in the good solvent benzene (B), medium solvent methylethylketone (M), and bad solvent decalin (D); all at 25°C [6]. Left: logarithmn of relative viscosity η/η_1 as a function of the mass concentration c_2 of the polymer. For comparison, $\eta/\eta_1 = f(c_2)$ of hard spheres with $\rho_2 = 1$ g/mL. Right: Logarithm of specific viscosity, $\eta_{sp} = (\eta - \eta_1)/\eta_1$, as a function of the logarithm of the overlap parameter $c[\eta]$.

Theoretical exponents α are 1/2 for non-draining, unperturbed coils of flexible chains in theta solvents and 0.764 for non-draining perturbed coils in good solvents at sufficiently high molecular weights (see Fig. 11-1). Experimental values for flexible coils are often in the range $0.5 < \alpha < 0.764$ which is often caused by too small molecular weight ranges and systematic variations of the widths of molecular weight distributions with molecular weights. Values of $\alpha > 0.764$ are found for wormlike chains (see medium molecular weights of PBLG in Fig. 11-1) but these apply only to limited molecular weight ranges (PBLG and DNA in Fig. 11-1). See also Volume I, p. 100, and Volume III, p. 95 ff.

11.3.4   Polymer Coils in Concentrated Solutions

Single polymer coils occupy large volumes in the limit of infinite dilute solution, for example, 155 mL for 1 g of the poly(styrene) of Fig. 11-2 although the specific volume of the glassy polymer is only 0.862 mL/g. At higher polymer concentrations, polymer coils must therefore start to overlap and one can no longer expect η_{sp}/c to be a linear function of the polymer concentration c.

The concentration dependence of viscosities η or specific viscosities η_{sp} in this region is often described by empirical equations such as

(11-10) $\lg (\eta_{sp}/c) = \lg [\eta] + k_M[\eta]c$; **Martin equation (Bungenberg-de Jong eq.)**

(11-11) $\eta = K_\eta c^\gamma$; $\gamma = 2\text{-}4$

with the system-specific constants k_M, K_η, and γ. Both equations apply only to limited ranges of concentrations as one can see from plots of $\lg (\eta/\eta_1) = f(c)$ or those of $\lg \eta_{sp} = f(\lg (c[\eta]))$ (Fig. 11-2).

The greater the molecular weight, the more space is needed by a coil molecule. At a certain **overlap concentration** c^*, coils therefore start to overlap and the polymer concentration changes from "dilute" to "semi-dilute" or "semi-concentrated," respectively. However, coils do not have a solid surface and one therefore has to use one of the characteristic coil dimensions to describe c^*: the radius of gyration s, the end-to-end distance r, or one of the hydrodynamic radii R_h, i.e., $R_{D,\Theta}$ (diffusion) or $R_{V,\Theta}$ (viscosity).

The overlap concentration c_s^* with respect to the radius of gyration can be calculated from the molecular mass, $m_{mol} = M/N_A$, the molecular volume, $V_{mol} = [(4 \pi s^3)/3][N_A/M]$, and the molecular weight dependence of the radii of gyration, $s = K_s M^v$ (Volume III, p. 110) if one approximates coils as equivalent spheres and takes the radius of gyration as the radius of the equivalent sphere:

$$(11\text{-}12) \qquad c^* = (3\ M)/(4\ \pi\ N_A s^3) = [3/(4\ \pi\ K_s^3 N_A)]M^{1-3v} = K_s^* M^{1-3v}$$

In theta solvents, one has $s = s_0 = 1.28\ R_{D,\Theta} = 1.28 \cdot 1.07\ R_{V,\Theta}$ (Volume III, p. 183), $R_{V,\Theta} = 1.28 \cdot 1.07 \cdot [(3\ [\eta]_\Theta M_2)/(4 \cdot 2.5\ \pi N_A)]^{1/3}$, and therefore

$$(11\text{-}13) \qquad c_{s,\Theta}^* \approx 0.973/[\eta]_\Theta \approx 1/[\eta]_\Theta$$

The critical overlap concentration of flexible coils in good solvents is *defined* by analogy as

$$(11\text{-}14) \qquad c^* \equiv 1/[\eta]$$

Segment concentrations are considerably larger above the overlap concentration c^* than below. Coils must therefore contract in good solvents until they reach a second critical concentration c^{**} at which they adopt their unperturbed dimensions. This concentration for the change from "semi-concentrated" to "concentrated" solutions is controlled by the molecular weight M^{**} and the corresponding intrinsic viscosity $[\eta]^{**}$, respectively, at which intrinsic viscosities $[\eta]$ in good solvents start to deviate from $[\eta]_\Theta$ of theta solvents. The second critical concentration is approximately

$$(11\text{-}15) \qquad c^{**} = 0.477/[\eta]^{**}$$

The corresponding value of $[\eta]^{**}$ for poly(styrene) in benzene at 25°C is ca. 6 ml/g (Fig. 11-1) so that for this system $c^{**} \approx 0.08$ g/mL.

In plots of $\lg \eta_{sp} = \lg [(\eta - \eta_1)/\eta_1] = f(\lg c[\eta])$, one observes curves that have a linear part with a slope of unity at low values of $\lg c[\eta]$ (Fig. 11-2, right, and Fig. 11-3, left) and another linear part with a larger slope at high values of $\lg c[\eta]$ (Fig. 11-3). The slope of unity at low polymer concentrations arises because η_{sp} here is directly proportional to $c[\eta]$ (Eq.(11-8)) for non-self-associating polymers. For self-associating polymers, functions $\eta_{sp}/c = f(c)$ are more complicated than Eq.(11-8), depending on the type of self-association (see Section 11.3.6), and η_{sp} may now be proportional to $(c[\eta])^x$ with $x > 1$. Such a behavior was found for synthetic amyloses (Fig. 11-3, right).

At large values of $c[\eta]$, one always finds $\eta_{sp} \propto (c[\eta])^q$ with $q > 1$. Since $[\eta] \propto M^\alpha$ (Eq.(11-9)) and at high c also $\eta_{sp} = (\eta/\eta_1) - 1 \approx \eta/\eta_1$, one obtains

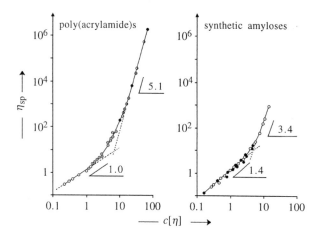

Fig. 11-3 Logarithm of specific viscosity, $\eta_{sp} = (\eta - \eta_1)/\eta_1$, at zero shear rate as a function of the logarithm of the product $c[\eta]$.
  Left: poly(acrylamide)s in water at concentrations between 0.1 and 5 % [7].
  Right: various concentrations of synthetic amyloses with degrees of polymerization between 300 and 1100 (48 600 < M < 330 000) in water or dimethylsulfoxide [8].

(11-16) $\eta/\eta_1 \propto c^q M^{\alpha q}$

Above the second critical concentration c^{**}, concentrated polymer solutions behave like melts with a melt viscosity $\eta \propto M^{3.4} = M^{\alpha q}$ (p. 88), i.e., q = 3.4/α. Since α equals 1/2 for unperturbed coils and 0.764 for perturbed ones, one obtains

(11-17) $\eta/\eta_1 \propto (c[\eta])^{6.8}$ (theta solvents)

(11-18) $\eta/\eta_1 \propto (c[\eta])^{4.45}$ (good solvents)

According to theory, only values of q = 6.8 (theta solvents) or q = 4.45 (good solvents) should be found. Such exponents are indeed found for several types of non-self-associating, flexible macromolecules in various solvents, including hyaluronates in water (Volume III, Fig. 12-5).

Deviations from these standard values are observed for some water-soluble macro-molecules (Fig. 11-3). The value of q = 3.4 for amyloses points to an exponent α = 1 in $[\eta] = f(M)$ (Eq.(11-9)) which seems likely for the narrow range of molecular weights of these wormlike macromolecules (see the slopes of lg $[\eta]$ = f(lg M) of wormlike DNA and PBLG molecules in the same molecular weight range (Fig. 11-1). The reason for the value of q = 5.1 for poly(acrylamide)s is unclear (association?).

11.3.5 Wormlike and Rigid Macromolecules

The constancy of bond angles between chain atoms, the steric hindrance and/or intra-molecular attractions between adjacent substituents leads to a certain **persistence** of chains (Volume III, p. 98) which is small for flexible chains but prominent for polymers in helicogenic solvents and for liquid-crystalline polymers.

Examples of stiffer, wormlike chains are poly(γ-benzyl-L-glutamate)s (PBLG) in *N,N*-dimethylformamide with various degrees of polymerization, *X*. These molecules need eight monomer units to generate a helix and nine units to form a stable one, but that helix is so short that the ratio of helix length to helix diameter is almost unity. Such molecules are spheroidal.

The axial ratio of helical PBLG molecules increases with increasing molecular weight and molecules becomes more rodlike, which causes α in Eq.(11-9) to increase from 0.5 at $M \approx 4400$ to 1.2 at $M \approx 80\ 000$ (Fig. 11-1). However, α never reaches the value of 2 for infinitely long, truly rigid rods since helices maintain a certain flexibility because of fluctuations about conformational and bond angles. At very high molecular weights, PBLG molecules rather behave like perturbed random coils ($\alpha = 0.764$) similar to very long double helices of DNA (Fig. 11-1). The function lg $[\eta] = f(\lg M)$ thus adopts an S-like shape for the whole range of molecular weights of wormlike molecules.

11.3.6   Associating Macromolecules

Intramolecular and Intermolecular Associations

Viscosities of polymer solutions are strongly affected by associations and aggregations. **Association** is defined in this book as reversible formation of physical bonds between chemical groups and **aggregation** as the corresponding irreversible process. Literature often uses "association" and "aggregation" as synonyms.

Associations may be between different types of molecules, for example, between a group of a polymer molecule and solvent molecules (so-called solvation) or between polymer groups themselves, either intramolecular (such as in the formation of helices) or intermolecular between groups of different polymer molecules (so-called **self-associations**). Physical bonding may be via hydrogen bridges, dipole-dipole interactions, π-π interactions, or via so-called entropic (= hydrophobic) bonds (see Fig. 11-4).

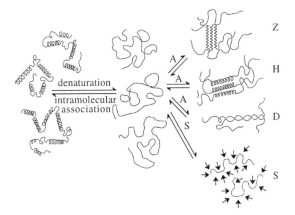

Fig. 11-4 Schematic representation of intramolecular (left) and intermolecular associations (right).

Intramolecular association can lead to partial helices (shown) or intramolecular networks (not shown); the dissolution of such protein structures is known as denaturation.

Intermolecular associations (A) may be via endgroups (not shown), by lateral combination of zig-zag chains (Z) or helices (H) or by formation of double helices (D) or triple helices (not shown). Association of amphoteric substances (S) prevents self-associations.

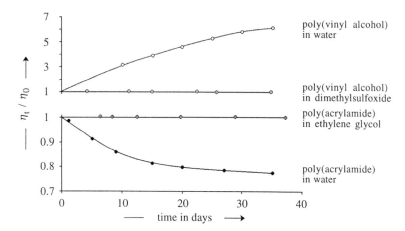

Fig. 11-5 Time dependence of the ratio η/η_0 of dynamic viscosities η at time t and η_0 at time zero of a poly(vinyl alcohol) and a poly(acrylamide), at constant concentration [9].

Poly(vinyl alcohol) (PVAL) does not associate in dimethylsulfoxide ($\eta/\eta_0 \neq f(t)$). The viscosity of its aqueous solution increases with time and so does the apparent molecular weight (not shown): in water, PVAL self-associates intermolecularly.

Poly(acrylamide) dissolves molecularly in ethylene glycol and the solution viscosity does not vary with time. The viscosity of its aqueous solution decreases with time although the apparent molecular weight remains constant: poly(acrylamide) associates intramolecularly.

Intermolecular associations are always accompanied by intramolecular ones but whether intramolecular ones also lead to intermolecular ones depends on many factors that are often not understood in detail ("wrong" groups in polymers, competing interactions polymer-polymer and polymer-solvent, cooperative effects (helix formation, generation of crystallites, etc.).

Spectroscopic methods usually detect and determine both intramolecular and intermolecular associations. Molecular weight methods sometimes allow one to distinguish between open and closed self-associations (see below) whereas viscosity measurements are not specific and provide relatively little information.

Viscometry can deliver information about the type of *intramolecular* association if intermolecular ones can be reasonably excluded. Short-range forces lead to physical bonds between neighboring groups along the chain which stiffens and shortens the chain, for example, by converting coil-like segments to helical ones. Long-range forces between groups that are separated from each other by many chain segments lead to intramolecular physical crosslinks. In both cases, molecules become more compact while maintaining a coil-like structure and the viscosity decreases with time (poly(acrylamide) in water, Fig. 11-5).

Intramolecular association does not change the concentration of viscometrically active species: the number of chemical molecules equals the number of physical ones. *Intermolecular* association, on the other hand, combines chemical molecules to fewer (but larger) physical ones. Viscosities will increase with time during such intermolecular self-associations since viscosities are affected more by the mass of the viscometrically active species than by their number (poly(vinyl alcohol) in water, Fig. 11-5). At higher concentrations, intermolecular associations may lead to highly swollen physical networks (gels; Section 11.6) or to precipitation.

Open and Closed Associations

Intermolecular asscociations comprise two borderline cases. In **open associations**, **unimers** (i.e., non-associated molecules of molecular weight M_I) are in thermodynamic equilibrium with their dimers, trimers, ..., N-mers with double, triple, ..., N-times the molecular weights of M_I, i.e., M_{II}, M_{III}, ..., M_N. The association is "open" since there is no upper limit for the molecular weight of the associates:

$$(11\text{-}19) \quad
\begin{aligned}
M_1 + M_1 &\rightleftarrows M_{II} \\
M_{II} + M_I &\rightleftarrows M_{III} \\
M_{III} + M_I &\rightleftarrows M_{IV} \rightleftarrows 2\,M_{II} \rightleftarrows 4\,M_I \quad \text{etc.} \\
\cdots\cdots\cdots\cdots\cdots\cdots & \\
M_N + M_I &\rightleftarrows M_{N+1} \quad \text{etc.}
\end{aligned}$$

The greater the mass concentration of the polymer, the greater the number of larger associates and the greater the apparent molecular weight, M_{app} (apparent molecular weights are measured at finite concentrations but calculated from experimental data by using equations for zero concentrations). In conventional plots of $\overline{M}_n/\overline{M}_{n,app} = f(c)$, ordinate values decrease with increasing polymer concentration c if the polymers self-associate in certain solvents while they increase for the same polymer in other solvents (Fig. 11-6). The strong initial decrease of $\overline{M}_n/\overline{M}_{n,app}$ with c is caused by the increasing size and proportion of associates and not by negative second virial coefficients.

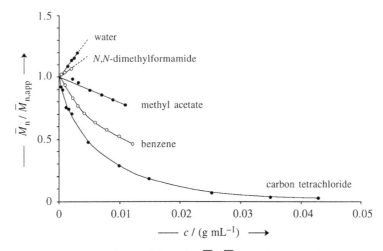

Fig. 11-6 Concentration dependence of the ratio $\overline{M}_n/\overline{M}_{n,app}$ of true number-average molecular weight and apparent number-average molecular weight as a function of the concentration c of a poly-(ethylene glycol), $HO(CH_2CH_2O)_{n-1}CH_2CH_2OH$ ($\overline{M}_n$ = 594 g/mol), in different solvents at 25°C [10]. By kind permission of the Schweizerische Chemische Gesellschaft, Zurich, Switzerland.

The polymer does not self-associate in water and DMF but does so via its OH endgroups in methyl acetate, benzene, and carbon tetrachloride. The negative initial slopes for the last three solvents are *not* caused by negative virial coefficients but by the equilibrium constants of self-association. Linearization of data for poly(ethylene glycol)s of different molecular weights (Volume III, p. 340 ff.) showed that equilibrium constants of self-association per OH endgroup are independent of the molecular weight of the polymers. At low polymer concentrations, the effect of second virial coefficients is vastly overshadowed by the effect of equilibrium constants, which contribute significantly to $\overline{M}_n/\overline{M}_{n,app}$ only at higher polymer concentrations.

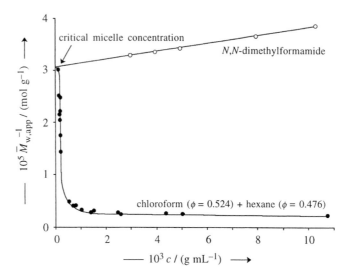

Fig. 11-7 Concentration dependence of inverse apparent weight-average molecular weights, $\overline{M}_{w,app}$, of a poly(γ-benzyl-L-glutamate) ($\overline{M}_w$ = 33 400 g/mol, $\overline{M}_n$ = 31 500 g/mol) [11].
The polymer is molecularly dissolved in DMF at 70°C but selfassociates at 25°C in a mixture of chloroform and hexane. The association is of the closed type with a calculated degree of association of N = 12 and an equilibrium constant of ^NK_c = 5·10^{55} (L/mol)^{11}.

In contrast to open associations with different species with molecular weights M_I, M_{II}, ..., M_N, only two species exist in equilibrium for **closed associations**:

(11-20) $N\,M_I \rightleftarrows M_N$

Similarly to open associations, more and more N-mers are formed with increasing polymer concentration. However, the concentration of these N-mers is so small below a so-called **critical micelle concentration** (CMC) that N-mers ("micelles") usually cannot be detected experimentally. The concentration of N-mers increases dramatically at concentrations above CMC, which causes inverse apparent molecular weights to drop dramatically (Fig. 11-7).

Similarly to open associations, functions $1/\overline{M}_{n,app}$ = f(c) and $1/\overline{M}_{w,app}$ = f(c) can be modified at higher concentrations by virial coefficients. Since polymer-solvent interactions relate to all monomeric units but polymer-polymer interactions only to certain groups or segments, self-association is possible in thermodynamically good solvents.

Viscosities of Associating Systems
Viscosities and reduced viscosities of solutions of self-associating polymers depend on many parameters: concentration, molecular weight, molecular weight distribution, association (open, closed, mixed types), equilibrium constants and degrees N of association, polymer-solvent interactions, shapes of unimers and associates, temperature, shear rates, time effects, etc. Expectedly, these effects can cause very different concentration dependencies of viscosities (compare the viscosity of various plant gums (Volume II, p. 418)).

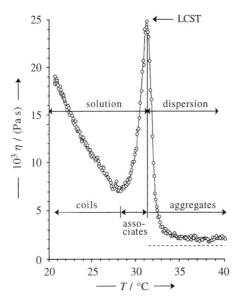

Fig. 11-8 Temperature dependence of the viscosity η of a 0.15 wt% aqueous solution of a poly(N-isopropyl acrylamide) in water at a shear rate of $\dot{\gamma} = 5.81$ s^{-1}, measured at a temperature increase of 1 K/min [12]. The lower critical solution temperature (LCST) is 31°C. The dotted line (- - -) respresents the viscosity of a 0.15 wt% dispersion of hard spheres (equals approximately the viscosity of water).

Especially strong effects can be expected at temperatures near critical solution temperatures. Exothermic systems phase separate with increasing temperature at a lower critical solution temperature (LCST, p. 422); an example is poly(N-isopropyl acrylamide) (PIAA) in water (Fig. 11-8).

At low temperatures, molecules of this polymer exist as random coils. With increasing temperature, hydrophobic interactions between isopropyl groups become stronger. The molecules become more compact and the viscosity decreases. On approach to the LCST, PIAA molecules associate intermolecularly to form loose associates which causes the viscosity to increase steeply. At the LCST, an entropically controlled phase separation sets in, the intermolecular associates cluster to more compact dispersed particles, and the viscosity drops sharply and considerably.

The dispersed aggregates are spheroidal and fairly dense. The viscosity of the dispersion is less than twice that of a dispersion of hard spheres and, at higher temperatures, practically independent of the temperature, as predicted by theory.

11.3.7 Polyelectrolytes

Many commercially used polysaccharides and other polymers are polyelectrolytes, i.e., polymers that carry many ionic groups per molecule. These groups can dissociate or associate, depending on their type, environment, and temperature.

Weakly charged groups of the same type repel each other in dilute solution, which leads to a stiffening of polymer chains. Effects of oppositely charged counterions become important at higher polyelectrolyte concentrations in dilute solution and in solutions of strongly charged polymer molecules. In these cases, triple ions $\oplus\ominus\oplus$ and $\ominus\oplus\ominus$ are more stable than ion pairs $\oplus\ominus$ or single ions $\oplus$ and $\ominus$ as estimated from potential energies (Volume III, p. 352). Higher ion aggregates are even more stable, which is the reason why NaCl forms stable ion crystals.

Reduced viscosities therefore show the following concentration dependence. In very dilute solution, polyelectrolyte molecules are substantially dissociated according to the mass action law. The many equal charges repel each other, which leads to a stiffening of chains. The resulting rodlike nature of polymer segments produces large reduced viscosities (Fig. 11-9). The dissociation of ion pairs such as $\sim M^{\oplus}_{\sim} /A^{\ominus}$ decreases with increasing polyelectrolyte concentration, which leads to more compact polyelectrolyte molecules and thus to smaller reduced viscosities.

Addition of a salt reduces the ion clouds around the chains and decreases the dissociation further. Polyelectrolytes thus show smaller reduced viscosities in salt solutions than in salt-free solvents. At high concentrations of added salt, many triple ions are formed, which increases the concentration of counterions in the interior of the coil relative to the exterior. The resulting osmotic effect causes additional water to enter the interior of the polymer coil, which extends at high concentrations of added electrolytes: reduced viscosities increase with increasing polyelectrolyte concentration (Fig. 11-9, right).

The dependence of reduced viscosities on polyelectrolyte concentration can be described by the often used empirical **Fuoss equation** which gives the inverse reduced viscosity as a function of the square root of the polyelectrolyte concentration:

$$(11\text{-}21) \qquad c/\eta_{sp} = A_{FS} + K_{FS}c^{1/2} \quad ; \quad A_{FS} \neq 1/[\eta]$$

Contrary to general belief, the intercept at $c^{1/2} \rightarrow 0$ does not deliver the inverse intrinsic viscosity, mainly because of strong effects of shear rates (see Volume III, p. 402).

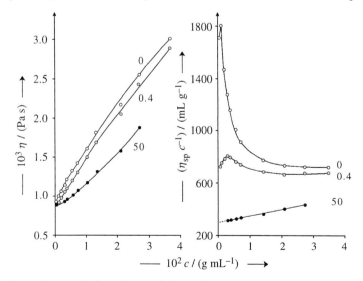

Fig. 11-9 Dependence of viscosities η (left) and reduced viscosities, η_{sp}/c, (right) on the concentration c of a sodium pectinate in water (0) and in aqueous sodium chloride solutions of [NaCl] = 0.4 mmol NaCl per liter and 50 mmol NaCl per liter, respectively, at 27°C [13].

11.3.8   Non-Newtonian Solutions

At higher shear rates, viscosities of melts and moderately and highly concentrated polymer solutions are no longer independent of the shear rate (p. 89). This non-Newtonian behavior sets in above a certain critical value $\gamma_{e,crit}$ of the elastic shear deformation γ_e. At $\gamma_e > \gamma_{e,crit}$, one can describe experimental data by a function

$$(11\text{-}22) \qquad \lg (\Theta/\Theta_0) = \lg A - \lg (\dot{\gamma}\Theta_0) \quad ; \quad A = \text{const.}$$

where $\Theta = \eta/G = \gamma_e/\dot{\gamma}$ is the ratio of viscosity η and shear modulus G which equals the ratio of elastic shear deformation γ_e and shear rate $\dot{\gamma}$, respectively, and the subscript 0 indicates the Newtonian state.

However, shear moduli and elastic shear deformations are not known in most cases. Eq.(11-22) is therefore transformed with $\Theta = \eta/G$ and $\Theta_0 = \eta_0/G_0$, respectively, to

$$(11\text{-}23) \qquad \eta/\eta_0 = AG/(\eta_0\dot{\gamma}) \; ; \; \lg (\eta/\eta_0) = \lg (AG) - \lg (\eta_0\dot{\gamma}) \quad \text{(non-Newtonian range)}$$

and the logarithm of the reduced viscosity, η/η_0, is plotted as a function of the logarithm of the product $\eta_0\dot{\gamma}$ of Newtonian shear viscosity η_0 and shear rate $\dot{\gamma}$. However, the abscissa of such plots is distorted because the shear modulus G is not constant. Some authors therefore plot $\lg (\eta/\eta_0) = f[\lg (\dot{\gamma}/\dot{\gamma}_{0,1})]$ where $\dot{\gamma}_{0,1}$ is the shear rate at which the viscosity η has dropped to 10 % of the Newtonian viscosity (Fig. 11-10). The resulting master curve was independent of the type and concentration of various polysaccharides.

A similar master curve was obtained for the whole Newtonian and non-Newtonian range by using the empirical function $\lg (\eta/\eta_0) = (-1/b) \lg [1 + (\sigma_{21,crit}^{-1}\eta_0\dot{\gamma})]^{by}]$ with y = −2/3 (Volume III, p. 507). The resulting curve was independent of temperature as well as the constitution and molecular weight of unbranched polymers.

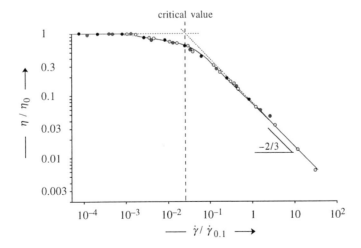

Fig. 11-10 Dependence of the ratio of apparent viscosity η and Newtonian viscosity η_0 on the ratio $\dot{\gamma}/\dot{\gamma}_{0,1}$ of shear rates for various concentrated aqueous solutions of varius polysaccharides that are used as thickeners (alginates, carageens, guarans, locust bean gum) [14]. $\dot{\gamma}_{0,1}$ is the shear rate at which the apparent viscosity η is 1/10 of the Newtonian viscosity η_0.

11.3.9   Lyotropic Liquid Crystals

Several polymers are lyotropic, i.e., they form liquid-crystalline domains in suitable solvents. The type and extent of their liquid-crystalline domains (nematic, smectic, cholesteric) (Volume III, p. 239 ff.) is controlled by the temperature, the type and interaction of polymer and solvent, the polymer concentration, and the applied shear rate.

For example, the polysaccharide xanthan in 0.1 mol/L aqueous solutions at 25°C forms relatively rigid, wormlike chains with persistence lengths 125 nm (Volume III, page 107). At a very low shear rate, the viscosity of such a solution increases strongly with increasing polymer concentration, passes through a sharp maximum, decreases, passes through a minimum, and increases again (Fig. 11-11). The greater the shear rate, the less pronounced are the initial increases of η and the subsequent maxima and minima. The difference between maxima and minima becomes smaller until it finally disappears at high shear rates.

This viscosity behavior reflects the physical structure of polymer molecules and their domains. In very dilute solutions at rest, xanthan molecules are present as isolated wormlike chains. With increasing polymer concentration, more and more of the rigid segments of these chains arrange themselves parallel to each other and form intermolecular contacts. The greater the shear rate at a given polymer concentration, the more the segments will align in the flow direction and the smaller will be the viscosity.

The solution phase separates at a certain polymer concentration that is independent of the shear rate. In this 2-phasic region, liquid-crystalline domains LC reside in an isotropic sea I at lower concentrations whereas isotropic solution droplets I exist in a liquid crystal matrix at higher concentrations (Fig. 11-11). At still higher concentrations, the liquid becomes 1-phase anisotropic.

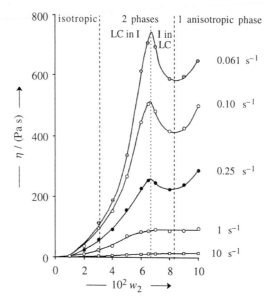

Fig. 11-11 Dependence of the shear viscosity η on the mass fraction w_2 of a xanthan in an aqueous solution of 0.1 mol/L NaCl at 25°C at various shear rates [15].
By kind permission of the Dr. Dietrich Steinkopff Verlag, Darmstadt, Germany.

11.4   Polymer Solutions

11.4.1   Oil Recovery

Conventional oil production retrieves only 15-30 % of the crude oil from reservoirs (primary oil recovery; Volume II, p. 46). Higher yields are obtained by secondary oil recovery, i.e., by reinjection of the gas that is produced as a byproduct, by pumping in CO_2 or water vapor, by vaporization of pumped-in liquid water at the high temperature of the reservoirs, or by partial burning of oil in the reservoir. Even larger oil yields of up to 45 % result from tertiary oil recovery in which aqueous solutions of alkali, detergents, or polymers are injected. These solutions should detach the highly viscous oil from the pore walls of the porous rocks and push it to the pump head.

Polymers for tertiary oil recovery must fulfill several requirements. Small proportions of these polymers must produce large additional volumes of oil at very low cost. The polymers must also be thermally stable and should not degrade by shearing or by other additives. They should not produce microgels that may clog pores of rocks. No polymer should be lost by adsorption on rocks, inclusion in pores, or complexation with multivalent metal ions. The viscosity of polymer solutions should also not be reduced by interaction with added surfactants or on contact with brine that is present in reservoirs.

These requirements are fulfilled by very few polymers. Commercially used are aqueous solutions of the polysaccharide xanthan (Volume II, p. 407) or of partially hydrolyzed poly(acrylamide)s in concentrations of 250-1500 ppm (= 0.025-0.150 %). Laboratory tests evaluated the usefulness of poly(ethylene glycol)s, poly(vinyl pyrrolidone)s, polyglucans [$\rightarrow$3β-DG$_p$1-3β-DG$_p$1-3β-DG$_p$*1$\rightarrow$] with branches via the 6 position of glucose units G$_p$*, guarans, hydroxyethyl celluloses, cellulose sulfates, and block copolymers from styrene and poly(2-vinyl pyridine) that was quaternized with $-CH_2CH_2CH_2SO_3^{\ominus}$. None of these polymers were successful in the field.

Poly(acrylamide)s and xanthans with molecular weights of several millions are useful because their aqueous solutions have high viscosities at small polymer concentrations. For example, a non-associating polymer with [η] = 1500 mL/g in a good solvent (k_H = 0.3) has a relative viscosity of $\eta/\eta_1 \approx 3.2$ at a polymer concentration of 1000 ppm ($\approx$ 0.001 g/mL) according to Eq.(11-8), where higher terms can be neglected. Viscosities can be even larger if coils associate or overlap. For example, a 0.048 g/mL solution of a poly(acrylamide) of M = 8·10^6 g/mol in an aqueous solution of 0.1 mol/L Na_2SO_4 was reported to have a Newtonian viscosity of 1580 Pa s, i.e., a relative viscosity of 1.6·10^6.

Xanthans and poly(acrylamide)s differ in their chemical structure and physical properties. Xanthans have helical segments that associate and form cholesteric mesophases at higher xanthan concentrations (see Fig. 11-11). Since helical structures are fairly stable, viscosities of xanthan solutions are influenced much less by salts in reservoirs than those of partially hydrolyzed poly(acrylamide)s. Xanthan chains also do not degrade easily on strong shearing because their helical segments are predominantly ordered laterally in liquid-crystalline domains. On the other hand, glycosidic bonds between the monomeric units of xanthans can be split relatively easily by water at the high temperatures in oil reservoirs. The resulting lower-molecular weight xanthans have drastically reduced efficiencies. Xanthans also tend to plug reservoirs with low permeabilities, probably by forming domains with liquid-crystalline structures.

Poly(acrylamide)s, on the other hand, form very loose random coils that overlap at fairly low concentrations. The resulting entanglements cannot disentangle fast enough on extensional flow which causes polymer chains to tear (Volume III, p. 513). Like xanthans, poly(acrylamide)s also hydrolyze under oil field conditions. However, their hydrolysis does not degrade chains but converts $-CONH_2$ groups into $-COOH$ groups. The resulting carboxyl groups coordinate with calcium and magnesium salts of brine in reservoirs which causes poly(acrylamide)s to precipitate. At elevated temperatures, $-CONH_2$ and $>CH-$ groups of poly(acrylamide)s can also be oxidized.

On balance of all their good and bad properties, poly(acrylamide)s fare somewhat better than xanthans, especially since poly(acrylamide)s are less expensive and can be injected more easily into oil fields than xanthans.

11.4.2 Drag Reducers

Fast-flowing liquids become turbulent above a critical Reynolds number (Fig. 11-12). The resulting eddies detach from the walls and increase the resistance to flow as measured by the ratio λ of dissipated frictional energy and applied kinetic energy per unit volume. The additional friction consumes much more energy than laminar flow.

Up to 80 % of this energy loss can be avoided by adding so-called drag reducers to flowing liquids. For example, addition of drag reducers to water for fire-fighting produces a jet with greater reach, causes more water to flow per unit time, and/or allows one to use smaller hose diameters. An 80 % reduction of friction reduces the pressure loss to 1/5 so that the required pressure can be reduced to 2 bar from 10 bar after application of drag reducers.

Injection of drag reducers also increases flow velocities of waste water during sudden downpours. Drag reducers are also added to high-pressure water jets for cutting or chopping solids. The flow of oil in pipelines is improved by adding some ppm of guar gum about every 100 km ($\approx$ 60 miles), which decreases the pumping power required.

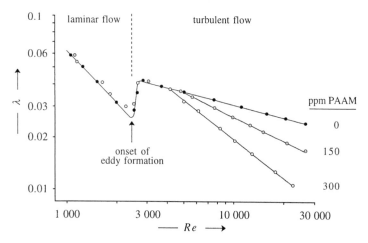

Fig. 11-12 Friction ratio λ of water and solutions of poly(acrylamide) (PAAM) in 0.5 mol/kg aqueous NaCl as a function of the Reynolds number $Re = \rho v L / \eta$ [16] where ρ = density, v = speed, L = length, and η = viscosity.

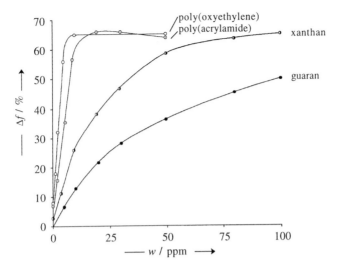

Fig. 11-13 Dependence of drag reduction Δf on the mass concentration w in parts per million (ppm) for a poly(ethylene oxide), a poly(acrylamide), a xanthan, and a guaran [17].

Drag reducers for hydrocarbon liquids are methacrylate polymers or aluminum soaps and for water mainly poly(ethylene oxide)s (PEOX) or poly(acrylamide)s (PAAM). Xanthan, guaran, and other polysccharides are less effective (Fig. 11-13). High-molecular weight PEOX and PAAM act in concentrations as low as a few parts per million. Their molecules thicken laminar layers of liquids near walls of tubes which reduces the formation of eddies. The flow in the center of the current remains turbulent.

Eddies leaving the wall sometimes dissolve abruptly, which subjects polymer molecules to a temporary elongational flow that stretches coil-like molecules and orients rod-like ones in the flow direction. The molecules break in the center if the elongational flow reaches a critical value (Volume III, p. 513). This critical value is lower for rod-like segments and molecules than for coiled ones because the former are more easily oriented in the flow direction than the latter where elastic (entropic) forces always try to reconstitute the thermodynamically more favorable coil structure. Rodlike molecules are therefore much more efficient drag reducers than the coil-like ones of the same degree of polymerization, X. However, the rodlike xanthan of Fig. 11-13 had a lower degree of polymerization than the poly(ethylene oxide) and poly(acrylamide), which made it less effective despite its rodlike characters.

11.4.3 Viscosity Improvers

With the exception of fully synthetic motor oils (engine oils), all modern engine oils for internal combustion engines contain 0.5-4 % polymers, which should improve the viscosity, the so-called viscosity index (VI), and/or the pour point (see below). Such polymers are methacrylate copolymers, ethylene-propylene bipolymers, ethylene-propylene-diene terpolymers, hydrogenated styrene-isoprene block polymers, and some hydrogenated star-like poly(isoprene)s. These polymers are sold as concentrates in oil which also includedetergents, antioxidants, and other additives.

Pour Point

Most motor oils contain paraffin waxes that have not been removed during oil refining. These waxes crystallize on cooling and the resulting crystallites form a physical network that increases the pour point of the oil. The pour point is the temperature at which an oil, fat, lubricant, etc., drops from a thermometer under its own weight.

Addition of 0.01-0.2 wt% methacrylate copolymers modifies the structure of paraffin crystallites, which increases the flow and reduces the pour point. Such polymers are bipolymers of methyl methacrylate and alkyl methacrylates with C_{12}-C_{14} alkyl groups.

Viscosity Modifiers

Oil refining also delivers light oils (Volume II, p. 51) that are not viscous enough at the temperatures at which engines (motors) operate. These oils are thus thickened by addition of polymers. For example, addition of $c = 0.01$ g/mL polymer with $[\eta] = 100$ mL/g and $k_H = 0.3$ to the good solvent base oil increases η/η_1 by a factor of 2.3 (see Eq.(11-8)). Such a viscosity increase does not seem very much, but engine oils with viscosities of, say, 3.5 Pa s and (3.5/2.3) Pa s behave completely differently with respect to temperature changes. In the past, poly(isobutylene)s were used as such viscosity modifiers of motor oils for the then fairly slow engines of American cars. However, these polymers degrade rapidly in the much faster running European motors (engines) and are therefore practically no longer used.

VI Improvers

More important is the improvement of the temperature dependence of viscosities of motor oils. Car engines run at high temperatures. The kinematic viscosity v of oils at these temperatures should therefore be neither too high nor too low. However, a medium oil viscosity at high temperatures usually means a very high dynamic viscosity η at normal temperature: the motor would not start because the cylinders would not move. Different oils are thus used for the various types of climates and motors (Table 11-3).

Table 11-3 Single-grade motor oils according to SAE (Society of Automotive Engineers).

Grade	$\dfrac{\eta_{max}}{Pa\,s}$ (start)	$\dfrac{\eta}{Pa\,s}$ (pumpability)	$\dfrac{v}{mm^2\,s^{-1}}$ (at 100°C)	$\dfrac{\eta_{max}}{Pa\,s}$ (150°C; $\dot{\gamma} = 10^6\,s^{-1}$)
SAE 0W	6.2 at –35°C	60.0 at –40°C	>3.8	
SAE 5W	6.6 at –30°C	60.0 at –35°C	>3.6	
SAE 10W	7.0 at –25°C	60.0 at –30°C	>4.1	
SAE 15W	7.5 at –20°C	60.0 at –25°C	>5.6	
SAE 20W	9.5 at –15°C	60.0 at –20°C	>5.6	
SAE 25W	13.0 at –10°C	60.0 at –15°C	>9.3	
SAE 20			5.6 - 9.3	2.6
SAE 30			9.3 - 12.5	2.9
SAE 40			12.5 - 16.3	2.9 (+0W to 10W); 3.7 (+15W to 25W)
SAE 50			16.3 - 21.9	3.7
SAE 60			21.9 - 26.1	3.7

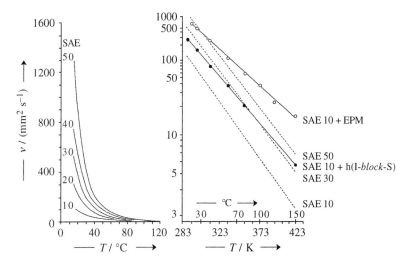

Fig. 11-14 Left: Temperature dependence of the kinematic viscosities v of single-grade motor oils SAE 50, 40, 30, 20, and 10 [18]. Right: Walther diagram with ASTM slopes for the same single-grade oils SAE 10, 30, and 50 and for the oil SAE 10 plus 2 wt% of a hydrogenated isoprene-styrene diblock polymer h(I-*block*-S) (●) or 2 wt% of an ethylene-propylene copolymer EPM (○).

Oils for continuous use are characterized by single numbers and those for the winter by an additional W. Single-grade oils such as SAE 20 are suitable for fairly narrow ranges of temperature. Multigrade oils carry two numbers. For example, the dynamic viscosity of an SAE 10W-40 oil should be smaller than 3.5 Pa s at –20°C (10W) and its kinematic viscosity at 100°C should be between 12.5 and 16.3 mm²/s (40) (Table 11-3).

The viscosities of motor oils decrease strongly with increasing temperature (Fig. 11-14). In a limited temperature range, the temperature variation can be described by the empirical **Walther function**, Eq.(11-24), where the kinematic viscosity, $v = \eta/\rho$, is given in centistokes. If dynamic viscosities were used instead, linearity would be lost.

(11-24) $\lg \lg (v + 0.8) = f(\lg T)$

The slopes of the resulting straight lines are known as ASTM slopes. The lines should be as flat as possible, which can be achieved by adding polymers to single-grade oils that then become multigrade oils (Fig. 11-14).

The efficiency of the added polymer P is measured by the ratio Q of the specific kinematic viscosities, $v_{sp} \equiv (v - v_0)/v_0$, at 100°C (formerly: 210°F = 98.9°C) and 40°C (formerly: 100°F = 37.8°C)

(11-25) $Q = v_{sp,100}/v_{sp,40}$ (c_P = constant)

The temperature dependence of the kinematic viscosity of a motor oil is described by its so-called **viscosity index (VI)** which was introduced in 1929 by the US Society of Automotive Engineers (SAE). The viscosity index is based on those two standard oils that had at the time of the introduction of the SAE scale the largest (L) and the smallest (S) temperature coefficients of kinematic viscosities.

Table 11-4 Kinematic viscosities v, Q values, and viscosity indices VI of motor oils and their blends with different VI improvers [20, 21]. PAMA = poly(methyl methacrylate-*co*-alkyl methacrylate), PIB = poly(isobutylene), SEP = styrene-ethylene-propylene-copolymer, hSB = hydrogenated styrene-butadiene block polymer, EPM = ethylene-propylene copolymer. 1 mm^2 s^{-1} = 1 centistokes.

| Oil or Blend | v/(mm^2 s^{-1}) | | | Q | VI | SAE |
	0°F	100°F	210°F			oil
Base oil 1		33.3	5.25			
PAMA concentrate		33.5	5.34			
Blend with 5.4 % PAMA	1526	55.2	9.48	1.225	152	10W20
Blend with 10 % PAMA	1777	76.6	13.50	1.215	146	10W40
Blend with 20 % PAMA	3010	164.8	29.30	1.160	142	20W60
Base oil 1		33.3	5.25			
PIB concentrate		32.6	5.30			
Blend with 10 % PIB	3270	74.0	10.75	0.859	129	20W30
Blend with 20 % PIB	5690	74.0	19.32	0.817	134	20W50
Base oil 2 + 0.8 % SEP				0.91	130	
Base oil 2 + 2.0 % SEP				0.85	136	
Base oil 2 + 0.8 % hSB				0.87	124	
Base oil 2 + 2.0 % hSB				0.85	132	
Base oil 2 + 0.8 % EPM				0.87	133	
Base oil 2 + 2.0 % EPM				0.78	133	

Standard oil L is a paraffin-rich oil (Pennsylvania oil) (VI defined as 100) whereas standard oil S is a Texas oil from the Gulf of Mexico that is rich in aromatics (VI defined as 0). Viscosities of L depend relatively little on temperature but those of S do so very much.

For the calculation of the viscosity index of a test oil (blend or unknown oil), one first measures the kinematic viscosity of this oil at 100°C. From the series of standards one then selects those two fractions 1 and 2 that have the same kinematic viscosity at 100°C as the test oil. After measuring the two kinematic viscosities v_1 and v_2 of these standards at 40°C, fractions 1 and 2 are assigned the viscosity indices 0 and 100. The viscosity index, VI = $100 (v_1 - v)/(v_1 - v_2)$, is then calculated from the kinematic viscosities v at 40°C and 100°C of the test oil.

All polymer-containing motor oils have VI values above 100 (Table 11-4), i.e., their kinematic viscosities are far less temperature dependent than those of their polymer-free base oils. For Q values, only oils containing alkyl methacrylate copolymers have values of $Q > 1$ whereas oils with added polyolefins always have $Q < 1$. This difference is caused by the difference in the temperature dependence of kinematic viscosity increments which is a measure of the change of coil dimensions with temperature.

An added polymer always increases oil viscosities. However, it is not the absolute viscosity increase that is important but the *relative* increase with temperature. Lower class single-grade oils have a relatively good viscosity-temperature profile as compared to higher classes. For example, a temperature rise from 15°C to 150°C decreased the kinematic viscosity v/(mm^2/s) of an SAE 10 oil from 118 to 2.29 after but that of an SAE 50 oil decreased more dramatically from 1305 to 5.97, which corresponds to viscosity differences of 116 mm^2/s *versus* 1299 mm^2/s and to viscosity ratios of 52 *versus* 219.

Addition of 2 wt% of a hydrogenated styrene-isoprene block polymer h(I-*block*-S) to SAE 10 increased $v/(mm^2/s)$ at 15°C from 118 to 266 and at 150°C from 2.29 to 4.9. The viscosity difference changed to 261 (*versus* 116 for the base oil) and the viscosity ratio to 54 (*versus* 52 for the base oil). These numbers indicate that the polymer addition worsened the temperature behavior of the SAE 10 oil. However, the numbers are still better than those for the higher class SAE 50 oil: 261 *versus* 1299 mm^2/s for the viscosity difference and 54 *versus* 219 for the viscosity ratio. Furthermore, the motor oil from oil SAE 10 plus polymer is now in a higher SAE class because of its larger viscosity.

Because of the double-logarithmic plot *and* the added 0.8 (Eq.(11-24)), one now obtains in the Walther diagram a less steep ASTM slope for the SAE 10 + h(I-*block*-S) oil than for the SAE 10 (Fig. 11-14, right). This plot thus overrates small absolute viscosity increases at high temperatures whereas large absolute viscosity increases show up only as small increases in the values of lg lg (v + 0.8).

The improved viscosity-temperature behavior of polymer-containing motor oils was long thought to come solely from an increase of polymer coil dimensions with increasing temperature. For example, mineral oil is a bad solvent for poly(methyl methacrylate-*co*-alkyl methacrylate)s PAMA at low temperatures but a good one at higher ones. The increasing coil dimensions lead to an increase of specific kinematic viscosities (Fig. 11-15) that compensates in part for the simultaneous decrease of the viscosity of the base oil: ASTM slopes become less steep (Fig. 11-14).

However, a temperature actuated coil expansion cannot be the sole reason for improved viscosity indices. Kinematic viscosity increments, $(v - v_0)/v_0$, of a PAMA-containing oil do increase linearly with temperature up to ca. 150°C but then pass through a maximum (Fig. 11-15) whereas those of the same oil with EPM decrease and those of the oil with h(I-*block*-S) polymer decrease, pass through a maximum, and decrease again. The latter behavior is caused by micellization of the polymer.

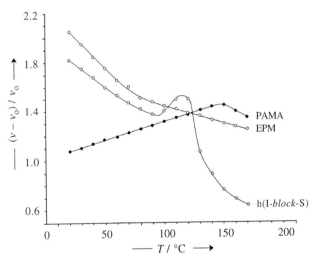

Fig. 11-15 Temperature dependence of the kinematic viscosity increments $(v - v_0)/v_0$ of base oil solutions of a poly(alkyl methacrylate) copolymer [PAMA], an ethylene-propylene copolymer [EPM], and a hydrogenated isoprene-styrene block polymer [h(I-*block*-S)] [20]. Polymer concentrations c were not reported but are usually somewhat above the critical concentrations c^* for coil overlap. $c < c^*$ would not give good thickening (no entanglements) and $c > c^*$ would lead to severe shear degradation.

Both base oils and VI index improvers are degraded thermally and oxidatively at the high temperatures of car engines. The shearing action of engine cylinders also subjects oil films to shear rates of more than 10^6 s^{-1} which leads to two effects. First, polymer-containing motor oils show shear thinning (Figs. 4-10 and 11-10), a reversible process that decreases viscosities (p. 89). Second, polymer molecules are irreversibly degraded to smaller molecular weights by extensional flow (Volume III, p. 513). For the latter reason, molecular weights of VI improvers are limited to 40 000-100 000 for ethylene-propylene-(diene) copolymers, 100 000-200 000 for hydrogented styrene-isoprene block polymers, and 100 000-350 000 for methacrylate copolymers. Polymers with narrow molecular weight distributions are better VI improvers than those with broad ones since the former have proportionally less high-molecular weight molecules.

Synthetic Motor Oils

Modern high-performance car engines need special engine oils which has led to the development of so-called synthetic oils. At present, the American Petroleum Institute (API) categorizes base oil stocks into five groups:

Group I base stocks comprise fractions from the fractional distillation of petroleum that are further treated to remove sulfur compounds, waxes, etc.

Group II comprises base stocks from the hydrocracking of fractionally distilled petroleums. Group I and II base stocks are known as **mineral oils** in the trade.

Group III is similar to Group II but has higher VI values. Group III base stocks are known as "synthetic oils" in the United States, a term that cannot be used in Germany and Japan because these oils are not really synthetic but refined mineral oils.

Group IV base stocks are truly synthetic oils consisting of poly(α-olefin)s.

Group V comprises all other synthetic motor oils such as poly(alkylene glycol)s, perfluorinated poly(alkyl ether)s, and polyol esters.

In the United States, oil changes for American cars with common American engine oils are recommended at intervals of 5000-7000 miles. Standards of European car manufacturers (original equipment manufacturers (OEM)) vary from manufacturer to manufacturer, depending on the engines. Some of the special synthetic oils for gasoline driven cars now last 2 years or 30 000 km (ca. 19 000 miles). Some of these oils contain up to 20 % additives of 10 different types such as VI improvers, detergents, dispersants, metal deactivators, defoaming agents, etc.

11.5 Polymer Dispersions

11.5.1 Overview

Dispersions are systems containing a continuous phase (the dispersion medium) and at least one discontinuous phase, the dispersed phase (L: *dis* = in different directions, *spargere* = scatter, distribute). The dispersion of the discontinuous phase may also be aided by dispersing molecules (dispersants). Dispersion media are always fluid (liquid or gas) whereas dispersed phases are always condensed (solids or liquids).

Dispersed phases always have defined surfaces with special surface properties that differ from those of the interior of the dispersed entity. In order to qualify as a "dispersion," dispersed materials must therefore have a minimum size which is about 5 nm for the diameter of a hard sphere.

Dispersions are subdivided into **aerosols** (= solids or liquids dispersed in gases), **emulsions** (liquids dispersed in liquids), and **suspensions** (solids dispersed in liquids). These older definitions of colloid chemistry differ from the use of the same words in terms like "dispersion polymerization," "emulsion polymerization," and "suspension polymerization" in polymer science (see below and Volume II, p. 156).

Examples of aerosols are fog (= water droplets in air) or smoke (solid particles in air). Emulsions usually consist of water and oils or fats, either as O/W emulsions of oil/fats in water (examples: milk, mayonnaise) or as W/O emulsions of water in fats/oils (examples: butter, ointments). Water is also important as a dispersion medium for suspensions, for example, of particles of clay or other minerals.

Polymer dispersions are always emulsions or suspensions of polymer particles in liquids, predominantly in water. A **polymer latex**, usually just called **latex**, is an aqueous dispersion of polymer particles with a glass temperature T_G below the application temperature. The term "latex" originally referred to the aqueous dispersion of natural rubber (p. 250). Polymers of **latex paints** also have glass temperatures below application temperatures because no paint film would be formed otherwise (Section 12.3.5).

Polymer latices mostly contain 40-70 wt% polymer particles with diameters of 0.05-5 μm, most often less than 1 μm. For stabilization, latices contain surfactantss in proportions of up to 10 %. Surfactantss are usually anionic molecules but there are also cationic and nonionic ones (Section 11.5.2).

Polymer dispersions are produced by two groups of methods. The main group involves a concomitant polymerization of monomers and dispersion of the resulting polymer particles by dispersion, emulsion, or suspension polymerization that is aided by dispergating agents (Volume II, p. 156 ff.). "Dispersion polymerization" is the term for a polymerizing water-in-oil *emulsion*. "Emulsion polymerization" and "suspension polymerization" do *not* refer to the *polymerization* process itself because both polymerization methods are special cases of polymerizing oil-in-water emulsions. Instead, the terms rather indicate the *properties* of the resulting dispersed phases: emulsion polymerization leads to emulsified polymer particles (example: styrene/butadiene copolymer particles with $T_G < T$) whereas suspension polymerization delivers suspended polymer particles (example: poly(styrene) particles with $T_G > T$). Polymers can also be dispersed *after* the polymerization process but this method is less frequently used.

11.5.2   Stabilization

Dispersions do not result spontaneously from the mixing of their components. They are rather thermodynamically unstable mixtures that need to by stabilized by **dispersion agents (dispersants)** that provide a *kinetic* barrier for coagulation. After some time, all dispersed materials precipitate or flocculate.

The inherent instability of dispersions is caused by the Brownian movement of the dispersed particles. As the particles approach each other, long-reaching ($\approx$ 10 nm) dis-

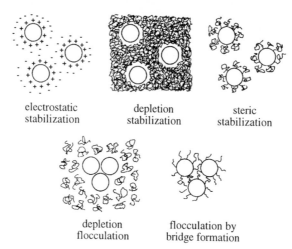

electrostatic   depletion   steric
stabilization   stabilization   stabilization

depletion   flocculation by
flocculation   bridge formation

Fig. 11-16 Stabilization and flocculation of dispersed particles.

persion forces (van der Waals interactions) induce dipoles and cause correlations in the positions of electrons relative to the nuclei of their atoms. Hence, more stable dispersions are obtained if particles repel each other. Such stabilizations can be caused by electrostatic forces between dispersed particles, by steric hindrances of adsorbed molecules, or by combinations thereof (Fig. 11-16).

Electrostatic stabilization consists of the adsorption of ionic dispersants on the dispersed particles. As a results, all particles now carry charges of the same sign (for example, positive charges in Fig. 11-16, top left) and repel each other since van der Waals attraction forces between dispersed particles are no longer effective.

Electrostatic stabilizations can only be used for aqueous dispersions whereas stabilizations by added polymers are also possible for polymers dispersed in non-aqeous solvents. Stabilizations by nonionic polymers are furthermore insensitive to added electrolytes. Two types of stabilization by polymers can be distinguished: depletion stabilization and steric stabilization.

In **depletion stabilizations**, dispersions are provided with fairly *large* concentrations of well soluble protecting colloids (for example, poly(vinyl alcohol)) or insoluble, finely dispersed inorganics (Pickering emulsifiers, for example, $BaSO_4$) (Fig. 11-16, top center). In contrast, *small* concentrations of polymers may actually promote flocculations if water is a thermodynamically bad solvent for these polymers (Fig. 11-16, bottom left).

These added substances should not adsorb on the dispersed particles. Instead, their presence should rather decrease the probability for collisions between dispersed particles and thus also that of precipitation. The stabilizers act by one or more than one of the following mechanisms. They may increase the viscosity of the dispersion medium and/or the particle/water surface tension or they may decrease the density difference between the dispersed polymer and the dispersion medium

In **steric stabilizations**, added polymers are actually adsorbed on the surface of the dispersed particles (Fig. 11-16, top right). Steric stabilizers therefore need to be amphiphilic: their hydrophobic parts adsorb on the surface of the dispersed hydrophobic particles whereas their hydrophilic parts form a protective layer that faces the hydrophilic dispersion medium (for adsorption, see Section 12.2.1).

In principle, homopolymers can do the job. However, copolymers are much more effective, especially graft and block polymers. The longer the hydrophobic sequences of these polymers, the more stable is their adsorption on the dispersed hydrophobic particles. The efficiency of steric stabilizers thus increases in the order homopolymers < random copolymers < graft polymers < diblock polymers. Triblock polymers are less effective than diblock polymers, especially if they are of the type hydrophobic–hydrophilic–hydrophobic because the two hydrophobic end-blocks may reside on different dispersed particles, which leads to precipitation (Fig. 11-16, bottom right).

Steric stabilization is caused by at least three effects. Combinatorial effects always counteract flocculations since these effects arise from the entropy of mixing of the solvent molecules of the dispersion medium with the medium-facing chain segments of the adsorbed polymeric stabilizer molecules. A second contribution stems from the difference in the free volumes of the low-molecular weight molecules of dispersants and the adsorbed macromolecules of the stabilizer; this contribution may have both entropic and enthalpic parts. Enthalpic and/or entropic contributions may also come from specific interactions between the molecules of stabilizers and the dispersion media.

The relative contribution of these effects is difficult to assess. Thermodynamically, **flocculations** may occur by cooling as well as by heating of dispersions, depending on the signs of the enthalpic and entropic terms (Table 11-5; see also next Section):

• Flocculation occurs on heating, if changes of enthalpy (ΔH) and entropy (ΔS) are both positive and the magnitude of $|\Delta H/T\Delta S|$ is greater than 1.

• Dispersions flocculate on cooling if ΔH and ΔS are both negative and $|\Delta H/T\Delta S|$ is smaller than 1.

In both cases, flocculation temperatures of the dispersions correlate strongly with the theta temperatures of free stabilizer molecules in the same dispersing medium (Table 11-5). Flocculations can thus be prevented if the dispersing medium is a thermodynamically good solvent for those segments of polymeric stabilizer molecules that face the dispersing medium and form the protective layer.

Table 11-5 Comparison of flocculation temperatures T_f of latices that have been sterically stabilized with polymers of molecular weight M with theta temperatures Θ of the same stabilizers in the same dispersing media [21]. Flocculation occurred on heating (↑) or cooling (↓) of dispersions.

Polymeric stabilizer	M	Dispersing medium		$T_f/°C$	$\Theta/°C$
Poly(oxyethylene)	10 000	0.39 mol/L MgSO$_4$ in H$_2$O	↑	45	42
Poly(oxyethylene)	96 000	0.39 mol/L MgSO$_4$ in H$_2$O	↑	43	42
Poly(oxyethylene)	1 000 000	0.39 mol/L MgSO$_4$ in H$_2$O	↑	44	42
Poly(vinyl alcohol)	57 000	2.0 mol/L NaCl in H$_2$O	↑	28	27
Poly(acrylic acid)	9 800	0.2 mol/L HCl in H$_2$O	↓	14	14
Poly(dimethylsiloxane)	3 200	heptane/ethanol (51:49 = V/V)	↓	67	67
Poly(dimethylsiloxane)	48 000	heptane/ethanol (51:49 = V/V)	↓	65	67
Poly(α-methylstyrene)	9 400	butyl chloride	↓	254	263
Poly(α-methylstyrene)	9 400	butyl chloride	↑	130	139

11.5.3   Flocculation

Flocculants are added to suspensions of fine particles in order to speed up the sedimentation or filtration of the particles. They are used to clear water for household and industrial uses as well as waste water that is produced by mining of minerals, in paper production, during commercial food preparation, etc. On a per-weight basis, more inorganic flocculants are used than organic ones; on a per-cost basis, it is just the opposite..

Inorganic flocculants comprise aluminum derivates such as aluminum sulfate hydrate $Al_2(SO_4)_3 \cdot 18 \, H_2O$, sodium aluminates $Na_2Al_2O_3$ and $NaAlO_2$, poly(aluminum chloride) $Al(OH)_{1.5}(SO_4)_{0.125}Cl_{1.2}$, various iron and silicon compounds as well as several types of lime from CaO to $Ca(OH)_2 \cdot Mg(OH)_2$.

Organic flocculants are water-soluble natural or synthetic polymers. Natural flocculants are mainly polysaccharides such as starch, guar gum, xanthan, alginic acid, sodium carboxycellulose, and tannins; they cost less than synthetic organic flocculants but require higher dosages. Most technically used synthetic organic flocculants are based on copolymers of acrylamide with various proportions of comonomers. About 45 % of these acrylamide copolymers are polyanions, 45 % are polycations, and about 10 % are nonionic. Other cationic organic polymeric flocculants are based on polymers with nitrogen chain atoms such as poly(ethylene imine), poly(vinyl imidazoline), etc.; the nitrogen atoms of these polymers are usually quaternized.

Anionic organic polymeric flocculants comprise sodium and ammonium salts of poly(acrylic acid) as well as the sodium salt of poly(styrene sulfonic acid). Nonionic organic-polymeric flocculants include poly(acrylamide) and for special purposes also poly(ethylene oxide) and poly(N-vinyl pyrrolidone). All of these polymers have molecular weights between 10^3 and 10^7.

The mechanism of flocculation is the mirror image of that of stabilization (Section 11.5.2). Polyelectrolytes as flocculants neutralize electrical charges on the surfaces of charged particles, which then no longer repel each other and flocculate. Nonionic flocculants act on sterically stabilized dispersions and form either bridges between particles or precipitate dispersed particles by a so-called depletion flocculation (Fig. 11-16).

Flocculations by **bridge formation** occur if one flocculant molecule connects two dispersed particles; this type of flocculation is thus a physical crosslinking. Such bridges can be formed by triblock polymers $A_nB_mA_n$ where the two end-blocks are adsorbed on different particles. Diblocks A_nB_m are stabilizers (see previous page).

Depletion flocculation is caused by small concentrations of polymers. It occurs when the distances between the surfaces of dispersed particles become smaller than the diameters of added polymer coils, i.e., if the concentration of dispersed particles is large. In this case, the space between dispersed particles is depleted of polymer molecules and filled instead with the pure dispersing medium. The mixing of the (almost) pure dispersing medium (the solvent) with the polymer solution decreases the Gibbs energy of mixing, which leads to flocculation. Depletion flocculations can thus be expected for small concentrations of high-molecular weight polymers, thermodynamically bad solvents for these polymers, and large concentrations of dispersed particles.

Conversely, addition of a highly concentrated solution of high-molecular weight coil-forming macromolecules to less concentrated particle dispersions will lead to an expansion of coil sizes and thus to a depletion stabilization (see Fig. 11-16).

11.6   Polymer Gels

11.6.1   Introduction

Gels are relatively stiff but soft masses consisting of small proportions of a gel-forming material and a large proportion of a liquid. They can easily be deformed by shearing but are fairly resistant against hydrostatic pressure.

The name "gel" is derived from gelatin, the degradation product of collagen. Warm solutions of gelatin congeal on cooling to gels (L: *gelare* = to freeze, to congeal). **Gellants** or **thickeners** are gelation-causing compounds such as certain polymers and some association colloids like metal soaps. **Hydrogels** are formed from aqueous solutions of such polymers. **Jellies** are foods based on hydrogels from gelatin or pectin. The term "jelly" is also used for matter with the consistency of a jelly such as a petroleum ointment. **Slimes** are naturally occurring thin gels. Gels also comprise creams, ointments, and pastes which are stiff emulsions that are not necessarily based on polymers. **Creams** have a higher water contents than **ointments**; **pastes** also contain dispersed solid particles.

Gels result from the formation of intermolecular chemical or physical bonds between the molecules of the gel-forming substance. These bonds act as crosslinking sites. If their concentration is small, i.e., lengths of crosslinks between junctions are long, and the interaction between the molecules of the liquid and the gel-forming substance is large, then the gel will swell and hold the liquid quite firmly.

Chemical gels form from certain multifunctional monomers above a critical monomer conversion, either by chain polymerization (Volume I, p. 342) or by polycondensation or polyaddition (Volume I, p. 491). The crosslinks of these gels are true chemical bonds, most commonly, covalent ones. Chemical gels do not dissolve or melt.

Physical gelation (gelling) is caused by the formation of physical bonds between segments of preformed polymer molecules. It is observed most often for water-soluble polymers, especially polysaccharides. However, several other polymers also gel, for example, atactic linear poly(styrene) in a series of organic solvents.

Physical gelation of solutions occurs on changes of temperature or solute concentrations. Like chemical gel formation, conversions solution → gel are sharp, which points to cooperative effects. In many cases, the formation of physical gels is thermoreversible. Other formations of physical gels are thermally irreversible because they are caused by the formation of crystallites that melt only at much higher temperatures.

Some physical gel formations require only very small polymer concentrations, for example, 0.6 wt% gelatin or 0.2 wt% agar, both in water. Gelling leads to strong "thickening" effects which are used in many industries, especially in the food industry.

11.6.2   Chemically Crosslinked Gels

Polymers

Gels of covalently crosslinked polymers are used for many purposes. Examples are column fillings for the separation of water-soluble polymers such as dextrans in gel chromatography, poly(acrylamide)s in electrophoresis, and ionic group-containing poly(styrene)s for ion exchange.

Crosslinked polymers from the copolymerization of 2-hydroxyethyl methacrylate with small proportions of ethylene glycol dimethacrylate or methylene bisacrylamide are used as soft contact lenses. Crosslinked silicones serve as implantations in cosmetic surgery, for example, in breast implants. Other hydrogels from crosslinked synthetic polymers are used as wound dressings for burns, coatings for synthetic organs, and reservoirs for the delivery of pharmaceuticals.

Starch phosphates are starches that are crosslinked by phosphoryl chloride or sodium trimetaphosphate. Their hydrogels are thickeners, for example, for pie fillings.

Starches grafted by acrylonitrile and poly(acrylamide)s grafted by acrylic acid absorb a hundred to a thousand times their weight of water and up to fifty times their weight of urine. One of the main uses is for disposable diapers. The world production of these **superabsorbents** (SAP; super slurpers) is more than a million tons per year.

Properties

The absorption of solvents by weakly crosslinked polymers causes swelling which starts at the surface and progresses to the interior. The speed of this process is not controlled by the diffusion coefficient of the swelling agent but by that of the segments of the polymer. The further the swelling progresses, the more it is counteracted by the elastic retraction forces of the network.

In the resulting swelling equilibrium, the Gibbs energy of mixing of polymer and solvent equals the Gibbs energy of elasticity of the gel. Thermodynamically good solvents for uncrosslinked polymers are therefore also good swelling agents for crosslinked polymers; they swell more than thermodynamically bad solvents.

The stiffness of chemically crosslinked gels increases with decreasing **degree of swelling**, $Q = V/V_0$, where V, V_0 are the volumes of the swollen and dry gel, respectively. The smaller the degree of swelling, the greater is the mass concentration $c = 1/(Qv_0)$ of the polymer in the gel where v_0 = specific volume of the dry gel.

The stiffness of a gel is given by its modulus of elasticity, E, which can be obtained by measuring the uniaxial compression of the gel. This modulus is proportional to the number concentration C_{el} of elastic network chains, $E = K''C_{el}$, where C_{el} is inversely proportional to the 3rd power of the average distance L_x between crosslinking points or, more generally, to the dimensionality d so that $E = K''C_{el} = K'/L_x{}^d$.

A crosslinked polymer swollen by a thermodynamically good solvent can be treated as a moderately concentrated solution of a polymer with an infinitely large molecular weight. In such a "solution," the average distance L_x of junctions equals the **screening length** (= **correlation length** (Volume III, p. 187)) L_{cl} of polymer solutions. This quantity got its name because the intermolecular effects of excluded volumes are screened by the much more effective influence of the junctions.

The screening length is inversely proportional to the bth power of the polymer concentration c, $L_{cl} = Kc^{-b}$. Scaling theory (Volume III, p. 124) predicts for the exponent a value of $v/(vd - 1)$ where d = dimensionality of the system and v = exponent in the dependence of the radius of gyration s on molecular weight M, $\langle s^2 \rangle^{1/2} = K_s M^v$ The modulus E should thus depend on a power z of the polymer concentration in the gel:

$$(11\text{-}26) \qquad E = K''C_{el} = K'L_{cl}{}^{-d} = (K'/K^d)c^{db} \equiv K^*c^{db} = K^*c^{vd/(vd-1)} = K^*c^z$$

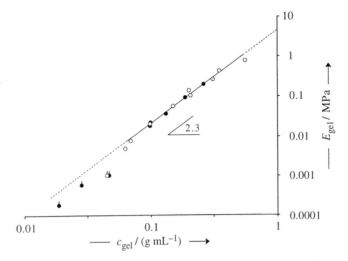

Fig. 11-17 Logarithm of the modulus of elasticity E_{gel} from uniaxial compression as a function of the logarithm of the mass concentration c of the polymer in the gel for crosslinked poly(styrene)s in benzene [22]. Styrene was crossliked with ethylene dimethacrylate (O) or divinyl benzene (●) as co-monomer. Values marked with ' were obtained at low monomer concentrations. They indicate defects in the network structure. Networks from higher monomer conversions are far more homogeneous.

Covalently crosslinked gels have dimensionalities of d = 3. Since ν = 0.588 ≈ 3/5 for polymers in good solvents, moduli of elasticity of gels swollen in good solvents should increase with increasing polymer concentration c according to $E = K^* c^{2.309} \approx K^* c^{9/4}$. The predicted exponent of ca. 2.3 was indeed found (Fig. 11-17).

11.6.3 Physically Crosslinked Gels

Gelation

Gelation to chemically crosslinked gels depends on the functionality of the parent monomers and on the type of polymerization but not on the system itself. In contrast, gelations to physically crosslinked gels are system specific, i.e., the formation and the resulting physical structure of gels depend strongly on the chemical structure of the parent polymers, the interaction of these polymers with the solvent, and the type of physical crosslinks. In order to form such crosslinks, polymer molecules must have segments that can associate at the appropriate conditions (type of solvent, polymer concentration, temperature) with segments of other polymer molecules.

Gel-forming segments must have a minimum length but should not be too long since "infinitely long" ones would assemble to long-range ordered structures that would not accommodate solvent molecules. Gel-forming polymer segments must therefore be separated from each other along the chain by longer non-gel-forming segments, i.e., polymer molecules must have a block character. Blocks can be fixed or virtual but junctions may be disordered or ordered (double helices, bundles of helices, sequences of hydrogen bond bridges, and the like (Fig. 11-4)).

In poly(vinyl alcohol)s and poly(vinyl chloride)s, for example, the length of gel-forming blocks is controlled by the number of monomeric units that form a segment with the same microconfiguration per unit, the steric sequence length. In polysaccharides, lengths of segments are often controlled by "wrong" monomeric units or branch points. Segment lengths may also be virtual if the initially formed physical crosslinking sites fixate the spatial structure of the network in such a way that other, potentially gel-forming segments can no longer diffuse to the initially formed junctions. Such gelations occur in concentrated solutions of amylose (Volume II, p. 383).

Polymer solutions gel if two or more segments of the same macromolecule reside in junctions together with segments of other macromolecules. Since a chain contains many gellable segments, critical polymer concentrations for gelation are independent of polymer molecular weights above a certain minimum value of M. For example, gelling of aqueous solutions of synthetic *amyloses* of degrees of polymerization of $300 < X < 2800$ always occurs at polymer concentrations above ca. 1 %. Gelations are also independent of overlap concentrations of polymer coils, which are in this case 7.2 % for $X = 300$ but 2.35 % for $X = 2800$. Gelation here is rather caused by an impeded crystallization. In water, small amylose molecules ($X < 100$) precipitate but larger ones ($X > 1100$) predominantly gel.

For technically used *poly(vinyl alcohol)s* (PVAL), the relevant quantity is the sequence length of syndiotactic blocks. In the trans conformation of isotactic sequences, all OH groups reside "on the same side" (Volume I, p. 124). This allows formation of intramolecular hydrogen bridges between adjacent monomeric units, which is the reason for the all-trans conformation of crystalline isotactic poly(vinyl alcohol).

Positions of OH groups alternate in the all-trans conformation of syndiotactic poly-(vinyl alcohol) chains. Hydrogen bridges between adjacent groups are not possible and the polymer chain forms a helix. However, crystallites cannot form if helical sequences are too short, i.e., if sequence lengths of syndiotactic units are not long enough. Examples are poly(vinyl alcohol)s from conventional poly(vinyl acetate)s.

The gelation of sufficiently concentrated aqueous solutions of poly(vinyl alkohol)s proceeds as follows. Heating desolvates the polymers and the solution demixes, for example, at the lower critical solution temperature (Fig. 10-4, II) of 60°C of a 0.04 g/mL solution of a PVAL of $M = 140\ 000$ containing 7 mol% of vinyl acetate units. Above this temperature, two liquid phases are formed, one rich in polymer molecules and the other one not so rich. The polymer molecules in the polymer-rich phase form short helical segments that bundle intermolecularly to small crystallites which form a network with the crystallites as junctions (Fig. 11-4). Similar crystallites are also formed on gelation of poly(vinyl chloride) in certain plasticizers.

Gelling of aqueous polymer solutions requires delicate balances between hydrophilic and hydrophobic groups of the polymer and additional processes such as formation of double or triple helices and/or crystallizations. The hydrophilic/hydrophobic balance can be shifted by addition of other chemical compounds. An example is the gelling of pectins which are naturally occurring homopolymers or copolymers of *galacturonic acid* (Volume II, p. 416). The gelling of these polymers is promoted by addition of sugars whose hydroxyl groups compete with the OH and COOH groups of pectins for the solvating water molecules. The resulting "naked" pectin molecules form hydrophobic bonds and gel.

The effect is used in the cooling of jams, for example, strawberry jams, where one uses at least the same mass of common sugar as strawberries. Since these berries contain only ca. 10 % pectins, one has about 20 times more OH groups from the sugar than from the pectin. Addition of acids such as citric acid, $HOCH(COOH)(CH_2COOH)_2$, shifts the dissociation equilibrium of galacturonic acid units which, in turn, allows to lower the proportion of sugar. The resulting uncharged COOH groups bind far less water than $COO^{\ominus}$ groups, which in turn makes the pectin more hydrophobic and promotes gelling.

Self-association of polymer chains can be pushed back by offering hydrophobic polymer groups lipophilic partners. Spaghetti foams on cooking in water because proteins (up to 14 wt% in wheat flour) denature and aggregate. The native macroconformations of these proteins possess hydrogen bridges that dissociate at higher temperatures. The spheroidal protein molecules thus open to coils, which exposes many hydrophobic groups that associate intermolecularly. The resulting physical network is insoluble and precipitates.

Addition of vegetable oils to the cooking spaghetti leads to solvations of hydrophobic protein groups of the spaghetti by the lipophilic groups of the oils. The hydrophilic groups of the attached oil molecules face the water which maintains the solubility of the protein molecules and decreases the foaming.

Properties

Physically crosslinked gels are usually characterized by their shear moduli G and not by their tensile moduli E which are related to G by $E = 2 G(1 + \mu)$ in the case of isotropic bodies. The tensile moduli of these gels are 3 times as large as the shear moduli since Poisson's ratios μ of the gels are 1/2, like those of water or natural rubber (Volume III, p. 523). Highly swollen gels should furthermore maintain their volume on deformation.

Hystereses on dynamic deformation lead to complex shear moduli, $G^* = G' + iG''$, (Volume III, p. 596 ff.). The real modulus G' (shear-storage modulus, in-phase modulus, elastic modulus) is a measure of the stiffness and shape stability of the specimen. The imaginary modulus G'' (shear loss modulus, out-of-phase modulus, viscous modulus) indicates the loss of usable mechanical energy by dissipation as heat.

Chemically and physically crosslinked gels do not differ in the dependence of moduli on the polymer concentration in the gel. In both types, moduli depend on the degree of swelling which is controlled by the concentration of junctions which in turn is a function of the polymer concentration c in the gel. These dependences lead in Eq.(11-26) (see below) to the exponent z of the concentration c of the polymer in the gel below where z = vd/(vd – 1), d = dimensionality, and v = exponent in the function $s = f(M)$, a measure of the goodness of the solvent.

However, chemically and physically crosslinked networks do differ in the dependence of moduli on degrees of polymerization. Chemical networks have an "infinitely large" degree of polymerization that remains constant on shearing except for very high shear rates that destroy the network irreversibly. In contrast, shearing constantly severs and reforms the weaker physical junctions of physical networks.

Shear moduli of physical networks thus depend on the probability of finding a segment in a junction. This probability is proportional to the square of the degree of polymerization of the polymer molecules because junctions can only be formed by addition of a segment to the segment of another molecule. Eq.(11-26) for the concentration dependence of the moduli of chemical networks has therefore to be modified for physical networks by introducing the square of the degree of polymerization of polymers:

$$(11\text{-}27) \qquad G' = KX^2c^z = K(cX^{2/z})^z$$

The shear storage modulus G' of a physically crosslinked gel should therefore depend on the power z of the variable $cX^{2/z}$ with $z = vd/(vd - 1)$. Examples are amyloses in water that self-associate at low concentrations, precipitate from more highly concentrated solutions of low-molecular weight polymers (p. 483), and physically crosslink to gels at higher molecular weights (p. 507). Since water is obviously a thermodynamically bad solvent for amyloses (otherwise they would not precipitate from water), one can set $v = 1/2$. The dimensionality d of physically crosslinked gels is between 2 and 3 (Volume III, p. 127) and can therefore be assumed to be $d = 5/2$. The function $lg\ G' = f(lg\ cX^{2/5})$ should therefore have a slope of $z = 5$ which is indeed observed (Fig. 11-18).

In good solvents with $v = 0.588 \approx 3/5$, one should expect a slope of 3.13 (≈ 3.0) and in moderately good solvents slopes of $3.13 \le z \le 5.0$.

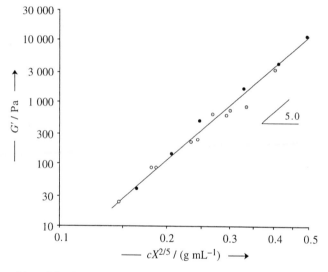

Fig. 11-18 Logarithm of the shear storage modulus G' as a function of the product of mass concentration c and effective degree of polymerization, $X^{2/5}$, of highly swollen aqueous gels of synthetic amyloses with degrees of polymerization of $X = 300$ (o), 660 (o), and 1100 (●) [23].

Literature to Chapter 11

11.1 OVERVIEW

R.L.Davidson, Ed., Handbook of Water-Soluble Gums and Resins, McGraw-Hill, New York, 3rd ed. 1980

Y.L.Meltzer, Water-Soluble Polymers, Noyes Publ., Park Ridge (NJ) 1981

E.Bekturov, E.A.Bakauova, Synthetic Water-Soluble Polymers in Solution (in Russian), Izdatelstvo "Nauka" Kazachskoj SSR, Alma Ata (1981); Synthetic Water-Soluble Polymers in Solution, Hüthig and Wepf, Heidelberg 1986

P.Molyneux, Water-Soluble Synthetic Polymers: Properties and Behavior, CRC Press, Boca Raton (FL), Vol. I (1983), Vol. II (1984); Water-Soluble Synthetic Polymers II, Macrophile Associates, London 1990

C.A.Finch, Ed., Chemistry and Technology of Water-Soluble Polymers, Plenum, New York 1983

E.W.Flick, Water-Soluble Resins, Noyes Publ., Park Ridge (NJ), 2nd ed. 1991

C.A.Finch, Ed., Industrial Water-Soluble Polymers, Royal Soc.Chem., London 1996

R.M.Fitch, Polymer Colloids, Academic Press, San Diego (CA) 1997

11.2 SOLUBILITY OF POLYMERS

C.M.Hansen, The Three-Dimensional Solubility-Parameter and Solvent Diffusion Coefficient. Their Importance in Surface Coating Formulation, Danish Technical Press, Copenhagen 1967

O.Fuchs, Gnamm-Fuchs: Lösungs- und Weichmachungsmittel, Wissenschaftl. Verlagsgesellschaft, Stuttgart 1980, 3 vols. (solvents and plasticizers; extensive list of solvnts for polymers)

A.F.M.Barton, CRC Handbook of Polymer-Liquid Interaction Parameters and Solubility Parameters, CRC Press, Boca Raton (FL) 1983; 2nd edition as: CRC Handbook of Solubility Parameters and Other Cohesion Parameters, CRC Press, Boca Raton (FL), 1991

A.F.M.Barton, Polymer-Liquid Interaction Parameters and Solubility Parameters, CRC Press, Boca Raton (FL) 1990

Ch.Wohlfahrt, CRC Handbook of Thermodynamic Data of Aqueous Polymer Solutions, CRC Press, Boca Raton (FL) 2004

Ch.Wohlfahrt, CRC Handbook of Thermodynamic Data of Polymer Solutions, CRC Press, Boca Raton (FL) 2006

Ch.Wohlfahrt, CRC Handbook of Enthalpy Data of Polymer-Solvent Systems, CRC Press, Boca Raton (FL) 2008 (3 vols.)

11.3 VISCOSITY OF POLYMER SOLUTIONS

C.W.Macosko, Rheology. Principles, Measurements, and Applications, Wiley, New York 1984

D.N.Schulz, J.E.Glass, Ed., Polymers as Rheology Modifiers (ACS Symp.Ser. No. 462), ACS, Washington (DC) 1991

R.Lapasin, S.Priel, Rheology of Industrial Polysaccharides. Theory and Applications, Blackie, London 1995

R.G.Larson, The Structure and Rheology of Complex Fluids, Oxford Univ.Press, New York 1999

F.A.Morrison, Understanding Rheology, Oxford Univ.Press, New York 2001

J.Furukawa, Physical Chemistry of Polymer Rheology, Springer, Berlin 2003

–, Handbook of Rheology Modifiers, Synapse Information Resources, Endicott (NY) 2006

11.4.1 OIL RECOVERY

K.S.Sorbie, Polymers in Improved Oil Recovery, CRC Press, Boca Raton (FL) 1990

11.4.2 DRAG REDUCERS

R.H.J.Sellin, R.T.Moses, Ed., Proceedings of the 3rd International Conference on Drag Reduction, University of Bristol, Bristol, UK 1984

W.-M.Kulicke, M.Kötter, H.Gräger, Drag Reduction Phenomena with Special Emphasis on Homogeneous Polymer Solutions, Adv.Polym.Sci. **89** (1989) 1

11.4.3 VISCOSITY IMPROVERS

D.Klamann, Lubricants and Related Products, Verlag Chemie, Weinheim 1984

E.R.Booser, Ed., CRC Handbook of Lubrication (Theory and Practice of Tribology), CRC Press, Boca Raton (FL) 1984

11.5 POLYMER DISPERSIONS

T.Sato, R.J.Ruch, Stabilization of Colloidal Dispersions by Polymer Adsorption, Dekker, New York 1980

K.O.Calvert, Polymer Latices and Their Applications, Macmillan, New York 1982

J.W.Goodwin, Ed., Colloidal Dispersions, Royal Soc.Chem., London 1982

Th.F.Tadros, The Effect of Polymers on Dispersion Properties, Academic Press, London 1982

P.Becher, Ed., Encyclopedia of Emulsion Technology, Vol. 1, Basic Theory, Dekker, New York 1983

R.D.Vold, M.J.Vold, Colloid and Interface Chemistry, Addison-Wesley, Reading (MA) 1983

D.Napper, Polymeric Stabilization of Colloidal Dispersions, Academic Press, London 1983

R.Buscall, T.Corner, J.F.Stageman, Ed., Science and Technology of Polymer Colloids, Elsevier, New York 1985

S.E.Friberg, P.Bothorel, Ed., Microemulsions: Structure and Dynamics, CRC Press, Boca Raton (FL) 1987

I.Piirma, Polymeric Surfactants, Dekker, New York 1992

R.B.McKay, Ed., Technological Applications of Dispersions, Dekker, New York 1994

M.Bourrel, R.S.Schechter, Microemulsions and Related Systems, Dekker, New York 1998

D.Distler, Ed., Wässrige Polymerdispersionen, Wiley-VCH, Weinheim 1998

J.C.T.Kwak, Polymer-Surfactant Systems, Dekker, New York 1998

K.Esumi, Ed., Polymer Interfaces and Emulsions, Dekker, New York 1999

E.Kissa, Dispersions. Characterization, Testing, and Measurement, Dekker, New York 1999

I.D.Morrison, S.Ross, Colloidal Dispersions. Suspensions, Emulsions, and Foams, Wiley, Hoboken (NJ), 2nd ed. 2002

D.Urban, K.Takamura, Eds., Polymer Dispersions and Their Industrial Applications, Wiley-VCH, Weinheim 2002

T.F.Tadros, Ed., Colloids and Interface Science, Wiley-VCH, Weinheim. Volume 1: Colloid Stability. The Role of Surface Forces, Part I (2006); Volume 2: Colloid Stability. The Role of Surface Forces, Part II (2006); Volume 3: Colloid Stability and Application in Pharmacy (2007); Volume 4: Colloids in Cosmetics and Personal Care (2007); Volumes 5 and 6 in preparation

11.5.3 FLOCCULATION

K.J.Ives, Ed., The Scientific Basis of Flocculation, Sijtjoff and Noordhoff, Alpen aan den Rijh, Netherlands, 1978

J.Bratby, Coagulation and Flocculation, Croydon, UK, 1980

G.Kaluza, Flocculation of Pigments in Paints–Effects and Cause, Progr.Org.Coatings **10** (1982) 289

B.Dobiáš, Coagulation and Flocculation, Dekker, New York 1993

J./Bratby, Coagulation and Flocculation in Water and Wastewater Treatment, IWA publ., Seattle (WA) 2006

11.6 GELS

G.O.Phillips et al., Gums and Stabilizers for the Food Industry, Elsevier, London, Volumes **3** (1986),
   4 (1988), **5** (1990)

J.R.Mitchell, D.A.Ledward, Ed., Functional Properties of Food Macromolecules, Elsevier, Amsterdam 1986

A.H.Clark, S.B.Ross-Murphy, Structural and Mechanical Properties of Biopolymer Gels, Adv.Polym.Sci. **83** (1987) 57

W.Burchard, S.B.Ross-Murphy, Ed., Physical Networks: Polymers and Gels, Elsevier Appl.Sci., London 1990

J.Guenet, Thermoreversible Gelation of Polymers and Biopolymers, Academic Press, London 1992

J.Cohen-Addad, Ed., Physical Properties of Polymeric Gels, Wiley, New York 1996

Y.Osada, K.Kajiwara, Gels Handbook, Academic Press, San Diego 2000 (4 vols.)

Y.Osada, A.R.Khokhlov, Polymer Gels and Networks, Dekker, New York 2002

A.Steinbüchel, S.K.Rhee, Eds., Polysaccharides and Polyamides in the Food Industry, Wiley-VCH, Weinheim 2005

11.6.2 CHEMICALLY CROSSLINKED GELS

F.L.Buchholz, A.T.Graham, Ed., Modern Superabsorbent Polymer Technology, Wiley, New York 1997

11.6.3 PHYSICALLY CROSSLINKED GELS

J.-M.Guenet, Thermoreversible Gelation of Polymers and Biopolymers, Academic Press, London 1992

K. te Nijenhuis, Thermoreversible Networks, Adv.Polym.Sci. **130** (1997) 1

References to Chapter 11

[1]   J.Eigner, P.Doty, J.Mol.Biol. **12** (1965) 549 (compilation of literature data)

[2]   P.Rohrer, H.-G.Elias, unpublished data

[3]   F.Abe, Y.Einaga, H.Yamakawa, Macromolecules **26** (1993) 1891

[4]   S.M.Aharoni, N.S.Murthy, Polym.Commun. **24** (1983) 132, recalculated from Table 1

[5]   R.Geddes, J.D.Harvey, P.R.Wills, Biochem.J. **163** (1977) 201

[6]   D.J.Streeter, R.F.Boyer, Ind.Engng.Chem. **43** (1951) 1790

[7]   W.-M.Kulicke, R.Kniewske, J.Klein, Progr.Polym.Sci. **8** (1982) 373, Fig. 31

[8]   M.J.Gidley, Macromolecules **22** (1989) 351, Fig. 1

[9]   W.-M.Kulicke, M.Kötter, H.Greiger, Adv.Polym.Sci. **89** (1989) 1, Fig. 16

[10] H.-G.Elias, R.Bareiss, Chimia **21** (1967) 53, Fig. 1

[11] H.-G.Elias, J.Gerber, Makromol.Chem. **112** (1968) 122, Fig. 2; J.Gerber, H.-G. Elias, Makromol.Chem. **112** (1968) 142, Fig. 1

[12] K.C.Tam, X.Y.Wu, R.H.Pelton, Polymer **33** (1992) 436, Fig. 3

[13] D.T.F.Pals, J.J.Hermans, Rec. Trav. **71** (1952) 433

[14] E.R.Morris, A.N.Cutler, S.B.Ross-Murphy, D.A.Rees, J.Price, Carbohydr.Polym. **1** (1981) 5, Fig. 32

[15] R.Oertel, W.-M.Kulicke, Rheol.Acta **30** (1991) 140, Fig. 5

[16] W.Interthal, Angew.Makromol.Chem. **123/124** (1984) 387, Fig. 3

[17] P.R.Kenis, J.Appl.Polym.Sci. **15** (1971) 607, Fig. 1

[18] H.G.Müller, Tribology International (June 1978), Fig. 1; Angew.Makromol.Chem. **67** (1978) 61, Figs. 5 and 7

[19] T.W.Selby, ASLE Transact. **1/1** (1957) 68, Tables 1 and 2

[20] G.Ver Strate, M.J.Struglinski, Polym.Mat.: Sci.Eng. **61** (1989) 252, Fig. 3

[21] D.Napper, Polymeric Stabilization of Colloidal Dispersions, Academic Press, Londom 1983

[22] G.Hild, R.Okasha, M.Macret, Y.Gnanou, Makromol.Chem. **187** (1986) 2271, various tables

[23] Calculated from data of A.H.Clark, M.J.Gidley, R.K.Richardson, S.B.Ross-Murphy, Macromolecules **22** (1989) 346, Fig. 5 (data: personal communication by M.J.Gidley, 30 August1991)

12   Coatings and Adhesives

12.1   Introduction

Polymers are used not only as plastics, fibers, and elastomers (Chapters 6-10) but also in large amounts in adhesives and coatings as adherents, binders of particulates, and thickeners. **Adhesives** (Section 12.3) include **sealants** whereas **coatings** (Section 12.4) also comprise paints and lacquers. In a wider sense, "coating" is also related to **printing inks** (Section 12.5) as well as to **depots** for active substances (Section 12.6). In the vast majority of applications, adhesives, sealants, coatings, printing inks, and depots are obtained from solutions; the use of dispersions and melts is not so prominent.

In 2000, adhesives consumed $4 \cdot 10^6$ t/a of synthetic polymers, dispersion-based paints $7.5 \cdot 10^6$ t/a, and organic solvent-based paints and lacquers $15.5 \cdot 10^6$ t/a,. For costs and ecology, organic solvent-based adhesives and paints are slowly being closed out.

Coatings are now based mainly on synthetic polymers whereas one-half of the consumed adhesives stems from natural polymers and the other half from synthetic ones. Coatings use predominantly thermoplastics and thermosets; rubbers and thermoplastic elastomers play a minor rôle.

The market for adhesives and coatings is split among very different polymers, which contrasts with the situation for fibers, rubbers, thermoplastics, and thermosetting resins where *four* polymers each control the markets: thermoplastic polyesters, polyamides, acrylics, and poly(olefin)s constitute 95 % of the market for synthetic fibers; polyurethanes, amino resins, phenolics, and unsaturated polyesters govern 90 % of the use of thermosetting resins; poly(ethylene)s, poly(vinyl chloride), poly(propylene), and poly-(styrene) account for 85 % of thermoplastics; and styrene-butadiene, butadiene, ethylene-propylene, and solid acrylonitrile-butadiene rubbers make up 79 % of synthetic rubbers. In contrast, just 50 % of the adhesives and coatings market are controlled by the *eight* largest polymer groups. This market is spread much more evenly among various polymers, which reflects the much more diverse demands on adhesives and coatings.

The market for permanent adhesives belongs almost exclusively to synthetics since older natural glues such as collagen-based bone glue (Volume II, p. 544) have fallen out of favor and starch-based sizes are only temporary adhesives. Nature does produce **bioadhesives** with excellent adhesive powers but there are problems with availability, harvesting, and work-up of these materials. At present, bioadhesives are being extensively researched because of health/safety and environmental concerns about synthetics.

Bioadhesives are biological polymers with many different chemical compositions. Most of them are proteins or polysaccharides. A bacterial exopolysaccharide (= microbial polysaccharide (Volume II, p. 372)) has been tested recently as a wood adhesive. Other examples are the fiber-reinforced protein adhesives of mollusks where the matrix consists of proteins that are rich in 3,4-dihydroxyphenyl-L-alanine (DOPA). The fibers of these **marine cements** vary with the species: collagen-resembling compounds in mussels, silk protein types in reef-forming worms, and β-chitin in tunicates. Mussel adhesive proteins are used to attach cells to plastics in tissue culture experiments.

Marine cements are not to be confused with **marine glues** which were used to close gaps between wooden planks of ships. These man-made materials consisted of asphalt, rubber, or shellac as a binder and petroleum or carbon disulfide as a solvent.

12.2 Polymers in and at Surfaces

Polymers are the essential components of coatings and adhesives but not necessarily the main ones. As adhesives, they connect the surfaces of two materials, and as binders they attach pigments of coatings to surfaces. The action of these polymers depends on the chemical structures of polymer molecules as well as on their physical structures near the surfaces (polymers, metals, glass, wood) which, in the case of polymer surfaces, is affected by the top surface regions of polymeric materials.

12.2.1 Polymer Surfaces

In common language, "surface" denotes both the *outer boundary* of an object as well as the material *layer* that constitutes such a boundary to other matter. In science, "surface" also refers to the boundary and the upper layer of condensed matter (solid, liquid) but usually only to a surrounding gas. "Surfaces" are thus distinguished from "interfaces" between two solids, two non-miscible liquids, or a solid and a non-dissolving liquid.

Polymer molecules in the surface of polymeric melts or solids usually have quite different physical structures than the same molecules in the interior of melts or solids (Volume III, p. 273). The physical structure of such surface regions depends on processing conditions and often also on the contacting environment (gas, liquid). It may also change with time because polymer segments near surfaces may be somewhat mobile even below their glass temperature. Polymer surfaces can also be slowly plasticized in contact with the atmosphere. The composition of an interface layer may be furthermore affected by the contacting layer since the layer attempts to have the lowest possible Gibbs interfacial energy.

Surfaces (and interfaces) are usually also "rough", i.e., the ratio of true (micro and submicro) surface to geometric (macro) surface is larger than unity. In general, surfaces and interfaces are not clean because they contain adsorbed, absorbed, and/or adherent substances such as oxygen, water, and fats.

12.2.2 Adsorption of Macromolecules

"Adsorption" is the physical binding of "alien" molecules (the adsorbate or adsorptive) to a solid surface (the adsorbent) (L: *ad* = to, toward; *sorbere* = to drink in, suck).

Adsorption of macromolecules from their melts or solutions on solid surfaces differs characteristically from that of dissolved or gaseous low-molecular weight substances (Volume III, p. 286 ff.). In a first approximation, small molecules are more or less spheroidal; their diffusion to and adsorption on a solid surface is fast. After a fairly short time, the adsorbed amount becomes constant and the physical structure of the adsorbed layer no longer changes with time.

The mass of the adsorbed molecules is proportional to the area of the covered surface. According to Langmuir, the fraction f_O of the covered surface of the adsorbent is a function of the mass concentration c of the low-molecular weight adsorbate:

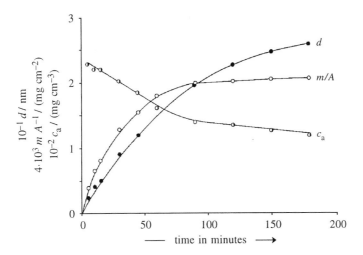

Fig. 12-1 Thickness d, adsorbed mass m per area A, and concentration c_a in the adsorbed layer as a function of time during the adsorption of a poly(styrene) PS ($\overline{M}_n$ = 176 000 g/mol) on a chromium surface from a 5 mg/mL solution of PS in the thermodynamically bad solvent cyclohexane [1].

(12-1) $f_O/(1 - f_O) = K \cdot c$

The adsorbed mass per area increases with time and becomes constant, which is also observed for dissolved polymer molecules as adsorptives (Fig. 12-1). However, this constancy of the adsorbed amount does not signal an adsorption equilibrium. According to ellipsometry (the measurement of the state of polarization of reflected light), thicknesses of adsorbed layers continue to increase with time whereas the concentration of polymer molecules in this layer decreases further.

This time dependency is caused by two effects. The poly(styrene) of Fig. 12-1 has a molecular weight distribution from which the faster diffusing small molecules are adsorbed first. These molecules are later replaced by slower diffusing larger molecules which have more contact points per molecule. If the adsorbed molecules form unperturbed random coils in solution (such as for poly(styrene) molecules in cyclohexane at 35°C) and if the random coil structure remains practically unchanged on adsorption, then the *thickness d* of the adsorbed layer must increase with the *number* concentration of monomeric units. However, since high-molecular weight chain molecules have smaller coil densities than low-molecular weight ones (Volume III, p. 102), polymer *weight concentrations* c_a in the adsorbed layer will decrease with time (Fig. 12-1).

Physical structures of adsorbed chain molecules may also change with time since they can be adsorbed in many different ways, which contrasts with the behavior of gas molecules, which can be approximated as spheres for many purposes. For example, the adsorption of coil molecules via their endgroups leads to polymer brushes (Fig. 12-2, left). The mass as well as the thickness of the adsorbed layer here are directly proportional to the molecular weight of the polymer (Volume III, p. 290 ff.). At high concentrations of contact points, chain molecules are very much stretched (Fig. 12-2, left), whereas at low concentrations of contact points they maintain their coil structure except, of course, for the polymer segments near the contact point. In both limiting cases for adsorptions via endgroups (brush and coil), Langmuir adsorption isotherms are observed.

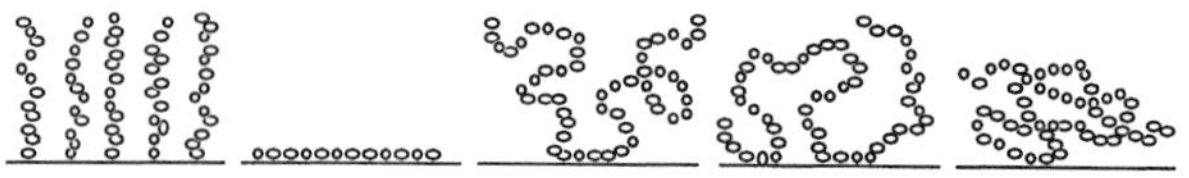

	brushes	two-dimensional coils	three-dimensional coils	loops and trains	spheres or collapsed coils
m_a	$\sim M$	$\neq f(M)$	$\neq f(M)$	$\sim M^a$	$\sim M^{1/3}$
d_a	$\sim M$	$\neq f(M)$	$\sim M^{1/2}$	$\neq f(M)$	$\sim M^{1/3}$

$$0 \leq a \leq 1/2$$

Fig. 12-2 Types of adsorptions of chain molecules on solid surfaces. m_a = adsorbed mass, d_a = thickness of adsorbed layer, M = molecular weight of adsorbate.

In the other limiting case, all monomeric units carry adsorbable groups and all these groups are adsorbed. This scenario requires isolated polymer coils (i.e., highly diluted polymer solutions) and a large adsorption energy per monomeric unit. In this case, adsorbed polymer chains will lie flat on the surface of the adsorbent where they will form two-dimensional coils (Fig. 12-2, second from left). Neither the adsorbed mass nor the thickness of the adsorbed layer will be affected by the molecular weight of the polymer. Again, the adsorption will follow Langmuir's adsorption isotherm.

Depending on the polymer concentration and the thermodynamic goodness of the solvent, several possibilitiess exist if adsorptions proceed via monomeric units but adsorption energies are not very large (Fig. 12-2). Coil molecules in theta solvents adopt their unperturbed dimensions so that the radius of gyration, s_0, is proportional to the square root of the molecular weight, $s_0 \sim M^{1/2}$. If unperturbed coils are adsorbed more or less intact in one molecule thick layers, maximum thicknesses d_a of adsorbed layers should be twice the radius of gyration so that $2 s_0 \sim d_a \sim M^{1/2}$ (Fig. 12-2, center). In this case, the adsorbed mass per area, $m_a = \rho_{mol} V_a$, should be independent of the molecular weight since the density ρ_{mol} of the molecule is proportional to $1/s_0^3$ (which in turn is proportional to $1/M^{3/2}$) and the volume V_a of the adsorbed molecule is proportional to $s_0^3 \sim M^{3/2}$.

The predicted function $s \sim M^{1/2}$ is indeed approximately found for the radii of gyration of poly(styrene) molecules that were adsorbed from a theta solvent (Fig. 12-3). At higher molecular weights, radii of gyration become less proportional to $M^{1/2}$, which points toward an adsorption via loops and trains (Fig. 12-2, second from right). Such adsorptions can be expected for greater polymer concentrations and higher molecular weights because chains begin to overlap and coils begin to interpenetrate each other.

Melts of polymer chains should behave like theta solutions since in both cases chain molecules are present as unperturbed coils. Entanglements of high-molecular weight chain molecules lead to layers that are several molecule diameters thick.

The adsorption of polymers from solution is also affected by the thermodynamic goodness of the solvent. In a thermodynamically bad solvent (theta solvent), contacts between polymer segments and the adsorbent are energetically preferred over contacts with segments of other polymer molecules. Chromium surfaces therefore adsorb poly(styrene) molecules from their solutions in the theta solvent cyclohexane at $\Theta = 35.4°C$ but not from their thermodynamically good solvent 1,4-dioxane. The more polar the polymer molecule, the more contacts it can make to polar surfaces, which leads to loops and trains (Fig. 12-2, second from right) and a more compact layer.

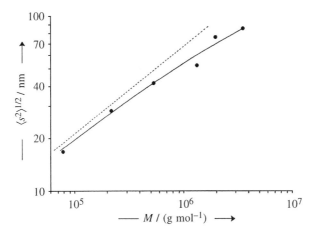

Fig. 12-3 Molecular weight dependence of the square root of the mean square unperturbed radius of gyration of adsorbed polymer molecules in the adsorption of poly(styrene)s on chromium surfaces from their solutions in cyclohexane at 36°C (theta solution) [1b].
● Experimental data, - - - theoretical function $\langle s^2 \rangle_0^{1/2} = f(M)$ for unperturbed poly(styrene) coils in cyclohexane solutions, —— empirical function $\langle s^2 \rangle^{1/2} = f(M)$ for molecules in the adsorbed layer.

12.2.3 Wettability

The wetting of solid surfaces S by liquids L (solvents or melts) is an adsorption of molecules that is controlled by the **interfacial tension** γ_{SL} between the solid and the liquid. According to the **Young equation**,

$$(12-2) \qquad \gamma_{SV} = \gamma_{SL} + \gamma_{LV} \cos \vartheta$$

interfacial tensions can be obtained from the surface tensions γ_{SV} between solids S and solvent vapors V and γ_{LV} between liquid solvents L and their vapors V and the cosine of the **contact angle** ϑ between the liquid and its solid support (see Volume III, p. 275 ff., for details). The cosine of the contact angle, $\cos \vartheta$, describes the wettability of a solid surface by a liquid (Fig. 12-4).

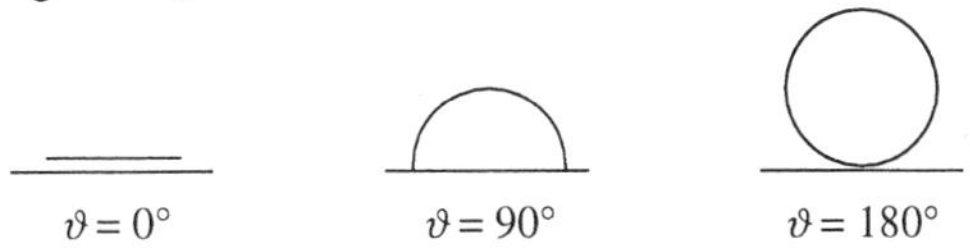

$$\vartheta = 0° \qquad\qquad \vartheta = 90° \qquad\qquad \vartheta = 180°$$

Fig. 12-4 Contact angles ϑ of liquids on a solid surface. The surface is completely wettable if $\cos \vartheta = 1$ (left) and completely unwettable if $\cos \vartheta = -1$ (right).

The quantities of Eq.(12-2) are not easy to measure and one therefore often determines a so-called **critical surface tension** γ_{crit} of the solid from the dependence of contact angles ϑ of various liquids on this solid on the surface tension γ_{LV}. In the **Zisman method**, one plots γ_{LV} as a function of $(1 - \cos \vartheta)$ and extrapolates the values of γ_{LV} to $\cos \vartheta \to 1$ (see Volume III, p. 281 ff.):

$$(12-3) \qquad \gamma_{LV} = \gamma_{crit} - K (1 - \cos \vartheta)$$

The critical surface tension γ_{crit} of many polymers equals approximately the interfacial tension γ_{SV}. Examples are γ_{crit} values of 18.5 mN/m for poly(tetrafluoroethylene), 33 mN/m for poly(ethylene), 39 mN/m for poly(vinyl chloride), and 46 mN/m for polyamide 6.6. In general, values of γ_{crit} increase with the polarity of the polymers.

All these polymers are not wetted by water which has a surface tension of 72 mN/m. Likewise, poly(tetrafluoroethylene) (γ_{crit} = 18.5 mN/m) is not wetted by oils and fats (γ_{LV} = 20-30 mN/m) which makes it useful for coatings on frying pans since it also has a high melting temperature of 327°C. Because this kind of teflon (see p. 345) is difficult to process and may also cause health problems on overheating, many cooking utensils are now coated with poly(*p*-phenylenesulfide) which apparently has a similar γ_{crit}.

Critical surface tensions of polymers are difficult to correlate quantitatively with measures of polymer polarities such as dipole moments. One of the problems is that chemical compositions of surface regions of solid polymers differ from those of the polymer interiors (see below). These compositions can be probed by various methods, for example, by secondary ion mass spectroscopy (SIMS) to depths of less than 1 nm and by electron spectroscopy for chemical analysis (ESCA) to depths of less than 5 nm.

Polymer surface regions are often rich in certain polymer groups. For example, the surface regions of solid aromatic polyimides tend to be richer in polar imide carbonyls than in other groups. In contrast, surface regions of poly(vinyl pyridine)s have a greater concentration of apolar CH_2 and CH groups than the interior. The 1-10 nm thick surface region of styrene-ethylene oxide block polymers S_i–EO_j is rich in styrene units: according to ESCA, it contains 30 % S units if the polymer composition is S_{10}–EO_{90} and 83 % S units if it is S_{30}–EO_{70}. The deviation of the average composition of the surface region from that of the interior depends on the history of the specimen, including the contact materials or the solvents used in processing.

Surface regions can be modified by several methods: mechanical roughening, swelling by solvents, chemical reactions, corona discharges, and treatment with flames or a gas plasma in nitrogen. These procedures introduce new chemical groups, modify surface structures, and/or roughen surfaces, all of which change wettabilities and the adhesive strength of joints, depending on the substrate and the treatment (Table 12-1).

Table 12-1 Effect of the type of surface treatment of substrates on the force F that is required to fracture a material consisting of a polymer that is glued to a substrate by an epoxide/versamide [2]. Fracturing was by overlap shearing (see Fig 12-5, left). Fracture occurred in the polymer itself (P), in the glue layer (G), or in aluminum as the substrate (Al).
Al = aluminum, HDPE = high-density poly(ethylene), NR = natural rubber, PC = polycarbonate, PET = poly(ethylene terephthalate), PP = poly(propylene), PVC = poly(vinyl alcohol), Q = silicone.

Polymer	Substrate	Required force F in Newton for fracture after treatment by						
		corona	flame	oxidation	solvent	no treatm.	roughening	plasma
PVC	Al					1 P	10 P	22 P
Q	Al				4 P		6 P	27 P
NR	NR					6 G	48 G	34 P
PP	PP	49 G	137 G	1380 P				430 P
PC	PC				590 G		1250 G	690 G
PET	Al			52 P	790 P	800 P	990 P/Al	1410 Al
HDPE	HDPE	550 G	640 G	680 G				1620 P

Fig.12-5 Schematic representation of spreading on a rough surface. Left: incomplete wetting; center: complete wetting; right: wetting with contraction of the coating.

For good adhesion, adhesive and substrate should have many strong bonds between their surface groups, which requires (a) large contact areas, (b) good wetting of the substrate by the adhesive, and (c) formation of covalent or at least very polar physical bonds between adhesive molecules and the surface groups of the substrate.

Surface treatment can change some or all of these conditions to various extents so that very different results are obtained (Table 12-1; see also Section 12.3). Some treatments increase the contact area of the adherent and therefore the **roughness** of surfaces, i.e., the ratio of true (microscopic and submicroscopic) surface area and geometric surface area. Freshly cleaved mica has a roughness of 1 whereas polished metal surfaces have roughnesses of 1.5-2. A good wetting agent will bind more strongly on a rough surface (Fig. 12-5) whereas a bad wetting agent will do better on a smooth one, everything else being equal. Of course, roughening may introduce new chemical groups besides changing microscopic and submicroscopic surface structures. Hydrophilicity is improved by providing the surface with more effective hydrophilic groups.

The Holy Grail of hydrophobicity is contact angles of $\vartheta_c = 180°$ for advancing and receding water droplets ($\vartheta_{adv}/\vartheta_{rec} = 180°/180°$). These superhydrophobicities are often called **Lotus effects** because of the especially hydrophobic leaves of the South Asian aquatic Lotus plant *Nelumbo nucifera* (not the Mediterranean shrub *Zizyphus lotus*).

Lotus effects can be obtained from porous surfaces composed of a polymer with an areal fraction $a_P = 1 - a_A$ and a contact angle ϑ_P and air with a_A and $\vartheta_A = 180°$. In this case, **Cassie's law**, $\cos \vartheta_c = a_P \cos \vartheta_P + a_A \cos \vartheta_A$, reduces to $\cos \vartheta_c = a_P(1 + \cos \vartheta_P) - 1$. The Lotus effect will thus be greater, the smaller a_P and the more ϑ_P approaches 180°.

For example, water droplets on a plane surface of hydrophobic isotactic poly(propylene) have a contact angle of 116°. Dissolution of the polymer in hot *p*-xylene, casting the solution on a substrate, cooling, and subsequent evaporation of the solvent leads first to a gel and then to a coating that, on a nanoscale, looks like a lotus plant. A water droplet on this coating had a contact angle of $\vartheta_c = 160°$ which corresponds to $a_P \approx 10.7 \%$.

A "perfect" hydrophobic surface with $\vartheta_{adv}/\vartheta_{rec} \approx 180°/180°$ was recently achieved by treating silicon wafers with toluene solutions of CH_3SiCl_3 that contained a trace of water. The resulting coating consisted of a nano-network with ca. 40 nm thick stems which, in turn, were composed of a molecular network with $>Si(CH_3)-$ junctions that were connected via $-O-$ units to either other $>Si(CH_3)-$ units or the surface of the silicon wafer.

Discussions of wettabilities usually assume that the wetting agent neither contracts nor expands during appliction. But such volume changes do happen during many practical applications. Evaporating solvents from drying paints or solution and dispersion adhesives leave behind polymers that occupy less space than the paint or adhesive. Solidification of polymer melts leads to contraction, which is more severe if the polymer crystallizes. Hardening of adhesives or coatings by polymerization usually also results in contraction (Volume I, p. 186).

Such effects influence wettabilities greatly. Liquid water does not wet poly(tetrafluoroethylene) and runs off PTFE surfaces as pearls. However, water expands on freezing. The resulting ice sticks stubbornly in the crevices of rough PTFE surfaces (Fig. 12-5, center). For this reason, airplanes must be carefully deiced although their wings are coated with teflon.

A few monomers polymerize with expansion because the total number of bonds is reduced: two σ bonds are replaced by one shorter π bond (the special case in Eq.(6-77), Volume I), and/or crystalline monomers are converted to less dense amorphous polymers. Such monomers lead to well-adhering coatings. However, these coatings lack certain other properties that are important for coatings and adhesives.

12.3   Adhesives

12.3.1   Overview

Adhesives are non-metallic materials that bind two parts, the adhere**nts** (L: *adhaerere* = to stick to, from *ad* = toward, *haerere* = to stick). They are commonly known as glues because glues were the first class of adhesives known to mankind (see below). Adhesives must have many properties such as tack = stickiness, rapid bonding to the adherents, strong bonding to adherents after "setting" or drying, sufficient inherent strength, and also often durability. In general, one or more of these qualities dominate. Adhesives are therefore often subdivided according to these qualities. An example is so-called **structural adhesives** with strong bonding to adherents.

The structure of adherents should not change upon application of the adhesive. Adhesives are therefore distinguished from **solvent cements** (shortened to "**solvents**") that penetrate and soften the surfaces of plastics, which allows the polymer molecules of one part to self-diffuse into the other part (see below). "Adhesives" also do not include **solders**, i.e., neither low-melting metal solders (low-melting eutectics) nor solder glasses.

"Adhesive" is a general term that includes glues, pastes, latices, cements, etc. The adhesion-active components (**binders**) of adhesives are predominantly polymers or oligomers; monomers are fairly rare.

Adhesives are subdivided according to the type of their adhesive components, their physical states, or their types of application. These categories are historical and by no means systematic.

Adhesives are subdivided according to their *physical states* into liquid, disperse, paste, powder, and dry film.

According to the type of *binders*, adhesives are divided into natural and synthetic ones based on thermoplastics, elastomers, and thermosets.

Natural adhesives comprise those from animals, vegetable, or natural gums. They are sold as solutions or dry powders and mainly used for glues and pastes (see below). Natural adhesives have the lowest strength of all adhesives but they are inexpensive, have long shelf lifes, and are easy to apply. Applications include mainly paper, cardboard, and light wood constructions.

Elastomeric adhesives are based on natural rubber and synthetic rubbers and sold as aqueous latex and as dispersions in organic solvents. Some of them are modified with resins to give "**adhesive alloys.**" Elastomeric adhesives have high flexibilities but low strengths.

Adhesives based on thermoplastics are subdivided into **hot melts** (see below) and so-called **thermoplastic adhesives** which are thermoplastics (or asphalt or oleoresins!) that are either dissolved in organic solvents or emulsified in water. Thermoplastic adhesives are fairly soft and have poor creep strength but are not very costly. They are used for non-critical applications (see below).

Thermosetting adhesives need crosslinking ("curing"), usually by heat and/or pressure. Some of them are two-part materials that need to be mixed before application. Thermosetting adhesives have the strongest bonding strengths of all adhesives and also good creep strength, fair peel strength, and low impact strength. However, they are brittle and show little resilience.

So-called **alloy adhesives** are combinations of thermosets with either thermoplastics or rubbers. They have better flexibilities, toughnesses, and impact resistances than thermosetting adhesives themselves.

There are many types of *applications* of adhesives.

Glues were originally obtained from hides and bones of animals. In adhesive technology, "glues" are aqueous solutions of organic or inorganic, synthetic or natural polymers. Glues are used for postage stamps, envelopes, labels, and inexpensive plywood. Examples of glues are (for the chemistry, see polymer sections in Volume II):

- Protein-based glues: glutin glues (less pure than gelatine from the same sources) from collagen (fish glue is liquid; hide and bone glues are sold as dry flakes); albumin glues from blood proteins; glutens from wheat or soy beans; casein glues from milk proteins; casein-latex glues from mixtures of casein and natural or synthetic rubber latices for bonding wood, linoleum, or laminated plastics to metal.
- Polysaccharides: starches, dextrins, methyl celluloses, or plant gums (mucilages, which are reserve foods of plants); so-called **vegetable glue** is mostly tapioca paste.
- Water glass: sodium and potassium salts of silicic acid;
- Synthetic polymers: poly(vinyl alcohol), poly(vinyl methyl ether), poly(acrylic acid), phenolic resins, amino resins.

Glues are called **pastes** if they form highly viscous masses from which one cannot draw threads. Pastes result from the swelling of spheroidal particles in water (starch granules, flour, methyl cellulose, etc.). Industrial starch pastes contain 15-25 wt% swollen starch granules; they are used at 55-70°C. Methyl cellulose has a lower critical solution temperature at ca. 55°C above which polymer particles swell without dissolution. Cellulose based pastes can be used in concentrations as low as 2-5 %. Pastes are used for paper and paper board.

Adhesive dispersions are mostly oil-in-water (O/W) dispersions of polymers with glass temperatures below the application temperature. Examples are poly(isobutylene), poly(acrylic esters), poly(vinyl acetate), poly(vinyl ethers), natural rubber, and phenolic and amino resins. The dispersed polymers may also contain plasticizers, solvents, resins, and fillers. Also used are water-in-oil (W/O) dispersions of water in poly(acrylic esters).

Solution adhesives are solutions of organic polymers in organic solvents that either evaporate (**drying adhesives**) or polymerize on binding (**reactive adhesives**). The first group comprises polymers in evaporating solvents such as solutions of natural rubber, chlorinated natural rubber, poly(isobutylene), poly(vinyl chloride), poly(vinyl acetate), poly(acrylate)s, shellac, colophony, copal, dammar, etc. The second group consists of polymers in polymerizing monomers such as unsaturated polyesters in styrene.

Bonding cements (for short: **cements** (not to be confused with inorganic cements (Volume II, p. 569)) are malleable adhesives containing fillers but no solvents or only volatile ones. These bonding cements are subdivided according to their main ingredients (starch, rubber, water glass, magnesia, etc.), the materials to be bonded (metal, wood, etc.), or the type of processing (filling in, smoothing, etc.). Bonding cements serve simultaneously as adhesives and fillers of gaps, for example, in lap joints. True gap-filling materials for the filling or sealing of holes, etc., do not need adhesive strengths.

Cements include

- **Rubber cements** = rubbers dissolved in organic solvents. Strong rubber cements are compounded with gums, resins, or synthetic polymers. Conventional rubber cements detoriate on exposure because the rubber is uncured. Curing cements are crosslinkable by crosslinking agents (curing agents) and/or heat and pressure. Rubber cements are based on natural, chloroprene, and styrene–butadiene rubbers.
- **Latex cements** are solutions of rubber latices in organic solvents. They have excellent tack and bond strongly to paper, leather, and fabrics.
- **Pyroxylin cements** based on cellulose nitrate are used as household cements.

Hot melt adhesives (for short: **hot melts** or **hot adhesives**) are melts of thermoplastics that act adhesively at temperatures above glass temperatures. Examples are so-called atactic poly(propylene), aliphatic polyamides, poly(isobutylene)s, poly(vinyl butyral), and poly(vinyl acetate). They are often compounded with fillers, waxes, and bitumen. Hot melts are also applied as **film glues** (**film adhesives**), both with and without carrier (substrate). They are used in carpeting, laminating, packaging, and book binding.

Contact adhesives (**pressure-sensitive adhesives**, **impact adhesives**, etc.) are applied to both surfaces. They bind these surfaces after applying some pressure. For **close contact glues**, distances between parts to be bonded should be less than 1/20 of an inch.

Adhesives consist not only of a polymer (the **binder**) but also many other ingredients with names that often designate other properties than the same names in other fields:

- **fillers**: solid, non-volatile, *non-adhesive* additives for the improvement of properties or the lowering of costs;
- **extenders** (**adulterating agents**): non-volatile, *adhesive* additives for lowering costs;
- **solvents**: volatile liquids that dissolve ingredients *physically*;
- **dispersants**: volatile liquids that disperse ingredients;
- **diluents**: solvents or dispersants that lower the concentration of adhesives and/or decrease viscosities;
- **liquefying agents**: compounds that lower the solidification temperature;
- **hardeners**: compounds that promote chemical curing or physical drying;
- **retarders**: compounds that slow down the setting of adhesives and thus increase their shelf life and/or their pot life.

Adhesives are also subdivided according to many other criteria such as:

- the base material (phenolic resin, starch paste, water glass cement, etc.);
- the type of delivery (liquid, powder, self-adhesive film, etc.);
- the processing (cold setting, hot setting, cold pressing, etc.);
- the type of hardening (melt, solution, reactive, etc.);
- the state after application (solid, soft); and
- the application (wood glue, wallpaper paste, etc.).

Adhesive bonding offers many advantages compared to other types of joining (nailing, screwing, welding, solution welding, etc.). It leads to better stress distributions over greater areas, increases dimensional stabilities, acts as a barrier against humidity or electrical charges, allows the joining of very thin sheets, often permits fast processing, and is usually cost-saving. Adhesives and putties are therefore used in considerable amounts. For example, the United States, annually uses ca. 5 million tons of organic and inorganic polymers for these purposes, half of which are natural products, especially starch. Other important materials are water glass, natural rubber and glutins. Of the synthetic polymers, 80 % are vinyl polymers, phenolic resins, synthetic rubbers, and urea-formaldehyde resins. About 50 % of synthetic polymers for adhesives are used in construction, about 33 % in packaging, and the reminder for textiles, automotive goods, furniture, and electrical and electronic applications.

12.3.2   Adhesion

Adhesion is not only important for adhesive joints but also for the reinforcement of plastics by fibers (Section 9.3.1), the granulation of plastics, the inherent tack of rubbers (Section 7.2.2), the static electrification of surfaces by dust (Section 14.2.7), the interaction between substrates and biological cells, and many other phenomena. This section discusses only adhesive joints.

The strength of an adhesive joint is controlled by the adhesion between the adhesive and adherent (the "support") and by the cohesion of the adherents. Science defines "adhesion" as the physical and/or chemical interaction between the matter of a solid surface and another contacting materials. The strength of adhesion is controlled by the number of contacts per unit area and the strength of the contacts. In some cases, contacts may be chemical bonds. However, physical bonds are much more prevalent, i.e., hydrogen bridges, dipole-dipole interactions, and van der Waals interactions. In a few cases, charge transfer is important; a technical example is xerography (Section 14.2.8).

In most cases, adhesion is caused by attractions between electrically neutral entities, which has led to microscopic and macroscopic theories. Microscopic theories are based on interactions between small molecules, for example, between two hydrogen molecules. The corresponding equations are then extended to greater collectives of molecules. Macroscopic theories are based on the electric field between two dielectrics.

All of these theories consider the interaction adhesive–adherent as utmostly important and neglect the interactions adherent–adherent and adhesive–adhesive. In this case, the strength of an adhesive bond is given by the force per unit area between **adsorbent** (here the adherent) and **adsorbate** (here the adhesive). The strength of such bonds can be calculated from the forces acting between the chemical groups of the adherent and adhesive and the area requirements of these groups, which leads to strengths of 500-2500 MPa for covalent bonds, 200-800 MPa for hydrogen bonds, and 80-200 MPa for van der Waals bonds if the surface area of the adherent is totally covered by adhesive groups.

However, strengths of adhesive joints are much smaller, only ca. 20 MPa or less, because of various factors. (1) Roughness is neglected because the force is calculated per geometric area. (2) Only a fraction of the adhesive groups are active since the surface area of the adherent is usually not completely wetted and only a few groups per molecule of the polymeric adhesive actually adhere to the surface of the substrate (Fig. 12-2). (3) Adsorbed polymer molecules may furthermore entangle with ones that are not adsorbed. (4) Conventional tests of adhesive strength also measure not only the force between adherent and adhesive but also the deformations and cohesions of all partners (see Section 12.3.4). Technical tests of adhesion are therefore affected by many more factors than the scientific definition of "adhesion" suggests.

Adhesion is thus controlled by more than just interactions between chemical groups. An important factor is the physical state of adherent and adhesive. Metals and glasses are used as adherents at temperatures far below their transformation temperatures and the adhesion occurs here only between their surface groups and the groups of the adhesive. Wood, on the other hand, is a porous material, which can be penetrated by some adhesives. Polymers as adherents may or may not be entered by adhesive oligomers or polymers, depending on glass temperatures and miscibilities.

These differences in the accessibility of surface regions of the adherent and the interactions between the chemical groups of adhesive and adherent determine the strength of the adhesive bond. For the optimal strength of an adhesive joint for a particular demand (shearing, bending, etc.), each adherent requires a tailor-made adhesive as indicated by the large number of marketed adhesives. There is no universal adhesive, marketing notwithstanding.

Three basic types exist: adherent and adhesive are both above their glass temperatures (E/E), both are below their glass temperatures (G/G), or one is below the glass temperature and the other one above (G/E).

In the E/E type, segments of both the adherent and the adhesive are mobile and can diffuse into each other if they are thermodynamically miscible. Autohesion is observed if adherent and adhesive are chemically the same (Section 7.2.2). Autohesion is promoted by crystallization which increases cohesion between groups.

G/G types are the other extreme. Since both adherent and adhesive are below their glass temperatures, segments are not very mobile (self-diffusion coefficients have been theoretically estimated as ca. 10^{-21} cm^2/s). G/G types do exist in adhesive joints but their formation obviously needs a fluid adhesive (melt or solution).

Polymer molecules of E type adhesives cannot diffuse into a G type adherent since polymers of the latter are below their glass temperatures. An exception exists if the adhesive contains a good solvent for the adherent and the polymer molecules of adherent and adhesive are thermodynamically miscible. In any case, chain ends of polymer molecules of G type adhesive are somewhat mobile. They can fill crevices of rough surfaces of adherents, especially under pressure. Adhesion of G/E type joints is therefore promoted by roughening surfaces of adherents. Also important is the adsorption of polymer chains of the adhesive on the surface of the adherent. However, relative importance of adsorption and diffusion is difficult to estimate since they both depend in approximately the same way on time and temperature.

Pressure effects are utilized in **self-adhesive** tapes, all of which contain rubbers or thermoplastic elastomers such as natural rubber, ethylene–vinyl acetate copolymers, poly(vinyl methyl ether), or styrene-isoprene-styrene triblock polymers. Hydrocarbon-based self-adhesives usually contain **tackifiers** such as colophony or pinene resins. Despite the name, tackifiers do not increase the tack but rather the glass temperatures and elastic moduli of adhesives (see also p. 261).

Table 12-2 Polymers in adhesives for various applications, excluding paper and household.

Polymer	Construction	Packaging	Textiles	Automotive	Furniture	Electrical
Elastomers						
Natural rubber			+	+		
Styrene-butadiene rubbers	+	+	+	+		+
Acrylonitrile-butadiene rubber	+	+	+			
Poly(chloroprene) rubber	+		+	+	+	+
Butyl rubber		+		+		
Polyurethanes	+	+	+	+	+	
SBS-triblock polymers	+					
Thermoplastics						
Poly(ethylene)	+	+	+			
Poly(propylene), atactic		+				
Poly(vinyl chloride)	+		+	+		+
Poly(acryl ester)s	+					
Ethylene-vinyl acetate copm.		+	+		+	
Poly(vinyl acetate)			+			
Poly(vinyl alcohol)		+				+
Poly(vinyl butyral)				+		+
Polyamides, aliphatic		+	+			
Cyanoacrylates						+
Cellulose ethers	+					
Carboxymethyl celluloses	+					
Thermosetting polymers						
Alkyd resins				+		
Phenolics	+			+	+	
Amino resins	+					
Urea resins					+	
Furan resins	+					
Unsaturated polyesters	+	+		+	+	+
Epoxy resins	+	+			+	+
Polysulfides				+		
Silicones				+		+

12.3.3   Action of Adhesives

The world production of synthetic polymers for adhesives is ca. $4 \cdot 10^6$ t/a (2000). About 43 % of these polymers are used for water-based adhesives, 21 % for reactive adhesives, 13 % for melt adhesives, and 23 % for other types.

Because of the many different types of adherents and the many different types of applications, a great variety of adhesives are marketed. Table 12-2 shows polymers that are used as or in adhesives in six major groups of industries. The table does not include natural polymers except natural rubber.

Hot melt adhesives are amorphous polymers that are applied at temperatures above their softening temperature (the "melting temperature"). On deposition on the adherent, the viscosity of the melt should not be so low that the adhesive flows away and not so high that wetting becomes difficult. Adhesion is caused by the solidification of the melt. Best results are obtained with viscosities of 10-10 000 Pa s.

Optimal adhesion is obtained at a certain melt viscosity η and hence at a certain molecular weight M since $\eta = f(M^a)$. Low-viscosity (= low-molecular weight) melts can penetrate porous adherents; an example is the glueing of interlinings to gentlemen's suits. High-molecular weight melts lead to stronger bonding because of the greater concentration of contact points (Section 12.2) but they are also more difficult to process.

A small proportion of branches per molecule decrease the melt viscosity and therefore the diffusion rate. Very strongly branched polymer molecules provide less contact points than linear or lightly branched ones so that adhesion passes through a maximum with increasing degree of branching.

For adhesive action, **solvent-based adhesives** (glues, dispersions, solutions) require removal of liquids by evaporation or heating. The liquid is commonly water since organic liquids are expensive, toxic, environmentally damaging, and often flammable. Most of these adhesives are used to glue porous cellulose-based substrates (paper, cardboard, wood) because they act as wicks that siphon the water fast from the adhesive into the substrate.

The low viscosity of solvent-based adhesives eases processing. The liquids of these adhesives may also swell the adherents if the solubility parameters of the liquids match those of the adherents. Swelling promotes the interdiffusion of the polymer component of the adhesive into the adherent (shift from the G/E type to the E/E type).

The solvents should plasticize the adherents and the adhesive. After application, solvents should diffuse fast out of the adhesive; solvents must therefore be volatile.

Reaction adhesives harden by polymerization of liquid monomers to yield solid binders. Such adhesives may be 1-component or 2-component systems. Common one-component systems harden by irradiation (UV light, electron beams) or by exposure to the environment. Cyanoacrylates, $CH_2=C(CN)(COOR)$, are known as **super glue** or **crazy glue**. They remain liquid in closed containers but polymerize in air because the weak base H_2O initiates their anionic polymerizations. Other 1-component systems are the so-called PUR-1 K systems based on isocyanates.

On the other hand, acrylic esters of the type $CH_2=C(CH_3)-CO-X-OC-C(CH_3)=CH_2$ with $X = OCH_2O$ or $O(CH_2)_xO$ and $x = 3$ or 4 remain liquid in the presence of air, for example, in transport containers of poly(ethylene) that are permeable to air. These compounds react with oxygen from air and form stable hydroperoxides. The hydroperoxides decompose in the absence of air, i.e., after acrylic esters are placed between two adherents. The resulting radicals initiate the polymerization of the carbon-carbon double bonds of the acrylic ester. These **anaerobic adhesives** are 1-component adhesives that are activated by amines. They are used to glue steel threads.

A special case of reaction adhesives is **structural adhesives** which are either thermosetting resins or alloy adhesives (see above). These adhesives must be cured at temperatures of ca. 150°C. However, ABS components of alloy adhesives soften at 90°C and it will therefore take some time before these adhesives become solid and have sufficient strength. The cooling time slows down production processes, for example, of cars. The remedy is usually an additional mechanical fastening, which increases costs.

A new curing method overcomes this difficulty by providing hot-curing alloy adhesives with 5 wt% of ca. 22 nm large single crystals of iron oxide that are covered with SiO_2 (MagSilica®). Application of a high-frequency magnetic field causes the iron oxide molecules to vibrate, which generates heat. Because of the large surface area of the particles, the heat is immediately transferred to the curing agent which reacts within seconds. In this process, adherents are not heated and therefore also not deformed. For recycling, adhesive joints are again subjected to the same high-frequency magnetic field but with greater intensity.

Most reactive adhesives are, however, materials consisting of two liquid components such as epoxides + amines or isocyanates + hydroxy compounds, usually formulated with plasticizers and fillers. Both components are mixed immediately before application. Alternatively, each component can also be microencapsulated (Section 12.6.1). Application of pressure to the mixtures of capsules liberates the contents which start to react and form the adhesive.

12.3.4   Adhesive Joining

In general, surfaces of adherents must be prepared before adhesives can be applied. First, they must be cleaned to remove surface contaminants, which prevent good contact between adhesive and surface and may impede wetting by the adhesive. If necessary, contact areas are increased by roughening surfaces, either mechanically or chemically. Inert surfaces may also be etched, oxidized, etc., in order to generate groups which interact more easily with the groups of the adhesive. Examples are hydroxyl groups, that may form hydrogen bonds or react with other reactive groups, or free radicals, for example, from the treatment of poly(tetrafluoroethylene) with sodium.

Strengths of adhesive jonts are tested by very many methods (Fig. 12-6). All of these tests are affected not only by the strength of the adhesion but also by the cohesion of the adherents and the deformation of adhesive and adherent. Results also depend on the testing speed since deformations are viscoelastic effects.

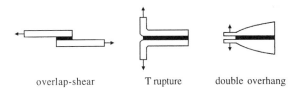

overlap-shear   T rupture   double overhang

Fig. 12-6 Schematic representation of three of many testing methods for adhesive joints.

Application of stress should deform adhesives but not adherents. A deformation of adhesives would lead to a uniform distribution of stress within the adhesive layer; this is most likely going to happen if adherents are rigid. A weak adhesive layer between strongly deformable adherents such as soft plastics will be deformed more strongly at the ends of the adhesive joint as shown in Fig. 12-7 for the shearing of an overlap joint. The uneven deformation leads to stress peaks, which renders the adhesive joint weak although the adhesion itself is strong.

Conversely, adhesive joining of metal parts by good wetting adhesives leads to strong adhesions because thick metal parts do not deform easily. Adhesive joining of thin metal foils is much more difficult. In this case, adhesives must be provided with plasticizers so that the adhesive layer becomes more deformable than the foils.

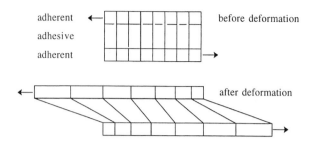

Fig. 12-7 Schematic representation of an adhesive joint in which the adherent is more deformable than the adhesive.

On loading, adhesive joints may debond at various locations. A crack may develop in the middle of the adhesive layer at one end of the joint and then spread through the whole layer (Fig. 12-8, left). After debonding, surfaces of both adherents will be covered by fractured adhesive in this "cohesive failure."

The adhesive bond may also fail directly at the interface adherent-adhesive (Fig. 12-8, second from left). This "adhesive failure" or "interfacial fracture" usually occurs if the adhesive has an insufficient fracture toughness (for fracture toughness, see Volume III, p. 629 ff.).

Tensile pre-stresses in the adhesive layer may also cause fractures to alternate between the two surfaces ("fracture jumping"). The resulting sections of the now "free" interfaces may be small or large (Fig. 12-8, second from right). This type of failure often occurs if one tries to remove price labels.

Fractures may also initiate at the end of the adhesive layer and spread through the adherent if the adherent is less tough than the adhesive (Fig. 12-7, right).

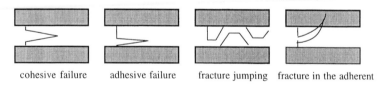

cohesive failure   adhesive failure   fracture jumping   fracture in the adherent

Fig. 12-8 Some types of failure of adhesive joints.

12.4 Paints and Coatings

12.4.1 Overview

Surfaces of many goods are covered with layers of other materials. These layers should protect the goods against the environment or wear and tear, embellish it by color and/or gloss, or lend it certain special properties such as reflection, electrical insulation, water repellance, and the like. The layers are usually polymers brought upon the surface by painting or coating (Sections 12.4.2-12.4.8; see also Section 4.8.1), by printing (Section 12.5), or by encapsulation (Section 12.6). Metallizing (Section 4.8.2) and oxide coating (Section 4.8.3) are far less often used.

Surface coatings (usually just "**coatings**") are thin layers of 0.001-0.1 mm thickness that adhere strongly to surfaces. They can be applied to surfaces by various coating and spray processes (Section 4.4.2) from solutions, dispersions, or melts. Special processes produce polymer coatings from powders (Section 12.4.8) or *in situ* from monomers by polymerization (Section 12.4.7) or by so-called plasma polymerizations (Section 4.8.4).

Surface coatings are most often applied as **paints**, sometimes called **lacquers**. Most paints consist of **binders** (polymers or prepolymers), solvents or diluents, dyestuffs (pigments or dyes, Section 3.2.1), fillers, and other adjuvants (catalysts, stabilizers, etc.). **Clear lacquers** (**clears**) are paints without dyestuffs and fillers; if made with drying oils, they are known as **varnishes**. All these terms are often used interchangeably.

The word **lacquer** is derived from the French word *laque* (Italian *lacca*, Arabic *lakk*, Persian *lak*) which has its roots in the Sanskrit word *laksha* = hundred thousand because very many ("hundred thousand") insects are needed to produce one ounce of shellac (from "shell" and "lac"), one of the earliest clear lacquers (Volume II, p. 79). *Lak* is also the root of the word **lake** which is a name for any of the bright pigments from the combination of an organic dye with metal salts or tannin. The word **varnish** is derived from the Latin *veronix* which has been traced to the Greek *berenike,* the old Greek name of the city of Bengasi (Libya) in which varnish was used for the first time in the Western world. Varnishes have been known in Japan and China since ca. 1500 BC and in Egypt since ca. 1400 BC.

Paints (lacquers) are historically subdivided according to the

type of binder	(asphalt, copal, alkyd, polyurethane, etc.),
content of pigments	(clear, translucent, etc.),
type of processing	(brush, dip, spray, etc.),
processing sequence	(base, top coat, etc.),
type of drying	(air, oven, 4 hours, etc.),
use properties	(insulation, rust protection, etc.),
application	(exterior, interior, furniture, leather, etc.).

The world-wide consumption of paints including dispersions, powders, etc., is ca. $23 \cdot 10^6$ t/a (2000) which amounts annually to ca. 4 kg per capita whereas the per capita consumption in highly industrialized countries is up to 20 kg per capita per year. Uses of paints vary from country to country (Table 12-3) but solvent-based paints are still the largest group as the global statistics for the modern classes of paints (2000) reveal:

solvent-based paints (binders dissolved in organic solvents)	$10.0 \cdot 10^6$ t/a
latex paints (aqueous dispersions of binders)	$7.5 \cdot 10^6$ t/a
water-borne coatings (water-soluble binders)	$3.1 \cdot 10^6$ t/a
high solids (solvent-poor or solvent-free systems)	$1.6 \cdot 10^6$ t/a
non-aqueous dispersions (NAD)	$< 0.1 \cdot 10^6$ t/a
powder lacquers	$0.7 \cdot 10^6$ t/a

Table 12-3 Types of paints used in various countries (1997).

Type	Percent of use in the region		
	Western Europe	USA	Japan
Organic solvent-based paints	57	55	83
Paints with water-soluble binders	12	12	11
High solids paints	15	27	1
Powder lacquers	12	5	4
Paints curing by radiation	4	1	1

12.4.2   Structure

Composition

Coats are produced from the binders of paints and coatings. Like the base materials of adhesives (Section 12.3), binders are polymers or prepolymers with many different chemical structures. Molecular weights of binders vary with the type of coating:

10^2-10^3 for drying oils and systems that harden by UV light or electron beams;

10^3-10^4 for alkyd resins, water-soluble binders, polymers for electrodeposition, high solids, and powder lacquers;

10^4-10^5 for solvent-based paints based on acrylic or polyester;

10^5-10^6 for non-aqueous dispersions and water-based dispersions of poly(vinyl acetate), poly(vinyl chloride), or cellulose derivatives;

10^6-10^7 for latex paints based on acrylates, styrene, or butadiene.

During binding, prepolymers polymerize. Polymers from these paints as well as polymers from all other types of paints must adhere to the substrate and form a pore-free film of ca. 10 μm thickness. The film does not prevent the access of oxygen or water to the substrate but should delay and slow down the uptake of O_2 and H_2O as much as possible. This is achieved by the greatest possible proportion of pigment and/or filler particles in the paint, the so-called critical pigment volume concentration (CPVC, see below). Pigments and fillers must be well bound by the polymers that are therefore called **binders**.

Coatings are predominantly applied from solutions and dispersions, less from melts and powders (Table 12-3). The liquids in solutions and dispersions are just part of the **vehicles** for application, a term that also comprises other volatile components, plasticizers, and even binders. At best, vehicles have no effect on the properties of coatings and at worst, a bad one. The trade also distinguishes between solvents and diluents. "**Solvents**" are non-aqueous liquids that are already present in commercial paints and coatings. "**Diluents**" are added by the end user in order to obtain the desired consistency.

Liquids in paints and coatings aid in the putting on of pigments, fillers, and binders. The transport of these substances to the substrates as well as the quality of the resulting surface film depends strongly on the viscosity of the paint as well as on its dependence on the shear rate. Solvents and diluents that lead to low viscosities are called "good solvents" by the trade which contrasts with thermodynamics where a "good solvent" leads to expanded polymer coils and thus to high solution viscosities.

Liquid paints should not be too "thin" and not too "thick." They should run smoothly and generate smooth liquid layers without brush marks before surface films are formed. Solvent-based and water-based paints should thus be thixotropic so that they do not drip when taken up by the brush but become less viscous at the high shear rates of brushing (up to 20 000 s^{-1}!). This should allow initil leveling of the coating and be followed by a viscosity increase that prevents sagging. Such viscosity considerations are less problematic for dispersions and latices which are usually Newtonian liquids because their binders are present as spheres or spheroids.

Dyestuffs for paints and coatings are mostly **pigments** but sometimes **dyes** (p. 25). **Fillers** (p. 30 ff.) do not act as pigments but as extenders that lower costs. They may also increase the adhesion of the paint and improve the strength of the coating. Dyestuffs, fillers, and binders make up the **solids content** of the paint.

Dyestuffs come in a great variety of types. Even air may serve as a "pigment" if water droplets are encapsulated by polymers. The capsules are mixed into the paint. On drying, capsule walls solidify while water escapes and is replaced by air. The large difference in refractive indices of binders (n = 1.3-1.5) and air ($n \approx 1$) leads to a high hiding power (see Section 15.2.3). Air as a whitening agent can replace as much as 5-10 % of the white pigment titanium dioxide.

Paints may also contain various other adjuvants. **Driers** promote "drying", i.e., the crosslinking by oxidation of film-forming binders through formation of free radicals that accelerate polymerization. Driers are *not* **dehydrating agents (dehumidifiers)**. Examples of driers are heavy metal salts of carboxylic acids such as cobalt octoate or lead naphthenate. Dissolved driers are called **siccatives** (L: *siccus* = dry). **Hardening accelerators** are catalysts; an example is *p*-toluenesulfonic acid for the proton-catalyzed crosslinking reaction of unsaturated polyester resins and melamine resins.

Anti-skinning agents are usually radical catchers that slow down the oxidation of siccative containing, oxidatively drying paints and prevent the formation of skins and wrinkles. An example of a non-volatile anti-skinning agent is hydroquinone. Another group of such agents are oximes that disappear with the evaporating solvents and thus do not extend the drying time.

Leveling agents generate smooth surfaces. They consist of small proportions of not very volatile liquids such as cyclohexanone or butylene glycol or oligomeric resinous substances. On evaporation of volatile liquids, these agents become concentrated in the drying coating, which keeps it fluid for a longer time, which in turn prevents wrinkles that form on fast drying.

Wetting agents promote the dispersion of solids and prevent the flocculation which preserves gloss, hiding power, and the uniformity of hue.

Floating inhibitors prevent the flooding (horizontal) and floating (vertical) of pigments with different densities and as well as surface reactions of paints during storage and drying. These inhibitors are surface active substances such as ionic surfactants and silicon oils. Floating in dispersions can also be prevented by adding substances that increase viscosities (see Section 12.4.5).

Critical Pigment-Volume Concentration

Each paint has a critical pigment-volume concentration (**CPVC**) above which properties of coatings change drastically. For example, fracture strengths and flexibilities decrease while the tendency to form bubbles increases (Fig. 12-9). These property changes occur above the CPVC because there is no longer enough binder to coat pigment and filler particles. However, the CPVC is lower than the maximum packing density of particles, for example, lower then 74.05 % for hexagonal close-packed unimodal spheres (p. 77) since particles are always surrounded by some binder.

Pigment-volume concentrations of the same pigment/filler vary with the type of paints. For example, latex paints and non-aqueous dispersions (NADs) have smaller CPVCs than solvent-based or water-based paints. During drying, binder molecules from the latter paints pack between pigment particles according to their macroscopic density. This is not possible for binder molecules of latices and NADs because they are in dispersed particles which remain as individuals during the early drying stage.

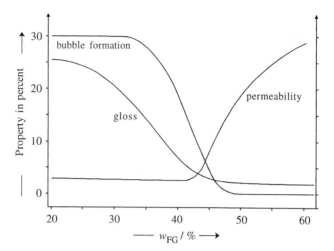

Fig. 12-9 Schematic representation of the variation of bubble formation, gloss, and permeability of a paint with the weight fraction w_{PG} of the pigment/filler [3].

The packaging of fairly rigid latex particles is rather controlled by the ratio of diameters of pigment particles and latex particles; during film formation, the individuality of latex particles is largely maintained (Section 12.4.5). The larger the diameter of latex particles L, the fewer of them can coat a pigment particle PG. Because of the proportionality $d^3 \sim V$ of diameters d and volumes V, ratios N_{PG}/N_L of numbers of pigment and latex particles at the CPVC should be proportional to the 3rd power of the ratio of their diameters, i.e., $N_{PG}/N_L = K(d_L/d_{PG})^3$. If so, all latex paints and non-aqueous dispersions should have the same CPVC. However, the experimentally found exponent is only 8/3 instead of 9/3, regardless of the type and diameter of the latex particles (Fig. 12-10)

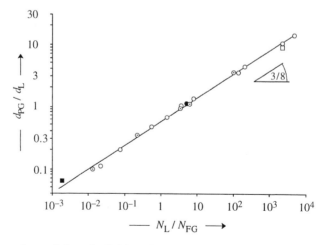

Fig. 12-10 Dependence of the ratio N_L/N_{PG} of the numbers of latex or hydrosols (L) and pigment (PG) particles on the ratio of the diameters d_L and d_{PG} of these particles at the critical pigment-volume concentration, plotted as $d_L/d_{PG} = f(N_L/N_{PG})$ [4].
The diameters of latices were 0.18 µm (○), 0.20 µm (○), and 0.35 µm (●), respectively, and those of hydrosols 0.2 µm (□) or 26 µm (■). The solid line corresponds to $N_L/N_{PG} = 4.25 \, (d_L/d_{PG})^{8/3}$.
By permission of the American Chemical Society, Washington (DC)

An exponent of $z = 8/3$ in $N_{PG}/N_L = K(d_L/d_{PG})^{8/3}$ indicates a variation of critical pigment-volume concentrations with the size of the latex particles: the larger the latex particles, the smaller the CPVC. Since $d_L/d_{PG} = 1$ for spherical latex and pigment particles with the same diameter, one also has $N_{PG}/N_L = 4.25$ and therefore a value of CPVC $\equiv$ $V_{PG}/(V_{PG} + V_L) = 1/((N_{PG}/N_L) + 1) = 1/(1 + 4.25) \approx 0.19$. Hence, the critical pigment-volume concentration is very low, only 19 %.

The function $N_{PG}/N_L = K(d_L/d_{PG})^{8/3}$ also applies to latex particles that are much larger than the pigment particles. The exponent 8/3 can therefore not result simply from a packing of latex particles into interstitial spaces between closely packed pigment particles. The molecular meaning of the fractal exponent 8/3 does not seem to be known.

The CPVC depends on the method that is used for its determination (see Fig. 12-9). The most sensitive property seems to be the ability of the coating to mask the substrate. This **hiding power** increases with the difference between the refractive indices n of pigments and polymers ($n_P \approx 1.3$-1.6, rarely up to 1.8 (Table 15-1)). Paints should therefore have pigments with refractive indices $n_{PG} > 2$.

At the CPVC, not enough polymer (binder) is present in order to coat pigment and filler particles completely. The paint layer also contains trapped air or even air as a filler (see above) with $n \approx 1$ which adds to the hiding power. The hiding power thus increases dramatically at pigment concentrations above the CPVC.

Pigments and fillers must be well dispersed in paints and paint films. Since all dispersions are thermodynamically unstable (Section 11.5.2), *all* dispersed particles must be stabilized. Stabilization is especially important if several types of dispersed particles are present, for example, not only pigments and fillers but also latex particles.

Dispersion of pigment particles is increased signficantly if these particles bind strongly to binders, either by chemical bonds or by adsorption or if the paint system contains surfactants, i.e., surface active chemical compounds. The best dispersions are obtained if pigments and surfactants have the same **HLB values**.

HLB values describe the *hydrophilic-lipophilic balance* of *neutral, low-molecular weight* surfactants on a scale from 1 to 20. They are calculated from HLB = 20 M_{seg}/M where M = molecular weight of the surfactant and M_{seg} = molecular weight of the hydrophilic parts of that molecule. The arbitrary factor 20 was assigned to the most hydrophilic molecule that was known at the time the HLB method was proposed (1954). Modern lipophilic molecules may have HLB values up to 40.

HLB values do not apply to ionic surfactants (phosphates, sulfates, sulfonates) and also not to associating polymers such as segmented bipolymers from propylene oxide and ethylene oxide. HLB values near 1 describe lipophilic compounds and those near 20 hydrophilic ones. Typical values are 4-6 for W/O emulsifiers, 8-18 for O/W emulsifiers, and 13-15 for detergents.

The stability of pigment dispersions is also strongly affected by the polymeric binders. In principle, the same rules apply as for the stabilization of latices (Section 11.5.2). An example is the stability of TiO_2 dispersions which settle in toluene within seconds but in toluene solutions of poly(butadiene) in minutes, and in toluene solutions of fatty acids in about one hour. The dispersions remain stable for hours and days if the added poly(butadiene) contains COOH groups that are statistically distributed along the chain and even months if COOH groups are end groups.

Film Formation and Adhesion

After application, solvent-free high solids and powdered lacquers need to form good strong surface films on the substrates which requires temperatures that are somewhat

Table 12-4 Glass temperatures T_G, deformation temperatures T_D (bending under the same load), and temperatures T_F for film formation (= temperature for the formation of a mechanically stable film).

Polymer	$T_G/°C$	$T_D/°C$	$T_F/°C$
Poly(butyl acrylate)	−54	−48	<5
Poly(ethyl acrylate)	−24	−18	<5
Poly(vinyl propionate)	10	5	5
Poly(methyl acrylate)	10	12	18
Poly(vinyl chloride-*co*-methyl acrylate) (80:20)	?	63	70

higher than the glass temperature of the binders (Table 12-4). This condition considerably reduces the number of possible film-building thermoplastics but is less critical for thermosetting materials.

Film formation proceeds by evaporation of liquids from water and organic solvent-based solutions of thermoplastics and by crosslinking polymerization for liquid thermosetting prepolymers. Polymer dispersions present special problems (Section 12.4.5).

Generally, paints and coatings cannot be applied directly to surfaces, in part, because surfaces are too rough (would give uneven surface film or would need too much paint) and, in part, because the adhesion of the surface film is insufficient. Therefore, surfaces of substrates need to be prepared before paint is applied.

Holes and larger irregularities of surfaces of substrates are patched or smoothed over with **fillers** (or **stoppers**) that are rich in solids. Filled or unfilled surfaces are then coated with **primers** which promote adhesion, prevent an uptake of the so-called undercoat (if any, see below) by the pores of the substrate (for example, wood), or increase the resistance against corrosion (in the case of metallic substrates). A typical **wood primer** consists of a poly(vinyl butyral) with ca. 43 % butyral, 50 % acetate, and 7 % hydroxy groups in a mixed solvent from ethanol and butanol. These primers often contain a second resin, such as melamine or epoxy.

Primers for metals usually consist of two components that react which each other and the metal surface; they are therefore also called **reaction primers**. Such primers consist, for example, of 9 wt% poly(vinyl butyral), 8.6 wt% basic zinc chromate, 1.3 wt% magnesium silicate, and 0.1 wt% furnace black in 61 wt% ethanol and 20 wt% butanol, and solution B with 18 wt% phosphoric acid, 16 wt% water, and 66 wt% ethanol. These primers are also called **wash primers** because they were "washed" with scrubbers upon the steel decks of warships in order to passivate the steel against corrosion and rust formation by phosphating the steel, i.e., putting on a phosphate layer. **Shop primers** are reaction primers for metals that do not emit unhealthy gases if primed metals are welded.

Primers for metals do not prevent corrosion *per se* because their layers let oxygen and water pass through as all polymer layers do after some time. Instead, they rather prevent the access of ions that would accelerate corrosion.

Substrates and primed substrates are also provided with one or more than one additional layer of **surfacers** (called **undercoats** for house and wall paints). These coatings are highly pigmented and rich in extenders and serve to smoothen smaller irregularities of substrate surfaces. Surfacers and undercoats can also be sanded.

One sometimes also applies **sealers** which prevent the transfer of substances from the substrate to the coating layers, or between the various coating layers, including the **top** coat (**finish**). Most top coats are pigmented but the automotive industry also uses **clear-coats** on top of the finish. Clearcoats consists of polyols with various contents of aromatic groups that are crosslinked by either melamine-formaldehyde resins (**1K systems**) or polyisocyanurates (**2K systems**).

Paints are usually applied to surfaces of metals, wood, stones, gypsum, etc., but rarely to those of plastics themselves because pigmenting of plastics is more durable and cost effective than painting finished parts. The big exceptions are certain automotive parts such as paintable side panels or bumpers which consist of rubber-modified thermoplastics (p. 454). The rubber component of these plastics accommodates the paint.

12.4.3 Solvent-borne Coatings

Classical coatings are solvent-borne, i.e., they contain binders, pigments, etc., in *organic* solvents, called **thinners, reducers**, or **carrying agents**. Thinners are usually mixtures because no single solvent can dissolve all film-forming ingredients and lead to the required properties. For example, a slower-evaporating solvent increases the leveling of the coating and also the gloss but a too slow one can cause sagging. Paints have viscosities of 6-100 Pa s and are applied with shear rates between 10^2-10^4 s^{-1} (brushing) and 10^3-10^6 s^{-1} (spraying) or 10^4-10^6 (rolling).

During the last decades, production of solvent-borne coatings has steadily decreased in Western countries because of the relatively large costs of ingredients, non-optimal properties of coatings, high cost of solvent recovery, and protection of work places (fire), workers (health), and the environment. Since many of the other paint/coating systems (p. 530) demand high investment costs for machinery, "high solids" have been developed (Section 12.4.7) that require lesser proportions of organic solvents.

Solvent-poor systems have been known for many centuries, for example, the **"drying" oil paints** based on linseed oil, wood oils (tung oil, walnut oil), or soybean oil. The unsaturated fatty acids of the glycerol esters of these oils react with oxygen from air and form peroxides. Decomposition of these peroxides generates free radicals that initiate the crosslinking polymerization of the double bonds (Volume II, p. 80).

Drying oils are characterized by the **iodine number** (**iodine value**, IV) which indicates the mass of iodine that is bound by 100 g of oil, or by a **drying index** (DI) that is calculated from the percentage of linoleic acid + 2 times the percentage of linolenic acid in the oil. Drying oils have IV > 170 and DI > 70; semidrying oils, IV = 100-170 and DI = 65-70.

Oil paints form tough-hard and relatively weather-resistant coatings. They penetrate rust and are therefore used as a primer on iron surfaces but they harden only slowly and yellow and saponify easily. Newer reacting solvent-based coatings are dominated by epoxies and polyurethanes as binders. Poly(butadiene) oils, oligoacrylates, and oil-free unsaturated polyester/melamine–formaldehyde resins are also important.

Most solvent-based coatings form paint films by evaporation of organic solvents. Binders of these paints include nitrocelluloses, cellulose esters (2 1/2-acetate, acetopropionate, acetobutyrate), chlororubbers (chlorinated natural rubber, chlorinated poly(ethylene), chlorinated poly(propylene)), polyolefins, acrylics, polyvinyl compounds (poly-(vinyl acetate), poly(vinyl chloride), poly(vinylidene chloride), poly(vinyl fluoride), etc.), alkyd resins, saturated and unsaturated polyesters, polyurethanes, epoxies, silicone resins, asphalt, bitumen, and urea, melamine, and phenolic resins.

Varnish is a special type of a solvent-based coating (for the word origin, see p. 531). It consists of a drying oil (see above), a resin (see also Volume II, p. 78), and a thinner (= solvent) but does not contain pigments. It is mainly used for wood where it delivers a transparent, hard, protective coating.

Enamel varnish (enamel paint) is a highly glossy, extremely hard-surfaced (by paint standards) varnish. It got its name from the similarly glossy but much harder true enamels (**vitreous enamels**) from low-melting mixtures of silicates, borates, and fluorides of Na, K, Pb and Al that form chemically resistant but impact-sensitive coatings, mostly on iron or cast iron surfaces but sometimes also on glasses or earthenware.

Lacquer is now often used as a general term instead of "paint" but originally referred to a paint containing a varnish resin from the tree *Toxicodendron vernicifluum* (varnish tree) of China and Japan. The active ingredient of this tree is urushiol, a mixture of various phenols and proteins.

12.4.4   Water-borne Coatings

Water-borne coatings contain water-soluble binders which are either copolymers with hydrophilic comonomeric units or oligomers and polymers in which hydrophilic groups have been introduced by post-reactions. Hydrophilic groups are mostly carboxyl, ammonium, or ethylene oxide groups. Typical polymers are maleinated oils and epoxides, alkyd resins, polyesters, acrylics, and poly(butadiene)s with molecular weights between 5000 and 10 000. Thickeners are mainly polyurethanes or cellulose derivatives.

Water-borne coatings are applied conventionally by brushing, dipping, spraying, etc. Their advantage is the low cost of non-toxic water; their disadvantage is the large evaporation energy of water which causes high drying temperatures and/or long drying times.

Water-borne coatings with polyelectrolytes as binders are used for surface coatings by **electrodeposition** of metals in which the metal forms one electrode and the container wall of the dipping bath the other. The deposition does not occur by direct discharge of polyelectrolyte at the electrode; for example, for polyanions *not* according to the reaction $(P^{z\ominus} - z\,e^{\ominus}) \rightarrow [P] \downarrow$. Instead, polyanions react with protons from the reaction $H_2O \rightarrow (1/2)\,O_2 + 2\,H^{\oplus} + 2\,e^{\ominus}$ which leads to a **electrocoagulation** according to

$$(12\text{-}4) \qquad P^{z\ominus} + z\,H^{\oplus} \rightarrow [PH_z] \downarrow$$

Cationic resins with quaternary ammonium groups also electrocoagulate. All other cationic resins react with hydroxyl ions according to

$$(12\text{-}5) \qquad PH_z^{z\oplus} + z\,OH^{\ominus} \rightarrow [P] \downarrow + z\,H_2O$$

Cationic depositions are advantageous because they do not lead to dissolution of metal from the cathodes and they do not attack the phosphate layer from the treatment of surfaces with wash primers (p. 536). For these reasons, industry mostly uses 10 % aqueous solutions of polycations, often with acetate counterions. However, electrodepositions stop after one thin layer is formed. Additional layers can be deposited only if the first layer is porous or electrically conducting.

12.4.5 Latex Paints

A **latex** is an aqueous **dispersion** of a water-insoluble polymer that acts as a binder in paints. The polymer is **emulsified**, i.e., above its glass temperature ($T > T_G$), and not **suspended**, i.e., below its glass temperature ($T < T_G$). Latex droplets are usually spheroids with diameters between 0.05 μm and 5 μm. Latices are therefore a special type of **hydrosols** which are aqueous dispersions of finely divided solids or liquids.

Most polymer latices are produced by **emulsion polymerization** of an emulsified monomer to an emulsified polymer (p. 501). Emulsifying of polymer melts or polymer solutions is less often used. The emulsified polymers form spheres or spheroids, which leads to very low viscosities of very highly concentrated dispersions (p. 480). Polymer contents of latices are usually 60-70 wt%, i.e., near the limit for the maximum hexagonal close packing of spheres (74.05 vol%, p. 77).

Latices are not very costly. Water as a fluid agent is also neither toxic nor flammable but it evaporates only slowly at room temperatures. Acceleration of evaporation by heating requires much energy and is costly but is unavoidable for mass production. Water should not be retained by the resulting coating because of its adverse actions such as plasticization and hydrolysis of the binder.

All dispersions are thermodynamically unstable, especially, if they contain both emulsified monomer droplets and suspended pigment particles as latex paints do. In solvent-based paints, in contrast, dispersions of pigments and fillers can be improved by adsorptive or chemical bonding of binders to pigment particles or by using reactive groups on the surface of pigments as polymerization initiators for the creation of polymer layers around pigment particles.

Like most other types of coatings, latex paints contain not only pigments and binders but also other components. Latex structures are beneficially affected by water that contains poly(acrylamide). Foaming is prevented by added silicone oils. Flow properties are improved by hydroxyethylcellulose or plant gums. Addition of bentonite, triethanolamine aluminate, or zirconium carbonate leads to thixotropy, i.e., a temporry reduction of apparent viscosity with shearing (p. 90).

Film formation from latices proceeds in several steps. In the first step, water evaporates and latex particles become less mobile until they finally pack closest (Fig. 12-11). In the ideal case of hexagonally densest packing, each latex particle is surrounded by 12 other ones (p. 77).

On further evaporation, interfacial tensions water/air, water/polymer, and polymer/air promote the **coalescence** of latex particles to a coherent film. The rate of coalescence is affected by many factors, for example, the temperature, the glass temperature of the binder, the presence of tensides, etc. In the simplest case, surfactants are absent and polymers are not plasticized and have glass temperatures that are higher than the drying temperature.

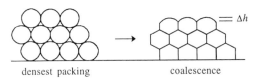

densest packing   coalescence

Fig. 12-10 Packing of latex particles and coalescence to a film [5a]. Δh = wrinkle height.

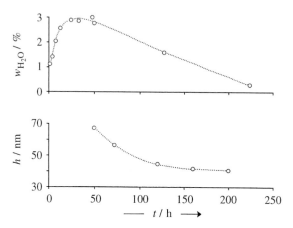

Fig. 12-12 Top: Time dependence of water content of a film from a latex of a poly(i-butyl methacryl-ate) ($M \approx 10^5$, $T_G = 65°C$) at 50°C. Bottom: decrease of wrinkle height with time [5b].
Wrinkle height is the distance between the top of a latex particle and the point of first contact with another latex particle (see Fig. 12-11). The original diameter of latex particles was 310 nm.

An example is poly(isobutyl methacrylate) (PiBMA) with a glass temperature of $T_G = 65°C$. At 50°C, interstitial volumes between latex particles of a dry latex film of this polymer take up water because the chemical potential of the condensed water is smaller than that of the humid atmosphere if the wetting angle is smaller than 90°. However, the moisture content of the latex film passes through a maximum and then decreases at a constant rate (Fig. 12-12). According to atomic force microscopy, wrinkle heights decrease continuously during dehydration; the height becomes constant after ca. 200 h.

This decrease is caused by a compression stress, i.e., the negative capillary pressure that is generated by the interface water–air. Although the main proportion of moisture evaporates fairly fast during film formation from a *latex*, dehumidification of the latex *film* is not complete as long as the film is in equilibrium with atmospheric water vapor. The capillary pressure is ca. 1 MPa for a wetting angle of 20°C and a moisture content $\phi_{H_2O} = 10^{-3}$ of the latex film. It increases to ca. 240 MPa for $\phi_{H_2O} = 10^{-9}$.

Capillary pressure is the decisive factor for film formations from latices. Film formation from *dry* latex particles *above* their glass temperature T_G is slower by a factor of 10 because it is driven only by the surface tension γ of the polymer. Such films from dry particles are formed faster from small particles than from large ones because the driving stress is inversely proportional to the radius R of particles according to $\sigma = 2\,\gamma R^{-1}$.

Films from latices are fairly brittle if they are dried below T_G. If such films are annealed at temperatures far above their glass temperature, particles are deformed and polymer segments from different particles begin to interpenetrate each other: the film coalesces (L: *co* = together, *alescere* = to grow, from *alere* = to nourish). The interpenetration is reflected by the dependence of the wrinkle height h on the square root of time, $h = K \cdot t^{-1/2}$ (Fig. 12-12, bottom), which indicates a diffusion process. Since the polymer molecules are present as coils and the radius of gyration of such coils varies with the square root of the molecular weight, one can expect that the diffusion rate, and therefore the rate of coalescence, decreases with increasing molecular weight. In general, one can expect two molecular weight regions, one for unentangled coil molecules and one for entangled ones. For crossover molecular weights, see Volume III, p. 605.

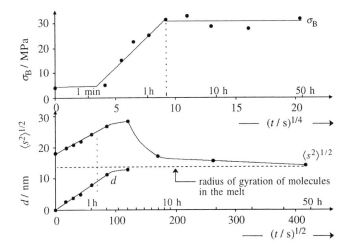

Fig. 12-13 Time dependence of properties of a sintered film from a poly(styrene) latex with a particle radius of 27.5 nm [6].

Top: Wood diagram. Fracture strength of the film as a function of the 4th root of time [6a].

Center: Radii of gyration of entities (see text) as a function of the square root of time [6b]. The broken horizontal line indicates the radius of gyration of unperturbed coils in the melt.

Bottom: Penetration depth d of polymer molecules as a function of the square root of time [6b].

The latex was prepared by free-radical emulsion polymerization of styrene using a nonionic surfactant. The resulting poly(styrene) had a molecular weight of 250 000 and a glass temperature of 100°C. The tenside was removed, and the latex particles were isolated and then pressed to a void-free film for 40 min at 110°C at a pressure of 9 MPa before measuring penetration depths, radii of gyration, and fracture strengths. Dotted vertical lines indicate the same time.

The penetration depth d of latex particles of poly(styrene) molecules increases linearly with the logarithm of time (Fig. 12-13, bottom) until it becomes constant at approximately the same time at which the radius of gyration of *latex particles* has reached a maximum (Fig. 12-13, center). The particles become larger because their interpenetration leads to recognizable aggregates.

After ca. 4 hours, interpenetration is complete. From then on, the measurable radius of gyration is an average over the gyration radii of both particles and polymer molecules. As the boundaries between particles disappear, radii of gyration decrease until they finally become the radius of gyration of unperturbed poly(styrene) molecules as it exists in the melt.

After a short induction period of 1-2 minutes, fracture strengths of the polymer film increase with the fourth root of time (Fig. 12-13, top) as predicted by the **reptation theory** of self-diffusion of polymer molecules in melts. This theory assumes that polymer chains with molecular weights above the critical molecular weight for entanglement move through a kind of tube formed by other segments like a reptile through underbrush (Volume III, p. 477). The fracture strength becomes independent of time after the interpenetration of segments is complete.

Latex paints are especially important in construction where one prefers acrylics for exteriors and copolymers of vinyl acetate with vinyl chloride or ethylene for interiors. In the United States, interior coatings also use styrene-butadiene copolymer latices with heat-crosslinkable groups. In Europe, such polymers play no rôle for this application since they are not light stable.

12.4.6   Non-aqueous Dispersions

Non-aqueous dispersions (**NAD**s) are dispersions of polymers in organic solvents, mostly of acrylic resins in gasoline. Their annual production is less than 100 000 t/a, mainly because they present the same problems as solvent-borne paints.

12.4.7   High Solids

Solvent-borne coatings contain **volatile organic compounds (VOC)** that are defined by the EPA as organic compounds that participate in a photochemical reaction. Many countries have therefore issued regulations for the maximum admissible emission of VOCs. In the United States, interior and exterior paints should emit no more than 250 g VOC per liter of paint (California: 50 g/L (2006)). Such VOC concentrations are difficult to achieve by conventional solvent-borne coatings. Replacement by other systems such as water-borne systems, latex paints, and powder lacquers is often not feasible for industrial production because of high investment costs for new processing lines.

The solution to this problem has been the development of solvent-poor systems (**high solids, HS**) that satisfy governmental regulations about VOCs. These paints contain not more than 20-30 % volatile organic compounds. Furthermore, conventional organic solvents are replaced by liquid "monomers" with low VOC values that act as diluents and sometimes also as crosslinking agents. Inert polymeric binders are replaced by reactive "oligomers." On curing, "monomers" and "oligomers" polymerize by crosslinking.

"*Monomers*" are mostly acrylates and methacrylates but some are also allyl derivatives, N-vinyl compounds, acetoacetates, and low-molecular weight epoxides. These compounds may be "monofunctional" such as lauryl methacrylate, "bifunctional" like the dimethacrylic ester of 1,6-hexanediol, "trifunctional" such as the trimethacrylate of 1,1,1-trimethylolpropane, $CH_3CH_2C(CH_2OH)_3$, etc.

"Monofunctional" compounds are of course bifunctional with respect to polymerization where they produce long chain segments between the oligomers that act as crosslinking junctions. "Bifunctional monomers" and all higher functional monomers are crosslinking agents themselves. The higher the functionality of monomers, the faster the curing (hardening), the crosslinking density, the hardness of the coating and its resistance against solvents. Flexibilities of paint films can be increased by using monomers with oligo(oxyethylene) groups such as the ethoxylated trimethylolpropane triacrylate, $CH_3CH_2C[CH_2O(CH_2CH_2O)_nOCCH=CH_2]_3$, where n may adopt a different value for each of the three substituents.

"*Oligomers*" serve as binders. They are usually (meth)acrylated epoxides and urethanes (M = 1000-30 000) but sometimes also polyesters and oligoacrylates.

acrylated epoxide

acrylated urethane

Z_1 = various groups, Z_p = difunctional polyol segment

Monomer-oligomer mixtures are applied to substrates and cured *in-situ*. Curing is by electron beams (EB) which generate free radicals that initiate the crosslinking copolymerization of monomers and oligomers. Hardening occurs in less than 1 second.

Alternatively, one can also cure with ultraviolet light, using a photoinitiator (Volume I, p. 320). Such photoinitiators are benzophenone, benzildimethylketal, or 2-hydroxy-2-methyl-1-phenyl-1-propanone. UV/EB curing is used for the coating of metal and wood furniture, clear coats on plastic and metal containers, and circuit boards.

Special acrylic esters are also "thermally cured" by free-radical initiators at temperatures down to 0°C. Such curing does not need the special equipment that is required for EB and UV hardenings.

Initiators for thermal curing are azo compounds such as *N,N*'-azobisisobutyronitrile or peroxides such as methylethylketone peroxide, *t*-butyl benzoate, cumene hydroperoxide, or 1,1-di(*t*-butylperoxy)-3,3,5-trimethylcyclohexane (see Volume I, p. 314). The formation of radicals from these initiators is thermal at elevatd temperatures but activated at lower temperatures by **metal driers** which are salts of Al, Mg, Mn, Sn, Co, etc. (but not Na or K) of higher fatty acids, naphthenic acids, or resin acids (rosin acids).

The curing is aerobic on the surface but anaerobic in the interior. It is slower at the surface since the oxygen of air is a good radical scavenger. In order to obtain a fast, homogeneous, and complete curing throughout the specimen, oxygen in and at the surface must be excluded either by working under a nitrogen blanket or by addition of radical scavengers such as metal soaps or allyl compounds.

Monofunctional and multifunctional acrylic esters are also used as reactive diluents and modifiers, respectively, for 2-component epoxides. The curing of these systems is usually by aliphatic amines such as diethylene triamine and triethylene tetramine but sometimes also by cycloaliphatic amines, amidoamines, or polyamides.

Amines react with acrylic esters in a *Michael addition* to a secondary amine which then reacts with another acrylic ester molecule (not shown) or by crosslinking with the epoxide:

(12-6)

$$
\begin{array}{ccc}
CH_2 = CH & + \ R'NH_2 & R'NH - CH_2 - CH_2 \\
\quad | & \xrightarrow{\hspace{1.5cm}} & | \\
COOR & & COOR
\end{array}
$$

A recent addition is completely solvent-free paints that contain an aspartic ester as the second component besides multifunctional isocyanates. These systems can be applied with spray guns because they have the consistency of easy flowing oils.

12.4.8 Powder Lacquers

Powder lacquers are powders of thermoplastic or thermosetting resins. The powders should be crumbly or should crumble easily on grinding since powder particles that do not break easily would melt or sinter by the heat that is generated by grinding. Lumping of particles can be prevented during grinding at room temperature if the thermosetting resins have glass temperatures greater than ca. 40-50°C.

Thermoplastic powder lacquers comprise polyamides, ethylene-vinyl acetate copolymers, poly(vinyl chloride), poly(ethylene)s, polyesters, and polyepoxides, which are usually applied by fluidized bed coating (p. 107). Films are formed at 140-190°C. Because of these temperatures, thermoplastic powder lacquers cannot be applied to wood or plastic and also not to soldered metals.

Thermosetting resins are mostly applied by electrostatic powder spraying. The resins are dry blends of pigments, binders and curing agents (= crosslinking agents). Binders are usually thermosetting resins such as epoxides which are used as such or combined with polyesters. Other important types are polyurethanes, polyesters combined with triglycidyl isocyanurate, and acrylates combined with oxazolines. Molecular weights are 1000 to 10 000. Hardening should not be too fast because a fast viscosity increase would lead to the formation of wrinkles.

Powder lacquers save energy, do not produce much waste, and do not significantly pollute the air. However, they are difficult to formulate, cannot easily match a targeted color, and require large investment costs for their manufacture and processing.

12.5 Printing Inks

The first printing processes were mechanical processes in which a **setting copy** (text, picture) was transferred to a **printing plate** (flat or roll). Printing ink was applied to the plate and then transferred to the substrate (paper, cardboard, fabric, etc.) by pressing the printing plate against the substrate. Depending on the relative heights of the printing and non-printing areas of the printing plate, one distinguishes between various types of printing (Section 12.5.1) which require different types of printing inks (Section 12.5.2).

Most mechanical printing processes are fast but need large investments for machinery and costly printing plates. They are therefore used only for high-volume printing.

Electronic printing processes do not require printing plates but are only suitable for small-volume printing. These printing processes include electrophotography (Section 14.2.8), microcapsule printing (Section 12.6.1), and ink-jet printing.

In **ink-jet printing**, very small electrically charged ink droplets are directed to moving substrates by computer-controlled nozzles. The ink dries by evaporation of the carrier. Substrates are mostly paper and cardboard but may also be plastic films and containers as well as textiles. Papyrus cannot be printed by this method.

12.5.1 Printing Processes

Relief Printing

In **relief printing**, printing inks are applied to raised printing areas (letters, etc.) of printing plates. The plates are then pressed against the printable substrate. Relief printing comprises letterpress and flexography.

Letter press uses rigid printing plates, for example, woodcuts or metal letters arranged on a plate. Originally, whole plates were carved from wood; modern printing began with movable metal letters (p. 236). Letters were poured from lead one at a time (monotype) or in whole lines (for example, linotype). Typesetting is now by computer.

Flexography uses flexible printing plates, originally from rubber, for type forms or gravures. Type forms for both flexography and letterpress are now produced mainly by **photopolymerization** (Volume I, p. 374). In this process, a photopolymerizable mass is put on a plastic or metal film and then irradiated with ultraviolet light through a mask with the desired pattern. The irradiated sections polymerize to solids; the non-irradiated sections are washed out (see Section 14.2.1). The resulting relief is the printing plate.

Gravure

In gravure, non-printing parts of printing plates are raised and printing parts are engraved. Printing plates are copper for copper engraving or steel for steel engraving. For small editions, plane plates are used and for large numbers of copies, rotogravure.

Rotogravure uses a steel cylinder which is covered by a thin copper layer. This layer is etched according to the print design with up to 8000 little indentations per square centimeter. The cylinder is then chromium plated in order to provide a hard surface.

The cylinder rotates half-submerged in a bath with the printing ink. The deepenings take up the paint; surplus paint is wiped off. By pressing the rotating cylinder against the substrate (paper, etc.), the paint in the deepenings is transferred to the substrate where it produces a tiny spot of color. The lightness (intensity) of the spot varies with the depth of the indentation which allows one to obtain halftones. For color prints, each ink (cyan, magenta, yellow, black) requires a separate printing cylinder.

Intaglio printing (from Italian *intagliare* = to engrave) with pastelike paints (Table 12-5) and deep gravures produces raised patterns; it is used for US bank notes. Embossing by intaglio printing is often simulated on letter heads and calling cards by sintering powdered resins on freshly printed, still wet letterpress or with lithographic printing inks.

Planography

In planography, printing and non-printing areas are on the same level. Planographic printing methods make use of the fact that oil and water repel each other.

The oldest planographic process is **lithography** (G: *lithos* = stone; *graphein* = to write) which was invented in 1798 by Johann Alois Senefelder who used the fine-pored, dense, easily cleavable "Sonthofener Schiefer" (Sonthofen limestone) as a printing plate (*Schiefer* is German for "slate" but this is a look-alike limestone). The original process used fat to cover the non-printing parts and aqueous printing inks for the printing ones. Later, this was reversed by using water-soluble materials (e.g., gum arabic and then poly-(vinyl alcohol)) for the non-printing parts and fat-containing printing inks for the printing ones. By pressing, the ink from the printing parts was transferred to the substrate. A later refinement of this process first transferred the pattern from the printing plate to a special paper that was treated with starch and then to a metal plate as a printing stock.

In **offset printing (offset lithography)**, printing is not from a stone as in the original lithographic process but from a thin sheet of paper for small numbers of copies and from aluminum, steel, zinc, or plastics sheets for large ones. These sheets are the carriers of a polymer film that takes up the printing ink and acts as the true printing stock. The paint from the printing stock is then transferred to a rubber blanket on a cylinder where the pattern appears *offset* (i.e., left and right sides exchanged). Printing from the rubber blanket to the paper then produces the correct pattern.

The pattern on the polymeric printing stock is produced by photopolymerization of a masked light-sensitive layer in a process called photolithography (see Section 14.2.11). Early layers consisted of egg white, gum arabic, or poly(vinyl alcohol) that were provided with chromium salts as sensitizers and crosslinked by the action of visible light.

Present-day *negative* processes use aryl azides and diazonium compounds. On irradiation of the unmasked parts of the pattern by near-ultraviolet light, aryl azides decompose and form nitrenes which free-radically crosslink binders such as cyclized poly(isoprene). The masked parts of this polymer remain uncrosslinked and are dissolved by an organic solvent. The crosslinked parts remain as a picture of the mask.

A newer process uses a poly(dimethylsiloxane) (PDMS) coat on a titanium-coated poly(ethylene terephthalate) film. The image is impressed upon the PDMS coat by infrared radiation which penetrates PDMS and heats the titanium beneath. The titanium breaks through the PDMS and creates grooves with hydrophobic PDMS walls and a hydrophilic bottom that readily and precisely accepts hydrophilic inks.

Positive processes employ naphthoquinone diazides which form difficultly soluble complexes with novolacs as binders. On irradiation of unmasked parts, naphthoquinone diazides convert to indene carboxylic acids that are dissolved because they have much higher solubilities. The masked parts form a positive pattern.

Screen Printing

In screen printing, printing parts are materials that let the paint through. In the original **silk-screen printing**, silk fabrics were used. Present materials are synthetic fabrics or wire gauze. The surfaces of these screens are provided with a photopolymer which is then irradiated according to the desired pattern. The irradiated areas harden (crosslink) and the non-irradiated ones are washed out. The resulting printing template is pressed against the substrate and the printing ink pressed through the openings.

12.5.2 Printing Inks

The different printing methods demand different pigment contents and different viscosities of printing inks (Table 12-5). Newspaper printing, a relief printing, is fast and therefore demands a low viscosity printing ink with small contents of pigments. In the US, binder contents of news inks are lower than the ones in Europe. Book printing, on the other hand, demands a much greater print quality and therefore printing inks with greater viscosities and larger pigment contents.

Intaglio printing demands printing inks with particularly large viscosities. All other printing inks lead to "coatings" that are far thinner than those produced by paints and other surface coatings (Table 12-5).

Conventional printing inks dry by evaporation of solvents, a process that can be accelerated by use of polymerizable binders such as acrylates (acrylate esters, acrylate carbamates, epoxy acrylates). The polymerization of these compounds is initiated by free radicals from the decomposition of photo initiators that is caused by ultraviolet light or electron beams (Volume I, p. 372). UV curable printing inks consist of 10-20 % reactive monomers, 10-15 % photo initiators, 20-40 % oligomers as thickeners, 15-20 % pig-

Table 12-5 Pigment content w_{PG} and viscosity η of printing inks and the resulting thicknesses d of films from mechanical printing methods compared to typical surface coatings. [1] Additional printimg methods: + 5 % digital printing, 11 % all other types of printing listed in this table.

Process	% use (US) [1]	w_{PG}/%	η/(Pa s)	d/μm
Relief printing				
Letterpress	2	20 - 80	1 - 50	3 - 5
News ink		8 - 12	0.2 - 1	
Flexography	22	10 - 40	0.1 - 10	2 - 4
Lithographic processes	43			
Lithography		20 - 80	10 - 80	2
Offset lithography			3	
Gravure	17			
Steel engraving				
Rotogravure		10 - 30	0.05 - 1	5 - 12
Intaglio			<200	5 - 75
Screen printing				12 - 25
Surface coatings				12 - 150

ments and fillers, and 5-10 % adjuvants (lubricants, flow improvers). The most often used photo initiator is a mixture of benzophenone and Michler's ketone.

Other curable printing inks are based on epoxides that are cationically polymerized by Lewis acids from the decomposition of diazonium salts. There are also printing inks based on thiols that harden by a free-radical initiated addition of –SH groups to ethylenic double bonds ($\sim$R–SH + CH$_2$=CH$\sim$ $\rightarrow$ $\sim$R–S–CH$_2$–CH$_2\sim$).

12.6 Depots

Depots are very small containers for the protective coating of small to miniscule amounts of substances such as solid or liquid pigments, adhesives, pharmaceuticals, herbicides, fungicides, or fertilizers (Table 12-6). These substances are set free for immediate action by sudden destruction of depot walls or for successive release by slow erosion of the protective coating.

12.6.1 Microcapsules

Microcapsules are spheroidal bodies of 1-5000 μm diameter with a polymeric wall that surrounds the protected goods. Wall thicknesses are ca. 1 μm for capsule diameters of less than 10 μm and ca. 50 μm for those of 3000 μm. Early microcapsules consisted of gelatin. They were used for the encapsulation of carbon black particles of non-dirtying carbon paper or single-use typewriter ribbons.

Microcapsules can be obtained from very many different polymers by a variety of methods. They are usually obtained in three steps: preparation of an emulsion or a suspension, encapsulation of substances, and isolation of capsules. The walls of capsules can

Table 12-6 Commercial microcapsules.

Wall materials	Encapsulation by	Application
Amino resins	*in situ* polycondensation	carbon paper
Ethyl cellulose	coacervation	pharmaceuticals
Gelatin	spray drying	vitamins
Maltodextrins	extrusion	food aromas
Polyureas	interfacial polycondensation	pesticides
Waxes	spray drying	fertilizers

be produced by a variety of methods such as *in-situ* polymerization, interfacial polycondensation, fluidized bed coating, dipping, centrifugal processes, spray drying, coacervation, electrostatic deposition, etc.

For *spray drying*, the goods to be encapsulated are emulsified or suspended in a solvent that contains a film-forming polymer. The dispersion is then sprayed into a hot stream of an inert gas where the solvent evaporates and leaves a firm polymeric coat around the goods. The resulting microcapsules are free flowing powders.

Alternatively, the goods can also be moved around in a tumbling mixer, vibrating trough, or rotary barrel mixer while being sprayed with the molten or dissolved depot material. The resulting coat solidifies from the melt, dries from the solution, or solidifies by polymerization of the monomeric coating material.

For coating in *fluidized beds*, goods are floated in an ascending vertical current of air while being sprayed with a solution of the coating material. The wall layer is formed by evaporation of the solvent.

In *dipping processes*, the goods pass through a thin layer of the liquid wall material. The resulting liquid coat solidifies in a hardening bath.

In *centrifugal coating*, liquid wall material resides as a thin layer on the porous exterior surface of a rotating drum. The goods are sprayed onto this layer by a rotating disc atomizer, pass though the layer in the pores, and exit as encapsulated goods into the interior of the drum where they are continuously removed. The thin layer on the drum is constantly renewed.

In *nozzle spraying*, the goods exit from a nozzle that is surrounded by a circular nozzle through which the wall material is sprayed against the exiting goods. The jet stream breaks by capillary wave fracture (Fig. 6-11) and the resulting droplets solidify.

Phase separations serve in many encapsulations. In these processes, the goods are dispersed in a polymer solution. The polymer is then deposited on the goods as a gel phase, either by temperature change, by addition of a precipitant, or by a solution of another polymer that does not mix with the first one.

The dispersed goods thus act as a kind of nucleating agent for the emerging phase separation by preventing the deposited gel from running together. The gel phase is a concentrated solution of the deposited polymer (a **coacervate** (L: *co* = together, *acervare* = to heap)). In phase separations of solutions by added non-solvents, coacervates consist of the polymer, the solvent, and the non-solvent (volume III, p. 317). A special case is *complex coacervation* in which the phase separation is caused by formation of a complex from the originally present, charged polymer and an added polymer with the opposite charge.

Continuous stirring prevents the gel-coated microencapsulates from sticking together. The gel coat is then solidified; for example, by cooling the gel below its glass temperature or by chemical crosslinking.

The encapsulated goods are set free at once by pressing the microcapsule. The release is slower by melting, dissolution, or chemical degradation of the wall. Encapsulated goods may also diffuse slowly through the wall (Section 12.6.2). Diffusion times agree well with the times calculated from permeability coefficients (Section 13.2). They depend on both the nature of the wall as well as the goods. For example, the half-life for the per- meation of water through a 25 μm thick wall of poly(butadiene) rubber is only about 2 hours, for a poly(ethylene) of the same thickness 42 days, and for a half as thick poly(trifluorochloroethylene) more than 2 1/2 years.

12.6.2 Other Depots

In many cases, depots should not be released at once as is required for adhesives and certain paints but their release rates should be controlled over longer time spans. Examples are pharmaceuticals, herbicides, fungicides, and long acting fertilizers. The release can be diffusion controlled or chemically controlled. In most cases, one is interested in zeroth order processes because they would set free the same amount per unit time over a long time period in such **time-release systems**.

In *diffusion-controlled systems*, active agents are released from a reservoir or a matrix. In **reservoir systems,** a core with the agent is surrounded by a polymeric hull which can be the wall of a capsule or microcapsule, a membrane, a hollow fiber, a liposome, or a dendrimer. Pharmaceuticals usually employ membranes as the hull.

The flow $J = A^{-1}(dm/dt)$ of the mass m of the agent per unit time through the area A of the hull is controlled by the first Fickian diffusion law, $J = -D \cdot (dc/dx)$, according to which the flow is given by the diffusion coefficient D and the concentration gradient dc/dx. Since the thickness dx of the wall is constant, one only needs to produce a constant concentration difference between the interior and the exterior of the reservoir in order to maintain a release rate dm/dt.

A constant concentration gradient can be obtained if a reservoir contains the active agent as a powder in amounts far above its solubility limit. A liquid entering the reservoir will then always dissolve the same amount per volume so that $dc = const.$ and also $dc/dt = const.$: the release is a zeroth order process in which the same amount is always released per unit time (Fig. 12-14, top left).

In ideal **matrix systems**, active agents are evenly distributed within a solid polymer. Agents in the center of the matrix therefore have to diffuse larger distances to the wall than those at the periphery. Hence, the exit of the active agent is not a zeroth order process but proceeds according to diffusion laws. In most cases, it is a first order process in which the exiting amount is proportional to the square root of time (Fig. 12-14, bottom right). Matrix systems are far less costly to prepare than reservoir systems; unfortunately, they are not suited for many applications.

The release of active agents can also be *chemically controlled* in various ways. Active agents can be either bound covalently, by hydrogen bonds, or by dipole-dipole interactions to the polymer chains of the matrix or just physically dispersed or dissolved in the matrix.

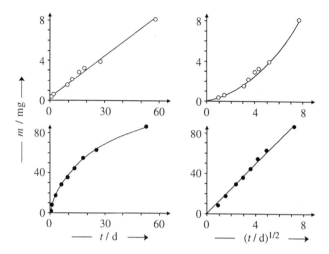

Fig. 12-14 Total mass *m* of agents that are released from polymers as a function of time *t* (left) or as a function of the square root of time (right) [7]. Top: a reservoir system containing the fungicide tri-butyltin fluoride ($m_0 = 41$ mg) in a reservoir from ethylene-vinyl acetate copolymer. Bottom: a matrix system with the herbicide 2,4-dichlorophenoxy acetic acid ($m_0 = 253$ mg) in a high-density poly-(ethylene) containing iron oxide particles.

Covalently bound agents can be released from the matrix hydrolytically or enzymatically and then transported in various ways. The release from the reservoir will be controlled by diffusion of the agent if the agent is bound to a solid matrix. However, if the polymer matrix with the bound agent is molecularly dissolved in the reservoir, then the rate of release will be controlled by the rate of the cleavage reaction.

In pharmaceutical applications, it is often preferred that reservoir walls are either permeable or are able to break down so that the agent-carrying polymer chains may exit reservoirs intact and travel to their place of action, provided they can pass through biological cell membranes. The latter systems are especially effective since they may consist of up to 80 % of the active agent as compared to 10-30 % for reservoir and matrix systems.

The active agent may also be dispersed or dissolved in a solid matrix but not be bound to it. It can then be released by a chemical or enzymatic breakdown of the matrix. A release by erosion of the solid matrix surface will be of zeroth order if the surface area remains constant. Of course, that is not possible for spherical matrices but can be achieved by large, thin disks because the reduction of thickness can be neglected.

Literature to Chapter 12

12.2.1 POLYMER SURFACES
D.T.Clark, W.J.Feasts, Eds., Surfaces, Wiley, New York 1979
B.W.Cherry, Polymer Surfaces, Cambridge University Press, Cambridge, UK, 1981
S.R.Culler, H.Ishida, J.L.Koenig, The Use of Infrared Methods to Study Polymer Interfaces,
    Ann.Rev.Mater.Sci. **13** (1983) 363
F.Garbassi, M.Morra, E.Occhiello, Polymer Surfaces. From Physics to Technology, Wiley,
    New York 2000
P.Chen, Ed., Molecular Interfacial Phenomena of Polymers and Biopolymers, CRC Press, Boca
    Raton (FL) 2006

12.2.2 ADSORPTION OF MACROMOLECULES
A.Takahashi, M.Kawaguchi, The Structure of Macromolecules Adsorbed on Interfaces,
    Adv.Polym.Sci. **46** (1982) 1
R.H.Ottewill, C.H.Rochester, A.L.Smith, Eds., Adsorption from Solution, Academic Press,
    New York 1983
G.D.Parfitt, C.H.Rochester, Eds., Adsorption from Solution at the Solid/Liquid Interface,
    Academic Press, London 1983

12.2.3 WETTABILITY
Yu.S.Lipatov, A.E.Feinerman, Surface Tension and Surface Free Energy of Polymers, Adv.
    Colloid Interface Sci. **11** (1979) 195
L.Gao, T.J.McCarthy, Langmuir **22** (2006) 5998; –, J.Am.Chem.Soc. **128** (2006) 9052 (Lotus
    Effect)
V.M..Starov, M.G.Velarde, C.J.Radke, Wetting and Spreading Dynamics, CRC Press, Boca Raton
    (FL) 2007

12.3 ADHESIVES
N.J.DeLollis, Adhesives, Adherents, Adhesion, Krieger, Huntington (NY), 2nd ed. 1980
W.C.Wake, Adhesion and the Formulation of Adhesives, Appl.Sci.Publ., Barking, Essex, UK,
    2nd ed. 1982
A.Pizzi, Ed., Wood Adhesives, Dekker, New York, Vol. 1 (1983), Vol. 2 (1989)
P.M.Hergenrother, High-Temperature Adhesives, ChemTech **14** (1984) 496
R.D.Adams, W.C.Wake, Structural Adhesive Joints in Engineering, Elsevier, New York 1984
S.R.Hartshorn, Ed., Structural Adhesives: Chemistry and Technology, Plenum, New York 1986
T.E..Lipatova, Medical Polymer Adhesives, Adv.Polym.Sci. **79** (1986) 85
M.A.Krenceski, J.F.Johnson, S.C.Temin, Chemical and Physical Factors Affecting Performance of
    Pressure-Sensitive Adhesives, J.Macromol.Sci.-Revs.Macromol.Chem.Phys. **C 26** (1986) 143
A.J.Kinloch, Adhesion and Adhesives. Science and Technology, Chapman and Hall, New York 1988
K.Booth, Industrial Packaging Adhesives, CRC Press, Boca Raton (FL) 1990
– Desk Top Data Bank, Adhesives, International Plastics Selector, San Diego, CA (no date given)

12.3.2 ADHESION
D.H.Kaeble, Physical Chemistry of Adhesion, Wiley, New York 1971
N.J. DeLollis, Adhesives, Adherends, Adhesion, Krieger, Malabar (FL) 1980
S.Wu, Polymer Interface and Adhesion, Dekker, New York 1982
D.M.Brewer, D.Briggs, Industrial Adhesion Problems, Wiley, New York 1986
I.M.Skeist, Ed., Handbook of Adhesives, Van Nostrand Reinhold, New York, 2nd ed. 1990
L.-H.Lee, Ed., Fundamentals of Adhesion, Plenum, New York 1991
L.-H.Lee, Ed., Adhesive Bonding, Plenum, New York 1991
V.L.Vakula, L.M.Pritykin, Polymer Adhesives: Basic Physicochemical Properties, Prentice-Hall,
    Englewood Cliffs (NJ) 1992
D.E.Packham, Ed., Handbook of Adhesion, Wiley, New York 1993
A.V.Pocius, Adhesion and Adhesives Technology, Hanser, Munich 1996
J.Comyn, Adhesion Science, Royal Soc.Chem., London 1997
R.J.Hussey, J.Wilson, Structural Adhesives Directory and Datebook, Chapman and Hall, London
    1997

K.L.Mittal, A.Pizzi, Eds., Adhesion Promotion Techniques, Dekker, New York 1999
A.V.Pocius, Adhesion and Adhesives Technology, Hanser Gardner, Cincinnati (OH), 2nd ed. 2002
M.Ash, I.Ash, Handbook of Adhesive Chemicals and Compounding Ingredients, Synapse Information
    Resources, Endicott (NY) 2004
W.Possart, Ed., Adhesion. Current Research and Applications, Wiley-VCH, Weinheim 2005
D.E.Packham, Ed., Handbook of Adhesion, Wiley, Hoboken (NJ), 2nd ed. 2005

12.4 PAINTS AND COATINGS

S.Paul, Surface Coatings. Science and Technology, Wiley, New York 1985
G.L.Schneberger, Understanding Paint and Painting Processes, Hitchcock, Wheaton (IL), 3rd ed. 1985
A.D.Wilson, J.W.Nicholson, H.J.Prosser, Eds., Surface Coatings, Elsevier, London 1988
L.J.Calbo, Ed., Handbook of Coatings Additives, Dekker, New York, Vol. I (1987), Vol. II (1992)
G.P.A.Turner, Introduction to Paint Chemistry and Principles of Paint Technology, Chapman and
    Hall, London, 3rd ed. (1988)
E.W.Flick, Ed., Handbook of Paint Raw Materials, Noyes, Park Ridge (NJ), 2nd ed. 1989
W.H.Morgans, Outlines of Paint Technology, Wiley, New York, 3rd ed. 1990
D.Satas, Ed., Coatings Technology Handbook, Dekker, New York 1991
E.D.Cohen, E.B.Gutoff, Eds., Modern Coating and Drying Technology, VCH, New York 1992
R.Lambourne, Paint and Surface Coatings. Theory and Practice, Ellis Horwood, Chichester 1993
A.R.Morrison, Ed., The Chemistry and Physics of Coatings, Royal Soc.Chem. London 1994
D.Stoye, Ed., Paints, Coatings and Solvents, VCH, Weinheim 1994
C.H.Harc, Protective Coatings. Fundamentals of Chemistry and Composition, Technomic,
    Pittsburgh (PA) 1994
S.Paul, Surface Coatings. Science and Technology, Wiley, New York, 2nd ed. 1995
T.C.Patton, Paint Flow and Pigment Dispersion, Wiley-Interscience, New York, 2nd ed. 1995
D.Stoye, W.Freitag, Eds., Lackharze; -, Resins for Coatings, Hanser, Munich 1996
A.R.Marrion, Ed., The Chemistry and Physics of Coatings, Royal Soc.Chem., London 1996
W.Freitag, D.Stoye, Eds., Paints, Coatings and Solvents, Wiley-VCH, Weinheim 1998
J.Bieleman, Lackadditive, Wiley-VCH, Weinheim 1998; -, English ed. 1999
R.Lambourne, T.A.Strivens, Paints and Surface Coatings, W.Andrew Publ., Norwich (NY),
    2nd ed. 1999
D.Satas, A.Tracton, Eds., Coatings Technology Handbook, Dekker, New York, 2nd ed. 2000
J.H.Bieleman, Ed., Additives for Coatings, Wiley-VCH, Weinheim 2000
R.A.Ryntz, Ed., Plastics and Coatings: Durability - Stabilization - Testing, Hanser, Munich 2001
D.G.Weldon, Failure Analysis of Paints and Coatings, Wiley, New York 2001
J.H.Bieleman, Driers, Chimia **56** (2002) 184
A.Goldschmidt, H.J.Streitberger, (BASF Handbook on) Basics of Coating Technology, W.Andrew,
    Norwich (NY) 2003
P.K.T.Oldring, Resins for Surface Coatings, Wiley, New York 2002 (3 vols.)
M.Ash, I.Ash, Handbook of Paint and Coating Raw Materials, Synapse Information Resources,
    Endicott (NY), 2nd ed. 2003
K.Tauer, Aqueous Polymer Dispersions, Springer, Berlin 2004
D.R.Karsa, Surfactants in Polymers, Coatings, Inks, and Adhesives, CRC Press, Boca Raton (FL)
    2004
J.J.Florio, D.J.Miller, Handbook of Coatings Additives, Dekker, New York, 2nd ed. 2004
B.Müller, U.Poth, Coatings Formulation, W.Andre, Norwich (NY) 2006
S.K.Ghosh, Ed., Functional Coatings by Polymer Microencapsulation, Wiley-VCH, Weinheim 2006
Z.W.Wicks, Jr., F.N.Jones, S.P.Pappas, D.A.Wicks, Organic Coatings. Science and Technology,
    Wiley, Hoboken (NJ), 3rd ed. 2007
A.A.Tracton, Ed., Coatings Materials and Surface Coatings, CRC Press, Boca Raton (FL) 2007

12.4.2.a CRITICAL PIGMENT-VOLUME CONCENTRATION

G.P.Bierwagen, T.K.Hay, The Reduced Pigment Volume Concentration as an Important Parameter
    in Interpreting and Predicting the Properties of Organic Coatings, Progr.Org.Coat. **3** (1975) 281
T.C.Patton, Paint Flow and Pigment Dispersion, Wiley-Interscience, New York, 2nd ed. 1979
D.Urban, K.Takamura, Eds., Polymer Dispersions and Their Industrial Uses, Wiley-VCH, Weinheim
    2002

12.4.2.b PROPERTIES OF PAINTS

K.Sato, The Hardness of Coating Films, Progr.Org.Coat. **8** (1980) 1

L.O.Kornum, H.K.Raaschou Nielsen, Surface Defects in Drying Paint Films, Progr.Org.Coat. **8** (1980) 275

M.Cremer, Evaluation of Hiding Power in Organic Coatings - Measurement, Methods and Their Limitations, Progr.Org.Coatings **9** (1981) 241

G.Kaluza, Flocculation of Pigments in Paints–Effects and Cause, Progr.Org.Coatings **10** (1982) 289

T.C.Patton, Paint Flow and Pigment Dispersion, Wiley-Interscience, New York, 2nd ed. 1995

E.W.Ott, Performance Enhancement in Coatings, Hanser, Munich 1998

D.G.Weldon, Failure Analysis of Paints and Coatings, Wiley, New York 2001

E.B.Gutoff, E.D.Cohen, Coating and Drying Defects, Wiley, New York, 2nd ed. 2006

12.4.2.c RADIATION CURING

S.P.Pappas, Ed., UV Curing: Science and Technology, Technology Marketing Corp., Stamford (CT), Vol. 1 (1978), Vol. 2 (1985)

C.G.Roffey, Photopolymerization of Surface Coatings, Wiley, Chichester 1982

R.Dowbenko, C.Friedlander, G.Gruber, P.Prucnal, M.Wismer, Radiation Curing of Organic Coatings, Progr.Org.Coat. **11** (1983) 71

G.A.Senich, R.E.Florin, Radiation Curing of Coatings, J.Macromol.Sci.-Rev.Macromol.Chem. Phys. **C 24** (1984) 239

P.K.T.Oldring, Ed., Chemistry and Technology of UV and EB Formulation of Coatings, Inks and Paints, SITA Technol., London 1991 (6 vols.)

A.Singh, J.Silverman, Radiation Processing of Polymers, Hanser, Munich 1991

N.Inagaki, Plasma Surface Modification and Plasma Polymerization, Technomic, Lancaster (PA) 1996

12.4.4 WATER-BORNE COATINGS

E.W.Flick, Industrial Water-Based Paint Formulations, Noyes, Park Ridge (NJ) 1988

K.Dören, W.Freitag, D.Stoye, Waterborne Coatings, Hanser, Munich 1994

D.R.Karsa, Ed., Waterborne Coatings and Additives, Royal Soc.Chem., London 1996

12.4.5 LATEX PAINTS

K.Calvert, Ed., Polymer Latices and Their Applications, Hanser, Munich 1981

12.4.7 HIGH SOLIDS

M.Takahashi, Recent Advances in High Solids Coatings, Polym.-Plast.Technol.Eng. **15** (1980) 1

12.4.8 POWDER LACQUERS

M.W.Ranney, Powder Coatings Technology, Noyes, Park Ridge (NJ) 1975

T.A.Misev, Powder Coatings, Wiley, New York 1991

12.4.x SPECIAL COATINGS

K.L.Mittal, Ed., Metallized Plastics, Dekker, New York 1997

L.W.McKeen, Fluorinated Coatings and Finishes Handbook, W.Andrew, Norwich (NY) 2006

12.6 DEPOTS

J.E.Vandegaer, Ed., Microencapsulation. Processes and Applications, Plenum, New York 1974

C.Tanquary, R.E.Lacey, Eds., Controlled Release of Biologically Active Agents, Plenum, New York 1974

J.R.Nixon, Ed., Microencapsulation, Dekker, New York 1976

A.Kondo, Microcapsule Processing and Technology, Dekker, New York 1979

A.F.Kydonieus, Ed., Controlled Release Technologies, CRC Press, Boca Raton (FL) 1980 (2 vols.)

R.Langer, N.Peppas, Chemical and Physical Structure of Polymers as Carriers for Controlled Release of Bioactive Agents: A Review, J.Macromol.Sci.-Revs.Macromol.Chem.Phys. **C 23** (1983) 61

T.J.Roseman, S.Z.Mansdorf, Eds., Controlled Release Delivery Systems, Dekker, New York 1983

P.B.Deasy, Microencapsulation and Related Drug Processes, Dekker, New York 1984

M.Rosoff, Ed., Controlled Release of Drugs: Polymers and Aggregate Systems, VCH, Weinheim 1989

P.J.Tarcha, Polymers for Controlled Drug Delivery, CRC Press, Boca Raton (FL) 1990
M.Rosoff, Ed., Vesicles, Dekker, New York 1996
E.Mathiowiz, Ed., Encyclopedia of Controlled Drug Delivery, Wiley, New York, Vol. 1 (1999),
    Vol. 2 (2000)
R.M.Ottenbrite, S.W.Kim, Polymeric Drugs and Drug Delivery Systems, Technomic, Lancaster (PA)
    2001
I.F.Uchegbu, A.G.Schätzlein, Eds., Polymers in Drug Delivery, CRC Press, Boca Raton (FL) 2006
S.K.Ghosh, Ed., Functional Coatings by Polymer Microencapsulation, Wiley-VCH, Weinheim 2006

References to Chapter 12

[1]   E.Killmann, J.Eisenlauer, M.Korn, J.Polym.Sci.-Polym.Symp. **61** (1977) 413, (a) Figs. 1 a-c,
        (b) data of Fig. 3
[2]   L.Perrone, Plastics Eng., (May 1980) 51, Table 3
[3]   W.K.Asbeck, M.Van Loo, Ind.Eng.Chem. **41** (1949) 1470, data of Fig. 1
[4]   S.T.Bowell, in J.K.Craver, R.W.Tess, Eds., Applied Polymer Science, Amer.Chem.Soc.
        Washington (DC) 1975, p. 597, Fig. 1
[5]   F.Lin, D.J.Meier, Polym.Mater.Sci.Eng. **73** (1995) 93, (a) Fig. 1, (b) Fig. 4
[6]   J.N.Yoo, L.H.Sperling, C.J.Glinka, A.Klein, Macromolecules **23** (1990) 3962, (a) Fig. 7,
        (b) Table III
[7]   L.R.Sherman, J.Appl.Polym.Sci. **27** (1982) 997, Figs. 3 and 4

13   Packaging Materials

13.1   Introduction

Pack(ag)ing materials enclose goods and active substances for transport and temporary protection but are removed before goods are used. They differ in this respect from coatings (Section 12.4) which are permanent.

Packing materials are used for packs (containers, boxes, barrels, bottles, cans, tubes, etc.), wrappings (sacks, bags, packets, etc.), and paddings. Large packs such as containers are usually made from wood or metals, medium-sized ones such as bottles or cans from glass, metals, or plastics, and small ones such as tubes from plastics. Wrappings are produced from cellulose (paper, cardboard, various fabrics based on plant fibers), aluminum, and many different plastics. Paddings consist of straw, wood-wool, and foamed rubbers or plastics. Industrialized countries consume about 25 kg packing materials per capita each year wherefrom ca. 40 % are paper and cardboard, 30 % glass, 15 % plastics, 10 % metals, and 5 % wood.

The annual world consumption of plastics for packing materials is about 15 % of the total plastics production and climbing. Containers and films show the strongest growth because plastics can economically replace glass and metals. Plastics-based packing materials show an above-average economic growth because plastics are easy to process and have a good combination of chemical and physical properties, especially low densities. Their energy and ecological balances are favorable (Section 8.6) and their prices are low. Still, large parts of the population view plastics as environmentally damaging because throw-away plastics articles are very visible and do not decompose easily (see Section 8.6). However, modern economies cannot do without plastic packing materials which save transportation costs (lighter weight) and protect goods. In third world countries, about 50 % of foodstuffs spoil because of lack of packing materials, absence of warehouses, and insufficient distribution systems; in Western countries, this is only ca. 1 %.

High-density poly(ethylene) still dominates the market for containers, but poly(ethylene terephthalate) is making inroads into the bottle market. Low-density poly(ethylene) is preferred for films because it can be stretched and thermally sealed. Poly(ethylene) has also mainly replaced plasticized poly(vinyl chloride) for carrying bags. Good market chances are predicted for multilayer films for cans for food and carbonated drinks. Isotactic poly(propylene) is the favorite for lids which have to be rigid and tough. In the United States, considerable amounts of foamed poly(styrene) are used for throw-away drinking cups for coffee and carbonated drinks.

13.2   Permeation

13.2.1   Fundamentals

Packing materials should be inert to packed goods and packed goods inert to packing materials. This demand is seldom fulfilled 100 %, not only not for plastics but also for the classical packing materials paper, wood, glass, and metals. For example, lead can be leached from lead glasses by acids in wines. Tin cans are actually steel cans that are coated with a thin layer of tin so that the steel cannot be attacked by acids. Modern steel cans are protected by coatings, the binders of which are thermoplastics of course.

In contrast to glass and metals, gases and liquids may permeate through plastics albeit often only after long times. A permeation from the interior to the exterior of a container leads to a loss of some contents, for example, carbon dioxide from carbonated drinks, aromas of coffee beans, or liquids from alcoholic beverages. Conversely, permeation of oxygen from the exterior into a container may oxidize the contents. Additives in the packing material may also migrate into the packed goods and change their appearance, composition, taste, etc.

Plastic walls may be permeated by very many different chemical compounds. The following discussion is concerned with those gases and liquids that can be dissolved in the packing material but do not interact so strongly with the material that the packing material is dissolved. The permeating substance will be called a **permeant** (L: *per* = through, *meare* = to go, pass); it can be a gas, a liquid, or a dissolved solid. The permeated material is referred to as a **matrix**; it may be a packing material, a membrane, etc.

Permeation begins with the uptake of gases and liquids by the matrix which causes the matrix to swell somewhat. This **sorption** is described by a parameter S. In the simplest case of *permanent gases*, S is a **distribution coefficient** which indicates the difference Δu of the proportions of the permeant on each side of the matrix (sheet, film, membrane, etc.) as a result of a constant pressure difference Δp:

(13-1) $\Delta u = S \cdot \Delta p$ (**Henry's law** for gases)

Δu may be the difference Δw of mass fractions w, $\Delta \phi$ of volume fractions ϕ, or Δx of mole fractions of the permeant. In this case, the solubility coefficient has the physical unit of an inverse pressure while its numerical value depends of the choice of the dimensionless quantity Δu.

For *liquids*, the driving force is not a pressure difference Δp but a concentration difference Δc_m on each side of the matrix. The permeated quantity is usually also measured as a concentration difference Δc so that the parameter S is now the Ostwald **solubility coefficient** with the physical unit of unity:

(13-2) $\Delta c = S \cdot \Delta c_m$

In practice, quantities Δc and Δc_m are not always given in congruent physical units but in different ones; for example, Δc in mol and Δc_m in mol/L, so that S may have different physical units; in this particular case, liter.

At the beginning of a measurement, a permeant will be completely on one side of the matrix and not at all on the other. The matrix will then first become saturated with the permeant before the permeant exits on the other side of the matrix, i.e., there will be a time lag (Fig. 13-1).

After the time lag, the permeated amount Δu, for example, the volume difference ΔV or the mass difference Δm, will increase linearly with time if the reservoir on the start side is large enough (Fig. 13-1). The time lag t_1 is determined by extrapolation of the straight line to $\Delta m = 0$. Theory defines the time lag as $t_1 = L_m^2/(6\ D)$ where L_m is the thickness of the matrix and D is the diffusion coefficient of the permeant in the matrix.

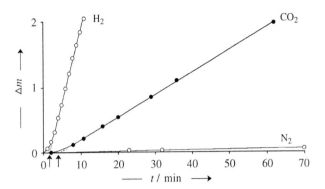

Fig. 13-1 Time dependence of the permeation of gases through a thin sheet of a styrene copolymer at 25°C [1]. The difference of quantities of gases on each side of the matrix is given in arbitrary units. ↑ indicates the time lag t_1 for H_2 and CO_2, respectively. The time lag could not be determined for nitrogen.

The time dependence of permeation is calculated by Fick's first law (Volume III, p. 367) according to which the diffusion flow $J_d = -D(\partial c/\partial r)$ is proportional to the diffusion coefficient D and the concentration gradient $\partial c/\partial r$. The diffusion flow from the greater concentration through the matrix to the smaller one is given by the change of mass m with time t per unit area A_m, $J_d = -(\partial m/\partial t)/A_m$. Replacing differentials by differences, equating, transforming, use of the matrix thickness $L_m = \Delta r$, and introduction of Eq.(13-2) leads to

$$(13\text{-}3) \qquad \Delta m = DS(A_m/L_m)\Delta c_m \Delta t = P(A_m/L_m)\Delta c_m(t - t_1)$$

where $P \equiv DS$ is the **permeability coefficient** with the physical unit length²/time if S is dimensionless and congruent physical units are used for Δm and Δc_m, for example, Δm as mass and Δc_m as mass per volume. The permeability coefficient P is thus

$$(13\text{-}4) \qquad P = \frac{\Delta m\, L_m}{A_m(\Delta c_m)(t - t_1)} \quad ; \quad \frac{L^2}{T} = \frac{ML}{L^2(ML^{-3})T} = \frac{NL}{L^2(NL^{-3})T} = \frac{VL}{L^2(1)T}$$

where the equations on the right side of the semicolon show the interrelationships between the length (L), area ($A = L^2$), volume ($V = L^3$), mass (M), amount-of substance (N), and time (T). If Δm is the amount-of-substance (e.g., N in mol), then Δc_m should be the amount concentration [C] with $N/V = N/L^3$ (e.g., in mol/cm³). If Δm is a volume with unit V, then Δc_m should be a volume fraction ϕ with the unit "1", etc.

In industry, one often hears that one should not combine all physical units (such as ML/[L²(M L⁻³)T] to L²/T because that would no longer allow one to immediately see the effect of changing one of the parameters Δm, L_m, A_m, or Δc_m. Of course, such effects can be seen immediately from Eq.(13-4): doubling of L_m doubles P, etc., even if one does not care to give the physical units.

There is a problem if one does not use congruent units but employs "practical" ones. In the Anglo-Saxon technical literature, permeation coefficients of gases are often reported according to ASTM in **BU units** which are called either **Barrer units** after their inventor or **barrier units** since the membrane, etc., does present a barrier to the gas (or if one thinks that "Barrier" must be a misprint?). In this approach, treatment of units of variables is incoherent and physical quantities are expressed in unconventional, outdated, or Anglo-Saxon units. Δm is treated as a volume and measured in cm^3 but Δc_m is not introduced correspondingly as a volume fraction but as a pressure in atmospheres, often without specification whether this is the outdated physical atmosphere (1 atm = 760 torr = 0.101 325 MPa) or the outdated technical atmosphere (1 at = 1 kp/cm^2 = 0,098 065 MPa). The thickness L_m is given in mils (1 mil = 10^{-3} inch) and the area A_m in square inches whereas the time difference $t - t_1$ is measured in days. The whole expression is then divided by 100 to give

(13-5) $$1\,BU = \frac{1\,cm^3 \cdot 1\,mil}{100\,in^2 \cdot 1\,atm \cdot 1\,day} \approx 4.497 \cdot 10^{-10}\,cm^2\,s^{-1}\,MPa^{-1}$$

   In the technical literature, one finds approximately 30 different units for the permeability P. Some of the equations used for the calculation of P leave out important parameters, others employ outdated physical units, and many of them report the physical unit(s) of P wrongly or not at all. In addition, a number of other measures are also used:

- The **oxygen transmission rate (OTR)** with the physical unit cm^3/(m$^2 \times$ 24 h) as a measure for the barrier action against oxygen, usually at 23°C and 50 % relative humidity. Inverse OTR values are additive if a membrane, film, etc., consists of three layers A, B and C: (1/OTR) = (1/OTR-A) + (1/OTR-B) + (1/OTR-C);

- the **gas permeability**, i.e., the OTR value per pressure differential with the unit cm^3/(m$^2 \times$ 24 h $\times$ bar) (1 bar = 10^5 Pa); and

- the **oxygen permeability** with the unit (cm$^3 \times$ 25 µm)/(m$^2 \times$ 24 h $\times$ bar) for a 25 µm thick film at (usually) 23°C and 50 % relative humidity.

13.2.2 Permeation of Gases

   Permeability coefficients of gases in polymers vary widely (Table 13-1). For example, the permeability of oxygen through cellulose 2 1/2-acetate is 400 times greater than that of cellulose itself, that of ethyl cellulose 5000 times, and that of poly(dimethylsiloxane) even 300 000 times. In all cases, these permeabilities are much smaller than those of membranes of Vycor glass with pore widths of ca. 5 nm.

   Permeability coefficients of H$_2$ and O$_2$ through nanoporous glass vary only by a factor 3.85 whereas those of these so-called permanent gases through poly(ethylene terephthalate) (PET) differ by a factor 1800. These differences between gases are caused by the "pore structure" of the matrix (called "membrane" if for separation, "wall" if for containment, etc.). Vycor glass, a chemically leached borosilicate glass, is a **pore membrane** whereas polymeric membranes are **solubility membranes**. **Pores** are defined here (but see p. 564) as channels with diameters that are considerably larger than the diameters of permeating molecules (for helium in Vycor glass: d_{pore} = 5 nm *versus* d_{mol} = 0.145 nm). Hence, interactions between pore walls and permeating molecules can be neglected in pore membranes. In solubility membranes, channels are either absent or so small that the permeating gas molecules cannot avoid interactions with the matrix material.

Table 13-1 Permeability coefficients P of gases at 0°C and a pressure of 760 torr $\approx$ 1.013 bar = 0.1013 MPa of gas separation membranes [2], arranged according to descending values for oxygen. Values in parentheses below the name of the gas indicate the molar van der Waals volume V_{mol} of the gas [3] and those in square brackets the molecule diameter d_{mol} [4].
     Vycor glass has a pore diameter of 5 nm. "PVC" is a plasticized polymer containing 30 wt% di(2-ethylhexyl)phthalate (DOP). Original data were reported in $(cm^3 \, cm)/(cm^2 \, s \, torr \, 10)$. The "10" probably has to be read as 10^{10} since coefficients would be much too high otherwise.

Membrane	$10^{14} \, P/(cm^2 \, s^{-1} \, Pa^{-1})$						
	H_2	He	O_2	N_2	CH_4	CO_2	H_2O
$V_{mol}/(L/mol)$	(0.0266)	(0.0237)	(0.0318)	(0.0391)	(0.0428)	(0.0427)	(0.0305)
d_{mol}/nm	[0.114]	[0.145]	[0.150]	[0.160]	[0.160]		[0.140]
Vycor glass	15 000 000	11 000 000	3 900 000	4 400 000	6 000 000	5 000 000	
PDMS	33 000	17 000	45 000	20 000	44 000	24 000	3 000 000
P4MP			2 400			7 000	
IR	3 800	2 700	1 700	750		9 800	23 000
PPE			1 100	230		5 600	300 000
Ethyl cellulose	230	2 300	750	230	450	3 000	1 800 000
FEP	1 100	3 800	450	160	110	130	2 500
HDPE	600	520	450	150		2 100	6 800
PC	1 100	5 000	110	38	270	600	110 000
LDPE			83	23		320	11 000
Cellulose acetate	600	1 800	60	23	23	180	750 000
PVC	1 000	1 100	45	15	150	280	44 000
PET	50	800	2.2	0.45	0.45	11	13 000
PVDC	6		0.38	0.075		2.3	75
Cellophane			0.15	0.038		0.45	980 000

The permeability coefficient $P = DS$ for the permeation through a pore membrane decreases therefore in a first approximation with increasing molecule diameter of the gas since it is controlled only by the diffusion coefficient of the gas. However, permeability coefficients of gases in solubility membranes are controlled by both the diffusion coefficient D and the solubility coefficient S of the gas which depends on the interaction between the gas and the matrix material of the membrane and thus on the polarizabilities of gas molecules and monomeric units of matrix molecules.

For these reasons, solubility coefficients of hydrogen and nitrogen are normally small, greater for oxygen, and the greatest for carbon dioxide. Because of the strong effect of S on P, permeability coefficients of CO_2 are ca. 5 times greater than those of O_2 for many polymers (Fig. 13-2) although the diffusion coefficient of CO_2 should be smaller than that of O_2 because of the greater molecular weight (44 versus 32) and the greater the van der Waals volume (0.0427 versus 0.0318 L/mol) of the former. The exceptionally small ratio $P_{CO_2}/P_{O_2} = 0.0011$ for poly(vinyl alcohol) (PVAL) is probably caused by large solubility coefficients from interactions between oxygen molecules and the hydroxyl groups of poly(vinyl alcohol).

Permeations of gases through polymeric membranes usually do not depend on the molecular weights M of polymers because both macroconformations of coil molecules in amorphous polymers and morphologies of semicrystalline polymers are independent of M if M is greater than ca. 5000.

The more flexible the polymer chain, the smaller is the activation energy of diffusion and the larger is the diffusion coefficient. Rubbers and elastomers have glass tempera-

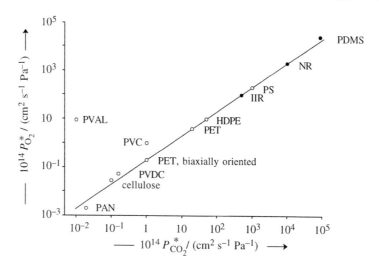

Fig. 13-2 Permeability coefficient P^* of oxygen as a function of the permeability coefficient of carbon dioxide for permeation of these gases through rubbers (●) and thermoplastics (○) at 30°C [5]. The straight line indicates $P^*_{CO_2}/P^*_{O_2} = 5$. At normal pressure p, a value of $P^* = 1 \cdot 10^{-14}$ cm^2 s^{-1} Pa^{-1} corresponds to $P = 1 \cdot 10^{-9}$ cm^2 s^{-1} since $p = 1 \cdot 10^5$ Pa. Note that data for polymers in this graph differ from those in Table 13-1 since specimens were prepared differently.

Polymers for Coca Cola® bottles should have permeability coefficients $P^*/(10^{-14}$ cm^2 s^{-1} Pa$^{-1}) < 1$ for O_2 and <0.5 for CO_2. For beer, the corresponding values are 0.05 (O_2) and 0.5 (CO_2).

tures much lower than room temperature and therefore usually much greater permeability coefficients than thermoplastics (see Fig. 13-2 and poly(dimethylsiloxane) (PDMS) in Table 13-1). Of elastomers, vulcanized butyl rubber (IIR) has an especially low permeability coefficient for oxygen and nitrogen and is therefore used for inner tubes of car tires (p. 283).

Low gas permeabilities of *thermoplastics* can be obtained by either a suitable chemical structure of the matrix or a special morphology of the system. Strongly polar monomeric units decrease the permeability because they lower the solubility of gases (example: poly(acrylonitrile) (PAN)). However, PAN is unusable as barrier resin for membranes and bottles because it cannot be melt processed. Typical **barrier resins** are therefore melt-processable, acrylonitrile-containing copolymers such as copolymers from styrene and acrylonitrile units (Lopac®; 70 % AN) and graft copolymers of methyl acrylate and acrylonitrile (26:74) on 10 % butadiene rubber (Barex®). Other barrier resins contain vinyl alcohol and ethylene units (for example, EVAL®).

Gas permeabilities of polymers are reduced by obstacles because they reduce both the diffusibility and solubility of gases; the former by increasing the path of permeating molecules and the latter by reducing the available space for absorption. Such obstacles are filler particles, crystalline polymer regions, and oriented chain segments. Platelets as fillers are more effective than rods and rods more than spheres. Examples of the influence of drawing are the permeabilities of unoriented and bi-oriented specimens of the same poly(ethylene terephthalate) in Fig. 13-2 and the effect of comonomer units and stretching on the permeability of ethylene polymers (Fig. 13-3).

Drawing increases the orientation of chain segments of polymers in general and the crystallinity of semicrystalline polymers in particular once a certain critical draw ratio

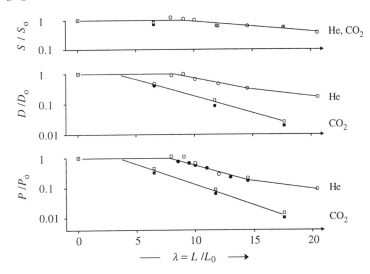

Fig. 13-3 Logarithms of relative solubility coefficients S/S_0, diffusion coefficients D/D_0, and permeability coefficients P/P_0 of helium (O,●) and carbon dioxide (□,■) at 25°C as functions of the draw ratio, $\lambda = L/L_0$ of films of a poly(ethylene) (O,□; Hizex 7000F) and a poly(ethylene-*co*-1-hexene) with a few percent of hexene units (●,■; Rigidex 002-55) [6,7].

L/L_0 is surpassed (Fig. 13-3). For both a poly(ethylene) and a copolymer of ethylene and 1-hexene, this ratio is ca. 8 for helium and only 4 for the larger CO_2 molecules. Diffusion coefficients are much more strongly affected than solubility coefficients so that permeabilities are more reduced by the former than by the latter.

Permeability coefficients P commonly increase with increasing temperature or, less commonly, decrease. In general, one finds that lg P = f($1/T$) (Fig. 13-4). This function shows two linear parts that intersect at the glass temperature T_G where chain segments become more mobile for $T > T_G$. The slopes are practically independent of the type of gas for $1/T > 1/T_G$, i.e., below T_G, but are slightly more variable above the glass temperature.

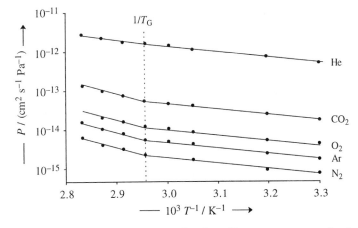

Fig. 13-4 Permeability coefficients of gases as a function of inverse temperature for the permeation of various gases through Barex® (see p. 560) [8].

The temperature dependence of permeability coefficients is controlled by the activation energy $\Delta E^{\ddagger}$ of diffusion and the enthalpy of dissolution, ΔH,

(13-6) $D = D_{\infty} \exp(-\Delta E^{\ddagger}/RT)$

(13-7) $S = S_{\infty} \exp(-\Delta H/RT)$

where D_{∞} = diffusion coefficient and S_{∞} = solubility coefficient at infinite temperature. Diffusion coefficients always increase with increasing temperature whereas solubility coefficients either decrease or increase, depending on thermodynamics. The plots of lg P = lg DS as a function of inverse temperature (Fig. 13-4) thus indicate also that diffusion is more important for permeability than solubility (see also Fig. 13-3).

Gas permeabilities are also strongly affected by the moisture uptake of polymer matrices if the polymers contain polar groups, especially hydrophilic ones. Such matrices must be conditioned before measurement which may take hours to days for thin membranes but weeks to months for thick walls at ambient temperatures.

It is not surprising that permeability coefficients P of polymers with hydrophilic groups increase with the contents of these groups and that they are greater at higher relative humidities RH (Fig. 13-5). However, it is remarkable that there are two linear sections for P = f(RH) at low and at high relative humidities and that the change from one to the other is always at the same % RH.

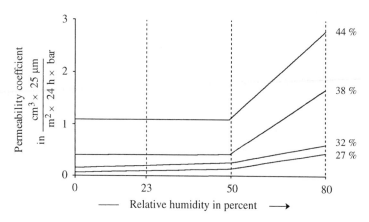

Fig. 13-5 Effect of relative humidity on the permeability coefficient of oxygen in films of ethylene-vinyl alcohol copolymers with various percentages of vinyl alcohol units [9]. Data at 50 % relative humidity and 23°C. 1 $(cm^3 \cdot 25~\mu m)/(m^2 \cdot 24~h \cdot bar) = 2.894 \cdot 10^{-11}~cm^2~s^{-1}~MPa^{-1}$.

13.2.3 Diffusion of Liquids in Polymers

The penetration of polymers by liquids and the subsequent permeation of these liquids is controlled by the glass temperature and the swellability of the polymers (Fig. 13-6). The time dependence of the permeated mass m of the liquid can be described by

(13-8) $m/m_{\infty} = KA_m t^n$

where m_∞ = permeated mass at infinite time, K = system constant, and A_m = cross-sectional area of the specimen. The exponent n of the time t is a function of the **Deborah number** *De*, i.e., the ratio of the diffusion time of the liquid and the relaxation time of the swollen polymer (Volume III, p. 486).

Polymer segments are mobile at temperatures above the glass temperature T_G of the polymer. The mobility is increased somewhat if the polymer is slightly swollen but the diffusion rate of the solvent is still small so that the mobility of the permeant is smaller than the relaxation time of polymer segments: the Deborah number is smaller than unity ($De < 1$). The movement of a permeant molecule in the swollen matrix causes "instantaneous" conformational changes of polymer segments because they have to "get out of the way."

In this so-called ideal Case I, both permeant molecules and polymer segments behave like viscous liquids and the diffusion rate can be described by the ideal Fick's law of diffusion with concentration-independent diffusion coefficients. The quantities in Eq.(13-8) become n = 1/2 and $KA_m = 4\,D/\pi$. Case I is typical for the permeation of gases through polymers as well as that of low-molecular weight, non-swelling liquids through elastomers.

Swellable polymers are softened by the advancing front of the solvent. The diffusion of the liquid in the polymer becomes faster and is now described by the generalized Fick's second law with n = 1/2 and a concentration-dependent diffusion coefficient D, for example, $D = D_0 \exp(-kc)$.

Permeations through polymers below the glass temperature are described by Case II in which the mobility of the permeant is greater than the relaxation time of polymer segments ($De > 1$). This case is characterized by a change of the physical structure of the polymer during the permeation process.

Physical structures of polymers do not change in the limiting case of $De \rightarrow \infty$ at temperatures far below the glass temperature. This permeation is characterized by a sharp boundary between the glassy core and the swollen zone that advances with constant speed. The permeated mass is directly proportional to the time (n = 1).

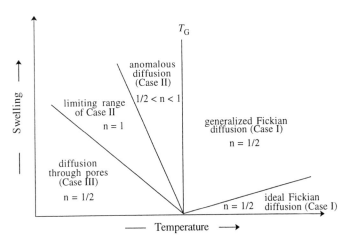

Fig. 13-6 Schematic representation of the temperature dependence of the permeation of liquids through polymers [10].

Relative movements of permeant molecules and conformational changes of polymer segments proceed more and more in unison in Case II the more the experimental temperatures approach the glass temperature of the dry polymer. This so-called anomalous diffusion is characterized by exponents $1/2 < n < 1$.

Case III is observed for the permeation of permeants with small molecule diameters through polymers with scarcely mobile segments, i.e., very little swelling of the polymer and temperatures far below the glass temperature of the polymer. In this case, permeant molecules will diffuse through "pores" of the matrix. These pores are micropores or interstitial volumes. The diffusion obeys Fick's first law (Case III of Fig. 13-6).

An official classification of "pores" according to their diameter d does not exist. For example, suggested terms are "macropores" when $d \geq 50$ nm, "mesopores" when 50 nm $\geq d \geq 2$ nm, "micropores" when $d < 2$ nm, and "supermicro" (0.7-2 nm), "ultramicro" (<0.7 nm), and "ultrapore" (<0.35 nm). Note that this use of "micro" does not conform to SI nomenclature ($\mu = 10^{-6}$; 1 μm $= 10^{-6}$ m $= 1000$ nm). All these pores are certainly "nanopores" (1 nm $=10^{-9}$ m). It is also very questionable whether an "ultrapore" with $d < 0.35$ nm should be called a "pore" at all because such distances are in the range of the length of chemical and physical bonds.

13.2.4   Migration of Adjuvants and Other Components

In the simplest case, migrations of additives such as adjuvants into polymers, low-molecular weight components of foodstuffs (aromas, etc.) through container walls, etc., is controlled by the two diffusion coefficients of the additive in the packing material and in the packed goods (for example, food), respectively, and the distribution coefficient of the additive for the goods and the packing material (see also Section 3.1.4). The additive can permeate from the packing material into the goods (see Fig. 13-7) or *vice versa*.

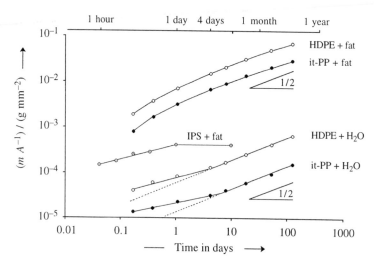

Fig. 13-7 Time dependence of the mass m of an antioxidant that migrated at 40°C through an area A of a high-density poly(ethylene) (HDPE), a poly(propylene) (PP) or an impact-resistant poly(styrene) (IPS) into either water or a synthetic fat [11,12]. The antioxidant was Irganox® 1076 (= octadecyl-3-(3,5-t-butyl-4-hydroxyphenyl)propionate) with weight fractions $w = 2.87 \cdot 10^{-3}$ in HDPE and PP and $w = 1 \cdot 10^{-3}$ in IPS. The synthetic fat was a mixture of synthetic triglycerides (HB 307). Slopes of 1/2 indicate diffusions according to Fick's law.

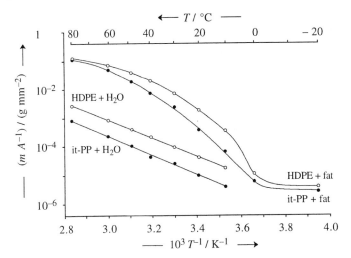

Fig. 13-8 Effect of temperature on the dependence of mass per area, m/A, of the antioxidant of Fig. 13-7 that migrated in 10 days from HDPE or it-PP into the synthetic triglyceride HB 307 [11].

In the first case, health problems may arise, for example, from the permeation of monomers (such as bisphenol A), plasticizers (such as certain phthalates), or antioxidants into foodstuffs. In the second case, packed goods loose one of their qualities, for example, aromas. An example of the first case is the permeation of antioxidants from polymeric pack materials into a synthetic fat or water as packed goods (Fig. 13-7).

The antioxidant of Fig. 13-7 dissolves very little in the polymer or in water, which leads to a slow migration from the polymer into the water. At short times, a time lag is observed. At long times, the migration follows Fick's diffusion law ($m/A = f(t^{1/2})$).

The solubility of this antioxidant in the hydrophobic polymers HDPE, PP and IPS and in the fat is much greater. During the first hours after contact, migration rates were large but then became smaller with time and finally followed Fick's law (HDPE, PP) or even became constant after one day (IPS), indicating a distribution equilibrium.

The migration of antioxidants from polymers into water is completely controlled by the diffusion out of the polymer. lg (m/A) decreases correspondingly with increasing $1/T$ (Fig. 13-8). Accordingly, slopes of the plots lg (m/A) = f($1/T$) are the same for HDPE–H_2O and for it-PP–H_2O. The migration from the polymers into the fat is more difficult to interpret because several effects overlap.

13.3 Applications

Differences in the permeation rates of gases and liquids and in migration rates of adjuvants are decisive for the choice of a polymer as packing material (containers, films) or as a membrane for the separation of gases, liquids, or ions. In some cases, one wants very small rates of permeation (for example, polymers for soft drink bottles) and in other cases, one wants great differences (for example, gas separations, fresh foods). For lack of space, only applications related to packing materials are discussed here.

13.3.1 Films

Thin planar materials are called **foil** (L: *folium* = leaf) if they are made from metals but **film** in common language if they consist of other substances (oil film, plastics film). Very thin films consisting of layers that are only one molecule thick are known as **monomolecular films**. Thin films for the separation of substances are said to be **membranes** (L: *membrana* = little skin); they may be solid or liquid (especially in biology).

Films may be self-supporting or not. Examples of non-self-supporting films are oils on water and dried paints or lacquers on solid surfaces. In polymer technology, the latter films are a special type of a **surface coating**, usually shortened to **coating** (p. 530 ff.). However, "coating" is not restricted to films obtained from liquids or dispersions but comprises also those prepared by powder lacquering or other methods.

According to ISO, a **film** is a thin plane product (hence, a solid!) with a thickness that is small compared to its length and width. The limits on thicknesses are arbitrary and vary from country to country. ASTM defines films accordingly but restricts the thickness to 40-250 μm and the tensile modulus to less than 70 MPa. Thicker solid specimens with thicknesses between 250 μm and 1.5 mm are named **sheets** or **sheetings** but the latter term is often used only for endless sheets. Thicker sheets are called **boards.**

Films and sheets are produced by many processes (see also Chapter 4): pouring of polymer solutions on flat surfaces and subsequent drying, pouring or extrusion of melts and solidification, calendering of solids, or *in-situ* polymerization of monomers. Their main application is as packaging material but they are also used for electrical insulation (Chapter 14), adhesive tapes (Chapter 12), and as carriers of information (photographic films, magnetic tapes, floppy disks, CDs). Other special applications are for solar panels and packing of active substances such as pharmaceuticals. Properties vary widely with the chemical structure of their constituing polymers and with processing conditions. The data in Table 13-2 are therefore only guide values.

Table 13-2 Guide values for properties of uncoated, stretched packaging films at 20°C and a relative humidity of 65 % (exception: permeability for water vapor at 40°C and relative humidity of 90 %). At a pressure of $1 \cdot 10^5$ Pa = 1 bar, a permeability for water vapor of $1 \cdot 10^{-9}$ cm^2 s^{-1} corresponds to a permeability of gases of $1 \cdot 10^{-14}$ cm^2 s^{-1} Pa^{-1}. Data in the draw direction (‖) or perpendicular to it (⊥).

Property		Physical unit	Cello-phane	EVAL	PVDC	LLDPE	LDPE	PET
Thickness		μm	20	15	30	38	38	12
Tensile modulus	‖	GPa	3.1	2.0	0.32		0.21	2.2
	⊥	GPa	2.1	1.9	0.33		0.25	1.9
Fracture strength	‖	MPa	135	80	84	43	24	174
	⊥	MPa	82	52	100	35	19	141
Elongation at break	‖	%	15	160	38	580	310	100
	⊥	%	42	200	52	700	580	58
Permeability								
water vapor		10^{-9} cm^2 s^{-1}	1.6	0.041	0.009		0.016	0.021
oxygen		10^{-14} cm^2 s^{-1} Pa^{-1}	320	0.28	7.1		3850	34
Haziness		%	1.5	1.8	6.0	14	5.0	2.5
Moisture content		%	15	3.0	0.1		0.1	0.1

Table 13-3 Effect of relative humidity on the permeability of packaging films for oxygen at 23°C. Saran and cellophane are free names in the United States but registered trademarks in Europe. Saran is a copolymer of vinylidene chloride that contains 15-20 % vinyl chloride for films and 13 % vinyl chloride and 2 % acrylonitrile for coatings.

Film from		$10^{14} \, P/(cm^2 \, s^{-1} \, Pa^{-1})$ at a relative humidity of				
		0 %	50 %	75 %	90 %	95 %
Poly(vinyl alcohol)	(PVAL)	45	-	-	-	>113 000
Poly((vinyl alcohol)$_{70}$-*co*-(ethylene)$_{30}$)	(EVAL)	77	86	158	13 500	17 100
Saran® 5233	(PVDC)	680	-	680	450	-
Cellophane	(= cellulose hydrate)	770	>22 500	-	-	-
Polyamide 6, biaxially oriented	(PA 6-BO)	6 800	-	-	-	22 500

In general, one prefers films that let through only a little moisture or none at all. Permeability of oxygen should also be low. For packaging cheese and meat products, Saran is the preferred material in the United States whereas Europe mainly uses multi-layer films (composite films). Low-density poly(ethylene) is chosen for bread and high-density poly(ethylene) for milk.

In contrast, packaging films for "breathing" food products such as fish, fresh meat, and vegetables should be well permeable to oxygen The material of choice for such films is plasticized poly(vinyl chloride) but this is criticized because of the use of DOP (DEHP) as plasticizer (p. 42) and the release of HCl on burning PVC-P (p. 362).

Permeabilities of films from polar polymers depend strongly on relative humidities (Table 13-3; for poly(vinyl alcohol-*co*-ethylene)s, see also Fig. 13-5). The higher the relative humidity, the more moisture is taken up by the films. The absorbed water acts as a plasticizer which makes polymer segments more mobile and increases the permeation of oxygen. This undesired effect can be counteracted by drawing of films (Table 13-4) or by use of multiple layer films (Section 13.3.2).

Table 13-4 Effect of drawing on properties of a 25 μm thick poly(propylene) films at 23°C and 75 % relative humidity (exception: 85 % relative humidity for the permeability of water vapor). At a pressure of $1 \cdot 10^5$ Pa = 1 bar, a permeability of $1 \cdot 10^{-9}$ cm^2 s^{-1} for water vapor corresponds to a gas permeability of $1 \cdot 10^{-14}$ cm^2 s^{-1} Pa^{-1}. Properties in the draw direction (‖) or perpendicular to it (⊥).

Property		Physical unit	not drawn	Properties of films drawn		
				mono-axially	biaxially (simultaneous)	biaxially (2 steps)
Total draw ratio		-	-	5.5	10	10
Tensile modulus	‖	MPa	500	2500	3000	2500
	⊥	MPa	900	-	3000	4000
Fracture strength	‖	MPa	50	250	200	140
	⊥	MPa	40	40	200	270
Elongation at fracture	‖	%	430	10	80	140
	⊥	%	540	700	80	40
Tear propagation strength	‖	N	7.6	-	-	0.25
	⊥	N	12	-	-	0.45
Permeability, oxygen		10^{-14} cm^2 s^{-1} Pa^{-1}	7.0	-	2.2	2.8
water vapor		10^{-9} cm^2 s^{-1}	0.018	-	0.0056	0.01

Table 13-5 Guide values for mechanical properties of some shrink films.

Property	Physical unit	PET	PP	LDPE	LDPE	PVC
Tensile strength	MPa	207	179	62	73	80
Fracture elongation	%	130	75	120	115	140
Tear resistance	mN/m	13.5	1.9	3.1	2.9	variable
Glass temperature	°C	98	− 24	− 118		80
Shrinkage temperature	°C	75-150	120-165	65-115	75-120	65-150
Melting temperature	°C	255	163	110		>310
Shrinkage stress	MPa	7.6	4.1	2.3	2.8	1.6

In general, films must be drawn in order to obtain the desired properties (Figure 13-3 and Table 13-4). In uniaxial drawing, crystalline domains of semicrystalline polymers are reoriented in the draw direction. Such films have increased tensile and fracture strengths in the draw direction (longitudinal direction (LD), machine direction (MD), ‖) whereas fracture strengths do not change or even decrease perpendicular to the draw direction (transverse direction (TD), ⊥). Greater improvements are obtained by biaxial drawing which may be in two steps or simultaneous in a quadratic or an octagonal way.

Film formation may freeze-in stresses, which is taken advantage of in the manufacture of **shrink films** (Table 13-5). Shrink films are produced by unidirectional or bidirectional orientation (PP, PET) or by injection blow molding (LDPE). High-density poly-(ethylene) is not suited for shrink films because of its high crystallization rate but shrink films can be obtained from 70:30 HDPE-LDPE blends. Frozen-in stresses are dissolved if films are rewarmed, which causes the film to shrink and to wrap goods tightly. Shrink films are especially of interest if shrinking can by done in hot-water baths.

Very thin films from solution casting, dipping, or chemical processes are mostly not self-supporting. A new wet process exploits the spontaneous formation of very thin films (Fig. 13-9) on liquid surfaces, albeit so far apparently not on a commercial scale. In this process, a water droplet is placed on a solution of a polymer in an organic solvent. Within a second, the droplet is covered by a thin layer of the polymer solution if the organic solvent has a high vapor pressure, the organic polymer solution has a greater density than the water, and the gas phase does not become saturated with the vapor of the organic solvent.

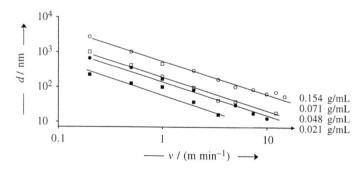

Fig. 13-9 Thickness d of films of various concentrations of a polyetherimide as a function of the haul-off speed v [13]. By kind permission of Hanser Publishers, Munich.

Evaporation of the organic solvent leads to the formation of a thin polymer film that is removed with constant speed. The polymer film is thinner, the lower the concentration of the polymer solution and the greater the haul-off speed (Fig. 13-9). Polymer solution and water are constantly replenished.

This process produced films from poly(styrene), polycarbonate, poly(methyl methacrylate), poly(hydroxybutyrate-co-valerate), or a polyetherimide that are 20-60 nm thick and several hundred meters long. The thicknesses of the thinnest of these films are in the range of diameters of unperturbed polymer coils (an unperturbed coil of a poly-(styrene) molecule of molecular weight 10^5 has a diameter of gyration of 17.4 nm).

13.3.2   Composite Films

The protective action of films can be improved by lacquering, laminating, or vacuum metallizing. Laminating of a polymer film by another polymer film or by more than one film (all called **doubling**) decreases considerably the transport of permeants through micropores of the base film because it is not very probable that micropores of the base film (carrier film, substrate, etc.) and the protective film are exactly on top of each other. Protective films may also be selective with respect to permeants that permeate easily through the carrier.

Classic cellophane (from *cell*ulose, *-o-* and G: dia*phan*is = transparent) is producedsimilarly to rayon (viscose) (p. 167) but it is then plasticized with glycerol, propylene glycol, ethylene glycol, or urea. Cellophane films are more glossy and more rigid than plastics films. Their main disadvantage is their permeability to water vapor. They are therefore either protected with nitro lacquers (dissolved cellulose nitrates) or, most often, with lacquers or a dispersion of Saran. Cellophane is now also laminated with poly(ethylene) films, using urea–formaldehyde resins as adhesive. The cellulose layer of this composite film protects aromas of packed goods whereas the poly(ethylene) layer prevents the permeation of water vapor

Modern composite films are produced by laminating solid films or extrusion where at least one film is added as a warm melt. **Laminating** comprises many different processes. In roll laminating (Fig. 13-10, left), two films A and B are fed to the machine and then pressed together by hot compression rolls (hcr); alternatively, they may also be glued together. A related laminating process is quilting where a wadding is placed between, for example, a vinyl sheet and a nonwoven fabric (Fig. 13-10, center).

In **extrusion** processes, films are extruded and joined while still warm, either with other extruding films (coextrusion) or with one or more ready sheets as in extrusion laminating (Fig. 13-10, right).

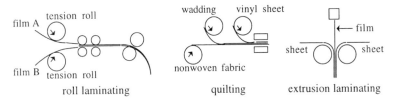

Fig. 13-10 Some processes for the manufacture of composite films.

Table 13-6 Composition, properties, and applications of some composite films. A = adhesive, EVAL = ethylene-vinyl alcohol copolymer, PA = polyamide, PE = poly(ethylene), PP = poly(propylene).

Composite film from	Important properties	Applications
LDPE + LDPE	void-free, multicolor	bags
LDPE + LLDPE	strength, elasticity	shrink films
LDPE + HDPE + LDPE	stiffness, moisture barrier	breakfast bags
PP + A + EVAL + A + PP	gas and aroma barrier	precooked meals
PA + ionomers	gas and aroma barrier	meat and cheese
PA + EVAL + A + PE	gas and aroma barrier	vacuum packaging

Extrusions with manifolds allow the use of polymers with very different viscosities for different films but are restricted to only a few layers. Extrusions with feed blocks (distributor blocks), on the other hand, lets one coextrude an almost unlimited number of layers albeit only from melts with adjusted viscosities.

Depending on the application (Table 13-6), commercial extruded composite films are composed of up to seven layers of two or more different polymers including those of adhesives (if any); one layer may also consist of aluminum. So-called **duplo films** from identical polymers may have even many more layers; a commercial iridescent film is made up of 100 layers (Section 14.1.7).

The simplest composite films consist of an outer carrier (base) and an inner hot sealed layer which may or may not enclose an adhesive layer, a barrier layer, etc. (see Table 13-7 for some packaging films). The base film controls the rigidity and strength of the composite film, acts as a barrier to gases and aromas, and should be printable or metallizable. Base films are most often cellophane, poly(styrene), poly(propylene), poly(ethylene terephthalate), polyamide 6, and high-density poly(ethylene). Poly(styrene) is favored because of its high surface gloss; it is mostly a colored impact type.

Table 13-7 Permeabilities P of various composite films for gases and water vapor at the indicated relative humidities RH. At a pressure of $1 \cdot 10^5$ Pa = 1 bar, a permeability of $1 \cdot 10^{-9}$ cm^2 s^{-1} for water vapor corresponds to a gas permeability of $1 \cdot 10^{-14}$ cm^2 s^{-1} Pa^{-1}.
Al = aluminium, C = cellophane, CN = cellulose nitrate ("nitrocellulose"), PE = poly(ethylene), PET = poly(ethylene terephthalate), PP = poly(propylene), PVDC = poly(vinylidene chloride) (copolymer!); BO = biaxially oriented; nm = not measurable.

Film consisting of (exterior → interior)	Thickness in μm	RH in % gas	RH in % H$_2$O	$10^{14}\,P/$(cm^2 s^{-1} Pa^{-1}) O$_2$	CO$_2$	N$_2$	$10^9\,P/$(cm^2 s^{-1}) H$_2$O
PP	20	0	-	2900	-	-	-
PP + PE	20/40	0	85	2100	-	-	0.0090
PP-BO + PE	20/40	0	85	1400	4050	290	0.0056
PP-BO + Al + PE	20/9/75	0	85	<0.1	nm	nm	<0.001
PET	12	0	85	1800	-	-	0.036
PET + PE	12/75	0	85	12	520	1.2	0.012
PET + PVDC + PE	12/3/75	0	85	12	64	1.2	0.0072
PET + Al + PE	12/9/75	0	85	<0.1	nm	nm	0.0011
CN + C + CN + PE	3/23/3/50	75	85	140	870	45	0.023
PVDC + C + PVDC + PE	3/23/3/50	75	85	12	45	3.5	0.0091

The hot sealing layer allows the closure of packages; its melting temperature must therefore be lower than that of the carrier. For extrusion laminating and extrusion coating, low-melting, low-density poly(ethylene)s are used ($T_M \approx 115°C$). Medium-density poly(ethylene) may be boiled and can be used for deep frozen goods because of its low glass temperature of less than $-80°C$. High-density poly(ethylene) with its higher melting temperature of $T_M \approx 135°C$ can be heat sterilized up to 121°C. Isotactic poly(propylene) ($T_M \approx 176°C$) can also be heat sterilized but is not usable for deep frezing because of is glass temperature of $T_G \approx -15°C$. Other polymers are used for special applications; examples are ionomers and ethylene-vinyl alcohol copolymers (Table 13-6).

Additional barrier layers are used if the barrier properties of carriers and hot sealing layers are insufficient. Such barrier layers consist of vinylidene chloride copolymers, poly(vinyl alcohol), vinyl alcohol-ethylene polymers, acrylonitrile copolymers, and aluminum (see Table 13-7).

13.3.3 Containers, Cans, and Bottles

Bottles and containers are mostly used for the packing of foods, drinks, and bottled water: in the United States, ca. 40 wide-mouth containers, 400 cans, and an uncounted number of bottles are used per capita and year. Most containers still consist of glass, aluminum, or steel, but plastics have made considerable inroads during the last years because they are light-weight and easy to process.

Cans and bottles for CO_2 containing drinks should not lose more than 15 % of the CO_2 during the sometimes considerable storage times. They should also allow only very little oxygen to permeate into the bottle or can (Table 13-8). Large bottles lose less CO_2 and gain less O_2 during storage than smaller ones because of their lower ratio of surface to volume: permeation is proportional to the surface whereas the gas content is proportional to the volume.

For this reason, Coca Cola® could be sold initially only in large plastics bottles and not in small ones. For example, 2-liter PET bottles containing 8 liters of CO_2 in 2 liters of water lost 15 vol% of the CO_2 in 110 days whereas the same loss of 15 vol% occurred in 60 days from 0.5 liter bottles with 2 liters of CO_2 in 0.5 liters of water.

Containers, cans, and bottles from plastics are produced either from relatively thick walled sheets or directly from the melt by blow forming (stretch forming) (p. 123) or extrusion or injection blow molding (p. 117). Stretch forming leads to greater rigidities, tensile strengths, and tensile impact strengths as well as to smaller permeabilities for oxygen, carbon dioxide, and water vapor than blow molding (Table 13-9).

Table 13-8 Targeted maximum permeabilities of gases through can and bottle walls.

Content	$\dfrac{10^{14}\ P(O_2)}{cm^2\ s^{-1}\ Pa^{-1}}$	$\dfrac{10^{14}\ P(CO_2)}{cm^2\ s^{-1}\ Pa^{-1}}$	$\dfrac{10^{10}\ P(H_2O)}{g\ cm^{-1}\ s^{-1}}$
Cola drinks	1	1.5	1.5
Fruit juices	0.5	1.5	1.5
Beer	0.05	1.5	1.5

Table 13-9 Influence of processing by blow molding (B, p. 117) or biaxial stretch blow molding (S) on tensile moduli E, tensile strengths σ, fracture elongations ε_B, tensile impact strengths F, turbidity, and permeabilities P of oxygen (P_O), carbon dioxide (P_C), and water vapor (P_W).

PET = poly(ethylene terephthalate), PS = crystal poly(styrene), SAN = styrene-acrylonitrile co-polymer (Barex (p. 560)), PP = poly(propylene), PVC-I = impact resistant poly(vinyl chloride).

Property		Physical unit	PET	PS	Properties of SAN	Barex®	PP	PVC-I
E	B	GPa	-	1.29	1.37	1.61	0.74	1.34
	S	GPa	-	1.72	1.72	1.87	1.32	1.78
σ	B	MPa	52	26	62	57	27	50
	S	MPa	-	69	75	63	47	56
ε	B	%	12	3	5	33	>50	200
	S	%	130	7	7	65	>50	84
F (vertical)	B	kJ/m²	-	-	3.8	15.8	8.6	18.1
	S	kJ/m²	-	-	10.3	33.4	39.7	33.6
F (horizontal)	B	kJ/m²	-	-	5.0	13.5	7.6	18.1
	S	kJ/m²	-	-	10.7	24.3	35.1	40.4
Turbidity	B	%	-	4.8	6.7	28	>30	-
	S	%	-	2.0	2.2	14	4	-
P_O ($\cdot 10^{14}$)	B	cm² s⁻¹ Pa⁻¹	0.50	-	-	0.041	9.8	0.075
	S	cm² s⁻¹ Pa⁻¹	0.31	-	-	0.023	4.6	0.038
P_C ($\cdot 10^{14}$)	B	cm² s⁻¹ Pa⁻¹	1.2	-	-	0.081	36	1.21
	S	cm² s⁻¹ Pa⁻¹	0.75	-	-	0.052	20	0.96
P_W ($\cdot 10^{9}$)	B	cm² s⁻¹	0.34	-	-	0.075	0.014	0.095
	S	cm² s⁻¹	0.16	-	-	0.046	0.0064	0.069

Food containers must be fairly stable at elevated temperatures. Preserves are filled into containers at 85-96°C; other foods are sterilized at 120-130°C for up to 45 minutes. Suitable food containers consist of poly(propylene)-adhesive-Saran-adhesive-poly-(propylene) or PET-adhesive-aluminum-adhesive-poly(propylene). These containers should not be heated above ca. 230°C because that would lead to relaxations of frozen-in stresses and thus to shrinkages. On the other hand, preheating to these temperatures is useful for compacting discarded containers.

An example is the recycling of the older type of 2-liter cola bottles which contained 63 g PET, 22 g HDPE in the bottom of the bottle, and 5 g paper label and adhesive.. For recycling, bottles were compacted and ground, paper was removed by air sifting (air classifying), and PET and HDPE were separated by flotation. The resulting raw PET is molten, solidified, and granulated. Recycled PET cannot be used for containers, films, etc., that will be in contact with food because of its increased content of acetaldehyde.

Metal cans for CO_2-containing soft drinks are more expensive than their contents! Plastic cans cost only about half as much as those of aluminum. However, their recycling is a problem: washing discarded cans is expensive, pollutes the environment, and may not be hygienic because of substances that have migrated into the plastics. The polymer blend resulting from simple melting of old plastic cans cannot be used directly after grinding and homogenization. However, this scrap can be used in specially developed composite films in proportions up to 50 wt%. For example, such a seven-layer film may consist of (1) 20-30 % carrier (HIPS, HDPE, PP), (2) 15-25 % scrap, (3) 2.5 % adhesive, (4) 5-10 % Saran or PE, (5) 2.5 % adhesive, (6) 15-25 % scrap, and (7) 20-30 % HIPS, HDPE, or PP.

Literature to Chapter 13

13.2 PERMEATION
H.B.Hopfenberg, Ed., Permeability of Plastic Films and Coatings, Plenum Press, New York 1974
H.T.Hwang, C.K.Choi, K.Kammermeyer, Separation Sci. **9**/6 (1974) 461 (extensive lists)
H.Yasuda, Units of Permeability Constants, J.Appl.Polym.Sci. **19** (1975) 2529
T.R.Crompton, Additive Migration from Plastics into Food, Pergamon Press, Oxford 1979
M.B.Huglin, M.B.Zakaria, Comments on Expressing the Permeability of Polymers to Gases,
    Angew.Makromol.Chem. **117** (1983) 1
J.Comyn, Ed., Polymer Permeability, Elsevier Appl.Sci.Publ., London 1985
W.R.Vieth, Diffusion in and Through Polymers, Hanser, Munich 1991
J.Stastna, D.De Kee, Transport Properties in Polymers, Technomic, Lancaster (PA) 1995
L.K.Massey, Permeability Properties of Plastics and Elastomers, W.Andrew, Norwich (NY), 2nd. ed.
    2003

13.3a APPLICATIONS: FILMS
O.J.Sweeting, The Science and Technology of Polymeric Films, Interscience, New York 1968 (2 vols.)
W.R.R.Park, Plastics Film Technology, Van Nostrand-Reinhold, London 1970
C.R.Oswin, Plastic Films and Packaging, Appl.Sci.Publ., Barking (Essex) 1975
D.H.Solomon, The Chemistry of Organic Film Formers, Krieger, Huntington (NY) 1977
C.J.Benning, Plastic Films for Packaging, Technomic, Lancaster (PA) 1983
J.H.Briston, Plastics Films, Longman, Harlow, England, 3rd ed. 1988
-, Desk Top Data Bank, Films, Sheets and Laminates, International Plastics Selector, San Diego
(CA)
J.H.Briston, Plastic Films, Harlow, Essex, 2nd ed. 1989
I.L.Gomez, High Nitrile Polymers for Beverage Container Applications, Technomic, Lancaster
    (PA) 1990
K.R.Osborn, W.A.Jenkins, Plastics Films: Technology and Packaging Applications, Technomic,
    Lancaster (PA) 1992
T.Kanai, G.A.Campbell, Eds., Film Processing, Hanser, Munich (OH) 1999
L.K.Massey, Film Properties of Plastics and Elastomers, W.Andrew, Norwich (NY), 2nd ed. 2004

13.3b APPLICATIONS: PACKAGING
W.S.Simms, Ed., Packaging Encyclopedia, Cahners Publ., Des Plaines (IL) 1985
J.Stepek, V.Duchacek, D.Curda, J.Horacek, M.Sipek, Polymers as Materials for Packaging,
    Ellis Horwood, Chichester 1988
G.M.Levy, Ed., Packaging and the Environment, Chapman and Hall, London 1993
S.E.M.Selke, Packaging and the Environment, Technomic, Lancaster (PA) 1994
A.L.Beady, K.S.Marsh, Eds., The Wiley Encyclopedia of Packaging Technology, Wiley,
    New York, 2nd ed. 1997
S.E.M.Selke, Understanding Plastics Packaging Technology, Hanser, Munich 1997
D.Twede, R.Goddard, Packaging Materials, Technomic, Lancaster (PA), 2nd ed. 1999
R.Hernandez, S.E.M.Selke, J.Culter, Plastics Packaging. Properties, Processing, Applications,
    Regulations, Hanser, Munich 2000
R.Coles et al., Food Packaging Technology, CRC Press, Boca Raton (FL) 2003

13.3c APPLICATIONS: MEMBRANES
J.H.Fendler, Membrane Mimetic Chemistry, Wiley, New York 1982
S.Sourirajan, T.Matsuura, Reverse Osmosis/Ultrafiltration Process Principles, National Research
    Council, Ottawa, Canada 1985
R.E.Kesting, Synthetic Polymeric Membranes–A Structural Perspective, Wiley, New York,
    2nd ed. 1985
N.Toshima, Polymers for Gas Separation, Wiley, New York 1992
R.E.Kesting, A.K.Fritzsche, Polymeric Gas Separation Membranes, Wiley, New York 1993
D.R.Paul, Y.Yampol'skii, Eds., Polymeric Gas Separation Membranes, CRC Press, Boca Raton
    (FL) 1994
T.Matsuura, Synthetic Membranes and Membrane Separation Processes, CRC Press, Boca Raton
    (FL) 1994

K.Mulder, Basic Principles of Membrane Technology, Kluwer Academic, Dordrecht, Netherlands 2nd ed. 1996

D.Muraviev, V.Gorshikov, A.Warshawsky, Eds., Ion Exchange, Dekker, New York 1999

T.Sata, Ion Exchange Membranes; Preparation, Characterisation, Modification and Application, Royal Soc.Chem., Cambridge, UK, 2004

B.Freeman, Y.Yampolskii, I.Pinnau, Eds., Materials Science of Membranes for Gas and Vapor Separation, Wiley, Hoboken (NJ) 2006

S.Pereira Nunes, K.-V.Peinemann, Eds., Membrane Technology in the Chemical Industry, Wiley-VCH, Weinheim. 2nd ed. 2006

Yu.Yampolskii, I.Pinnau, B.D.Freeman, Eds., Materials Science of Membranes for Gas and Vapor Separation, Wiley, Hoboken (NJ) 2006

A.B.Koltuniewicz, E.Drioli, Eds., Membranes in Clean Technologies, Wiley-VCH, Weinheim 2007 (2 vols.)

R.M.A.Roque-Malherbe, Adsorption and Diffusion in Nanoporous Materials, CRC Press, Boca Raton (FL) 2007

K.-V.Peinemann, S.Pereira Nunes, Eds., Membrane Technology, Wiley-VCH, Weinheim; Volume 1: Membranes for Life Science (2007); Volume 2: Membranes for Energy Conversion (2008)

A.B.Koltuniewicz, E.Drioll, Membranes in Clean Technologies, Wiley-VCH, Weinheim 2008

13.3d APPLICATIONS: POLYMER ABSORBENTS

F.L.Buchholz, A.T.Graham, Eds., Modern Superabsorbent Polymer Technology, Wiley, New York 1997

References to Chapter 13

[1]   P.Goeldi, H.-G.Elias, unpublished
[2]   W.Pusch, A. Walch, Angew.Chem. **94** (1982) 670, Table 11
[3]   R.C.Weast, Ed., Handbook of Chemistry and Physics, The Chemical Rubber Co., Cleveland (OH), 46th ed. (1965), D-92
[4]   O.A.Neumüller, Römpps Chemie-Lexikon, Frankh'sche Verlagsbuchhandlung, Stuttgart, 8th ed. (1979-1988), Volume IV, p. 2643
[5]   H.-G.Elias, Makromoleküle, Wiley-VCH, Weinheim, 6th ed., Vol. 2 (2001), Table 14-3
[6]   P.S.Holden, G.A.J.Orchard, I.M.Ward, J.Polym.Sci.-Polym.Phys.Ed. **23** (1985) 709
[7]   J.A.Webb, D.I.Bower, I.M.Ward, P.T.Cardew, J.Polym.Sci. **B 31** (1993) 743
[8]   H.Yasuda, T.Hirotsu, J.Appl.Polym.Sci. **21** (1977) 105, Fig. 2
[9]   H.Schenck, J.André, Kunststoffe **89**/4 (1999) 106, selected curves of Fig. 4
[10]   -, Chem.Eng.News **43**/41 (1965) 64. The diagram in this paper is based on the work of T.Alfrey, Jr., E.F.Gurnee, and W.C.Lloyd (Dow Chemical) (handwritten notes; personal communication of Turner Alfrey, Jr.).
[11]   J.Klahn, K.Figge, W.Freytag, Dtsch. Lebensmittel-Rdsch. **78**/7 (1982) 241 (various tables)
[12]   J.Klahn, K.Figge, F.Meier, Angew.Makromol.Chem. **131** (1985) 73, Table 5
[13]   G.Hinrichsen, A.Hoffmann, T.Schleeh, Ch.Macht, Kunststoffe-plast europe **99** (2000) 100, Fig. 6

14 Electrical and Electronic Uses of Polymers

14.1 Introduction

14.1.1 Overview

The commercial use of the earliest fully synthetic polymer, phenolic resin, was marginal until its usefulness as an electrical insulator was discovered in the early 1900s (Volume II, p. 261). Without phenolic resins, the young electrical industry could not have prospered since all other electrical insulators known at that time were expensive, difficult to manufacture and process, and/or not satisfactory for many applications. Early electrical insulators based on natural resins (amber, shellac) or natural polymers (gutta percha, silk) are no longer used. Porcelain, ceramics, and fire-proof glass serve today only for very special electrical purposes.

Modern synthetic polymers serve not only as insulators and protective materials (Section 14.2) but also as so-called photoresists in the production of electronic parts (Section 14.2.11), as photoconductors and binders in Xerography® (Section 14.2.8), and electrically conducting materials in batteries and many other applications (Section 14.3).

The world consumption of polymers for electrical/electronic applications is not known. In the United States, about 1 million tons of polymers are used annually for electrical applications, exclusive of the unknown consumptions by the automotive industry and for household appliances. About 20 % each are poly(styrene)s, poly(vinyl chloride)s, and low-density poly(ethylene)s and about 5 % each ABS, other poly(ethylene)s, and phenolic resins. The remaining 25 % are contributed by many other polymers, especially epoxy resins, polyamides, polycarbonates, urea-formaldehyde resins, poly(propylene), and poly(ethylene terephthalate), and several elastomers. A vast tonnage of these polymers is used as electrical insulators or for other protective functions, a smaller tonnage is provided with additives for electrical conduction, and an even smaller tonnage is intrinsically electrically conductive.

14.1.2 Types

Matter behaves differently on application of electrical fields. Some chemical compounds do not conduct electricity; they are **electrical insulators (dielectrics)** (L: *insula* = island). Other chemical substances are **electrical conductors** or even **superconductors** whereas still others are **semiconductors** with electrical conductivities that are midway between those of insulators and conductors (L: *com* = together, *ducere* = to lead).

Electrical conductivities are reported in siemens S per centimeter which is the inverse of the (volume) resistivity ρ in ohm Ω times centimeter (for names of physical quantities, see Section 14.1.3):

Superconductors	$\sigma \approx 10^{20}$	S/cm $\approx 10^{-20}$	Ω cm
Conductors	$\sigma \geq 10^{3}$	S/cm $\leq 10^{-3}$	Ω cm
Semiconductors	$\sigma = 10^{-9} - 10^{2}$	S/cm $= 10^{-2} - 10^{9}$	Ω cm
Insulators	$\sigma = 10^{-22} - 10^{-10}$	S/cm $= 10^{10} - 10^{22}$	Ω cm

Most commercial polymers are insulators (Section 14.2). Resistivities of cured phenolic resins are ca. 10^{10} Ω cm and those of poly(ethylene)s for cable sheatings ca. 10^{17} Ω cm. Dipole containing polymers are less insulating: poly(methyl methacrylate)s have resistivities of ca. 10^{16} Ω cm, epoxy rsesins of ca. 10^{15} Ω cm, and polyamides of ca. 10^{12} Ω cm. All of these polymers are fairly easily electrically charged (Section 14.2.7).

14.1.3 Electrical Quantities

Literature uses terms for physical quantities that are either based on modern or older standardization attempts (ISO, DIN, IEC, IUPAC, IUPAP) or on historic terminology, sometimes side-by-side. The following review is based on IUPAP-IUPAC terminology for names of physical quantities, ISO nomenclature for physical units of these quantities, and IUPAC recommendations for symbols of physical quantities. Outdated names and quantities are shown in quotation marks.

Application of an **electrical potential difference** ("voltage drop," "voltage") U (or ΔV or $\Delta\phi$) in volt V causes matter to resist the **electric current** I in ampere A. The **electrical resistance** $R = U/I$ is given in ohm, Ω = V A^{-1}, and is obtained according to $R = a/(\sigma db)$ from the conductivity σ (also as κ or γ) of the matter, the thickness d of the specimen, and the side lengths a and b of the rectangular specimen or the width a and distance b of electrodes, respectively. As resistance of the total *measuring system*, R is an "electrical *surface* resistance" which cannot be converted into the **resistivity** of *matter* (see below). For $a = b$, the resistance becomes $R = 1/(\sigma d) \equiv R_\square$, the "quadratic surface resistance."

The inverse of the resistance is the **conductance**, $G = 1/R$, which is measured in siemens, S = $1/\Omega$ = A/V. Since the physical unit siemens is the reciprocal of the physical unit ohm, literature often uses the backwards written ohm, i.e., mho instead of ohm$^{-1} \equiv$ S but mho is not a recognized physical unit.

The **electric current**, $I = dQ/dt$, describes the change of the **quantity of electricity** Q (= the **electric charge**) with time t (L: *currere* = to run). The **electric power**, $P = UI$, is reported in Watt, W = VA = J s^{-1}, i.e., as radiant flux of electricity.

The electric charge Q (= quantity of electricity) is measured in coulomb, C = A s. The electric charge per electrical potential difference, $Q/U = C$, is called **capacitance** and reported in farad, F = C/V.

The **resistivity** ρ is defined as the electrical resistance across opposite sides of a cube; it was therefore also called "volume resistivity" or "volume resistance." It is proportional to the resistance R and the cross-sectional area A and inversely proportional to the length L that is passed through by electrons, $\rho = R(A/L)$, and therefore given in Ω m = S^{-1} m.

In contrast to the electrical resistance, the resistivity depends only on the matter and not on the geometric dimensions of the measuring system (see above). It is therefore "specific" for a material. For this reason, it was formerly also called "specific volume resistivity," "specific resistivity," or "specific insulation resistance." This use of "specific" does not comply with the IUPAC definition of the word, which refers to a physical property divided by the mass (Section 16.4).

The reciprocal value of the (volume) resistivity ρ is the (electrical) **conductivity** σ (also: κ or γ), measured in S m^{-1}. For anisotropic materials, the conductivity is a tensor like the resistivity. The old name for conductivity is "specific conductance."

14.1.4 Testing

Electrical properties of insulators are commonly called **dielectric properties** (G: *di(a)* = through, *elektron* = amber (which becomes electrically charged on rubbing)). These properties comprise the relative permittivity (Section 14.2.2), the resistivity (Section 14.2.3), the electrical resistance (Setion 14.2.4), the dissipation factor (Section 14.2.5), the electric field strength (Section 14.2.6), and the static electrification (Section 14.2.7). All of these properties are obtained from standardized tests, for example, those of the CAMPUS® system (Table 14-1) which relies on IEC standards to which DIN also adheres. ASTM, CENELEC, and other standardization organizations use test conditions that deviate in part from that of IEC and CAMPUS®.

Table 14-1 Testing conditions for dielectric properties (all at 23°C and 50 % relative humidity) according to CAMPUS®. The table includes the numbers of standards as issued by the International Electrotechnical Commission (IEC). ASTM standards deviate in part from IEC standards. T = thickness, TO = transformer oil.

Property	Test condition	Physical unit	IEC	Size of specimen
Relative permittivity	50 Hz	1	250	1 ± 0.1 mm
	1 MHz	1	250	1 ± 0.1 mm
Dissipation factor	50 Hz	1	250	1 ± 0.1 mm
	1 MHz	1	250	1 ± 0.1 mm
Resistivity	contact electrodes	Ω cm	93	1 ± 0.1 mm
Electrical resistance	contact electrodes	Ω	93	1 ± 0.1 mm
Breakdown field strength	P25/P75	kV/mm	243-1	in TO (IEC 296)
Tracking resistance	test solution A	-	112	$\geq(15 \cdot 15 \cdot 4)$ mm^3
CTI-100 drop value	test solution A	-	112	$\geq(15 \cdot 15 \cdot 4)$ mm^3
Tracking resistance	M, test solution B	-	112	$\geq(15 \cdot 15 \cdot 4)$ mm^3
CTI M-100 drop value	M, test solution B	-	112	$\geq(15 \cdot 15 \cdot 4)$ mm^3
Electrolytic corrosion	-	-	426	$(30 \cdot 10 \cdot 4)$ mm^3

14.2 Insulators

Most polymers strongly resist the flow of an electrical current and are therefore valuable insulators for electrically conducting parts. Insulators must have certain minimum or maximum electrical properties and sometimes also thermal properties; their mechanical properties are usually less important. Some of these polymers are also used for protection (casings, etc.) where their mechanical properties are of prime concern.

The importance of the various electric properties varies with the sector of the electrical industry. In telecommunications, one is mainly interested in dielectric properties whereas in electric power engineering, resistivities, dissipation factors, and breakdown field strengths are of concern. In telecommunications, the behavior of polymers at high electric frequencies is in the forefront but in electric power engineering it is the behavior at high voltage, and for cables and electrical controls it is the electrical behavior at long times. Polymers for electrical purposes should be resistant against heat, humidity, and stress cracking and often also against mechanical deformation.

14.2.1 Polarizability

The electrical conductivities of metals and polymeric dielectrics differ in their mechanisms. Atoms of metals have many orbiting electrons. Application of an electric field causes a difference in the electric potential and electrons begin to flow. Since these electrons are shared by all atomic nuclei, electric conductivities, $\lambda/(\text{S/cm})$, of metals are not much affected by the chemical nature of metals: 613 900 (Ag), $\geq$580 000 (Cu), 365 900 (Al), and 100 200 (Fe), all at 300 K. Even constantans, a group of metallic alloys with fairly temperature-independent, "high" electric resistivities have electric conductivities of ca. 20 000 S/cm (* 40-55 % Ni, 45-60 % Cu, traces of Fe and Mn).

The situation is different for polymeric insulators. Atoms consist of positively charged nuclei surrounded by orbiting negatively charged electrons. The center of gravity of both types of charges is in the center of the atom. There are no electrical poles in an atom; atoms therefore have no permanent **electric dipole moment**, $\mu = Qr$, a tensor given by the product of electric charge, Q (in Coulomb C), and the shift, r, of electrons (in meter m), i.e., C m. For convenience, dipole moments are often reported in Debye units, D (the Debye is a non-SI unit with a value of $D = 3.335\ 64 \cdot 10^{-30}$ C m). Similarly, molecules composed of like atoms or those with different atoms but symmetric structures have also no electric dipole moment, i.e., D = 0 for H_2, CH_4, CCl_4, C_6H_6, etc.

Centers of gravity of negative and positive charges differ in non-symmetric molecules composed of different types of atoms. These molecules have **permanent electric dipole moments**, μ_p, which vary for small molecules between 0 and ca. 10 Debye units, for example, 0.11 for CO, 1.70 for CH_3OH, and 10.42 for CsCl. Static electric fields shift the center of gravity of the electron cloud relative to the center of gravity of the nuclei, which creates an **induced dipole moment**, μ_i.

The electric dipole moment μ has the physical unit C m. The **electric dipole moment per volume** (also called **dielectric polarization**), $P = \mu/V$, has therefore the physical unit C m^{-2}. The dielectric polarization, $P = (\varepsilon - \varepsilon_0)E$, is the product of the difference of **permittivities** of matter and in vacuum, $\Delta\varepsilon = \varepsilon - \varepsilon_0$, and the **electric field strength** E. Permittivities are measured in F m^{-1} = C V^{-1} m^{-1} and electric field strengths in V m^{-1}.

14.2.2 Relative Permittivity

Permittivities ε are difficult to measure. They are usually reported as **relative permittivities** (formerly: **dielectric constants**), $\varepsilon_r = \varepsilon/\varepsilon_0$, where ε_0 = permittivity of vacuum. Data are also reported as **electric susceptibilities**, $\chi_e = \varepsilon_r - 1$. The permittivity of vacuum is a natural constant with the value $\varepsilon_0 = (8.854\ 187\ 817\ ...) \cdot 10^{-12}$ F m^{-1}.

The relative permittivity of vacuum is unity; air at 20°C and 50 Hz has a value of $\varepsilon_r =$ 1.000 58. The greater the polarizability of a polymer, the larger its relative permittivity: poly(tetrafluoroethylene) has a value of $\varepsilon_r = 2.15$, poly(ethylene) 2.28, poly(oxymethylene) 3.6, poly(chloroprene) 9.0, and carbon black-filled, vulcanized *cis*-1,4-poly-(isoprene), 13.0. Water has a value of $\varepsilon_r = 81$ (see also Table 14-2).

Foamed ("expanded") thermoplastics and elastomers are "composites" of a polymer and air or other gases; they therefore have relatively low relative permittivities of 1-2. "Porous" polymers with electron microscopically invisible pores and relative permittivities of 1.8-2 are of high interest as interlayer dielectrics for future computer chips.

Table 14-2 Guide values for the water absorption Δw (at 50 % relative humidity), relative permittivity ε_r (at 1 MHz), resistivity ρ, electrical resistance R, loss tangent $\tan \delta$ = (at 1 MHz), electric strength S, and tracking force U (methods KA, KB, KC (p. 588)) of plastics [1]. All data at 23°C.

Polymer	$\dfrac{\Delta w}{\%}$	ε_r	$\dfrac{\rho}{\Omega\ \text{cm}}$	$\dfrac{R}{\Omega}$	$\tan \delta$	$\dfrac{S}{\text{kV mm}^{-1}}$	U/V KA	KB	KC
Perfluoralkoxy copolymer	0.03	2.1	10^{15}		0.0002	20			
Poly(tetrafluoroethylene)	0	2.15	10^{18}		0.0001	40	3b	>600	>600
Poly(propylene), it-		2.25	10^{17}		0.0002	90	3c	>600	>600
Poly(ethylene), LD	0.05	2.28	10^{17}	10^{13}	0.0002	110	3b	>600	>600
, HD		2.35	10^{17}		0.0001	130	3c	>600	>600
Poly(styrene), at-	0.1	2.5	10^{17}	10^{15}	0.0002	140	2	200	500
Poly(methyl methacrylate), at-		3.5	10^{17}		0.0500	50	3c	>600	>600
Styrene-butadiene copm (SB)		2.5	10^{16}		0.0004	110	2	200	550
Poly(N-vinyl carbazole)		3.0	10^{16}	10^{14}	0.0006	50	3b	600	600
Polycarbonate, bisphenol A	0.15	3	10^{16}		0.0050	50	1	175	225
Styrene-acrylonitrile copolymer		3	10^{16}		0.0100	70	2	275	500
Poly(butylene terephthalate)		3.3	10^{16}		0.0100	80	3a	450	>600
Poly(vinyl chloride), at-	<1.8	3.4	10^{16}	10^{13}	0.0200	40	3b	>600	>600
ABS		3.2	10^{15}	10^{13}	0.0100	120	3a	300	>600
Poly(oxymethylene)		3.6	10^{15}		0.0040	60	3b	>600	>600
Polyamide 6, dry	0	3.8	10^{15}		0.030	<150			600
, conditioned	9.5	7	10^{9}		0.30	80	3b	>600	>600
Unsaturated polyester resin		3.4	10^{14}	10^{12}	0.0100	50	3c	500	>600
ASA		3.4	10^{14}		0.0100	80	3a	300	>600
Epoxy resin		3.5	10^{14}		0.0050	50	3c	>600	>600
Cellulose acetobutyrate		4	10^{15}		0.0150	40	3b	>600	>600
Polyurethane		4	10^{13}		0.0500	30		>600	
Cellulose acetate	4.7	5.8	10^{13}		0.03	35			
Poly(vinyl fluoride)		7.4	10^{13}		0.08	140			
Poly(vinylidene fluoride)		6.43	10^{14}		0.17	10			
Polyamide 66, conditioned		4	10^{12}		0.3	90	3b	>600	>600
Phenolic resin, unfilled		7	10^{12}		0.300	10	1	175	150
, inorganic filler			$<10^{12}$		>0.02	>80	1		<150
, organic filler			$<10^{11}$		>0.05	>50	1		125
Urea resin		6	10^{11}		0.100	10	3a	>600	>600
Melamine resin, unfilled		7	10^{10}		0.300	10	3b	>600	>600
, inorganic filler			$<10^{9}$		≤ 0.5	<150	3c		>600
, organic filler			$<10^{11}$		≤ 0.4	<150	3c		>600
Poly(isoprene), 1,4-cis-		2.6	10^{14}		0.0002	23			
, vulcanized		3	10^{14}		0.002				
, vulc., carbon black	>15		10		0.1				
, vulc., metal	<18 000								

At temperatures below their melting temperatures, relative permittivities of semicrystalline, apolar polymers decrease slightly with increasing temperature; for example, from 2.15 at 25°C to 1.9 at 250°C for poly(tetrafluoroethylene) (T_M = 327°C) at 1 kHz. Another example is low-density poly(ethylene) (T_M = 115°C): 2.3 at –30°C but 2.18 at +90°C. Relative permittivities of crystalline polymers *de*crease because increasing temperatures lead to stronger segmental movements which cause the polymer to expand. As a result, distances, d, between atoms of different segments increase and therefore also attractions between those atoms. These attractions arise from relatively weak dispersion forces (London forces) which are proportional to d^{-6}.

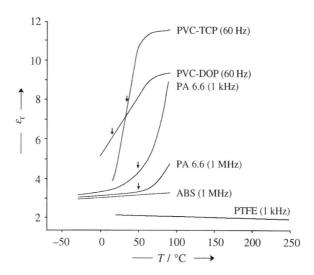

Fig. 14-1 Temperature dependence of relative permittivities, ε_r, of poly(tetrafluoroethylene), acrylonitrile-butadiene-styrene polymers (ABS), dry polyamide 6.6 at two frequencies, and a poly(vinyl chloride) containing 43 wt% dioctyl phthalate (DOP) or tricresyl phosphate (TCP). ↓ Dynamic glass temperatures of plasticized PVCs (at 60 Hz) or static glass temperatures of PA 6.6 and ABS.

Relative permittivities of polar polymers *in*crease with increasing temperature because of the increasing mobility of dipoles (see ABS in Fig. 14-1). This increase is especially pronounced once the glass temperature is surpassed (PA 6.6 in Fig. 14-1). The increase of relative permittivities is stronger, the smaller the applied frequency since dipoles can follow the electrical field more easily if the alternating current is smaller.

Plasticized polymers have strongly increased segmental mobilities and therefore also especially strong increases of relative permittivities with temperature (plasticized PVC in Fig. 14-1). For the same reason and especially because of their moisture content (water: $\varepsilon_r = 81$), conditioned poly(amide)s have much larger relative permittivities than their dried parent polymers.

Polymers containing (polar) catalyst residues, emulsifiers, etc., also show strongly increased relative permittivities. Not surprisingly, these effects are especially pronounced for polymers containing carbon black or metal powders: the relative permittivity of vulcanized *cis*-1,4-poly(isoprene) (NR; $\varepsilon_r = 3$) increases to more than 15 for carbon black filled NRs and up to 18 000 for metal-filled ones.

14.2.3   Resistivity

Application of small electric fields to dielectric substances causes dipoles to orient. Larger electical fields separate electrons from atoms and the material becomes ionized. It is these ions that carry the electric current and not electrons as in metals. Since ionization follows polarization, resistivities (= " volume resistances") decrease with increasing relative permittivity, at first strongly and then less strongly (Fig. 14-2). Resistivities are decreased by the same factors that increase relative permittivities: large mobilities of chain segments, polar groups in polymer segments and additives, and higher temperatures.

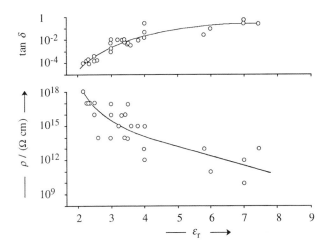

Fig. 14-2 Logarithm of resistivity ρ and loss tangent, tan δ, as a function of the relative permittivity ε_r of various plastics (data of Table 14-2).

Poly(tetrafluoroethylene) is practically not polarizable because of its strongly bound fluorine substituents; its resistivity is therefore large (Table 14-2). The greater the proportion of polar groups in a polymer and the larger the mobility of the chain segments, the more rapidly the resistivity decreases with increasing temperature (Fig. 14-3 and the series PS-SAN-ABS in Table 14-2).

Absorbed water decreases resistivities (see polyamide 6 in Table 14-2) and so do polar fillers and emulsifier residues. Resistivities of polymers decrease with increasing temperature because of the increased mobility of dipoles whereas electrical resistivities of metals increase slightly because they conduct electric currents by electrons and not by ions (Fig. 14-3).

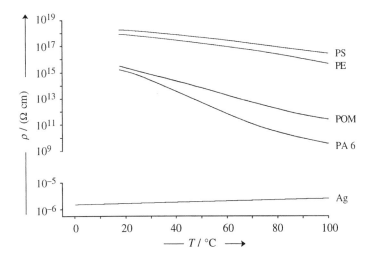

Fig. 14-3 Temperature dependence of resistivities ρ of poly(styrene) (PS), poly(ethylene) (PE), poly-(oxymethylene) (POM), poly(ε-caprolactam) (PA 6), and silver (Ag).

14.2.4   Electrical Resistance

Electrical resistances R are "electrical surface resistivities" of the measuring system and not "volume resistivities" of matter like electric resistivities ρ (see Section 14.1.3). Electrical resistances are therefore much more affected by external factors than are electric resistivities. An especially important factor is humidity because large humidities may lead to considerable ionic conductivities even if the concentration of impurities on surfaces is small. For this reason, electrical resistances are usually 2-3 decades smaller than resistivities (Table 14-2).

Carrier materials for printed circuits must therefore have especially large electrical resistances. Their polymers should not have polar groups and should also not attract moisture. Conversely, this effect is helpful for the prevention of static electrifications (Section 14.2.7) such as wiping surfaces with wet fabrics.

14.2.5   Loss Tangent

Fundamentals

The application of a *static electric field* to a dielectric material causes molecular dipoles in this material to try to orient themselves in the direction of the field. This orientation proceeds by random processes, i.e., by jumps or by diffusion. The applied electric field does not cause these movements but it rather affects the average degree of orientation of dipoles. Since molecular dipole moments reflect the orientation of molecules and molecule groups, the times required to establish macroscopic orientations correspond approximately to those that are required to reorient groups and molecules, respectively.

In *alternating electric fields*, dipoles of a dielectric try to follow the field which is less easy the faster the field changes direction, i.e., the higher the frequency (see the analogous situation for mechanical alternating fields (Volume III, p. 596 ff.)). The longer the time lag of orientation, the larger the consumed energy for this process. The lost energy per time is therefore a **power loss** P_L since this energy is converted to heat and therefore lost for possible use as work. Correspondingly, the **available power** P_A is reduced.

The power loss is higher the greater the frequency ω of the alternating field, the square of the electrical potential difference U, the capacitance C_0 in the absence of the dielectric, the relative permittivity ε_r, and the loss tangent, tan δ:

(14-1) $P_L = 2 \pi \omega U^2 C_0 \varepsilon_r \tan \delta$

The relative proportions of power loss and available power are controlled by the phase shift which is measured by the angle at which the phase angle ϕ of sinusoidal alternating electrical fields deviates from 90°, $\phi = 90° - \delta$, where $\delta = $ **loss angle**. The power loss is zero if the loss angle is 90°; all energy can then be converted to power. If current and voltage are in phase, $\delta = 0°$, all electrical energy will be converted to heat and the available power is zero.

The ratio of power loss P_L to available power P_A is the **loss tangent**, tan δ, also called "**dissipation factor**" or "**loss factor**" (but see below!):

(14-2) $\dfrac{P_L}{P_A} = \dfrac{UI\cos(90°-\delta)}{UI\sin(90°-\delta)} = \dfrac{\sin\delta}{\cos\delta} \equiv \tan\delta$

The sine of the loss angle, $\sin\delta$, is the **power factor**. It indicates the ratio of usable and apparent work.

The product of relative permittivity and loss tangent is defined as **dielectric loss index**, $\varepsilon_r \tan\delta$, formerly known as "loss factor" (the present (IUPAP, IUPAC) loss factor is $\tan\delta$ and not $\varepsilon_r \tan\delta$). Polymers with high dielectric loss indices can be heated by high electric frequencies, which allows electric welding. $\varepsilon_r \tan\delta$ has therefore also been called **welding factor**. Such a polymer is poly(vinyl chloride) with a welding factor of 0.068 as compared to the electrically non-weldable poly(styrene) with a welding factor of 0.0005. Conversely, poly(vinyl chloride) cannot be used as an insulator for high-frequency conducting materials. Such electrical insulators must rather be apolar polymers with small relative permittivities and low dielectric loss tangents.

Complex Permittivity

Sinusoidal alternating electrical fields cause **electric field strengths E** and **electric displacements, $D = \varepsilon E$**, to depend on the time t and the **angular frequency** ω of the field, which can be written as $E = E_0 \exp(i\omega t)$ and $D = D_0 \exp\{i(\omega t - \delta)\}$. The general equation with τ = relaxation time, δ = loss angle, and i = imaginary unit (i = $(-1)^{1/2}$),

(14-3) $\qquad D + \tau(dD/dt) = \varepsilon_0 E + \tau\varepsilon_\infty(dE/dt)$

leads with the expressions for E and D to the **complex permittivity, ε^***,

(14-4) $\qquad \varepsilon^* = \varepsilon_\infty + \dfrac{\varepsilon_0 - \varepsilon_\infty}{1 + i\omega\tau}$

where $\varepsilon_0 - \varepsilon_\infty$ = **dielectric increment**, ε_0 = permittivity in the relaxed state (i.e., at low angular frequencies), and ε_∞ = permittivity in the unrelaxed state (i.e., at a high angular frequency, usually a frequency below the infrared range). The complex permittivity is known by many names such as "complex dielectric constant," "dynamic electric constant," and "complex electric inductive capacity."

According to P. Debye, the complex permittivity can be split into a real part ε' (= **real permittivity**) and an imaginary part ε'' (= **imaginary permittivity**),

(14-5) $\qquad \varepsilon' = \varepsilon_\infty + \dfrac{\varepsilon_0 - \varepsilon_\infty}{1 + \omega^2\tau^2} \qquad ; \qquad \varepsilon'' = \dfrac{(\varepsilon_0 - \varepsilon_\infty)}{1 + \omega^2\tau^2}\,\omega\tau$

if the process has only one single relaxation time (Fig. 14-4).

In the past, and in most publications even today, the real permittivity is also called "real dielectric constant," "in-phase dielectric constant," "relative dielectric constant," and even "dielectric constant." The imaginary permittivity is known as "imaginary dielectric constant," "imaginary dielectric number," "out-of-phase dielectric constant," "dielectric absorption constant," and strangely also as "dielectric loss factor" (see above).

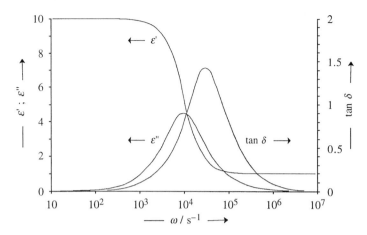

Fig. 14-4 Dependence of real permittivity ε', imaginary permittivity ε'', and loss tangent tan δ on the circular (angular) frequency $\omega = 2\,\pi v$ as calculated by the Debye equations, Eqs.(14-5) and (14-6), for a process with a single relaxation time and values of $\varepsilon_0 = 10$, $\varepsilon_\infty = 1$, and $\tau = 1 \cdot 10^{-4}$. v = frequency.

The dielectric loss tangent, tan δ, is then the ratio of the imaginary permittivity ε'' to the real permittivity ε':

$$(14\text{-}6)\qquad \tan \delta = \frac{\varepsilon''}{\varepsilon'} = \frac{(\varepsilon_0 - \varepsilon_\infty)\,\omega^2 \tau^2}{\varepsilon_0 + \varepsilon_\infty \omega^2 \tau^2} \qquad\qquad \varepsilon^* = \varepsilon' - i\varepsilon'' = \varepsilon' - (-1)^{1/2}\,\varepsilon''$$

Real permittivities ε' decrease with increasing angular velocities from large values in the relaxed state of polymers to zero in the unrelaxed state (Fig. 14-4). In contrast, imaginary permittivities ε'' and dielectric loss tangents, tan δ, both pass through maxima.

Dielectric Spectroscopy

Real and imaginary permittivities are interconnected by the **Kramers-Kronig equations**, for example, Eq.(14-7). These equations allow one to calculate the imaginary part, $\varepsilon''(\omega_0)$, at low frequencies ω_0 from the corresponding real part, $\varepsilon'(\omega_0)$, the permittivity ε_0 of vacuum, and the electrical conductivity σ at the circular frequency ω:

$$(14\text{-}7)\qquad \varepsilon''(\omega_0) = \frac{\sigma}{\varepsilon_0 \omega_0} + \frac{2}{\pi}\int_0^\infty \varepsilon'(\omega_0)\,\frac{\omega}{\omega^2 - \omega_0^2}\,d\omega$$

Both $\varepsilon' = f(\omega)$ and $\varepsilon'' = f(\omega)$ therefore contain the same information. However, absorption curves, $\varepsilon'' = f(\omega)$, are usually much more detailed than dispersion curves, $\varepsilon' = f(\omega)$, (Fig. 14-5), which is the reason why usually only imaginary parts are analyzed and discussed and not the real parts.

Maxima in the function $\varepsilon'' = f(\omega)$ indicate energy-absorbing transformations and relaxations, respectively; for example, dynamic glass temperatures in Fig. 14-5, top right. These maxima shift to higher frequencies at higher measuring temperatures, i.e., dynamic glass temperatures are higher at higher frequencies (Volume III, p. 466) because

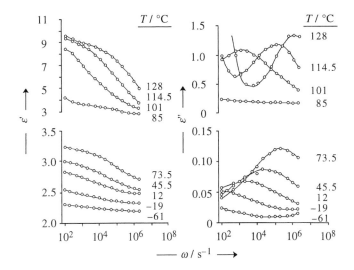

Fig. 14-5 Frequency dependence of real permittivities ε' (left) and imaginary permittivities ε'' (right) of a slightly crystalline poly(vinyl chloride) at various temperatures [2]. Top: β relaxation (glass transformation); bottom: γ relaxation. The α process (not shown) here is the melting temperature.
    By kind permission of Springer-Verlag, Berlin.

chain segments cannot follow the rapidly changing alternating electrical field fast enough. Other energy-absorbing relaxations at low temperatures are rotations of side groups, movements of short-chain segments, or boat-chair conversions of cyclohexane rings. The processes are characterized by upright Greek letters, starting with α_a for the glass temperature of amorphous polymers and α_c for the melting temperature of semi-crystalline ones, respectively. With decreasing temperature, designations follow the Greek alphabet in descending order (α, β, γ, δ, ε, ...).

Absorption maxima are observed if the angular frequency of the electric field corresponds to the natural relaxation frequency of the oriented dipoles, i.e., if the angular frequency equals the reciprocal relaxation time, $\omega = 1/\tau$. The maximum of the imaginary part is then $(\varepsilon'')_{max} = (\varepsilon_0 - \varepsilon_\infty)/2$. At the maximum of ε'', the real part adopts a value of $\varepsilon' = (\varepsilon_0 + \varepsilon_\infty)/2$.

Maxima in the function $\varepsilon'' = f(\omega)$ are fairly broad, which indicates a distribution of relaxation times. An example is the relaxation of chain segments of various lengths at the glass transformation. Literature describes the spectrum of relaxation times by several mathematical functions, most of which are based on the **Scaife equation**, Eq.(14-8):

(14-8) $$\frac{\varepsilon^* - \varepsilon_\infty}{\varepsilon_0 - \varepsilon_\infty} = [1 + (i\omega\,\tau)^{1-a}]^{-b}$$

This equation reduces to the **Cole-Cole equation** for $b = 1$, to the **Davidson-Cole equation** for $a = 0$, and to the **Havriliak-Negami equation** for $1 - a = c$, where a, b and c are empirical constants. For a process with a single relaxation time, Eq.(14-8) reduces to Eq.(14-4) if $a = 0$ and $b = 1$.

After introduction of $\omega\tau = \exp[\ln \omega\tau] = \exp k$, the Cole-Cole equation, Eq.(14-8) with $b = 1$, can be split into the real and the imaginary part:

(14-9) $\varepsilon' = \varepsilon_\infty + \dfrac{\varepsilon_0 - \varepsilon_\infty}{2}\left(1 - \dfrac{\sinh(1-a)k}{\cosh(1-a)k + \sin(a\pi/2)}\right)$

(14-10) $\varepsilon'' = \dfrac{\varepsilon_0 - \varepsilon_\infty}{2}\left(1 - \dfrac{\cos(a\pi/2)}{\cosh(1-a)k + \sin(a\pi/2)}\right)$

After eliminating k, one obtains

(14-11) $\left(\varepsilon' + \dfrac{\varepsilon_0 + \varepsilon_\infty}{2}\right)^2 + \left(\varepsilon'' + \dfrac{\varepsilon_0 - \varepsilon_\infty}{2}\tan\dfrac{a\pi}{2}\right)^2 = \left(\dfrac{\varepsilon_0 - \varepsilon_\infty}{2\cos(a\pi/2)}\right)^2$

For processes with a single relaxation time ($a = 0$), a plot of $\varepsilon'' = f(\varepsilon')$ (**Cole-Cole diagram**) shows above the ε' axis a semicircle with the center on the ε' axis (Fig. 14-6). The curve is no longer an exact semicircle for a distribution of relaxation times ($\alpha > 0$). It deviates from the shape of a semicircle, and its center is then below the ε' axis.

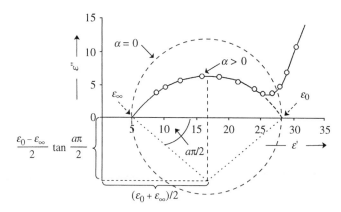

Fig. 14-6 Cole-Cole diagram of a poly(acrylonitrile) at 143.5°C. –o– Experimental data of [3]; – – – semicircle for a single relaxation time.

Effect of Polymer Constitution

The dielectric loss tangent is controlled by the polarizability, which in turn depends on the polarity of the dipoles and their proportion per molecule. Since common polymers are composed of the same chemical elements C, H, N, O, Cl, etc., and the polarizability controls both the relative permittivity and the loss tangent, one observes at *constant* temperature and constant frequency an increase of loss tangents with increasing relative permittivities that is independent of the chemical constitution (Fig. 14-2, top).

Apolar polymers such as poly(ethylene), poly(styrene), and poly(N-vinyl carbazole) have no electric dipoles or only weak ones and therefore small relative permittivites and loss tangents (low-density poly(ethylene) in Fig. 14-7). These polymers are used as insulators in high-frequency electrical engineering.

Loss tangents of conditioned low-density poly(ethylene) decrease slightly with increasing frequency, pass though a shallow minimum, and then increase (Fig. 14-7). The initial higher values are caused by moisture that is taken up by residues from polymer-

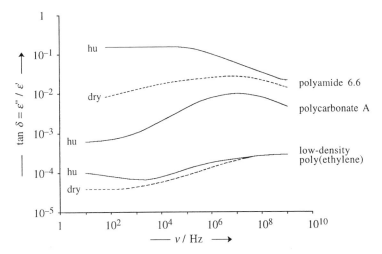

Fig. 14-7 Frequency dependence of dielectric loss tangents at 30°C. - - - Dry state, ———— conditioned at 23°C and 65 % relative humidity. hu = humid (moisture containing).

ization catalysts and other impurities. Loss tangents then decrease in the frequency range 10-10^3 Hz because water molecules can no longer follow the electric field. Such a decrease is also found for conditioned polyamide 6.6 albeit at higher frequencies because of hydrogen bridges between –NH–CO– and H_2O.

The subsequent increases of loss tangents of dry LDPE, dry PC, and dry PA 6.6 result from the imaginary part of relative permittivities (see Fig. 14-5) and are caused by the onset of energy-absorbing transformation or relaxation phenomena such as glass transformations and melting processes. At still higher frequencies, loss tangents may show maxima (dry PC in Fig. 14-7) and also minima because real relative permittivities decrease steadily but not linearly with increasing frequency (Fig. 14-5, left) whereas imaginary relative permittivities are absorption phenomena and may therefore lead to maxima and minima in ε' (Fig. 14-5, right) and therefore also in tan δ.

These absorption phenomena in alternating electric fields allow one to determine and examine glass transformations and other relaxation phenomena of polar polymers. These processes differ for chemical groups in main chains and in side chains. For example, the main-chain dipoles of solid poly(oxyethylene), $+O–CH_2–CH_2+_n$, can orient themselves only above the glass temperature because only then are segments sufficiently mobile. Side-chain dipoles of poly(vinyl methyl ether), $+CH_2–CH(OCH_3)+_n$, on the other hand, orient by either segmental movements of main chains or by aligning side groups. At temperatures above the glass transformation, these polymers exhibit two dispersion ranges: at low frequencies because of the mobility of chain segments and at high frequencies because of orientations of side groups. Below the glass temperature, only the effect of the side groups is observed.

14.2.6 Breakdown Field Strength and Tracking Resistance

The imaginary part of the relative permittivity may arise either from polar groups of the polymer itself or from ions from impurities. The continuous reorientation of dipoles by the alternating electric field generates heat which is difficult to dissipate because polymers have low heat conductivities. The resulting heat build-up leads to a rapid increase of the electrical conductivity because of the strong temperature dependence of the dissociation of ions. Finally, a breakdown occurs.

The **electric strength** S (**breakdown field strength**; formerly: dielectric strength) depends on the presence of polar groups and the mobility of dipoles (see the series PS-SAN-ABS and the data for dry and conditioned PA 6 in Table 14-2). It decreases with increasing temperature and/or frequency. Data for different types of polymers are difficult to compare because electric strengths are affected strongly by moisture contents and impurities from polymerization processes.

A **breakdown** is caused by an avalanche type increase of ion conductity in the *interior* of the specimen. However, a **short circuit** can also occur because of **tracking (track)** of electric currents on the *surface* of a specimen, i.e., **surface leakage**. Such creeping currents are caused by electrically conducting surface impurities, i.e., by **deposit tracking**. The burn-in of surface currents in organic polymers leads to tracks of charcoal which have much smaller resistivities (graphite: $\rho \approx 10^{-2}$ Ω cm!).

The **tracking resistance** is measured by the number of drops of a salt solution as "standard impurity" that are required to produce a breakdown at standard conditions. The test solution is dropped steadily into a defined gap between two electrodes on the surface of the test specimen; the electrodes are under constant voltage. The resulting arc can burn in and generate a track that then leads to a breakdown. The number of drops until the breakdown is a measure of the tracking resistance (ASTM). Alternatively, one can also keep the number of drops constant and change the voltage level (ISO, IEC, DIN, VDE, CAMPUS®).

Standard impurities with added wetting agents (method KB) lead to lower tracking resistances than those without wetting agents (method KC). In Europe, method KB is preferred whereas method KC is common in the United States.

A polymer is tracking resistant if electric arcs cannot generate carbon particles. Such particles are not formed if the action of the arc depolymerizes the polymer since the evaporating monomer prevents the deposition of salt. An example of such a monomer is poly(methyl methacrylate).

The same effect is obtained if the degradation of polymers leads to other volatile degradation products, for example, volatile hydrocarbons from poly(ethylene)s or a variety of chemical compounds from aliphatic polyamides. However, poly(N-vinyl carbazole) does not generate volatile decomposition products and therefore has a bad tracking resistance although it is a good electrical insulator.

14.2.7 Static Electrification

Static electrification causes accumulations of electric charges on insulating or ground surfaces, i.e., a surplus or a deficit of electrons. It is produced by a surface in contact with ionized air or by rubbing two surfaces against each other (**triboelectric charging**). Matter can be charged triboelectrically if its conductivity is less than 10^{-8} S/cm and the relative humidity is less than ca. 70 %. The earliest known electrically chargeable material was amber, a fossilized resin whose Greek name *elektron* became the Godfather to the negatively charged elementary particle "electron" and the science of electricity.

Table 14-3 Triboelectric series of polymers, (mass) charge densities q_m, and half-lifes t_{50} for the discharge of positive or negative charges. Industry uses mass-related charge entities (no official IUPAP symbol) whereas ISO and IUPAP define the (volume) charge density as $\rho = Q/V$ in C m^{-3} and the surface charge density as $\sigma = Q/A$ in C m^{-2} where Q = quantity of electricity (electric charge), V = volume, and A = surface area. Note that each partner carries both positive and negative charges.

Triboelectric series	Code	q_m/C g^{-1})	t_{50}/s (positive)	t_{50}/s (negative)
Positive end				
Silicone rubber	MQ	−0.18		
Phenol-formaldehyde thermoset	PF	−13.9		
Sheep's wool	Wo		2.5	1.6
Polyamide 6.6	PA		940	720
Cellulose, cellophane	C		0.3	0.3
cotton	CO		3.6	4.8
Poly(styrene)	PS	0.37		
Poly(ethylene terephthalate)	PET			
Epoxide thermoset	EP	−2.13		
Poly(acrylonitrile)	PAN		670	690
Poly(ethylene)	PE			
Poly(vinylidene chloride)	PVDC			
Poly(trifluorochloroethylene)	ECTFE	8.22		
Negative end				

Materials can be arranged qualitatively in a triboelectric series (Table 14-3, left column) in which a material is called positive if its ratio of positive/negative charges is greater than that of its rubbing partner, which is then negative relative to its partner (examples: MQ-PF, PF-PS). The resulting series is not absolute but may change somewhat according to experimental conditions, time, and environment. The triboelectric series also says nothing about sign and magnitude of charge densities and discharging times.

The sign and extent of triboelectric charging (electrostatic charging) depends on the partner and the duration of rubbing. For example, the first rubbing of poly(oxymethylene) against polyamide leads to an electric field strength of 360 V/cm, the tenth rubbing to one of 1400 V/m, and many more rubbings to a limit of 3000 V/cm. An antistatic containing ABS polymer had a limit of 120 V/cm against poly(acrylonitrile) but one of −1700 V/cm against polyamide 6.

Charge densities of triboelectrically charged surfaces are not uniform but merely represent averages. On surfaces, one may rather have "islands" with positive charges in a "sea" of negative ones or *vice versa* as one can see if the surface is sprinkled with different powdered dyes or pigments. In general, one type of charge dominates.

Charges from triboelectric charging disappear only slowly because surface conductivities of most polymers are low. Half-times for discharges differ for positive and negative charges. In a first approximation, the discharge time t_{dis} is proportional to the electrical resistance against discharging, $R = U/I$, and the capacitance, $C = Q/U$:

(14-12) $t_{dis} = kRC = kR(Q/U)$; k = dimensionless constant

For polymers, capacitances C or, alternatively, quantities of electricity Q and electrical potential differences U are difficult to measure. The discharge time t_{dis} can therefore be reduced only by lowering the electrical resistance R against discharging. In turn, this quantity depends on the resistance (= "surface resistivity") and the resistivity (= "volume resistivity") of the bulk polymer as well as on the resistivity of the surrounding air. The smallest of these quantities controls the constant k and thus the discharge time t_{dis}.

Half-times of discharge of polymers are often tens of minutes (PA 6.6, PAN (Table 14-3)) which is not only a nuisance but sometimes outright dangerous. Examples are the electrical charging of spinning rolls in spinning processes, the dusting of household articles made from plastics, the charging of electrical and electronic equipment such as computer casings; the "stickiness" of poly(styrene) "peanuts" for packaging, polymeric wrapping films, or freshly dry-cleaned polyester pants against clean woolen socks; the escape of gasoline vapors from unearthed rubber hoses during refueling of cars (danger of explosion); the warehousing of powders, grain, etc., in silos (dust explosions); etc.

Electrostatic charging can be prevented in different ways. In fiber spinning, deflecting rollers are neutralized by ionized air. At gas stations, rubber hoses are wrapped with metal sleeves or spirals. Alternatively, one can provide polymers with antistatic agents. For example, addition of 30 wt% carbon black to ethylene-propylene copolymers allows these polymers to keep all their mechanical properties but will raise electrical conductivities to ca. 10^{-2} S/cm from ca. 10^{-16} S/cm. Instead of internal antistatic treatments, one can also improve electrostatic behavior by external means which change (surface) resistances R but not (volume) resistivities ρ. An example is the surface treatment of polymers with compounds that bind moisture or reduce friction by addition of lubricants or coating with teflon. However, external antistatic treatments need to be renewed from time to time.

Conversely, electrostatic charging can be exploited for useful applications. Examples are electrostatic paint spraying, the flocking of goods to produce velvety surfaces, and electrophotography.

14.2.8   Electrophotography

Electrophotography is the term for electrostatic copying processes that use the generation of electrical surface charges to produce copies. It includes the **zinc oxide process** with negative electrostatic charging of zinc oxide as sensitizer and **xerography** with positive electrocharges (G: *xeros* = dry, *graphein* = to write).

In these processes, a photoconducting material is deposited on a metal cylinder and subsequently negatively charged in the dark by a corona discharge. In early xerography, the photoconductor was amorphous selenium; presently, He-Ne lasers still use As_2Se_3. The next generation of electrophotographic processes used poly(N-vinyl carbazole) which absorbs ultraviolet light and generates an exciton (an electron-hole pair (Volume I, p. 372)) that is ionized in an electric field. Poly(N-vinyl carbazole) behaves as an insulator in visible light but can be sensitized with certain electron acceptors such as 2,4,7-trinitro-9-fluorenone with which it forms charge-transfer complexes. Most copying machines and xerographic laser printers now use solid solutions of transport-active molecules in polymers, for example, triphenylamine in bisphenol A polycarbonate.

On projection of the image to be copied on the thus treated surface, interaction of light with the charge-transfer complex of the receptor and sensitizer causes the lighter areas to discolor whereas the darker ones remain negatively charged. The resulting latent

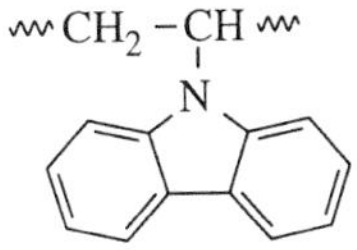

poly(N-vinylcarbazole)

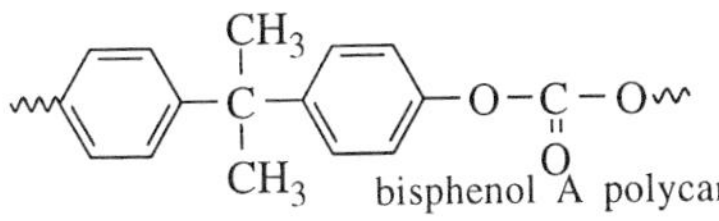

bisphenol A polycarbonate

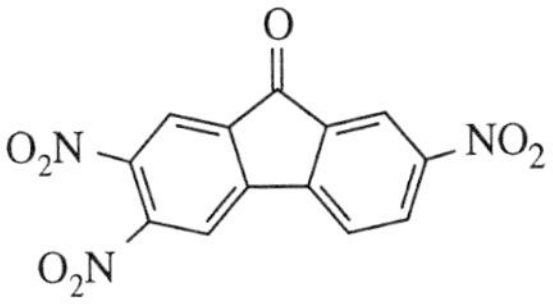

2,4,7-trinitro-9-fluorenone

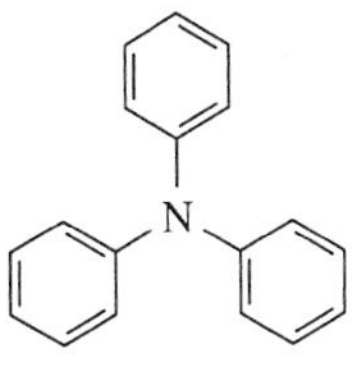

triphenylamine

image is then sprayed with resin-coated, positive developer particles which are deposited on the negatively charged, darker areas but not on the lighter ones thus creating a negative electrophotograph. On contact with negatively charged paper, the positively charged developer on the electrophotographic negative is transferred to the paper. Subsequent heating of the paper causes the resin to sinter, which fixates the positive copy.

Charge carriers of photoconducting polymers are generated by the interaction of charge transfer complexes with light. The conductivity of a polymer is given by the number concentrations N^*/V and P^*/V, respectively, of excited negative and positive charge carriers, their mobilities $(\mu_-)^*$ and $(\mu_+)^*$, respectively, in the excited states, and the electric charge, e:

(14-13) $\sigma_{light} = (N^*/V)(\mu_-)^* \, e \; + \; (P^*/V)(\mu_+)^* \, e$

Mobilities μ^* of carriers in some excited polymer systems approach that of amorphous selenium (Table 14-4).

Table 14-4 Carrier mobilities at 25°C and electrical fields of ca. 10^5 V/cm.

Material	$\mu^*/(cm^2\ V^{-1}\ s^{-1})$ positive	negative
Selenium, amorphous	10^{-1}	10^{-2}
Poly(isopropyl-N-vinyl carbazole)	10^{-1}	10^{-1}
Poly(N-vinyl carbazole)	$10^{-6} - 10^{-7}$	-
Poly(N-vinyl carbazole) + 2,4,7-trinitro-9-fluorenone (1:1)	10^{-8}	10^{-6}

14.2.9   Electrets

Electrets are dielectrics that can keep an electric field for long time spans; they are the electric analogs of magnets. Electrets are formed from materials with low electrical conductivities such as carnauba wax and polymers such as poly(propylene), poly(styrene), poly(methyl methacrylate), polyurethanes, and some polyamides.

Electrets are produced by two processes. In one process, the polymer is heated above its glass temperature T_G. After applying a strong electric field (e.g., 25 kV/cm), the polymer is cooled to $T < T_G$ while the field is maintained. The optimal charging temperature seems to be $T_G + 37°C$. In the second process, the polymer is cooled while flowing under pressure. This process requires a higher charging temperature of $T_G + 57°C$.

The charging orients dipoles. The material remains charged positive on one side and negative on the other after the electric field is removed. The charge difference declines slowly over months.

Reasons for the formation of electrets are not well known. It seems that both volume and surface polarizations are generated. At electric fields below ca. 10 kV/cm, one obtains volume polarizations. If such an electret is broken parallel to the charged surfaces, two new electrets are formed. However, a breakthrough occurs at electric fields above ca. 10 kV/cm which leads to surface polarizations.

This interpretation is supported by the polarizations that occur on charging at different electric field strengths. At small field strengths, fields generate charges with signs opposite to that of the field that may be caused by ionic impurities. At temperatures above the glass temperature of the polymer, distances between like charges can be increased and then frozen at $T < T_G$. At large field strengths, air breaks through and the surface of the electret has now the same polarization as the electrodes.

14.2.10 Pyro, Piezo, and Ferroelectrics

The heating of matter causes volume changes and, in crystals with polar axes, also orientations of internal dipoles. Surface charges change if ferromagnetic materials are heated to temperatures below the **Curie temperature**, which is the temperature at which the material becomes paramagnetic. On surfaces, heating or cooling generates an electric potential difference, the so-called **pyroelectricity** (G: *pyr* = fire) which was first observed on tourmaline by Pierre Curie.

In non-centrosymmetric crystal structures, electrical polarizations can also be produced by mechanical stresses (strain, pressure, etc.). The resulting change of polarization leads to a voltage between opposite sides of the specimen, so-called **direct piezoelectric effect** (G: *piezein* = to press, squeeze). Such piezoelectrics are used as pressure sensors.

Application of an electric potential on opposite sides of a piezoelectric material leads to expansion, contraction, or shear deformation (**converse piezoelectric effects**), which is used for actuators.

Piezoelectric materials thus allow one to follow mechanical or thermal signals by electrical signals or *vice versa*, which is used in loudspeakers and microphones as well as in alarm systems that are sensitive to movements or temperature changes.

In **direct piezoelectric effects**, polarizations p_i are directly proportional to mechanical stresses σ_j,

(14-14) $p_i = d_{ij}\sigma_j$

where i = 1, 2, 3 and j = 1, 2, 3 are orthogonal directions within the specimen. Direction 3 is usually the direction of polarization (direction of the electric field) and direction 1 is the direction of the mechanical stress (draw direction).

In **converse piezoelectric effects**, mechanical deformations S (elongations, shearings, etc.) are correspondingly proportional to the applied electric field E_j:

(14-15) $S_i = d_{ji}E_j$

In some cases, easy-to-process but less effective polymers have replaced the older ceramic materials. For example, barium titanate has a piezoelectric coefficient of $d_{31} = 190 \cdot 10^{-12}$ m/V whereas the strongest polymeric piezoelectric, the β modification of poly-(vinylidene fluoride), only has $d_{31} = 30 \cdot 10^{-12}$ m/V and $d_{32} = 2/3$ m/V, respectively. The γ phase of polyamide 11 has $d_{31} = 1.5 \cdot 10^{-12}$ m/V. Polyamides 5, 7, 9, and 11 become piezoelectric if their melts are quenched by ice water.

Pyroelectric crystals are called **ferroelectric** if their polar axis can be oriented reversibly by electric fields. Ferroelectrics show dielectric hysteresis on reversal of the fields. The name "ferroelectrics" was chosen in analogy to "ferromagnetics" (Fe, Co, Ni, CrO_2, etc.); it has nothing to do with iron.

Ferroelectric materials have small domains with reversible orientations of dipoles. These orientations are either parallel (**ferroelectrics**) or antiparallel (**ferrielectrics, antiferroelectrics**). The β modification of crystalline poly(vinylidene fluoride) is ferroelectric and so also is amorphous poly(vinylidene cyanide-*co*-vinyl acetate). Liquid crystalline polymers have been studied as ordered matrices for low-molecular weight ferromagnetic organic molecules.

14.2.11 Optical Lithography

Optical lithography is the name for processes that use ultraviolet light (hence "optical" (G: *optos* = light)) to create geometric patterns (hence "graphy" (G: *graphein* = to write) on a substrate (but no longer a stone (G: *lithos*)) using photolacquer techniques. It was originally used in **offset lithography** for manufacturing printing plates (p. 545), was then improved as **photolithography** for producing printed wiring boards (chips, printed circuit boards), and finally developed further as **microlithography** for patterns <10 μm and then as **nanolithography** (patterns <100 nm) for fabricating integrated circuits (**IC**) for microprocessors. X-rays (λ_0 = 1-0.01 nm) and electron beams (λ_0 = 0.02 nm at 3600 V) deliver even smaller patterns (**Next Generation Lithography (NGL)**).

These fields use the same principle albeit with increasing miniaturization, sophistication, and number of required manufacturing steps. Fig. 14-8 shows schematically the two possible variations of these techniques for the production of so-called *photoresists* for microprocessors from wafers and photolacquers.

Procedures

Wafers are practically always silicon single crystals because they can be produced with few impurities ($\approx 10^{-9}$ %!) as thin sheets (**wafers**) with exceptional planarity. In very special cases, synthetic sapphire is used. Before use, wafers must be thoroughly cleaned.

Polymers for optical lithography are **photolacquers**, i.e., solutions of polymers that deliver clear coatings (p. 531). In view of the enormous economic importance of the field, very many different polymers have been suggested and tried for use in optical lithography. Only a few became commercially interesting and even fewer survived the stringent application conditions and the rapid progress of the field.

Photolacquers are usually applied by spin-coating at 1200-4800 rpm (step II in Fig. 14-8) which delivers spin-coats of 0.5-2.8 μm thickness with extraordinarily small variations of thicknesses of ±5-10 nm. Photolacquers must be resistant against subsequent

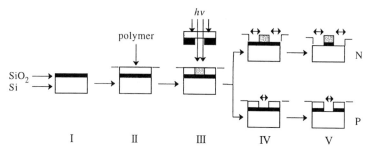

Fig. 14-8 Production of negative (N) and positive (P) resists for microprocessors by irradiating polymers. A plane silicon sheet (**wafer**; from Middle German *wafel* = waffle) is coated with a thin layer of
SiO_2 (I) and then with a polymer layer from a so-called photolacquer (II). Upon irradiation of the polymer through a mask (III), polymer layers in the non-covered, irradiated area (gray) become *less* soluble
(**negative (tone) process** (N)), leading to a **negative resist**; example: crosslinking) or *more* soluble (**positive (tone) process** (P), leading to a **positive resist**; example: degradation). The uncrosslinked polymer from the uncovered parts is dissolved by a solvent ("developer") (IV, top). The
SiO_2 of the wafer is then wet or dry etched (IV→V) at positions that are not covered by the polymer.
The resulting free silicon areas are subsequently doped (Section 14.3.3).

Manufacture of chips and printed wiring boards is similar, except for the following differences. Instead of a silicon wafer, a copper layer on a supporting board is used. The board consists of phenolic
resin and paper (FR-1, FR-2), epoxy resin and paper (FR-3), or epoxy resin and glass fibers (FR-4,
FR-5); FR = flame retardant. Etching is by aqueous solutions of $FeCl_3$, $Na_2S_2O_8$, or HCl.

chemical etching and insensitive to heat; for this reason, they are also called **photoresists**
or simply **resists**. After application, resists are dried ("prebaked") at 90-100°C.

The dried photolacquer is then irradiated (step III; for details, see next subsection)
and subsequently baked at higher temperatures. The irradiation causes a photochemical
reaction which renders the irradiated sections more soluble (**positive (tone) resist**) or less
soluble (**negative (tone) resist**). The uncrosslinked polymer of IV-N and the degraded
polymer of IV-P are removed by dissolution with a suitable solvent. Finally, the exposed
SiO_2 cover is removed by *wet etching* with buffered HF. The exposed Si is then doped
(Section 14.3.3). The production of modern memory chips may require 30 lithography
steps, 10 oxidations, 20 etching procedures, 10 doping steps, etc.

Printing

The polymers are then exposed to intense, uniform light (step III in Fig. 14-8) which
produces the desired pattern (N or P in Fig. 14-8). Several systems are used in this
"*printing*" process. A photomask in direct contact with the wafer offers the best resolution (minimum loss of resolution by light diffraction) but *contact printing* cannot be
used for mass production (damage on contact, 1:1 transfer of defects). Better is
proximity printing where the mask is positioned 10-50 μm above the photolacquer. For
very large scale processes, patterns are not transferred on a 1:1 scale but reduced to 4:1
or 5:1 and projected by a lens system. The resolution of this *projection printing* can be
improved by placing a refractive medium such as ultrapure water between projection
lens and photolacquer (*immersion lithography*).

The smaller the patterns (the gray area in Fig. 14-8, III), the better the resolution, the
more information can be packed on the microchip, and the smaller can be the electronic
device at the same or greater functionality. The smallest possible size is called the

critical dimension CD (or *target design rule*). For projection printing, this dimension is given by $CD = K_{proc}(\lambda_0/A_n)$ where K_{proc} = a process-related quantity (often 0.4), λ_0 = wavelength of incident light, and $A_n = (n_P^2 - n_M^2)^{1/2}/n_0$ = critical aperture of the lens (see also Section 15.1.5) where n_P = refractive index of the polymer, $n_0 = 1$ = refractive index of vacuum, and n_M = refractive index of the medium between the projection lens and the photolacquer. A polymer with $n_P = 2$ would thus have $A_n \approx 1.73$ in air which would lead to $CD = 0.231\ \lambda_0$ if $K_{proc} = 0.4$. For these conditions, the smallest possible dimension would thus be $CD \approx 57$ nm if a wavelength of $\lambda_0 = 248$ nm is used. The refractive index of polymers should therefore be as high as possible.

Microlithography has delivered resolutions of 0.35-3 μm if mid-UV (MUV; $\lambda_0 =$ 350-450 μm) from mercury lamps was used. This light is filtered and a single spectral line is selected, either the "g-line" (435.8 μm) or the "i-line" (365 nm). Resolutions of ca. 70 nm were obtained using so-called deep-UV (DUV) from an KrF excimer laser ($\lambda_0 =$ 248 nm) with a phase-shift mask. Nanolithography also applies DUV but employs an ArF excimer laser ($\lambda_0 = 193$ nm) which allows one to obtain resolutions of 65 nm if so-called phase masks are used and even resolutions of 40 nm by immersion lithography.

Early Negative Processes

Optical lithography began in 1826 with photographic experiments by Joseph Nicéphore Nièpce who noticed that asphalt hardened at sections that were irradiated with visible light. This technique was later combined with the even older technique of lithography (p. 545). In the resulting asphalt lithography, stone slabs were coated with asphalt and irradiated through a mask for hours. The non-hardened parts were then dissolved by turpentine oil. In modern parlance, this is a negative process with wet etching.

"Asphalt" is internationally the term for a natural or synthetic mixture of bituminous substances and inorganic minerals (Volume II, p. 58). "Bitumens" are mixtures of many different types of organic molecules; soluble fractions have molecular weights of 300-2000. The hardening of asphalt by irradiation is therefore a crosslinking of oligomeric and polymeric parts of bituminous substances.

The recognition of this process as a crosslinking photopolymerization led to the development of other, more practical, crosslinking systems. All systems must lead to highly crosslinked polymers that do not swell and resist undercutting on wet etching. Polymers must also adhere well to SiO_2, especially on wet etching.

The earliest successors of asphalt were water-soluble polymers such as gelatin, egg albumin, or gum arabic that were provided with a photosensitive dichromate salt. Irradiation triggered the photoinitiated oxidation of polymers to crosslinked materials. Natural polymers were later replaced by poly(vinyl alcohol).

These systems were followed in 1953 by the KPR process (*Kodak photoresist*) which used poly(vinyl cinnamate), $+CH_2-CH(O-CO-CH=CHC_6H_5)+_n$. On irradiation, cinnamate units crosslink intermolecularly to cyclodimer units. The process was first used for letter printing and later for the manufacture of circuit boards.

(14-16)

I   IIa   IIb   IIc

The next system used in microelectronics was based on cyclized poly(isoprene) (**RUI** = *r*ubber *u*nits *i*somerized) (Volume II, p. 249) which is a soluble polymer with molecular weights of 2000-10 000 that contains 50-90 % mono-, di-, and tricyclic units (IIa-c) besides remaining isoprene units I.

RUI is applied as a solution that contains a small proportion of an aromatic bisazide, for example, $Z[C_6H_4N^{\ominus}-N^{\oplus}\equiv N]_2$. Irradiation of the dried mixture decomposes bisazides to nitrogen and highly reactive dinitrenes III with two unpaired electrons per nitrogen. The dinitrenes crosslink poly(isoprene) by either adding to the >C=C< double bond and forming aziridine structures IV, Eq.(14-17), or by inserting themselves into >(C–H)– bonds. However, problems arose because mass production required proximity or projection printing (see previous page): crosslinking was reduced because nitrenes reacted with dissolved O_2 in films instead of with isoprene units, and the now less tightly crosslinked polymer was swollen by the developer, which distorted the image.

The Riston® series of DuPont avoids these problems. The photolacquers of this series consist of multifunctional acrylic monomers such as $C[CH_2OOC-CH=CH_2]_4$ and a photosensitizer, for example, an anthraquinone derivative, as initiator of the photopolymerization of the monomer to a crosslinked polymer. The Riston® series of compounds comprises many formulations for the manufacture of printed wiring boards.

(14-17)   III   IV

Positive Processes

Novolacs are soluble oligomers from the reaction of formaldehyde with phenols (Volume II, p. 261), containing predominantly cresol units (V). They dissolve in dilute caustic soda solution albeit very slowly. The dissolution rate is reduced by addition of small proportions of diazoquinones such as DNQ (VI). On irradiation, DNQ undergoes a Wolff transformation (VI → VII) which yieldss a highly soluble compound VII. For unknown reasons, the rate of dissolution of the irradiated regions (V + VII) by aqueous NaOH increases by orders of magnitude whereas the dissolution rate of the unirradiated regions remains very small. Hence, the novolac–DNQ system leads to a positive process.

V SO_2—O—R   VI SO_2—O—R   VII

The novolac-DNQ system offers many advantages: there is no problem with oxygen; the unirradiated regions swell only a little or not at all; the developer is water-based. The system was first used for chips and printed circuit boards and then for other electronics.

There are several other positive processes. Poly(2-butene-*alt*-sulfone), developed by Bell Labs, has a low ceiling temperature and depolymerizes easily into 2-butene and SO_2 on irradiation with electron beams. It is used for the preparation of photomasks.

Photomasks are also prepared from poly(methyl methacrylate) which depolymerizes to 100 % monomer on irradiation with electron beams (see also Volume I, p. 591).

Modern Processes

The novolac-DNQ system turned out to be impractical for the smaller structures desired for integrated circuits. These smaller structures are obtained by irradiation of photoresists with smaller wavelengths such as those from the 248 nm KrF excimer laser. At this wavelength and below, the DNQ-novolac system absorbs strongly, which attenuates the beam and makes the photolayer non-uniform.

Photoresists for smaller wavelengths are often based on poly(*p*-hydroxystyrene) derivatives (PHOST), for example, poly(*t*-butoxyoxycarbonylstyrene) (VIII), in combination with small proportions of photosensitizers consisting of onium salts such as the sulfonium salt $(C_6H_5)_3S^\oplus[AsF_6]^\ominus$. On irradiation of the unmasked areas of the photoresist, the photosensitizer decomposes into many products which react further and deliver a host of compounds, including the Brønsted acid $H^\oplus[AsF_6]^\ominus$. The proton of this acid attacks the non-polar VIII to give the polar poly(vinyl phenol) plus the gases CO_2 and C_4H_8, which leave the coating (Eq.(14-18)).

The reaction also regenerates $H^\oplus$ so that one proton from the Brønsted acid fragments ca. 1000 *t*-butoxy groups. The onium salt thus acts as a *photochemical acid regenerator*. Since one photon triggers so many reactions, only very low exposure doses are needed, i.e., the quantum yield is much greater than unity. For this reason, photoresists of this type became known as *chemically amplified photoresists*.

$$(14\text{-}18)$$

Poly(vinyl phenol) (IX) resulting from this reaction is acidic and dissolves easily in polar solvents such as aqueous NaOH. The system thus leads to a negative resist. However, systems with VIII (or related compounds) plus onium salts can also be used as positive resists with suitable developers such as anisole, $C_6H_5OCH_3$. In this solvent, the irradiated resist (i.e., IX) is insoluble whereas the original polymer VIII is soluble and can be washed away.

Polymers such as VIII cannot be used for the so-called 193 nm resists from ArF excimer lasers (p. 595) because the aromatic groups of VIII strongly absorb light at 193 nm. Therefore, various copolymers have been developed that do not absorb light in this region. These polymers do not contain aromatic groups but a variety of other groups.

Examples are terpolymers of methyl methacrylate (X) where methyl methacrylate units (R = CH$_3$) provide the polymer layer with mechanical stability and low shrinkage; methacrylic acid units (R = H) make the terpolymer water-soluble so that aqueous developers can be used; and *t*-butyl units (R = C(CH$_3$)$_3$) improve the etch resistance.

$$\text{X} \qquad\qquad \text{XI} \qquad\qquad \text{XII} \qquad\qquad \text{XIII}$$

Polymers for plasma etching may have bicyclo units such as XI with substitutents R' such as –COOC(CH$_3$)$_3$, –CH$_2$O–CO–O–C$_2$H$_5$, –COOH, etc. Dissolution of these compounds is improved by incorporation of maleic anhydride units XII.

So-called 157 nm resists are used for the irradiation of fluorine-containing monomers by a molecular fluorine excimer laser. An example is a methacrylate-type polymer X where the –CH$_3$ group has been replaced by –CF$_3$ and the substituent –R by the group –C(CF$_3$)$_2$–(1,4-C$_6$H$_{10}$)–C(CF$_3$)$_2$–OH. Another example is polymer XIII.

14.2.12 Low-κ Dielectrics

In digital circuits, conducting components such as transistors and interconnecting wires must be insulated from each other both horizontally and vertically (through interlayers) in order to avoid the transport of energy from one component to the other. This inadvertant energy transfer leads to "crosstalk", i.e., the unwanted phenomenon of a signal from one electronic component being affected by a signal from another one.

The energy transfer may be through a physical contact (conductive coupling; involving heat transfer), a shared magnetic field (inductive coupling), or an electrical field (capacitive coupling). In integrated circuits (IC), coupling is usually capacitive.

The capacitance *C* of a body, measured in farad F, is its ability to hold a quantity of electricity (= the electric charge *Q* in coulomb) per voltage (= the electrical potential difference *U* in volt), C = Q/U (L: *capere* = to hold, contain, take). The capacitance can be calculated from the permittivity and the geometry of the system, for example, for a plane dielectric material between two plane electrically conducting layers,

(14-19) $C = \varepsilon_r \varepsilon_0 A/d = \kappa \varepsilon_0 A/d$

where ε_r = relative permittivity, $\varepsilon_0 = 8.854 \cdot 10^{-12}$ F/m = permittivity of vacuum, *A* = contact area between dielectric and conductor, and *d* = thickness of dielectric.

The relative permittivity ε_r is traditionally called "dielectric constant" but is not a constant in the meaning of physics where "constant" designates a property that has the same value for all types of matter (example: Boltzmann constant). A "dielectric constant" is rather a property of a *specific material* and neither universal for all matter nor for all dielectrics. For unkown reasons, the microelectronics industry also uses the symbol κ (*kappa*), spoken as "kay", instead of the scientific symbol ε_r.

As electronic components of integrated circuits became smaller and smaller in microelectronics, distances *d* became smaller, too, which increases the likelihood of capacitive coupling and charge build-up. This problem prompted the search for so-called low-κ materials which are materials with dielectric constants smaller than the $\kappa = 3.9$ of SiO_2, the insulating material used in silicon chips. The desired low-κ materials need to combine many outstanding electrical, mechanical, chemical, and thermal properties, for example, very high electric field strengths (= dielectric strengths of industry); high tensile moduli, high fracture toughnesses, and good adhesion to both copper or aluminum as conductors and SiO_2 as insulator; high chemical resistances and etch selectivities, low solubilities in water and small moisture absorptions; and high thermal conductivities and stabilities as well as low dimensional changes on cooling or heating.

Only a few materials have these combinations of properties; all of them are polymers. *Inorganic polymers* for low-κ materials are all based on –Si–O– group-containing chemical structures. Silicon dioxide, $[SiO_2]_n$, is a crystalline or amorphous tectopolymer (Volume II, p. 577). Its κ value is lowered from 3.9 by doping: with fluorine to a fluorinated silica glass with $\kappa = 3.5$ and with carbon to $\kappa = 2.9$ (Table 14-5). Since air is an excellent dielectric ($\kappa \approx 1.0005$), porous silica glasses have even lower κ values: 2.5 for porous carbon doped SiO_2, 2.29 for a xerogel with 77 % porosity (Nanoglass®), and 2.0 for another porous SiO_2 with 65 % porosity. Properties of porous materials depend not only on the percentage of empty space but also on the size, shape, and interconnection of pores. However, SiO_2-based materials with very large porosities have lower mechanical strengths and are difficult to integrate into the production process.

Silicon dioxide is deposited on copper by chemical gas-phase reactions (chemical vapor phase deposition, CVD) of silicon at 800-1200°C, either by a dry process ($Si + O_2 \rightarrow SiO_2$), or a wet process ($Si + H_2O \rightarrow SiO_2 + 2\ H_2$), as well as from plasma-enhanced CVD (PECVD) of dichlorosilane at 900°C ($H_2SiCl_2 + 2\ N_2O \rightarrow SiO_2 + 2\ N_2 + 2\ HCl$) or of silane at 300-500°C ($SiH_4 + O_2 \rightarrow SiO_2 + 2\ H_2$). CVD is also used to deposit SiO_2 by thermal decomposition of tetraethoxysilane = tetraethylorthosilicate (TEOS) at 600°C ($Si(OC_2H_5)_4 \rightarrow SiO_2 + 2\ (C_2H_5)_2O$).

An SiO_2 with $\kappa = 2.9$ is the so-called carbon-doped SiO_2 (CDO) from the PECVD of trimethylsilane, $HSi(CH_3)_3$, with oxygen to porous $Si_iC_jH_kO_l$ with 10-25 % C-substituted silicon (Applied Materials). Nanoglass® from the reaction $Si(O_2H_5)_4 + 2\ H_2O \rightarrow SiO_2 + 4\ C_2H_5OH$ has an even greater porosity and lower κ (Table 14-5).

Table 14-5 Dielectric constants κ, glass temperatures T_G, tensile moduli E, thermal conductivities λ, and moisture absorptions Δw of low-κ materials. For abbreviations and structures, see next page.

Material	Trade name	κ	T_G °C	E GPa	λ W m^{-1} K^{-1}	Δw %
SiO_2		3.9			1.2	
CDO, 40 % porosity	Aurora® 2.7	2.5		>9	<0.4	
SiO_2, 77 % porosity	Nanoglass®	2.29			0.065	
HSQ	FOx®	2.9		4.7		
HSQ, 59 % porous	XLK®	2.2		2.5		
MSQ	HOSP®	2.65	>500			
MSQ, 56 % porous	LKD 5109	2.25		4.3		
PTFE AF	Teflon AF-2400	1.89	160, 240	1.3	0.19	0.12
PPX	Parylene N®	2.7	450	2.4		
PNB		2.2	365	7.3		<0.1
BCB, porous		2.65	≥350	2.9	0.24	<0.2
PPPh, porous	SiLK®	2.65	>450	2.45	0.18	0.01

Nanoglass® (AlliedSignal) from sol-gel technology is not put on chips by chemical vapor deposition but by a centrifugal method (spin-on) that produces from solutions very smooth films with thicknesses in the low micrometer range and minimal variations of thickness. The technology is also used to obtain low-κ *inorganic polymers* from semi-inorganic compounds such as poly(hydrogen silsesquioxane)s (HSQ; R = H) from the polycondensation of 5-10 % trialkoxysilanes, $HSi(OR)_3$, with H_2O in a mixed solvent containing more than 60 wt% methylisobutylketone (FOx® (*Fluid Oxide*); XLK® (*extra low k*); both DowCorning). Sol-gel technology also serves for the syntheses of *semi-inorganic* poly(methyl silsesquioxane)s (MSQ; R = CH_3) from the corresponding reaction of methyltrialkoxysilanes, $CH_3Si(OR)_3$ (for example, HOSP® (*hybrid organosiloxane polymer* (Honeywell)) and LKD (*low k dielectric* (JSR)). Porous films from these polymers have considerably lower κ values than non-porous ones (Table 14-5).

poly(sesquisiloxanes), R =H, CH_3 Teflon AF   Cyclotene

Even smaller dielectric constants κ can be expected for organic polymers which mostly have refractive indices in the range n = 1.29-1.73 (Section 15.1.1). Since Maxwell's law, $\kappa = n^2$, applies for the high frequencies at which microelectronic devices operate, this translates into κ values of 1.66-2.99 for nonporous organopolymeric dielectrics and 1.46-2.39 for these with 30 vol% air-filled pores.

An example is the bipolymer Teflon AF 2400 (DuPont) with tetrafluoroethylene and 2,2-bis(trifluoromethyl)–4,5-difluoro-1,3-dioxole units which has at visible light a refractive index of n =1.29 at 300°C and a κ = 1.89 at a frequency of 10^{10} Hz. This polymer is used for mechanical surface coatings and membranes and not as a dielectric because its two transformation temperatures (as T_G in Table 14-5) are fairly low and its dynamic modulus of E = 1.3 GPa at 25°C drops to very small values at 300°C. There are other grades of the Teflon AF family that are (or have been?) investigated for dielectrics.

Two other candidates for dielectrics were (or are?) poly(*p*-xylylene) (PPX) and poly(norbornene) (PNB) (Table 14-5). PPX is obtained by polymerizing [2.2]-*p*-cyclophane (= di-*p*-xylylene) (Volume II, p. 260). Norbornene (NB) yields two types of poly(norbornene)s (PN) and it is not clear which one was explored for dielectrics, the one from ring-opening metathesis polymerization (ROMP) or the one from Ziegler polymerization (ZN) (Volume II, p. 252 ff.).

(14-20)

Benzocyclobutene polymers (BCB; Dow Chemical) are used for integrated circuit interconnects. BCB is a spin-on dielectric that is sold as a 15-20 % xylene solution of the prepolymer from Cyclotene (p. 600). The monomer and prepolymer, respectively, cyclodimerize to the polymer at temperatures above 170°C. BCB has a $\kappa = 2.65$, no weight loss at 300°C, and a glass temperature of at least 350°C which is too low for chip production.

SiLK® (*silicon application low-K* material); Dow Chemical) is sold as a 15-20 % solution of a prepolymer in *N*-methylpyrrolidone. The prepolymer results from a Diels-Alder reaction of a bifunctional monomer I (Ph = phenyl) with a trifunctional monomer III to a trifunctional intermediate IM, Eq.(14-21), that on curing reacts to the branched prepolymer and then to a crosslinked, closed pores ($d < 2$ nm) containing polymer. SiLK® has excellent properties (Table 14-5) and is thermally stable at temperatures up to at least 450°C. These properties come at a price, though: $2000 buys 1 liter of the prepolymer solution from which ca. 95 % gets lost during the spin-on process!

(14-21)

14.3 Electrical Conductors

14.3.1 Electrical Conductivity

The electrical conductivity σ of a *material* is controlled by the number N of charge carriers per volume V, the elementary charge e, and the mobility μ of the carriers:

(14-22) $\sigma = (N/V)\,\mu\,e$

In *metals*, electricity is carried by (quasi)free electrons which are present in concentrations of 10^{21}-10^{22} electrons/cm^3 with mobilities of 10-10^6 cm^2/(V s). Electrical conductivities of metals are ca. 10^2-10^4 S/cm (Fig. 14-9); they decrease with increasing temperature.

In contrast, electrical conductivities of *semiconductors* (Fig. 14-9) increase with increasing temperature. Their carriers of electrical conductivity are either ions or electrons. However, the conductivity of semiconductors by electrons differs from that of metals where electrons are (quasi)free and can jump from the valence band into the conduction band without being activated. Semiconductors need an activation energy to do so; the crossing also creates "**holes**" in the valence band. These holes are treated mathematically as if they are additional, positive charge carriers, called **defect electrons**. Hence, electrical conductivities of semiconductors arise from electrons (*n*egative charge carriers; n-conductivity) or from defect electrons (*p*ositive charge carriers; p-conductivity).

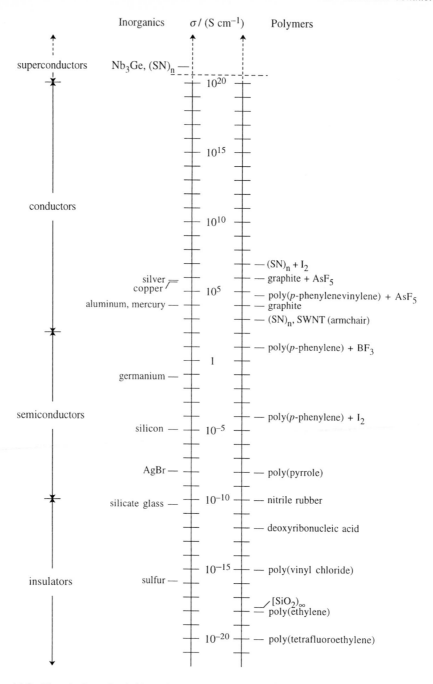

Fig. 14-9 Electrical conductivities of some elements, inorganic compounds, and inorganic and organic polymers at 20°C. Triniobium germanide Nb_3Ge and poly(azasulfene) $(SN)_n$ become superconducting with $\sigma > 10^{20}$ S/cm at temperatures near zero kelvin (near −273.15°C) whereas $Tl_2Ca_2Ba_2Cu_3O_{10+x}$ is superconducting at temperatures below −148°C. SWNT = single-wall carbon nanotube. Quartz = $[SiO_2]_\infty$, is a tecto polymer. At room temperature, sulfur consists mainly of cyclic octasulfur. Nitrile rubber is a copolymer from the free-radical copolymerization of acrylonitrile and butadiene.

Mobilities of charge carriers, $\mu/(cm^2\ V^{-1}\ s^{-1})$ in semiconductors correspond to those of metals. For n-conductors, mobilities vary between 150 in gallium phosphide (GaP) and 77 000 in indium antimonide (InSb). They are somewhat lower for p-conductors: between 60 in silicon carbide (SiC) and 1900 in germanium (Ge).

However, concentrations of charge carriers in semiconductors are much smaller than those in metals since the carriers are either equal amounts of electrons and defect elec- trons (high activation energies) or crystal defects (low activation energies). These defects are either "naturally" present or are generated artificially by doping semiconductors with foreign atoms.

In metals and semimetals such as germanium, atoms are packed fairly densely and charges are relatively easy to transfer. However, distances are fairly large between low- molecular weight organic molecules in crystal lattices. Furthermore, these lattices are only held together by weak van der Waals forces.

In polymer chains, strong covalent bonds exist in the chain direction but only weak van der Waals forces in all other directions. In addition, electrons in single bonds are all delocalized. Therefore, substances consisting of common polymer molecules have only very low electrical conductivities of less than ca. 10^{-10} S/cm: they are insulators. How- ever, there are some polymers that are semiconductors or conductors or become (semi)- conducting on doping (Section 14.3.3).

14.3.2   Electrically Conducting Polymers

Conventional polymers can be made electrially conducting by addition of metal pow- ders or carbon blacks. Such treatments prevent electrical charging (Section 14.2.7) and electromagnetic interferences.

For years, graphite was the only known polymer with an intrinsic electrical conducti- vity. Graphite consists of stacked layers of graphenes (see next page) which are large two-dimensional arrays of an(n)ellated 6-membered carbon rings with conjugated double bonds (L: *anellus* = small ring, from *anus* = ring (*annus* = year)). Perpendicular to the graphene plane, graphenes are bonded by van der Waals forces. Because of this difference in bonding, the electrical conductivity of graphite in the lattice plane is ca. 10^4 S/cm (similar to mercury) but only 1 S/cm perpendicular to it.

Similarly, single-wall carbon nanotubes (SWNT) also have an electrical conductivity of 10^4 S/cm if their carbon rings are in the so-called armchair arrangement (Fig. 14-10). However, the electrical conductivity drops to 10^2–10^{-9} S/cm, depending on the diameter of the carbon tubes, if the carbon rings are in the so-called zigzag arrangement (Volume II, p. 208). There are also SWNTs where the carbon rings are arranged at other angles, carbon nanotubes with more than one wall, etc.

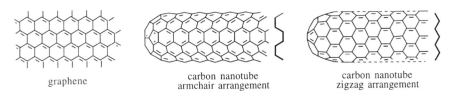

| graphene | carbon nanotube armchair arrangement | carbon nanotube zigzag arrangement |

Fig. 14-10 Graphene and carbon nanotubes (CNT).

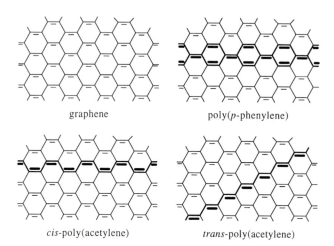

graphene   poly(*p*-phenylene)

cis-poly(acetylene)   *trans*-poly(acetylene)

Fig. 14-11 Top left: lattice structure of graphene (= monolayer of graphite molecules) (Volume II, p. 205). All others: projections of some conjugated polymer molecules on the graphene plane.

The name "poly(acetylene)" of the compounds of Fig. 14-11 refers to the origin of these polymers (= polymers of acetylene, HC≡CH). The constitution of the poly(acetylene)s of Fig. 14-11 is that of a poly(vinylene) with the monomeric unit –CH=CH–; the monomeric unit of a true poly(acetylene) would be –C≡C–. The *cis*-poly(acetylene) of Fig. 14-11 is the Z-trans form ("cis"-transoid); the *trans*-poly(acetylene) is the E-trans form ("trans"-transoid). In the "trans"-cisoid from (= E-cis form) (not shown), chains are arranged on the graphene lattice like the chains of *cis*-poly(acetylene) except that the double bonds are on the inclined graphene bonds and not on the horizontal ones.

Polymer chains with conjugated double bonds can be viewed as partial graphenes; examples are *cis* and *trans*-poly(acetylene) and poly(*p*-phenylene) (Fig. 14-11). Indeed, films of trans-poly(acetylene) have a black-silvery shine just like graphite. In infinitely long polymer chains with conjugated π-bonds, charges could be evenly distributed, which should result in a very mobile π electron cloud.

Very many different polymers have been synthesized in the search for electrically conducting polymers; simple examples are shown in Figs. 14-12 and 14-13 (see also examples in Figs. 15-12 to 15-14, and pp. 641-642). Most of these polymers turned out to be electrical conductors or semiconductors only after doping (Table 14-6; see Section 14.3.3).

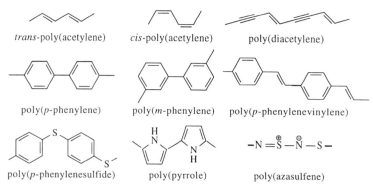

trans-poly(acetylene)   *cis*-poly(acetylene)   poly(diacetylene)

poly(*p*-phenylene)   poly(*m*-phenylene)   poly(*p*-phenylenevinylene)

poly(*p*-phenylenesulfide)   poly(pyrrole)   poly(azasulfene)

Fig. 14-12 Examples of dimeric units of unsubstituted polymers. According to X-ray studies, rings of poly(pyrrole) are arranged alternately.

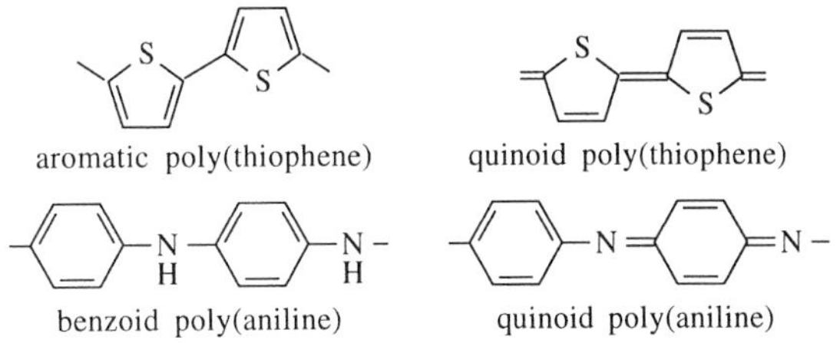

aromatic poly(thiophene) quinoid poly(thiophene)

benzoid poly(aniline) quinoid poly(aniline)

Fig. 14-13 Dimeric units of poly(thiophene) (PTP) and of poly(aniline) (PANI).

It is not always clear whether the monomeric or dimeric units of all of the polymers of Fig. 14-12 have indeed all the ideal structures ascribed to them or whether there are constitutional mistakes along the chain. Some polymers have distinct isomeric units. Poly(thiophene) can exist in an aromatic or a quinoid form with the same chemical composition (Fig. 14-13). Oxidation of benzoid poly(aniline) (B) leads to a quinoid form (Q), so that there is leucoemeraldine with the composition $(BB)_n$, emeraldine base with $(BQ)_n$, and pernigraniline with $(QQ)_n$ (see also Volume II, p. 450 ff.).

According to the theory of electrical conductivity, linear poly(*p*-phenylene) and polyenes such as the poly(acetylene)s are compounds with filled valence bands. In neutral polymers, charge carriers can therefore be generated only by considerable activation energies. Correspondingly, *trans*-poly(acetylene) is only a semiconductor ($\sigma \approx 10^{-9}$ S/cm) and poly(*p*-phenylene) even an electrical insulator ($\sigma \approx 10^{-15}$ S/cm) (Table 14-6).

The situation changes on introduction of large proportions (up to ca. 20 wt%) of positive or negative charge carriers, called **dopants** (see examples in Table 14-6). An example is the series of poly(aniline)s where emeraldine hydrochloride at pH = 5-8 is an electrical insulator with an electrical conductivity of ca. 10^{-10} S/cm. However, the conductivity rises 10 decades with increasing acidity to ca. 1 S/cm at pH = 1. The emeraldine salt of (±)-camphor sulfonic acid is even a weak conductor ($\sigma = 200$ S/cm). Such doped poly(aniline)s serve for many electronic applications including light-emitting diodes, rechargeable batteries, biosensors, etc.

Table 14-6 Guide values for electrical conductivities of some unsubstituted polymers at 25°C before and after doping. LiNp = lithium naphthalide; ⊕ positive charge carrier (p-dopant); ⊖ = negative charge carrier (n-dopant). For chemical constitutions of polymers, see Figs. 14-12 and 14-13. Literature data usually do not indicate dopant concentrations (for the strong effect of concentration, see Fig. 14-14).

| Polymer | | undoped | $\sigma/(S\ cm^{-1})$ after doping with | | | |
			AsF$_5$ ⊕	I$_2$ ⊕	BF$_3$ ⊕	LiNp ⊖
PPS	poly(*p*-phenylene sulfide)	10^{-16}	10	0.0001		
PPP	poly(*p*-phenylene)	10^{-15}	500	0.0001	10	5
PMP	poly(*m*-phenylene)		0.001			
PTP	poly(thiophene)	10^{-11}	0.02	6		
PAC	*trans*-poly(acetylene)	10^{-5}-10^{-4}	1200	500	100	200
PAC	*cis*-poly(acetylene)	10^{-10}-10^{-9}	1200	160		
PPY	poly(pyrrole)	10^{-8}	100	100	100	
PAS	poly(azasulfene)	$2\cdot10^3$		40 000		
C	graphite	10^4	1 000 000			

A high intrinsic electrical conductivity without any doping is shown only by poly(azasulfene), $+N=S^{\oplus}-N^{\ominus}-S+_n$, also known as poly(sulfurnitride), poly(thiazyl), or SNX polymer (Volume II, p. 585). In this polymer, every other $-SN-$ unit is intrinsically charged. In contrast to other polymers, electrical conductivities of this polymer increase with decreasing temperature until it becomes a superconductor at a temperature of ca. 0.3 K. However, these high electrical conductivities are observed only for single crystals of $+SN+_n$ which have the bad habit of exploding suddenly.

14.3.3 Doping

As mentioned above for emeraldine hydrochloride, electrical conductivities σ of conjugated polymers such as poly(acetylene), poly(thiophene), poly(pyrrole), etc., increase considerably if they are treated with dopants (Table 14-6). σ increases with increasing temperature and proportion of the dopant, first strongly and then more slowly (Fig. 14-14). At room temperature and fairly large concentrations of dopants, the resulting electrical conductivities are in the same range as those of good inorganic semiconductors.

In contrast to inorganic semiconductors, dopants do not produce lattice defects in conjugated polymers. Doping rather reduces polymers (adds an electron; negative charge (n-doping)) or oxidizes them (removes an electron; positive charge (p-doping)). p-Dopants are I_2, AsF_5, BF_3, PF_5, $FeCl_3$, CF_3COOH, $(CF_3)_2SO_4$, and H_2SO_4; lithium naphthalide is an n-dopant. n-Doping is thermodynamically favorable; it proceeds spontaneously. p-Doping requires energy input.

Even non-conjugated polymers can be doped. Treating non-conjugated *cis*-1,4-poly-(isoprene), $+CH_2-C(CH_3)=CH-CH_2+_n$, with iodine causes the polymer to become black and its electrical conductivity to increase from ca. 10^{-13} S/cm to 10^{-2}-10^{-1} S/cm. Raman spectroscopy indicates the presence of large proportions of $[I_3]^{\ominus}$, probably cyclic iodonium ions. The electrical conductivity of iodine-doped poly(isoprene) may result from electron or ion transport via poly(iodine) chains similar to those in amylose-iodine complexes where they cause the blue color.

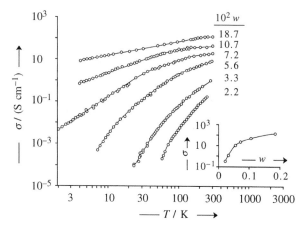

Fig. 14-14 Temperature dependence of the electrical conductivity σ of iodine-doped *cis*-poly(acetylene) [4]. Insert: conductivity at 300 K as a function of the weight fraction w of iodine.
By kind permission of Hanser Publishers, Munich.

Upon doping, polymers change their structure. For example, doping of the vinylene groups of poly(vinylene) with AsF_5 leads to carbenium ions,

$$(14\text{-}23) \qquad \text{wCH=CH—CH=CHw} \xrightarrow[- AsF_3, - 2 [AsF_6]^{\ominus}]{+ 3 AsF_5} \text{wCH}^{\ominus}\text{—CH=CH—CH}^{\ominus}\text{w}$$

whereas the reaction of poly(thiophene) with AsF_5 results in benzothiophene groups. The occurrence of chemical reactions explains why large proportions of dopants have to be used for polymeric semiconductors, in contrast to those for inorganic ones. Equilibria are obtained only after long times because diffusions of dopants into polymers are slow processes. For example, it took four days for AsF_5 to diffuse into poly(*p*-phenylene sulfide) before the final value of the electrical conductivity was reached.

14.3.4 Solitons

According to electron spin resonance spectroscopy, undoped *trans*-poly(acetylene)s contain ca. 1 quasi-free electron per 3000 carbon atoms. Doping provides far higher concentrations of charged entities; for example, carbenium ions in poly(acetylene), Eq.(14-23). However, the resulting electrically conducting polymers do not show **Curie paramagnetism** (caused by localized charge carriers) or **Pauli paramagnetism** (electrons delocalized over the whole system). These experimental results led to the hypothesis that the conductivity of doped poly(acetylene) is caused by solitons.

In physics and mathematics, a **soliton** is a self-reinforcing solitary wave that maintains its shape while it travels at constant speed. This feature was discovered about 150 years ago when the English engineer John Russell Scott observed a sole wave which rolled several kilometers in a narrow channel without changing its wave nature (L: *solus* = single, alone). The curious stability of the shape of this wave was caused by the peculiar movement itself; it is mathematically described by a set of non-linear differential equations. An (in)famous example of a soliton is a tsunami.

In polymer materials science, a **soliton** is a cation, radical, or anion at a topological **kink** that separates a state A from a state B with a reverse bond alternation (Fig 14-15).

positively charged soliton   neutral soliton   negatively charged soliton

Fig. 14-15 Solitons with positive charge (cation), neutral charge (radical), and negative charge (anion). A soliton is permanent, localized, and able to interact with other solitons.

Solitons can be formed only by polymers with degenerate ground states, i.e., two identical resonance structures with the same energy. The only known polymer with this structure is *trans*-poly(acetylene) (Fig. 14-16). For poly(pyrrole), see next page.

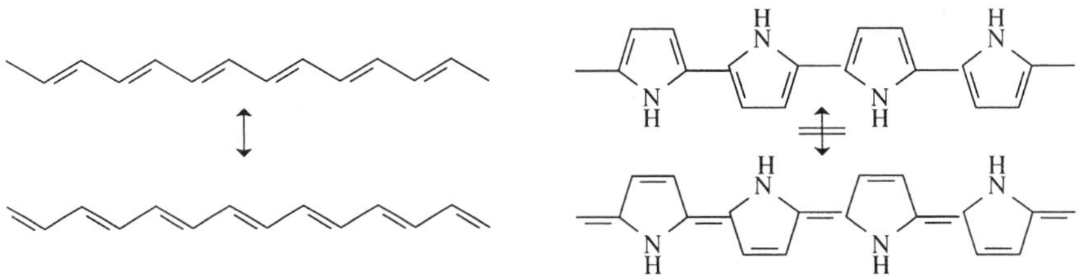

Fig. 14-16 Resonance structures of *trans*-poly(acetylene) (left) and poly(pyrrole) (right).

Poly(pyrrole) and other polymers with conjugated double bonds (except poly(acetylene)) do not form solitons because the two possible resonance structures are not identical. Other examples are poly(*p*-phenylene) (compare Figs. 14-11 and 14-12), poly(thiophene) (Fig. 14-13), and poly(aniline)s (Fig. 14-13). Polymers without identical resonance structures rather form polarons and bipolarons (Section 14.3.5).

Neutral solitons are free radicals which appear as defects during the isomerization of the polymers. Charged solitons are produced by doping; they may be carbenium ions or carbanions. Formations of solitons disturb planar structures, i.e., they produce kinks.

Kinks are small (Fig. 14-15). They are very mobile and may therefore show up anywhere in the chain. The high mobility of the resulting defects allows them to travel along the chain as a solitary wave, i.e., a soliton (Fig. 14-17).

Fig. 14-17 Travel of a neutral soliton along a *trans*-poly(acetylene) chain.

The soliton travels only along the chain but does not jump to other chains. Its movement is therefore anisotropic; there is no tunnel effect between the chains. Diffusion coefficients of solitons are thus one-dimensional; in *trans*-poly(acetylene), they are $3.8 \cdot 10^{14} \, s^{-1}$ in the chain direction but only $1.1 \cdot 10^{9} \, s^{-1}$ perpendicular to the chain. Similar effects are also found for the two-dimensional electrical conductivity of graphite: the ratio $\sigma_{\parallel}/\sigma_{\perp}$ of the conductivities in the graphite plane and perpendicular to the plane is ca. 10^4 before doping and 10^6 after doping with SbF_5.

Figs. 14-15 and 14-17 show solitons as localized entities. However, solitons may be delocalized over several bonds but not over the entire chain. According to theoretical calculations, a soliton in *trans*-poly(acetylene) extends over ca. 14 chain bonds. The charges are delocalized: 85 % of the charges are distributed over 15 % of the atoms. A molecule with an electron cloud that is evenly distributed over the whole chain would be unstable (**Peierls distortion**).

14.3.5 Polarons and Bipolarons

In crystal physics, a polaron is a quasiparticle consisting of an ion and an electron. Polarons are formed by introducing an electron into the lattice of a perfect ion crystal and applying an electric field. The electric charge of the traveling electron distorts the lattice by pushing back the neighboring lattice electrons, which in turn leads to an attraction between the negative charge of the introduced electron and the positive charges of the surrounding atomic nuclei. The resulting polarization cloud makes the introduced electron heavier so that it can be treated theoretically as a quasiparticle, i.e., a polaron.

This concept has been adopted by polymer materials science for processes that occur if conjugated polymers are doped with charge carriers (p. 606). On oxidation of a phenylene group of a neutral conjugated polymer such as poly(*p*-phenylene) by a p-dopant, an electron is split off and a radical-cation is formed. Because of the resulting distortion of the polymer chain, radical and cation do not remain in close contact on one phenylene ring but separate along the chain, for example, as shown in Fig. 14-18 for

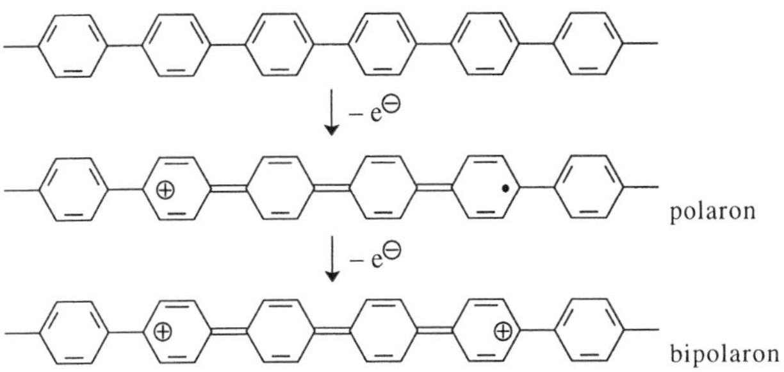

Fig. 14-18 Formation of polarons and bipolarons in oxidized poly(*p*-phenylene).

four phenylene rings. This "extended" radical-cation forms an extended resonance structure and can therefore be treated as one quasi-particle called a **polaron**.

Further oxidation of the *polaron* removes a second electron and similarly results in a
dication with two charges that are coupled over a section of the phenylene chain. These
coupled charges can be in the center of the chain as shown in Fig. 14-17 or at the ends
of the chain. Such entities are called **bipolarons**. Removal of another electron from the
uncharged polymer leads to two polarons (not shown) and not to a bipolaron.

Similarly, treatment of poly(pyrrole) with low concentrations of p-dopants leads to
paramagnetic polarons ("separated" radical-cations in an extended resonance structure).
High concentrations of p-dopants convert polarons to dications (bipolarons) which comprise about four pyrrole rings (Fig. 14-19) according to spectroscopy. This charged segment is able to travel through the whole chain by rearranging double and single bonds.

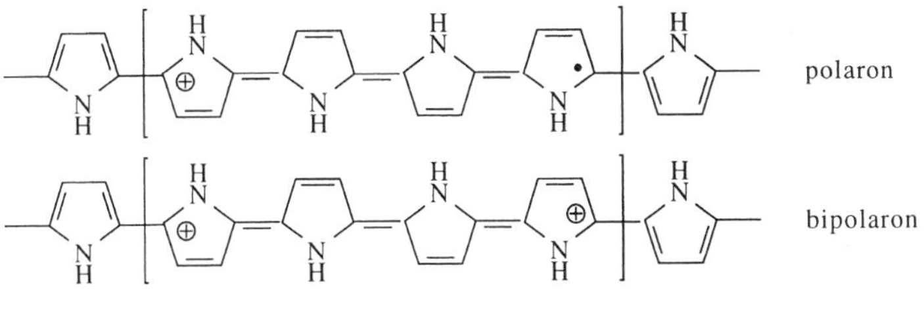

Fig. 14-19 Polarons (here a radical cation) and bipolarons (here a dication) of poly(pyrrole).

Bipolarons can also be formed from two polarons that share the same distortion of the
molecule. In this case, the two "free" electrons of the two polarons combine to one electron pair, which lowers the energy and leads to a dication, the bipolaron (Fig. 14-20).

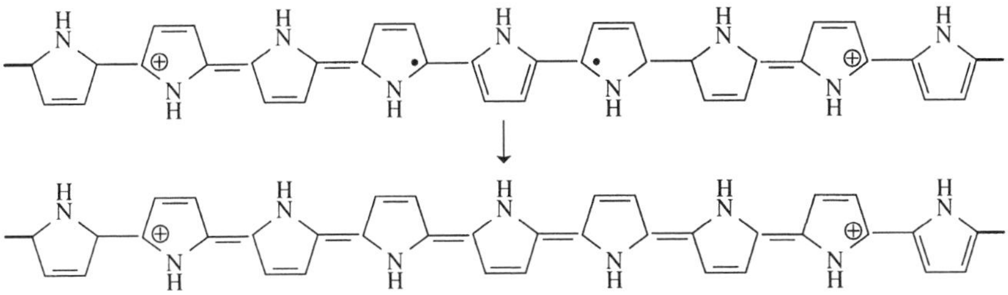

Fig. 14-20 Generation of a bipolaron (bottom) from two polarons (top) of a poly(pyrrole) molecule.

14.3.6 Organic Metals

On lowering temperature, electrical conductivities of undoped or lightly doped organic semiconductors such as poly(acetylene) decrease (Fig. 14-14) whereas those of very highly doped ones such as poly(acetylene)/$[ClO_4]^{\ominus}$ increase (not shown). Since such a behavior is typical for metals, highly doped semiconductors have been called **organic metals, synthetic metals,** or **low-dimensional conductors**.

Organic metals are not only formed by doped conjugated polymers but also by charge-transfer (CT) complexes of low-molecular weight molecules. An example is the CT complex from 7,7,8,8-*tetra*cya*no*quinodimethane (TCNQ) and *tetra*thia*ful*valene (TTF; also TFV) where alternate TCNQ and TTF molecules are stacked upon each other in a sandwich-like manner. These TCNQ/TTF complexes possess at 50-60 K the same electrical conductivity as copper at room temperature.

A system of conjugated double bonds is often a necessary but not a sufficient condition for high electrical conductivity. Poly(acetylene) and similarly structured and doped chemical compounds like poly(*p*-phenylene) and poly(*p*-phenylene sulfide) all have linear and "planar" chains. However, the phenylene rings of poly(*p*-phenylene) are at an angle of 23° to each other and poly(*p*-phenylene sulfide), $-\!\!+\!S\!-\!(p\text{-}C_6H_4)\!-\!\!+\!\!_n$, has no conjugated double bonds at all. Its phenylene rings are arranged at 45° to the planar zigzag chain of the sulfur atoms, which results in a 90° angle between phenylene rings. Despite the non-planar structure, poly(*p*-phenylene sulfide) is a fairly good electrical conductor.

The most important condition for organic polymers to conduct electricity seems to be the ability of orbitals to overlap. The p orbitals of the C=C double bonds of poly(acetylene) overlap because the polymer is planar. In poly(*p*-phenylene sulfide), the p and d orbitals of sulfur probably interact with the π system of phenylene rings. Indicative of this effect is the dramatic drop of electrical conductivity if S atoms are replaced by CH_2 groups. The same effect of chemical constitution can also be seen in the electrical conductivity of doped homo and regio copolymers (Table 14-7).

Electrically conducting polymers may replace many of the presently known batteries. For some years, a complex of poly(2-vinyl pyridine) and iodine with a conductivity of 10^{-3} S/cm has been used as the cathode in Li/I_2 batteries for pacemakers. These batteries have greater energy densities than lead acid batteries and a lifetime of ca. 10 years.

Batteries of poly(acetylene)/Li (n) and poly(acetylene)/AsF_5 (p) have greater electrical potential differences than conventional lead/lead oxide batteries (4.4 V *versus* 2.0 V) and

Table 14-7 Electrical conductivities of non-oriented, AsF_5 doped polymers and copolymers.

Monomeric units		$\sigma/(S\ cm^{-1})$ for			
–A–	–B–	$(A)_n$	$(B)_n$	$(A\text{–}B)_n$	$(A\text{–}B\text{–}B)_n$
–CH=CH–	–(p-C_6H_4)–	1200	500	3	
–(p-C_6H_4)–S–	–(p-C_6H_4)–	500	1	0.3	0.02
–(p-C_6H_4)–S–	–(p-C_6H_4)–O–	1	0.001	0.0001	0.000 001

Table 14-8 Theoretical and experimental gravimetric charge densities ρ_m, experimental gravimetric energy densities ε_m, and life cycles N_c of various systems.

System	$\rho_m/(A\ h\ kg^{-1})$		$\varepsilon_m/(W\ h\ kg^{-1})$	N_c
	Theory	Experiment	Experiment	Experiment
Graphite oxide	650	≈30		
Poly(aniline)	300	≈100	8	500
Lead dioxide	224			
Carbon black	170			
Poly(p-phenylene)	106	70		
Poly(pyrrole)	91	85		20 000
Zinc/NiOOH			<75	600

their capacity is ca. 40 times greater at ca. 1/10 of the weight. They also do not contain lead and do not emit toxic fumes. Other possible uses of conductive polymers are as capacitors and sensors, in integrated circuits and solar cells, as electrolysis membranes, for electromagnetic interference shielding, in non-linear optics, as light-emitting diodes, nanowires, components of conductive adhesives and inks, etc. However, gravimetric charge densities of doped polymers are not very large (Table 14-8) since they can take up only 1 charge per 10 chain atoms. Low polymer densities also lead to small energy densities.

A main problem for technical applications of polymeric synthetic metals is often their lack of environmental stability. Doped films of poly(pyrrole) maintain their electrical conductivity for 200 hours at 80°C but lose it in a humid atmosphere at a rate which increases with increasing temperature (Fig. 14-21). The remedy is to embed poly(pyrrole) in a glass fiber-reinforced, cured epoxy resin where its conductivity stays constant even after 44 days at 70°C and 95 % relative humidity.

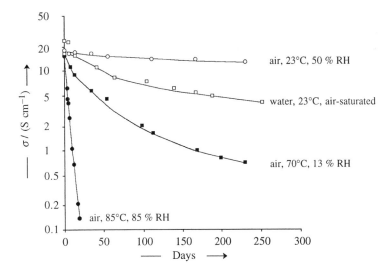

Fig. 14-21 Time dependence of the logarithm of electrical conductivity of films of doped poly(pyrrole) in air or under water at various temperatures and relative humidities [5].
By kind permission of Hanser Publishers, Munich.

Historical Notes to Chapter 14

First highly semiconducting ($\sigma = 1$ S/cm) polymer, an iodine doped and oxidized poly(pyrrole). Three papers by D.E.Weiss et al., Electronic Conduction in Polymers, Austral.J.Chem. **16**/6 (1963): Part I: R.McNeill, R.Siudk, J.H.Wardlaw, D.E.Weiss, p. 1056; Part II: B.A.Bolto, D.E.Weiss, p. 1076; Pt. III: B.A.Bolto, R.McNeill, D.E.Weiss, p. 1090

First highly conductive organic charge-transfer complex from tetrathiafulvalene (TTF) and tetracyanoquinodimethane (TCNQ): P.W.Anderson, P.A.Lee, M.Saitoh, Remarks on Giant Conductivity in TTF-TCNQ, Solid State Comm. **13** (1973) 595

The paper on iodine-doped *trans*-poly(acetylene), the first organic polymer with an electrical conductivity that approached the conductivity of silver and earned its three principal authors (Heeger, McDiarmid, Shirakawa) the Nobel Prize in 2000: C.K.Chiang, M.A.Druy, S.C.Gau, A.J.Heeger, E.J.Louis, A.G.McDiarmid, Y.W.Park, H.Shirakawa, Synthesis of Highly Conductive Films of Derivatives of Polyacetylene, (CH)x, J.Am.Chem. Soc. **100** (1978) 1013

See also: A.G.McDiarmid, Twenty-Five Years of Conducting Polymers, Chem.Commun. (2003) 1

Literature to Chapter 14

14.1 OVERVIEWS
A.R.Blythe, Electrical Properties of Polymers, Cambridge University Press, Cambridge 1979
D.A.Seanor, Ed., Electrical Properties of Polymers, Academic Press, New York 1982
W.Tiller Shugg, Handbook of Electrical and Electronic Materials, Van Nostrand Reinhold, New York 1986
C.C.Ku, R.Liepins, Electrical Properties of Polymers. Chemical Principles, Hanser, Munich 1987
J.I.Kroschwitz, Ed., Electrical and Electronic Properties of Polymers (= reprints of sections in Encyclopedia of Polymer Science and Engineering), Wiley, New York, 2nd ed. 1988
E.Riande, R.Diaz-Calleja, Electrical Properties of Polymers, Dekker, New York 2004

14.2 INSULATORS
P.Hedvig, Dielectric Spectroscopy of Polymers, Halsted, New York 1977
A.Bradwell, Ed., Electrical Insulation, Peregrinus, Stevenage, Herts, United Kingdom 1983
G.Heinicke, Tribochemistry, Hanser, Munich 1984
E.Riande, E.Saíz, Dipole Moments and Birefringence of Polymers, Prentice Hall, Englewood Cliffs (NJ) 1992
L.A.Dissado, J.C.Fothergill, Electrical Degradation and Breakdown in Polymers, The Institution of Electrical Engineers, Stevenage (UK) 1992; Peregrinus, Stevenage, Herts (UK) 1992
S.J.Havriliak, S.Havriliak, Jr., Dielectric and Mechanical Relaxation in Materials, Hanser, Munich 1996
J.P.Runt, J.F.Fitzgerald, Ed., Dielectric Spectroscopy of Polymeric Materials: Fundamentals of Application, Am.Chem.Soc., Washington (DC) 1997
B.Bhushan, Modern Tribology Handbook, CRC Press, Boca Raton (FL) 2000

14.2.8 ELECTROPHOTOGRAPHY
R.M.Schaffert, Electrophotography, Focal Press, London, 2nd ed. 1975
M.Stolka, D.M.Pai, Polymers with Photoconductive Properties, Adv.Polym.Sci. **29** (1978) 1
M.E.Scharfe, Electrophotography. Principles and Optimization. Research Studies Press, Letchwood, UK 1984
E.M.Williams, The Physics and Technology of Xerographic Processes, Wiley, New York 1984
M.Biswas, T.Uryu, Recent Advances in Photoconductive and Photosensitive Polymers, J.Macromol.Sci.-Revs.Macromol.Chem.Phys. **C 26** (1986) 249
L.B.Schein, Electrophotography and Development Physics, Springer, Berlin 1987
W.Gerhartz, Ed., Imaging and Information Storage Technology, VCH, Weinheim 1992

14.2.9 ELECTRETS
M.M.Perlman, Ed., Electrets, Charge Storage, and Transport in Dielectrics, Electrochem.Soc., Princeton (NJ) 1973
G.M.Sessler, Ed., Electrets (= Topics in Applied Physics **33**), Springer, Heidelberg 1987

14.2.10 PYRO, PIEZO, AND FERROELECTRICS
R.G.Kepler, Piezoelectricity, Pyroelectricity, and Ferroelectricity in Organic Materials, Ann.Rev.Phys.Chem. **29** (1979) 497
R.G.Kepler, R.A.Anderson, Piezoelectricity in Polymers, CRC Crit.Revs.Solid State and Materials Sci. **9** (1980) 399
G.W.Taylor, J.J.Gagnepain, T.R.Meeker, T.Nakamura, L.A.Shuvalov, Ed., Piezoelectricity, Gordon and Breach, New York 1985
J.M.Herbert, T.T.Wang, A.M.Glass, Eds., Applications of Ferroelectric Polymers, Blackie, Glasgow 1988
T.Ikeda, Fundamentals of Piezoelectricity, Oxford Univ.Press, Oxford 1990
H.S.Nalwa, Ferroelectric Polymers. Chemistry, Physics and Applications, Dekker, New York 1995

14.2.11 OPTICAL LITHOGRAPHY
S.S.DeForest, Photoresists. Materials and Processes, McGraw-Hill, New York 1975
D.J.Elliot, Integrated Circuit Fabrication Technology, McGraw-Hill, New York 1982
W.Schnabel, H.Sotobayashi, Polymers in Electron Beam and X-Ray Lithography, Progr.Polym. Sci. **9** (1983) 297
M.T.Goosey, Ed., Plastics for Electronics, Elsevier, London 1985
F.A.Voellenbroek, E.J.Spiertz, Photoresist Systems for Microlithography, Adv.Polym.Sci. **84** (1988) 85
W.M.Moreau, Semiconductor Lithography, Plenum, New York 1988
J.H.Lai, Polymers for Electronic Applications, CRC Press, Boca Raton (FL), 1989
M.S.Htoo, Ed., Microelectronic Polymers, Dekker, New York 1989
A.Reiser, Photoreactive Polymers, The Science and Technology of Resists, Wiley, New York 1989
R.R.Tummala, E.J.Rymaszewski, Ed., Microelectronics Packaging Handbook, Van Nostrand Reinhold, New York 1989
D.S.Soane, Z.Martynenko, Polymers in Microelectronics: Fundamentals and Applications, Elsevier, New York 1989
L.T.Manzione, Plastic Packaging of Microelectronic Devices, Van Nostrand Reinhold, New York 1990
Y.Tabata, I.Mita, S.Nonogaki, K.Horie, S.Tagawa, Polymers for Microelectronics - Science and Technology, Kodansha, Tokyo 1990; VCH, Weinheim 1990
C.P.Wong, Ed., Polymers for Electronic and Photonic Applications, Academic Press, San Diego (CA) 1992
R.Dammel, Diazonaphthoquinone-based Resists, Optical Eng.Press, Bellingham (WA) 1993
L.F.Thompson, C.Grant Willson, M.J.Bowden, Eds., Introduction to Microlithography, Am.Chem.Soc., Washington (DC), 2nd ed. 1994
L.F.Thompson, C.Grant Wilson, S.Tagawa, Eds., Polymers for Microelectronics, Resists and Dielectrics, Am.Chem.Soc., Washington (DC) 1994
P.Rai-Choudhury, Ed., Handbook of Microlithography, Micromachining and Microfabrication, Optical Eng.Press, Bellingham (WA) 1997
R.C.Jaeger, Lithography, Introduction to Microelectronic Fabrication, Prentice Hall, Upper Saddle River (NJ) 2002
H.-B.Sun, S.Kawata, Two-Photon Photopolymerization and 3D Lithographic Microfabrication, Adv.Polym.Sci. **170** (2004) 169
H.Ito, Chemical Amplification Resists for Microlithography, Adv.Polym.Sci. **172** (2005) 37

14.2.12 LOW κ DIELECTRICS
R.C.Jaeger, Thermal Oxidation of Silicon, Prentice Hall, Upper Saddle River (NJ) 2001
D.M.Dobkin, M.K.Zuraw, Principles of Chemical Vapor Deposition, Kluwer, Norwell (MA) 2003
S.P.Murarka, M.Eizenberg, A.K.Sinha, Eds., Interlayer Dielectrics for Semiconductor Technologies, Elsevier/Academic Press, Amsterdam 2003
M.R.Baklanov, K.Maex, M.Green, Eds., Dielectric Films for Advanced Microelectronics, Wiley, Hoboken (NJ) 2007

14.3 ELECTRICAL CONDUCTORS

W.E.Hatfield, Ed., Molecular Metals, Plenum, New York 1980

J.Mort, G.Pfister, Eds., Electronic Properties of Polymers, Wiley, New York 1981

H.Kuzmany, M.Mehring, S.Roth, Electronic Properties of Polymers and Related Compounds, Springer, Berlin 1985

M.T.Goosey, Ed., Plastics for Electronics, Elsevier, London 1985

R.R.Chance, D.Bloor, Eds., Polydiacetylenes, Nijhoff, Netherlands 1985

J.M.Margolis, Ed., Conductive Polymers and Plastics, Chapman and Hall, New York 1989

N.C.Billingham, P.D.Calvert, Electrically Conducting Polymers. A Polymer Science Viewpoint, Adv.Polym.Sci. **90** (1989) 1

W.R.Salaneck, I.Lundstrøm, B.Rånby, Eds., Conjugated Polymers and Related Materials, Oxford Univ.Press, New York 1993

J.A.Chilton, M.T.Goosey, Eds., Special Polymers for Electronics and Optoelectronics, Chapman and Hall, London 1995

S.Roth, One-Dimensional Metals–Physics and Materials Science, VCH, Weinheim 1995 (includes diamonds, graphite, poly(diacetylene), poly(cumulene))

G.Schopf, G.Kossmehl, Polythiophenes–Electrically Conductive Polymers, Adv.Polym.Sci. **129** (1996) 1

G.G.Wallace, G.M.Spinks, P.R.Teasdale, Conductive Electroactive Polymers, Intelligent Materials Systems, Technomic, Lancaster (PA) 1996 (emphasis on poly(pyrrole)s, poly(aniline)s, and poly(thiophenes))

H.S.Nalwa, Handbook of Organic Conductive Molecules and Polymers, Wiley, New York 1997 (4 vols.)

K.Müllen, G.Wegner, Eds., Electronic Materials: The Oligomer Approach, Wiley-VCH, Weinheim 1998

D.L.Wise, G.E.Wnek, D.J.Trantolo, T.M.Cooper, J.D.Gresser, Eds., Electrical and Optical Polymer Systems. Fundamentals, Methods, and Applications, Dekker, New York 1998

L.Rupprecht, Conductive Polymers and Plastics in Industrial Applications, Plastic Design Library, Norwich (NY) 2000

B.D.Malhotra, Handbook of Polymers in Electronics, RAPRA, Shawbury, Shrewsbury, Shropshire, UK 2002

W.R.Salaneck, K.Seki, A.Kahn, J.-J.Pireaux, Eds., Conjugated Polymer and Molecular Interfaces: Science and Technology for Photonic and Optoelectronic Application, Dekker, New York 2002

D.R.Gamota, P.Brazis, K.Kalyanasundaram, J.Zhang, Eds., Printed Organic and Molecular Electronics, Kluwer, New York 2004

Th.Dauxois, M.Peyrard, Physics of Solitons, Cambridge Univ.Press, New York 2006

T.B.Singh, N.S.Sariciftci, Progress in Plastic Electronic Devices, Ann.Rev.Mater.Res. **36** (2006) 199

F.Cataldo, Ed., Polyynes. Synthesis, Properties, and Applications, CRC Press, Boca Raton (FL) 2006

Anon., Emissive Materials - Nanomaterials, Adv.Polym.Sci. **199** (2006)

G.Hadziioannou, G.G.Malliaras, Eds., Semiconducting Polymers. Chemistry, Physics and Engineering, Wiley-Interscience, Weinheim, 2nd ed., Vol. 1 (2007)

T.A.Skotheim, J.R.Reynolds, Eds., Handbook of Conducting Polymers, CRC Press, Boca Raton (FL), 3rd ed. 2007 (2 vols.)

M.S.Freund, B.A.Deore, Self-Doped Conducting Polymers, Wiley, Hoboken (NJ) 2007

U.Scherf, D.Neher, Eds., Polyfluorenes, Adv.Polym.Sci. **212**, Springer, Berlin 2008

W.Barford, Electronic and Optical Properties of Conjugated Polymers, Oxford University Press, New York, 1st ed. (2005), 2nd ed. (2009)

References to Chapter 14

[1] H.-J.Mair, G.Zieschank, G.Hegemann, H.Janssen, Ullmann's Enzyklopädie der technischen Chemie, VCH, Weinheim, 4th ed., Vol. **15** (1978), expanded Table 1 (p. 451-452) and Table 8 (p. 479)

[2] Y.Ishida, M.Matsuo, K.Yamafuji, Kolloid-Z. **180** (1962) 108, Figs. 1-6

[3] Y.Ishida, O.Amano, M.Takayanagi, Kolloid-Z. **172** (1960) 129, data of Fig. 5

[4] S.Roth, K.Menke, Kunststoffe **73** (1983) 521, Fig. 4

[5] W.Sauerer, Kunststoffe **81** (1991) 694, Fig. 2

15 Polymers in Optics and Optoelectronics

Optics originally studied the generation, transmission, and detection of visible light (G: *optos* = visible) but includes now also ultraviolet and infrared light, i.e., electromagnetic radiation with wavelengths λ greater than X-rays and smaller than microwaves (Fig. 15-1). **Geometrical optics** treats light as linearly traveling rays whereas **physical optics** studies phenomena that depend on the wave nature of light which has a frequency of v and a speed of $c = c_0/n$ where c_0 = speed in vacuum and n = refractive index.

Two main groups of optical properties can be distinguished if light beams encounter matter and materials: those that depend on averages of molecular properties and those that arise from the deviation of local properties from these averages. The first group leads to phenomena that are described by geometrical optics: refraction, reflection, gloss, absorption, transparency, haze, birefringence, and irisdescence (Section 15.1). The second group is studied by physical optics; it comprises diffraction, interference, and polarization. Since this book is not concerned with matter *per se* but with properties of polymeric *materials*, only opacity and hiding power will be discussed (Section 15.2).

Light has the dual character of waves and particles, called photons. **Quantum optics** studies the basic science of photons whereas **photonics** refers to applied research and the development of photonic devices. **Optoelectronics** (Section 15.3) is a subfield of photonics that is concerned with the use of photons in electronic devices. Hence, it is also a subfield of **electro-optics** which studies interactions of light and *all* electric fields.

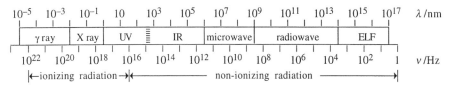

Fig. 15-1 Wavelengths λ and frequencies $v = c/\lambda$ of electromagnetic radiation. UV = ultraviolet, IR = infrared, ELF = extra-low frequency (overhead power lines). The range ≣ of visible light depends on the individual; it is usually 400-750 nm but sometimes 380-800 nm. $c_0 = 299\ 792\ 458$ m/s (exactly).

15.1 Geometrical Optical Properties

15.1.1 Refractive Index

A light beam, traveling with a speed v_A in a transparent medium A with a refractive index n_A, changes its speed to v_B and its direction to an angle of refraction β upon entering a transparent medium B with $n_B/n_A > 1$ by an angle of incidence α (Fig. 15-2). The ratio of the sine of these two angles is inversely proportional to the ratio of refractive indices of the two media (**Snell's law**; discovered by the Dutchman Willebrord van Roijen Snel (Snellius)). The **absolute refractive index** of medium B, n_B, is the ratio of the speed of the electromagnetic radiation *in vacuo*, v_0, to the speed v_B in the medium B. Relative refractive indices refer to two media. In general for media A and B (see Fig. 15-2):

(15-1) $v_A/v_B = \sin \alpha/\sin \beta = n_B/n_A$; $\alpha = \alpha'$, $\beta = \beta'$ ($n_A \equiv 1$ for $v_A = v_0$)

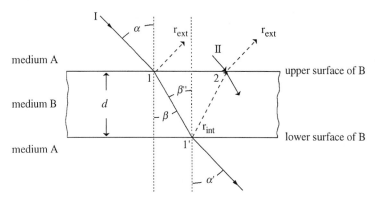

Fig. 15-2 Definition of angle of incidence, α, and angle of refraction, β, of a light beam I falling on a plane-parallel sheet (medium B) of thickness d in vacuum (medium A) with refractive index $n_A = 1$. —— Paths of refracted light (this Section), – – – paths of reflected light (see Section 15.1.7).

The variation of refractive indices n_B of a material B with the wavelength of incident light is often described by the **Abbé** number v (Abbé dispersion) that is calculated from the refractive indices of B at wavelengths of 656.3 nm, 589.3 nm, and 486.1 nm:

(15-2) $v = (n_{589} - 1)/(n_{486} - n_{656})$

Materials can separate the colors of light more easily the smaller the refractive index. For carbon chains, refractive indices $n_{D,20}$ at a wavelength of $\lambda = 589.3$ nm (D line in the visible spectrum of sodium) and 20°C vary practically linearly with the Abbé number (Fig. 15-3). Cellulose propionate and poly(dimethylsiloxane) deviate from this line.

According to the **Lorentz-Lorenz relation**, Eq.(15-3), refractive indices n of mono-meric units of molar mass M_u and density ρ depend on the induced polarizability of these units and therefore also on the induced dipole moment μ_i, that is produced by an electric field with the field strength E_i (Eq.(15-3)).

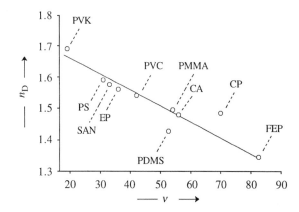

Fig. 15-3 Interrelationship between refractive index n_D at $\lambda = 589.3$ nm (D line) and the Abbé dispersion v for CA = cellulose acetate, CP = cellulose propionate, EP = epoxy resin, FEP = copolymer from tetrafluoroethylene and hexafluoropropylene, PDMS = poly(dimethylsiloxane), PMMA = poly-(methyl methacrylate), PS = poly(styrene), PVC = poly(vinyl chloride), PVK = poly(N-vinyl carb-azole), SAN = copolymer from styrene and acrylonitrile.

Table 15-1 Densities ρ and refractive indices n (at 589.3 nm) of various polymers at 25°C.
Indices: a = completely amorphous, c = completely crystalline, c,‖ = crystalline in the chain direction, c,⊥ = crystalline perpendicular to the chain direction, k = conventional plastics.

Polymer	$\dfrac{\rho_a}{\text{g cm}^{-3}}$	$\dfrac{\rho_k}{\text{g cm}^{-3}}$	$\dfrac{\rho_c}{\text{g cm}^{-3}}$	n_a	n_k	$n_{c,\perp}$	n_c	$n_{c,\parallel}$
Poly(tetrafluoroethylene)	(2.00)	2.28	2.302		1.36	1.376		1.376
Poly(dimethylsiloxane)	0.96	0.97		1.404				
Poly(propylene), it-	0.850	0.92	0.943	1.471	1.49	1.496	1.522	1.53
Poly(ethylene)	0.855	0.95	1.000		1.480	1.52	1.562	1.582
Poly(methyl methacrylate), at-	1.188	1.18		1.489	1.491			
Poly(vinyl alcohol), at-	1.265		1.340		1.490	1.505		1.55
Cellulose (n_k for flax)		1.54			1.54	1.531		1.595
Polycarbonate, bisphenol A-		1.22			1.586			
Poly(ethylene terephthalate)	1.335	1.38	1.455	1.573			1.641	
Poly(styrene), at-	1.06	1.05		1.591	1.59			
Poly(vinylidene chloride)	1.67	1.775	1.96	1.603			1.611	
Poly(N-vinyl carbazole)	1.2			1.69				

$$(15\text{-}3) \qquad \frac{n^2 - 1}{n^2 + 2} = \frac{4\pi}{3}\frac{\rho N_A}{M_u}\alpha_i = \frac{4\pi}{3}\frac{\rho N_A}{M_u}\frac{\mu_i}{E_i}$$

The polarizability of a monomeric unit is greater, the more electrons are present in its atoms and the more mobile are these electrons. Carbon atoms therefore have much larger polarizabilities than hydrogen atoms. Because contributions by hydrogen atoms can be neglected, most polymers with carbon chains have approximately the same refractive index of ca. 1.5 (Table 15-1). Deviations from this "normal" value are expected for large side groups (example: poly(N-vinyl carbazole)), strong polarizabilities (example: fluorine-containing polymers), and inorganic polymers (example: poly(dialkyl-siloxane)s). The constitution of all polymers lets one estimate that refractive indices of all organic polymers can only be in the range of 1.29-1.73.

Segments of crystalline polymers are usually more densely packed than those of amorphous ones: densities increase with increasing crystallinity (Table 15-1). Crystalline chain polymers are always anisotropic and therefore have different polarizabilities and also different refractive indices, in the chain direction (‖) and perpendicular to it (⊥). Solid polymers with different refractive indices in the three spatial directions x, y, and z are characterized by a **planarity index**, $I_{pl} = (n_x/2) + (n_y/2) - n_z$.

15.1.2 Reflection

A part of the light falling on a transparent homogeneous body is reflected at the plane of entry (external reflection (reflexion (UK)); r_{ext} in Fig. 15-2) and another part at the plane of exit (internal reflection; r_{int} in Fig. 15-2). Accordig to Eq.(15-4), the ratio of intensities of reflected light (I_r) and incident light (I_0) depends on both the angle of incidence, α, and the angle of refraction, β. The **reflectivity** $R = I_r/I_0$, is small for small angles of incidence but increases strongly at large angles (Fig. 15-4). Reflectivity is the limit of **reflectance**; reflectance of thin specimens is affected by internal reflections.

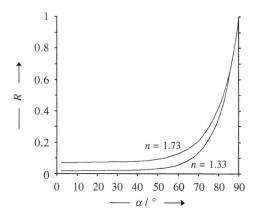

Fig. 15-4 Dependence of reflectivity R on the angle of incidence α for refractive indices of 1.33 and 1.73 according to the Fresnel equation, Eq.(15-4).

(15-4) $$R = \frac{I_r}{I_0} = \frac{1}{2}\left[\frac{\sin^2(\alpha-\beta)}{\sin^2(\alpha+\beta)} + \frac{\tan^2(\alpha-\beta)}{\tan^2(\alpha+\beta)}\right]$$ **(Fresnel equation)**

15.1.3   Gloss

Gloss, $G = R/R_{st}$, is defined as the ratio of reflectivity R of a body to that of a standard R_{st}; for example, a material with $n_D = 1.567$ in the paint industry. Because of Eq.(15-4) for R and R_{st}, gloss increases both with increasing refractive indices of specimens and with the angle of incidence (Fig. 15-5). The increase is stronger, the greater the angle of incidence: ideal-glossy surfaces reflect light preferentially in the direction of the observation angle whereas ideal-dull ones reflect light equally in all directions.

Real surfaces rarely are as glossy as calculated from Eq.(15-4). They are always somewhat rough (p. 521) and scatter light, which reduces the proportion of reflected light. An additional loss of light intensity is caused by light scattering from optical inhomogeneities in the zone directly below the surface. The relative proportion of these two types of scattered light depends on the angle of incidence. It can be obtained if the scattering is measured in air and then after immersion of the specimen in a medium with the same refractive index as the specimen itself. The difference of scattering from these two measurements gives the proportion of scattering from the surface.

Glittering occurs if different parts of the surface have different proportions of gloss. It is caused either by a stronger, directed reflection of light or by a contrast in light intensity or color between the glittering area and its environment. Glitter is sometimes desired and is produced in plastic films and shaped articles by adding metal particles or mica (p. 28, 36) to the polymer. In fibers, it can be obtained by providing the fiber with a triangular or trilobal cross-section (p. 168).

Shaped plastics articles sometimes show an undesired glitter from small spots. This effect is produced by small, uneven patches that are caused by evaporating volatile substances during injection molding. In industry, they are known as mica specks because they produce a similar effect to that caused by deliberately added mica.

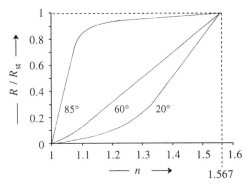

Fig. 15-5 Maximum theoretical gloss, R/R_{st}, as a function of the refractive index n of the specimen for three angles of incidence, α, as related to a standard with $n_D = 1.567$.

15.1.4 Transparency

About 10 % of all plastics are used for clear packaging films and transparent objects such as bottles and cans (Chapter 13). The most important transparent thermoplastic is probably atactic poly(styrene) (PS) and here especially the grade called crystal poly(styrene) (p. 325). Other important transparent plastics are statistical, atactic styrene copolymers with acrylonitrile (SAN), methyl methacrylate + acrylonitrile (MBS), or methyl methacrylate + acrylonitrile + butadiene (MABS); atactic block polymers of styrene and butadiene (SBS); atactic poly(methyl methacrylate (PMMA); polycarbonates (PC; mainly from bisphenol A); amorphous poly(ethylene terephthalate) (APET) and its copolyester with cyclohexane-1,4-dimethylol) (PETG); isotactic poly(propylene) (PP); atactic poly(vinyl chloride) (PVC), and isotactic poly(4-methyl-1-pentene) (PMP). Some other transparent polymers are used in considerably smaller amounts: certain types of fluorocopolymers, cycloolefin copolymers, polyamides, and thermoplastic polyurethanes. All of these polymers vary in their transmission, transparency, and haze (see below).

With the exception of clear lacquers (p. 531) and varnishes (p. 538), coatings should not be transparent at all, either for aesthetic reasons or for protection of the surface of the substrate against damage by light. Since binders of paints and coatings are usually transparent or translucent polymers, the desired hiding power is provided by suitable fillers (Section 12.2.3).

Ideal Transmission

Light falling on a body can be reflected, absorbed, scattered, and transmitted; the body may also show luminescence, i.e., the emission of light for any reason other than a rise in the temperature of the body. Since this section is concerned with transparency, luminescence (if any) is ignored and absorbance and scattering are treated as correction factors.

Light is both reflected and transmitted if it falls perpendicularly on an optically homogeneous, plane-parallel, non-absorbing, non-scattering, non-luminescing body. In this case, both the angle of incidence, α, and the angle of refraction, β, become zero, and the Fresnel equation, Eq.(15-4), reduces to

$$(15\text{-}5) \qquad R_0 = (n-1)^2 / (n+1)^2$$

The ideal (internal) transmission τ_{int} (L: *trans* = across, through, over beyond; *mittere* = to send) is the ratio of the intensity $I_{tr} = I_0 - I_r$ of the transmitted light to the intensity I_0 of the incident light where I_r = intensity of the reflected light:

$$(15\text{-}6) \qquad \tau_{int} = 1 - R_0 = \frac{I_{tr}}{I_0} = \frac{I_0 - I_r}{I_0} = 1 - \frac{I_r}{I_0}$$

According to Eq.(15-5), ideal transmissions of polymers cannot exceed values between 92.8 % (for $n = 1.73$) and 98.4 % (for $n = 1.29$) (p. 611). Polymers thus reflect between 1.6 % and 7.2 % of the incident light at the polymer–air surface.

Ideal transmission is almost never obtained since the specimen always scatters or absorbs some light. The most transparent polymer is poly(4-methyl-1-pentene) (PMP), with a refractive index of $n_D = 1.466$ (at 25°C), which should have an internal transmission of 96.4 % but has only one of ca. 90 %. Poly(methyl methacrylate) (PMMA), known as Plexiglas® or Lucite®, has a transmission of no more than 93 % instead of 96.1 % as calculated from $n_D = 1.492$ (25°C) in the range 430-1110 nm (Fig. 15-6).

Internal transmissions of PMMA decrease at wavelengths greater than ca. 1100 nm (not shown) because light is increasingly absorbed in the infrared region (780 ≤ λ_0/nm ≤ 500 000 nm). All polymers absorb light in some ultraviolet regions (5-400 nm).

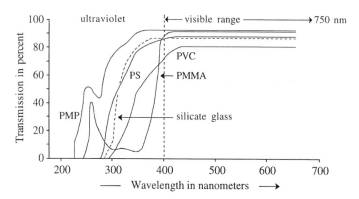

Fig. 15-6 Transmission of silicate glass, it-poly(4-methyl-1-pentene) (PMP), at-poly(methyl methacrylate) (PMMA), at-poly(styrene) (PS), and at-poly(vinyl chloride) (PVC) as a function of wavelength.

External Transmission

The brightness of light behind a transparent body (window, etc.), i.e., the **luminosity** (external transmission) of that body, is controlled not only by the ratio of the intensity I_{tr} of the transmitted light to the intensity I_0 of incident light but also by the intensity I_{fs} of light that is scattered at surfaces in the forward direction and by scattering centers within the body:

$$(15\text{-}7) \qquad \tau_{ext} = (I_{tr} + I_{fs})/I_0$$

The light behind a turbid pane with small intensity I_{tr} of transmitted light but large forward scattering intensity I_{fs} is therefore just as bright as the light behind a clear pane with large I_{tr} and small I_{fs}. However, objects such as pages of a book behind a turbid pane show less contrast and far less resolution.

Transparency and Translucency

In general, incident light with the intensity I_0 is not only transmitted with intensity I_{tr}, reflected with intensity I_r, absorbed with intensity I_a, and forward scattered with intensity I_{fs} but is also backward scattered with intensity I_{bs}. The **transparency** T (L: *trans* = through, beyond, across, over; *parere* = to show) of an object is therefore calculated as

$$(15\text{-}8) \qquad T = \frac{I_{tr}}{I_0 - I_r - I_a} = \frac{I_{tr}}{I_{tr} + I_{bs} + I_{fs}}$$

Technically, one distinguishes between transparency and **translucency** (L: *lucere* = to shine). Transparent bodies allow the passage of more than 90 % of the incident light; they appear clear even at great thicknesses. Objects with transparencies of 90 % or less are called translucent; they are clear only at small thicknesses. Such bodies are also called **contact clear** since they look turbid only in air ($n \approx 1$) but not in contact with liquids or solids ($n > 1$), for example, a shrink film in contact with packaged goods.

Haze, Milkiness, and Clarity

The translucency of a transparent material is caused by a non-negligible proportion of forward-scattered light. This **haze** H is calculated from the intensities of transmitted light (I_{tr}) and forward-scattered light (I_{fs}) as

$$(15\text{-}9) \qquad H = \frac{I_{fs}}{I_{tr} + I_{fs}} \qquad ; \qquad M = \frac{I_{fs} + I_{bs}}{I_{tr} + I_{fs} + I_{bs}}$$

In the United States, "haze" relates only to angles greater than 2.5°. The loss of clarity by the combined effects of forward and backward scattering is called **milkiness** M. The **clarity** is judged by viewing standardized, printed scales through the transparent objects (films, sheets, etc.).

Transparency of Plastics

Unfilled *amorphous* plastics are usually clear (Tables 15-2 to 15-4). Amorphous atactic poly(styrene) does not absorb visible light (Fig. 15-6) and is therefore clear if it does not contain residues from the polymerization (suspending agents, emulsifiers, etc.). Such a glass-clear polymer is the amorphous, so-called crystal poly(styrene) from the thermal polymerization of styrene.

However, the appearance of transparent polymers may be affected by light absorption. For example, the transparent, non-crystalline, atactic copolymer from styrene and acrylonitrile (SAN) absorbs light in the violet range (380-450 nm) and therefore looks

Table 15-2 Guide values for properties of rigid, transparent polymers at 23°C. PS = crystal poly-(styrene), SAN = styrene-acrylonitrile copolymer, PMMA = poly(methyl methacrylate), SMMA = styrene-methyl methacrylate copolymer, PMP = isotactic poly(4-methyl-1-pentene). Processing by extrusion (E) or injection molding (I).

Property	Physical unit	PS	SAN	PMMA	SMMA	PMP
Color, transparency	-	clear	yellowish	clear	clear	clear
Refractive index (D line)	1	1.590	1.575	1.491	1.562	1.466
Abbé number	1	30.9	35.3	57.2	34.7	56.4
Transmission, ideal	%	94.8	95.0	96.1	95.2	96.4
Transmission, actual	%	94	90	92		90
Density	g/cm^3	1.05	1.08	1.18	1.14	0.83
Linear thermal expansion coefficient	10^{-6} K^{-1}	70	70	70	65	120
Melting temperature (DSC)	°C	-	-	-	-	240
Glass temperature (DSC)	°C	100	115	113		40
Vicat temperature B	°C	90	100	105	105	179
Heat distortion temperature (1.82 MPa)	°C	80	88	100		41
Continued service temperature	°C	70	85	100		180
Tensile modulus	MPa	3200	3700	3300	3500	1800
Fracture strength	MPa	50	75	72	77	18
Fracture elongation	%	2	3	4		30
Impact strength (Charpy)	kJ/m^2	18	18	20	16	20
Notched impact strength (Izod, 3.1 mm)	J/m	93	27	24		150
(Charpy)	kJ/m^2	2	3.5	2		
Moisture uptake (24 h, 23°C, 50 % RH)	%	0.1		0.3		0.01
Processing	-	E, I	I	E, I	I	I

slightly yellowish. For the same reason, poly(vinyl chloride) (PVC) also has a yellowish tint. Since it is ca. 5 % X-ray crystalline, its light transmission is reduced from the ideal value of 95.5 % (at 25°C: n_D = 1.539) to ca. 80-85 %.

Semicrystallinity does not always reduce transparency. For example, isotactic poly(4-methyl-1-pentene) has degrees of crystallinity of 40-65 % which are caused by its lamellar structure but it is still glass clear (see also Section 15.2.2). It is even possible to convert opaque, semicrystalline poly(ethylene)s to clear films by carefully adjusting quenching and orientation.

Most transparent polymers of Tables 15-2 to 15-4 take up only a little moisture after 24 hours at 23°C and 50 % relative humidity and even after prolonged times. An exception is polyamide 6-3-T (p. 336) which contains 3.5 wt% moisture after 4 months at 23°C and a relative humidity of 50 % and 7.5 % water after being stored under water. The water acts as a plasticizer and makes the polymer more elastic: the fracture elongation doubles; Charpy tests of such notched specimens do not lead to fracture.

Some industrial applications require plastics that are not only rigid (Table 15-2) but also tough (Table 15-4). In some of these thermoplastics, impact strengths are increased by incorporation of rubber domains (PVC, PMMA, MABS) whereas in others improved impact strengths are produced by a special physical structure (example: SBS triblock polymers). However, such modified polymers are not as transparent as their matrix polymers. They are also somewhat hazy, which is especially noticeable in thicker parts (see Section 15.2.2).

Table 15-3 Guide values for properties of impact-resistant transparent polymers at 23°C. APET = amorphous poly(ethylene terephthalate), CAP = cellulose acetopropionate, PA 6-3-T = poly((2,2,4;2, 4,4)-trimethylhexamethylene terephthalamide), PC BPA = bisphenol A polycarbonate, PVC = poly-(vinyl chloride). nf = no fracture. RH = relative humidity. For processing: see Table 15-4.

Property	Physical unit	PC BPA	APET	PVC hard	CAP	PA 6-3-T dry
Color, transparency	-	clear	clear			clear
Refractive index (D line)	1	1.586	1.575	1.539	1.48	1.566
Abbé number	1	29.8				
Transmission, ideal	1	94.9	95.0	95.5	96.3	95.1
Transmission, actual	%	92			90	≤90
Density	g/cm^3	1.22	1.34	1.39	1.21	1.12
Linear thermal expansion coefficient	10^{-6} K^{-1}	68	80	75		80
Melting temperature (DSC)	°C	(230)	(255)			
Glass temperature (DSC)	°C	150	78	80		150
Vicat temperature B	°C	158	73	80		145
Heat distortion temperature (1.82 MPa)	°C	142	72			120
Continuous service temperature	°C					100
Tensile modulus	MPa	2500	2200	3000	< 2100	2800
Fracture strength	MPa	72	38	60		
Fracture elongation	%	150	300	50		70
Impact strength (Izod, 3.1 mm)	J/m	740		nf		nf
(Charpy)	kJ/m^2	nf	nf	70-nf	nf	
Notched impact strength (Izod, 3.1 mm)	J/m	960	90	4		360
Moisture uptake (24 h at 50 % RH)	%	0.15	0.16	0.25	0.15	0.26
Processing	-	E, I	B, E	E, I	I	E, I

Table 15-4 Guide values for properties of impact-modified transparent polymers at 23°C. MABS = polymer from methyl methacrylate, acrylonitrile, butadiene, and styrene (so-called transparent ABS); MBS = polymer from methyl methacrylate, butadiene, and styrene; PMMA = poly(methyl methacrylate), SBS = poly(styrene)-*block*-poly(butadiene)-*block*-poly(styrene), PVC = poly(vinyl chloride). nf =no fracture. B = blow molding, C = calendering, E = extrusion, I = injection molding.

Property	Physical unit	PVC impact	PMMA impact	SBS	MBS	MABS
Color, transparency	-					clear
Refractive index (D line)	1			1.54		1.53
Transmission, ideal	%					95.6
Transmission, actual	%					90
Density	g/cm^3	1.36	1.15	1.1	1.11	1.08
Linear thermal expansion coefficient	10^{-6} K^{-1}	70				93
Vicat temperature B	°C	82	≤103	≤68	≤95	90
Heat distortion temperature (1.82 MPa)	°C					85
Continuous service temperature	°C					85
Tensile modulus	MPa	2600	≤ 2900	≤ 1900	≤ 2600	2100
Fracture elongation	%					3.2
Impact strength (Charpy)	kJ/m^2	nf	25-115	25-nf	60-nf	80-nf
Notched impact strength (Charpy)	kJ/m^2	5-40	2-14	2-5	4-7	7-10
Processing	-	B, C, E, I	E, I	E, I	I	I

Contact Lenses

A contact lens is a thin corrective lense that is fitted over the cornea of the human eye. The lenses consist of either hard or soft polymers and must be constructed in such a way that a sufficiently thick cushion of tears can exist between the lens on one hand and the iris and the pupil on the other. This film enables the lens to float on the eye and provides the cornea with the necessary oxygen. Contact lenses should not be breeding grounds for bacteria and must therefore be either easy to clean or be throw-away articles.

The first contact lenses were produced in 1887 by the German glass blower F.E.Müller. In 1888, the German physiologist A.E.Fick (of Fick's diffusion laws fame) constructed the first fitted contact lens from glass based on molds from impressions of the eye. Today, contact lenses are made to fit from rigid or soft polymers after the curvature of the eye has been measured. These lenses are usually not worn overnight.

Rigid contact lenses are produced by injection molding of methyl methacrylate. Contact lenses from poly(methyl methacrylate) (PMMA) provide the wearer with better vision than soft ones. They also last longer and are better to clean but are less comfortable. Similarly to the flexible hydrophobic contact lenses from poly(dimethylsiloxane) or cellulose acetobutyrates, PMMA contact lenses require a fairly thick cushion of tears. Their main disadvantage is the lack of permeability to oxygen.

Contact lenses from hydrophilic polymers are soft and comfortable but do not last as long as rigid or flexible hydrophobic ones. The so-called disposable ones last only 1-2 weeks whereas the extended wear ones can be used 3-4 weeks albeit with some cleaning. These lenses can be produced by various processes, for example, by centrifugal casting of a polymerizing system of 2-hydroxyethyl methacrylate (HEMA) with some glycol dimethacrylate as crosslinking agent, a process that was invented in 1955-1961 by the Czech scientists O.Wichterle and D.Lim (Volume II, p. 297). The 0.8-3.5 nm wide "pores" of these polymers cannot be entered by the ca. 200 nm wide bacteria. However, proteins can still be deposited and, because they are good breeding grounds for bacteria, such contact lenses must still be cleaned periodically.

The uptake of water by HEMA polymers is 37 % which is surpassed by silicone hydrogels which have been available since 1999. The latter polymers have extremely high permeabilities to oxygen (see also p. 559).

15.1.5 Fiber Optics

According to Snell's law (p. 615), light traveling in a medium B of higher refractive index n_B is refracted at the interface with a medium A of lower refractive index n_A, i.e., bent towards B. The light is totally reflected ($\alpha = 90°$ in Fig. 15-2) if the angle of impingement β is less than a critical angle β_{crit}. For this condition and the vacuum, $n_B/n_A = \sin \alpha/\sin \beta$ (Eq.(15-1)) converts to $n_B = 1/\sin \beta_{crit}$. The critical angle of a material with $n_B = 1.5$ in vacuum ($n_A \equiv 1 = n_0$) is therefore $\beta_{crit} \approx 41.8°$. Such a total internal reflection allows one to transmit light over long distances without loss of intensity. Based on these principles, **fiber optics** transports information from a source to a receiver by so-called **optical fibers** that are embedded in hollow tubes (**wave guides**).

Fiber-optical systems consist of four components: emitters (= sources of radiation), optical fibers, connectors (of optical fibers) and detectors (= recipients of radiation). An

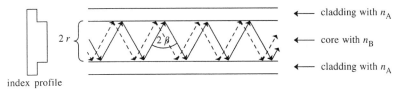

index profile

Fig. 15-7 Schematic representation of a step-index optical fiber with two modes. r = radius of core.

emitter consists of a laser and a diode which emits up to 1 billion digital impulses (as "0" or "1") per second. For telephone lines, impulses are transmitted by bundles of 10^5-10^6 optical fibers with diameters of ca. 50 µm per fiber, the ends of which are coupled by connectors. The detectors are usually diodes.

There are several types of optical fibers. Graded-index optical fibers consist of con-centric layers of glass in which from the interior to the exterior each more outward layer has a slightly lower refractive index than its more inward neighbor, i.e., these optical fibers have a graded index profile. A combination of refraction and total internal re-flection prevents the escape of light from the fiber.

Step-index optical fibers have rectangular index profiles (Fig. 15-7). They consist of an impulse-propagating cylindrical core that is surrounded by a tight-fitting coaxial mantle (**cladding**) with a lower refractive index than the core (Fig. 15-7). The core as well as the cladding may be a glass or a polymer. Light travelling through the core is to-tally reflected at the inner surface of the core if the opening angle β is smaller than the critical angle β_{crit} (see above). For cladded cores, the reflected proportion depends on the difference between the refractive indices n_B of the core material and n_A of the surrounding cladding. This difference should be as large as possible since it determines the **numerical aperture**, $A_n = \sin \beta_{crit} = (n_B^2 - n_A^2)^{1/2}/n_0$ (with β_{crit} = half-angle of the maximum core of light), at which the entering light can be guided in the fiber.

Step-index optical fibers can be of the single-mode or multimode type. The number of modes that can travel simultaneously in a fiber is related to the so-called **normalized frequency**, $V = (2 \pi r/\lambda_0)A_n$ where r = radius of the core, λ_0 = wavelength of light in vacuum, and A_n = numerical aperture. Core diameters are ca. 10-250 µm for multimode optical fibers and less than 10 µm for single-mode ones.

In principle, air is a good cladding "material" since it has the lowest refractive index of all matter. However, the surface of a glass core in air is not protected and can be easily scratched or covered by dust which leads to light scattering and a loss of light, hence the protective cladding which is usually 10-150 µm thick.

Optical glass (similar to lead crystal) combines high clarity with a high refractive index ($n_{D,20°C}$ = 1.650). It is therefore an excellent core for wave guides. Transparent polymers are less refractive than glass (see Table 15-2) but are more flexible and there-fore allow thicker fibers ($\leq$ 1000 µm *vs.* 125 µm) and a greater throughput of informa-tion per fiber. Cores and claddings of polymers can also be extruded together whereas the polymer cladding of glass must be put on in a separate step. Wave guides with poly-meric cores are therefore less expensive than those with glass cores.

For example, a system for visible light has a core of poly(methyl methacrylate) (n_D = 1.489 at 25°C) and a cladding of a partially fluorinated polymer ($A_n \approx 0.5$ for this system). For ultraviolet light, a system may consist of a silicate glass core and a cladding of poly(tetrafluoroethylene-*co*-hexafluoropropylene) ($A_n \approx 0.14$). Claddings often consist of two layers: a soft inner one as protection against microbending and an outer one of a tough, abrasion-resistant material.

Rigid optical fibers serve for the transmittal of data and acoustic signals (telephone, etc.). Flexible optical fibers allows light to be guided around corners. They serve in medicine for inspections of internal organs (bladder, colon, etc.), in transportation for tail lights of cars, etc.

15.1.6 Optical Birefringence

In general, polymeric materials are isotropic because their optical elements are statistically distributed throughout the specimen: segment axes in amorphous polymers and crystallites in semicrystalline ones. Anisotropies occur if these optical elements are preferentially oriented, for example, by uniaxial deformation or by stress. Since refractive indices usually differ in the three spatial directions, an optical birfringence is observed albeit only in transparent or translucent materials (see Volume III, Section 7.6.2).

15.1.7 Iridescence

Very thin transparent layers B with parallel surfaces such as oil films on water, soap bubbles in air, etc., show intense glossy colors in reflected or transmitted white light. These colors arise because light waves from the same source are not only refracted but also reflected at the two boundaries of the layers.

A light beam I (solid line in Fig. 15-2) that is refracted at point 1 of the top surface of layer (film) B of thickness d is not only refracted again at point 1' of the bottom surface of the layer but is also reflected to some extent at both the external top surface and the inner bottom surface (see broken lines in Fig. 15-2). The latter light is then again refracted and reflected at a point 2, etc. The path of a light beam I via point 1 to point 1' and then to point 2 is obviously much longer than the path of a light beam II from the same light source. The geometric shift Δ_{geo} is calculated from geometric considerations and Snell's law, Eq.(15-1). There is also a shift, $\Delta_{ph} = \lambda/2$, from the phase jump that the light wave experiences from the reflection at the more dense medium. The total shift is

(15-10) $\Delta = \Delta_{geo} + \Delta_{ph} = 2\ d[n^2 - \sin^2 \alpha]^{1/2} + \lambda/2$

Light waves are intensified if this shift Δ is an even multiple of λ [setting $\Delta = 2\ z(\lambda/2)$] and extinguished if the shift Δ is an odd multiple of λ [setting $\Delta = (2\ z + 1)(\lambda/2)$] where z = 1, 2, 3, ... in both cases. For light falling perpendicularly on the film ($\alpha = 0$), light is intensified at wavelengths $\lambda_1 = 4\ nd$, $\lambda_2 = (4\ nd)/3$, $\lambda_3 = (4\ nd)/5$, etc., and extinguished at wavelengths $\lambda_1' = 2\ nd$, $\lambda_2' = nd$, $\lambda_3' = (2/3)\ nd$, etc.

For example, a 250 nm thick film with a refractive index of 1.5 would have strong reflections at $\lambda_1 = 1500$ nm, $\lambda_2 = 500$ nm, $\lambda_3 = 300$ nm, etc., and no reflections at $\lambda_1' = 750$ nm, $\lambda_2' = 375$ nm, $\lambda_3' = 250$ nm, etc. Such a film reflects in the near infrared ($\lambda = 1500$ nm) and in the blue-green ($\lambda = 500$ nm).

White light consists of the whole spectrum of wavelengths, which would be intensified at the λ_i values (i = 1, 2, 3, etc.) and wiped out at λ_i' values. Wavelengths next to each λ_i would be somewhat less intensified and those next to λ_i' not completely wiped out but reduced in intensity, etc. By variation of the number of stacked layers in films, the thicknesses of layers with increasing position number of the layers as well as the refractive indices of polymers, one can produce selective colors or the whole rainbow. For example, a film with 1000 alternating nanolayers of polycarbonate and poly(methyl methacrylate) has a metallic shine.

15.2 Effects of Light Scattering

Media act in two ways on electromagnetic waves that pass through optically inhomogeneous systems. On one hand, amplitudes and phases are changed, i.e., the wave front will be distorted. In the language of optics, this results in a "loss of resolution" whereas the plastics industry calls this a "loss of clarity."

On the other hand, electromagnetic waves also lose a part of their energy because of scattering effects (Section 15.2.2). Forward scattering reduces the contrast and the specimen becomes hazy (p. 615). The loss of contrast from both forward and backward scattering makes the specimen milky (p. 621).

This section discusses two aspects of scattering phenomena: opacity (Section 15.2.1) and hiding power (Section 15.2.2). Other scattering phenomena are beyond the scope of this book, which is concerned with applications of polymers and not with fundamental aspects such as the determination of molecular parameters by light scattering (Volume III, Chapter 5).

15.2.1 Opacity

A body appears opaque if it scatters light because (a) refractive indices or densities vary locally and/or (b) it contains variations in the orientation of anisotropic volume elements that are larger than one-half of the wavelength of incident light.

Scattering of polarized light at small angles allows one to distinguish between these two sources of scattering. A horizontally polarized scattering (H_v) arising from incident vertically polarized light is caused by an anisotropy of scattering elements. A vertically polarized scattering (V_v) caused by incident vertically polarized light depends on both the anisotropy of scattering elements and on local differences of refractive indices.

However, local fluctuations of refractive indices result in opacity only if the specimen contains structures (such as microphases) that are larger than the wavelength of incident light. These structures cannot be too large because an infinitely large single crystal does not scatter light. The clarity of a specimen can therefore be increased considerably if the dimensions of scattering elements are reduced. Gains in clarity are less pronounced if differences between the refractive indices of both phases are reduced.

The appearance of heterogeneous specimens is controlled by the difference in refractive indices and by the absolute and relative size of phases. Examples are blends of poly(vinyl chloride) (PVC) and graft copolymers of styrene + acrylonitrile on butadiene rubbers (ABS) where the refractive index of PVC ($n_{D,20}$= 1.539) is greater than that of ABS. These materials look milky-yellowish if ABS is dispersed in PVC but milky-bluish if PVC is dispersed in ABS.

In order to improve the appearance of impact-modified polymers, one therefore tries to reduce the particle size of the dispersed rubber phase as well as the difference in refractive indices of the two phases. However, one can reduce the particle size of the dispersed phase only to an optimal diameter below which the impact strength will decrease.

The effect of the differing refractive indices of dispersed rubber particles (n_R) and surrounding thermoplastic matrix (n_M) can be reduced by the **core-shell technology** in which the dispersed particles with a refractive index $n_{R,core} > n_M$ are wrapped in a shell with a lower refractive index $n_{shell} < n_M$. The impact-modified thermoplastic is transparent if the average refractive index of the dispersed core-shell particles matches that of the matrix.

However, refractive indices can be matched only at a certain temperature, usually the use temperature, since refractive indices vary with temperature. At the higher processing temperatures, transparencies will be smaller and hazes larger (Fig. 15-8). The specimen will become increasingly more clear on cooling.

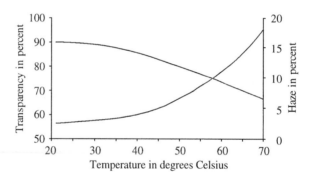

Fig. 15-8 Temperature dependence of transparency and haze of an impact-modified poly(methylmethacrylate) (Lucryl KR 2006/1) [1]. By kind permission of Hanser-Verlag, Munich.

15.2.2 Hiding Power of Coatings

A coating is applied to the surface of a substrate in order to protect the substrate and/or improve its appearance by hiding the surface. The **hiding power** of a coating is controlled by absorption, reflection, and scattering. It denotes the surface area in square meters that can be completely hidden by one liter of paint, i.e., $m^2/L = mm^{-1}$, or in the US, square feet (or square yards) covered by one gallon.

Hiding powers depend on reflection and scattering. The contribution of reflection to the hiding power of a coating can be estimated from Eq.(15-5) for light perpendicular to the substrate. In this equation, n is now the refractive index n_F of the filler (i.e., the pigment) and "1" (the refractive index of vacuum) is replaced by the refractive index n_S of the substrate:

(15-11) $R_0 = (n_F - n_S)^2/(n_F + n_S)^2$

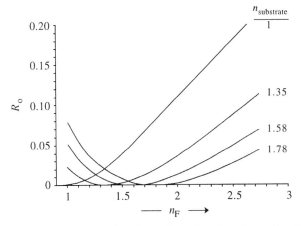

Fig. 15-9 Reflection R_0 of light normal to the surface of a substrate as a function of the refractive index n_F of fillers or pigments in clear coatings. Substrates with refractive indices n_S of 1.78 (for example, polyimide), 1.58 (poly(styrene)), 1.35 (poly(tetrafluoroethylene)), and 1.00 (vacuum). Calculations with Eq.(15-11) for an angle of incidence, $\alpha = 0$, and an angle of refraction, $\beta = 0$.

At constant pigment sizes, reflections R_0 decrease with increasing refractive index n_F, become zero at $n_F = n_S$, and increase again (Fig. 15-9). Because of its high refractive index, synthetic rutile with $n_F = 2.73$ is the the most often used white pigment (p. 27). Refractive indices of most other white pigments are considerably lower (Table 15-5).

Air with $n_F \approx 1.00$ is an excellent "white pigment" since it has a relative hiding power of 77.4 % of that of rutile (Table 15-5). For this reason, microporous pigments and fillers have better hiding powers than macroporous and compact ones since liquid binders cannot enter micropores so that air becomes trapped. However, air has very bad mechanical properties since its moduli, strengths, etc., are zero. Paints with microporous fillers can therefore only be used for purely decorative purposes (no abrasion).

Actual hiding powers depend not only on reflection but also on light scattering. The larger a particle, the more scattering centers it will have: initially, the scattering increases with the 3rd power of the particle diameter. However, the larger the particle, the smaller will be the proportion of backward scattering (Volume III, Fig. 5-5). Because of this effect, the scattering decreases with further increase of the particle diameter.

Table 15-5 Relative hiding power of various white pigments for poly(styrene) with ($n_D = 1.590$) as related to manufactured rutile, the white tetragonal crystal modification of TiO_2 with $R_0 \equiv 1$. Calculations with Eq.(15-11) for pure reflection.

White pigment	Refractive index	Relative hiding power Calculated	Experiment
Rutile (a tetragonal TiO_2 modification)	2.73	1.00	1.00
Anatase (a tetragonal TiO_2 modification)	2.50	0.71	0.78
Zinc sulfide	2.37	0.56	0.39
Zinc oxide	2.00	0.21	0.14
Lithopone (28 % ZnS + 72 % $BaSO_4$)	1.84	0.08	0.18
Heavy spar ($BaSO_4$)	1.64	0.005	
Air	1.00029	0.744	

With increasing particle diameter d_F, hiding power thus increases proportionally to d_F^3, passes through a maximum, and then decreases about proportionally to $1/d_F$. For this reason, some experimental hiding powers are greater and some are smaller than the theoretical ones from reflections without considering scattering (Table 15-5).

Hiding power also depends on the particle concentration. An increase of the particle concentration will increase the scattering intensity but only to a certain point above which the light scattered by one particle will hit another particle and then become scattered again, etc. This multiple scattering decreases the relative scattering intensity. The decrease of hiding power by this effect becomes noticeable if the distances between particles become smaller than about three times the particle diameters.

15.3 Optoelectronics and Photonics

15.3.1 Fundamentals

A **photon** is the energy quantum of electromagnetic radiation (G: *phos* = light). It has no electrical charge and zero mass while at rest. Its counterpart in the magnetic field is the **graviton**. On interaction with matter, photons appear either as particles (example: photoconductivity) or as waves (examples: diffraction, reflection). Like electrons, photons can be used to collect, store, transport, and process information. In analogy to electronics, this technology is thus called **photonics**.

An **energy quantum** (L: *quantum* = neuter of *quantus* = how great) is the smallest indivisible unit of energy E of **photons** in electromagnetic radiation (visible light, UV, IR, γ radiation, etc.) and of **phonons** in acoustical waves, lattice vibrations, and material waves. Quantum theory began in 1900 with the insight of Max Planck that energy in thermal radiation is not a continuous spectrum but can be changed only in discrete amounts (quanta), $E = Nh\nu$, where N = integer, h = Planck constant, and ν = frequency ($h = 6.626\ 075\ 5\ (40) \times 10^{-34}$ J s). The mass m_q of one quantum is given by the Einstein equation, $m_q = E/c^2$, where c = speed of light (in vacuum: $c_0 = 299\ 782\ 458$ m s^{-1} (exactly)). The mass of a quantum at rest is zero since $m_q = 0$ if $c = 0$. Moving quanta possess masses and may therefore also be measured in amounts-of-substance (e.g., in moles). The energy of a quantum is very small; for visible light it is only ca. 10^{-19} J.

In contrast to electrons, photons do not interact strongly with other photons or with external fields. Hence, photons transport information faster and with less energy loss over greater distances than electrons. This behavior of photons is advantageous for the use in transmission lines and high-density circuitry. The weak interactions between photons do not allow easy manipulations, however. For this reason, the hybrid technology of **optoelectronics** has been established where photons interact with electrons.

Materials for optoelectronics must possess two essential properties: transparency to light and high mobility of electrons. Metals are electrical *conductors* because the electrons of their atoms are not firmly bound to one atom but are shared by other atoms of the metal. In the language of **band theory**, the gap between the valence band and the conductance band is infinitesimal (Fig. 15-10, left). Application of a steady electric field thus produces a steady current of electrons. However, metals are not transparent.

Conversely, some *insulators* may be transparent but valence bands of all insulators are separated from conductance bands by fairly large energy gaps (Fig. 15-10, right). Electrons of atoms of dielectrics are firmly bound to atomic nuclei; they can surpass the gap (forbidden zone) only when subjected to large external energies.

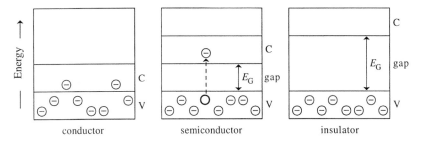

conductor   semiconductor   insulator

Fig. 15-10 Relative positions of valence bands V and conductance bands C in inorganic conductors, semiconductors, and insulators at ambient temperatures according to band theory. The **band gap** (vertical double arrow) is defined as the energy difference E_G between the highest occupied state and the lowest empty state (organic chemistry: between the *h*ighest *o*ccupied *m*olecular *o*rbital (HOMO) in the ground state and the *l*owest *u*noccupied *m*olecular *o*rbital (LUMO) in the excited state). In semiconductors such as Si, In, and Ga, valence electrons ⊖ may enter the low-laying conductance band leaving behind a hole O, called a defect electron. Application of a current causes electrons to travel and fill neighboring holes leaving behind holes that therefore *seem* to be traveling, too (Table 15-6).

Energies of band gaps are very small if presented in SI units such as joule. For convenience, they are therefore usually given in the traditional unit of electronvolt; 1 eV $\approx$ 1.602 177 33 (49) $\times$ 10^{-19} J. The value of the electronvolt in joule is not exact because it depends on the experimentally determined value of the elementary charge.

Solid conductors and insulators differ correspondingly in the **Fermi level** which indicates the energy in a solid at which one-half of the quantum states are occupied, i.e., where the average number of particles per quantum state is 1/2. The Fermi level of conductors is in the conductance band and that of insulators in the valence band.

Semiconductors have electrical conductivities between those of conductors and insulators (p. 575). Semiconductivity may be intrinsic, i.e., caused by crystal defects or thermal excitations, or extrinsic, i.e., caused by "foreign" atoms such as arsenic in indium (Table 15-6) or added dopants such as AsF_5 in conjugated polymers (p. 606).

In intrinsic semiconductors, the number of electrons in the conductance band equals the number of holes in the valence band. Extrinsic semiconductors may be doped with electron donors or electron acceptors leading to a large surplus of electrons (negative charges) in the former case (n-type) or to a large surplus of holes (positive charges) in the latter case (p-type).

Per definition, band gaps E_G of semiconductors are smaller than ca. 3-3.5 eV; they range in inorganics from 0.36 eV in InAs to 3.5 eV in GaN. Band gaps can be fine-tuned by dopant levels in inorganic and organic semiconductors or, in inorganic semiconductors, by alloying two materials in various proportions; for example, HgCdTe with $0 < E_G/eV < 1.5$ from CdTe (E_G = 1.5 eV) and HgTe (E_G = 0 eV).

The Fermi level of semiconductors is in the gap between the valence band and the conductance band. It is close to the conductance band in n-type semiconductors and close to the valence band in p-types.

Simple "inorganic" band theory fails to explain certain features of organic semiconductors (such as the absence of spin in charge carriers) which is the reason why the concepts of solitons, polarons, and bipolarons were introduced (p. 607). p-doping produces a delocalized radical cation (p. 608, bottom), i.e., a polaron, which has an energy that corresponds to the middle of the band gap (between C and V in Fig. 15-10, center). Removal of another electron from the polaron by further doping generates a bipolaron.

Table 15-6 Types, examples, and properties of an insulator (IN) and some semiconductors (SC). E_G = energy gap in electronvolts, μ = mobility of electrons (μ_n) or holes (μ_p), both in $cm^2/(V\ s)$.

Type	Example	E_G	μ_n	μ_p	Remarks
IN	C	5.45		1200	
SC, intrinsic, inorganic	Si	1.14	1750	440	
	Ge	0.67	3800	1850	
SC, extrinsic, inorganic	GaAs	1.35	8500	400	
	InAs	0.36	28 000	280	
SC, extrinsic, polymeric	poly(pyrrole)		0.022		polaron
	poly(aniline)		0.000035		bipolaron
	poly(*p*-phenylenevinylene)			0.0004	
	poly(3-hexylthiophene)			0.00005	

In molecules with conjugated double bonds, polarons and bipolarons extend over whole molecules (low-molecular weight organics) or segments of several monomeric units (organic polymers). Application of an electric field causes polarons and bipolarons to travel intermolecularly in both small and large molecules and additionally also intramolecularly along the chain in polymers (p. 608 ff.).

However, they can move along the whole polymer chain only if the chain is free of chemical and physical defects such as head-to-head or tail-to-tail connections of monomeric units, cis connections in trans structures, branches, chain ends, kinks in chains, and the like. Such defects will cause charge carriers to transfer to other chains. Similar intermolecular movements are the only possible ones for charge carriers in low-molecular weight organics, which therefore must have their molecules well oriented with respect to their axes by poling. As a result, mobilities of charge carriers in organic polymeric semiconductors are much smaller than those of inorganic semiconductors (Table 15-6).

In this respect, inorganic semiconductors based on silicon or compounds of indium, gallium, etc., are advantageous because their "molecules" can relatively easily be coaxed into (nearly) perfect crystal lattices, which eases the intermolecular hopping of charge carriers. Indeed, inorganic semiconductors can be prepared with high purities and high crystallinities but processing requires high temperatures and is costly.

In addition, there may be a problem with the future supply of indium which is presently (2007) consumed at a rate of ca. 1130 t/a (ca. 650 t/a from recycling, 480 t/a from zinc ores) whereas world reserves are estimated as only 11 000 t. The situation is less critical for other inorganic semiconductors such as Ga and Si: the anual consumption of gallium is ca. 184 t (100 t/a from bauxite which contains ca. 60 g of Ga per of ton bauxite; 84 t/a from scrap recycling) but gallium reserves are estimated as 1 million tons whereas silicon is practically inexhaustible thanks to the abundance of SiO_2.

For all these reasons, there is a tremendous interest in the synthesis and properties of polymeric semiconductors. At present, there is only a small industrial production of polymeric light-emitting diodes (Section 15.3.2), industrial application of polymeric semiconductors in solar cells is just beginning (Section 15.3.3), and polymers for non-linear optics (Section 15.3.4), field-effect transistors, sensors, and the like are at the research stage. However, non-conducting polymers have been used in supporting roles in optical applications such as claddings in fiber optics (Section 15.1.5) or embedding materials for low-molecular weight semiconducting organics.

15.3.2 Light-emitting Diodes

Electroluminescence is the optoelectronic phenomenon that a material emits infrared, visible, or ultraviolet light when it is excited by electrons from a strong electric field or by an electric current passing through (L: *lumen* = light; *escents*, past participle suffix of *escere* which indicates the beginning of a process). The effect is utilized in **light-emitting diodes (LEDs)**. In principle, LEDs consist of a semiconducting material, the *emitter layer* (EL), which is sandwiched between an electron-injecting cathode and a hole-injecting anode. The application of a voltage causes electrons and holes (= defect electrons) to move toward each other and meet in the emitter layer where they (re)combine to an electroneutral quasi-particle called an **exciton** (L: *excitare* = to excite, to arouse; from *ex* = out, *ciere* = to put in motion). The exciton carries excitation energy which it loses if it becomes trapped, for example, by crystal defects. The decay leads to the emission of light (fluorescence if it decays in less than 10 ns) or other electromagnetic radiation.

Radii of excitons can be much larger than atomic radii, for example, 11.5 nm in germanium (covalent radius of a Ge atom: 0.122 nm), which corresponds to small binding energies for electron-hole pairs. For example, the binding energy of an exciton in Ge is only 4 meV which is 3400 times smaller than that of a hydrogen atom (13.6 eV). In conjugated polymers, Coulombic attractions between electrons and holes are much larger so that the binding energies of excitons are much greater, usually ca. 100-1400 meV, which makes these excitons very stable.

Excitons differ from excimers and exciplexes which are excited complexes of *molecules*. The combination of two like molecules, $A^* + A \rightarrow [A{\cdots}A]^*$, leads to an **excimer** (*exci*ted di*mer*) whereas that of two unlike ones, $A^* + B \rightarrow [A{\cdots}B]^*$, delivers an **exciplex** (*exci*ted com*plex*).

The emitter layers of inorganic light-emitting diodes (**ILEDs**) are semiconducting alloys such as GaAs for the infrared, GaP for green light (555 nm), and AlN for the ultraviolet. In general, such emissive layers are doped.

Emitter layers can also be semiconducting low-molecular weight molecules or polymer molecules. Both lead to **organic light-emitting diodes (OLEDs)** but OLED is often used to indicate only small molecules. SMOLED for *sm*all *o*rganic molecule LEDs is uncommon. **Polymeric light-emitting diodes** are known as **PLEDs**.

Small organic molecules for OLEDs form semiconducting charge-transfer complexes (for example, TCNQ-TTF, p. 610). The first OLED (1953) employed the charge transfer complex of acridine orange and quinacrine; at present (2007), ca. 95 % of all carbon compound-based LEDs are (SM)OLEDs, mainly charge-transfer complexes of various substituted perylenes and differently substituted quinacridones. These small molecules are incorporated into displays by vapor deposition (see also p. 599).

acridine orange

perylene

quinacrine (mepacrine)

quinacridone

OLEDs and PLEDs are more advantageous than ILEDs since they do not require the expensive manufacture of single crystals. OLEDs and PLEDs differ with respect to charge transfer which is always intermolecular in OLEDs but both intermolecular and intramolecular in PLEDs. Small molecules for OLEDS can be synthesized with exactly defined constitutions whereas polymer molecules for PLEDS may have constitutional errors (see p. 633). However, polymers can be processed to thin films by solution-based processes similar to ink-jet printing (p. 544) which is one of the reasons why it is also proposed to embed light-emitting small organic molecules in non-light-emitting yet semiconducting polymeric layers.

PLEDs are composed of many layers of materials with thicknesses in the hundreds of nanometers, in the simplest case, a support (substrate), an electron (= radical anion (polaron)) providing cathode (ca. 200 nm), a hole (= radical cation (polaron)) delivering anode, layers for the transport of electrons and holes, and the emitter layer (ca. 80 nm) where electrons and holes meet and form excitons (Fig. 15-11). The emitter layer consists of conjugated chains of semiconducting polymer molecules (see p. 604 ff.).

The support is usually a transparent silicate glass. The anode and the hole transport layer must be (at least) semi-transparent for the luminescence that is emitted by the decay of excitons in the emitter layer. The cathode provides the electrons and the anode the holes. In the ideal case, the yield of excitons would be maximized if electrons and holes arrive at the emitter layer in equal proportions. In semiconductig polymers, however, mobilities of holes are greater than mobilities of electrons. The resultant imbalance can be remedied in part by choosing an anode that produces holes at a smaller rate than the cathode, i.e., an anode with a larger **work function** than that of the cathode.

The work function ϕ is the smallest energy that is needed to transport an electron from the Fermi level to the vacuum, i.e., from the solid to a point just outside the surface of the solid. Its magnitude depends strongly on the structure of the surface (roughness, adsorbed gas, oxidation, etc.). Work functions ϕ/eV of common polymers are 4.08 (PA 6.6), 4.22 (PS), 4.26 (PC, PTFE), 4.36 (Kapton), and 4.85 (PET).

Anodes often consist of *indium tin oxide* (ITO), an alloy of ca. 90 wt% In_2O_3 and ca. 10 wt% SnO_2 that is applied by spin coating. ITO is chosen because it not only has a large work function of 4.7 eV but also a rare combination of properties that are useful for LEDs: high transparency in the visible region, good electrical conductivity, and excellent adherence to substrates such as glass. However, it is also expensive, fragile, and inflexible, which is the reason why one tries to replace it by flexible, transparent semiconducting polymers (see below).

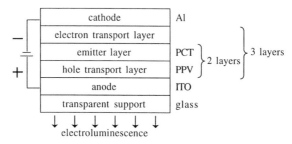

Fig. 15-11 Structure of a simple three-layer PLED. The example on the right shows the first *two-layer* PLED (no electron transport layer) that reached an internal electroluminescent efficiency of 4 % [2]. In *three-layer* PLEDs, electron transport layers may be LiF or CN-PPV (p. 635). Modern PLEDs achieve the same efficiencies as incandescent light bulbs (ca. 10 % of applied electrical energy).

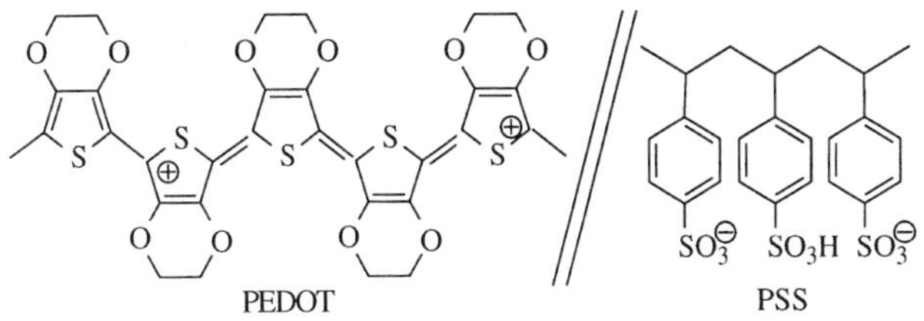

Fig. 15-12 Sections of polymer chains of poly(3,4-ethylenedioxythiophene) (PEDOT; shown as bipolaron) and poly(styrene sulfonate) (PSS) that form the polyelectrolyte PEDOT-PSS.

In the example of Fig. 15-11, the work function of the anode (ITO) was increased by spin coating ITO with a very thin layer (ca. 100 nm) of the semiconducting polymer poly(p-phenylene vinylene) (PPV, p. 604) which does not emit light. Later PLEDs used ITO-free semiconducting polymers with large work functions as anode, for example, PEDOT, PEDOT-PSS (Fig. 15-12; $\phi = 5.2$ eV), carbon ($\phi \approx 5$), or poly(aniline) (PANI) (p. 605), all as thicker films of 150-200 nm thickness.

PEDOT and PEDOT-PSS (= salt of PEDOT with PSS) are used not only for LEDs but also for solar cells (Section 15.3.3) and other photovoltaic devices. They are available in various grades; for example, PEDOT-PSS as Baytron P®, grade EL, and grade P. Electrical conductivities of PEDOT vary with the type of polymerization, $\sigma = 0.3$ S/cm from chemical polymerization *versus* 80 S/cm from electropolymerization, and can be increased by doping, annealing (by two decades), or addition of N-methyl pyrrolidone (by 3 decades). PEDOT-PSS is delivered as a 1.3-1.7 % dispersion in water from which it can be applied. In the visible range of 450-830 nm, PEDOT-PSS has a light transmission of ca. 80 %. However, ultraviolet light degrades PEDOT.

Cathodes should have low work functions, which is the reason why metals such as barium ($\phi = 2.5$-2.7 eV) or calcium ($\phi = 2.87$ eV) have been used. However, these metals are aggressive and need to be protected. Aluminum is better in this respect but has an undesirable greater work function ($\phi = 4.06$-4.26 eV). This, in turn, can be remedied in the device of Fig. 15-11 by using emitter layers with greater electron affinities, for example, CN-PPV instead of PPV. The electron injection from the cathode into the active layer is also improved by placing a 0.3-1 nm thin layer of LiF as the electron-transport layer between the cathode and the active layer (see Fig. 15-11).

The structure and properties of these and other semiconducting, conjugated polymers (examples: Figs. 15-13 and 15-14) can be changed over a wide range by introducing side groups and/or doping with various dopants to different levels. As a result, a wide variety of PLEDS are now available for practically all of the colors of the visible spectrum. White light is obtained by mixing blue, green, and red. Some examples of PLEDs are shown in Fig. 15-14.

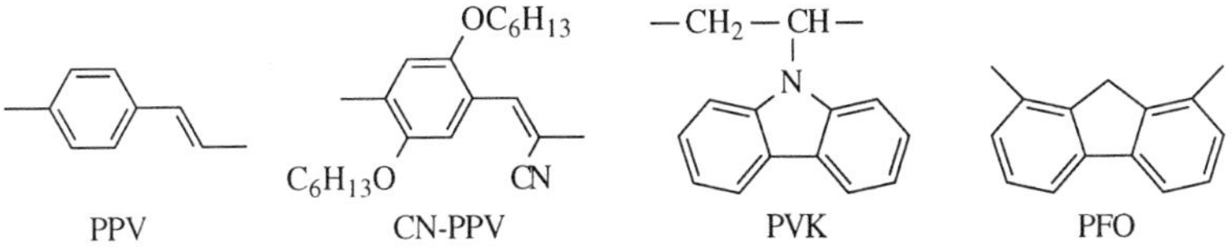

Fig. 15-13 Examples of monomeric units of polymers for PLEDs. PPV = poly(p-phenylenevinylene); CN-PPV = poly(2,5-dihexyloxy-1,4-phenylene-2-cyanovinylene), commonly known as poly-(cyanoterephthalylidene); PVK = poly(N-vinyl carbazole); PFO = poly(2,7-fluorene).

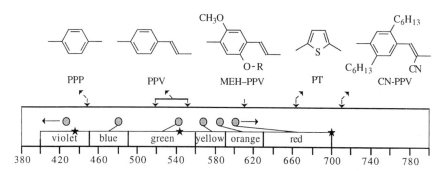

Fig. 15-14 Electroluminescence in the visible light region from some neutral (undoped) conjugated polymers shown as their monomeric units. Stars ★ indicate CIE primary colors for color matching tests: blue (435.8 nm), green (546.1 nm), red (700 nm) (CIE = *Commission International de l'Eclairage* = International Commission on Illumination).

Color designations vary widely from country to country and in the scientific and popular literature. For example, names and widths of color ranges in this graph are those of the English language Wikipedia. ◎ gives the positions of absorption maxima for these colors according to the German language Wikipedia where ⟵ indicates that the absorption maximum is at $\lambda < 425$ nm and ⟶ that the absorption maximum is at $\lambda > 600$ nm.

Note that the technical literature reports the color of the light from PPP as blue, the two absorptions of PPV as green (520 nm) and yellow (551 nm) but the color of the light as green, the light from MEH-PPV at 610 nm as red-orange, and the light from both PT and CN-PPV as red.

PPP = poly(*p*-phenylene), PPV = poly(*p*-phenylenevinylene), MEH-PPV = poly[2-methoxy-5-(2-ethylhexyloxy)-1,4-phenylene vinylene] with R = –O–CH$_2$–CH(C$_2$H$_5$)–C$_4$H$_9$, PT = poly(thiophene), CN-PPV = poly[*p*-(2,5-dihexylphenylene)-(2-cyanovinylene)].

15.3.3 Solar Cells

Photovoltaic cells produce electrical energy from light by the photovoltaic effect. The most important of these cells are solar cells that harvest sunlight. At present, practically all commercial solar cells are produced from inorganic materials, especially silicon, but also from other inorganic semiconductors. Efficiencies of Si cells range from 5-7 % for thin amorphous Si (a-Si) to >20 % for thick monocrystalline Si (c-Si). GaAs cells reach efficiencies near 30 %, CuInSe$_2$ cells (CIS) 12-15 %, and CdTe cells ca. 10 %.

Present solar cells (2008) from organic materials have efficiencies of less than 10 % and shorter lifetimes of up to 5000 hours. Organic dyestuffs presently deliver the best efficiencies whereas the first commercial polymer solar cells (2008) have efficiencies of only 3-5 %. However, polymer solar cells are attractive because they are flexible and transparent and can be produced as large panels at low cost.

The photovoltaic effect is the reverse of electroluminescence (Section 15.3.2). Technology therefore employs the same classes of materials for electroluminescence and photovoltaics. Correspondingly, LEDs and solar cells have similar designs (Fig. 15-15).

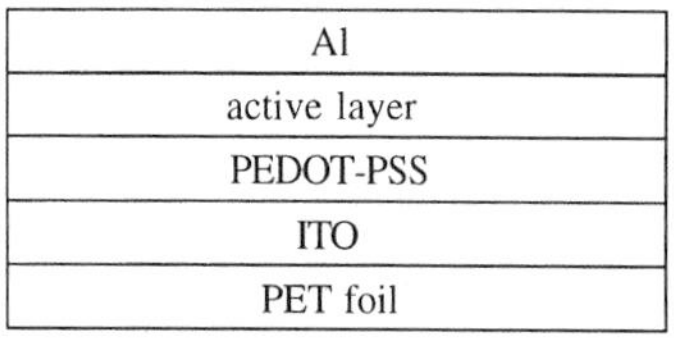

Al
active layer
PEDOT-PSS
ITO
PET foil

Fig. 15-15 Schematic drawing of a simple flexible polymer solar cell with conjugated polymers as active layers (see also Fig. 15-11). Light enters through the transparent PET foil. The thin ITO layer is the anode; aluminum is the cathode.

Like inorganic and polymer LEDs, inorganic and polymer solar cells also differ in their mechanisms. Inorganic solar cells have p-n junctions that generate large electric fields which separate the electrons and holes from the absorbed photons. The active layers of organic solar cells consist instead of a combination of electron-donating conjugated polymer chains with delocalized π electrons and electron-accepting molecules that may be polymeric or not. This combination can be a planar junction of a thin electron donor film in contact with a thin electron acceptor film or it can be a so-called bulk heterojunction (BHJ) consisting of a phase-separated mixture of electron donor and electron acceptor molecules.

Donor and acceptor molecules must be in close proximity for the following reason. Electron-donating polymers have conjugated chains with delocalized π electrons. These chains absorb photons with energies that exceed the energy gap between the lowest unoccupied and the highest occupied molecular orbitals. The absorption generates electroneutral excitons (p. 633) in which electrons (negative charge) and holes (positive charge) are strongly bound to each other by Coulomb forces with energies of 100-1400 meV.

Electrons and holes become separated at the interface between electron acceptors and electron donors but remain in close proximity when they try to travel to the electrodes with the opposite signs. However, mobilities of electrons and holes are small in conjugated polymers (see p. 632) and do not exceed 0.01-0.001 cm^2 V^{-1} s^{-1} in doped ones. Electrons and holes also have short diffusion lengths of only 3-10 nm. Therefore, long travel paths increase the probability that electrons and holes recombine, which reduces the efficiency of the process. Furthermore, if the hole is more mobile than the electron (or *vice versa*), a space charge is generated that hampers the travel of electrons or holes.

Distances for carrier diffusion must therefore be short, which is the reason that film layers in planar junctions of donor and acceptor layers are only several hundred nanometers thick. Bulk heterojunctions are better in this respect because "phase" regions are only several nanometers wide. The problem here is that phase structures are difficult to control and optimize. A possibly better way would be to embed the photovoltaic polymer in controlled pores of a thin substrate.

15.3.4 Nonlinear Optics

Fundamentals

A local electric field E creates a polarization p in molecules. In linearly polarized molecules, field and polarization are directly proportional, $p = \alpha E$, where α is the **polarizability** of the molecule. The polarization creates a dipole moment which in turn leads to a refractice index if the field is in the optical frequency range (Section 15.1.1).

Refractive indices are practically independent of field strengths if fields are small. At very high field strengths, however, practically all matter, including molecules, showa dependence of the refractive index on higher powers of the electric field, causing a **nonlinear response**, i.e., a **nonlinear optical** behavior (**NLO**). Such high electrical fields are generated, for example, by the very coherent, high-intensity light produced by lasers.

However, the power generated by most lasers is not large enough to produce a significant NLO effect in *all* materials. Only some substances have special structures which allow light beams to interact and produce notable NLO behavior. The non-linear optical effects arise if strong electric fields disturb the equilibrium positions of electrons, atoms, and molecules which in turn change the velocity (and thus the refractive index) or the frequency (and thus the wavelength) of the incident light.

The first effect is used for modulators, which are ultrafast light switches that change the intensity or the phase of light. The second effect doubles the frequency of light, for example, the light of an infrared laser with $\lambda = 1064$ nm ($\nu = c_0/\lambda \approx 2.82\cdot10^5$ s^{-1}) to green light ($\lambda = 532$ nm; $\nu \approx 5.64\cdot10^5$ s^{-1}) ($c_0 = 299\ 792\ 458$ m s^{-1} = speed of light in vacuum). This effect is desired for the storage of optical data because it allows one to quadruple the quantity of data that can be stored on optical discs (length × width!).

The non-linear variation of the polarization p of molecules and P of substances, respectively, with the applied electrical field can be described by power series. The various *local* fields F along the different molecule axes j, k, l induce a *molecule* polarization:

$$(15\text{-}12) \qquad p_i(\omega) = \alpha_{ij}(-\omega) \cdot F_j^{\omega}$$
$$+ \beta_{ijk}(-\omega; \omega_1, \omega_2) \cdot F_j^{\omega_1} F_k^{\omega_2}$$
$$+ \gamma_{ijkl}(-\omega; \omega_1, \omega_2, \omega_3) \cdot F_j^{\omega_1} F_k^{\omega_2} F_l^{\omega_3} + \dots$$

These local fields represent the differences between the applied electrical fields E and the fields experienced by the molecules. The external fields may have high frequencies ω (light waves) or low ones (electronic devices).

The proportionality constants in Eq.(15-12) are the **linear polarizability** α_{ij}, the **quadratic** (or **second**) **hyperpolarizability** β_{ijk}, and the **cubic** (or **third**) **hyperpolarizability** γ_{ijkl} (all also called **molecular susceptibilities**). The variables p and F are vectors because the interactions occur in all three spatial directions. The proportionality constants α_{ij}, β_{ijk} and γ_{ijkl} are tensors, however.

In analogy to Eq.(15-12), the dependence of the polarizability P of *materials* (bulk: substances, mixtures of substances, etc.) on the external field strength E can be written as

$$(15\text{-}13) \qquad P_I(\omega) = \chi_{IJ}^{(1)}(-\omega) \cdot E_J^{\omega}$$
$$+ \chi_{IJK}^{(2)}(-\omega; \omega_1, \omega_2) \cdot E_J^{\omega_1} E_K^{\omega_2}$$
$$+ \chi_{IJKL}^{(3)}(-\omega; \omega_1, \omega_2, \omega_3) \cdot E_J^{\omega_1} E_K^{\omega_2} E_L^{\omega_3} + \dots$$

where $\chi_{IJ}^{(1)}$, $\chi_{IJK}^{(2)}$, and $\chi_{IJKL}^{(3)}$ are the **first, second** and **third susceptibilities**.

Note that for each non-linear polarizability and susceptibility, respectively, both entrance and exit frequencies must be noted in parentheses (exit frequencies first). If, for example, two fundamental waves ω interact in the matter and generate the second harmonic wave with a frequency 2ω, then the first non-linear susceptibility (= quadratic susceptibility) is given as $\chi^{(2)}(-2\omega; \omega, \omega)$.

The **linear susceptibility** $\chi_{IJ}^{(1)}$ is a second-rank tensor that is related to the relative permittivity (aka: dielectric constant) ε_r by $\varepsilon_r = 1 + 4\pi\chi_{IJ}^{(1)}$. Because of the very large wavelengths, however, the **Maxwell equation** $\varepsilon_r = n^2$ applies. The linear susceptibility is thus also related to the refractive index, n. It is therefore responsible for refractions (Section 15.1), absorptions, and scattering effects (Section 15.2).

Molecules with sufficiently large **first non-linear** (**quadratic**) **hyperpolarizabilities** β_{ijk} and materials with the corresponding **first non-linear** (**quadratic**) **susceptibilities** $\chi_{IJK}^{(2)}$ (both third-rank tensors) are called **second-order NLO** materials. These values depend on the frequencies of E and P.

The values of β_{ijk}, γ_{ijkl}, $\gamma^{(2)}$ and $\gamma^{(3)}$ represent electric charges which are usually given in photonics and optoelectronics in so-called **electrostatic units** (esu, statcoulomb). However, the esu is not an SI unit. It is defined as the electric charge which acts on another, equal charge with an energy of 10^{-7} J if both charges are separated by 1 cm. According to ISO, the electrostatic unit should be replaced by the Coulomb (1 C = $3 \cdot 10^9$ esu).

Second-order NLOs allow one to add $(\omega_1 + \omega_2 \rightarrow \omega_+)$ or subtract $(\omega_1 - \omega_2 \rightarrow \omega_-)$ the entering optical frequencies ω_1 and ω_2. With $\omega + \omega \rightarrow 2\omega$, this leads to an optical **2nd harmonic generation (SHG)**. Second-order NLOs can also be used for **parametric frequency mixing** $(\omega_+ \rightarrow \omega_1 + \omega_2)$ and **optical rectification** $(\omega_1 - \omega_2 \rightarrow 0)$.

Generation of the $E_J^{\omega_1} E_K^{\omega_2}$ term of Eq.(15-13) by one optical and one electrical field (direct current and low frequency alternating current, respectively) gives rise to the **Pockels effect**. This **linear electrooptical effect** leads to a linear dependence of the refractive index on the electrical field strength. The resulting electrooptical coefficients R_{IJK} depend strongly on reorientations of molecules and movements of ions. They are usually much larger than the corresponding quadratic optical coefficients $\chi_{IJK}^{(2)}$ that depend mainly on electronic effects.

Molecules and substances with considerable **second non-linear hyperpolarizabilities** and **susceptibilities**, respectively, (both fourth-rank tensors), i.e., cubic coefficients γ_{ijkl} and $\chi_{IJKL}^{(3)}$, resp., are called **third-order NLOs**. They can be used to increase intensities.

In analogy to the linear electro-optical Pockels effect, a **quadratic electrooptical effect (Kerr effect)** is observed if one uses an optical field and the square of a direct current electrical field (d.c. field) or a low-frequency alternating electrical field (a.c.), respectively. Similarly to the Pockels effect, quadratic electrooptical coefficients S_{IJKL} are greater than the corresponding cubic optical coefficients $\chi_{IJKL}^{(3)}$.

The linear (Pockels) electrooptical coefficient R_{IJK} and the quadratic (Kerr) electro-optical coefficient S_{IJKL} are the proportionality coefficients in the relationship between the inverse square of the refractive index, n, and the field components E_K and E_L:

$$(15\text{-}14) \qquad \Delta (1/n^2)_{IJ} = R_{IJK}E_K + S_{IJKL}E_KE_L$$

Kerr coefficients S_{IJKL} can usually be neglected. The quadratic susceptibility $\chi_{IJK}^{(2)}$ can thus be calculated from the experimental Pockels coefficients R_{IJK}:

$$(15\text{-}15) \qquad \chi_{IJK}^{(2)}(-\omega,\omega,0) = -(1/2) [n_{ii}^2(\omega) \cdot n_{jj}^2(0)] R_{IJK}(-\omega,\omega,0)$$

Non-linear optical effects are shown by inorganic as well as low- and high-molecular weight organic compounds. However, organic materials have much shorter response times than inorganic materials (ca. 10^{-15} s *vs*. 10^{-8} s). In addition, organic polymers possess high mechanical strengths; they are also easy to prepare and process. Organic polymers with NLO properties are thus of high interest for applications in optoelectronics and photonics.

Like most young intermediate fields between established disciplines, photonics has developed its own jargon. An **"optical polymer"** here is not an optically *pure* polymer and the synonymous **"optically active polymer"** is not a polymer with optical activity in the organic chemical sense. In photonics, "optical polymers" rather consist of polymer molecules that carry groups with NLO properties in the main chain or in the side chain, or are mixtures of non-NLO polymers with low-molecular weight NLO compounds. Fibers from such "optical polymers" are called "polymer optical fibers" (POFs).

Second-order NLOs

The second term of the right-hand side of Eq.(15-13) becomes positive (βE^2) whether or not the field is positive ($+E$) or negative ($-E$). A negative field should however lead to a negative term ($-E^2$) if the molecule is centrosymmetric. It follows that the non-linear second (quadratic) hyperpolarizability β must be zero for centrosymmetric molecules. This limitation does not apply to the third (cubic) hyperpolarizability γ which describes the first non-linear term for centrosymmetric molecules.

The quadratic hyperpolarizability β does not equal zero for *molecules* with asymmetric electron distributions. Examples are molecules with strong electron donor-acceptor groups, for example, in the *para* position on the benzene ring:

(15-16) H_2N —⟨ ⟩— NO_2 ⟷ $H_2\overset{\ominus}{N}$ =⟨ ⟩= $\overset{\oplus}{N}O_2$

$$\beta = 34.5 \cdot 10^{30} \text{ esu}$$

$$\beta = 10.2 \cdot 10^{30} \text{ esu}$$

Correspondingly, quadratic hyperpolarizabilities β of neutral solitons are very small but those of positively or negatively charged ones are very large. β increases here according to $\beta \sim N^{3.05}$ with approximately the 3rd power of the number N of carbon atoms in the chain. Even stronger dependencies, i.e., $\gamma \sim N^q$, are predicted for the cubic hyperpolarizabilities of conjugated chains ($5.0 < q < 5.25$). According to these theoretical calculations, γ is negative for singly (positively or negatively) charged solitons but positive for doubly charged polarons (i.e., bipolarons).

Similar considerations apply to *assemblies of molecules*, for example, high- or low-molecular weight crystals, liquid crystals, or Langmuir-Blodgett films. An electronic asymmetry here can be generated by chiral groups in side-chain liquid crystal molecules. Such molecular assemblies interact with the entering light of frequency ω which produces new light waves with double, triple, etc., frequencies.

Corresponding effects are observed for electromagnetic waves with other than optical frequencies. Entry frequencies of 0 and ω generate exit frequencies of $-\omega$ in substances with quadratic susceptibilities. This Pockels effect is used for modulators (p. 639).

Substances with $\chi^{(3)}(-\omega; \omega,0,0)$ lead to quadratic electrooptical effects ($-\omega$ is the exit frequency and $\omega,0,0$ are the entry frequencies). This effect is used for liquid crystal displays. The response times here cover more than 15 decades; they are (10^{-15}-10^{-12}) s for electronically stimulated, non-resonating NLOs, (10^{-12}-1) s for photochemically generated effects, and (10^{-9}-10^{-3}) s for thermally induced effects.

In order to achieve large second-order susceptibilities $\chi^{(2)}$ in polymeric materials, it is not sufficient that the NLO carriers (NLO chains, NLO side groups, and/or low-molecular weight NLO molecules (such as dyestuffs) in non-NLO polymers) exhibit large quadratic hyperpolarizabilities β. The NLO carriers must also aggregate to macroscopic non-centrosymmetric regions so that contributions of individual tensor components become additive. In inorganic compounds such as lithium niobate $LiNbO_3$, this is achieved by a control of the crystal growth.

Statistically oriented segments of organic molecules can by aligned by (**electric field**) **poling** which is the application of an electrical field at temperatures above the glass transformation T_G. Electric fields with strengths of (10^5-10^6) V/cm are generated between two electrodes or by a corona discharge and held for some time. At these temper-

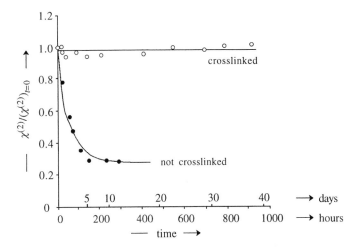

Fig. 15-16 Time dependence of the normalized quadratic susceptibility of a polymer before and after crosslinking [3]. The uncrosslinked polymer with the following repeating unit had a weight-average molecular weight of $\overline{M}_w$ = 7720 g/mol (corresponds to a weight-average degree of polymerization of repeating units of $\overline{X}_w \approx 13$) and a non-uniformity ("polydispersity") index of $\overline{M}_w / \overline{M}_n$ = 1.6.

$$-\underset{O}{\overset{O}{\overset{\|}{C}}}-NH(CH_2)_6NH-\underset{O}{\overset{O}{\overset{\|}{C}}}-O(CH_2)_6-\underset{O}{\overset{O}{\overset{\|}{S}}}-\!\!\!\bigcirc\!\!\!-N=N-\!\!\!\bigcirc\!\!\!-\underset{CH_3}{\overset{}{N}}-CH_2CH_2O-$$

The polymer was poled 15 min at 115°C. The non-optimized Pockels coefficient (SHG coefficient) was r_{33} = 40 pm/V, which led, according to Eq.(15-15) with a refractive index of $n \approx 1.42$, to a calculated linear optical activity of $\chi^{(2)}$ = 80 pm/V. By permission of Plenum Press, New York.

atures, $T > T_G$, molecule segments become mobile and polar segments orient themselves more or less in the field direction. The polymer is then quenched in order to lock-in the resulting non-centro-symmetric segment orientations. Finally, the field is removed.

However, molecule segments of non-crosslinked polymers may be somewhat mobile even below the glass temperature. As a consequence, the quadratic susceptibility $\chi^{(2)}$, normalized to the initial value of $(\chi^{(2)})_{t=0}$, decreases after poling and finally approaches a constant value (Fig. 15-16) The values of $\chi^{(2)}/(\chi^{(2)})_{t=0}$ stay constant if the mobility of segments is suppressed by crosslinking the polymer chains.

In another strategy, polymers with high glass temperatures are used so that a reorganization of the poled chain segments becomes unlikely. An example is the commercial polymer I with biphenyl groups and a $T_G > (300\text{-}400)°C$. Industry prefers polyimide chains because they are thermally stable at the required high processing temperatures. Another, commercial NLO polymer II (Optimer®) employs small, thermally stable 1,4-phenylene chain units that each carry a large thermally stable substituent in which a dialkylamine donor and a dicyanovinyl acceptor are bound to two thiophene units.

Table 15-7 Second non-linear susceptibilities $\chi^{(3)}$ of poly(diacetylene)s, $-\!\!+\!\!C\equiv C-CR=CR+\!\!-_n$, with various substituents R at different wavelengths λ_0 [4]. The letter codes were not explained by [4].

Polymer	Thickness d of specimen in μm	$10^9 \chi^{(3)}$/esu at λ_0/nm =		
		1830	1940	2100
Poly(BTFP)	0.27	0.64	0.32	0.20
Poly(PTS)	0.90	0.23	0.091	0.040
Poly(DFMP)	1.09	0.080	0.069	0.056

Third-order NLOs

Optoelectronics is based on the interaction of photons and electrons. However, much faster transfers and conversions of signals can be obtained by **photonics** where photons interact with photons and light is thus controlled by light. Such exclusively optical systems require materials with considerable second non-linear susceptibilities $\chi^{(3)}$, i.e., those with a third-order dependence of the refractive index on the field.

In contrast to the second-order NLO effects, third order NLO effects can be obtained from centrosymmetric systems. For example, such large effects are observed for polymer molecules with conjugated double bonds such as substituted poly(diacetylene)s (Table 15-7). The effects depend on the wavelength.

Table 15-8 Second non-linear susceptibilities $\chi^{(3)}$ of polymers (third-order NLOs) at wavelengths λ at which resonance (res.) or no resonance (n-res.) exists.
Polymers: I = Poly(acetylene), II = poly(1,4-phenylene vinylene), III = poly(2,5-thiophene), IV = poly(1,4-phenylene benzobisthiazole), V = with 4-BCMU-substituted poly(diacetylene) (red modification), VI = emeraldine base (p. 598), VII = poly(4-phenyl-2,6-quinoline).

Repeating unit	$10^9 \chi^{(3)}$ / esu		λ / nm	
	res.	n-res.	res.	n-res.
I —CH=CH—	1.1	0.10	620	950
II —CH=CH—	0.4		602	
III	0.4		602	
IV		0.45		602
V		0.4		602
VI	0.037		1830	
VII	0.011	0.0023	1200	2380

According to present understanding, third-order NLO effects arise if polymer molecules are conjugated over large distances, possess easily polarizable electrons, and have small optical band gaps. Non-resonating third-order NLO molecules do not absorb light; electrons here are only polarized by the interactions with very intense light beams. The larger the conjugated sections, the less firmly are electrons bound and the easier are polarizations. The polarization increases especially strongly if resonance is present at the entering wavelength (or the overtones) since this leads to an absorption of light (Table 15-8).

So far, poly(acetylene) (I) has shown the highest value of any third order NLO (Table 15-8). However, even this value is about two decades too low for practical applications. Poly(acetylene) is furthermore easily oxidized by oxygen from air.

Literature to Chapter 15

15.0 GENERAL REVIEWS
R.Ross, A.W.Birley, Optical Properties of Polymeric Materials and Their Measurements, J.Phys. D (Appl.Phys.) **6** (1973) 795
R.S.Hunter, The Measurement of Appearance, Wiley, New York 1975
G.H.Meeten, Ed., Optical Properties of Polymers, Elsevier, London 1986
W.Schnabel, Polymers and Light, Wiley-VCH, Weinheim 2007

15.1.5 FIBER OPTICS
L.A.Hornal, Ed., Polymers for Lightwave and Integrated Optics; Technology and Applications, Dekker, New York 1992
J.Marceau, Ed., Plastic Optical Fibers Practical Applications, Wiley, New York 1997
R.A.Shotwell, An Introduction to Fiber Optics, Prentice Hall, Columbus (OH) 1997
J.Hecht, Understanding Fiber Optics, Prentice Hall, Columbus (OH) 1999
H.S.Nalwa, Ed., Polymer Optical Fibers, Amer.Sci.Publ., Stevenson Ranch (CA) 2004
M.G.Kuzyk, Polymer Fiber Optics. Materials, Physics, and Applications, CRC Press, Boca Raton (FL) 2007
M.L.Calvo, V.Lakshminarayanan, Optical Waveguides. From Theory to Applied Technologies, CRC Press, Boca Raton (FL) 2007

15.1.6 OPTICAL BIREFRINGENCE
E.Riande, E.Saíz, Dipole Moments and Birefringence of Polymers, Prentice Hall, Englewood Cliffs (NJ) 1992

15.1.7 IRIDESCENCE
T.Alfrey, Jr., E.F.Gurnee, W.J.Schrenk, Physical Optics of Iridescent Multilayered Plastic Films, Polym.Eng.Sci. **9** (1969) 400

15.3 OPTOELECTRONICS AND PHOTONICS
Y.R.Shen, The Principles of Nonlinear Optics, Wiley, New York 1984
D.J.Williams, Organic Polymeric and Non-Polymeric Materials with Large Optical Nonlinearities, Angew.Chem. **96** (1984) 637; Angew.Chem.Int.Ed.Engl. **23** (1984) 690
D.S.Chemla, J.Zyss, Eds., Nonlinear Optical Properties of Organic Materials and Crystals, Academic Press, Orlando (FL) 1987
D.L.Wise, G.E.Wnek, D.J.Trantolo, T.M.Cooper, J.D.Gresser, Photonic Polymer Systems. Fundamentals, Methods, and Applications, Dekker, New York 1988
R.A.Hahn, D.Bloor, Eds., Organic Materials for Non-Linear Optics, Royal Soc. Chem., London 1989; CRC Press, Boca Raton (FL) 1989
J.Messier et al., Eds., Nonlinear Optical Effects in Organic Polymers, Kluwer Academic, Dordrecht, Netherlands, 1989

P.N.Prasad, D.J.Williams, Introduction to Nonlinear Optical Effects in Molecules and Polymers, Wiley, Chichester 1991

J.C.Brédas, R.Silky, Conjugated Polymers. The Novel Science and Technology of Highly Conducting and Nonlinear Optically Active Materials, Kluwer Academic, Dordrecht, Netherlands 1992

L.A.Hornak, Ed., Polymers for Lightwave and Integrated Optics, Dekker, New York 1992

J.Zyss, Ed., Molecular Nonlinear Optics, Academic Press, Orlando (FL) 1993

F.Agallo-Lopez, Electrooptics, Academic Press, San Diego 1994

J.A.Chilton, M.T.Goosey, Eds., Special Polymers for Electronics and Optoelectronics, Chapman and Hall, London, 1995

H.S.Nalwa, S.Miyata, Nonlinear Optics of Organic Molecules and Polymers, CRC Press, Boca Raton (FL) 1996

D.L.Wise, G.E.Wnek, D.J.Trantolo, T.M.Cooper, J.D.Gresser, Electrical and Optical Polymer Systems, Dekker, New York 1998

W.R.Salaneck, K.Seki, A.Kahn, J.-J.Pireaux, Eds., Conjugated Polymer and Molecular Interfaces, Dekker, New York 2001

W.Barford, Electronic and Optical Properties of Conjugated Polymers, Oxford Univeristy Press, New York 2005

J.P.Dakin, R.G.W.Brown, Eds., Handbook of Optoelectronics, CRC Press, Boca Raton (FL) 2006 (2 vols.)

M.C.Gupta, J.Ballato, The Handbook of Photonics, CRC Press, Boca Raton (FL), 2nd ed. 2007

A.Rogers, Essentials of Photonics, CRC Press, Boca Raton (FL), 2nd ed. 2008

S.-S.Sun, L.H.Dalton, Introduction to Organic Electronic and Optoelectronic Materials and Devices, CRC Press, Boca Raton (FL) 2008

H.S.Nalwa, Ed., Handbook of Organic Electronics and Photonics, Amer.Sci.Publ., Los Angeles (CA) 2008, 3 vols.

15.3.2 LIGHT-EMITTING DIODES

J.Shinar, Ed., Organic Light-Emitting Devices: A Survey, Springer. Berlin 2004

H.S.Nalwa, Ed., Handbook of Luminescence, Display Materials and Devices, Volume 1, Organic Light-Emitting Diodes, Amer.Sci.Publ., Los Angeles (CA) 2003

G.Müllen, U.Scherf, Eds., Organic Light Emitting Devices. Synthesis, Properties, and Applications, Wiley-VCH, Weinheim 2006

H.Yersin, Ed., Highly Efficient OLEDs with Phosphorescent Materials, Wiley-VCH, Weinheim 2008

15.3.3 SOLAR CELLS

A.Mayer, S.Scully, B.Hardin, M.Rowell, M.McGehee, Polymer-based Solar Cells, Materials Today **10** (2007) 28

15.3.4 NONLINEAR OPTICS

C.J.Brabec, V.Dyakonov, J.Parisi, N.S.Sariciftci, Eds., Organic Photovoltaics, Springer, Berlin 2003

S.-S.Sun, N.S.Sariciftci, Eds., Organic Photovoltaics. Mechanisms, Materials, and Devices (Optical Engineering), CRC Press, Boca Raton (FL) 2005

References to Chapter 15

[1] G.Lindenschmidt, N.Noiessner, Kunststoffe **85** (1995) 1066, Fig. 3

[2] A list of early work on OLEDs, starting with the 1953 papers of A.Bernanose et al. on OLEDs from acridine orange and quinacrine, can be found in http://en.wikipedia.org/wiki/OLED (accessed 2008-10-18). The PLED in Fig. 15-11 was designed by N.C.Greenham, S.C.Moratti, D.D.C.Bradley, R.H.Friend, A.B.Holmes, Nature **365** (1993) 628

[3] L.R.Dalton, C.Xu, B.Wu, A,W.Harper, in P.N.Prasad, Ed., Frontiers of Polymers and Advanced Materials, Plenum Press, New York 1994, p. 175, Fig. 8

[4] H.Matsuda, S.Okzda, H.Nakanishi, M.Kato, J.Photopolym.Sci.Tech. **2** (1989) 253, reported by J.Lee, S.Tripathy, H.Matsuda, S.Okada, H.Nakanishi, ACS Polym.Prepr. **31**/1 (1990) 414. The paper did not explain the meaning of BTFP, PTS, and DFMP.

16 Appendix

16.1 SI Physical Quantities and Units

According to **Maxwell**, many physical properties can be described in quantitative terms by *quantity calculus*. The value of a physical quantity (symbol in *italics*) equals the product of a numerical value (Roman (upright) type) and a physical unit (symbol in upright letter):

$$\text{physical quantity} = \text{numerical value} \times \text{physical unit}$$

This equation can be manipulated by the ordinary rules of algebra. If, for example, a certain item has a length L of 0.002 meters (m) = 2 millimeters (mm), then this may be written as

$$L = 0.002 \text{ m} \quad \text{or} \quad L = 2 \cdot 10^{-3} \text{ m} \quad \text{or} \quad 10^3 \, L/\text{m} = 2 \quad \text{or} \quad L = 2 \text{ mm} \quad \text{or} \quad L/\text{mm} = 2$$

but *not* as $10^{-3} \, L/\text{m} = 2$. For example, a column head $10^2 \, F/(\text{N m}^{-2})$ for a column entry of 7.35 thus indicates a force $F = 7.35 \cdot 10^{-2} \text{ N/m}^2$. Literature data often do not follow these SI (Systéme International) rules by the International Standardization Organization (ISO), which are adopted by IUPAP (International Union of Pure and Applied Physics), IUPAC (International Union of Pure and Applied Chemistry), etc. Instead one finds various nonrational notations such as F, N m^{-2} or F [N m^{-2}] for a column entry of $7.35 \cdot 10^{-2}$, often with wrong algebraic statements such as $10^{-2} \, F$, N m^{-2}.

The International System of Units (Système international; **SI system**) uses seven **base physical quantities** and seven **SI base units** (Table 16-1). It replaces the formerly used CGS and MKS systems; it is *not* "the metric system." The SI system also defines several **derived units** (Table 16-2) which now include the radian and steradian (not shown) that were formerly considered "supplementary units." Several older units may still be used for the time being (Table 16-4).

In order to avoid cumbersome numbers, symbols of units may be prefixed with SI prefixes (Table 16-5). The decimal sign between digits in a number is either a period (US) or a comma (e.g., Europe). No commas or periods should be used to separate groups of three digits for better reading. Instead, spaces should be applied (exception: groups of four digits may be written without space). Examples:

US: 1 234.567 8 or 1234.5678; Europe: 1234,5678 or 1 234,567 8; but *not* 1,234.5678.

Numbers and units are separated by a gap and groups of quantities or units by spaces or multiplication signs ($\cdot$ or $\times$).

All countries except the United States of America, Liberia, and Myanmar have adopted the SI system. In many countries, the SI system is the only system of weights and measures that can be lawfully used in trade and commerce. American technical literature and sometimes also American scientific literature still uses American and old Imperial British units although the metric system was introduced in 1896 by an act of Congress and by law all U.S. government units were supposed to convert to the SI system by the end of 1992. The following tables list names and symbols of base and derived SI units, temporarily allowed non-SI units, and SI prefixes. This book follows IUPAC/IUPAP recommendations; exceptions are noted in the list of symbols of physical quantities.

Table 16-1 Base quantities and base units.

| Physical quantity | | SI base unit | | SI symbol |
SI symbol	SI name	English name	American name	
l	length	metre	meter	m
m	mass	kilogramme	kilogram	kg
t	time	second	second	s
I	electric current	ampere	ampere	A
T	thermodynamic temperature	kelvin	kelvin	K
n	amount of substance	mole	mole	mol
I_v	luminous intensity	candela	candela	cd

Table 16-2 Some derived SI units and their symbols (recommendations by IUPAC and IUPAP) used in this volume. For a comprehensive list, see Volume III, Table 19-2.

Physical quantity		SI unit	
Symbol Name		SI name	SI symbol and unit(s)
P	power, radiant flux	watt [1]	W $= V\,A$ $= J\,s^{-1}$ $= m^2\,kg\,s^{-3}$
E	energy, work, heat	joule	J $= N\,m$ $= m^2\,kg\,s^{-2}$
$\boldsymbol{F}$ [2]	force	newton	N $= J\,m^{-1}$ $= m\,kg\,s^{-2}$
-	impact strength (US)	newton	$J\,m^{-1}$ $= m\,kg\,s^{-2}$
G	weight	newton	N $= J\,m^{-1}$ $= m\,kg\,s^{-2}$
-	impact strength (Europe)	-	$J\,m^{-2}$ $= kg\,s^{-2}$
γ	interfacial tension	-	$J\,m^{-2}$ $= N\,m^{-1}$ $= kg\,s^{-2}$
p, σ	pressure, stress	pascal	Pa $= J\,m^{-3}$ $= N\,m^{-2}$ $= m^{-1}\,kg\,s^{-2}$
	impulse, momentum	-	$N\,s$ $= m\,kg\,s^{-1}$
I	electric current	ampere	A $= C\,s^{-1}$
Q	electric charge	coulomb	C $= A\,s$
U	electric potential, electromotive force	volt	V $= W\,A^{-1}$ $= J\,C^{-1}$ $= m^2\,kg\,s^{-3}\,A^{-1}$
R	electric resistance	ohm	Ω $= V\,A^{-1}$ $= m^2\,kg\,s^{-3}\,A^{-2}$
G	electric conductance	siemens	S $= \Omega^{-1}$ $= m^{-2}\,kg^{-1}\,s^3\,A^2$
C	electric capacitance	farad	F $= C\,V^{-1}$ $= m^{-2}\,kg^{-1}\,s^4\,A^2$
ε	relative permittivity [3]	-	1
Φ_{v}	luminous flux	lumen	lm $= cd\,sr$
E_{v}	illuminance	lux	lx $= cd\,sr\,m^{-2}$
t, θ [4]	Celsius temperature	degree Celsius	$^{\circ}C$

[1] Names of units derived from persons' names are not capitalized. Exceptions are "degree Celsius" and two-letter symbols for dimensionless quantities (Reynolds number, Deborah number, etc.).
[2] Vectorial: symbols are in bold letters (directions: $\boldsymbol{u}, \boldsymbol{v}, \boldsymbol{w}$).
[3] Formerly: dielectric constant.
[4] This book uses T as a symbol for both the kelvin and the Celsius temperature (see Table 16-4).

Table 16-3 Fundamental constants used in this book (CODATA: 2006 adjustment).

Physical quantity	Symbol	= number × physical unit	IUPAC symbol
Boltzmann constant	k_{B}	$= 1.380\ 6504 \cdot 10^{-23}\ J\,K^{-1}$	k
Avogadro constant	N_{A}	$= 6.022\ 141\ 79 \cdot 10^{23}\ mol^{-1}$	N_{A} or L [1]
Loschmidt constant	$N_{\mathrm{A}}/V_{\mathrm{m}}$	$= 2.686\ 7774 \cdot 10^{25}\ m^{-3}$	n_{o} [2]
Molar gas constant	R	$= 8.314\ 472\ J\,mol^{-1}\,K^{-1}$	R
Atomic mass constant	m_{u}	$= 1.660\ 538\ 782 \cdot 10^{-27}\ kg\ (= m(^{12}C)/12)$	m_{u}
Permittivity of vacuum	ε_{o}	$= 1/(\mu_0 c_0^2) = (8.854\ 187\ 817...) \cdot 10^{-12}\ F\,m^{-1}$ (exact)	
Atomic mass constant [3]	m_{u}	$= 1.660\ 538\ 782 \cdot 10^{-27}\ kg\ (= m(^{12}C)/12)$	
Molar volume, ideal gas	V_{m}	$= 22.413\ 996 \cdot 10^{-3}\ m^3\,mol^{-1}\ (T = 273.15\ K, p = 101\ 325\ Pa)$	
Faraday constant	F	$= 96\ 485.3399\ C\,mol^{-1}$	

[1] The IUPAC symbol N_{A} is inconsistent with the IUPAC use of N as a symbol for a number because N_{A} has the unit of an inverse amount-of-substance. The first to determine the value of the "Avogadro constant" was Loschmidt, not Avogadro (hence the alternative symbol L). The old German term for N_{A} was indeed Loschmidt number, N_{L}.
[2] The "Loschmidt constant" was determined first by Avogadro, not by Loschmidt (hence the old German term "Avogadro number"). The IUPAC use of n_{o} as the symbol for an inverse volume is inconsistent with the other IUPAC use of n as the symbol for the amount-of-substance.
[3] Also called "unified atomic mass constant."

Table 16-4   Older physical units. * Non-SI units that may be used with SI prefixes or SI units.

Physical quantity	Physical unit Name	Physical unit Symbol	Physical unit Value in SI units	Notes
Time	minute	min	= 60 s	1, 2)
Time	hour	h	= 3600 s	1, 2)
Time	day	d	= 86 400 s	1, 2)
Time	month	(mo)	= 30 d (see Table 16-6)	2, 3)
Time	year	(yr, a)	= 365 d (see Table 16-6)	2, 4)
Length	ångstrøm	Å	$= 10^{-10}$ m = 0.1 nm	5, 6)
Volume	liter	l, L	$= 10^{-3}$ $m^3 \equiv$ 1 L	7)
Mass	ton(ne) *	t	$= 10^3$ kg	8)
Mass	unified atomic mass unit [4]	$u = m_a(^{12}C)/12$	$\approx 1.660\ 54\cdot10^{-27}$ kg	9)
Energy	electronvolt *	eV	$\approx 1.602\ 18\cdot10^{-19}$ J	10)
Pressure	bar *	bar	$= 10^5$ Pa	5)
Temperature	Celsius temperature	°C	$= \theta/°C = (T/K) - 273.15$	6.11)

[1]   IUPAC allows the use of the non-SI units "minute", "hour", and "day" in appropriate contexts although these three physical units are not part of the SI system.

[2]   These units should not be used with SI prefixes.

[3]   "Month" is not a scientific unit. In commercial data, the symbol for "month" is often "mo."

[4]   "Year" is not a scientific unit. In commercial data, the symbol for "year" is often "yr." For international use, "a" is preferred as the symbol (L: *annus* = year).

[5]   This unit is approved for "temporary use with SI units" in fields where it is presently used.

[6]   Names of units derived from person's names are not capitalized. Exceptions are "degree Celsius" and two-letter symbols for dimensionless quantities (Reynolds number, Deborah number, etc.).

[7]   The "liter" is not an SI unit but both "liter" and its symbols "l" and "L" may be used with SI prefixes such as deci (d), centi (c), milli (m), or micro (μ). The symbol "L" is preferred over the symbol "l" because the letter "l" (el) may be confused with the number "1."

[8]   IUPAC allows the use of the physical unit "ton(ne)" $\equiv$ 1000 kg (especially for technical and commercial data) which is, however, not an SI unit. The ton is often used with SI prefixes (kiloton, megaton). The "metric" ton (UK: tonne) with the symbol "ton" is not to be confused with "long ton" (1 long ton $\approx$ 1016.047 kg) or "short ton" (1 short ton $\approx$ 907.185 kg). These two "tons" are often used in commerce and statistics without the adjectives "long" and "short."

[9]   The value of this unit depends on the experimentally determined value of the Avogadro constant N_A; the value of the corresponding SI unit is therefore not exact. The unified atomic mass (physical unit: kg) is sometimes called the dalton (symbol Da) which is not an SI unit. In the biosciences, and recently also in polymer science, "dalton" has erroneously come to mean the relative molecular mass = molecular weight (physical unit: 1) or the molar mass (physical unit: g/mol)!

[10]   The value of this unit depends on the experimentally determined value of the elementary charge e; the value of the corresponding SI unit is therefore not exact.

[11]   The SI unit of the Celsius temperature *interval* is the degree Celsius (SI symbol of the unit: °C) which is equal to the kelvin (*not* "degree kelvin" or "degree Kelvin"). The *symbol* of the unit kelvin (small k!) is K, *not* °K. Celsius is always written with a capital C.

When quoting temperatures in kelvin, a space should be written between the numerical value and the symbol K as it is customary for all physical properties. However, if temperatures are given in degrees Celsius, no space should exist between the numerical value and the symbol. Thus: T = 298 K but also T = 25°C (this book).

IUPAC now recommends the symbol θ for the physical property "Celsius temperature." In polymer science, a capital theta (Θ) is the traditional symbol for the property "theta temperature." However, the symbols θ and Θ are easily mixed up, even by experienced polymer scientists. This book thus uses the symbol T for both thermodynamic temperatures and Celsius temperatures; the possibility of a mix-up is remote since physical units are always given.

Table 16-5 SI prefixes for SI units. The prefixes were introduced because two groups of Western countries used the same names of numbers for different numbers [1], for example, "billion."

Factor	Prefix [3]	Symbol [4]	Common name [1] United States	Europe	Origin of prefixes [2]
10^{24}	yotta [5]	Y	septillion	quadrillion	L: *octo* = eight [$10^{24} = (10^3)^8$]
10^{21}	zetta [6]	Z	sextillion	trilliard	L: *septem* = seven [$10^{21} = (10^3)^7$]
10^{18}	exa	E	quintillion	trillion	G: *hexa* = six [$10^{18} = (10^3)^6$]
10^{15}	peta	P	quadrillion	billiard	G: *penta* = five [$10^{15} = (10^3)^5$]
10^{12}	tera	T	trillion	billion	G: *teras* = monster
10^9	giga	G	billion	milliard	G: *gigas* = giant
10^6	mega	M	million	million	G: *megas* = big
10^3	kilo	k	thousand	thousand [7]	G: *khilioi* = thousand
10^2	hecto [8]	h	hundred	hundred	G: *hekaton* = hundred
10^1	deca [9]	da	ten	ten	G: *deka* = ten
10^{-1}	deci	d	one tenth	one tenth	L: *decima pars* = one tenth
10^{-2}	centi	c	one hundredth	one hundredth	L: *pars centesima* = one hundredth
10^{-3}	milli	m	one thousandth	one thousandth	L: *pars millesima* = one thousandth
10^{-6}	micro [10]	μ	one millionth	one millionth	G: *mikros* = small
10^{-9}	nano	n	one billionth	one milliardth	G: *nan(n)os* = dwarf
10^{-12}	pico	p	one trillionth	one billionth	I: *piccolo* = small
10^{-15}	femto	f	one quadrillionth	one billiardth	D, N: *femten* = fifteen
10^{-18}	atto	a	one quintillionth	one trillionth	D, N: *atten* = eighteen
10^{-21}	zepto [6]	z	one sextillionth	one trilliardth	L: *septem* = seven [$10^{-21} = (10^{-3})^7$]
10^{-24}	yocto [5]	y	one septillionth	one quadrillionth	L: *octo* = eight [$10^{-24} = [10^{-3})^8$]

[1] The names of numbers (= numerals) of most countries are based on the so-called long scale (LS) of numbers which uses "million" (= 10^6) as the base and multiplies the exponent 6 with the series of whole positive integers (1, 2, 3, 4, etc.). The names of the numbers are then given by prefixes of "illion" that are derived from Latin, starting with *bi* (2). Examples: 10^6 = million = $(10^6)^1$, $(10^6)^2$ = billion = 10^{12}, $(10^6)^3$ = trillion = 10^{18}, $(10^6)^4$ = quadrillion = 10^{24}, etc.

Numbers that are greater than these long scale numbers by factors of 10^3 are given the same prefixes as the preceding long scale numbers but the ending "illiard.". Examples: $10^6 \cdot 10^3 = 10^9$ = milliard, $10^{12} \cdot 10^3 = 10^{15}$ = billiard, $10^{18} \cdot 10^3 = 10^{21}$ = trilliard, etc. This series of numbers is therefore $(10^6)^{1.5}$ = milliard, $(10^6)^{2.5}$ = billiard, $(10^6)^{3.5}$ = trilliard. etc.

The United States, followed in 1974 by the United Kingdom and a number of other English-speaking countries, use the so-called short scale (SS) which is based on the series 10^{3x} with x = 2 (million), x = 3 (billion), x = 4 (trillion), etc. In this system, multipliers x no longer correspond to the Latin words since "billion" is now $10^{6.3}$ and not $10^{6.2}$, trillion is $10^{6.4}$ and not $10^{6.3}$, etc. Because of this inconsistency and the different meanings of numerals in the long and short scales of numbers, commercial publications often use expressions such as "thousand millions" for "billions" (SS) and "billiards" (LS), "million millions" instead of "trillions" (SS) and "billions" (LS), etc.

[2] D = Danish, G = Greek, I = Italian, L = Latin, N = Norwegian.

[3] Names of prefixes for numbers of 10^6 and greater always end with "a" and names of prefixes for numbers of 10^{-6} and smaller always with "o."

[4] Symbols for numbers of 10^6 and greater are always capital letters.

[5] ISO prefixed "otta" with y because a symbol "O" would be confusing. Ditto for "octo" (10^{-24}).

[6] ISO replaced "s" by "z" in order to avoid the use of "s." Ditto for "zepto" (10^{-21}).

[7] French: mille.

[8] NIST recommends "hekto" (etymologically correct, see right column) but ISO uses "hecto."

[9] NIST recommends "deka" (etymologically correct, see right column) but ISO uses "deca."

[10] USA: μ as the symbol for "micro" is neither known to the general public nor to newspapers and magazines; it is also not on typewriter and computer keyboards. The prefix "μ" is therefore sometimes replaced by the non-SI prefix "mc" (from "micro"; 1 mcg = 1 microgram ≡ 1 μg). The non-SI prefix "ml" is then substituted for the SI prefix "m" = "milli" (1 mlg = 1 milligram ≡ 1 mg).

16.2 Conversion of Units

Table 16-6 Conversion of old units.

Name	Old unit	= SI unit
Lengths		
Statute mile	1 mile	= 1609.344 m
Yard	1 yd = 3 ft	= 0.914 4 m (exact)
Foot	1 ft = 1' = 12"	= 0.304 8 m (exact)
Inch	1 in = 1"	= 2.54 cm (exact)
Mil or thou	1 mil = 0.001 in	= 25.4 μm (exact)
Micron	1 μ	= 10^{-6} m ≡ 1 μm
Millimicron	1 mμ	= 10^{-9} m ≡ 1 nm
Ångstrøm	1 Å	= 10^{-10} m ≡ 0.1 nm
Areas		
Square mile	1 sq. mile = 640 acres	= 2 589 988.110 m^2
Hectare (land)	1 ha = 100 a	= 10 000 m^2
Acre	1 acre = 4 roods	= 4046.856 m^2
	1 acre = 160 square rods	
	1 acre = 4840 square yards	
Square yard	1 sq. yd. = 9 sq. ft.	= 0.836 127 36 m^2
Square foot	1 sq. ft. = 144 sq. in.	= 0.092 030 4 m^2
Square inch	1 sq. in.	= 6.451 6·10^{-4} m^2
Barn	1 b	≡ 10^{-28} m^2

Volumes (= units of capacity)

The liter (symbol L or l) is now defined as $1·10^{-3}$ m^3 and can be used instead of the cubic meter. (1 m^3 ≡ 10^3 L). The liter (cgs) is 1 L = 1.000 028·10^{-3} m^3.

In the United Kingdom of Great Britain and North Ireland, the gallon (UK) and related units of measure were used for both liquid measures (fluid measures) as well as for dry measures.

In the United States of America, the gallon is still used for liquids. For dry goods, the US uses the bushel. The barrel is a commercial measure of liquids *and* weights. The actual measures of gallons, bushels, and barrels varies with the type of goods.

Cubic yard	1 cu. yd.	= 764.554 857 L
Barrel (UK)	1 barrel	= 163.61 L
Barrel (US)	1 barrel = 31.5 gallons	= 119.237 L
Barrel (US barrel alcohol)	50 US gallons	= 189.25 L
Barrel (US barrel petroleum)	1 bbl = 42 US gal	= 158.987 L
Bushel (UK)	1 bu (UK) = 8 gal (UK)	≈ 36.368 L
Bushel (US struck bushel)	1 bu = 4 pecks	= 35.239 L (US stricken bushel)
Cubic foot	1 cu. ft.	= 23.816 846 592 L (exact)
Gallon (UK, British or Imperial)	1 gal (UK)	= 4.545 96 L
Gallon (US dry)	1 dry gal = 4 dry qts.	= 4.405 L
Gallon (US fluid)	1 gal = 4 fluid quarts	= 3.785 412 L
Board foot	12·12·1 cu. in.	= 2.3597 L
Imperial quart (UK; liquid or dry)	1 qt.	= 1.136 522 L
Dry quart (US)	1 qt. = 67.2 cu.in.	= 1.101 211 L
Liquid quart (US)	1 qt. = 57.75 cu.in.	= 0.946 353 L
Reputed quart (UK; wine, spirits)	= 26.67 UK fluid oz.	= 0.757 775 L
Cubic inch	1 cu. in.	= 16.387 cm^3
Dry pint (US)	1 pt.	= 0.550 606 L

Table 16-6 Conversion of old units (continued).

Name	Old unit	= SI unit
Masses		
Long ton (UK)	1 ton = 2240 lbs	= 1016.046 909 kg
Ton, tonne (UK)	1 t	≡ 1000.000 000 kg
Short ton (US)	1 sh. ton = 2000 lbs	= 907.184 74 kg
Barrel cement (US)	376 lbs	= 170.551 kg
Barrel salt or beans (US)	280 lbs	= 127.006 kg
Barrel flour (US)	196 lbs	= 88.904 kg
Barrel gun powder (US)	25 lbs	= 11.34 kg
Long hundredweight (US)	1 long ctw = 112 lbs	= 50.802 3 kg
Hundredweight (UK)	1 cwt = 112 lbs	= 50.802 3 kg
Short hundredweight (US)	1 sh. cwt	= 45.359 2 kg
Pound (international)	1 lb = 7000 grains	= 453.592 37 g
Pound (avoirdupois) (US)	1 lb = 16 oz.	= 453.592 427 7 g
Pound (apothecaries' or troy)	1 lb = 5760 grains	= 373.242 g
	1 lb = 12 Troy oz	
Ounce (avoirdupois) (US)	1 oz.	= 28.349 52 g
Ounce (apothecaries')	1 oz ap = 24 scruples	= 31.103 5 g
Ounce (Troy)	1 Troy oz = 480 grains	= 31,103 5 g
	1 Troy oz = 20 dwt	= 31.103 5 g
Atomic mass constant	1 u	= $1.660\ 540\ 2 \cdot 10^{-27}$ kg
Mass of an electron	-	= $9.109\ 39 \cdot 10^{-31}$ kg
Time		
Light year	1 ly	= $9.4605 \cdot 10^{15}$ m
Year	1 a	= 365 days (statistics only)
Months	1 mo	= 30 days (statistics only)
Day	1 d	= 24 h = 86 400 s
Hour	1 h	= 60 min = 3600 s
Minute	1 min	= 60 s
Temperatures		
Degree Celsius	y°C – 273.15°C	= x K
Degree Fahrenheit	(z°F – 32°F)(5/9)	= y°C
Angles		
Degree of plane angle	1°	= (π/180) rad = $1.745\ 329\ 2 \cdot 10^{-2}$ rad
Minute of plane angle	1'	= (1/60)° = (π/10 800) rad
		= $2.908\ 882 \cdot 10^{-4}$ rad
Second of plane angle	1"	= (1/60)' = (π/648 000) rad
		= $4.848\ 136\ 6 \cdot 10^{-6}$ rad
Densities (1 kg m^{-3} = $1 \cdot 10^{-3}$ g cm^{-3})		
Specific gravity	1 lb/cu. in.	= 27.679 904 71 g cm^{-3}
	1 oz/cu. in.	= 1.729 993 853 g cm^{-3}
	1 lb/cu. ft.	= $1.601\ 846\ 337 \cdot 10^{-2}$ g cm^{-3}
	1 lb/gal (US)	= $7.489\ 150\ 454 \cdot 10^{-3}$ g cm^{-3}
Energy, Work, Heat (1 J = 1 N m = 1 W s)		
Coal unit (European hard coal)	1 ton SKE	≙ 29.31 GJ
Coal unit (US)	1 ton coal	≙ 27.92 GJ
Coal unit (UK)	1 ton coal	≙ 24.61 GJ
Short ton bituminous coal (US)	1 T coal	≙ 26.58 GJ

Table 16-6 Conversion of old units (continued).

Name	Old unit	= SI unit

Energy, Work, Heat (1 J = 1 N m = 1 W s) (continued)

Name	Old unit	= SI unit
Kilowatt-hour	1 kWh	= 3.6 MJ
Horsepower-hour	1 hph	= 2.685 MJ
Cubic foot-atmosphere	1 cu.ft.atm.	= 2.869 205 kJ
British thermal unit	1 BTU$_{mean}$	= 1.055 79 kJ
British thermal unit, interntional	1 BTU$_{IT}$	= 1.055 056 kJ
-	1 cu.ft.lb(wt)/sq.in.	= 195.237 8 J
Liter atmosphere (cgs)	1 L atm	= 101.325 0 J
-	1 m kgf	= 9.806 65 J
Calorie, international	1 cal$_{IT}$	= 4.186 8 J
Calorie, thermochemical	1 cal$_{th}$	= 4.184 J
-	1 ft-lbf	= 1.355 818 J
-	1 ft-pdl	= 4.215 384 J
-	1 erg	= $1 \cdot 10^{-7}$ J = 0.1 μJ
Electronvolt	1 eV	$\approx 1.602\ 177\ 33 \cdot 10^{-19}$ J

Force

Name	Old unit	= SI unit
-	1 ft-lbf/in. notch	= 53.378 64 N
Kilogram force	1 kgf	= 9.806 65 N
Pound-force	1 lbf	= 4.448 22 N
Ounce-force	1 oz.f.	= 0.2780 N
Poundal	1 pdl	= 0.138 255 N
Gram-force	1 gf	= $9.806\ 65 \cdot 10^{-3}$ N
Pond	1 p	= $9.806\ 65 \cdot 10^{-3}$ N
Dyne	1 dyne	= $1 \cdot 10^{-5}$ N

Length-related Force

Name	Old unit	= SI unit
-	1 kp/cm	= 980.665 N m^{-1}
-	1 lbf/ft	= 14.593 898 N m^{-1}
-	1 dyn/cm	= $1 \cdot 10^{-3}$ N m^{-1}

Area-related Force, Pressure, Mechanical Stress (1 MPa = 1 N mm^{-2})

Name	Old unit	= SI unit
Physical atmosphere	1 atm = 760 torr	= 0.101 325 MPa
-	1 bar	= 0.1 MPa
Technical atmosphere	1 at	= 0.098 065 MPa
-	1 kp/cm^2	= 0.098 065 MPa
-	1 kgf/cm^2	= 0.098 065 MPa
-	1 lbf/sq.in.	= $6.894\ 76 \cdot 10^{-3}$ MPa
Pound-force per square inch	1 psi	= $6.894\ 76 \cdot 10^{-3}$ MPa
Inch mercury (at 32°F)	1 in.Hg	= $3.386\ 388 \cdot 10^{-3}$ MPa
Inch water (at 39.2°F)	1 in.H$_2$O	= 249.1 Pa
Torr	1 torr	= (101 325/760) Pa $\approx$ 133.322 Pa
Millimeter mercury	1 mm Hg	= $13.5951 \cdot 9.806\ 65$ Pa $\approx$ 133.322 Pa
-	1 dyn/cm^2	= $1 \cdot 10^{-5}$ MPa
Millimeter water	1 mm H$_2$O	= $9.806\ 65 \cdot 10^{-6}$ MPa

Power (1 W = 1 J s^{-1})

Name	Old unit	= SI unit	
Horsepower (boiler)	1 hp	= 9810 W	
Horsepower (electric)	1 hp	= 746 W	
Horsepower (UK)	1 hp	= 745.700 W	
Horsepower* (metric)	1 PS	= 735.499 W	(* D: *Pferdestärke*)
British thermal unit per hour	1 BTU/h	= 0.293 275 W	
Calorie per hour	1 cal/h	= $1.162\ 222 \cdot 10^{-3}$ W	

Table 16-6 Conversion of old units (continued).

Name	Old unit	= SI unit
Thermal Conductivity		
-	1 cal/(cm s °C)	= 418.6 W m^{-1} K^{-1}
-	1 BTU/(ft h °F)	= 1.731 956 W m^{-1} K^{-1}
-	1 kcal/(m h °C)	= 1.162 78 W m^{-1} K^{-1}
Heat Transfer Coefficient		
	1 cal/(cm^2 s °C)	= 4.186 8·10^4 W m^{-2} K^{-1}
	1 BTU/(ft^2 h °F)	= 5.682 215 W m^{-2} K^{-1}
	1 kcal/(m^2 h °C)	= 1.163 W m^{-2} K^{-1}
Length-related Mass (Fineness = Titer = Linear Density)		
Tex (see p. 161)	1 tex	= 1·10^{-6} kg m^{-1}
Denier	1 den	= 0.111·10^{-6} kg m^{-1}
Fracture Length		
Gram per denier	1 g/den	= 9·10^3 m
Textile Strength		
Gram-force per denier	1 gf/den = gpd	= 0.082 599 N tex^{-1} = 0.082 599 m^2 s^{-2}
(gram-pound per denier)		= 98.06 MPa · (density in g cm^{-3})
Dynamic Viscosity		
Poise	1 P	= 0.1 Pa s
Centipoise	1 cP	= 1 mPa s
Kinematic Viscosity		
Stokes	1 St	= 1·10^{-4} m^2 s^{-1}
Heat Capacity		
Clausius	1 Cl	= 1 cal$_{th}$/K = 4.184 J K^{-1}
Molar Heat Capacity		
Entropy unit	1 e.u.	= 1 cal$_{th}$ K^{-1} mol^{-1} = 4.184 J K^{-1} mol^{-1}
Electrical Conductance		
Inverse ohm	1 mho	= 1 S
Electrical Field Strength		
-	1 V/mil	= 3.937 008·10^4 V m^{-1}

Concentrations

Science uses various measures of the abundance of one substance in all other substances present (see next page). For convenience, industry employs various other measures of abundance *after* mixing:

%	= 1 part per hundred parts = per cent (= percent)
‰	= 1 part per thousand parts = per mill (= per mil)
ppm	= 1 part per 1 million parts
ppb	= 1 part per 1 billion parts (to be avoided because of the various meanings of "billion")
ppt	= 1 part per 1 trillion parts (to be avoided because of the various meanings of "trillion")

The plastics and the rubber industry also use

phr   = 1 part per hundred parts of resin or rubber, i.e., *before* mixing

16.3 Concentrations

Concentrations measure the abundance of substance 1 in all substances i present; $i = 1, 2, 3, \ldots$

Mass fraction $= w_1 = m_1/\Sigma_i\, m_i = m_1/m = c_1/c$. Mass m_1 of substance 1 divided by the sum of masses m_i of all substances i. Since all masses reside in the same gravity field, a mass fraction can also be called a **weight fraction** ("weight" is not an accepted ISO term; it was formerly the name of a mass in a gravity field).

The value of $100\, w_1$ is called weight percent (wt%) and the value of $1000\, w_1$ is called weight pro-mille (wt-‰). The English language literature also uses part per hundred (1 pph = 1 %), part per million (1 ppm = 10^{-4} %), and part per (American) billion (1 ppb = 10^{-7} %), and part per (American) trillion (1 ppt = 10^{-10} %).

Volume fraction $= \phi_1 = V_1/(\Sigma_i\, V_i) = V_1/V$. Volume V_1 of substance 1 divided by the total volume of all substances i *before* the mixing process.

Mole fraction, amount fraction, number fraction $= x_1 = n_1/(\Sigma_i\, n_i) = n_1/n$. Amount n_1 of substance 1 divided by the sum of amounts n_i of all substances i. "Amount-of-substance" (short: "amount") is measured in moles, never in kilograms; it is not a mass! The amount-of-substance was (and still is) erroneously called "mole *number*" in the literature.

Mass concentration = mass density. In polymer science, generally as $c_1 = m_1/V$, i.e., as mass m_1 of substance 1 per volume V of mixture *after* mixing. IUPAC recommends the symbols γ_1 or ρ_1 instead of c_1 but ρ_1 may be confused with the same symbol for the mass density of a *neat* substance. Mass concentrations are usually called "concentrations" in the literature.

Number concentration = number density of entities. $C_1 = N_1/V$. Number N_1 of entities of type 1 (molecules, atoms, ions, etc.) per volume V of mixture *after* mixing. IUPAC recommends the name "concentration" for this physical quantity but this may be confused with the much more common "concentration" = mass concentration.

Amount-of-substance concentration = amount concentration. In polymer science, this is usually defined as $[1] = n_1/V$, i.e., amount-of-substance 1 per volume V of mixture *after* mixing. IUPAC recommends $c_1 = n_1/V$, which may be confused with the symbol c_1 for the mass concentration. The amount concentration is often called "**mole concentration**" or **molarity** and given the symbol M; the latter symbol is not recommended by IUPAC and should not be used with SI prefixes (i.e., not mM for an amount concentration of "millimole" per Liter).

Molality of a solute. $a_1 = n_1/m_2$, i.e., amount n_1 of substance 1 per mass m_2 of solvent 2. Molalities are often denoted by m, which should not be used as the symbol for the unit mol kg^{-1}.

16.4 Ratios of Physical Quantities

The terms "normalized", "relative", "specific", and "reduced" are sometimes used with different meanings although they are clearly defined.

Normalized requires that the quantities in the numerator and denominator are of the same kind. A normalized quantity is always a fraction (quantity of the subgroup divided by the quantity of the group); the sum of all normalized quantities equals unity.

Relative also refers to quantities that are of the same kind in the numerator and denominator but the quantity in the denominator may be in any defined state. Example 1: relative viscosity $\eta_r = \eta/\eta_2$ as the ratio of the viscosity η of a solution to the viscosity η_2 of the solvent 2. Example 2: relative humidity = ratio of moisture content of air to moisture content of air saturated with water (both at the same temperature and pressure).

Specific refers to a physical quantity divided by the mass. The symbol of a specific quantity is the lower case form of the symbol of the quantity itself. Example: specific heat capacity $c_p = C_p/m =$ = heat capacity (in heat per temperature) divided by mass.

The so-called specific viscosity $\eta_{sp} = (\eta - \eta_1)/\eta_1$ is *not* a *specific* quantity. Indices are italicized only if they refer to a constant physical quantity.

Reduced refers to a quantity that is divided by a specified other quantity. Example: reduced osmotic pressure Π/c = osmotic pressure Π divided by the mass concentration c.

Dimensionless quantity: a product or a ratio of two or more different physical quantities that are combined in such a way that the resulting physical quantity has the physical unit of unity (i.e., is "dimensionless").

16.5   Polymer Codes: General Aspects

Conventional names of polymers (plastics, rubbers, fibers, etc.) and their monomers are often abbreviated in the technical and scientific literature. The recommended abbreviations and acronyms vary; ISO, IUPAC, ASTM, BS, DIN, etc., sometimes use the same codes for different plastics or give the same plastic different codes. There are also many "other" abbreviations and acronyms that are not standardized or recommended by international, national, scientific, or industrial organizations.

Tables 16.7-16.14 contain lists of abbreviations and acronyms for fibers, rubbers, plastics, and some other chemicals. Synthetic polymers are listed in alphabetical order of the names of their constituting monomers and not in alphabetical order of polymer names. A selective list of abbreviations and acronyms in alphabetic order can be found in Volume II of this series of books.

Note that some abbreviations and acronyms are registered trademarks in either some or all countries of the world. Examples are EVAL® (worldwide), PAN® (in Europe (DIN uses PAS)), PPO® (worldwide)), and SAN® (Japan and the United States).

For **plastics**, industry in most countries now adheres to the *ISO system* of abbreviations and acronyms [ISO 1043-1986(ε)]. This system cites the abbreviation of the main component first, usually prefixed by P for homopolymeric thermoplastics, followed by codes for after-treatments or special properties. The ISO system is thus a hierachical system.

Abbreviations for comonomers are separated from the main component by a dash; they are not prefixed by P but may be followed by M (for *mo*nomer). The copolymer of ethene (E), propene (P), and a diene monomer (DM) is thus called EPDM.

After-treatments are also indicated by letter *after* the symbol for the parent polymer; the two symbols are separated by a dash. For example, acronyms are PE for poly(ethylene) and PE-C for chlorinated poly(ethylene).

Differences in properties are also indicated by letters or combinations thereof *after* the symbol for the polymer. An example is PE-HD for high-density poly(ethylene).

ASTM abbreviations for plastic names are less systematic (ASTM-D 1600-94a); they usually adhere to a "natural" system. They thus abbreviate the spoken language by indicating special characteristics, modifications, etc., by letters *in front* of that of the symbol for the parent polymer. Examples are HDPE = high-density poly(ethylene), CPE = chlorinated poly(ethylene), and BOPP = biaxially oriented poly(propylene).

IUPAC (Pure Appl.Chem. **59** (1987) 691) uses its own system that sometimes adheres to ISO, sometimes to ASTM, and sometimes to no other system.

The same *polymer* may carry different abbreviations for plastics, fibers, rubbers, and recyclables. An example is poly(ethylene terephthalate): PET (plastics), PES (fiber), PETE (recyclable plastics). Another example is poly(tetrafluoroethylene): PTFE (plastics) and PTF (PTFE as fiber raw material).

Recyclable plastics use numbers instead of letters for the different types.

Rubbers mostly have abbreviations and acronyms that end with an R or today also with an M; silicone elastomers being an exception (see below). Apparently, the same symbols are used for rubbers (in this book defined as uncrosslinked polymers) and elastomers (defined as chemically or physically crosslinked polymers with elastomeric properties; see p. 249).

Fibers, especially natural ones, have their own abbreviations and acronyms (see Section 16.6).

Silicones (polysiloxanes) are characterized by two different sets of abbreviations. ISO characterizes *siicone polymers* using Q as general symbol for silicones. Substituents are indicated by letters: F = fluoro, M = methyl, P = phenyl, V = vinyl. FMQ is thus a silicone, $+\text{SiRR}'\text{-O}+_n$, with fluorine and methyl substituents.

The *General Electric* terminology, on the other hand, characterizes siloxanes and polysiloxanes by the functionality of their *monomers* and *monomeric units*, respectively: M indicates monofunctional, D difunctional, T trifunctional, and Q quadrifunctional siloxane units with the implicit assumption that substituents are usually methyl groups. Substitution by groups other than methyl are indicated by a prime (M', D', T', Q') but the type of group is not specified.

Since M, D, T, and Q symbolize the units $(CH_3)_3SiO_{0.5}$, $(CH_3)_2SiO_{1.0}$, $(CH_3)SiO_{1.5}$, and SiO_2, respectively, symbols for *siloxanes* are obtained by adding the appropriate letters: MM indicates $(CH_3)_3SiOSi(CH_3)_3$ and MD_2M the linear tetramer $(CH_3)_3SiO[Si(CH_3)_2O]_2Si(CH_3)_3$ whereas the cyclic tetramer, $[(CH_3)_3SiO]_4$ has the symbol D_4. MM' may thus be $(CH_3)_3SiOSi(CH_3)_2(C_6H_5)$ or $(CH_3)_3SiOSi(CH_3)(C_6H_5)_2$, etc. The same symbols apply to the repeating units of *polysiloxanes*, i.e., D for $–(CH_3)_2SiO–$; D' for $–(CH_3)(C_6H_5)SiO–$, or $–(CH_3)(CH_2=CH)SiO–$, etc.

16.6 Names and Codes of Fibers

Names and codes of fibers are regulated by ISO Standard 2076 of the International Organization for Standardization (ISO), the European Textile Characterization Law, and for the commerce of the United States by the Federal Trade Commission (FTC). Names of and codes for man-made fibers have been proposed by the International Bureau for the Standardisation of Man-Made Fibres (Bureau International pour le Standardisation de la Rayonne et des Fibres Synthétiques (BISFA), Brussels, Belgium), DIN (Deutsches Institut für Normung (German Standardization Institute)), and ASTM (American Society for Testing and Materials). ASTM codes refer to polymers and not to fibers *per se*.

Table 16-7 Codes for fibers. As a rule, chemical names of fibers do not indicate the exact chemical composition of their base polymers. The old DIN code (indicated by #) has been replaced by the new DIN code. For specifications with respect to compositions, see Table 6-1 (BISFA and FTC) and Table 16-8 (FTC). [1] Proposed name, not yet official.

| Source or | Name | | | Codes | |
chemical name	ISO	FTC	DIN#	ASTM	DIN
Natural fibers: protein fibers (silks)					
Mulberry spinner			Ms		SE
Tussah moths			Ts		ST
Natural fibers: protein fibers (wools)					
Alpaca (sheep camel)			Ap		WP
Angora rabbit			Ak		WA
Beaver					WB
Cashmere			Kz		WS
Guanaco (lama)					WU
Llama			Lm		WL
Mohair (from angora goat)			Mo		WM
Otter					WT
Sheep			Wo		WO
-, virgin wool					WV
Vicuña			Vi		WG
Yak					WY
Natural fibers: protein fibers (hairs)					
Camel			Km		WK
Cow (Rind)			Ri		WR
Goat			Hz		
Hare			Hs		
Human					HA
Horse			Rh		
Rabbit			Kn		
Tibet goat			Tz		
Natural fibers: cellulose textile fibers					
Cotton			Bw		CO
Flax			Fl		LI
-, linen					LI
-, half-linen					HF
Ramie (China grass, rhea)			Ra		RA
Natural fibers: cellulose fibers (hard fibers and others)					
Alfalfa					AL
Broom					GI
Calotropis (akon fiber)			An		
Coir (coconut fiber)			Ko		CC
Esparto (alfa grass)			Ag		

Table 16-7 Codes for fibers (continued).

Source or chemical name	Name ISO/BISFA	FTC	Codes BISFA	DIN#	ASTM	DIN
Fique (Mauritius fiber)			Fi			
Hemp						HE
-, Bombay hemp, Madras hemp, Sunn			Sn			SN
-, Manila- (Abaca)			Ma			AB
Henequen (Agave)			He			
Jute			Ju			JU
Kapok (Bombax wool)			Kp			KP
Kenaf			Kf			KE
Maguey						MG
Moss, American			Tl			
Phormium (New Zealand flax)			Nf			
Rosella (Java jute)			Ro			
Sisal (sisal hemp)			Si			SI
Urena (Congo jute, cadillo, aramina)				Ur		
Regenerated Polysaccharides						
Alginic acid (metal salts)	alginate		ALG	AL		AG
Cellulose, regenerates						
- from cupro process	cupro	cupra (rayon)	CUP	CC		CU
- from viscose process	viscose	rayon	CV	CV		VI
- high strength	modal	rayon	CMD			MD
- from organic solvents	lyocell	rayon	CLY			
Cellulose 2 1/2-acetate	acetate	acetate	CA	CA	CA	AC
Cellulose triacetate	triacetate	triacetate	CTA	CT	CTA	TA
Regenerated Proteins						
Protein fiber	protein	azlon				PR
Ardein fiber				AR		
Casein fiber				KA		CS
Zein fiber				ZE		
Synthetic Fibers (for chemical compositions, see Table 6-1)						
Elastomer fiber, ester type	elastomultiester	-	EME			
Melamine-formaldehyde	melamine	-	MF		MF	
Phenolic resin	-	novoloid			PF	
Poly(acrylate)	-	anidex				
Poly(acrylonitrile)	acrylic	acrylic	PAN	PAC	PAN	PAN
Poly(acrylonitrile) CoPM	modacrylic	modacrylic	MAC	PAM		MA
Polyamide, aliphatic	polyamide	nylon	PA	PA	PA	PA
, aromatic	aramid	aramid	AR			
Poly(chloroprene)	-	rubber			CR	
Polydiene	elastodiene	rubber	ED	LA	EU	EL
Poly(diene-*co*-acrylonitrile)	-	lastrile			NBR	
Poly(ethylene)	polyethylene	olefin	PE	PE	PE	PE
Poly(ethylene terephthalate)	polyester	polyester	PES	PES	PET	PES
Polyimide, aromatic imides	polyimide	-	PI	PI	PI	
Poly(lactide)	synterra [1]	-	PLA			PL
Poly(1,4-phenylenesulfide)	-	sulfar			PPS	
Poly(propylene), isotactic	polypropylene	olefin	PP	PP	PP	PP
Poly(styrene) (>85 % sty)				PST		
Poly(tetrafluoroethylene)	fluorofiber	-	PTFE			FL
Polyurea	-	-		PUA		PB
Polyurethane	-	-		PUR	PUR	PU
-, segmented (>85 % PUR)	elastane	spandex	EL	PUE		EA

Table 16-7 Codes for fibers (continued).

Source or chemical name	Name ISO/BISFA	FTC	BISFA	DIN#	ASTM	DIN
Poly(vinyl alcohol), acetalized	vinylal	vinal	PVAL	PVA		VY
Poly(vinyl chloride)	chlorofiber	vinyon	CLF	PVC	PVC	
Poly(vinylidene chloride)	-	saran		PVD	PVDC	
Poly(vinylidene cyanide)	-	nytril				
Trivinyl fiber	-	-				TV
Inorganic Fibers						
Asbestos fiber				As		AS
Glass fiber (fiberglass)	glass	glass	GF	GL		GF
Carbon (fiber) (>90 % C)	carbon	carbon	CF			CF
Metal fiber	metal	metallic	MTF	MT		ME
Fibers coated with metals	metallized	metallic				
Mineral fiber (from melts)	-	metallic		ST		
Mineral fiber (from ceramics)	ceramic	-	CEF			
Slag wool				SL		
Other Fibers						
unspecified						AF
unknown or textile residue						TR

Table 16-8 Compositions of textile fibers based on § 303.7 of the United States Federal Trade Commission (FTC) (www.ftc.gov/os/statutes/textile/rr-textl.htm; accessed 2008-11-17. For the Standard 2076 of the International Standardization Organization (ISO) and the corresponding rules of BISFA (*Bureau International pour la Standardisation des Fibres Artificielles* = International Bureau for the Standardization of Man-Made Fibers), see also Table 6-1, p. 155.

Acetate. A manufactured fiber in which the fiber-forming substance is cellulose acetate. Where not less than 92 % of the hydroxyl groups are acetylated, the term **triacetate** may be used as a generic description of the fiber.

Acrylic. A mnufactured fiber in which the fiber-forming substance is any long-chain synthetic polymer composed of at least 85 wt% of acrylonitrile units, $-CH_2-CH(CN)-$.

Anidex. A manufactured fiber in which the fiber-forming substance is any long-chain synthetic polymer composed of at least 50 wt% of one or more esters of a monohydric alcohol and acrylic acid, $CH_2=CH-COOH$.

Aramid. A manufactured fiber in which the fiber-forming substance is a long-chain synthetic polyamide in which at least 85 % of the amide linkages $-NH-CO-$ are attached directly to two aromatic rings.

Azlon. A manufactured fiber in which the fiber-forming substance is composed of any regenerated naturally occuring protein.

Elastoester. A manufactured fiber in which the fiber-forming substance is a long-chain synthetic polymer composed of at least 50 wt% of aliphatic polyether and at least 35 wt% of polyester as defined in paragraph "Polyester."

Fluoropolymer. A manufactured fiber containing at least 95 % of a long-chain polymer synthesized from aliphatic fluorocarbon monomers.

Glass. A manufactured fiber in which the fiber-forming substance is glass.

Melamine. A manufactured fiber in which the fiber-forming substance is a synthetic polymer composed of at least 50 wt% of a crosslinked melamine polymer.

Metallic. A manufactured fiber composed of metal, plastic-coated metal, metal-coated plastic, or a core completely covered by metal.

Modacrylic. A manufactured fiber in which the fiber-forming substance is any long chain synthetic polymer composed of less than 85 wt% but at least 35 wt% of acrylonitrile units, $-CH_2-CH(CN)-$, except fibers qualifying under paragraphs "Anidex" or "Rubber (2)."

Table 16-8 Compositions of textile fibers based on § 303.7 of the United States Federal Trade Commission (FTC) (continued).

Novoloid. A manufactured fiber containing at least 85 wt% of a cross-linked novolac.

Nylon. A manufactured fiber in which the fiber forming substance is a long-chain synthetic polyamide in which more than 85 % of the amide-linkages, $-NHCO-$, are attached directly to aliphatic groups.

Nytril. A manufactured fiber containing at least 85 % of a long-chain polymer of vinylidene dinitrile, $-CH_2-C(CN)_2-$, where the vinylidene dinitrile content is no less than every other unit in the chain.

Olefin. A manufactured fiber in which the fiber-forming substance is any long-chain synthetic polymer composed of at least 85 wt% of ethylene, propylene, or other olefin units, except amorphous (non-crystalline) polyolefins qualifying under paragraph "Rubber (1)".

When the fiber-forming substance is a cross-linked synthetic polymer, with low but significant crystallinity, composed of at least 95 wt% of ethylene units and at least one other olefin unit, and the fiber is substantially elastic and heat resistant, the term **lastrol** may be used a a generic description of the fiber.

PBI. A manufactured fiber in which the fiber-forming substance is a long-chain aromatic polymer having reoccurring imidazole groups as an integral part of the polymer chain.

PLA. A manufactured fiber in which the fiber-forming substance is composed of at least 85 wt% of lactic acid ester units derived from naturally occurring sugars.

Polyester. Any long chain synthetic polymer composed of at least 85 wt% of an ester of a substituted aromatic carboxylic acid but not restricted to substituted terephthalate units, $-O-R-O-CO-p-C_6H_4-CO-$, or para-substituted hydroxybenzoate units, $-O-R-O-p-C_6H_4-CO-$.

When the fiber is formed by the interaction of two or more chemically distinct polymers (of which none exceeds 85 wt%), and contains ester groups as the dominant functional units (at least 85 wt% of the total polymer content of the fiber), and which, if stretched at least 100 %, durably and rapidly reverts substantially to its unstretched length when the tension is removed, the term **elasterell-p** may be used as a generic description of the fiber.

Rayon. A manufactured fiber composed of regenerated cellulose, as well as manufactured fibers composed of regenerated cellulose in which substituents have replaced not more than 15 % of the hydrogens of the hydroxyl groups. Where the fiber is composed of cellulose precipitated from an organic solution in which no substitution of the hydroxyl groups takes place and no chemical intermediates are formed, the term **lyocell** may be used as a generic description of the fiber.

Rubber. A manufactured fiber in which the fiber-forming substance is comprised of natural or synthetic rubber, including the following categories:

    (1) A manufactured fiber in which the fiber-forming substance is a hydrocarbon such as natural rubber, polyisoprene, polybutadiene, copolymers of dienes and hydrocarbons, or amorphous (non-crystalline) polyolefins.

    (2) A manufactured fiber in which the fiber-forming substance is a copolymer of acrylonitrile and a diene (such as butadiene) composed of not more than 50 wt% but at least 10 wt% of acrylonitrile units, $-CH_2-CH(CN)-$. The term **lastrile** may be used as a generic description for fibers falling into this category.

    (3) A manufactured fiber in which the fiber-forming substance is a polychloroprene or a copolymer of chloroprene in which at least 35 wt% of the fiber-forming substance is composed of chloroprene units, $-CH_2-CCl=CH-CH_2-$.

Saran. A manufactured fiber in which the fiber-forming substance is any long-chain synthetic polymer composed of at least 80 wt% of vinylidene chloride units, $-CH_2-CCl_2-$.

Spandex. A manufactured fiber in which the fiber-forming substance is a long-chain synthetic polymer comprised of at least 85 % of a segmented polyurethane.

Sulfar. A manufactured fiber in which the fiber-forming substance is a long-chain synthetic polysulfide in which at least 85 % of the sulfide linkages $-S-$ are attached directly to two aromatic rings.

Vinal. A manufactured fiber in which the fiber-forming substance is any long-chain synthetic polymer composed of at least 50 wt% of vinyl alcohol units, $-CH_2-CHOH-$, and in which the total of the vinyl alcohol units and any one or more of the various acetal units is at least 85 wt% of the fiber.

Vinyon. A manufactured fiber in which the fiber-forming substance is any long-chain synthetic polymer composed of at least 85 wt% of vinyl chloride units, $-CH_2-CHCl-$.

16.7 Codes for Rubbers

Table 16-9 Monomers and codes for rubbers. These codes are mostly not different from codes for fibers or plastics composed of identical or similar polymers. Rubbers are arranged in alphabetical order of the monomers that were used for their syntheses (not according to their main monomeric units). A PM after the names of monomers indicates specific chemical groups that were introduced during polymer synthesis or by an after-reaction of the polymer. "Other" indicates outdated or commonly used codes that are not sanctioned by ASTM, BS (British Standards), or ISO. PM = polymer.
  * IUPAC recommendation R 1629 (1980); ASTM-D 1418-76.

Monomers (plus polymer after-treatment, if any)	ASTM	BS	ISO	Other
Acrylic ester + acrylonitrile	ANM			
Acrylic ester + butadiene	ABR*			AR
Acrylic ester + 2-chlorethyl vinyl ether	ACM			
Acrylonitrile + butadiene	NBR*		NBR	GR-N
Acrylonitrile + butadiene, PM contains carboxyl	XNBR		NBR-X	
Acrylonitrile + butadiene, PM hydrogenated			NBR-H	HNBR
Acrylonitrile + butadiene + styrene, PM with COOH	XABS			
Acrylonitrile + chloroprene	NCR*			
Alkylene sulfide	ASR			
Allyl glycidyl ether + propylene oxide				GPO
Butadiene	BR*		BR	BP, CBR
Butadiene + styrene	SBR*	SBR	SBR	GR-S
Butadiene + styrene, PM contains carboxyl groups	XSBR			
Butadiene + styrene, triblock polymer (elastic)	YSBR			SBS
Butadiene + styrene, block copolymer with COOH	YXSBR			
Butadiene + styrene + 2-vinyl pyridine	PSBR			
Butadiene + 2-vinyl pyridine	PBR*			
Chloroprene	CR*	CR	CR	
Chloroprene + styrene	SCR*			
Cyclooctene (as trans polymer)				TPA, TPR
Chlorotrifluoroethylene	CF			
Chlorotrifluoroethylene + vinylidene fluoride	CFM			
Epichlorhydrin	CO*		CO	CHR
Epichlorhydrin + ethylene oxide	ECO			CHC
Ethylene (PM after-chlorinated)	CM			CPE
Ethylene (PM perchlorosulfonated)	CSM	CSPR		CSR
Ethylene + ethyl acrylate	EEA		E/EA	
Ethylene + methyl acrylate	EAM			AEM
Ethylene + propylene	EPM	EPR	EPM	AP
Ethylene + propylene + non-conjugated diene	EPDM	EPTR	EPDM	APT, EPT
Ethylene + tetrafluoroethylene				ETFE
Ethylene + vinyl acetate	EVM		EVAC	EVA, EVAC
Fluorinated carbon chains (fluoro-, perfluoroalkyl-, perfluoroalkoxy groups)	FKM		FPM	
Hexafluoropropylene + tetrafluoroethylene				FEP
Hexafluoropropylene + vinylidene fluoride	FPM			
Isobutylene	PIB*	PIB	PIB	PIS
Isobutylene + isoprene	IIR*	Butyl		GR-I, PIBI
Isobutylene + isoprene (PM after-brominated)	BIIR			
Isobutylene + isoprene (PM after-chlorinated)	CIIR			
Isoprene (natural rubber)	NR*			NK
Isoprene (natural rubber, easy processable)				SP
Isoprene (natural rubber, standardized Indonesian)				SIR
Isoprene (natural rubber, standardized Malaysian)				SMR

Table 16-9 Monomers and codes for rubbers (continued).

Monomers (plus polymer after-treatment, if any)	ASTM	BS	ISO	Other
Isoprene (natural rubber, technically classified)				TC
Isoprene (synthetic rubber)	IR*	IR		CPI, PIP
Isoprene (PM chlorinated)				RUC
Isoprene + acrylonitrile	NIR			
Isoprene + styrene	SIR* [a)]			
Isoprene + styrene (triblock polymer)				SIS
Isoprene + styrene (triblock polymer), hydrogenated				SEBS
Norbornene	PNR			PNR
Phosphornitrile chloride, with fluoro substituents				PNF
Propylene oxide	PO			OPR
Propylene oxide + allyl glycidyl ether	GPO			POR
Silicone, general	Q			
Silicone with fluoro substituents	FQ			
Silicone with fluoro + methyl substituents	FMQ		MFQ	FSI
Silicone with fluoro, methyl, vinyl substituents	FVMQ			
Silicone with methyl substituents	MQ			SI
Silicone with phenyl substituents	PQ			
Silicone with phenyl + methyl substituents	PMQ		MPQ	PS
Silicone with phenyl, vinyl, methyl substituents	PVMQ		MPVQ	
Silicone with vinyl substituents	VQ			
Silicone with vinyl + methyl substituents	VMQ		MVQ	
Sulfide (thioplastics)				T, TR, OT
Tetrafluoroethylene + trifluoronitrosomethane + nitrosoperfluorobutyric acid (nitroso rubber)	AFMU			CNR
Urethane elastomer	UE	UR		PUR
Urethane elastomer with polyester segments	AU		AU	
Urethane elastomer with polyether segments	EU		EU	
Urethane polyester elastomer, isocyanate crosslinkable	AU-I			
Urethane polyester elastomer, peroxide crosslinkable	AU-P			
Vinylidene fluoride + hexafluoropropylene	FKM		FPM	
Tetrafluoroethylene + perfluoroalkyl vinyl ether				PFA
Thermoplastic elastomer, general	TPEL		TPE	TPE, TPR
Thermoplastic elastomer, ester based				TPE-E
Thermoplastic elastomer, ether-amide based				TPE-A
Thermoplastic elastomer, ether-ester based	TEE			
Thermoplastic elastomer, olefin based	TEO			TPE-O, TPO
Thermoplastic elastomer, polyether type	PEBA			
Thermoplastic elastomer, styrene/diene based	TES			TPE-S
Thermoplastic elastomer, urethane based	TPUR			

Additional codes that are used in the trade:

ICR   = initial concentration rubber (= rubber from undiluted natural rubber latex),
OER   = oil-extended rubber,
RTV   = room temperature vulcanization.

[a)] Not to be confused with IR = "standardized Indonesian Natural Rubber."

16.8 Codes for Thermoplastics and Thermosets

Names of comonomeric units are arranged in alphabetical order and not according to their proportion in the polymer. ASTM D 1600-94a, DIN 7728, ISO 1043-1986(ε)], IUPAC [Pure Appl. Chem. **59** (1987) 691]; Other = frequently used conventional codes. PM = Polymer.

ISO: Special properties of plastics are indicated by up to four letters *after* the code that are separated from the code by a hyphen. ASTM places such letters *in front* of the code.

In general, the following letters are used: B =biaxial (oriented); C = chlorinated; D = density; E = expanded; F = flexible, or fluid; H = high; I = impact; L = linear or low; M = mass, medium, or molecular; N = normal or novolac; O = oriented; P = plasticizer content; R = raised or resol; U = ultra or unplasticized; V = very; W = weight; X = X-linked (cross-linked).

Table 16-10 Codes based on monomers. Monomer names are given in alphabetical order. For space reasons, the following abbreviations are used in the column "Polymer of": EPD = ethylene-propylene-diene; MMA methyl methacrylate; PE = poly(ethylene), PM = polymer.

Polymer of	ASTM	DIN	ISO	IUPAC	Other
Acrylonitrile	PAN	PAC		PAN	PAN®
Acrylonitrile + butadiene	PBAN				
Acrylonitrile + butadiene + MMA + styrene					MABS
Acrylonitrile + EPD + styrene		A/EPDM/S		AES	
Acrylonitrile + methyl methacrylate	AMMA	A/MMA	A/MMA		
Acrylonitrile + styrene	SAN	SAN	SAN	SAN	PSAN, AS
Acrylonitrile + styrene + chlorinated PE	ACS	A/PE-C/S			
Acrylic acid	PAA				PAS
Acrylic ester + acrylonitrile + butadiene	ABA	A/B/A			
Acrylic ester + acrylonitrile + styrene	ASA	ASA	A/S/A		
Acrylic ester + ethylene	EEA		E/EA		EAA
Acrylic ester + maleic anhydride					AMA
Adipic acid + hexamethylenediamine	PA 6.6	PA 66	PA 66	PA 6.6	
Adipic acid + tetramethylene glycol					PTMA
Allyl diglycol carbonate	PADC				
Aminobenzoic acid, *p*-					PAB
Aminotriazole					PAT
Aminoundecanoic acid, 11-		PA 11	PA 11		
Anilien					PAL
Arylamide	PARA				
Aryl sulfone					PAS
Azelaic acid anhydride					PAPA
Benzimidazole					PBI
Bisphenol A + phosgene (= polycarbonate)	PC	PC	PC	PC	
Bitumen + ethylene					ECB
Butadiene + methyl methacrylate + styrene	MBS	MBS	PBS		
Butadiene + styrene, thermoplastic	SB	S/B	S/B	S/B	PASB
Butadiene + styrene, polym. in emulsion					E-SBR
Butadiene + styrene, thermoplast. blockcopm.	YSBR				
Butene-1	PB	PB	PB		BT
Butyl acrylate	PBA	PBA	PBA	PBA	
Butylene glycol + terephthalic acid	PBT	PBTP	PBTP		PTMT
Butyl methacrylate					PBMA
Caprolactam, ε-	PA 6	PA 6	PA 6	PA 6	PA 6
Caprolactam, ε- + ω-dodecanolactam	PA 6.12		PA 612		
Chlorotrifluoroethylene	PCTFE	PCTFE	PCTFE	PCTFE	
Chlorotrifluoroethylene + ethylene					ECTFE
Cresol + formaldehyde	CF	CF	CF	CF	

Table 16-10 Codes based on monomers (continued).

Polymer of	ASTM	DIN	ISO	IUPAC	Other
Cyclohexanedimethylol, 1,4- + terephthalic acid					PCDT
Cycloolefin copolymers					COC
Diallylchlorendate (= diallyl 1,4,5,6,7,7-hexa- chlorobicyclo-(2,2,1)-5-ene dicarboxylate	PDAC				
Diallyl fumarate	PDAF				
Diallyl isophthalate	PDAIP				
Diallyl maleate	PDAM				
Diallyl phthalate	PDAP	PDAP	PDAP	PDAP	DAP
Dichlorobenzene, 1,4- + disodium sulfide	PPS	PPS			
Dimethylphenol, 2,6- (polyphenylene ether)	PPE		PPE		PPO™
Dodecanolactam (laurolactam)	PA 12	PA 12	PA 12	PA 12	
Ethyl acrylate					PEA
Ethylene	PE	PE	PE	PE	PE
Ethylene, PM impact resistant			PE-HI		
Ethylene, PM crosslinked		PE-X	PE-X		XLPE
Ethylene, PM with high desity	HDPE	PE-HD	PE-HD		
Ethylene, PM with low density	LDPE	PE-LD	PE-LD		
Ethylene, PM with low density, linear	LLDPE		PE-LLD		
Ethylene, PM with medium density	MDPE		PE-MD		
Ethylene, PM with very low density					VLDPE
Ethylene, PM with ultrahigh molecular weight	UHMW-PE		PE-UHMW		
Ethylene + chlorotrifluoroethylene					ECTFE
Ethylene + ethyl acrylate	EEA	E/EA			EEA
Ethylene + methacrylic acid	EMA	E/MA			
Ethylene + methyl acrylate + vinyl chloride		VC/E/MA	VC/E/MA		
Ethylene + methyl methacrylate		E/MMA	E/MMA		
Ethylene + propylene	EPM	E/P	E/P		PEP
Ethylene + propylene + diene	EPD	EPDM	EPDM		
Ethylene + tetrafluoroethylene	ETFE	E/TFE	E/TFE		
Ethylene + vinyl acetate	EVA	E/VA	EVAC		VAE
Ethylene + vinyl acetate + vinyl chloride		VC/E/VAC	VC/E/VAC		
Ethylene + vinyl chloride	VCE	VC/E	VC/E		
Ethyl acrylate + ethylene	EEA	E/EA	E/EA		
Ethylene glycol					PEG
Ethylene glycol + maleic anhydride	UP	UP	UP	UP	
Ethylene glycol + terephthalic acid (ester)	PET	PET	PETP	PETP	PETE
Ethylene glycol + terephthalic ester + another glycol	PETG				
-, PM fast crystallizing					CPET
-, PM oriented					OPET
Ethylene oxide	PEO	PEOX	PEOX	PEO	
Formaldehyde (or trioxane) (polyacetal)	POM	POM	POM	POM	
Formaldehyde + furan	FF		FF		
Formaldehyde + melamine	MF	MF	MF	MF	MF
Formaldehyde + melamine + phenol	MPF	MPF	MPF		MP
Formaldehyde + phenol	PF	PF	PF	PF	
Formaldehyde + urea	UF	UF	UF	UF	
Furfural + phenol	PFF				
Hexafluoropropylene + tetrafluoroethylene	FEP	FEP	FEP		
Hexamethylenediamine + sebacic acid	PA 6.10	PA 610	PA 610		PA 6.10
Hydroxybenzoic acid, p-	POB				PHB
Hydroxyethyl methacrylate, 2-					PHEMA
Isobutene	PIB	PIB	PIB	PIB	IM

Table 16-10 Codes based on monomers (continued).

Polymer of	ASTM	DIN	ISO	IUPAC	Other
Isocyanurate		PIR	PIR		
Laurolactam (dodecanolactam)	PA 12	PA 12	PA 12	PA 12	PA 12
Linseed oil, epoxidized		ELO	ELO		
Maleic anhydride + styrene	SMA	S/MA	S/MA		
Methacrylimide		PMI	PMI		
Methyl acrylate					PMA
Methyl acrylate + vinyl chloride	VCMA	VC/MA	VC/MA		
Methyl-α-chloromethacrylate	PMCA				
Methyl methacrylate	PMMA	PMMA	PMMA	PMMA	
Methyl methacrylate + vinyl chloride	VCMMA			VC/MMA	
Methyl pentene-1, 4-	PMP	PMP	PMP		TPX
Methyl styrene, α-	PMS	PMS	PMS		PAMS
Methyl styrene, α- + styrene	SMS	S/MS	S/MS		
Octyl acrylate + vinyl chloride	VCOA	VC/OA	VC/OA		
Perfluoroalkoxyalkane	PFA	PFA	PFA		
Propylene	PP	PP	PP	PP	
Propylene, PM oriented			PP-O		OPP
Propylene, PM biaxially oriented			PP-BO		BOPP
Propylene + tetrafluoroethylene (alternating)					TFE-P
Propylene oxide	PPOX	PPOX	PPOX		
Soy bean oil, epoxidized		ESO	ESO		
Styrene	PS	PS	PS	PS	PS
Styrene, polymer high impact		PSHI	PS-HI		HIPS
Styrene, polymer oriented film					OPS
Styrene, polymer expanded					EPS
Styrene, polymer impact	SRP	IPS			
Tetrafluoroethylene	PTFE	PTFE	PTFE	PTFE	
Tetrahydrofuran					PTHF
Tetrahydrofuran, PM with hydroxyl endgroups					PTMEG
Triallyl cyanurate	PTAC				
Trifluoroethylene					P3FE
Trioxane + comonomer	POM	POM	POM	POM	
Vinyl acetate	PVAC	PVAC	PVAC	PVAC	PVA
Vinyl acetate + vinyl chloride	PVCA	VC/VAC	VC/VAC	PVCA	
Vinyl carbazole, N-	PVK	PVK	PVK		
Vinyl chloride	PVC	PVC	PVC	PVC	PCU, V
-, PM as flexible film			PVC-F		F-PVC
-, PM as rigid film			PVC-R		RPVC
-, PM oriented			PVC-O		OPVC
-, polymerized in bulk			PVC-M		M-PVC
-, polymerized in emulsion			PVC-E		E-PVC
-, polymerized in suspension			PVC-S		S-PVC
Vinyl chloride + vinylidene chloride	VCVDC	VC/VDC	VC/VDC		
Vinyl fluoride	PVF	PVF	PVF	PVF	
Vinylidene chloride	PVDC	PVDC	PVDC	PVDC	
Vinylidene fluoride	PVDF	PVDF	PVDF	PVDF	
Vinyl methyl ether					PVME
Vinyl pyrrolidone, N-	PVP	PVP	PVP		

Table 16-11 Codes based on characteristic polymer groups. * Saran coated, ** metal coated

Polymer	ASTM	DIN	ISO	IUPAC	Other
Epoxide	EP	EP	EP	EP	
Epoxide, glass fiber reinforced					GEP
Epoxy ester			EPE		
Epoxy ester, aromatic	ARP				
Epoxy ester, saturated (thermoplastic)	TPES	SP	SP		
Epoxy ester, unsaturated	UP	UP	UP	UP	
Epoxy ester, unsaturated, glass fiber reinforced					FRP, GUP
Epoxy ester alkyd	PAK				
Epoxy ester ether, thermoplastic elastomer	TEEE				
Epoxy ester imide	PEI		PEI		
Epoxy ester urethane	PAUR		PEUR		
Polyamide, aliphatic	PA	PA	PA	PA	PA
Polyamide, aromatic	PARA				PAR
Polyamide, metal coated	PA**				
Polyamide, Saran coated	PA*				
Polyamide imide	PAI	PAI	PAI		
Polyarylamide	PARA				
Polyarylate	PAT				
Polyarylether	PAE				
Polyaryletherketone	PAEK				
Polyarylsulfone					PAS
Polybenzimidazole					PBI
Polybismaleimide					BMI
Polybisphenol A-sulfone					PSF
Polycarbodiimide					PCD
Polycarbonate, aromatic	PC	PC	PC	PC	
Polyether-block-amide	PEBA	PEBA	PEBA		
Polyetheretherketone	PEEK		PEEK		
Polyetherimide	PEI	PEI			
Polyetherketone	PEK				
Polyethersulfone	PES PESU	PES	PES		
Polyetherurethane	PEUR				
Polyimide	PI	PI	PI		
Polyimidesulfone	PISU				
Polyisocyanurate		PIR			
Polyparabanic acid					PPA
Polyphenylenether	PPE	PPE	PPE		PPO®
Poly(1,4-phenylenesulfide)	PPS				
Polyphenylenesulfone	PPSU	PPSU	PPSU		PSU
Polyphthalamide	PPA				
Polysulfone	PSU	PSU	PSU		
Polyurethane	PUR	PUR	PUR	PUR	
Polyurethane, thermoset	TSUR				
Polyurethane, thermoplastic	TPUR				TPU
Silicone (plastic)	Si	SI	SI		

Table 16-12 Codes based on polymer transformations.

Resulting polymer	ASTM	DIN	ISO	IUPAC	Other
Carboxymethylcellulose	CMC	CMC	CMC	CMC	
Carboxymethylhydroxyethylcellulose					CMHEC
Casein (denatured)	CS	CSF	CS	CS	
Cellulose based plastics, general	CE				
Cellophane	C				
Cellophane, Saran coated	C*				
Cellulose acetate	CA	CA	CA	CA	
Cellulose acetobutyrate	CAB	CAB	CAB	CAB	
Cellulose acetopropionate	CAP	CAP	CAP	CAP	
Cellulose nitrate	CN	CN	CN	CN	NC
Cellulose propionate	CP	CP	CP	CP	
Cellulose triacetate	CTA	CTA	CTA		TA
Ethyl cellulose	EC	EC	EC	EC	
Hydroxyethyl cellulose					HEC
Hydroxypropyl cellulose					HPC
Hydroxypropylmethyl cellulose					HPMC
Methyl cellulose		MC			
Poly(ethylene), chlorosulfonated	CSM				CSR
Poly(ethylene), after-chlorinated	CPE	PE-C	PEC		
Poly(ethylene-*co*-vinyl alcohol)	EVA	E/VAL			EVAL™
					EVOH
Poly(propylene), chlorinated		PP-C			
Poly(vinyl alcohol)	PVAL	PVAL	PVAL	PVAL	
Poly(vinyl butyral)	PVB	PVB	PVB		
Poly(vinyl chloride), after-chlorinated	CPVC	PVC-C	PVCC		PC, PeCe
Poly(vinyl formal)	PVFM	PVFM	PVFM	PVFM	

Table 16-13 Miscellaneous codes. Blends are not always physical mixtures because they may contain graft polymers that result from the blending process.

Polymer, blend, etc.	ASTM	DIN	ISO	IUPAC	Other
Blend from PC and ABS (example)		PC+ABS	PC+ABS		
Blend from PA 6 and PA 12 (example)		PA 6/12	PA 6/12		
Molding compound, sheets (see p. 108)					SMC
Molding compound, fibrous premix (see p. 108)					BMC
Molding compound, with high strength					HMC
Prepreg with high content of glass fibers					HMC
Plastic, reinforced by man-made fibers					CFK
Plastic, fiber reinforced					FRP
Plastic, glass fiber reinforced					GFK, GRP
Plastic, carbon fiber reinforced		KEK			
Plastic, metal fiber reinforced					MFK
Poly(acrylonitrile-*co*-styrene) and chlorinated PE					ACS
Poly(ethylene terephthalate), bi-oriented		PET-BO	PET-BO		
Polyester, film, metallized					MPE
Polymer, liquid-crystalline	LCP				
Polymer, interpenetrating					IPN
Polyolefin, amorphous					APO

16.9   Other Codes

Table 16-14 Codes for monomers, adjuvants, solvents, etc.

Chemical compound	DIN	ISO	IUPAC	Other
Azobisisobutyronitrile, *N,N'*-				AIBN
Benzylbutylphthalate	BBP	BBP		
Benzyloctyladipate (benzyl-2-ethylhexyladipate)		BOA		
Dibenzoylperoxide				BPO
Dibutyl phthalate	DBP	DBP	DBP	
Dicapryl phthalate	DCP	DCP	DCP	
Didecyl phthalate				DDP
Diethyl phthalate		DEP		
Diheptyl phthalate		DHP		
Dihexyl phthalate		DHXP		
Diisobutyl phthalate	DIBP	DIBP		
Diisodecyl adipate	DIDA	DIDA	DIDA	
Diisodecyl phthalate	DIDP	DIDP	DIDP	
Diisononyl adipate		DINA		
Diisononyl phthalate	DINP	DINP		
Diisooctyl adipate	DIOA	DIOA	DIOA	
Diisooctyl phthalate	DIOP	DIOP	DIOP	
Diisopentyl phthalate				DIPP
Diisotridecyl phthalate	DITP	DITDP		
Dimethylacetamide, *N,N*-				DMAC
Dimethylformamide, *N,N*-				DMF
Dimethylphthalate		DMP		
Dimethylsulfoxide				DMSO
Dimethylterephthalate				DMT
Dinonyl phthalate		DNP	DNP	
Dioctyl adipate (di-2-ethylhexyladipate)	DOA	DOA	DOA	
Dioctyl azelate (di-2-ethylhexylazelate)	DOZ	DOZ	DOZ	
Dioctyldecyl phthalate		DODP		
Dioctyl isophthalate (di-2-ethylhexylisophthalate)		DOIP	DOIP	
Dioctyl phthalate (di-2-ethylhexylphthalate)	DOP	DOP	DOP	
Dioctyl sebacate (di-2-ethylhexylsebacate)	DOS	DOS	DOS	
Dioctyl terephthalate (di-2-ethylhexylterephthalate)	DOTP	DOTP		
Diphenylcresyl phosphate		DPCF		
Diphenyloctyl phosphate		DPOF		
Diundecyl phthalate				DUP
Hexamethylphosphortriamide				HMPT
Methyl methacrylate				MMA
Tetrahydrofuran				THF
Tetramethylsilane				TMS
Tetraoctyl pyromellitate	TOPM	TOPM		
Toluene diisocyanate (tolylene d.; toluylene d.)				TDI
Trichloroethyl phosphate		TCEF		
Triethylenediamine				DABCO®
Triisooctyl trimellitate	TIOTM	TIOTM		
Tricresyl phosphate	TCF	TCF	TCP	TKP, TTP
Trioctyl mellitate	TOTM	TOTM		
Trioctyl phosphate (tri(2-ethylhexyl)phosphate)	TOF	TOF	TOP	
Triphenyl phosphate	TPF	TPF	TPP	
Vinyl acetate				VAC
Vinyl chloride				VC, VCM

Literature about Physical Measures and Units

R.A.Nelson, SI. The International System of Units, Amer.Assoc.Physics Teachers, College Park (MD), 2nd ed. 1982

H.G.Jerrard, D.B.McNeill, A Dictionary of Scientific Units, Chapman and Hall, London, 5th ed. 1986

I.Mills, T.Cvitaš, K.Homann, N.Kallay, K.Kuchitsu, Eds., (IUPAC Commission on Physicochemical Symbols, Terminology and Units), Quantities, Units, and Symbols in Physical Chemistry, Blackwell, Oxford, UK 1988 ("Green Book")

ASTM Committee E-43, Standards for the Use of the International System of Units (SI): The Modern Metric System, ASTM, West Conshohocken (PA) 1997

F.Cardarelli, Scientific Unit Conversion, Springer, Berlin 1997

17 Subject Index

Entries are listed in strict alphabetical order; they may consist of a single word, abbreviations, acronyms, or combinations thereof. For alphabetization, technical terms consisting of two nouns are considered to be one word, whether written as two words (example: acetal polymer), with a hyphen (for example, tension-thinning), or in parentheses, brackets, or braces (example: "Catalyst, def.", comes before "Catalyst efficiency")). Qualifying numbers and letters as well as hyphens, parentheses, brackets, and braces in names of chemical compounds such as 1-, 1,4-, α-, β-, o-, m-, p-, L-, D-, etc., also have been disregarded for alphabetization. Terms consisting of an adverb and a noun are arranged according to the noun (example: Molar mass → Mass, molar).

The following abbreviations are used: def. = definition; ff. = and following. For polymer codes, see also Appendix (Chapter 17).